Combinatorial Chemistry and Technologies

Methods and Applications

Second Edition

Combinatorial Chemistry and Technologies

Methods and Applications

Second Edition

Edited by

Giorgio Fassina

Stanislav Miertus

Boca Raton London New York Singapore

A CRC title, part of the Taylor & Francis imprint, a member of the Taylor & Francis Group, the academic division of T&F Informa plc.

Published in 2005 by
CRC Press
Taylor & Francis Group
6000 Broken Sound Parkway NW, Suite 300
Boca Raton, FL 33487-2742

Printed in the United States of America on acid-free paper
10 9 8 7 6 5 4 3 2 1

International Standard Book Number-10: 0-415-30830-5 (Hardcover)
International Standard Book Number-13: 978-0-415-30830-4 (Hardcover)

Library of Congress Cataloging-in-Publication Data

Combinatorial Chemistry and Technologies / edited by Stanislav Miertus and Giorgio Fassina. – 2nd ed.
p. cm.
Includes bibliographical references and index.
ISBN 0-8247-5837-4
1. Combinatorial chemistry. I. Miertus, Stanislav. II. Fassina, Giorgio, 1960-

RS419.C59 2004
615'.19–dc22 2004062069

Taylor & Francis Group
is the Academic Division of T&F Informa plc.

Visit the Taylor & Francis Web site at
http://www.taylorandfrancis.com

and the CRC Press Web site at
http://www.crcpress.com

Preface

Combinatorial methodologies have dramatically changed the drug discovery process in the pharmaceutical industry, offering an unlimited source of new molecular entities to be screened for activity. Innovative chemistries, softwares, hardwares, and advanced molecular biology protocols have been developed in the past few years to generate numerically complex and structurally diverse libraries of compounds, as well as new automated approaches for simultaneous screenings. This field not only has integrated, complemented and revitalized existing research activities, but also tremendously stimulated the evolution of other disciplines and technologies, such as solid phase chemistry, molecular modeling, *in vitro* screening of biological activities, miniaturization and automation, to satisfy the demand to design, produce high yield and purity, and rapidly screen huge numbers of molecules. Combinatorial technologies have merged probability with chemistry, molecular biology, rational design and automation. The rapidly growing interest in combinatorial technologies all over the world has been not only due to the possibility of identifying new drugs to treat human diseases, but also to their broad applicability to other fields, such as the diagnostic, new materials, and catalysis sectors. Many books have been published so far on combinatorial techniques, but offering just a limited perspective of the field, focusing on selected aspects without covering all the different approaches and integrated technologies involved.

Therefore, Marcel Dekker agreed to publish the book on combinatorial chemistry and technology, covering a broad spectrum of methodologies and applications. The first edition appeared in 1999 and this book provided a comprehensive coverage of methodologies employed for the design, synthesis and screening of molecular libraries. Major topics included generation of molecular diversity by chemical methods using solution and solid phase chemistries, biological approaches for the production and screening of peptide, antibody and oligonucleotide libraries, the application of computer assisted approaches to guide library synthesis, the use of high throughput screening methodologies to accelerate lead discovery, development of automation and robotics, and economic and legal issues. The book received a wide acknowledgement and interest from the international scientific community. Due to the very dynamic development of methods and fields of applications of combinatorial chemistry and technologies, it has become evident that there is a need to revise and update the content of individual chapters, as well as to extend it to include other important topics, such as catalysis and application in biotechnology. Therefore, a second extended and revised edition is being published.

The book is intended for scientists as well as doctoral fellows willing to be introduced to the topic but without prior detailed knowledge of the field. All the basic approaches and methodologies are reviewed in sufficient detail to provide university level teachers with a useful textbook for courses in combinatorial technologies.

At the same time, the book can be useful to technologists and industrial R&D researchers as an introductory survey on various aspects of combinatorial chemistry and combinatorial technology.

We would like to thank all the authors who contributed to this book, for their expertise and highly qualified collaboration.

Giorgio Fassina **Stanislav Miertus**

Editors

Dr. Giorgio Fassina has a Ph.D in bioorganic chemistry, and started his scientific career as a visiting associate at the National Cancer Institute, National Institutes of Health (Bethesda, MD, USA). Later, he joined the biotechnology sector holding management positions and in 2001 founded XEPTAGEN S.p.A., a biotechnology company exploiting combinatorial technologies for biomarker discovery in oncology, where he is currently the managing and scientific director. A member of the boards of several consulting firms in the biotechnology sector, he also has been adjunct professor of molecular biology at the University of Bologna since 1995, and an international advisor for the United Nations Industrial Development Organization (UNIDO) for technology transfer in the field of combinatorial technologies. His research interests have been focused on ligand design and synthesis for biopharmaceutical applications and biomarkers discovery in oncology. The recipient of several awards for his scientific accomplishments, including the Federchimica Prize, Dr. Fassina is the author of more than 140 papers in peer reviewed journals and the main inventor of more than 25 international patent applications.

Professor Stanislav Miertus graduated in chemistry in 1971, received his Ph.D. in physical chemistry in 1975 and he has been full professor at the Slovak Technical University since 1989. Since 1997, he is the area director for pure and applied chemistry of the International Centre for Science and High Technology of the United Nations Industrial Development Organization (ICS-UNIDO), in Trieste, Italy, covering different fields of chemistry and technology, including catalysis and green chemistry, environmentally degradable plastics and remediation technologies, combinatorial chemistry and combinatorial technologies. Among his professional experiences he was project leader at Poly Tech Trieste, Italy; professor at Faculty of Pharmacy, Trieste; visiting scientist at the University of Pisa; visiting professor at the City University of New York; Researcher at the Institut de Chimie Biophysique, Paris, and others. His research activities have been focused on computational and physical chemistry, especially elucidation of structure and properties of bioactive compounds and biomacromolecules and computer-assisted combinatorial chemistry. He coauthored approximately 200 papers and 3 patents. He is coauthor and coeditor of more than 20 scientific books, books of proceedings and technical compendia.

Contributors

Alessandro Weisz
Dipartimento di Patologia Generale
Seconda Università
degli Studi di Napoli
Napoli, Italy

Alfredo Paio
GlaxoWellcome
Medicines Research Centre
Verona, Italy

Andrea Missio
GlaxoWellcome
Medicines Research Centre
Verona, Italy

Andrew Bradbury
International School for Advanced Studies
Trieste, Italy

Antonio Fiordelisi
XEPTAGEN S.p.A.
Pozzuoli, Italy

Aris Persidis
Argonex, Inc.
Charlottesville, Virginia

Árpád Furka
Department of Organic Chemistry
Eötvös Loránd University
Budapest, Hungary

Claudio Scafoglio
Dipartimento di Patologia Generale
Seconda Università
degli Studi di Napoli
Napoli, Italy

Concetta Ambrosino
Dipartimento di Patologia Generale
Seconda Università
degli Studi di Napoli
Napoli, Italy

Daniel O. Cicero
Department of Chemical Sciences
and Technologies
University of Rome Tor Vergata
Rome, Italy

Daniela G. Berta
Chemistry Research Department
Pharmacia Italy S.p.A.
Italy

Daniela Palomba
XEPTAGEN S.p.A.
Pozzuoli, Italy

Daniele Sblattero
International School for Advanced Studies
Trieste, Italy

Edith S. Monteagudo
IRBM, MRL Rome
Rome, Italy

Eduard R. Felder
Chemistry Research Department
Pharmacia Italy S.p.A.
Italy

Elisabetta de Magistris
GlaxoWellcome
Medicines Research Centre
Verona, Italy

Enrico Burello
Department of Biochemistry
Biophysics and Macromolecular
Chemistry
University of Trieste
Trieste, Italy

Giorgio Fassina
XEPTAGEN S.p.A.
Pozzuoli, Italy

Giovanna Palombo
TECNOGEN S.c.p.A.
Piana di Monte Verna, Italy

Hennie R. Hoogenboom
University Hospital Maastricht
Maastricht, The Netherlands

Henning Ulrich
Department of Biochemistry
Instituto de Química
Universidade de São Paulo
São Paulo, Brazil

J.M. Domínguez
Instituto Mexicano del Petróleo
Molecular Engineering Program
México city, Mexico

Larry Gold
NeXstar Pharmaceuticals
Boulder, Colorado

Luca Beneduce
XEPTAGEN S.p.A.
Pozzuoli, Italy

Lucia Altucci
Dipartimento di Patologia Generale
Seconda Università
degli Studi di Napoli
Napoli, Italy

Lucia Carrano
Biosearch Italia S.p.A.
Gerenzano, Italy

Luigi Cicatiello
Dipartimento di Patologia Generale
Seconda Università
degli Studi di Napoli
Napoli, Italy

Maria Dani
TECNOGEN S.c.p.A.
Piana di Monte Verna, Italy

Maria Marino
XEPTAGEN S.p.A.
Pozzuoli, Italy

Nikolai Sepetov
Nanoscale Combinatorial
Synthesis Inc.
Tucson, Arizona

Olga Issakova
Selectide – Aventis Pharmaceuticals
A Subsidiary of Hoechst Marion
Roussel
Tucson, Arizona

Pierfausto Seneci
GlaxoWellcome
Medicines Research Centre
Verona, Italy

Rafael Ferritto
GlaxoWellcome
Medicines Research Centre
Verona, Italy

Simon Hufton
University Hospital Maastricht
Maastricht, The Netherlands

Stanislav Miertus
Department of Biochemistry
Biophysics and Macromolecular
Chemistry
University of Trieste
Trieste, Italy

Stefano Donadio
Biosearch Italia S.p.A.
Gerenzano, Italy

Valerie J. Gillet
University of Sheffield
Sheffield, United Kingdom

Valery V. Antonenko
Affymax Research Institute
Santa Clara, California

Vladimir Frecer
International Centre for Science
and High Technology
Trieste, Italy

Table of Contents

1 Combinatorial Chemistry and Combinatorial Technologies: Principles and Applications

Giorgio Fassina and Stanislav Miertus

The time and cost needed for the development of new drugs have increased steadily during the past three decades. Estimated costs for introducing a new drug in the market now reach around $200–300 million U.S., and this process takes around 10–12 years after discovery. This increase in time and cost is mainly due to the extensive clinical studies of new chemical entities required by competent regulatory agencies, such as the U.S. Food and Drug Administration (FDA) and, to a lesser extent, to the increased costs associated with research. The time and cost required for clinical and preclinical evaluation of new drugs is not likely to decrease in the near future and, as a consequence, a key issue for pharmaceutical companies to stay in the market has been to increase the number of new drugs in the development pipeline. Drug discovery in the past has been based traditionally on the random screening of collections of chemically synthesized compounds, or extracts, derived from natural sources, such as microorganisms, bacteria, fungi, and plants, of terrestrial or marine origin, or by modifications of chemicals with known physiological activities. This approach has resulted in many important drugs, but the ratio of novel to previously discovered compounds has diminished with time. In addition, this process is very time consuming and expensive. A limiting factor was the restricted number of molecules available, or extract samples to be screened, since the success rate in obtaining useful lead candidates depends directly on the number of samples tested. Chemical synthesis of new chemical entities often is a very laborious task, and additional time is required for purification and chemical characterization. The average cost of creating a new molecular entity in a pharmaceutical company is around $7500 U.S./ compound [1]. Generation of natural extracts, while very often providing interesting new molecular structures endowed with biological properties, leads to mixtures of different compounds at different concentrations, thus making activity comparisons very difficult. In addition, once activity is found in a specific assay, the extract needs to be fractionated in order to identify the active component. Quite often, the chemical synthesis of natural compounds is extremely difficult, thus making the lead development into a new drug a very complex task. While the pharmaceutical industry was demanding more rapid and cost-effective approaches to lead discovery, the advent of new methodologies in molecular biology, biochemistry, and genetics, leading to the identification and production of an ever-increasing number of enzymes, proteins, and receptors involved in biological processes of pharmacological relevance, and good candidates for the development of screening assays, complicated this scenario even more. The introduction of combinatorial technologies provided an unlimited source of new compounds, capable of satisfying all these needs. This approach was so appealing and full of promise that many small companies started to flourish, financed by capital raised from private investors. Once combinatorial technologies clearly demonstrated the potential to identify new leads with a previously unknown speed, the majority of these companies were purchased by big pharmaceutical companies. Combinatorial approaches were originally based on the premise that the probability of finding a molecule in a random screening process is proportional to the number of molecules subjected to the screening process. In its earliest expression, the primary objective of combinatorial chemistry focused on the simultaneous generation of large numbers of molecules and on the simultaneous screening of their activity. Following this approach, the success rate of identifying new leads is greatly enhanced, while the time required is considerably reduced.

The development of new processes for the generation of collections of structurally related compounds (libraries), with the introduction of combinatorial approaches, has revitalized random screening as a paradigm for drug discovery, and has raised enormous excitement about the possibility of finding new and valuable drugs in short times and at reasonable costs. However, the advent of this new field in drug discovery did not obscure the importance of "classical" medicinal chemistry approaches, such as computer-aided rational drug design and QSAR, for example, but, instead, catalyzed their evolution to complement and integrate with combinatorial technologies.

The word "combinatorial" appeared in the scientific literature at the beginning of the 1990s, but the generation of the first combinatorial libraries can be dated back to the beginning of the 1980s. The first reports dealt with the simultaneous production of collections of chemically synthesized peptides, produced by solid-phase methods on solid supports [2–6]. Peptides were particularly suited for combinatorial synthesis, given the well-established synthetic protocols available, the great number of different molecules attainable, and the potential to generate leads of biological and pharmaceutical value. The use of peptide libraries was greatly accelerated by the introduction of biological methods for library preparation, and by the use of phage display technology, which provided interesting advantages over the synthetic counterpart [7, 8]. At the same time, the first papers on the generation of oligonucleotide libraries appeared in the literature [9, 10], thus suggesting the possibility of extending the applicability of combinatorial approaches to other classes of synthetic or natural oligomeric compounds, such as carbohydrates. There are many important biologically active glycoconjugate drugs whose carbohydrate constituents are associated with the molecular mechanism by which these drugs exhibit their effects. With these drugs, exploration of carbohydrate molecular diversity has the potential for identifying novel agents with enhanced potency. As a conformational rigid and functionally rich system, carbohydrates also provide valuable molecular scaffold systems, around which helps/facilities to generate primary screening libraries.

A broad variety of new synthesis and screening methods are currently grouped under the term "combinatorial." These methods include parallel chemical synthesis and testing of multiple individual compounds, or compound mixtures, in solution, synthesis and testing of compounds on solid supports, and biochemical or organism-based synthesis of biological oligomers coupled to selection and amplification strategies. All these different methods have expanded rapidly, each with its putative advantages, disadvantages, and proponents, and a broad coverage of all the diverse approaches used for library generation and screening is provided in the following chapters.

Many active compounds have been selected to date following combinatorial methodologies, and a considerable number of those have progressed to clinical trials. However, combinatorial chemistry (CC) and related technologies for producing and screening large numbers of molecules, also find useful applications in other industrial sectors not necessarily related to the pharmaceutical industry. Emerging fields of application of combinatorial technologies are the diagnostic, the downstream, processing, the catalysis, and the new-material sectors. In the first case, CC can be successfully applied to the identification of previously unknown epitopes, recognized by antibodies in biological fluids associated with pathological conditions. The selected epitopes can then be used for the development of diagnostic kits useful for the identification and quantification of the antibody of interest. In the downstream processing field, combinatorial chemistry finds application in the selection of ligands able to recognize specific macromolecules of biotechnological interest, such as proteins, antibodies, or nucleic acids. This is of great importance to industry, since the major costs associated with the production of recombinant molecules for therapy are associated with the purification of the desired target molecule from crude feedstocks. The availability of specific and selective ligands, such as monoclonal antibodies, to be used in affinity chromatography for the capture and concentration of the target from crude samples, will considerably reduce the costs of producing biopharmaceuticals [11]. Combinatorial technologies have also been applied to the identification of new macromolecules, endowed with catalytic activity, for reactions where natural enzymes are inactive. This application, even if still at an early stage, is drawing considerable attention from the industrial sector, since the availability of new enzymes may reduce the production costs of many chemicals.

The different technologies and strategies used in the production of combinatorial libraries are now so well developed, that it is easy to plan synthetic schemes for the generation of a huge number of compounds. Since the rate at which compounds can be screened constitutes a limitation to the use of combinatorial technologies, it is important to be selective about the compounds which are synthesized. Computational methods are very valuable from this point of view to assist in the design of combinatorial libraries. The main requirement for lead generation is often to maximize the range of structural types within the library, with the expectation that a broad range of activities will result. As a consequence, diversity analysis is an important aspect of library design. The diversity of libraries may be measured by the use of similarity or dissimilarity indexes, which make intermolecular comparisons possible. Measures of chemical similarity have been developed for similarity searching in chemical databases. The calculation of the similarity between two molecules involves characterization of the molecules by using chemical/structural descriptors, and then the application of similarity coefficients to quantify the similarity.

With the increased speed at which new drug entities are now synthesized and evaluated for pharmacological activity, a need has arisen to provide fundamental metabolism data at the early stages of drug discovery. Strategies are being developed to permit drug metabolism data to be an important part of early drug discovery. Many important properties of drugs related to metabolism could be the deciding factor in whether or not a compound is selected for clinical development. Some of these include the pharmacokinetic properties. Other related factors that can help discovery teams make decisions about which structures to pursue include measurements of metabolic stability, protein binding, P450 inhibition and absorption.

In combinatorial chemistry, due to the high number of chemical manipulations required to synthesize libraries of compounds, automation is unavoidable. Many research groups, in both academic and industrial settings, are developing automated instruments tailored specifically to these needs, and this technology field is acquiring an extremely important role for the development of combinatorial technologies for this millennium.

An important reason for the rapid development of this field is its unique patent history. There are no strategic patents in this fields, but many companies own patents that enable them to pursue unique chemistries; thus, new entrants have the peculiar opportunity to gain access to this field with no major limitations.

Combinatorial technology is a platform technology that is integrating very well with other technologies, such as functional genomics and proteomics, which are used to focus new lead generation. Combinatorial technology is evolving rapidly, and its value will increase with time because of this feature.

REFERENCES

1. Chabala JC, Pharmaceutical Manufacturers Association, *Drug Discovery Management Subsection*, Philadelphia, PA, September 19–21, 1993.
2. Furka A, Sebestyen F, Asgedom M, Dibo G, Abstr. 14th Int. Congr. *Biochem.*, Prague, Czechoslovakia, 1988, Vol. 5, p. 47.
3. Houghten RA, General method for the rapid solid-phase synthesis of large number of peptides: specificity of antigen-antibody interaction at the level of individual aminoacids, *Proc. Natl. Acad. Sci. USA*, 82:5131–5135, 1985.
4. Fassina G, Lebl M, Chaiken IM, Screening the recognition properties of peptide hormone sequence mutants by analytical high performance liquid affinity chromatography on immobilized neurophysin, *Coll. Czech. Chem. Commun.*, 53:2627–2636, 1988.
5. Lam KS, Salmon SE, Hersh EM, Hruby VJ, Kazmierski WM, Knapp RJ, A new type of synthetic peptide library for identifying Ligand binding activity, *Nature*, 354:82–84, 1991.
6. Hougthen RA, Pinilla C, Blondelle SE, Appel JR, Dooley CT, Cuervo JH, Generation and use of synthetic combinatorial libraries for basic research and drug discovery, *Nature*, 354:84–86, 1991.
7. Cwirla S, Peters EA, Barret RW, Dower WJ, Peptides on phage: a vast library of peptides for identifying ligands, *Proc. Natl. Acad. Sci. USA*, 87:6378–6382, 1990.
8. Scott JK, Smith GP, Searching for peptide ligands with an epitope library, *Science*, 249:386–390, 1990.
9. Tuerk C, Gold L, Systematic evolution of ligands by exponential enrichment: RNA ligands to bacteriophage T4 DNA polymerase, *Science*, 249:505–510, 1990.
10. Ellington AD, Szostak JW, In vitro selection of RNA molecules that bind specific ligands, *Nature*, 346:818–822, 1990.
11. Fassina G, Verdoliva A, Odierna MR, Ruvo M, Cassani G, A protein a mimetic peptide for the affinity purification of antibodies, *J. Mol. Recogn.*, 9:564–560, 1996.

2 Combinatorial Chemistry: From Split-Mix to Discrete

Árpád Furka

CONTENTS

I. INTRODUCTION

Pharmaceutical research has always been a slow and expensive process. Generally, thousands of new compounds need to be prepared to find a new drug. The conventional one-by-one synthesis of so many compounds, as well as their one-by-one testing, was very tedious and time-consuming. When one compares the production of new compounds—before the eighties—with the industrial manufacturing of different other items, for example automobiles, and the extensive application of production lines and automation, the backwardness of synthetic technologies becomes quite obvious. The advent of the new combinatorial synthetic methods introduced in the eighties dramatically changed this situation. These methods initiated a revolution: first in pharmaceutical research, and then in other fields devoted to discover other useful materials. The combinatorial methods induced a new way of thinking and radically changed our theory and practice in designing and preparing new substances for pharmaceutical research and other applications. The use of combinatorial methods expanded very fast and contributed to the foundation of a rapidly growing new scientific field, combinatorial chemistry.

The synthetic methods applied in combinatorial chemistry are used to prepare either large series of individual compounds or mixtures of large number of compounds. Both mixtures and series of compounds are termed libraries. The synthetic methods themselves can be classified into two categories:

- Real combinatorial synthetic methods
- Parallel synthetic methods.

The product of a parallel synthesis is a library containing a limited number of discrete compounds. The real combinatorial methods can produce mixtures containing a very large number of compounds, but can be used to make individual compounds, too.

In parallel synthesis, each product is usually prepared individually in a separate reaction vessel. The number of operations that have to be executed in the course of the synthetic process is practically the same as in the case of one-by one-preparation of the same compounds. The real gain in parallel synthesis is in the reaction time. Tens, or even hundreds, of compounds can be prepared within the time range devoted for making a single substance. As later will be shown, the real combinatorial synthesis is so organized to allow production of a very large number of compounds in a limited number of reaction vessels, and in very short time.

The combinatorial synthetic methods work according to the combinatorial principle, which means that they make possible to prepare compound libraries comprising all molecular structures that can be theoretically deduced from the monomers used in the synthesis as building blocks. These methods, at the same time, are very efficient.

One of these methods is the split-mix procedure, originally called "portioning-mixing" method, invented by Furka et al. [1–3]. This method was originally developed to enable the user to prepare millions of new peptides, but, later, was successfully used to synthesize organic libraries, too. The method is an embodiment of the combinatorial principle and the "combinatorial thinking", which constitutes the theoretical background of the method and proved to be so fruitful in other areas, too.

The biological methods of preparation of peptide libraries were introduced by three different groups in 1990 [4–6]. These methods, which are also based on combinatorial principles, can be used to prepare even billions of peptide sequences. These sequences appear attached to the end of protein chains. Only the natural L-amino acids can be used as monomers in composition of the libraries. Very effective screening methods are also developed to determine potential bioactive sequences.

Another combinatorial method, the light-directed, spatially addressable parallel chemical synthesis, was developed by Fodor and his colleagues in 1991 [7]. Arrays of about 1000 individual peptides were prepared on the surface of glass slides. The key technology applied in the synthesis was photolithography, which is regularly used in making computer chips. Although the synthesized array may be composed of any collection of individual sequences, the method is most effective when the library is constructed according to the combinatorial principles.

II. THE SPLIT-MIX SYNTHESIS

A combinatorial oligomer library, containing all sequences that can be deduced from 3 different (white, gray and black) monomers, could be synthesized by conventional methods, following the branches of the combinatorial tree demonstrated in Figure 2.1. The sequences of the library members can be read by going from the origin along the branches (e.g. white–white–white, black–black–black on the left, and right sides, respectively). Looking at the combinatorial tree, one can deduce a very important rule, the combinatorial distribution rule:

Each product formed in a given reaction step of a combinatorial process has to be distributed into samples, then react each sample with one of the monomers of the next reaction step.

The split-mix method realizes the combinatorial distribution rule by mixing the products after each reaction step, then distributing the mixture into equal portions.

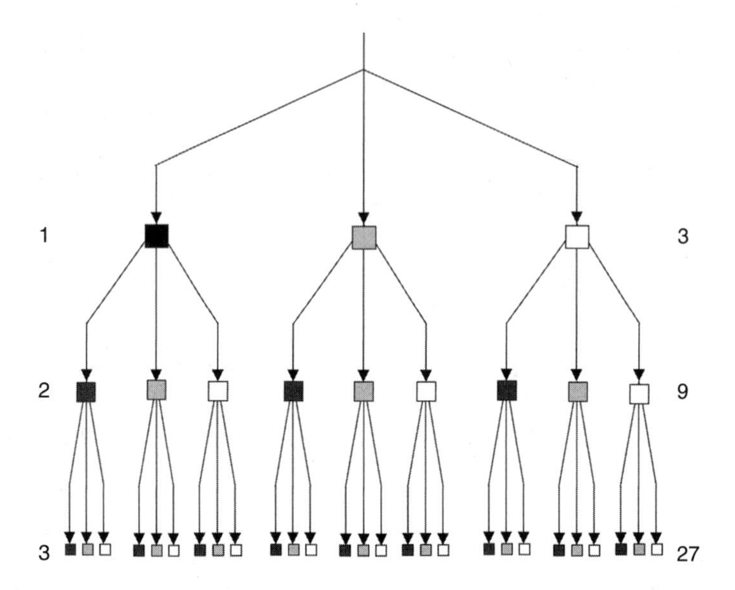

FIGURE 2.1 The combinatorial tree. Each square in the figure represents a reaction vessel, and their black, gray and white colors symbolize the monomers that are coupled in the vessels. The numbers on the left side indicate the number of the reaction step and the order of branches. The numbers on the right side show three things: number of reaction vessels used in the reaction step, the number of executed reaction cycles and the number of products.

The split-mix method was developed for preparing peptide libraries. The method is based on Merrifield's solid phase procedure, published in 1963 [8]. Each coupling cycle of the solid phase synthesis is replaced by the following simple operations:

Dividing the solid support into equal portions;
Coupling each portion individually with only one of the amino acids;
Mixing the portions.

The method is exemplified by the synthesis of a peptide library on solid support, using only three different amino acids. The scheme of the procedure is shown in Figure 2.2. The amino acids are represented by white, gray and black circles. In the first coupling cycle, the amino acids are coupled to equal portions of the resin. The product—after recombining and mixing the portions—is the mixture of the three amino acids bound to resin. In the second cycle, this mixture is again divided into three equal portions and the amino acids are individually coupled to these mixtures. In each coupling step, three different resin-bound dipeptides are formed. As a consequence, the end product—after mixing—is a mixture of nine dipeptides. The divergent, vertical and convergent arrows indicate portioning, coupling (with one kind of amino acid) and combining-mixing,

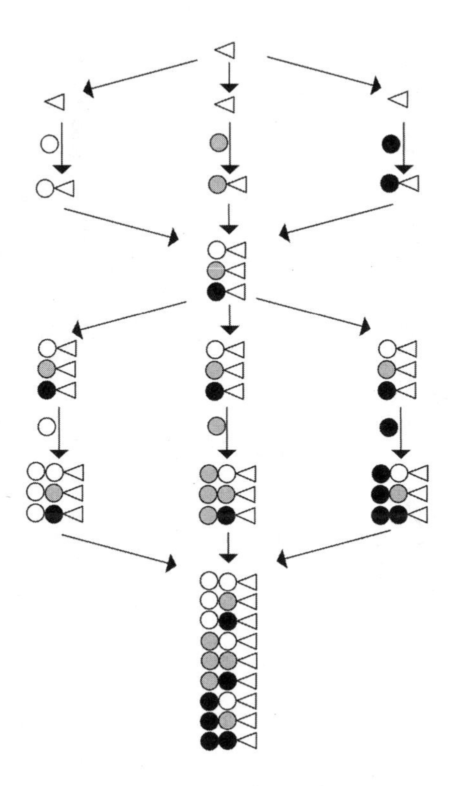

FIGURE 2.2 Scheme of the split-mix synthesis. The triangles represent the solid support. The white, gray and black circles are amino acids or other kinds of monomers.

respectively. If the synthesis is continued, after executing a further portioning, coupling and mixing step leads to the formation of a mixture of 27 resin-bound tripeptides (Figure 2.3A). And, if an additional coupling step is carried out, the end product comprises 81 tetramers (Figure 2.3B,C, and D).

A. Features of the Synthesis

The split-mix synthesis has several key features that are crucial in the utility of the method in drug discovery or other kinds of applications.

1. Formation of All Possible Sequences

By looking at the di-, tri- and tetrapeptide libraries in Figures 2.2 and 2.3, and examining their sequences, it becomes quite clear that these libraries contain all sequences that can be deduced from the three (white, gray and black) amino acids. No more sequence combinations of the white, gray and black circles can be deduced than those that are seen in the figures. That means: the consecutive execution of the three simple operations (portioning, coupling and mixing) ensures—with mathematical accuracy—the formation of all possible

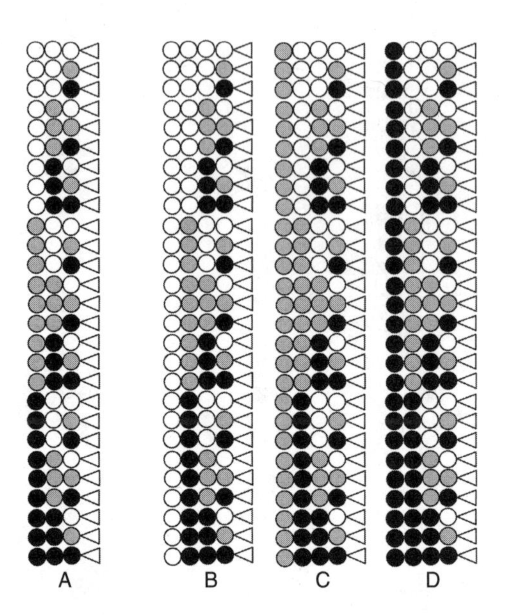

FIGURE 2.3 Trimer (A) and tetramer (B, C, D) libraries constructed from three different monomers. Triangles: solid support; white, gray and black circles: monomers.

sequence combinations of amino acids, or other kinds of building blocks, used in the synthesis. No extra care has to be taken to ensure this.

The combinatorial principle embodied in the synthesis captured the imagination of many researchers all over the world and had a profound effect on the development of the field. The combinatorial nature of the product of the split-mix synthesis is also reflected in its name, "combinatorial library."

2. Efficiency of the Synthesis

Figures 2.2 and 2.3 also show that, starting with a single substance, the resin, used as solid support, the number of compounds is tripled after each coupling step: first $3 \times 1 = 3$ ($= 3^1$) resin bound amino acids, then $3 \times 3 = 9$ ($= 3^2$) resin bound dipeptides, then $3 \times 9 = 27$ ($= 3^3$) resin bound tripeptides and, finally, $3 \times 27 = 81$ ($= 3^4$) tetrapeptides are formed. This means that the number of products increases exponentially as the synthesis proceeds. If, instead of three monomers, 20 different amino acids are used in the synthesis, the number of peptides in each coupling step is increased by a factor of 20. The total number of the synthesized peptides can be expressed by a simple formula 20^n, where n is the number of amino acid residues in the peptides.

If all operations are executed manually in the synthesis, one coupling step with each of the 20 amino acids can be easily realized every day. As a consequence, in 2, 3, 4, 5 or 6 days, 400, 8,000, 160,000, 3,200,000 or 64 million peptides, respectively, can be made. Or, to put it in another way, a chemist, working with the method, can produce, in a single week, more compounds than were made in the whole previous history of chemistry. Such efficiency had never been dreamed of before the introduction of the method, and this justifies the explosion-like development of the field of combinatorial chemistry.

3. Formation of Compounds in One-to-One Molar Ratio

In the area of pharmaceutical research, libraries are prepared in order to find biologically active substances among the new products. In the identification process, or screening, the goal is to find the biologically most effective component. Serious problems may arise in screening mixtures of compounds in which the components are not present in equal quantities. A low activity component, for example, if it is present in a large amount, may show a stronger effect than a highly active component present in lower quantity. It is important, therefore, to prepare libraries in which the constituents are present in equal molar quantities. The split-mix method was designed to comply with this requirement. Before each round of couplings, the resin is thoroughly mixed, then is divided into equal portions. Homogenization by mixing ensures that the previously formed peptides, or other compounds, are present in practically equimolar quantities in each portion. Since couplings with the different monomers are executed on spatially separated samples, it is possible to use appropriate chemistry to drive each coupling reaction to completion, regardless of the reactivity of the amino acids or other kinds of monomers. Since each previously formed peptide is quantitatively transformed into an elongated new one, in principle, both the numbers of peptides originally present and their equimolar ratio is preserved in every reaction vessel in each step. If the final products are cleaved from the support, a mixture is formed. For reasons outlined above, there is good reason to expect that the components of the mixture are present in equal molar quantities.

There are several reasons that may cause deviations from equimolarity of the compounds. One of these reasons is, for example, if some reactions remain uncompleted, even if the reagents are used in large excess and the couplings are repeated.

4. Individual Compounds on Beads

The split-mix procedure has another intrinsic feature that plays an important role in applications, and gives a unique character to the method: on any individual bead of the solid support, only one kind of peptide or other compound is formed. This may seem surprising at first glance, but becomes quite understandable upon closer examination. In Figure 2.4, the fate of a randomly selected bead is followed in a three-step coupling process. The bead in every coupling step meets only a single amino acid. This is the amino acid, which is used for coupling in the particular reaction vessel, to where the given bead was randomly transferred. Thus, only this single amino acid is coupled to all of its free sites. This amino acid in the first, second and third coupling

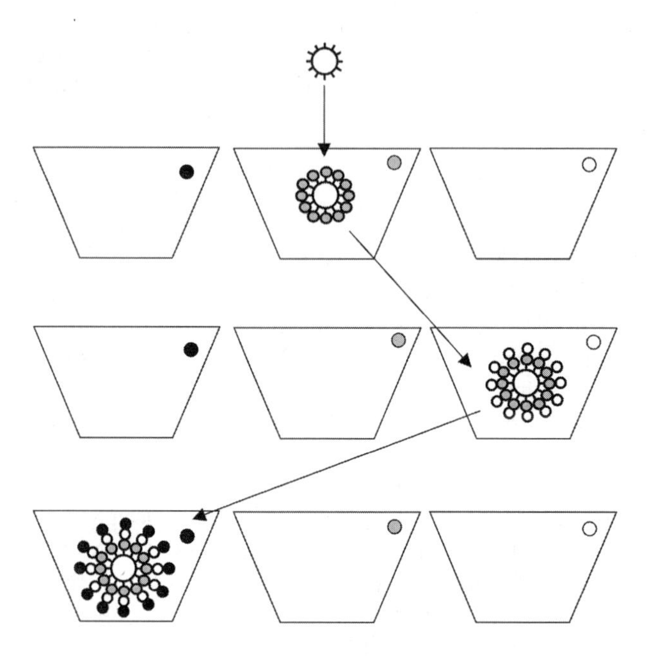

FIGURE 2.4 Formation of a single compound on each bead. Large circle: resin bead; smaller black, gray and white circles: monomers (e.g. amino acids).

step, is the gray, white and black one, respectively, so the bead ends up with the gray–white–black sequence (read in order of coupling).

In the split-mix synthesis, like in other solid phase procedures, the beads behave very much like tiny reaction vessels, which do not interchange their contents with the other ones. Each of the millions of these reaction vessels preserves its content until the end of the synthesis, when they become containers of a single substance. If peptides are produced, their identity can be determined by automatic sequencing [9]. It is sufficient to sacrifice a fraction of the total quantity for this purpose. All this means that the split-mix synthesis is, in fact, a parallel procedure, with unprecedented efficiency, however, leading to individual compounds. This feature of the split and mix synthesis allows screening the products in three different ways:

Doing binding experiments with the individual compounds uncleaved from the beads

Cleaving the product from a single bead and test it as an individual compound

Pooling the beads before cleavage, then carrying out screening with a solution of a mixture.

In the last case, a special deconvolution procedure has to be followed, in order to identify the compound responsible for the biological activity, or other useful property.

5. Applicability of the Split-Mix Procedure for the Synthesis of "Organic" Libraries

Peptides are not the only potential drug candidates. In most cases, other kinds of small organic molecules are preferred, because of their reduced susceptibility to enzymatic degradation. The split-mix method is fully applicable in the synthesis of organic libraries. Both sequential type and cyclic libraries can easily be prepared if the reaction conditions for solid phase are well developed. It has to be emphasized, however, that the advantages of the split-mix method can be fully exploited only in the case of multi-step synthetic procedures. For realization of the one-pot procedures suggested by Ugi [10], for example, the parallel procedures are better-suited.

The synthesized libraries are often screened as individual substances released from beads. Determination of the structure of the various organic compounds is not as simple as sequencing peptides. In order to circumvent this difficulty, in several laboratories encoding of the beads, is introduced. The molecular units of the encoding tags are attached to the beads in parallel with the coupling of the organic building blocks of the library. The attached encoding tags record the synthetic history of the beads. Thus, by determining the code of a bead, the structure of the expected product can be deduced.

Different encoding types were described in the literature like encoding with sequences Nielson et al. 1993 [11], Nikolaiev et al. 1993 [12], Kerr et al. 1993 [13], binary encoding Ohlmeyer et al. 1993 [14] and encoding with fluorescent colloids Battersby et al. 2000 [15]. When encoding by sequences, the encoding tags are either peptides or oligonucleotides. Their sequences encode both the identities of organic reagents coupled to the bead and the order of their coupling. In the binary encoding system the, coding units are halobenzenes carrying a varying length hydrocarbon chain attached to the beads through a cleavable spacer. It is characteristic for this labeling technique that the coding units do not form a sequence. It is simply their presence which codes for the organic building blocks and their position.

6. The Liquid Phase Split-Mix Synthesis

In 1995 a liquid phase variant of the split-mix method was described [16]. The synthesis is carried out on polyethylene glycol monomethyl ether (MeO-PEG) support. This polymer is soluble in the course of the reaction, and the homogeneous phase is advantageous for coupling, but can be precipitated, allowing washing out the excess of the reagents. This method has an additional advantage. The molar ratio of the products in not affected by statistics. The method has, however, a disadvantage, too. Only mixtures can be made. Since there are no beads, the one-bead-one-compound feature of the original method is completely lost.

B. Deconvolution Methods

Libraries prepared by application of real combinatorial synthetic methods are usually submitted to screening experiments, either as soluble mixtures or as unknown discrete compounds cleaved from, or tethered to individual beads of the solid support. The task in deconvolution is to identify the substance that has a desired property. The deconvolution methods can be classified into two groups: deconvolution of mixtures, cleaved from support and deconvolution of tethered libraries.

1. Mixtures Cleaved from the Solid Support

As already mentioned, components of libraries produced by the original split-mix method are discrete compounds. If the libraries are cleaved from the solid support, however, mixtures are formed. The products of the liquid phase synthesis are always mixtures. At the beginnings, finding an active component in a mixture of thousands, or millions of structurally related compounds seemed to be a task like finding a needle in a haystack. Later on, however, reliable methods have been developed to solve this problem. All these methods are based on preparation and screening of properly designed partial libraries.

a. The Iteration Method

The principle of the iteration method was first described by Furka in an unpublished, notarized document [17]. Experimental realization was first demonstrated in deconvolution of a multi-component peptide library prepared by Geysen and his colleagues [18]. Application to libraries prepared by the split-mix method was published in 1991 [19].

The principle of the method is outlined in Figure 2.5, demonstrating a simple example for determination of a bioactive sequence in three stages. First,

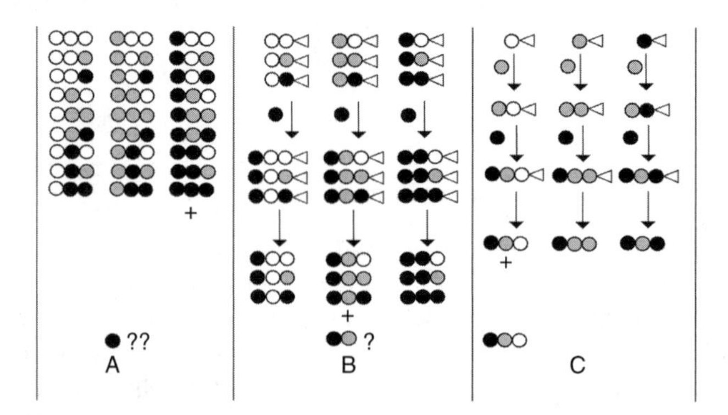

FIGURE 2.5 Deconvolution by iteration. The triangles represent the solid support, and the white, gray and black circles are amino acids. A: coupling position 3; B: coupling position 2; C: coupling position 1.

a tripeptide library is synthesized by applying the same three (white, gray and black) amino acids in all three coupling cycles. Before mixing the resin portions in the first and second coupling cycles, a small sample is taken out from all portions for later use.

In the first stage of the iteration process, the amino acid occupying the N-terminal position (coupling position 3) is determined. After the last coupling cycle of the library synthesis, the samples are not mixed. They are cleaved and screened separately (A). The components of the three samples are differing only in the amino acid occupying coupling position 3 (left) and this makes possible its identification. If, after cleavage, the sample marked by + shows activity in screening, it means that the position 3 (N-terminal) amino acid is the black one.

In the second stage of the process, the first determined black amino acid is coupled to each of the three samples taken before mixing in the dipeptide stage of the synthesis. The products of couplings differ only in the amino acid occupying coupling position 2 (B). The three products are cleaved from the resin, then submitted separately to screening, to identify the amino acid occupying coupling position 2 in the active peptide. If the sample marked by + carries the activity, this assigns the gray amino acid to the second position of the active tripeptide.

The third stage of iteration goes back to the samples taken after the first coupling cycle of the synthesis (C). To each of the unmixed samples, first the gray, then the black, amino acid is coupled, since it is already known that these two amino acids occupy the second and third positions, respectively, in the active peptide. Each of the products contains a single tripeptide. Determination of the activity of the samples after cleavage identifies the coupling position 1 (C-terminal) amino acid (white). The final sequence, starting from the N-terminus, is: black–gray–white.

Many successful experimental results show that a previously unthinkable task can be accomplished by iterative synthesis and screening of partial libraries. Even the need of iterative synthesis and screening can be eliminated by properly designed and pre-prepared sets of partial libraries (see below).

b. Positional Scanning

The sets of pre-prepared partial peptide libraries and their applicability in screening were introduced by two groups in 1993: Furka and his colleagues [20, 21] and Pinilla and coworkers [22]. In all peptides present in a special partial library, named first order sub-library [23], used in positional scanning, one position is occupied by the same amino acid, while, in all other positions, any other amino acid may occur. Such library can be prepared by omitting the portioning operation in one selected coupling position and using a single amino acid in coupling. In all other coupling positions, normal split-mix steps are executed.

Figure 2.6 shows all possible first order sub-libraries (B–J) that can be deduced from a full library (A). In B, C, D, for example, the first coupling

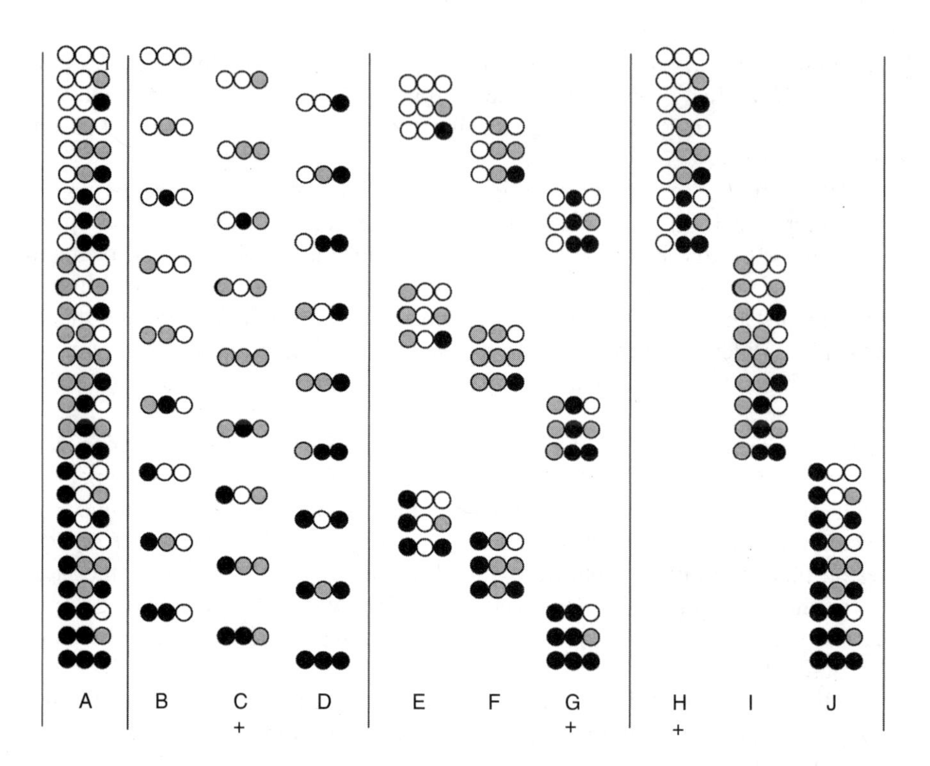

FIGURE 2.6 Positional scanning. A: full library; B–J: a full set of first order sub-libraries.

position (C-terminal, first from right) is occupied by the white, gray and black amino acid, respectively. In E, F, G the second coupling positions while in H, I and J the N-terminal positions are occupied by the white, gray and black amino acids, respectively. If the C-terminal position in the bioactive peptide happens to be also occupied by the gray residue, then the C sub-library is expected to show activity in screening. Since this sub-library comprises all possible sequences with gray C-terminal, the bioactive peptide must be present in the sub-library. If the bioactive peptide has black or white C-terminal, the C sub-library has to be inactive, since the bioactive peptide is not present in it.

In order to determine the sequence of the bioactive tripeptide, all the nine sub-libraries need to be tested. If those marked with + are the active ones in screening (H, G and C), then the sequence of the bioactive peptide (from N to C terminal) is white–black–gray. In deconvolution of a tripeptide library prepared from 20 amino acids, for example, 60 sub-libraries have to be synthesized and screened. For screening of hexapeptides, a set of 120 sub-libraries is needed. Once the set of sub-libraries is prepared in sufficient quantity, it can be used in many different screening experiments.

c. Determination of Amino Acid Composition of Bioactive Peptides. Omission Libraries

Omission libraries, developed by Câmpian and his colleagues in 1998 [24], make it possible to determine the amino acid composition of bioactive peptides and, as a consequence, to reduce the time and cost of deconvolution. Omission libraries can be prepared by omitting, in the library synthesis, one amino acid in all coupling positions. Composition of omission libraries (B, C, D), and their relation to the full library (A), is demonstrated in Figure 2.7. It can be seen that, from C, for example, all gray containing peptides are missing. If the gray amino acid happens to be present in the active component of the full library, the – gray (C) omission library is expected to be inactive in screening, since the active peptide is missing from the mixture. If an omission library proves to be active in screening, it means that the omitted amino acid is not present in the active peptide, or it is not essential for the activity. This feature makes it possible to determine the amino acid composition of bioactive peptides, by screening a full set of omission libraries. The number of omission libraries in a full set is the same as the number of amino acids used in the

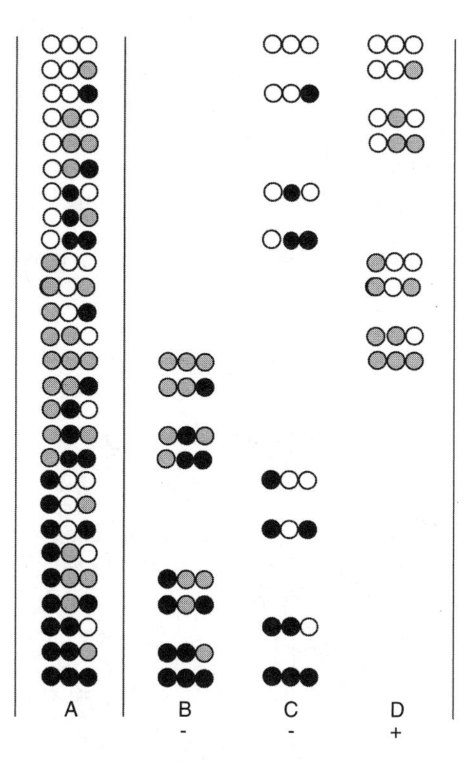

FIGURE 2.7 Omission libraries. A: full library; B, C and D: "white," "gray" and "black" omission libraries.

synthesis of the full library. If the B and C omission libraries are inactive, and only D is active, it means that the active tripeptide is composed from the white and gray amino acids. One of these amino acids, of course, occupies two positions in the sequence.

The advantages of using omission libraries are illustrated by a simple example. If the full tripeptide library is composed from 20 amino acids, it contains 8000 peptides. Only 20 omission libraries have to be prepared (or be bought) and tested (instead of the 60 sub-libraries needed in positional scanning) to determine the amino acid composition. By varying the three (or less) amino acids in all three positions, a library can be created containing only 27 peptides and still comprising the bioactive sequence. These peptides can even be prepared by parallel synthesis to identify the active one, or a positional scanning can be carried out with the nine first-order sub-libraries of this very simple library.

2. Split-Mix Libraries Screened as Individual Compounds

a. *Tethered Libraries*

Tethered libraries prepared by the split-mix method have an enormous advantage: they contain individual compounds, enclosed in beads as containers. Several deconvolution methods take advantage of this fact. The first such method was published in 1991 [9]. In this procedure, the screening test was carried out with peptide libraries tethered to the support. The beads were mixed with the solution of the target protein carrying a color label. The beads binding the target were distinguished by color (Figure 2.8). The colored beads were picked out manually and then sequenced after removal of the attached protein.

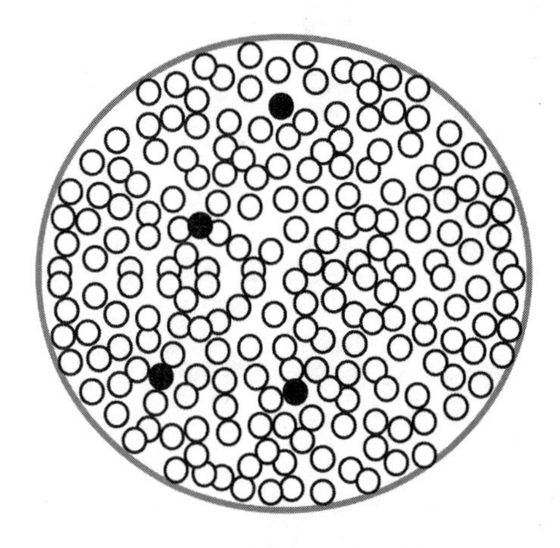

FIGURE 2.8 Binding test with tethered peptide library.

b. Compounds Cleaved from Individual Beads

A different approach in screening of the split-mix libraries is to cleave the compound from individual beads and use solutions of the individual compounds in the screening experiments. One of these procedures, developed by Ohlmeyer and coworkers [14], is used for deconvolution of libraries of small organic compounds. The libraries were prepared by applying the binary encoding technique and using a photolabile linker, which allows a two-stage release of the organic substance. After portions of beads were distributed into small vessels (Figure 2.9), the first portions of the substances were released by irradiation. The content of each vessel was then submitted to screening. If one of them proved to be active (marked by an arrow in the figure), the beads were re-distributed into vessels, each containing a single bead. After releasing the second portion of the substances, a second screening identified the bead that carried the active substance (marked by an arrow). Finally, the encoding molecules were released from the "active" bead and determined by electron capture gas chromatography, thus identifying the structure of the organic molecule responsible for the biological activity.

III. SYNTHESIS OF DISCRETE COMPOUNDS

The parallel synthetic methods, while being very slow and expensive when compared to the split-mix procedure, have two advantages:

- The products of the parallel synthesis are known individual compounds and no special deconvolution process is needed for their identification
- The products can be prepared in relatively large—multiple milligram—quantities.

There were successful efforts to modify the split-mix method, in order to eliminate the disadvantages while preserving the high efficiency.

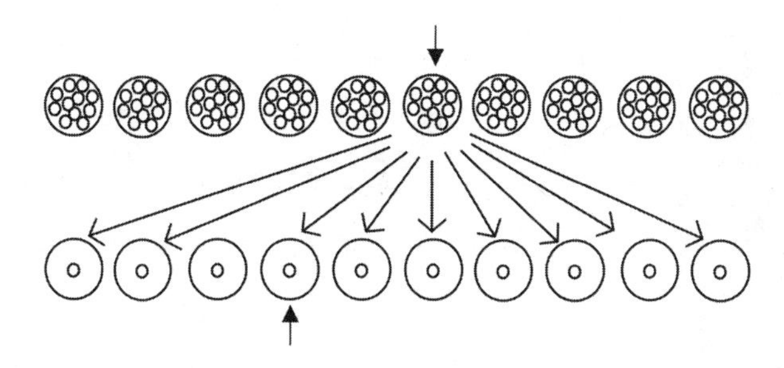

FIGURE 2.9 Two-stage identification of beads carrying active organic molecule. The arrow shows the vessel containing the active substance.

A. Split-Mix Synthesis Using Resin Enclosed in Radiofrequency-Encoded Capsules

This synthesis, suggested by two groups [25, 26], follows the split-mix pattern. The solid support units are permeable capsules, containing resin. Between two synthetic steps the units are pooled, then are sorted, one-by-one, according to the combinatorial principle. To enable proper sorting, capsules need to be encoded. In the radiofrequency encoding method suggested by the two groups, an electronic chip, a small bar, is enclosed, besides the resin, into the permeable capsules (Figure 2.10A). The method has been developed into a commercial product at IRORI. The key operation in the synthetic process is sorting which has been automated. Figure 2.10/B–G demonstrates the components of the automatic sorter. The capsules are pooled into a vibratory bowl (B), then they pass through a solenoid gate (C), where their code is read by an antenna (D) and, based on the code, the computer (F) determines into which vessel (G) the capsule must be delivered for the next synthetic step. The delivery is executed by the X–Y movement of the delivery mechanism. After sorting, the capsules collected in each vessel are reacted with a different monomer. The automatic sorter can sort 10,000 capsules in 10 hours. IRORI developed a manual version of sorting, too. There are other commercial devices that also apply radiofrequency encoding.

B. The String Synthesis

1. Principle

It has been shown by Furka and his colleagues [27] that, in split synthesis, encoding becomes unnecessary if the spatial relation to each other of the

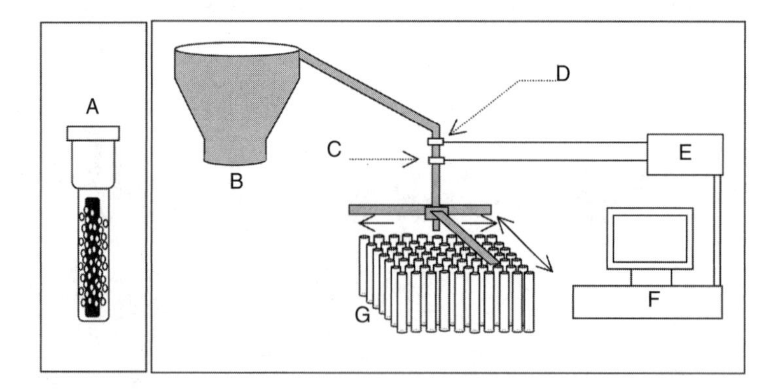

FIGURE 2.10 The IRORI automatic sorter. A: permeable capsule containing resin and a transpoder; B: vibratory bowl; C: solenoid gate; D: antenna; E: control unit; F: computer; G: vessel.

macroscopic solid support units can be followed in every phase of the synthetic process. This can be accomplished if:

The unlabeled support units are arranged into spatially-ordered groups.
The relative spatial arrangement of the units can be maintained during the chemical reactions.
The support units are redistributed between the reaction steps according to a predetermined pattern.
The synthetic history of each support unit is traced by a computer.

In this case, the products formed in the support units can be identified by computer prediction, based on the relative positions of the units occupied in the final groups. There are different possibilities for practical realization of the principles. In the string synthesis described below, the relative spatial arrangement of the units is maintained by stringing them.

2. Support Units and Strings

The support units used in the string synthesis were Chiron Mimotopes crowns, shown in Figure 2.11. They were used together with different color stems, also purchased from Chiron Mimotopes. Spatially ordered groups are formed from the crowns by stringing them together with a thread. The stems serve to string the crowns together and to facilitate redistribution. The stems were modified as shown in Figure 2.11. First, a hole was drilled to allow the string to be passed, and then the stem was carved to keep the holes parallel and facilitate threading while in the sorting device. The modified stems can be used repeatedly. SynPhase Lanterns, available at Mimotopes Pty. Ltd. (http://www.mimotopes.com), may also be used in place of the crowns.

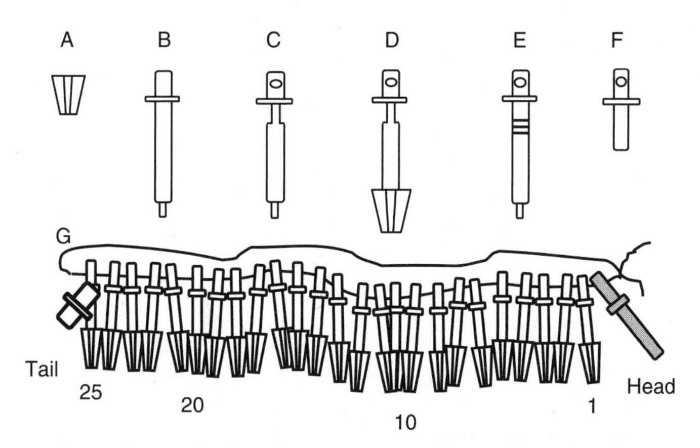

FIGURE 2.11 Crown and string. A: crown; B: stem; C: carved stem with hole; D: support unit; E: full-length stem labeling the head of the string. The scratches represent the string number; F: half stem that marks the tail of the string; G: Stringed crowns. Positions of the crowns are numbered from the head.

The string itself must be resistant to solvents and other reaction conditions occurring in the synthesis. In our example, a polyethylene fishing line was used. The stringed crowns are shown in Figure 2.11G. In order to be able to unequivocally define the position of the crowns, the two ends of the string must be distinguishable. The head is marked by an empty full-length stem (Figure 2.11E) and the tail is labeled by a stem cut in half (Figure 2.11F). The numbering of the positions of the crowns start at the head. The number of strings used in a reaction step depend on the number of building blocks used in that reaction step. A separate string is assigned to each block. As a consequence, the strings themselves must be numbered, or otherwise labeled. The simplest way to label the strings is to make visible scratches on the stem marking the head. Using colored stems is also a possibility. If lanterns are used as support units, it is worthwhile to take into account that they have a hole in their center, so they can be threaded without the need for stems.

The chemical reactions were carried out on crowns by coupling amino acids onto the respective strings. The number of crowns used in the synthesis is equal to the number of the expected products. If five amino acids are used in a coupling step, five different strings are assembled, each containing the same number of crowns. After threading the crowns, each string is placed into a reaction vessel and coupled with a different amino acid. All couplings were carried out in the five reaction vessels (Figure 2.12).

3. Manual Device for Redistribution

The strings coming from the reaction vessels after completed couplings are named source strings. Their crowns are then rearranged into the strings of the next reaction step, denoted as destination strings. Redistributions can be carried out using a very simple device that can be easily made by a machine shop. The device contains two identical pieces shown in Figure 2.13. Both pieces are metal plates with several numbered parallel slots and bent at the two edges. Before sorting, the crowns hang in the slots of the source tray, as shown in Figure 2.14. In the sorting process the crowns are pushed into the slots of the destination tray (Figure 2.15). It is important to place each string into the slot carrying the same number. Also, position the heads and tails of the source strings into the slots of the source tray as indicated in the figure, otherwise the software (see later) cannot be used. The destination strings should also be

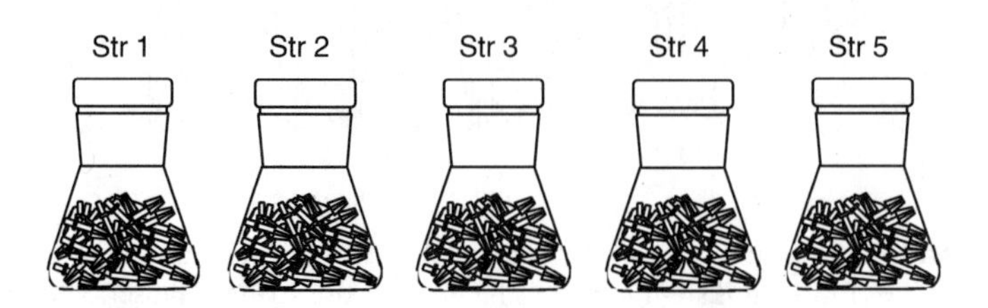

FIGURE 2.12 Strings in flasks.

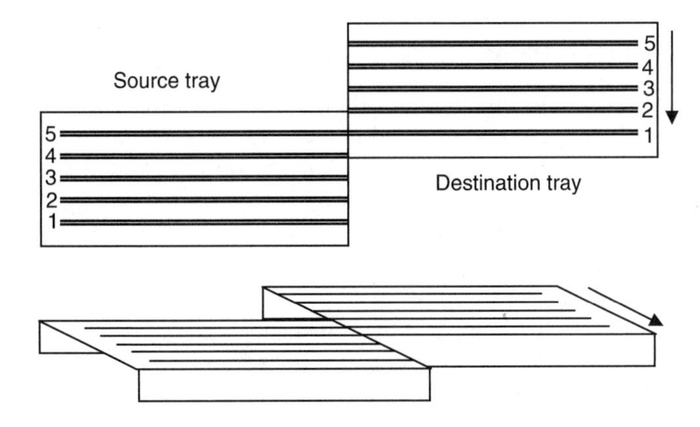

FIGURE 2.13 Sorting device. Top and side view.

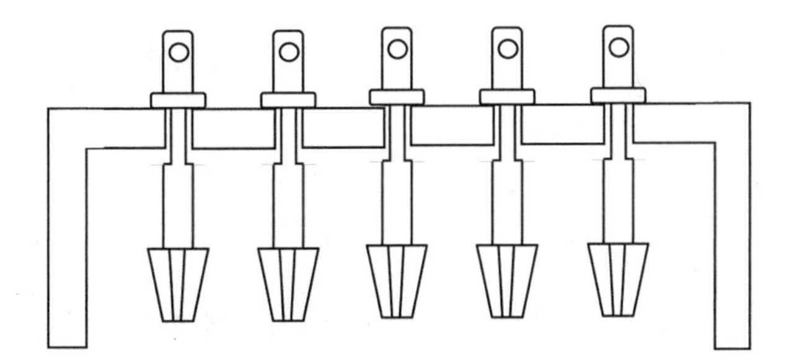

FIGURE 2.14 Crowns in the slots of the sorting device.

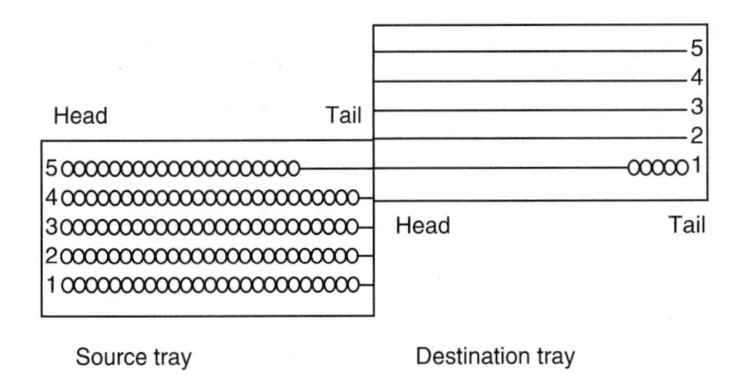

FIGURE 2.15 Sorting operation. Pushing crowns from source tray into destination one.

numbered according to the numbers of the destination slots, and render their heads and tails to the heads and tails of the destination slots. The crowns are loaded into the slots while still attached to the string, and then the string is cut and removed. The crowns are then sorted in string-free form, and then each group of crowns occupying a destination slot are restrung.

4. Redistribution Pattern

The series of redistributions made in the course of the procedure must ensure formation of all components expected in a combinatorial synthesis. This can be realized by following the combinatorial distribution rule: the crowns of a string containing the same product have to be evenly distributed among the strings of the next reaction step. Obeying this rule, there are still many possibilities to carry out the redistributions [27]. The semi-parallel sorting described below is designed to ensure fast redistribution.

The first column in Figure 2.16 demonstrates redistribution of 125 crowns in a single sorting cycle. The crowns are delivered from each source slot in

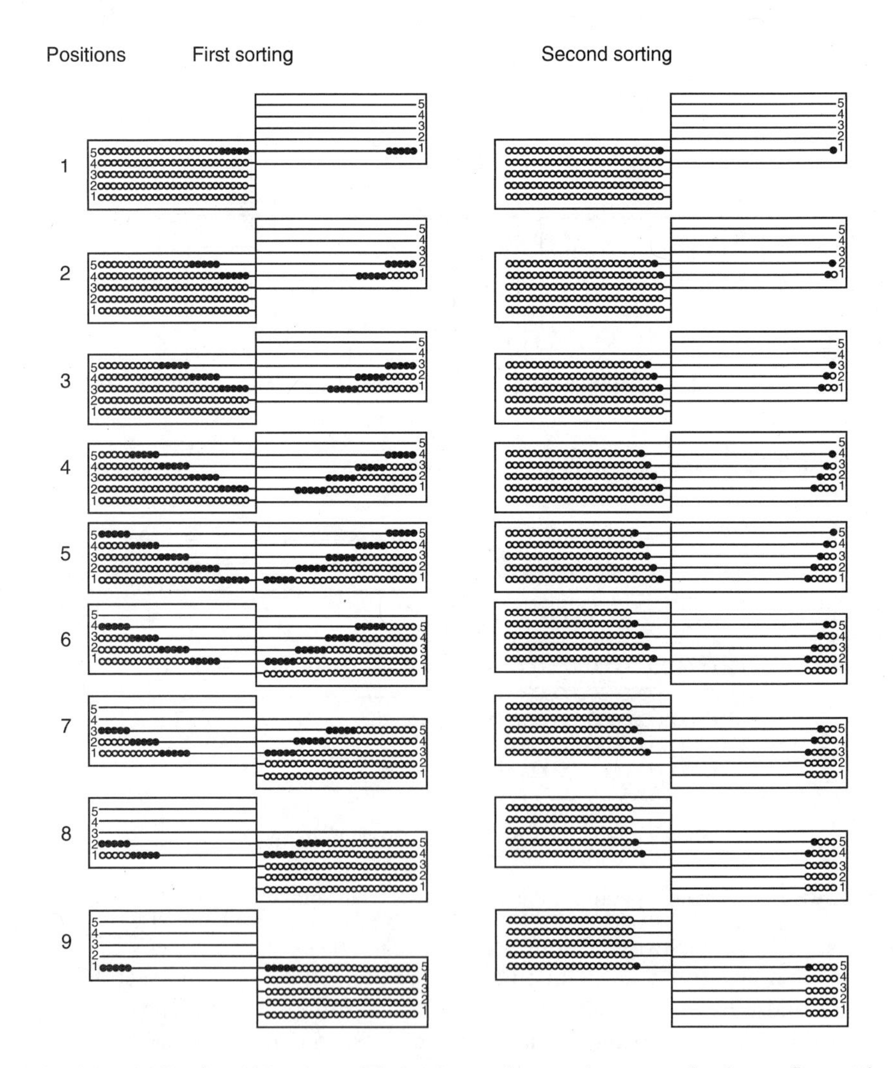

FIGURE 2.16 Sorting 125 crowns. First column: first sorting; second column: first cycle of the second sorting.

groups of five. Position 1 is the starting arrangement of the source and destination trays. The No. 5 source slot is in alignment with No. 1 destination slot. From this position, 5 crowns are pushed into the No. 1 destination slot, then the destination tray is repositioned. In position 2, No. 4 and No. 5 source slots are facing No. 1 and No. 2 destination slots, respectively. This makes it possible to push 5 crowns into both No.1 and No. 2 destination slots. The destination tray is moved step by step into positions 3 through 9, while in each position 5 crowns are pushed over into all destination slots that are in alignment with the slots of the source trays. Thus, in positions 3, 4, 5, 6, 7, 8 and 9, altogether 15, 20, 25, 20, 15, 10 and 5 crowns are pushed into destination slots, respectively. The nine positions represent a full redistribution cycle, since there are no more possible relative positions of the source and destination slots. Since the deliveries of crowns mostly occur in groups, the redistribution is fast.

5. Software

The software is written in visual basic and the data appear in excel sheets. The software can be downloaded *via* Internet from the following address:

http://szerves.chem.elte.hu/furka

by clicking on the title Excel Book appearing on the lower part of the main page. The software is available for those who have Excel installed in their computer.

The software can handle up to 1000 crowns, up to 20 monomers and up to 9 reaction steps. Figure 2.17 shows the datasheet of the Excel book where the starting data have to be entered. The starting data are: the number and symbols of the building blocks (monomers) used in the coupling steps. The symbols are single-letter abbreviations. In the case of peptide synthesis, the symbols correspond to the respective amino acids. The areas where data can be entered are yellow on the screen. Row 26 shows the string numbers. All monomers that are entered into a column are assigned to successively undergo coupling with the string appearing in the same column. Several data are instantly calculated and appear in the blue regions of the screen. Among these data are the total number of crowns needed in the synthesis and the number of coupling steps (column B). The number of source and destination slots used in the first and subsequent sorting steps (D and E), and the number of crowns occupying these slots (F and G), also appear, along with the number of crowns that contain the same product (H). The number of crowns in a group that has to be moved in every sorting cycle from a source to a destination slot can be seen in column I.

The program can be started by pushing Ctrl S. The result of calculations appears in sheets Sort #1 through Sort #9. The sheets show a block of products present in the crowns of the source slots and, below these, a block of products sorted into the destination slots. Position of the crowns are counted downward from the top. The number of sheets showing the results of couplings and sortings is equal to the number of sortings plus one. The last sheet contains the predicted product distribution on the final strings.

Starting data for semi-parallel sorting
Sorting: in every cycle delivery starts from the highest number (rightmost) source slot into the first (leftmost) destination slot
Run: Ctrl + S
The number of reagents and the symbols of monomers (A, B, C etc.) have to be entered
The number attached to the sequences only show the original position of the units
The number of units in a block to be delivered is indicated by red numbers
Do not delete blue cells! Enter data only into yellow cells!

Number of building blocks (maximum number is 52)		Sort number	Number of slots		Crowns in tubes		Identical crowns	Crowns to move
			Source	Destin.	Source	Destin.		
CP 1	5	1	5	5	25	25	25	5
CP 2	5	2	5	5	25	25	5	1
CP 3	5	3	0	0	0	0	1	0
CP 4		4	0	0	0	0	0	0
CP 5		5	0	0	0	0	0	0
CP 6		6	0	0	0	0	0	0
CP 7		7	0	0	0	0	0	0
CP 8		8	0	0	0	0	0	0
CP 9		9	0	0	0	0	0	0
CP 10								

Total number of crowns	125
Number of coupling steps	3

Maximum number of reagents 20
Maximum number of sorter slots 20

Maximum number of crowns 1,000

	Monomers in coupling							
	1	2	3	4	5	6	7	8
CP 1	I	F	L	V	G			
CP 2	E	F	W	Y	S			
CP 3	E	F	W	Y	S			
CP 4								
CP 5								
CP 6								
CP 7								
CP 8								
CP 9								
CP 10								

FIGURE 2.17 Input sheet.

6. Synthesis of a Library of 125 Tripeptides

The synthesis was carried out using 125 Chiron Mimotopes crowns, derivatized with Fmoc-Rink amide linker. The procedure was started with the formation of five strings by threading 25 crown units on Berkley Fire Line fishing line. Five Fmoc protected amino acids were used in each coupling position, as demonstrated in the flow diagram of the synthesis (Figure 2.18).

a. *Coupling*

Couplings were carried out with strings placed in 100-ml flasks. 10 ml 1:1 v/v DMF-piperidine was added to remove the protecting group and then placed on

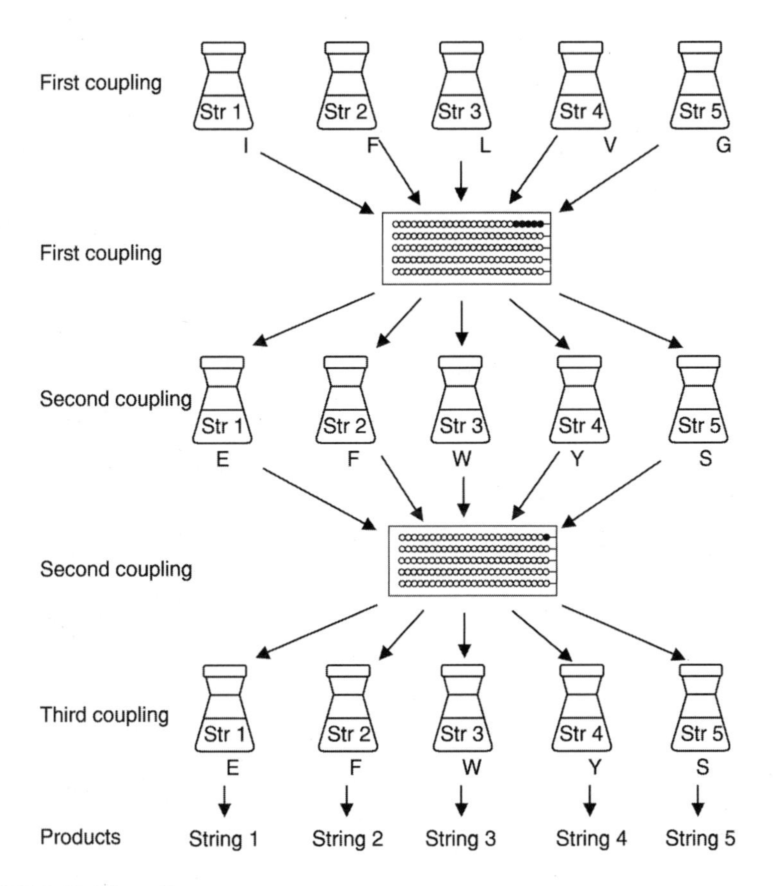

FIGURE 2.18 Flow diagram.

an orbital mixer for 30 minutes. Solutions were decanted from strings, then washed with 3 × 15 ml DMF, 15 ml DCM, 15 ml DMF, 15 ml DCM and 2 × 15 ml DMF. The deprotection operation was once more repeated, then finally washed with 2 × 15 ml DCM. After drying, 10 mmol Fmoc amino acid, 10 mmol HOBt and 15 mmol DIC was added in 10 ml NMP solution and placed on orbital mixer for 2 hours. The solution was then decanted and washed with 3 × 40 ml DMF, 40 ml DCM 2 × 40 ml DMF. The above coupling operation was once more repeated, then finally washed with 2 × 40 ml DCM. The crowns were dried in an oven, then the strings were removed for sorting.

b. Cleavage

In order to cleave the products from the support, the crowns were separately placed in a test tube, 1 ml 1:1 v/v piperidine-DMF was added and allowed to stand for 30 minutes, then filtered, washed with 3 × 2 ml DMF, 2 ml DCM, 2 ml DMF and 2 × 2 ml DCM. The crowns were placed back into test tubes, 1 ml 95% TFA/H_2O added, then allowed to stand for 30 minutes. The solutions

TABLE 2.1
Content of No. 1 strings (Str.) after couplings (Cpl.) and sortings (Sort) and position of products on the final strings. Amino acids are indicated with one letter symbols

Position	Cpl. 1 Str. 1	Sort 1 Str. 1	Cpl. 2 Str. 1	Sort 2 Str. 1	Str. 1 Products	Str. 2 Products	Str. 3 Products	Str. 4 Products	Str. 5 Products
1	I	I	EI	EI	EEI	FEI	WEI	YEI	SEI
2	I	I	EI	FI	EFI	FFI	WFI	YFI	SFI
3	I	I	EI	WI	EWI	FWI	WWI	YWI	SWI
4	I	I	EI	YI	EYI	FYI	WYI	YYI	SYI
5	I	I	EI	SI	ESI	FSI	WSI	YSI	SSI
6	I	F	EF	EF	EEF	FEF	WEF	YEF	SEF
7	I	F	EF	FF	EFF	FFF	WFF	YFF	SFF
8	I	F	EF	WF	EWF	FWF	WWF	YWF	SWF
9	I	F	EF	YF	EYF	FYF	WYF	YYF	SYF
10	I	F	EF	SF	ESF	FSF	WSF	YSF	SSF
11	I	L	EL	EL	EEL	FEL	WEL	YEL	SEL
12	I	L	EL	FL	EFL	FFL	WFL	YFL	SFL
13	I	L	EL	WL	EWL	FWL	WWL	YWL	SWL
14	I	L	EL	YL	EYL	FYL	WYL	YYL	SYL
15	I	L	EL	SL	ESL	FSL	WSL	YSL	SSL
16	I	V	EV	EV	EEV	FEV	WEV	YEV	SEV
17	I	V	EV	FV	EFV	FFV	WFV	YFV	SFV
18	I	V	EV	WV	EWV	FWV	WWV	YWV	SWV
19	I	V	EV	YV	EYV	FYV	WYV	YYV	SYV
20	I	V	EV	SV	ESV	FSV	WSV	YSV	SSV
21	I	G	EG	EG	EEG	FEG	WEG	YEG	SEG
22	I	G	EG	FG	EFG	FFG	WFG	YFG	SFG
23	I	G	EG	WG	EWG	FWG	WWG	YWG	SWG
24	I	G	EG	YG	EYG	FYG	WYG	YYG	SYG
25	I	G	EG	SG	ESG	FSG	WSG	YSG	SSG

were decanted into vials. The crowns were washed with 1 ml 95% TFA/H_2O and the solutions added to the same vials, then dried in a rotawap.

c. *Sorting*

Before beginning the sorting operations, the starting data were entered into a computer (Figure 2.17). As appeared in the last column of the datasheet, in the first sorting the crowns needed to be moved from each slot in groups of five. In the second sorting the crowns had to be moved one by one. The first sorting is demonstrated in the first column of Figure 2.16. Sorting of the 125 crowns was finished in the nine positions of a single cycle. The second sorting is shown in the second column of the figure. In this case, only a single crown was moved from each slot. In the nine positions of the first sorting cycle shown in the figure, altogether 25 crowns were delivered. The rest of the crowns were redistributed in additional four cycles (not shown in the figure).

d. Product Distribution

Prediction by the computer of product distribution on the strings appeared in sheets Sort #1, Sort #2 and Sort #3. Some of the predictions are summarized in Table 2.1. It can be seen that, after coupling 1 on string 1, as expected, all units contained I. After the first sorting, string 1 contained five products in groups of five crowns. The product distribution in the rest of the strings was exactly the same. After coupling 2, string 1 contained five dipeptides in groups of five crowns. After the second sorting, as the table shows, all products in crowns of string 1 were different. The product distribution in the rest of the strings—not shown in the table—were exactly the same. Positions of the formed tripeptides on the five strings, after the third coupling, are shown in the rest of the columns of Table 2.1.

In order to verify the product distribution, randomly selected sequences were independently synthesized and compared to those cleaved from the crowns. The data gathered by HPLC and MS data unequivocally proved the predicted product distribution.

ACKNOWLEDGEMENT

The authors thank OTKA T34868 and NKFP 1/047 for the grants.

REFERENCES

1. Furka Á, Sebestyén F, Asgedom M, Dibó G, Cornucopia of peptides by synthesis, *Highlights of Modern Biochemistry*, Proc. 14th Int. Congr. Biochem., VSP, The Netherlands, 1988, Vol. 5, p. 47.
2. Furka Á, Sebestyén F, Asgedom M, Dibó G, More peptides by less labour, Proc. 10th Int. Symp. Med. Chem., Budapest, 1988, p. 288.
3. Furka Á, Sebestyén F, Asgedom M, Dibó G, General method for rapid synthesis of multicomponent peptide mixtures, *Int. J. Peptide Prot. Res.*, 37:487–493, 1991.
4. Scott JK, Smith GP, Searching for peptide ligands with an epitope library, *Science*, 249:386–390, 1990.
5. Cwirla SE, Peters EA, Barrett RW, Dower WJ, Peptides on phage: A vast library of peptides for identifying ligands, *Proc. Natl. Acad. Sci. USA*, 87:6378–6382, 1990.
6. Devlin JJ, Panganiban LC, Devlin PE, Random peptide libraries: A source of specific protein binding molecules, *Science*, 249:404–406, 1990.
7. Fodor SPA, Read JL, Pirrung MC, Stryer L, Lu AT, Solas D, *Science*, 251:767–773, 1991.
8. Merrifield RB, Solid phase peptide synthesis, I—The synthesis of a tetrapeptide, *J. Am. Chem. Soc.*, 85:2149–2154, 1963.

9. Lam KS, Salmon SE, Hersh EM, Hruby VJ, Kazmierski WM, Knapp RJ, A new type of synthetic peptide library for identifying ligand-binding activity, *Nature*, 354:82–84, 1991. and its correction, *Nature*, 360:768, 1992.
10. Ugi I, *Isonitrile Chemistry*, New York, Academic Press, 1971, pp. 1–278.
11. Nielsen J, Brenner S, Janda KD, Synthetic methods for the implementation of encoded combinatorial chemistry, *J. Am. Chem. Soc.*, 115:9812–9813, 1993.
12. Nikolaiev V, Stierandova A, Krchnak V, Seligmann B, Lam KS, Salmon SE, Lebl M, Peptide-encoding for structure determination of nonsequenceable polymers within libraries synthesized and tested on solid-phase supports, *Peptide Res.*, 6:161–170, 1993.
13. Kerr JM, Banville SC, Zuckermann RN, Encoded combinatorial peptide libraries containing non-natural amino acids. *J. Amer. Chem. Soc.* 115:2529–2531, 1993.
14. Ohlmeyer MHJ, Swanson RN, Dillard LW, Reader JC, Asouline G, Kobayashi R, Wigler M, Still WC, Complex synthetic chemical libraries indexed with molecular tags, *Proc. Natl. Acad. Sci.* USA, 90:10922–10926, 1993.
15. Battersby BJ, Bryant D, Meutermans W, Matthews D, Smythe ML, Trau M, Toward larger chemical libraries: encoding with fluorescent colloids in combinatorial chemistry, *J. Am. Chem. Soc.*, 122:2138–2139, 2000.
16. Han H, Wolfe MM, Brenner S, Janda KD, Liquid-phase combinatorial synthesis. *Proc. Natl. Acad. Sci.* USA, 92:6419–6423, 1995.
17. Furka Á, Study on possibilities of searching for pharmaceutically useful peptides, 1982, http://szerves.chem.elte.hu/Furka/
18. Geysen HM, Rodda SJ, Mason TJ, *A priori* delineation of a peptide which mimics a discontinuous antigenic determinant, *Mol. Immunol.*, 23:709–715, 1986.
19. Houghten RA, Pinilla C, Blondelle SE, Appel JR, Dooley CT, Cuervo JH, Generation and use of synthetic peptide combinatorial libraries for basic research and drug discovery, *Nature*, 354:84–86, 1991.
20. Furka Á, Sebestyén F, Peptide sub-library kits, PCT application, WO 93/24517, 1993.
21. Sebestyén F, Dibó G, Furka Á, Efficiency and limitations of the "portioning-mixing" peptide synthesis, In Schneider CH, Eberle AN eds., Peptides 1992, ESCOM, Leiden, 1993, p. 63.
22. Pinilla C, Appel JR, Houghten RA, Positional scanning synthetic peptide combinatorial libraries, In Schneider CH, Eberle AN eds., Peptides 1992, ESCOM, Leiden, 1993, p. 65–66.
23. Furka Á, Sub-Library Composition of Peptide Libraries, Potential Application in Screening, *Drug Development Research*, 33:90–97, 1994.
24. Câmpian E, Peterson ML, Saneii HH, Furka Á, Deconvolution by omission libraries, *Bio-organic and Medicinal Chemistry Letters*, 8:2357–2362, 1998.
25. Moran EJ, Sarshar S, Cargill JF, Shahbaz M, Lio A, Mjalli AMM, Armstrong RW, Radiofrequency tog encoded combinatorial library method for the discovery of tripeptide-substituted cinnamic acid inhibitors of the protein tyrosine phosphatase PTPSB *J. Am. Chem. Soc.*, 117:10787–1788, 1995.
26. Nicolaou KC, Xiao X-Y, Parandoosh Z, Senyei A, Nova MP, Radiofrequency encoded combinatorial chemistry, *Angew. Chem. Intl. Ed. Engl.*, 34:2289–2291, 1995.
27. A Furka, JW Christensen, E Healy, HR Tanner, H Saneii. The string synthesis. A spatially addressable split procedure. *J. Comb. Chem.* 2:220–223, 2000.

3 Glycopeptide Libraries

Daniela Palomba and Giorgio Fassina

CONTENTS

Combinatorial and computational methodologies, for the design and the synthesis of peptide and peptidomimetic analogs, represent a powerful tool in the search of novel bioactive molecules for therapeutic intervention in disease. The rapid improvement of these technologies, including *in silico* rational molecular design, innovative solid and solution phase processes, and effective no time-consuming high throughput screening (HTS), has demanded the creation of a broad spectrum of new molecular entities for discovering new drugs or/and materials. Recent advances in organic synthesis, and in the area of combinatorial solid and solution chemistry, have given technological support for the creation of polyfunctional scaffolds and the construction of several unique libraries. Sugar amino acids constitute a relevant class of such scaffolds, and an effective set of peptidomimetic building blocks, providing high diversity and a wide font of natural molecules [1].

The relevant biological role of glycopeptides and the need to generate mimics of biopolymers, such as peptidomimetics, suggest the importance of these molecules as key synthetic targets.

I. BIOLOGICAL AND SYNTHETIC RELEVANCE OF GLYCOPEPTIDES

A. Glycopeptides as Relevant class of Peptidomimetics

The relative low cost of synthesis and the more restricted side effects of small peptides, as compared to other classes of bioactive molecules, including proteins, support their medicinal use. However, peptides derived from natural amino acids are proteases substrates and might be eliminated by the immune system as non-self molecules, limiting their use in medicinal chemistry.

Peptidomimetics, such as glycopeptides, represent a valid approach to pharmaceutical treatments [2]. They are peptido-organic molecules, containing non-peptidic structural elements, including sugar, nucleosidic or lipidic moieties, that are capable of mimicking or antagonizing the biological action(s) of natural parent peptide molecules, imitating the template peptides, but lacking the peptidic bonds. Therefore, peptidomimetics are not substrate of proteases, and are likely to be active *in vivo* for a longer period of time as compared to the template peptides. In addition, they might be less antigenic and might show an overall higher bioavailability.

Glycosylation of peptides is an efficient method to stabilize peptide by inhibition of enzymatic degradation, to conformationally modify peptide and usually to increase peptide solubility. Peptides glycosylation also creates additional diversity in peptide libraries since the high functionalities and variability of sugar moieties.

Moreover, glycopeptides may become, in the near future, important targets for medicinal chemists as drugs, such as novel antibiotic agents, or as specific markers for certain types of tumors, due to their key role played in important biological events, including inflammation, metastasis, cellular trafficking and cell surface recognition.

B. Glycopeptides Biosynthetic Pathway

Glycosylation is the most abundant post-translational modification of proteins or peptides. Carbohydrates are covalently linked to specific amino acids imparting an important structural diversity involved in biological recognition processes, and in events regulating activity in cell adhesion and infection [3]. The glycosylation reaction in the biosynthetic pathway of proteins is a post-translational step, due to a concerted action of different enzymes, including glycosyl hydrolases and transferases. Furthermore, biosynthetically created glycoproteins have a high degree of heterogeneity, a limit for structural and functional analysis. Therefore, glycoproteins are not easily obtained by gene technology, and organic synthesis provides an important alternative and a unique system to obtain homogeneous glycolconjugates.

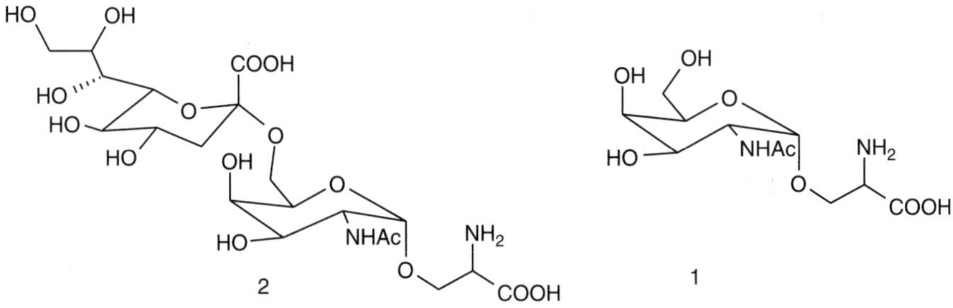

FIGURE 3.1 T_N (Gal NAC-serine/threonine) **1** and T_N sialyl **2** antigen-glycoamino acid building blocks.

Such class of homogeneous glycopeptides has been considered, in recent years a powerful source of candidates for a novel class of vaccines [4]. In particular, glycopeptides, conjugate to antigen motifs T_N, T, sialyl-T_N and 2, 3- sialyl-T, are putative candidates as tumour vaccines, due to their expression on mucins in epithelial breast, ovary, stomach, colon cancers. Chemical and chemoenzymatic synthesis of building blocks carrying these antigens have been designed. The chemical synthesis may be conducted via high regio- and stereoselective sialylation of galactopyranoside saccharides in order to produce acceptor and donor templates useful in solid phase synthesis as reported by Xia et al. [5]. On the other hand, George et al. [6] have performed chemoenzymatic strategies using recombinant sialyltransferases and N-acetylglucosaminyl transferase in solution phase for coupling with solid phase assembled T and T_N glycopeptide (Figure 3.1). Furthermore, conjugate glycopeptides have been employed as immunomodulators in many immunological investigations where synthetic glycopeptides, coupled to other proteins have led to identification of novel and powerful selective ligands or antigens [7, 8].

II. GLYCOPEPTIDES SYNTHESIS

Several combinatorial approaches for the synthesis of glyco-amino acids have been developed, due to the complexity of carbohydrate conjugates. The synthetic challenge is to get access to these building blocks by automated synthesis in order to generate glycopeptides library, to fully understand their biological roles and to widen their use in medicinal chemistry.

A. Synthetic Chemical Strategies

Many methods, merging peptide-chemistry with glycotechnology, have been improved, in order to introduce sugar moiety in peptide chain providing bases to create novel combinatorial libraries in automated or semi-automated manner.

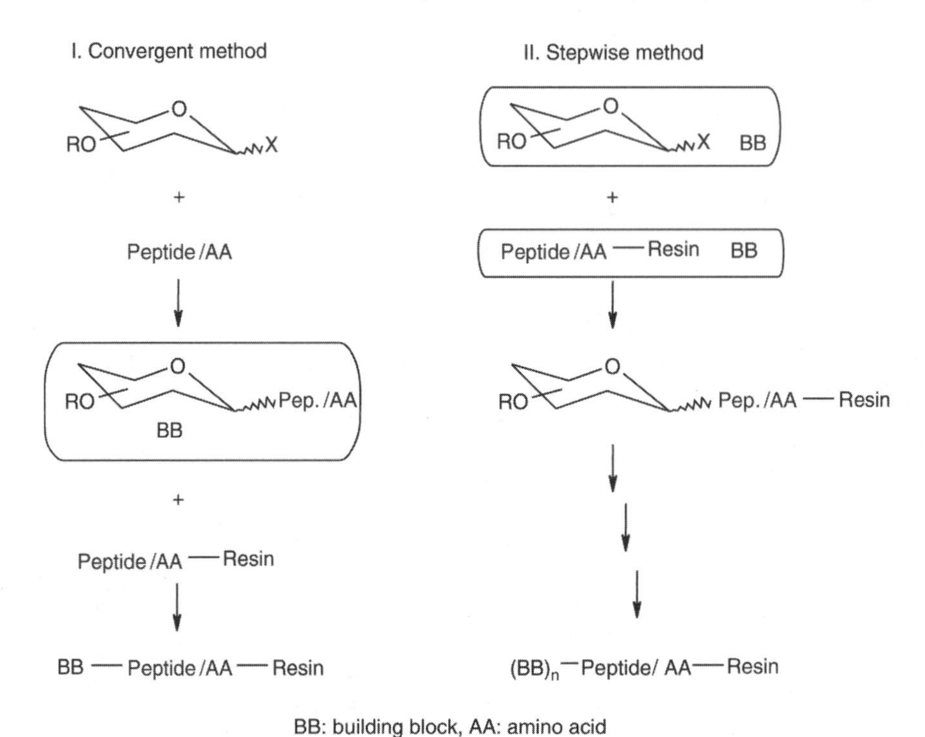

FIGURE 3.2 Convergent method and stepwise method.

The construction of glycopeptides can be performed by using two chemical strategies (Figure 3.2):

1. *Convergent method:* The peptide and sugar moieties are coupled to each other by using activation and orthogonal protecting groups, as required, from substrates, and then introduced in designed library as building blocks.
2. *Stepwise method:* The amino acids and sugar moieties are linked step by step as independent building blocks for solid phase synthesis of designed library; this approach requires orthogonal side-chain protecting groups for the glycosylation step.

Recent technological improvements have allowed the use of solid phase methods, such as Boc- and Fmoc-strategy, not only for peptides, but also for glycopeptides (Table 3.1). Fmoc-strategy is more used because it requires milder conditions than Boc strategy and for the wider selection of compatible protective groups.

In both, cases for sugar moieties the acetyl protective groups are very useful for stabilizing the O-glycosidic linkage against TFA or hydrogen fluoride

TABLE 3.1
Fmoc vs Boc α-amino group temporary protecting group strategy

Fmoc(9-fluorenylmethyloxycarbonyl) Strategy	Boc (*tert*-butoxycarbonyl) Strategy
Removal under mild basic conditions (Pip, DBU)	Removal by TFA
Basic-stable protecting groups for sugar hydroxyl and amino acids side chain functional groups	Acid-stable protecting groups for sugar hydroxyl and amino acids side chain functional groups
Es. Benzyl-, acetyl- groups for carbohydrates	Es. Benzyl-, acetyl- groups for carbohydrates
Cleavage from solid support and deprotection of amino acids side chain protecting groups by TFA and suitable scavenger	Cleavage from solid support and deprotection of amino acids side chain protecting groups by HF, super strong acid
Benzyl-, acetyl- groups for carbohydrates removed under basic or reductive conditions	Benzyl-, acetyl- groups for carbohydrates removed under basic conditions

treatment [9]. Acetyl- are usually preferred to benzyl- protection because acetate groups are removable under mild basic conditions preventing the β-elimination. Furthermore, benzyl-ethers are not always compatible with sulfur or aromatic containing amino acids, due to their deprotection *via* catalytic hydrogenation. Allyl esters, or allyloxycarbonyl templates are useful protective groups in glycopeptide libraries construction due to the mild neutral cleavage conditions via π-allyl palladium complex formation [7]. However, as reported by Otvos et al. [10], the coupling efficiency of glycosylamino acids is enhanced when sugar residue is unprotected, especially for solid phase glycosylation reaction, although the glycosidic linkage becomes more acid sensitive.

Due to such sensitivity of glycosidic linkage, various acid-labile linkers have been shaped in order to minimize glycosylamino acid moiety exposition to acid environment, including Wang, Rink, PAL, SASRIN and also the HYCRAM linker [11] which provides the releasing of protected glycopeptides under neutral conditions (Pd(0) salt) [12] (Table 3.2). These linkers are immobilized on resin solid supports including polystyrene and polyamide/polyethylene glycol copolymer (PEGA, [13]), useful for glycopeptides synthesis. However in stepwise method a careful selection of resins is necessary for the marked influence of support in glycosylation efficiency [14].

The synthetic manipulation of O-glycosylamino acid molecules is usually more complex because they may undergo the acid-catalysed anomerisation of glycosidic bonds (Figure 3.3) and they are also usually more sensitive to acid conditions, due to their aldolic nature (e.g. serine and threonine derivatives). However, the acid-sensitivity is modulated by saccharide and peptide nature.

TABLE 3.2
Linker for peptide solid phase synthesis

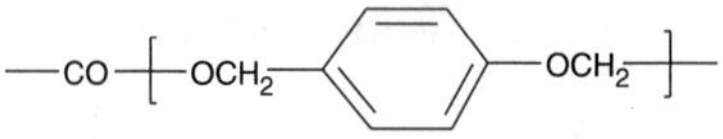	WANG
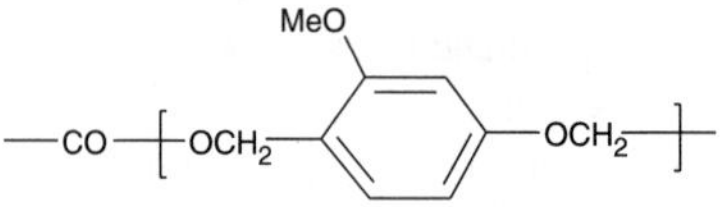	SASRIN
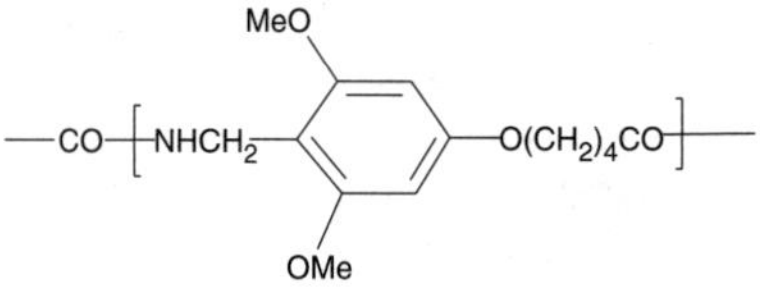	PAL
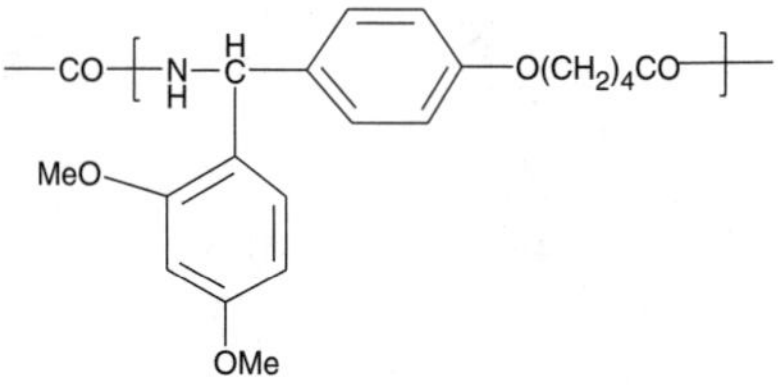	RINK
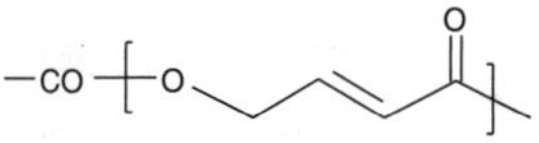	HYCRAM
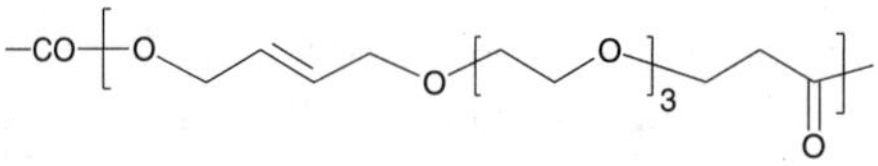	HYCRON
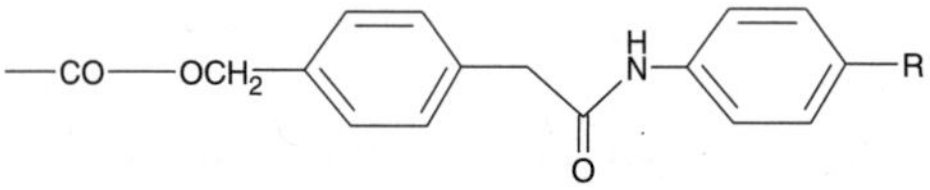	PAM (R: resin)
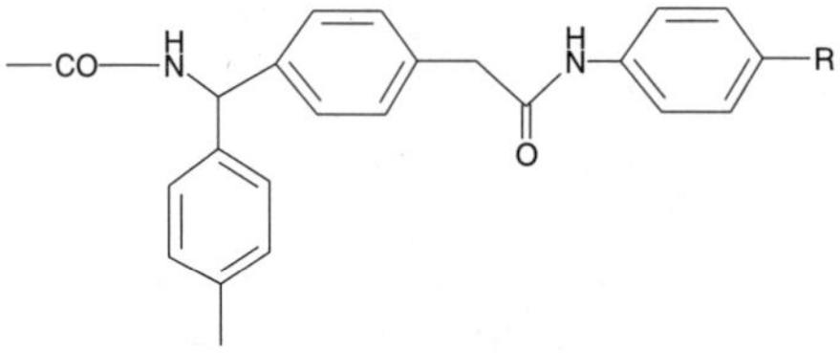	MBHA (R: resin)

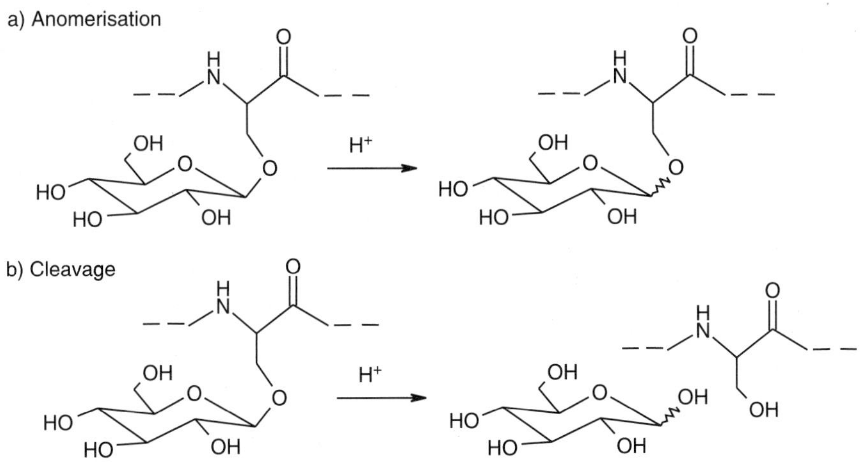

FIGURE 3.3 Acid sensitivity of O-glycopeptides.

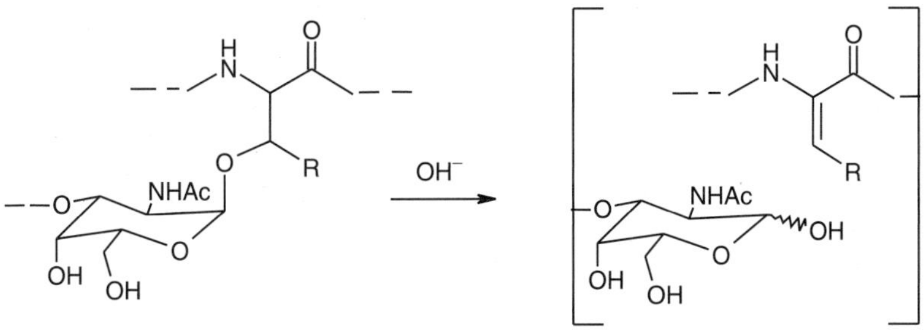

FIGURE 3.4 β-Elimination of O-glycopeptides under alkaline conditions.

In alkaline solution, usually employed in deblocking step of synthesis, O-glycopeptides easily afford β-elimination of glycan side chain (Figure 3.4). This base-catalysed degradation depends on the acidity of αCH group adjacent to the carboxylic group (i.e.: O-glycosyl serine/threonin esters > O-glycosyl serine amide > O-glycosyl free carboxylate.)

B. Chemoenzymatic Strategies

In chemoenzymatic synthesis, the glycosylamino acids are obtained by enzymic glycosylation and used as building blocks in combinatorial libraries creation [15]. This latter technology requires, first, the solution and implementation of several technical problems to be considered a general method-like chemical approach. Use of enzymatic methods can avoid some regio- and

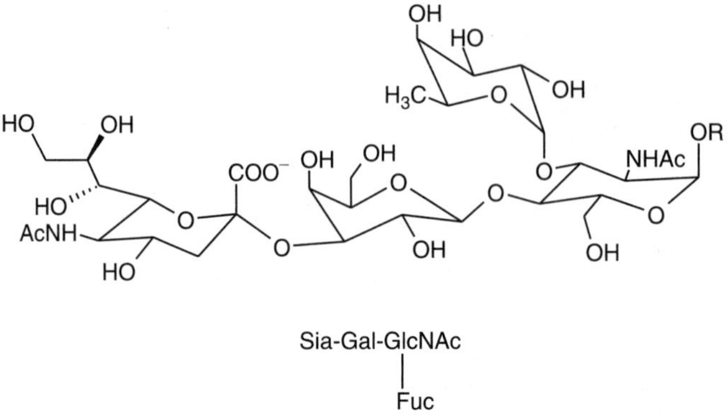

FIGURE 3.5 Sialyl Lewis^x (sLe^x).

stereo-selectivity problems, especially in the synthesis of glycopeptide containing complex oligosaccharide, and simplify the protective groups pattern. Valuable applications of enzymatic glycosyltransferase chemistry were published by Wong et al. [16] and Unverzagt et al. [17]. They combined glycosyltransferase reactions with an alkaline phosphatase, in order to obtain solid phase building blocks containing Sialyl Lewis^x tetrasaccharide template, for the investigation of cell adhesion phenomena (Figure 3.5).

III. GLYCOPEPTIDES LITERATURE REVIEW

The more recurrent motifs in glycopeptide libraries are those mimicking the structural core of the most important classes of biological relevant proteins [18, 19]. They can be grouped in two main classes:

- N-glycopeptides
- O-glycopeptides

A. N-Glycopeptides

Proteins, modified by the covalent attachment of a carbohydrate to an asparagine residue via a β-N-glycosidic linkage, constitute the major class of N-glycoproteins. This motif is found in hen egg albumin, human transferrin, membrane and virus glycoproteins (i.e. gp120 of HIV-1), important proteins involved in key processes of transport and molecular recognition in the eukariotic cellular compartment and on the cell membrane. The consensus sequence characterizing these glycoproteins is Asn-X-Ser/Thr, where X is any amino acid except proline. Sugar moieties usually is acetylglucosamine **1**, but in bacteria, glucosamine **2** or N-acetyl-galactosamine **3** have also been found (Figure 3.6). The glucosamine-asparagine linkage is the most frequently synthesized building block [20], because of its natural abundance and its stability towards acidic and basic conditions encountered in solid phase

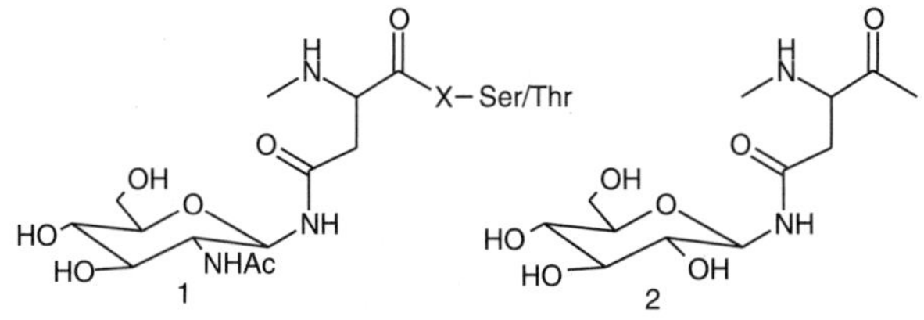

FIGURE 3.6 N-glycosyl asparagine linkage.

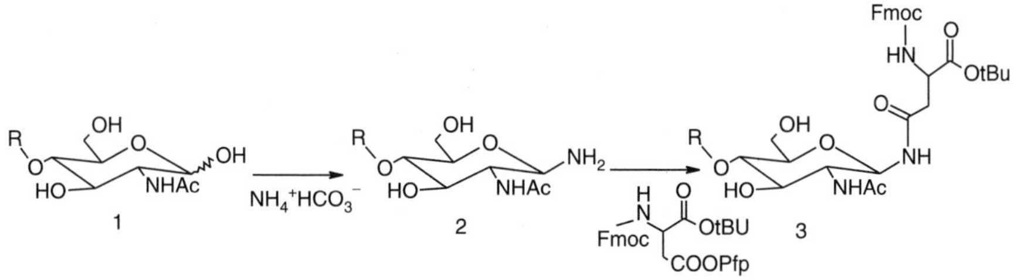

FIGURE 3.7 N-glycosyl asporagine linkage synthesis.

peptide synthesis. N-acetyl glucosamine moiety [21] could be obtained from hydrogenation of corresponding N-azides acetyl-protected sugar and then coupled with Z-Asp-OBzl *via* DCC activation as reported by Kunz and coworkers. Glycosyl azides allow a versatile manipulation of protective group pattern and represent a useful class of glycopeptide building blocks in contrast to glycosyl-amine, which are more sensitive templates. An alternative method [22] for the coupling between N-acetyl glucosamine unprotected moieties, obtained with saturated ammonium bicarbonate, and α-carboxy-pentafluorophenyl provides activated esters of NH-Fmoc-Asp-OtBu (Figure 3.7). This latter approach is very useful for coupling peptides with usually unprotected oligosaccharides isolated from natural sources. For instance, N-glycopeptides characterized by mannotrioside-di-N-acetyl-chitobiose core linked to Asn are used as models for studying interactions involved in adhesion processes. Shimon and co-workers have developed a convenient and general route to obtain these building blocks *via* a coupling procedure of glycosylamines to aspartic acid [23].

B. O-Glycopeptides

Most biologically relevant glycoproteins of mammals, including membrane proteins such as human glycophorin and mucins, present various sugars, such

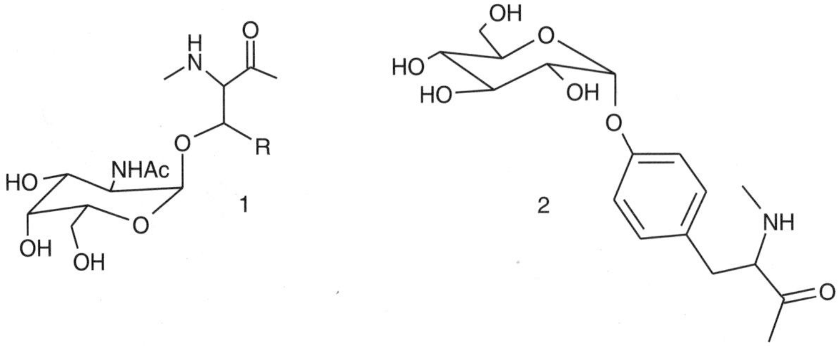

FIGURE 3.8 α-o-glyco-linkage (**1**) and α-o-glyco- and malto-sylated aromasic hydroxy side chain of Tyr-194 (**2**).

as N-acetyl-galactosamine attached to the side chain hydroxyl group of Ser, Thr, Tyr via α-O-glyco-linkage, (Figure 3.8), **1**).

Mucins glycoproteins play important biological roles such as protection and lubrification of respiratory and digestive tracts, they contain a basic structure of 2-acetamido-2-deoxy-D-galactose linked to hydroxy groups of Thr or Ser occurring in all saccharide side chains. During tumour transformation, the glycosilation pattern of mucins changes markedly suggesting their significant role in neoplastic transformations [4]. Libraries containing mucins core have been synthesized to better understand glycosylation profile of cell membrane and their adhesion processes [24].

Recently, another relevant class of O-linked glycoproteins occurring in glycogenin has been discovered. Glycogenin is 38 kDa protein involved in glycogen biosynthesis. The recurrent motif is the α-O-gluco- and malto-sylated aromatic hydroxy side chain of Tyr-194 (Figure 3.8, **2**). To confirm the configuration of glycosidic linkage and study the secondary structure of this important precursor of glycogen several glycotyrosine building blocks useful in libraries construction and screening have been synthesized [25]. The strategy employed provides direct glycosilation of Nα-Fmoc amino acid Pfp ester, in order to synthesize several building blocks, including N-, O- linked glycopeptide of Tyr, directly on solid phase Rink-PEGA-resin [26].

Another class of relevant glycoproteins are the selectin transmembrane glycoproteins. Selectins are expressed on the surface of leukocytes and recognize biomolecules containing the tetrasaccharide sialyl-Lewis^{x} (sialyl-CD15). This tetrasaccharide, composed of sialic acid, galactose, fucose, and N-acetyl-galactosamine, is found on all circulating myeloid cells. The carbohydrate recognition process between selectins and ligands, mediating the early steps of the adhesion cascade, is still unclear. Therefore, many glycosylamino acids building blocks providing the sialyl-Lewis^{x} core (Figure 3.5) have been synthesized [27].

Other naturally occurring glycosidic bonds have been reviewed from Hayes and Hart as putative motif of novel libraries [28].

IV. COMBINATORIAL GLYCOPEPTIDE LIBRARIES

Synthesis of glyco-amino acids building blocks, for convergent solid phase synthesis of gliycopeptide libraries has been the first approach pursued from many research groups, because of the intrinsic chemical complexity of a sugar moiety. The need of orthogonal protection groups strategies, the difficult stereo- and regio-chemical control, the possibility of several side reactions limit, for carbohydrates, the creation of a general synthetic procedure like, in peptide solid phase strategies.

In this section various synthetic approaches that have been applied in building blocks preparation for combinatorial libraries assembly will be discussed.

The therapeutic use of many polycyclic glycopeptide antibiotics, including vancomycin, ristocetin, characterized by central core of β-D-glucose attached to phenolic peptide side chain and coupled to unusual amino sugars, have stimulated the synthesis of glycoamino acids building blocks for creation of novel libraries in order to identify more potent antibiotics directed towards resistant bacteria [29]. For instance, the clinically used antibiotic vancomycin is a glycopeptide, active against several antibiotic-resistant strains of bacteria such as MRSA (methicillin-resistant Staphylococcus aureus). Many glycopeptide libraries have been generated in order to optimize this lead molecule [30, 31] for example dimers vancomycin libraries created by the application of two ligation methods (disulfide formation and olefin metathesis).

Synthesis strategies have also been developed to obtain glycopeptide libraries in order to facilitate the investigation of glycoprotein-mediated cell–cell interaction in biological recognition events or the identification and analysis of the interaction between carbohydrates and their receptors.

To achieve this goal, Meldal and St Hilaire [32] have pursued the approach to synthesize glycopeptides to mimic the activity of natural carbohydrate ligands, in order to enhance their affinity for proper receptor. In many cases, glycopeptides have been found to be superior ligands for several receptors, including L-, E-, P- selectins of the immune system, bacterial adhesins, and also to be more easily assembled in libraries displays than oligosaccharides. Sialopeptide libraries, sialic acid containing glycopeptides, were synthesized and screened in order to find a high affinity ligand for sialoadhesin (Siglec-1) and provide valuable information about the active site of receptors. Libraries, containing a sialic acid at the N-terminus of a peptide (lactamization), or holding sialic acid at the centre of peptide skeleton, were produced by using the Mix & Split method and the ladder synthesis via a photolabile linker to PEGA resin, in order to easily analyse libraries by MALDI-TOF mass spectrometry and to screen them for binding of receptors directly on solid phase [33].

Due to the growing interest in recognizing process of the innate immune system, the synthetic glycopeptides are becoming an invaluable tool to establish the fine specificity of immune response provided by carbohydrate-mediated B and T cells. For this purpose, glycosylamino-acids building blocks, containing N-α-Fmoc serine glycosylated *via* saccharides 1,2-trans peracetates and Lewis acid catalysis, have been used to generate libraries of molecules

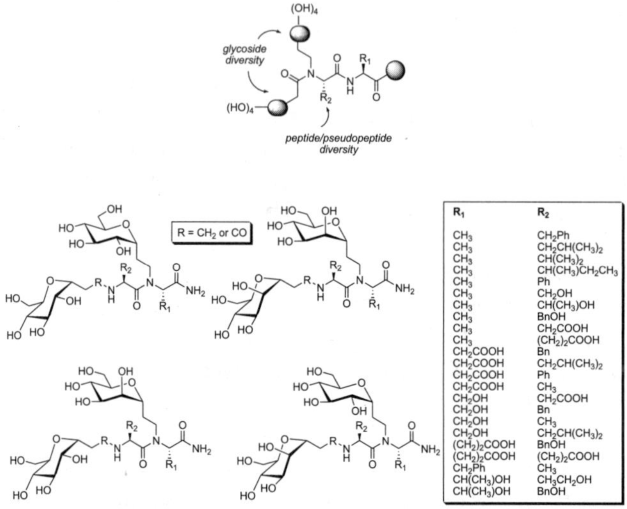

R_1	R_2
CH_3	CH_2Ph
CH_3	$CH_2CH(CH_3)_2$
CH_3	$CH(CH_3)_2$
CH_3	$CH(CH_3)CH_2CH_3$
CH_3	Ph
CH_3	CH_2OH
CH_3	$CH(CH_3)OH$
CH_3	BnOH
CH_3	CH_2COOH
CH_3	$(CH_2)_2COOH$
CH_2COOH	Bn
CH_2COOH	$CH_2CH(CH_3)_2$
CH_2COOH	Ph
CH_2COOH	CH_3
CH_2OH	CH_2COOH
CH_2OH	Bn
CH_2OH	CH_3
CH_2OH	$CH_2CH(CH_3)_2$
$(CH_2)_2COOH$	BnOH
$(CH_2)_2COOH$	$(CH_2)_2COOH$
CH_2Ph	CH_3
$CH(CH_3)OH$	CH_3CH_2OH
$CH(CH_3)OH$	BnOH

FIGURE 3.9 Libraries tested as inhibitors of enzymes involved in the biosynthesis of the mycobacterium cell wall.

mimicking Helper-T-cell stimulating peptides [34]. By using a different approach, Dukic-Stefanovic et al. [35] N-terminally acetylated gluco-dipeptide library to be used against AGE (advanced glycation end products in age-related diseases) antibodies to characterize their immunoreactivity (epitope mapping).

Several glycosylated oligopeptides have been synthesised with N- or O-glycosidic deviate from the naturally occurring linkage, in order to enhance enzycmatic stability. The introduction of sugar moiety in biologically active peptides has been performed also in order to modify conformation and properties of original peptide. Many glycosylation approaches have been applied, for these purposes, in solid-phase glycopeptides libraries synthesis including the use of glycosyl trichloroacetimidate donors for serine, threonine, tyrosine side chains glycosylation; glycosyl iodoacetamides approach, that provides access to glycopeptide alpha-thioesters [36] and the reductive amination approaches, via glycoside aldehyde derivatives, improved by Arya P. BarkLey A. et al. for the versatile creation of artificial glycopeptide libraries. These latter libraries, combining diversity in both peptides/pseudopeptides and glycoside moieties (Figure 3.9), were tested as inhibitors of enzymes (i.e. UDP-galactopyranose mutase) involved in the biosynthesis of the mycobacterium cell wall [37, 38].

V. MITSUNOBU LIGATION PROCESS FOR GLYCOPEPTIDE LIBRARIES GENERATION

Synthetic access by stepwise method to glycopeptides molecules requires flexible solution- and solid-phase combinatorial strategies. The selected

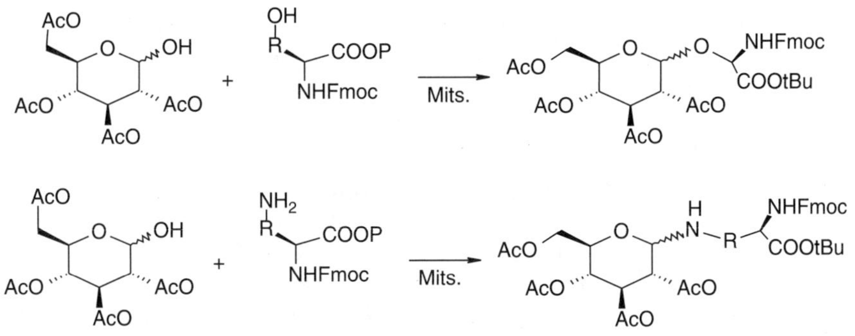

FIGURE 3.10 Mitsunobu glycosylation process R: side-chain amino acids; P: protective group or resin linker.

glycosylation method should proceed in high yield, with a known stereospecific pattern.

Xeptagen S.p.A. has improved a computer assisted combinatorial chemistry approach, based on an innovative application of Mitsunobu reaction, that could be used to design and synthesize new peptido-mimetic libraries, in order to find novel leads efficient in different biological targets (Figure 3.10).

The developed combinatorial synthetic process is based on Mitsunobu reaction [39] in order to introduce sugar in peptide skeleton by a regio- and stereo-selective manner. This methodology exploits the diversity provided by a monosaccharide, which is an ideal molecular scaffold, for his conformationally rigidity and functionally rich system, and provides access to novel glycopeptides molecules. In addition, the Mitsunobu reaction for the coupling of anomeric hydroxyl of carbohydrate and side-chain functional group of selected peptides, could be performed with high yields on a solid support. The plethora of carbohydrate moieties and amino acids from commercial sources provides the potential for vast chemical diversity when constructing such libraries.

A. *In Silico* Combinatorial Analysis of Building Blocks

The unique chemistry was adapted to use in combinatorial synthesis and was complemented by *in silico* technologies for the rational design of libraries, in order to optimize diversity of building blocks, obtain molecules with remarkable enzymatic stability, but also to get information from analysis of i.e. conformational constrains, molecular rigidity, size, charge, and hydrophobicity.

The *in silico* filters are used to design a subset of building blocks from the number of potentially useful molecules in order to rationally reduce the huge numbers of synthesizable compounds. The structural diversity of the glycopeptides library could be tailored to the target by selecting *in silico* side-chain mimetics, scaffolds, and substitution patterns, and by using specifically designed softwares able to detect similar shape and similar electronic distribution in different molecules, allowing the identification of the building blocks most suitable to construct the appropriate peptido-mimetic molecule. Computational methods for library design allow to tailor the library generation by the rapid and cost-effective building blocks elimination, by using the basic knowledge of medicinal chemistry merged to the chemical–physical proprieties of molecules [40].

As model library for building blocks *in silico* optimization we selected the SAX library, where S were sugar moieties, A were amino acids glycosysdable by Mitsunobu coupling process, and X were natural or non-natural amino acids (Table 3.3 a,b,c). The collection diversity can be evaluated by using many quantitative molecular descriptors [41]. Our sets have been screened using mainly topological descriptors, which differentiate the molecules according mostly to their size, degree of branching, flexibility, overall shape, and electronic descriptors including:

Topological (2D) descriptors: Zagreb, the Zagreb index, is defined as the sum of the squares of vertex valencies [42]; Balaban, JX and JY, is a highly discriminating descriptor, whose values do not substantially increase with molecule size and the number of rings present [43]; Kappa indices; PHI; SubgraphCount; Chi indices; Hosoya, the Hosoya

TABLE 3.3
SAX sets

1) S: sugar moieties

1,2,3,4,6- penta O-acetyl-α-D-Mannopyranose
1,2,3,4,6- penta O-acetyl-α- and β-D-Galattopyranose
1,2,3,4,6- penta O-acetyl-α- and β-D-Glucopyranose
1,2,3,5- tetra O-acetyl-β-D- and L-Ribofuranose
1-O-acetyl-2,3,5- tri-O-benzoyl-β-D-Ribofuranose
D Galactosamine HCl
D(+) Glucosamine HCl
D(+) Mannosamine HCl
D(+) Digitoxose
D-(−) Arabinose
β-D-Allose

2) A: natural and not-natural amino acids

3) X: amino acids glycosysdable by Mitsunobu

X	pka_3
His	9.17
Tyr	10.07
Lys	10.53
Cys	10.78
Asn	
Gln	

index, is the sum of all (nonzero) p(k); Wiener, the Wiener index, is the sum of the chemical bonds existing between all pairs of heavy atoms in the molecule [44].

Structural descriptors: Hbond donor, number of hydrogen bond donors; Hbond acceptor, number of hydrogen-bond acceptors; Rotlbonds, number of rotatable bonds in the molecule having rotations that are considered to be meaningful for molecular mechanics. All terminal H atoms are ignored (for example, methyl groups are not considered rotatable); MW, Molecular weight.

Electronic descriptors: F Charge, sum of formal Charge.

Thermodynamic descriptors: $AlogP_{98}$, logarithm of the octanol/water partition coefficient, atom-type value; the AlogP98 descriptor is an implementation of the atom-type-based AlogP method using the latest published set of parameters.

The diversity between structures has been calculated by the formula I as Euclidean Distance (D).

$$D_{a,b} = \sqrt{\sum_{j=1}^{n} \left(x_{ja} - x_{jb}\right)^2} \quad j = 1, 2, \ldots, N \text{ descriptors; } a, b \text{ models} \tag{I}$$

Finally, applying MaxMin algorithm II, the sets of molecules for which the Euclidean distance is maxima have been determined.

$$\text{MaxMin} = \max\left\{\min\left\{D^2_{a,b}\right\}\right\} \tag{II}$$

The sugar sets have been reported in 3D topological and chemical-physical space, obtaining six clusters, where differences between examined molecules are exclusively due to the chirality of skeleton carbon atoms (Figure 3.11)

The two amino acids sets have been grouped in 3D topological and chemical–physical space, on the basis, not only of topological descriptors and $AlogP_{98}$ (Figure 3.12), but also on the basis of structural, electronic and thermodynamic descriptors (Figure 3.13).

B. Combinatorial Synthesis Procedure

The Mitsunobu reaction leads to the alkylation of alcohols with various nucleophiles or acids (HA) *via* a redox system, composed by diethylazadicarboxylate (DEAD) and triphenylphosphine (TPP) (Figure 3.14). A limit in the application of Mitsunobu process is the pK_a value of the acid counterpart, that must be usually smaller than 11; therefore, many improved redox systems have been developed in order to solve this problem.

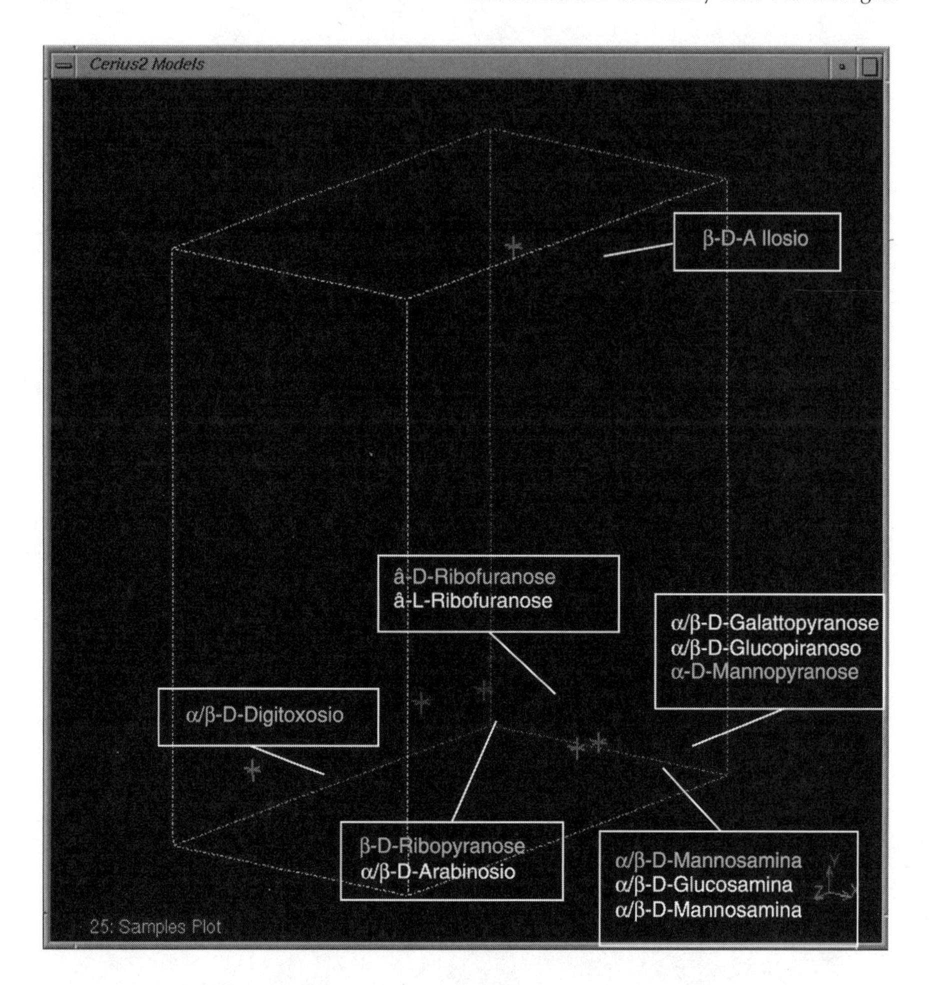

FIGURE 3.11 Sugar sets: 3D topological and chemical-physical space.

The Mitsunobu procedure for the incorporation of carbohydrate moiety into combinatorial peptide libraries has been executed on solution phase, by using gluco-tyrosine as glycoamino acid model (prepared as shown in Figure 3.15) and improved on a solid support for model peptide GlyGly-Tyr (β-tetra acetyl glucose)-GlyGly.

In solution, phase amino acid derived methyl ester N-Fmoc-Tyr-OMe **4** was successfully condensed under Mitsunobu conditions with 2,3,4,6-tetra-Oacetyl-D-glucose **2** to afford the fully protected glucosylamine **5**. The reaction provides products of analytical purity and predictable sterochemistry, as confirmed by $^{1}H/^{13}C$ NMR. Synthetic conditions and protective groups for glycopeptides synthesis have been investigated. Mitsunobu conditions employed in solution phase are based on the improved redox system 1,1′-azodicarbonyldipiperidine (ADDP)—tributylphosphine (TBP) in

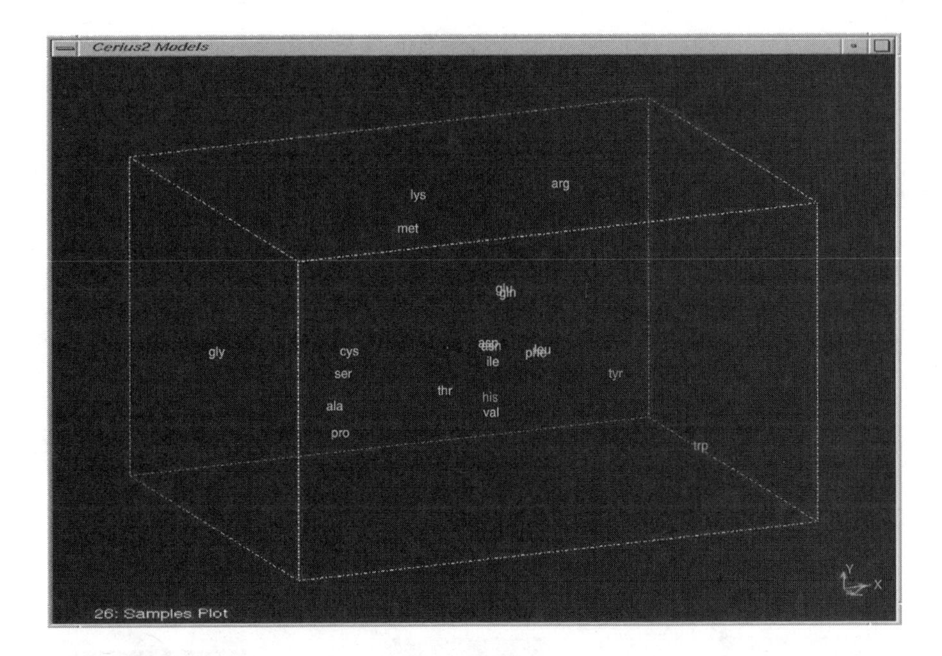

FIGURE 3.12 Amino acids sets: 3D topological and Alog P_{98} space.

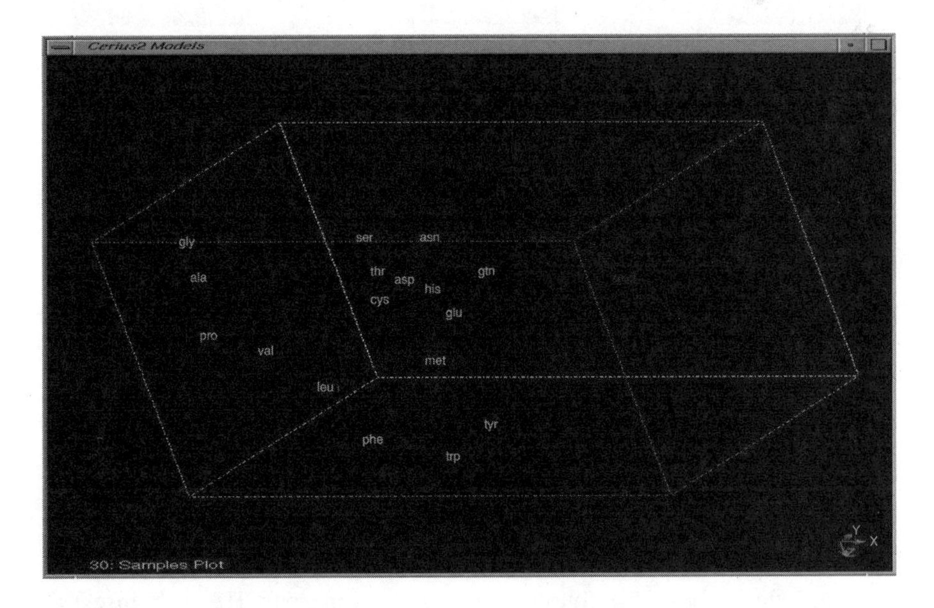

FIGURE 3.13 Amino acids set: 3D chemical-physical space.

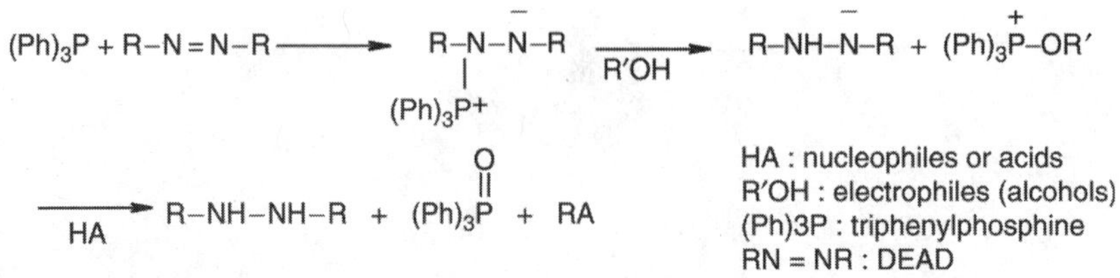

(Ph)3P + R–N=N–R → R–N–N–R (Ph)3P+ → (R'OH) R–NH–N–R + (Ph)3P–OR'

→ (HA) R–NH–NH–R + (Ph)3P=O + RA

HA : nucleophiles or acids
R'OH : electrophiles (alcohols)
(Ph)3P : triphenylphosphine
RN = NR : DEAD

FIGURE 3.14 Mitsunobu reaction's mechanism.

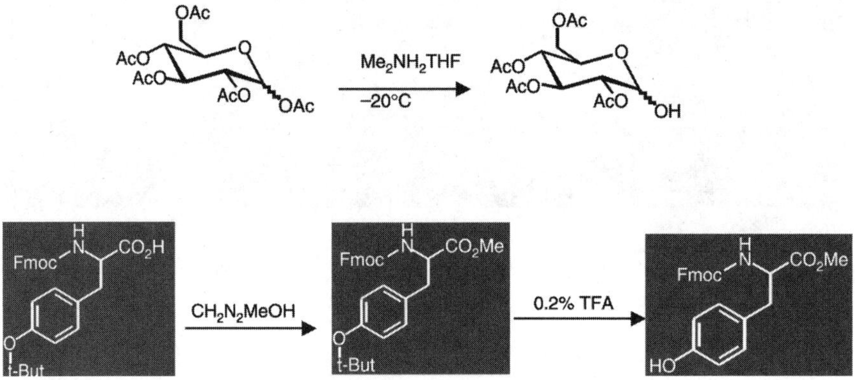

OAc AcO AcO AcO OAc O
Me2NH2THF
–20°C
OAc AcO AcO AcO OH O

H N Fmoc CO2H O t-But
CH2N2MeOH
H N Fmoc CO2Me O t-But
0.2% TFA
H N Fmoc CO2Me HO

FIGURE 3.15 Synthesis of gluco-tyrosine.

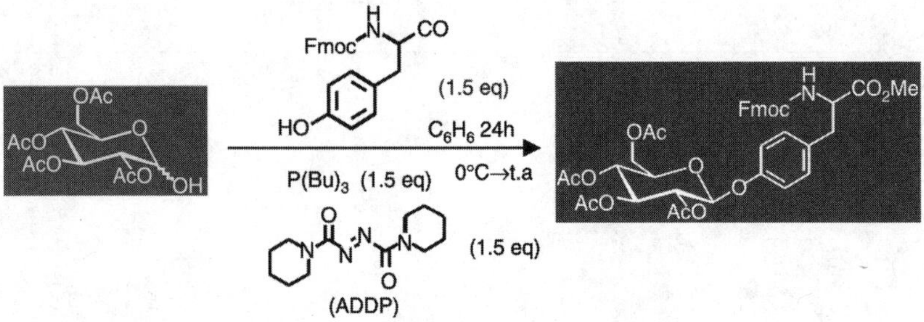

OAc AcO AcO AcO OH O
H N Fmoc CO HO
(1.5 eq)
C6H6 24h
P(Bu)3 (1.5 eq)
0°C→t.a
(ADDP) (1.5 eq)
H N Fmoc CO2Me OAc AcO AcO AcO O O

FIGURE 3.16 Mitsunobu ADDP-TBP redox system.

benzene, which allows the coupling with acid counterpart HA **4** (Figure 3.16) of pK_a larger than 11 (i.e. Tyr pK_a 10.5), as reported by Tsunoda [45].

The application of Mitsunobu method in solid phase has required the use of other solvents and redox systems due to the properties of resin support [46]. After the examinations of several Mitsunobu reagents and conditions, the coupling was optimized with TMAD (N,N,N′,N′-tetramethylazodicarboxamide) and tributylphosphine as redox system; the reaction was conducted in anhydrous THF/CH_2Cl_2 solvent mixture, which is compatible with resin solid support, at room temperature and with 10-fold excess of sugar moiety. The glycosylation step was not accompanied by side-reactions. As target compounds for our synthesis, we selected GlyGly-Tyr(β-tetra acetyl glucose)-GlyGly. The process has been designed to minimize the exposure of the glycosidic bonds to acid, required for resin cleavage or side chain deprotection step. The conditions used for Fmoc-deprotection and de-O-acetylation step of sugar moiety, including sodium carbonate, sodium methoxide and hydrazine-hydrate in methanol, were found not to cause β-elimination or racemisation of peptide, stereocenters. Final cleavage and side chain deprotection were performed with short TFA treatment. This latter procedure is not broadly applicable, due to glycosidic bonds acid instability of certain molecules, such as O-fucosidic-derivatives and non protected sugars; therefore TFA cleavage conditions must be modulated on the basis of substrate nature.

The adaptation of this procedure in solid phase allows expansion of the easy and automated access to novel glycosylamino acid building blocks required for chemical libraries construction, not only by convergent, but also by stepwise, methods.

VI. CONCLUSION

Novel technologies, including Mitsunobu ligation process, convergent chemical or chemoenzymatic synthesis, and other linking procedures, contribute to the goal of creating diverse glycopeptides moieties in combinatorial displays, in order to perform extensive biological high throughput screening of wider molecules repertoires: a great tool for drug discovery and an invaluable implement in basic research [47]. Recently, chemoselective synthesis, compatible with major sensitive glycosydic linkage, and chemoenzymatic [48, 49] procedures providing milder selective deprotection and coupling processes, are promising tools in glycopeptides building blocks assembly and solid phase synthesis. A challenge for the future is to fully merge combinatorial approaches, including computer-assisted design, automated chemical and/or enzymatic synthesis and automated screening, with genetic technologies, in order to cover the therapeutic range of glyco-proteins and -peptides from antimicrobial and anti-inflammatory agents to tumour vaccines [50, 51] and immunomodulators [52].

BIBLIOGRAPHY

1. Chakraborty TK, Ghosh S, Jayaprakash S, Sugar amino acids and their uses in designing bioactive molecules, *Curr. Med. Chem.*, 9(4):421–35, 2002.
2. Chakraborty TK, Ghosh S, Jayaprakash S, Sugar amino acid based scaffolds-novel peptidomimetics and their potential in combinatorial synthesis, *Chem. High Throughput Screen*, (5):373–87, 2002.

3. Seitz O, Glycopeptide synthesis and the effects of glycosylation on protein structure and activity, *Chembiochem*, 1(4):214–46, 2000.
4. Kunz H and Birnbach S, Synthesis of *O*-glycopeptides of the tumor-associated T_N and T antigen type and their binding to bovine serum albumin, *Angew. Chem. Int. Ed. Engl.*, 25:360–2, 1986.
5. Xia J, Alderfer JL, cf. Piskorz, Matta KL, Total synthesis of sialylated and sulfated oligosaccharide chains from respiratory mucins, *Chemistry*, 6(18): 3442–51, 2000.
6. George SK, Schwientek T, Holm B, Reis CA, Clausen H, Kihlberg J, Chemoenzymatic synthesis of sialylated glycopeptides derived from mucins and T-cell stimulating peptides, *J. Am. Chem. Soc.*, 123(45):11117–25, 2001.
7. Kunz H, Glycopeptides of biological interest: a challange for chemical synthesis, *Pure Appl. Chem.*, 65:1223–32, 1993.
8. Kurosaka A, Kitagawa H, Fukui S, et al., A monoclonal antibody that recognize a cluster of a disaccharide, NeuAcα2-6GalNAc, in mimic type glycoproteins, *J. Biol. Chem.*, 263:8724–6, 1998.
9. Otvos Jr. L, Wroblewsky K, Kollat E, Perczel A, Hollosi M, Fasaman GD, HCJ Ertl, J Thurin, Coupling strategies in solid-phase synthesis of glycopeptides, *Peptide Res.*, 2, 362–66, 1989.
10. Otvos Jr. L, Urge L, Hollosi M, Wroblewsky K et al., Automated solid-phase synthesis of glycopeptides, Incorporation of unprotected mono- and disaccharide units of N-glycoprotein Antennae into T cell epitopic peptides, *Tetrahedron Letters*, 31:5889–92, 1993.
11. Seitz O, Kunz H, HYCRON, an allylic anchor for high-efficiency solid phase synthesis of protected peptides and glycopeptides, *J. Org. Chem.*, 62:813–826, 1997.
12. Seitz O, Kunz H, A novel allylic anchor for solid-phase synthesis of protected and unprotected O-glycosylated mucin-type glycopeptides, *Angew. Chem. Int. Ed. Engl.*, 34:803–805, 1995.
13. Meldal M, Auzanneau FI, Hindsgaul O, Palcic MM, A PEGA resin for use in the solid phase chemical/enzymatic synthesis of glycopeptides, *J. Chem. Soc. Chem. Commun.*, 1849–50, 1994.
14. Paulsen H, Schleyer A, Mathieux N, Meldal M and Bock K, New solid-phase oligosaccharide synthesis on glycopeptides bound to solid phase, *J. Chem. Soc. Perkin Trans. I*, 281–93, 1997.
15. Köpper S, Polymer-supported enzymic synthesis on a preparative scale, *Carbohydrate Research*, 265:161–66, 1994.
16. Schuster M, Wang P, Paulson JC, Wong CH, Solid-phase chemical-enzymatic synthesis of glycopeptides and oligosaccharides, *J. Am. Chem. Soc.*, 116:1135–6, 1994.
17. Unverzagt C, Kunz H, Paulson JC, High-efficiency synthesis of syaliloligosaccharides and sialoglycopeptides, *J. Am. Chem. Soc.*, 112:9308–9, 1990.
18. Mizuno Mamoru, Recent trends in glycopeptide synthesis, *Trends in Glycoscience and Glycotechnology*, 13, 69:11–30, 2001.
19. Shin I, Lee J, Facile Synthesis of *N*- and *O*-Glycosylated α-Aminooxy Acids as Building Blocks for the Structural Studies of Glycosylated Pseudopeptides, *Synlett.*, 9:1297, 2000.
20. Arsequell G, Valencia G, Recent advances in the synthesis of complex N-glycopeptides, *Tetrahedron Asymmetry*, 10:3045–94, 1999.

21. Unverzagt C, Kunz H, Stereoselective synthesis of glycosides and anomeric azides of glucosamine, *J. Prakt. Chem.*, 334:570–8, 1992.
22. Meldal M, Bock K, Pentafluorophenyl esters for temporary carboxyl group protection in solid phase synthesis of N- linked glycopeptides, *Tetrahedron Lett.*, 31:6987–90, 1990.
23. Cohen-Anisfeld ST, Lansbury PT, A practical, convergent method for glycopeptide synthesis, *J. Am. Chem. Soc.*, 115:10531–37, 1993.
24. Schweizer F, Glycosamino acids: building blocks for combinatorial synthesis – Implications for drug discovery, *Angew. Chem. Int. Ed. Engl.*, 41:23053, 2002.
25. Jansson AM, Jensen KJ, Meldal M, Lomako J, Lomako WM, Olsen CE, Bock K, Solid-phase glycopeptide synthesis of tyrosine-glycosylated glycogenin fragments as substrates for glucosylation by glycogenin, *J. Chem. Soc. Perkin. Trans.*, 11001, 1996.
26. Jensen KJ, Meldal M, Bock K, Glycosilation of phenols: preparation of 1,2- cis and 1,2- trans glycosylated tyrosine derivatives to be used in solid-phase glycopeptide synthesis, *J. Chem. Soc. Perkin. Trans.*, 1, 2119, 1993.
27. Lowary T, Meldal M, Helmboldt A, Vasella A and Bock K, Novel type of rigid C-linked glycosylacetylene-phenylalanine building blocks for combinatorial synthesis of C-linked Glycopeptides, *J. Org. Chem.*, 63:9657, 1998.
28. Hayes BK, Hart GW, Novel forms of protein glycosylation. *Curr. Opin. Struct. Biol.*, 4:692–6, 1994.
29. Williams DH, Rajananda V, Williamson MP, Bojesen G, *Topics in Antibiotic Chemistry*, ed. P.G. Sammes, Wiley, New York, 1980, vol.5, p.119.
30. Nicolaou KC, Cho SY, Hughes R, Winssinger N, Smethurst C, Labischinski H, Endermann R, Solid and solution phase synthesis of vancomycin and vanxcomycin analogues with activity against vancomycin-resistant bacteria, *Chemistry*, 7(17):3798–823, 2001.
31. Nicolaou KC, Hughes R, Cho SY, Winssinger N, Labischinski H, Endermann R, Synthesis and biological evaluation of vancomycin dimers with potent activity: target accelerated combinatorial synthesis, *Chemistry*, 7(17):3824–43, 2001.
32. St Hilaire PM, Meldal M, Glycopeptide and oligosaccharide libraries, *Angew. Chem. Int. Ed. Engl.*, 39(7):1162–79, 2000.
33. Halkes KM, St Hilaire PM, Crocker PR, Meldal M, Glycopeptides as oligosaccharide mimics: high affinity sialopeptide ligands for sialoadhesin from combinatorial libraries, *J. Comb. Chem.*, 5(1):18–27, 2003.
34. Elofsson M, Roy S, Walse B, Kihlberg J, Solid-phase synthesis and conformational studies of glycosylated derivatives of helper-T-cell immunogenic peptides from hen-egg lysozyme., *Carbohydr. Res.*, 17, 246:89–103, 1993.
35. Dukic-Stefanovic S, Schicktanz D, Wong A, Palm D, Riederer P, Niwa T, Schnizel R, Munch G, Characterization of antibody affinities using an AGE-modified dipeptide spot library, *J. Immunol. Methods.*, 266(1–2):45–52, 2002.
36. Macmillan D, Daines AM, Bayrhuber M, Fitsch SL, Solid-phase synthesis of thioether-linked glycopeptide mimics for application to glycoprotein semi-synthesis, *Org. Lett.*, 4(9):1467–70, 2002.
37. Arya P, Barkley A, Randall KD, Automated High-Throughput Synthesis of Artificial Glycopeptides. Small-Molecule Probes for Chemical Glycobiology, *J. Comb. Chem.*, 4(3):193–8, 2002.

38. Arya P, Kutterer KM, Barkley A, Glycomimetics: A programmed approach toward neoglycopeptide libraries, *J. Comb. Chem.*, 2(2):120–6, 2000.
39. Mitsunobu O, The use of diethyl azodicarboxylate and triphenylphosphine in synthesis and transformation of natural products, *Synthesis,* 1, 1981.
40. Matter H, Baringhaus KH, Naumann T, Klaubunde T, Pirard B, Computational approaches towards the rational design of drug-like compound libraries, *Comb. Chem. High. Throughput Screen*, 4(6):453–75, 2001.
41. Todeschini R, Consonni V, Handbook of Molecular Descriptors, Wiley-VCH, Weinheim Germany, in *Methods and Principles in Medicinal Chemistry*, vol.11, pp. 667, 2000.
42. Balaban S, Filip PA, Ivanciuc O, Computer generation of acyclic graphs based on local vertex invariants and topological indices derived canonical labelling and coding of trees and alkanes, *J. Math. Chem.*, 11:79–105, 1992.
43. Bonchev D, Information Theoric Indices for Characterization of Chemical Structures Research Studies Press, Letchworth UK, p. 249, ISBN 0-471-90087-7, 1983.
44. Ivanciuc T, Klein DJ, Seitz WA, Balaban AT, Wiener index extension by counting even/odd graph distances, *J. Chem. Inf. Comput. Sci.*, 41(3), 2001.
45. Tsunoda T, Ozaki F, Ito S, Novel Reactivity of Stabilized Methylenetributylphosphorane: A new Mitsunobu Reagent, *Tetrahedron Letters*, 35(81):5081–82, 1994, TMAD.
46. Rano TA, Chapman KT, Solid phase synthesis of aryl ethers *via* the mitsunobu reaction, *Tetrahedron Letters*, 36:3789–92, 1995, Solid phase synthesis (aryl resin coupling) TMAD Bu_3P.
47. Schweizer F, Glycosamino acids: building blocks for combinatorial synthesis-implications for drug discovery, *Angew. Chem. Int. Ed. Engl.*, 41(2):231–53, 2002.
48. Schuster M, Wang P, Paulson JC, Chi-Huey WongSchuster. Solid-phase chemical-enzymic synthesis of glycopeptides and oligosaccharides, *J. Am. Chem. Soc.*, 116:1135–6, 1994.
49. Galli-Stampino L, Meinjohanns E, Frische K, Meldal M, Jensen T, Werdelin O, Mouritsen S, T-cell recognition of tumor-associated carbohydrates: the nature of the glycan moiety plays a decisive role in determining glycopeptide immunogenicity, *Cancer Research*, 57:3214–222, 1997.
50. Sames D, X-T Chen, Danishefsky SJ, A convergent total synthesis of an extended mucin motif, *Nature*, 389:587–91, 1997.
51. Braun P, Waldmann H, Kunz H, Chemoenzymatic synthesis of O-glycopeptides carryng the tumor associated TN-antigen structure, *Bioorg. Med. Chem.*, 1:197–207, 1993.
52. Feizi T, Carbohydrate-mediated recognition systems in innate immunity, *Immunol. Rev.*, 173:79–88, 2000.

4 Design of Protease Inhibitors by Computer-Assisted Combinatorial Chemistry

Vladimir Frecer, Enrico Burello and Stanislav Miertus

CONTENTS

I. INTRODUCTION

Proteases remain at the forefront of understanding and intervention in human biochemistry and pathological processes, such as degenerative diseases and malignant tumor development. They also play a key role in determining the infectiveness of various pathogens, ranging from retroviruses to fungi and protozoa. Therefore, proteases represent attractive targets for the design of specific inhibitors as potential therapeutic agents [1]. One protease family that has recently attracted attention is that of the aspartic proteases [2, 3]. This family also includes mammalian proteins, such as the digestive enzyme pepsin, renin that is involved in control of blood pressure and lysosomal cathepsin D. Thus, inhibition of the proteases has therapeutic value in conditions varying from hypertension, inflammation, tumor metastasis, and Alzheimer's diseases, to fungal or viral infections and malaria.

In the past decade we have witnessed the development of numerous inhibitors of the aspartic protease (PR) of human immunodeficiency virus (HIV) [4–6]. The genome of HIV encodes enzymes necessary for viral replication, which are translated as part of a larger polyprotein precursor, whose proteolytic processing during the virus assembly and maturation is performed by the PR. Therefore, PR is essential for the HIV life cycle and represents a suitable anti-HIV drug target. Several PR inhibitors, such as saquinavir, indinavir, nelfinavir, ritonavir, amprenavir, and lopinavir, were approved by FDA for the clinical use in patients with acquired immunodeficiency syndrome (AIDS). AIDS, typically, leads to opportunistic infections or malignancies associated with the underlying damage of the immune system by the viral particles characterized by progressive loss of CD4 helper T cells. Unfortunately, despite their initial success in viral load reduction, widespread use of these inhibitors resulted in the emergence of drug-resistant HIV mutants [7] that pose a continuing challenge for the design of new potent PR inhibitors active against wider spectrum of HIV PR mutants, including the drug-resistant HIV strains.

Combinatorial chemistry techniques have been developed to provide large numbers of drug candidates, via parallel synthesis and screening of libraries of compounds, to increase the rate and probability of generating lead compounds with desired potencies and specificities. Computational methods are being increasingly used to assist the design of combinatorial libraries, in order to reduce the number of compounds that have to be synthesized, without decreasing significantly the chemical diversity of the library and the probability of finding the leads, while increasing the drug-like character of the molecules and similarity to the known active substances [8–11]. The relevant computational procedures are described as virtual library, focusing or targeting and in silico or virtual screening. They comprise several basic strategies. Fragment-based and, analog-based, virtual library focusing, represent selection procedures that enable the choosing of sets of reagents (fragments) or analogs, so as to generate libraries that focus around the properties of an active compound—typically, a drug-like molecule. On the other hand, structure-based targeting uses known, or suspected, active site of a target receptor to select compounds, which are likely to bind within the defined active site.

In collaboration with University of Trieste, we have developed rational approaches for the design and synthesis of peptidomimetic and non-peptidic inhibitors of HIV PR, utilizing structure-based [12–15], as well as combinatorial, library design methods [16, 17]. In this paper, we survey computer-assisted studies on the design, focusing and in silico screening of virtual combinatorial libraries of peptidomimetics and cyclic ureas, as potential anti-HIV agents, that were carried out in our laboratory.

II. COMBINATORIAL DESIGN OF PEPTIDOMIMETIC INHIBITORS OF HIV PR

We summarize here the efforts to generate a virtual library of peptidomimetic pentameric inhibitors of the HIV PR, derived from ritonavir (Figure 4.1), developed by the Abbott Laboratories [18,19] using computer-assisted combinatorial chemistry methods [20].

The synthetic scheme of ritonavir-like C_2-symmetric peptide inhibitors permits the use of amino acid (AA) and carboxylic acid (CA) building blocks, that are reacted with the central diaminomonohydroxy core [21] to yield various peptidomimetic pentamers. Thus, a virtual library of pseudo pentapeptides can be generated by attaching appropriate "side chains" to the scaffold containing the core, to obtain analogs consisting of five building blocks, each carrying one side chain attachment site (R-group) (Figure 4.2). In practice, the central core is first synthesized from an amino acid AA_1 and an aldehyde ALD_4, while attaching amino acid AA_2 and the ending carboxylic acids CA_3 and CA_5 complete the inhibitor pentapeptide.

From several hundred amino acids, aldehydes, and carboxylic acids available in the commercial databases of suppliers of chemicals [22], we chose 473 CA_3 and CA_5 for the ending R_3 and R_5 positions, 86 and 60 natural

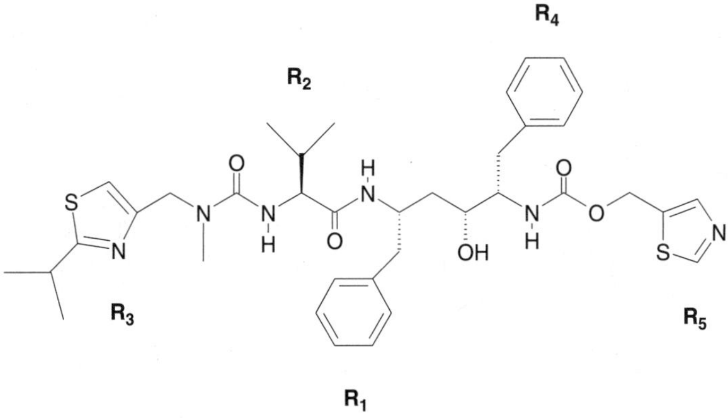

FIGURE 4.1 Ritonavir, potent peptidomimetic inhibitor of HIV PR [18, 19].

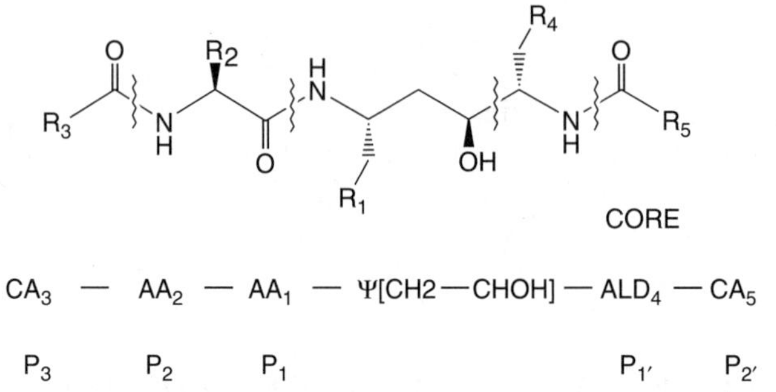

CORE

CA_3 — AA_2 — AA_1 — Ψ[CH2—CHOH] — ALD_4 — CA_5

P_3 P_2 P_1 $P_{1'}$ $P_{2'}$

FIGURE 4.2 Diaminomonohydroxy core based scaffold of peptidomimetic inhibitors of HIV PR. The nomenclature of Schechter and Berger used here [36] designates the substrate cleavage sites P_6–P_5–P_4–P_3– P_2–P_1 ... P'_1–P'_2–P'_3–P'_4, etc. with the scissile bond between P_1 and P'_1 and the C-terminus of the substrate on the prime site. The corresponding binding sites on the enzyme are indicated as S_6–S_5–S_4–S_3–S_2–S_1 ... S'_1–S'_2–S'_3–S'_4, etc.

and unusual AA_2 and AA_1 for positions R_2 and R_1, respectively, and 37 ALD_4 for the position R_4. We selected mainly fragments that contained a methylene link to the pseudo peptide backbone, to enhance the flexibility of the side chains of the generated analogs and improve their interaction with the specificity pockets of the binding site of HIV PR receptor.

The estimated size of a virtual library that contains all the initially selected building blocks (fragments) is prohibitive, even for a computational analysis:

$$473\,(CA_3) \times 86\,(AA_2) \times 60\,(AA_1) \times 37\,(ALD_4) \\ \times 473\,(CA_5) \sim 8.9 \times 10^{10} \text{ compounds}$$

Therefore, we introduced filters and penalties to reduce the number of fragments used and, subsequently, the analogs generated, rather than enumerated, the full library containing all available combinations of reagents.

A. Selection of Fragments

The strategy we employed to reduce the size of the virtual library was based on independent analysis of the properties of individual building blocks, in order to select subsets of suitable and diverse reactants from a pool of available chemicals.

First, we collected and analyzed a set of 14 crystal structures of known HIV PR inhibitors bound to the native PR, which included ritonavir, saquinavir, indinavir, nelfinavir, amprenavir, lopinavir, KNI-272, DMP-323, BMS-182193, GR-137615, L-700417, SB-204144, PNU-140690, and A-77003 [23, 24] and were taken from Protein Data Bank [25]. The aim of this analysis was to reveal optimum ranges of molecular properties of the potent tight binding transition-state-mimicking peptidomimetic and nonpeptidic HIV PR inhibitors, for both the whole inhibitor molecules and their fragments. To characterize the physico-chemical properties of the fragments and inhibitors, 13 different descriptors related to the shape, size, molecular field, and conformational flexibility, which are related to the accommodation of the reagents in the specificity pockets of the PR receptor binding site, were computed using the Cerius^2 program [20]. The optimum ranges of the properties were obtained by statistical analysis of the computed descriptors for 14 active PR inhibitors and their building blocks, in terms of upper and lower bounds and average values (Table 4.1).

Each of the AA_1, AA_2, CA_3, ALD_4 and CA_5 fragments to be used in the virtual library generation was then characterized by the same set of descriptors.

TABLE 4.1
Physicochemical and structural properties of known potent peptidomimetic and nonpeptidic HIV PR inhibitors and their building blocks used in the focusing of combinatorial libraries

	Descriptor Ranges[a]					
	Fragments[b]					
Descriptors	AA_1	AA_2	CA_3	ALD_4	CA_5	Inhibitors[c]
Dipole moment [debye]	0.1–2.7	0.0–3.7	0.1–2.6	0.1–2.9	0.0–3.7	0.8–23.0
Surface area [Å^2]	110–188	85–167	122–206	122–143	79–167	579–937
Principal moment of inertia [Da·Å^2]	24–119	13–131	59–230	57–113	13–131	2340–6398
Molecular volume [Å^3]	72–144	56–127	94–154	94–113	56–127	462–752
Molecular weight [Da]	58–139	44–134	92–170	92–124	44–134	505–795
No. of rotatable bonds	0–0	0–2	0–3	0–1	0–2	12–27
No. of hydrogen bond acceptors	0–2	0–1	0–3	0–1	0–3	5–9
No. of hydrogen bond donors	0–2	0–2	0–1	0–0	0–2	2–7
Lipophilicity (AlogP98)	−0.1–2.3	0.6–2.3	0.7–2.3	2.3–2.4	−0.8–2.0	1.8–7.4

[a]Descriptor ranges were derived by analyzing the properties of building blocks (residues) of 14 potent peptidomimetic and nonpeptidic HIV PR inhibitors, for which crystal structures of inhibitor : PR complex have been published in the Protein Data Bank [25].
[b]AA_1 and AA_2 represent amino acids at P_1 and P_2 positions, CA_3 and CA_5 mean carboxylic acids, constituting the P_3 and P'_2 positions, and ALD_4 is an aldehyde forming the P'_1 building block of the peptidomimetic analogs (Figure 4.2).
[c]Considered HIV PR inhibitors : ritonavir, amprenavir, indinavir, saquinavir, nelfinavir, lopinavir, A77003, BMS-182913, DMP-323, KNI-272, SB-204144, PNU-140690, GR-137615, and L-700417 [23, 24].

Next, a penalty function was assigned to the reagents, in order to discriminate those with property values lying outside the previously determined optimum ranges. Fragment filtering was then carried out by selecting only the reagents with suitable properties. At the same time, we applied the MaxMin distance-based diversity function to select the most diverse fragments that efficiently span the chemical space. From Monte Carlo optimization of molecular diversity under the property ranges, with a predetermined number of final fragments, we obtained the focused reagent subsets (Figure 4.3) that reduce the size of the virtual library to:

$$12\,(CA_3) \times 5\,(AA_2) \times 6\,(AA_1) \times 3\,(ALD_4) \times 10\,(CA_5) = 10{,}800 \text{ compounds}$$

The virtual library was then generated by attaching the selected fragments to the peptidomimetic scaffold (Figure 4.2).

B. Selection of Analogs

The size of the generated fragment-focused virtual library can be further reduced by applying additional library focusing methods to select a small subset of most diverse analogs. For this purpose, optimization of the analogs' geometry was performed with preserved pseudopeptide backbone in bound conformation, leaving only the side chain atoms flexible. Next, molecular physico-chemical descriptors were computed for the 10,800 generated analogs. Finally, an analog-filtering procedure was applied, which compared the properties of the analogs with the property ranges calculated for the set of known HIV PR inhibitors, which permitted the selection of only those analogs that complied with the optimal property ranges of potent PR inhibitors (Table 4.1). The property ranges were used to define a penalty function, which was then employed in the identification of 100 most diverse peptidomimetic analogs with suitable molecular properties (lowest penalty values).

C. Structure-Based Library Targeting

The size of the analog-focused diverse subset was further reduced, by applying structure-based targeting and *in silico* screening methods to select a small and highly restricted set of analogs—lead compounds. Docking of the analogs to the binding site of a receptor is routinely used to filter out compounds with low predicted binding affinities to their target. Typically, Monte Carlo methods are employed to search for analog structures and conformations that can fit into the binding site of the targeted enzyme [26]. Fitting of various conformers, generated for each analog into the binding site model, was carried out by comparing principal moments of inertia of both the binding site and the conformer.

Thus, the selected subset of 100 most diverse peptidomimetic analogs was docked to the model of the PR receptor (Figures 4.4 and 4.5), derived from the crystal structure of ritonavir : PR complex [18,19], using Monte Carlo

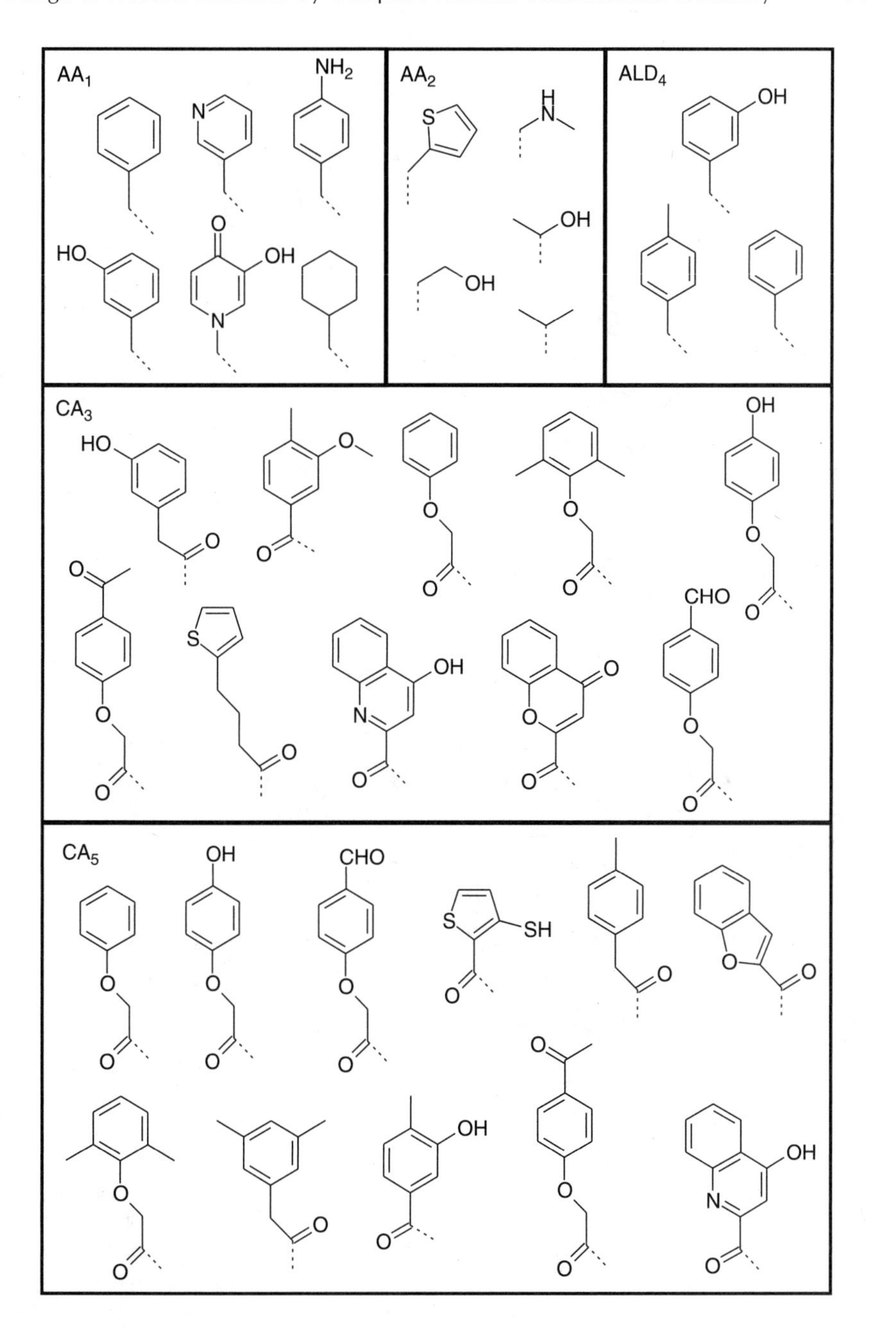

FIGURE 4.3 Selected fragments—"side chains" of amino acids (AA_1, AA_2), aldehydes (ALD_4), and carboxylic acids (CA_3, CA_5), building blocks of the peptidomimetic inhibitors of HIV PR, selected by fragment-based library focusing techniques.

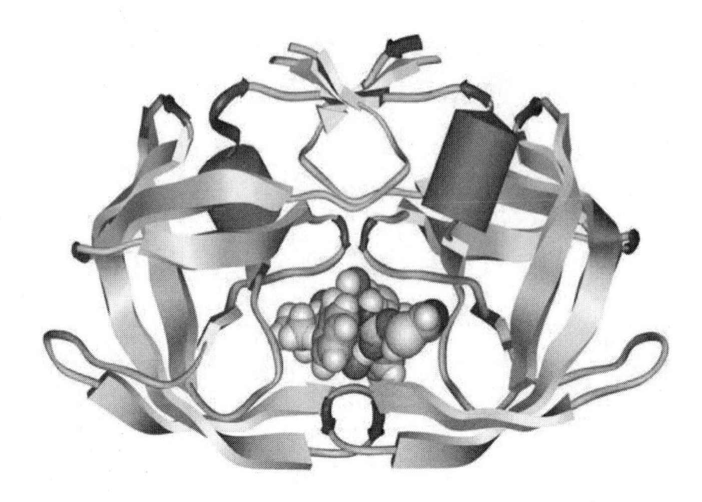

FIGURE 4.4 Crystal structure of inhibitor : enzyme complex ritonavir : PR [19].

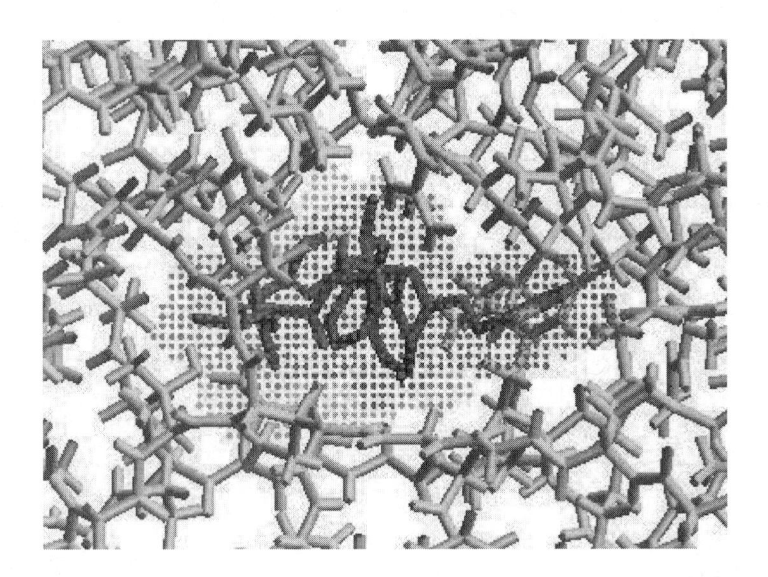

FIGURE 4.5 Close-up of the binding site of the ritonavir : PR complex (Figure 4.4) with ritonavir in cylinder representation (hydrogens were omitted for better clarity) occupying the binding site of HIV PR (green). The binding site model (red vertices) is filled with the docking grid, which encloses the space available for different conformations of analogs when docked to the HIV PR.

procedure with a flexible analog fitting algorithm. This procedure generated 20 best binding conformers per analog, which were clustered into conformational families, based on their mutual r.m.s. deviations, using the Jarvis-Patrick method [30]. The best conformer for each analog, which displayed the highest docking score computed using molecular mechanics and the cff91 force field [27], was then selected. Subsequently, a ligand fit scoring function, which estimated the inhibition (binding) constant (K_i) by combining ligand–receptor nonbonding interaction energy (E_{vdW}) with two additional terms related to the buried polar surface area between the ligand and the receptor (C_{pol}, T_{pol}) as: $pKi = a - b \cdot E_{vdW} + c \cdot C_{pol} - d \cdot T_{pol}^2$ [28], was computed. This virtual screening for inhibition activity of the designed library of analogs resulted in the selection of 48 most promising analogs, with predicted pK_i constants better than 3.0 ($K_i < 10^{-3}$ M).

D. QSAR Analysis

To relate the predicted inhibition constants (pK_i) derived from the predefined general ligand scoring function [28] with the in vitro inhibition activity against the HIV PR, we correlated the experimental K_i constants of the 14 known PR inhibitors with the ligand fit scoring function parameters of these inhibitors, which were submitted to the same docking and scoring procedure applied previously to the focused subset of peptidomimetics. The resulting regression equation (Figure 4.6) was used to predict the inhibition potencies against the HIV PR for the selected 48 analogs. The 16 best compounds with the highest predicted pK_is that combined high inhibition activity against the HIV PR with diversity criteria were selected as potential lead compound candidates (Figure 4.7).

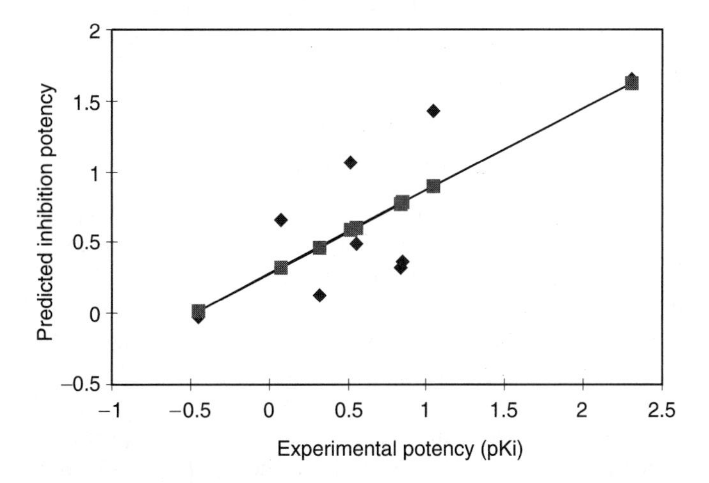

FIGURE 4.6 QSAR linear regression analysis of known HIV PR inhibitors that relates the inhibition constants K_i to molecular scoring function parameters [27] E_{vdW}, C_{pol}, and T_{pol}^2, which characterize binding of the inhibitors to the PR model (Figure 4.4):

$$pK_i = -1.374 - 0.009 \cdot E_{vdW} - 0.113 \cdot C_{pol} + 0.005 \cdot T_{pol}^2$$

($N = 12$, outliers = 2, $R^2 = 0.85$, $R_{xv}^2 = 0.61$, F-test = 11.23, $\alpha > 95\%$).

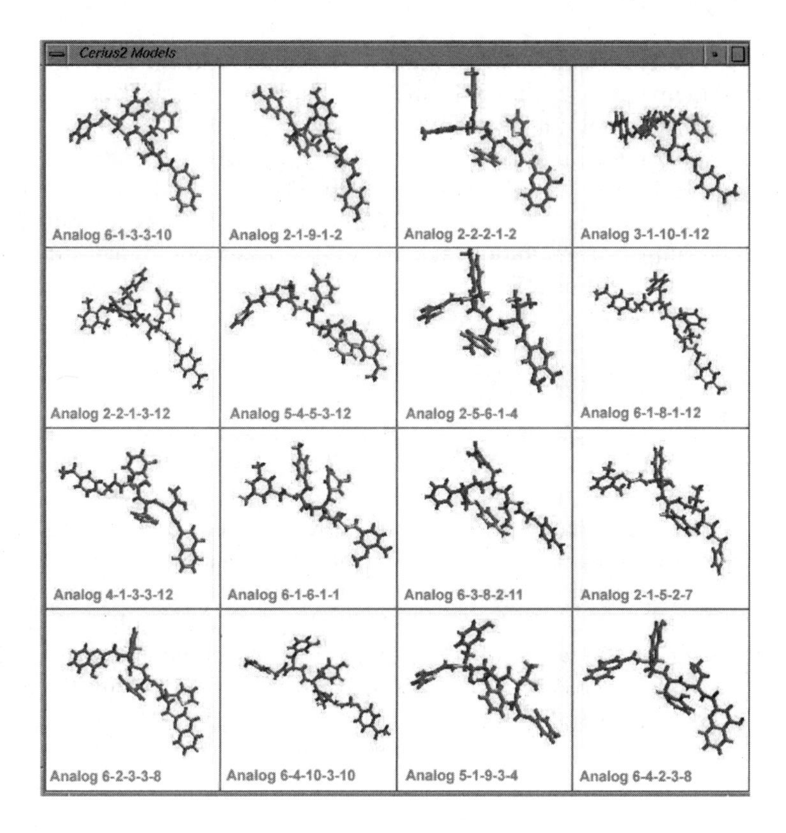

FIGURE 4.7 The 16 best predicted leads–peptidomimetic inhibitors of HIV PR with the lowest values of the predicted inhibition constant toward the PR, as displayed in Cerius^2 [20].

III. SIMILARITY ANALYSIS

Among these 16 best lead compound candidates, one exceptional analog (Analog-5-1-9-3-4) (Figure 4.8) with a predicted inhibition constant to the HIV PR of approximately 10nM, was identified. To optimize the properties of this analog we searched the full fragment-focused virtual library of 10,800 analogs for compounds topologically similar to the lead. To this end, we calculated a set of topological indexes, such as InfoContent, Balaban and Wiener [29], and clustered the library into more than 1400 clusters. The cluster, which contained the lead, also included 19 other topologically similar analogs. These molecules were docked to the PR receptor, then a ligand scoring function was calculated and their pK_i were predicted from the QSAR equation derived for known HIV-1 PR inhibitors (Figure 4.6). Out of them, 9 most potent lead variants, with predicted K_i between 10 and 100 nM, were selected for synthesis and experimental determination of the inhibitory activity (Table 4.2, Figures 4.9 and 4.10). The outcome of experimental testing will be published elsewhere.

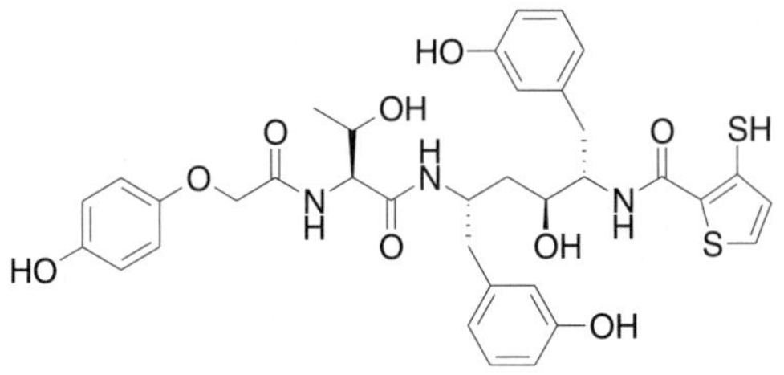

FIGURE 4.8 Lead compound—Analog 5-1-9-3-4 with exceptionally high predicted inhibition potency against HIV PR.

TABLE 4.2
Library subset containing 9 potent analogs, obtained by similarity search around the previously identified and targeted lead—the Analogue 5-3-9-1-4, with predicted inhibitory activities against HIV PR in the low nanomolar range

Analog	Docking Score[a] [kJ·mol^{-1}]	Minimized Energy[b] [kJ·mol^{-1}]	Ligand Score[c]	E_{vdW}[d] [kJ·mol^{-1}]	C_{pol}[e] [Å^2]	T^2_{pol}[f] [Å^4]	Predicted Activity, K_i[g] [nM]
Analog 2-5-8-2-4	−6220.7	−334.5	6.43	−345.4	21.1	502.5	13
Analog 2-4-9-3-4	−12140.3	−184.0	6.09	−316.2	20.8	464.8	35
Analog 5-3-9-1-4	−20526.9	−251.9	5.46	−282.3	18.7	377.2	112
Analog 4-1-8-1-4	−8699.4	−346.1	5.39	−317.0	15.1	262.3	78
Analog 4-5-8-1-4	−8747.4	−391.6	5.07	−296.9	14.7	287.7	80
Analog 3-1-1-1-4	−5416.8	−686.1	5.01	−278.4	15.9	329.2	101
Analog 2-1-8-2-4	−4551.5	−249.1	4.97	−325.3	11.1	122.9	109
Analog 5-4-9-1-4	−9642.1	−321.7	4.95	−293.8	13.9	246.1	111
Analog 5-1-9-3-4	−5467.4	−653.9	4.87	−262.3	16.3	348.9	125

[a]Docking score was derived by Monte Carlo assisted fitting of flexible ligand to rigid HIV PR receptor.
[b]Energy of the analog minimized in the binding site of the HIV PR using cff91 force field [27].
[c]Ligand scoring function that predicts binding affinity of the analogs to the HIV PR model, based on the computed parameters E_{vdW}, C_{pol} and T^2_{pol} [28].
[d]E_{vdW}—ligand–receptor nonbonding interaction energy.
[e]C_{pol}—buried polar surface area between a protein and ligand, which involves attractive protein–ligand interactions.
[f]T^2_{pol}—squared buried polar surface area that involves both attractive and repulsive protein–ligand interactions.
[g]Predicted inhibition constant of HIV PR, computed by the QSAR equation valid for known HIV PR inhibitors (Figure 4.6).

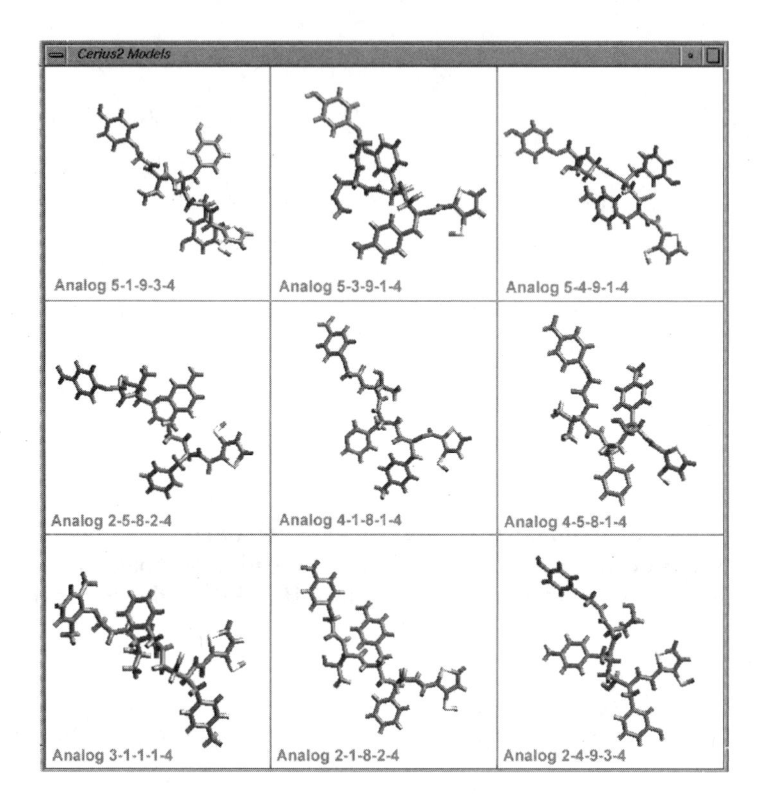

FIGURE 4.9 Molecular models of the nine potent lead variants derived by similarity analysis of the Analog 5-1-9-3-4, generated by the ligand fit docking algorithm of Cerius^2 [20].

IV. COMBINATORIAL DESIGN OF NONPEPTIDIC INHIBITORS OF HIV PR

In the second example, we summarize our efforts to design highly focused virtual library subset of nonpeptidic inhibitors of HIV PR, using structure-based computational approach [20].

Recently, cyclic ureas have been reported to constitute an entirely new class of potent ccand perspective nonpeptidic inhibitors of HIV PR [31, 32]. Design of virtual library of cyclic urea inhibitors (Figure 4.11) involves attachment of fragments to the central core—seven-membered urea ring—following a predefined synthetic scheme (Figure 4.12). The central core of the inhibitor is synthesized from amino acid (AA) and aldehyde (ALD) building blocks (R_1 and R_2), while the attachment of two flanking groups to the ring nitrogens, in the form of chlorinated cyclic hydrocarbon (RING), at positions R_3 and R_4, yields the final cyclic urea inhibitor.

The number of analogs that can be potentially synthesized in a combinatorial experiment, by using the cyclic amino acids, aldehydes and

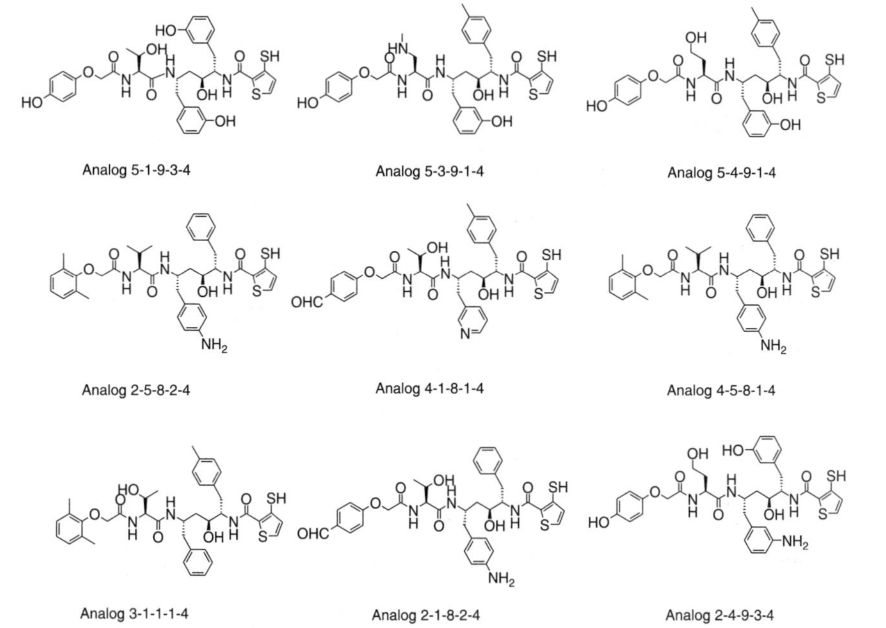

Analog 5-1-9-3-4   Analog 5-3-9-1-4   Analog 5-4-9-1-4

Analog 2-5-8-2-4   Analog 4-1-8-1-4   Analog 4-5-8-1-4

Analog 3-1-1-1-4   Analog 2-1-8-2-4   Analog 2-4-9-3-4

FIGURE 4.10 Chemical structures of the nine potent lead variants derived by similarity analysis of the Analog 5-1-9-3-4, with predicted K_i values against HIV PR in the low nanomolar range.

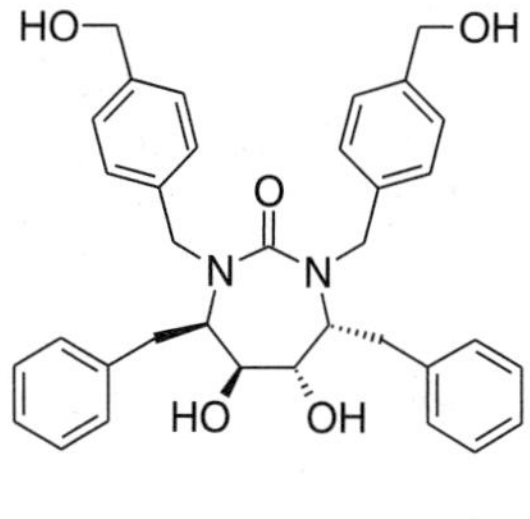

FIGURE 4.11 Cyclic urea inhibitor DMP-323, with inhibition constant against the wild type HIV PR $K_i = 340\,\text{pM}$ [31].

chlorinated cyclic hydrocarbons available in commercial databases of suppliers of chemicals [22], can easily exceed the capacity of any synthesis and screening apparatus:

$$850\,(\text{AA}) \times 735\,(\text{ALD}) \times 1350\,(\text{RING}) \times 1350\,(\text{RING}) \sim 1.1 \times 10^{12} \text{ compounds}$$

Therefore, we again introduced appropriate "structural" filters and penalties to reduce the number of fragments employed in the generation of virtual library of cyclic ureas.

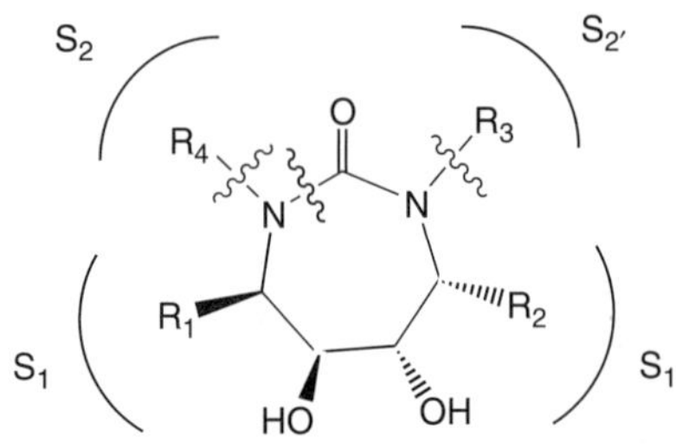

FIGURE 4.12 Synthetic scheme for cyclic ureas (see also Figure 4.2).

A. Selection of Fragments

We used the same strategy for the fragment-based library focusing as described above, which relies on the optimum ranges of molecular properties obtained by analyzing 14 tight binding HIV PR inhibitors (Table 4.1). The same properties and descriptors were then calculated for the pools of available reactants—amino acids, aldehydes, and chlorinated cyclic hydrocarbons building blocks. Penalty scores were assigned to fragments whose descriptor values laid outside the optimal ranges, in order to filter out those that differed in their properties. The diversity between the fragments was evaluated using the distance-based MaxMin function and the topological indices: Balaban, Hosoya, Wiener, and Zagreb [30]. Six amino acids and three aldehydes were selected as the subset of suitable and most diverse building blocks capable of covering the topological space explored (Figure 4.13).

By examining the structures of tight binding representative HIV PR inhibitors, we realized that a high structural variability is allowed at the fragments in positions R_3 and R_4, filling the S_2 and S'_2 specificity pockets of the substrate-binding site of the HIV PR, i.e. in the chlorinated cyclic hydrocarbons. Therefore, we decided to use a higher number of diverse fragments in the R_3 and R_4 positions, in order to cover in greater detail the available chemical space, namely, four five-membered hydrocarbon rings, six six-membered rings and eight condensed rings (in total 18 RING fragments) (Figure 4.13). The fragment-based library focusing thus reduced the size of the library of cyclic urea inhibitors to:

$$6\,(\mathrm{AA}) \times 3\,(\mathrm{ALD}) \times 18\,(\mathrm{RING}) \times 18\,(\mathrm{RING}) = 5832 \text{ compounds}$$

The virtual library was then enumerated using the selected sets of fragments.

B. Selection of Analogs

The size of the generated fragment-focused library of cyclic ureas was further reduced by applying analog-based focusing methods. Molecular properties of the generated analogs were computed for their "bound" conformations.

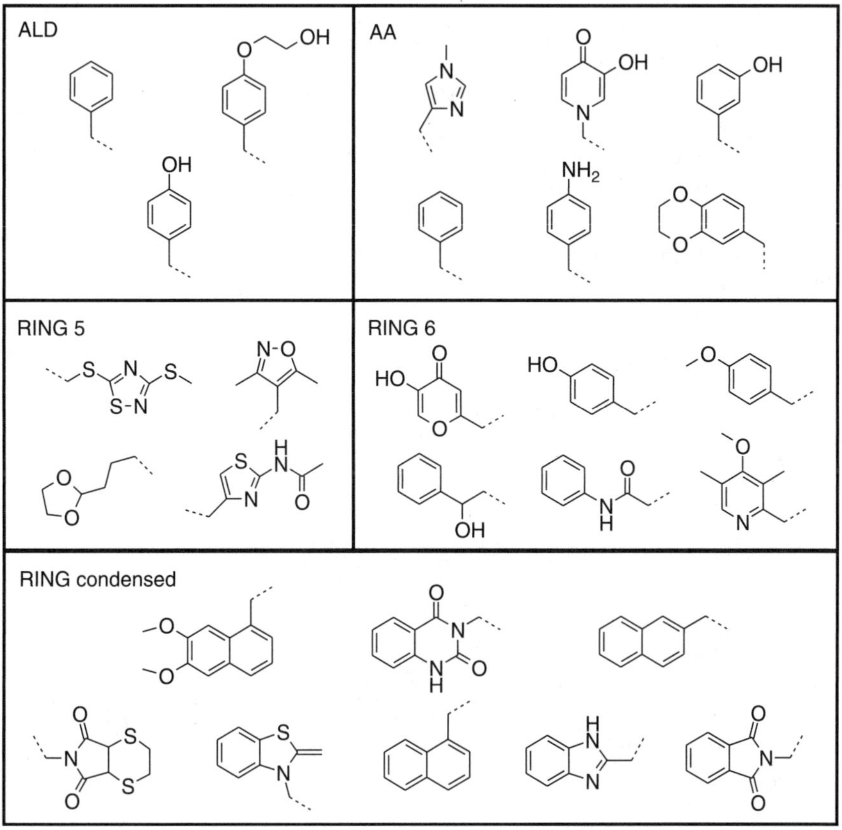

FIGURE 4.13 Selected fragments (side chains) of amino acids (AA), aldehydes (ALD), and chlorinated cyclic hydrocarbons (RING 6, RING 5, RING condensed) building blocks of the cyclic urea inhibitors of HIV PR, chosen by fragment-based library focusing techniques.

Analog-filtering procedure was applied, based on the molecular physico-chemical descriptors and optimum property ranges calculated for the set of known HIV PR inhibitors (Table 4.1), which permitted the selection of 100 most diverse cyclic urea analogs with suitable molecular properties.

C. Structure-Based Library Targeting

The size of the analog-focused library subset of 100 cyclic ureas was further reduced, by applying structure-based library targeting and *in silico* screening methods to select a highly restricted set of best binding analogs. The 100 analogs of the subset were docked to the model of the HIV PR receptor, derived from the crystal structure of DMP323 : PR complex [31], using the Monte Carlo docking algorithm of Cerius^2 [20]. The procedure yielded 20 conformers per analog, which were clustered into 5 conformational families, based on their mutual r.m.s. deviations. The best conformer for each analog which displayed the highest docking score, was then selected. The analogs were screened and ranked according to the ligand fit scoring function values [28]. This *in silico* screening for the highest inhibition activity resulted in the selection of nine highly targeted lead compounds (Figures 4.14 and 4.15), with predicted K_i constants estimated from the QSAR model of known HIV PR inhibitors (Figure 4.6) in the picomolar range (Table 4.3). This set of nine cyclic ureas was proposed for synthesis and experimental determination of the inhibitory activity to collaborating laboratories; the outcome of testing will be published elsewhere.

V. CONCLUDING REMARKS

In this paper we summarized two examples describing techniques of computer-assisted combinatorial chemistry, namely, the strategy of structure-based

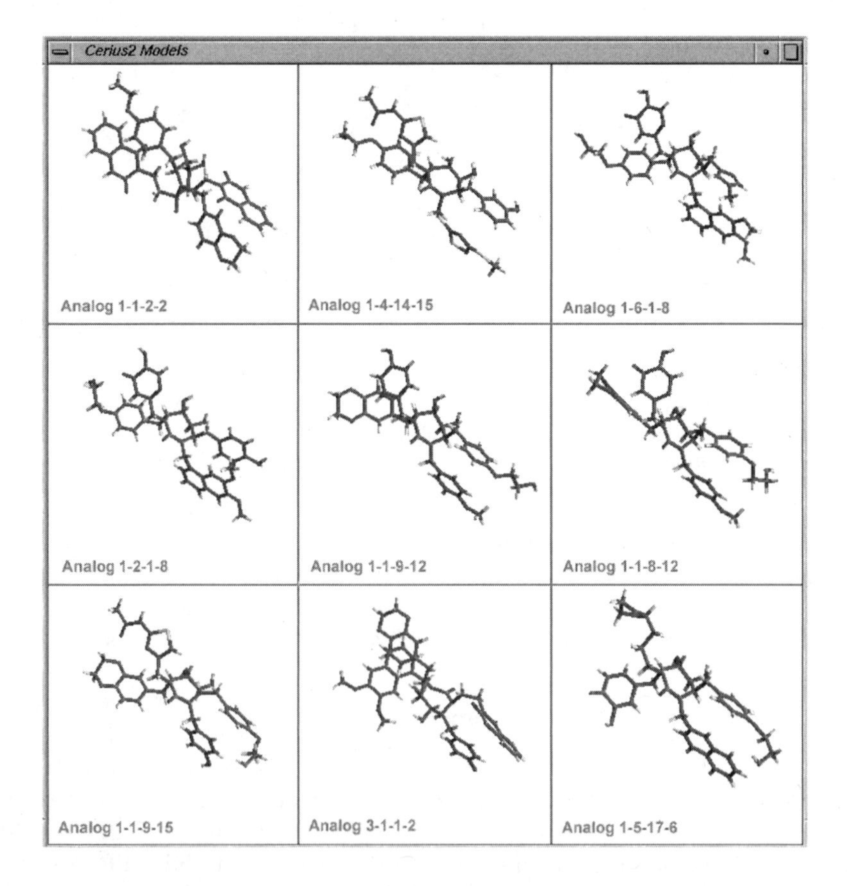

FIGURE 4.14 The nine best designed cyclic urea inhibitors of HIV PR obtained by structure-based library focusing, with predicted K_is against the HIV PR in the picomolar range, as displayed in Cerius^2 [20].

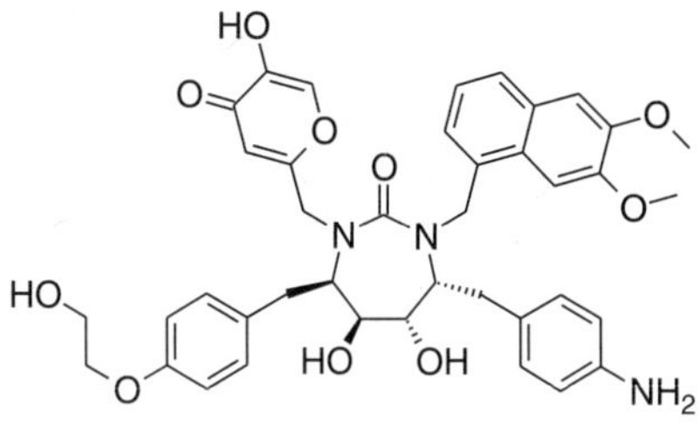

Analog 1-2-1-8

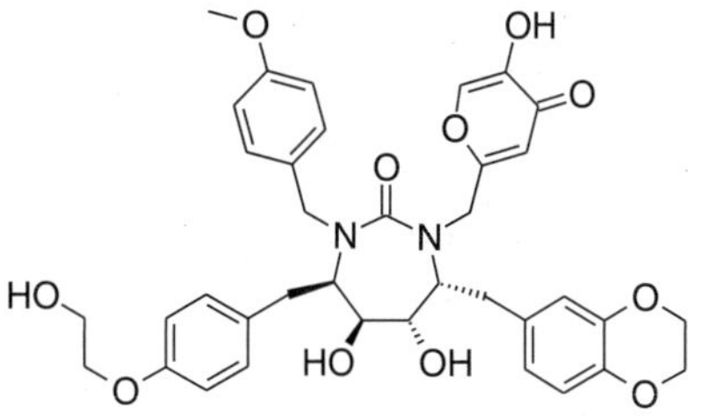

Analog 1-1-8-12

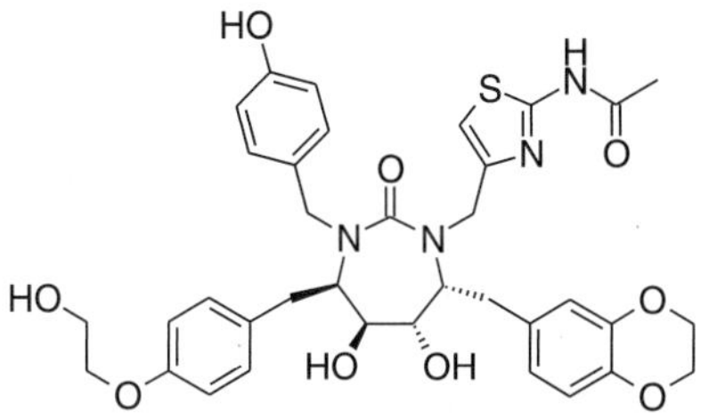

Analog 1-1-9-15

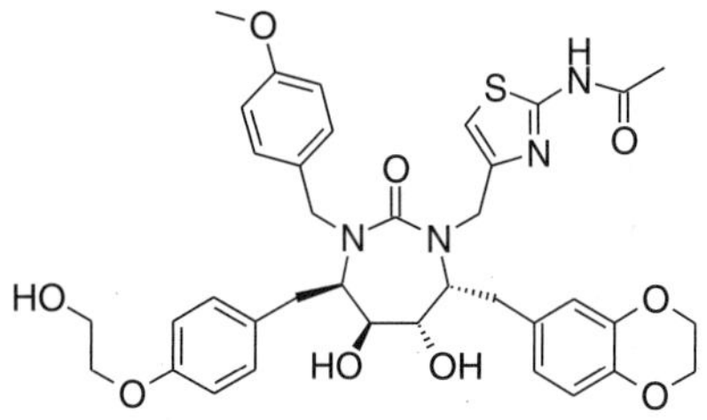

Analog 1-1-9-12

FIGURE 4.15 Chemical structures of the four most potent designed cyclic urea leads.

combinatorial library focusing, targeting and virtual screening. Computational approaches are becoming increasingly attractive, because they can significantly reduce the cost, time, and labor, required to synthesize and screen combinatorial libraries through addition of structural information about the targeted receptor. Addition of structural information (as well as other filters, such as drug-likeness, toxicity, and ADME characteristics) into the design phase of combinatorial experiment, significantly enhances the success rate in valuable lead generation [32–34]. We have illustrated one possible approach to focusing of virtual libraries of HIV PR inhibitors and the adjustment of their sizes to the capacity of an individual laboratory, by applying more and more stringent selection criteria, based on the structure and properties of the targeted receptor. This approach is sufficiently versatile to be applied in the design of agents' action on various therapeutic targets. We have to bear in mind, however, that the properties of the resulting highly focused library subsets depend strongly on the relevance of the descriptors, filters, penalty, scoring functions, and other selection criteria used. The presented examples yielded small, highly focused, subsets containing lead compounds—possible drug candidates—with strong predicted anti-HIV potencies against the wild type HIV PR. Addition of relevant structural information on the structure of most frequently occurring drug-resistant forms of the HIV PR may help to tackle also the problem of resistance.

TABLE 4.3
Targeted library subset containing nine potent cyclic urea inhibitors against the HIV PR drug target, generated via structure-based virtual library targeting and identified through *in silico* screening

Analog	Docking Score[a] $[kJ \cdot mol^{-1}]$	Minimized Energy[b] $[kJ \cdot mol^{-1}]$	Ligand Score[c]	E_{vdW}[d] $[kJ \cdot mol^{-1}]$	C_{pol}[e] $[Å^2]$	T^2_{pol}[f] $[Å^4]$	Predicted Activity, K_i[g] [pM]
Analog 1-5-17-6	−1193.3	−51.8	9.01	−384.5	44.4	2045.2	0.47
Analog 1-2-1-8	−1563.1	89.6	8.94	−352.0	52.8	2954.6	0.002
Analog 1-1-9-12	−981.7	−510.1	8.89	−374.4	45.84	2249.6	0.08
Analog 1-1-8-12	−278.9	−326.9	8.83	−359.5	48.1	2471.6	0.01
Analog 1-1-2-2	−205.3	−822.6	8.76	−395.3	38.4	1554.1	27.5
Analog 1-4-14-15	−1211.9	−784.2	8.74	−379.3	44.9	2323.6	0.02
Analog 1-6-1-8	−413.0	81.1	8.71	−393.9	39.6	1744.5	3.5
Analog 3-1-1-2	−334.4	−28.3	8.69	−393.5	41.8	2073.5	0.13
Analog 1-1-9-15	−911.2	−713.5	8.66	−376.1	43.7	2193.8	0.08

[a]Docking score was derived by Monte Carlo assisted fitting of flexible ligand to rigid HIV PR receptor.
[b]Energy of the analog minimized in the binding site of the HIV PR using cff91 force field [27].
[c]Ligand scoring function that predicts binding affinity of the analogs to the HIV PR model based on the computed parameters E_{vdW}, C_{pol} and T^2_{pol} [28].
[d]E_{vdW}—ligand–receptor nonbonding interaction energy.
[e]C_{pol}—buried polar surface area between a protein and ligand, which involves attractive protein–ligand interactions.
[f]T^2_{pol}—squared buried polar surface area that involves both attractive and repulsive protein–ligand interactions.
[g]Predicted inhibition constant of HIV PR, computed by the QSAR equation valid for known HIV PR inhibitors (Figure 4.6).

REFERENCES

1. Deadman J, Proteinase inhibitors and activators strategic targets for therapeutic intervention, *J. Pept. Sci.*, 6:421–431, 2000.
2. Cooper JB, Aspartic proteinases in disease: a structural perspective, *Curr. Drug Targets*, 3:155–173, 2002.
3. Leung D, Abbenante G, Fairlie DP, Protease inhibitors: current status and future prospects, *J. Med. Chem.*, 43:305–341, 2000.
4. Hoegl L, Korting HC, Klebe G, Inhibitors of aspartic proteases in human diseases: molecular modeling comes of age, *Pharmazie,* 54:319–329, 1999.
5. Wlodawer A, Gustchina A, Structural and biochemical studies of retroviral proteases, *Biochem. Biophys. Acta.*, 1477:16–34, 2000.
6. De Clercq E, New anti-HIV agents and targets, *Med. Res. Rev.*, 22:531–565, 2002.
7. Boden D, Markowitz M, Resistance to human immunodeficiency virus type 1 protease inhibitors, *Antimicrob. Agents Chemother.*, 42:2775–2783, 1998.

8. Martin YC, Diverse viewpoints on computational aspects of molecular diversity, *J. Comb. Chem.*, 3:231–250, 2001.
9. Schneider G, Trends in virtual combinatorial library design, *Curr. Med. Chem.*, 23:2095–2101, 2002
10. Beavers MP, Chen Z, Structure-based combinatorial library design: methodologies and applications, *J. Mol. Graph Model*, 20:463–468, 2002.
11. Hobbs DW, Guo T, Library design concepts and implementation strategies, *J. Recept. Signal Transduct. Res.*, 21:311–356, 2001.
12. Miertus S, Furlan M, Tossi A, Romeo D, Design of new inhibitors of HIV-1 aspartic protease, *Chem. Phys.*, 204:173–180, 1996.
13. Frecer V, Miertus S, Tossi A, Romeo D, Rational design of inhibitors for drug-resistant HIV-1 aspartic protease mutants, *Drug Des. Disc.*, 15:211–231, 1998.
14. Tossi A, Bonin I, Antcheva N, Norbedo S, Benedetti F, Miertus S, Nair AC, Maliar T, Dal Bello F, Palu G, Romeo D, Aspartic protease inhibitors—An integrated approach for the design and synthesis of diaminodiol-based peptidomimetics, *Eur. J. Biochem.*, 267:1715–1722, 2000.
15. Frecer V, Miertus S, Interactions of ligands with macromolecules: rational design of specific inhibitors of aspartic protease of HIV-1, *Macromol. Chem. Phys.*, 203:1650–1657, 2002.
16. Burello E, Bologa C, Frecer V, Miertus S, Application of computer-assisted combinatorial chemistry in antiviral, antimalarial and anticancer agents design, *Mol. Phys.*, 100:3187–3198, 2002.
17. Burello E., Frecer V, Miertus S, Combinatorial design and focusing of library of cyclic urea inhibitors of aspartic protease of HIV-1, *chem. Biochem.* submitted.
18. Sham HL, Kempf DJ, Molla A, Marsh KC, Kumar GI, Chen CM, Kati WM, Stewart K, Lal R, Hsu A, Betebenner D, Korneyeva M, Vasavanoda S, McDonald E, Saldivar A, Wideburg N, Chen X, Niu P, Park C, Jayanti V, Grabowski B, Granneman GR, Sun E, Japour AJ, Leonard JM, Plattner JJ, Norbeck DW, ABT-378, a highly potent inhibitor of the human immunodeficiency virus protease, *Antimic. Agents Chemother.*, 42:3218–3224, 1998.
19. Kempf DJ, Sham HL, Marsh KC, Flentge CA, Betebenner D, Green BE, McDonald E, Vasavanonda S, Saldivar A, Wideburg NE, Kati WM, Ruiz L, Zhao C, Fino L, Patterson J, Molla A, Plattner JJ, Norbeck DW, Discovery of ritonavir, a potent inhibitor of HIV protease with high oral bioavailability and clinical efficacy, *J. Med. Chem.*, 41:602–617, 1998.
20. *Cerius^{2}Life Sciences*, version 4.5, MSI Inc., San Diego, CA, 2000.
21. Benedetti F, Magnan M, Miertus S, Norbedo S, Parat D, Tossi A, Stereoselective synthesis of non symmetric dihydroxyethylene dipeptide isosteres *via* epoxy alcohols derived from alpha-amino acids. *Bioorg. Med. Chem. Lett.*, 9:3027–3030, 1999.
22. *Available Chemicals Directory*, Version 95.1, MDL Informations Systems Inc., San Leandro, CA. http://cds3.dl.ac.uk/cds/cds.html. 2002.
23. Kempf DJ, Sham HL, Protease inhibitors HIV, *Curr. Pharm. Design*, 2:225–246, 1996.
24. Wlodawer A, Vondrasek J, Inhibitors of HIV-1 protease: a major success of structure-assisted drug design, *Annu. Rev. Biophys. Biomol. Struct.*, 27:249–284, 1998.

25. Berman HM, Westbrook J, Feng Z, Gilliland G, Bhat TN, Weissig H, Shindyalov IN, Bourne PE, The Protein Data Bank, *Nucl. Acids Res.*, 28:235–242, 2000. http://www.pdb.org
26. Peters KP, Fauck J, Frommel C, The automatic search for ligand binding sites in proteins of known three-dimensional structure using only geometric criteria, *J. Mol. Biol.*, 256:201–213, 1996.
27. Maple JR, Hwang MJ, Stockfish TP, Dinur U, Waldman M, Ewing CS, Hagler AT, Derivation of class II force fields. I. Methodology and quantum force field for the alkyl functional group and alkane molecules, *J. Comput. Chem.*, 15:162–182, 1994.
28. Böhm HJ, The development of a simple empirical scoring function to estimate the binding constant for a protein–ligand complex of known three-dimensional structure, *J. Comp. Aided Molec. Design*, 8:243–256, 1994.
29. Kier LB, Hall LH. Molecular connectivity in chemistry and drug research. In: G de Stevens, ed. *Medicinal Chemistry*, New York, Academic Press, 1976, pp. 14–27.
30. Willett P, Similarity-searching and clustering algorithms for processing databases of two-dimensional and three-dimensional chemical structures, In: PM Dean, ed. *Molecular Similarity in Drug Design*, Glasgow, Chapman and Hall, 1994, pp. 110–137.
31. Lam PYS, Jadhav PK, Eyermann CJ, Hodge CN, Ru Y, Bacheler LT, Meek JL, Otto MJ, Rayner MM, Wong YN, Chang CH, Weber PC, Jackson DA, Sharpe TR, Erickson-Viitanen S, Rational design of potent, bioavailable, nonpeptide cyclic ureas as HIV protease inhibitors, *Science*, 263:380–384, 1994.
32. Amazaki TY, Hinck AP, Wang XY, Nicholson LK, Torchia DA, Wingfield P, Stahl SJ, Kaufman JD, Chang CH, Domaille PJ, Lam PY, Three-dimensional solution structure of the HIV-1 protease complexed with DMP323, a novel cyclic urea-type inhibitor, determined by nuclear magnetic resonance spectroscopy, *Protein Sci.*, 5:495–506, 1996.
33. Bohm HJ, Stahl M, Structure-based library design: molecular modelling merges with combinatorial chemistry, *Curr. Opin. Chem. Biol.*, 4:283–286, 2000.
34. Mason JS, Good AC, Martin EJ, 3-D pharmacophores in drug discovery, *Curr. Pharm. Des.*, 7:567–597, 2001.
35. Darvas F, Dorman G, Papp A, Diversity measures for enhancing ADME admissibility of combinatorial libraries, *J. Chem. Inf. Comput. Sci.*, 40:314–322, 2000.

5 Solid-Phase Synthesis of Organic Libraries

Eduard R. Felder and Daniela G. Berta

A number of valid reasons have prompted the research-oriented chemical industry to invest substantial resources into the effort to dramatically increase the evaluation efficiency of new chemical entities. The objective is to optimize rapidity and cost in the early phases of the process, leading to the identification of compounds with promising properties for further development into commercial products (typically pharmaceuticals or agrochemicals). In drug discovery, the wealth of new molecular targets with therapeutic potential is bound to increase, due to the efforts to understand the mechanisms of diseases and due to the data retrieved from genomics [1, 2].

The situation is not uncommon that biological assays are initially run without a valid lead compound (besides the positive controls), triggering the desired response. Long years of experience in molecular modeling have revealed the insufficient predictive capability of de novo ligand design, unless extensive information is known about both the target and already existing ligands. Without precise knowledge of the structural and physico-chemical parameters involved in a particular ligand–target interaction, the intellectual efforts are better spent on planning an efficient process of exhaustive exploration of the space available at the target site, based on experimental approaches which benefit from the recently developed combinatorial technologies. The added value of combinatorial libraries consists of their systematic organization and design, which support high-throughput synthesis, as well as testing of new entities, but which may equally take into account such pragmatic factors as ease of synthesis and limitation of molecular weight in the range of potential oral availability.

Without the constraints imposed by pre-existing lead structures, identified from the conventional compound sources, like corporate collections or natural extracts, the design of combinatorial libraries for lead finding (*discovery libraries*) may concentrate on the generation of highly diverse molecules with low molecular weight and favorable stability. The ability to produce large numbers of *small molecule libraries* on a variety of drug-like templates (rather than "chain-like" oligomeric structures) and to make them available to the high-throughput screening, will steadily decrease the likelihood that unattractive, complex, lead structures need to be dealt with. This will pave the way for more efficient and comprehensive follow-up activities in the lead optimization phase. Computational studies assessing diversity measures of library components have indicated that one single low-molecular-weight template (with multiple substituent sites) may be used to generate highly diverse compounds, but that the common template structure also introduces a residual similarity. It is, therefore, important to develop the capability to work with a variety of templates. An organization's performance in generating molecular diversity (and testing it) will turn out to be a key success factor, with an impact, not only on the speed of new product developments, but also on the quality of the compounds brought forward. While there seems to be a general consensus in this respect, the means, in terms of strategies and technologies, to positively influence this success factor, are much more debatable. Evidently, the output of new molecular entities, meeting the criteria of diversity, stability, and molecular weight range, depends, not only on the chemical creativity and know-how in place, but also on aspects of technological sophistication, automation and logistics. In addition, there are important considerations to be made on acceptable compromises in how rigorously the synthetic processes should adhere to the standards common for individual syntheses. Depending on the size of a library, and the complexity of a synthetic pathway, one needs to decide on the most sensible level of processing for each sample. This ranges from (ideally) the purification of all intermediates and final products, to the simplest work-up procedures of removing volatile reagents from crude products after cleavage from the solid support. In most cases, the ambition is to broadly validate protocols, to such an extent, that simple, automatable, work-up delivers crude materials with sufficient purity for biological testing. The acceptable purity level usually depends on the type of assay employed, i.e. on its sensitivity and robustness.

Before touching on more specific aspects of combinatorial chemistry, the peculiarities of solid-phase synthesis should be kept in mind (see Figure 5.1), since they are particularly advantageous in this context. The automation of synthetic processes greatly benefits from the uniform and predictable physical properties of intermediates grafted on the solid support, as well as from the ease of removing excess reagents by simple washes and filtration.

As of today, apart from straightforward one-step derivatizations, the majority of "analoging" procedures by combinatorial synthesis are carried out on solid support. This overview focuses on organic chemistry adapted to the operation mode of solid-phase synthesis. The "translation" of protocols from solution chemistry to validated solid-phase procedures is a nontrivial task, often requiring substantial development time. This is particularly the case when broad validation for a large variety of *building blocks* is sought. In general,

Solid phase synthesis

- Products are grafted to polymer beads during assembly
- Multi-step reactions without purification/solution of intermediates
- Large excess of reagents drives the equilibrium to more quantitative conversions
- "Automation friendly"
- *"One bead one sequence" approach* allows the preparation of simultaneously millions of physically seperate individual compounds on solid microcarriers (beads)

FIGURE 5.1 Solid-phase synthesis.

diversity schemes combine large sets of building blocks (reactants) with each other, to such an extent, that it is not possible to apply individually tailored reaction conditions for each combination. Suitable conditions with general validity need, therefore, to be identified in a more or less laborious and extensive *scope and limitations study*, where a representative set of combinations are synthesized under systematically varied conditions.

In this chapter, we concentrate on the issues involved in the design of (generic) diversity systems for *lead finding* (discovery libraries) and illustrate these considerations with a small selection of suitable examples. The cases were chosen on the basis of their potential to be utilized for the efficient preparation of highly diverse combinatorial libraries. On the other hand, it is important to point out that there is a whole area of activities, concerned with the preparation of focused thematic libraries, where the synthetic pathway needs to be customized to an already existing lead structure, originally identified from sources other than combinatorial synthesis (see the representative example on taxol derivatives in Ref. [31]). The purpose of such efforts, usually, is lead *optimization* by "analoging". The problems in this area reside in the time investment needed to adapt the solution synthesis scheme to a suitable protocol for the solid phase. This may be a very laborious and inefficient process, particularly in cases where insoluble reagents (e.g. catalysts) are involved, or where precise dosage of reactants is crucial (e.g. 1:1 ratio of equivalents). To some extent, these aspects will be de-emphasized with the rapid ongoing progress of general organic synthesis on solid phase. Alternatively, the industry also engages in major efforts to increase speed and automation of solution-phase chemistry, which is the naturally more straightforward way to produce analogs of leads originally synthesized in solution.

The relevant factors for successful combinatorial chemistry contributions to the drug discovery effort are summarized in Figure 5.2.

In assessing the value of a combinatorial synthesis scheme, with respect to its potential for *lead finding* applications, a number of considerations should be taken into account. The main factors of importance are listed in Figure 5.3.

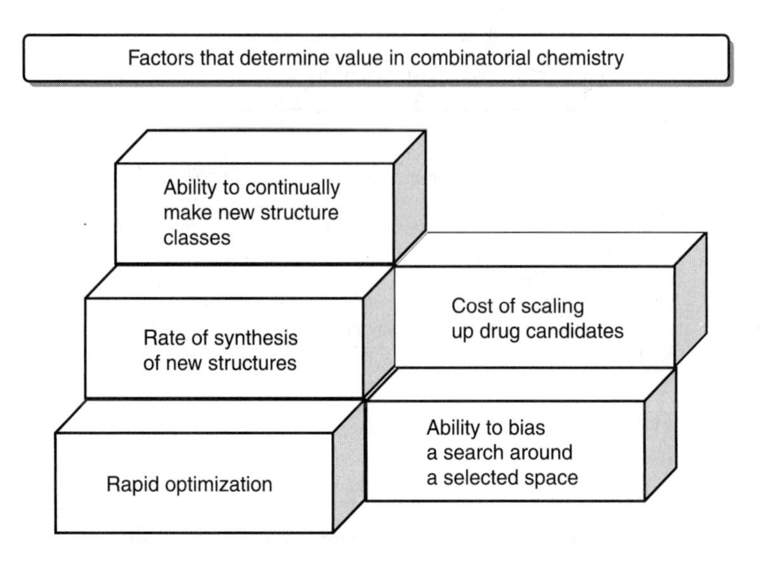

FIGURE 5.2 Factors that determine value in combinatorial chemistry.

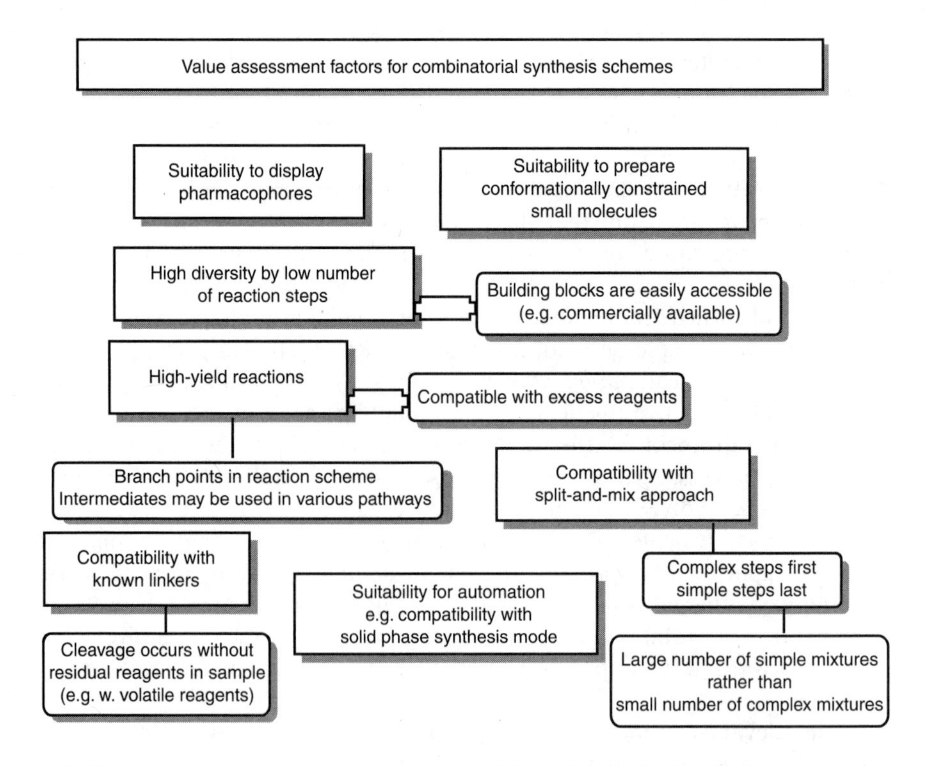

FIGURE 5.3 Value assessment factors for combinatorial synthesis schemes.

The difficulty of designing a sequence of compatible reactions, refining the experimental conditions and reaching near-quantitative yields over multistep procedures, is reflected in the proportionally modest number of mature chemical diversity systems, compared to the abundant number of organic reactions described on solid phase.

The following list of combinatorial synthesis schemes (Figure 5.4) is a collection of examples which fulfil the criteria of broad applicability. More

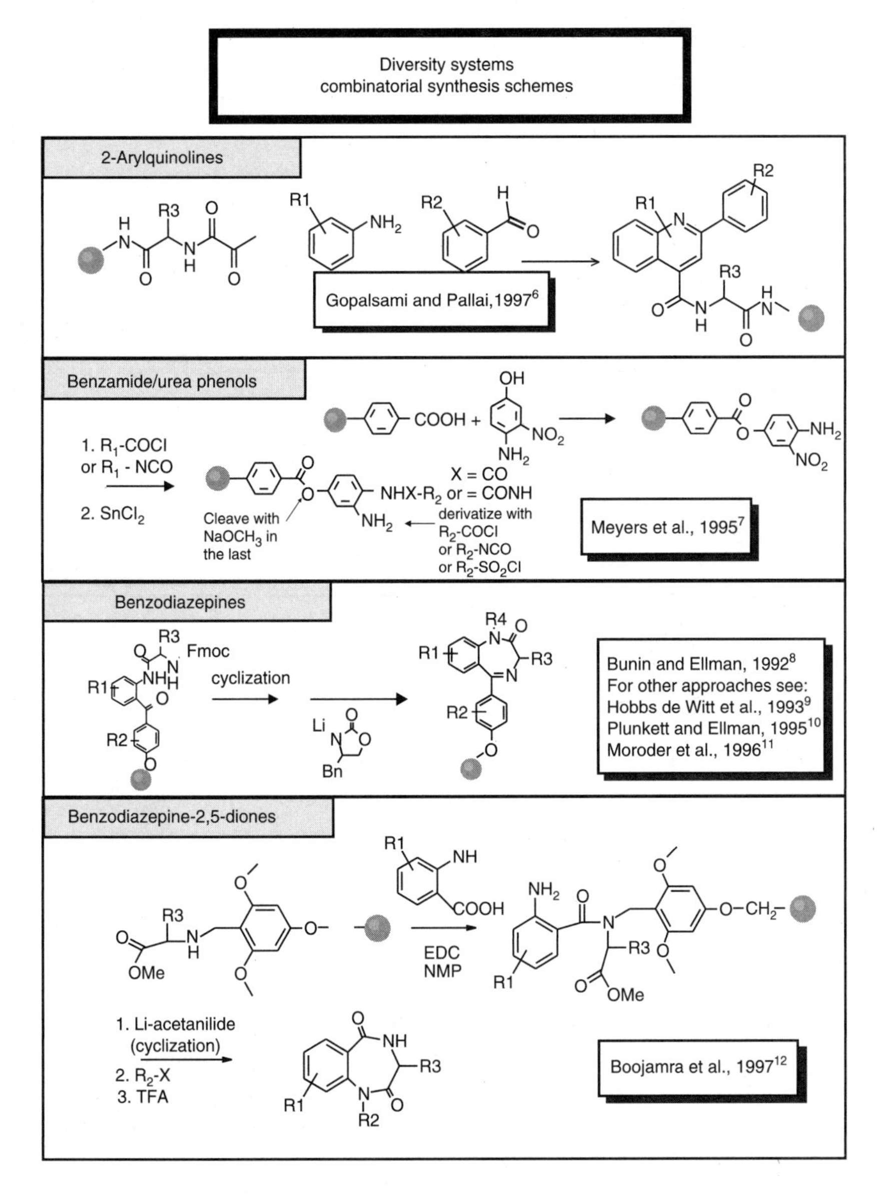

FIGURE 5.4 Diversity systems combinatorial synthesis schemes.

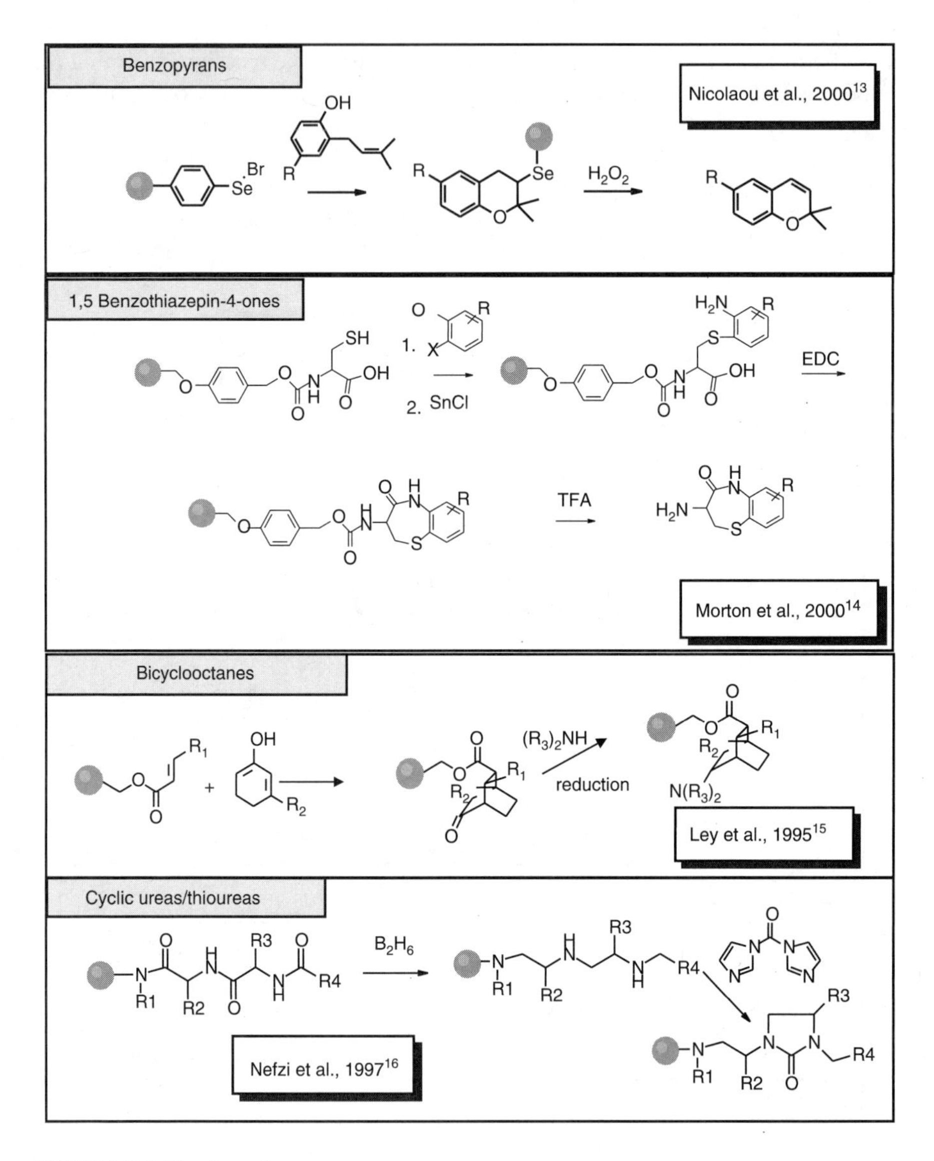

FIGURE 5.4 Continued.

comprehensive compilations were published in specific reviews [3–5]. For subscribers, electronic databases such as SPORE (Molecular Design Ltd., San Leandro, USA) and SPS (Synopsys Scientific Systems Ltd., Leeds, UK) are excellent vehicles to keep up-to-date with the latest developments in solid-phase synthesis and to carry out structure searches.

The area of diversity generation from simple building blocks, by means of ingenious combinatorial synthesis schemes, remains a controversial field,

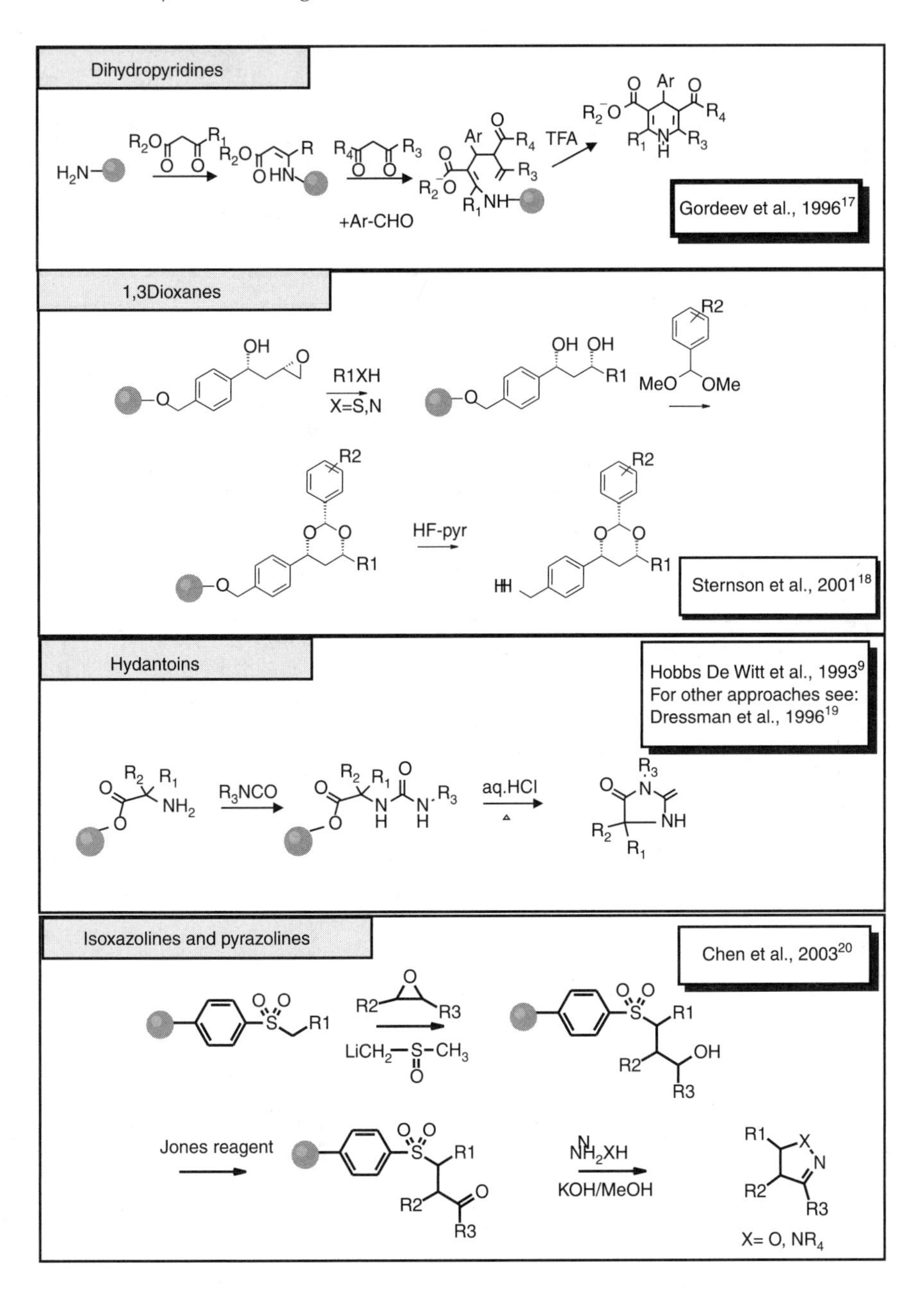

FIGURE 5.4 Continued.

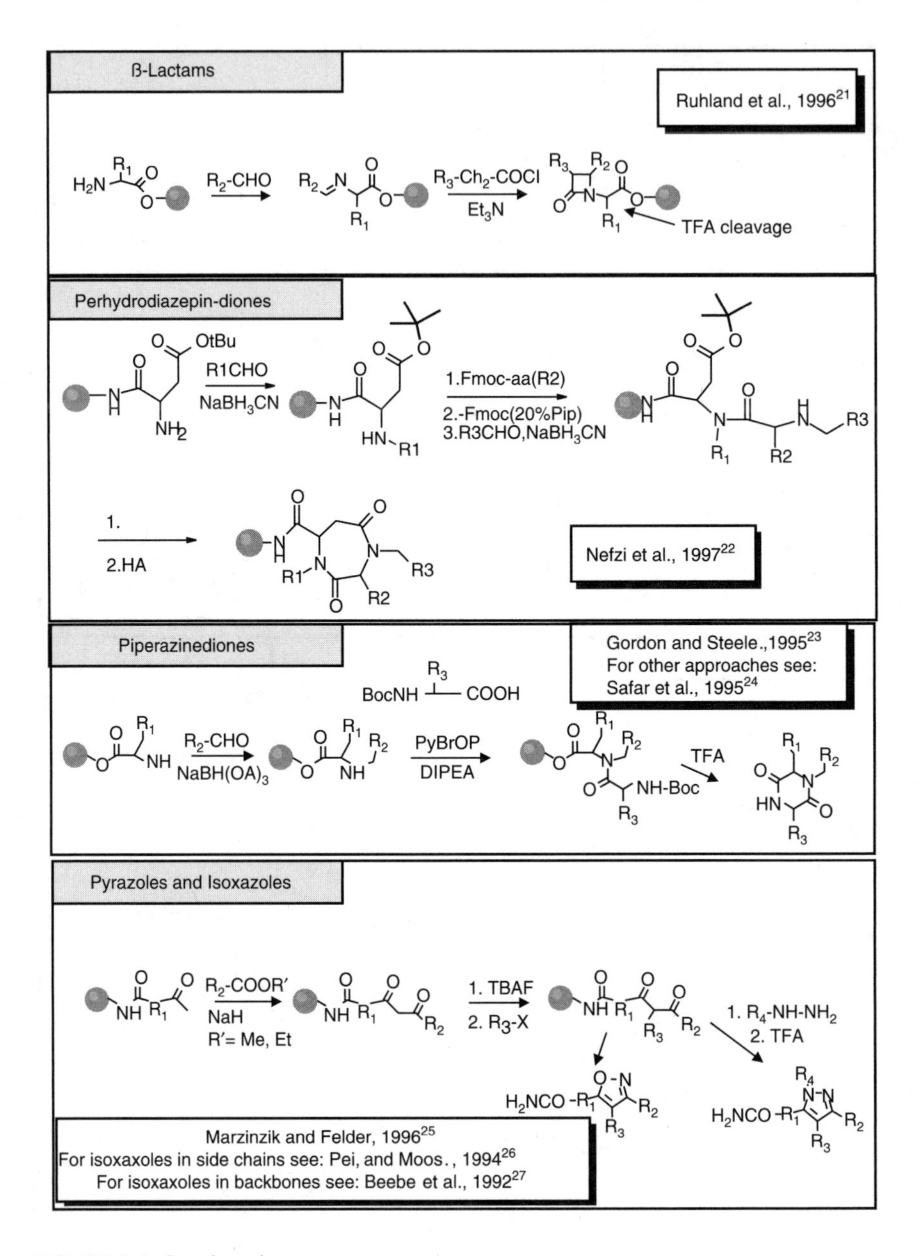

FIGURE 5.4 Continued.

which has yet to carry conviction for the general organic chemist. The chemists' deeply rooted ambition to demonstrate the art of synthesizing any given structure proposed by nature or design, tends to draw the efforts toward lead optimization and "analoging", rather than toward *de novo* lead discovery.

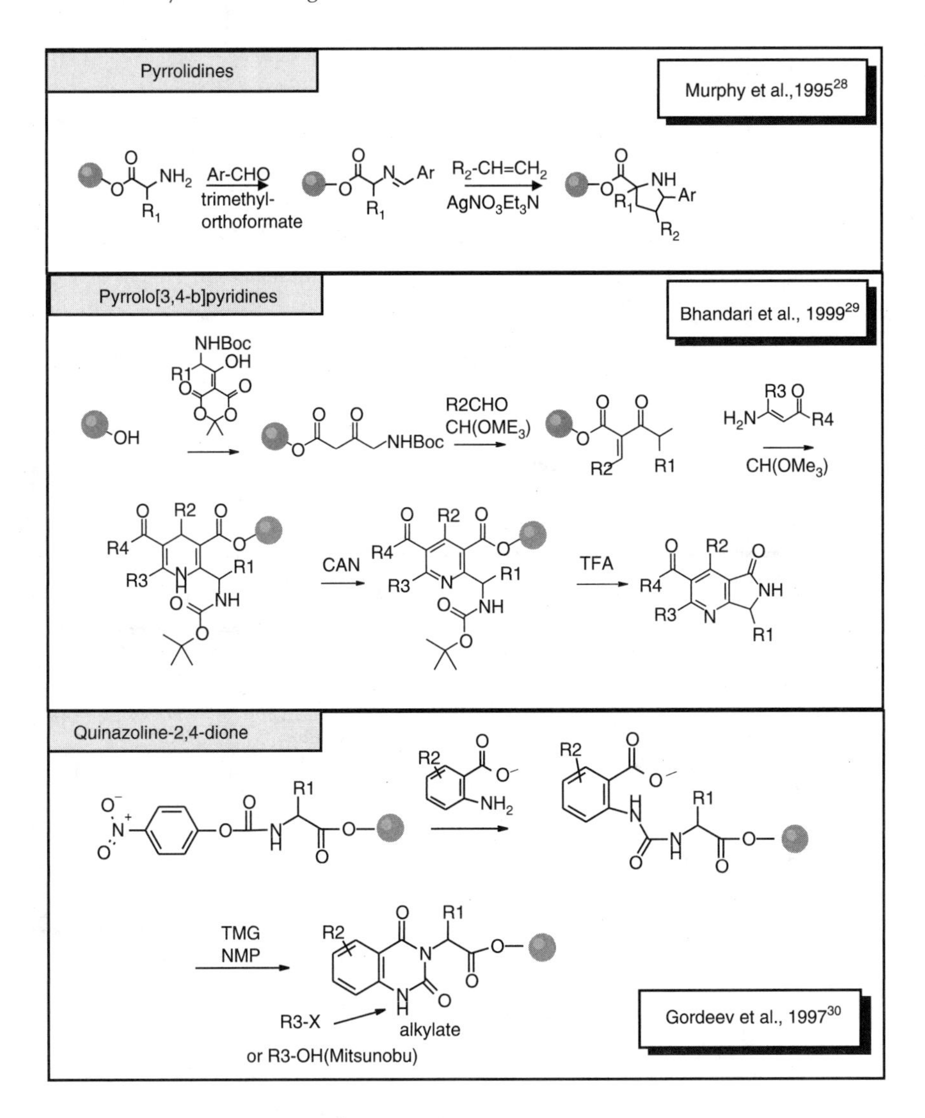

FIGURE 5.4 Continued.

A change in this respect will depend on the success in demonstrating the added value of combinatorial leads for drug development. The incremental improvements brought about by increasing the efficiency of optimizing "conventional" leads with laboratory automation alone, will not suffice to translate into the "paradigm shift" proclaimed a few years ago. A deeper impact on the quality of discovered primary leads is necessary, in terms of their simplicity and suitability for further modular derivatizations.

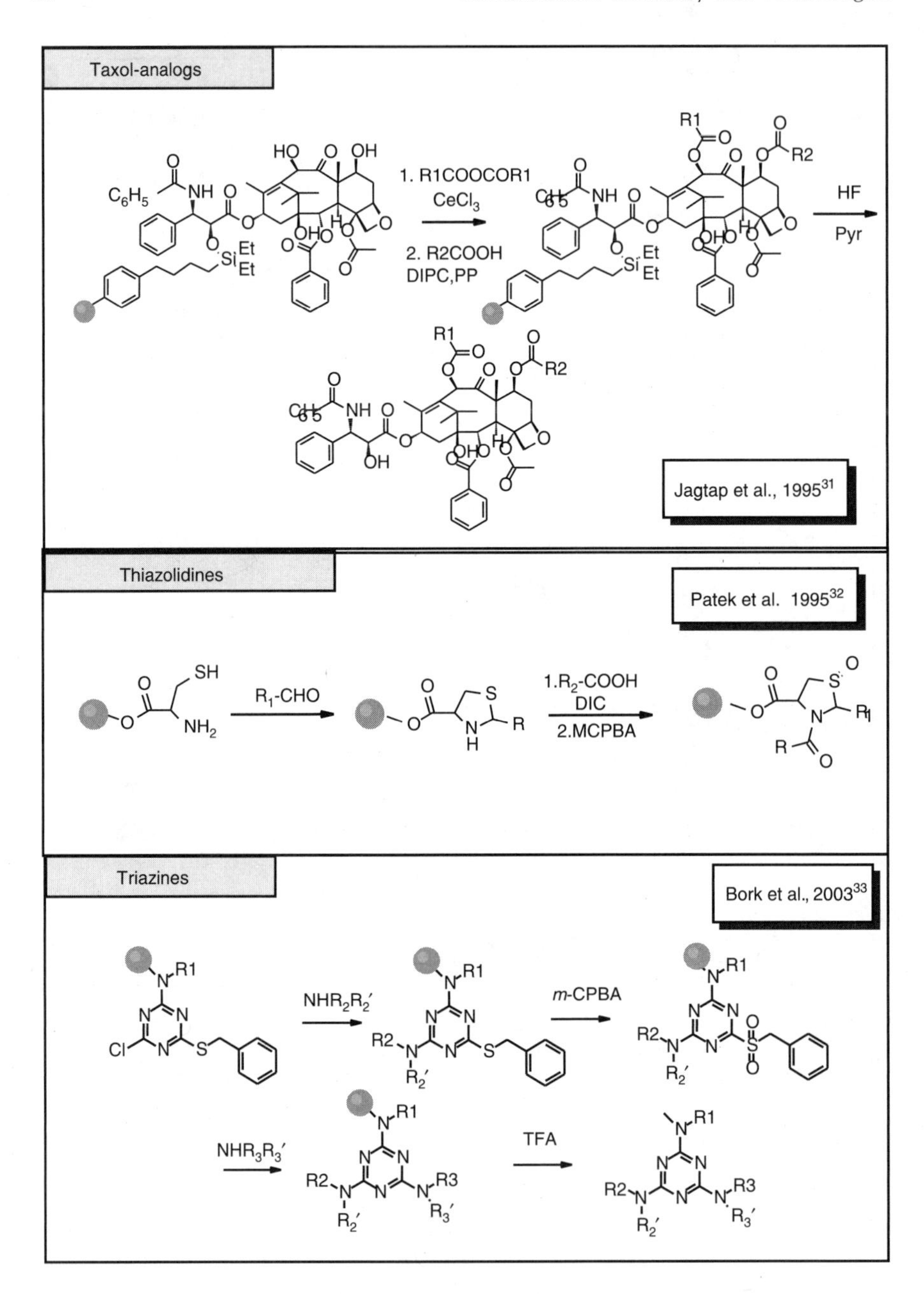

FIGURE 5.4 Continued.

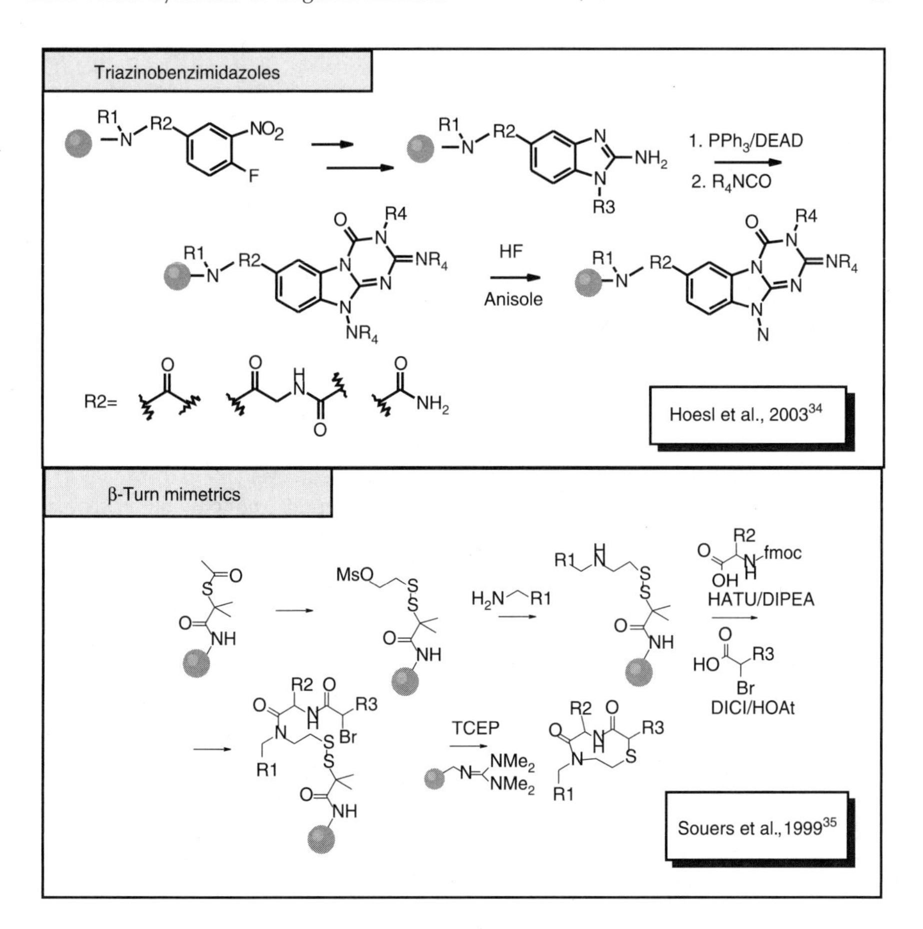

FIGURE 5.4 Continued.

REFERENCES

1. (a) Evans WE, Johnson JA, Pharmacogenomics: the inherited basis for interindividual differences in drug response, *Annu. Rev. Genom. Human. Genet.*, 2:9–39, 2001. (b) Burley SK, Bonanno JB, Structuring the universe of proteins, *Annu. Rev. Genom. Human. Genet.*, 3:243–262, 2002.
2. Rubin EM, Barsh GS, Biological insights through genomics—mouse to man, *J. Clin. Invest.*, 97:275–280, 1996.
3. (a) Dolle RE, Comprehensive survey of combinatorial library synthesis: 2000, *J. Comb. Chem.*, 3:477–517, 2001. (b) Dolle RE, Comprehensive survey of combinatorial library synthesis: 1999, *J. Comb. Chem.*, 2:383–433, 2000. (c) Dolle RE, Nelson Jr KH, Comprehensive survey of combinatorial library synthesis: 1998, *J. Comb. Chem.*, 3:235–282, 1999. (d) Dolle RE, Comprehensive survey of chemical libraries yielding enzyme inhibitors, receptor agonists and antagonists, and other biologically active agents: 1992–1997, *Mol. Diversity*, 3:199–233, 1998.

4. Balkenhohl F, Ch von dem Bussche-Huennefeld, Lansky A, Zechel Ch, Kombinatorische synthese niedermolekularer organischer verbindungen, *Angew. Chem.*, 108:2436–2488, 1996.
5. Terrett NK, Gardner M, Gordon DW, Kobylecki RJ, Steele J, Combinatorial synthesis—the design of compound libraries and their application to drug discovery, *Tetrahedron*, 51:8135–8173, 1995.
6. Gopalsamy A, Pallai PV, Combinatorial synthesis of heterocycles: solid phase synthesis of 2-arylquinoline-4-carboxylic acid derivatives, *Tetrahedron Lett.*, 38:907–910, 1997.
7. Meyers HV, Dilley GJ, Durgin TL, Powers TS, Winssinger NA, Zhu H, Pavia MR, Multiple simultaneous synthesis of phenolic libraries, *Mol. Diversity*, 1:13–20, 1995.
8. Bunin BA, Ellman JA, A general and expedient method for the synthesis of 1,4-benzodiazepine derivatives, *J. Am. Chem. Soc.*, 114:10997–10998, 1992.
9. Hobbs De Witt S, Kiely JS, Stankovic CJ, Schroeder MC, Reynolds Cody DM, Pavia MR, Diversomers: an approach to nonpeptide, nonoligomeric chemical diversity, *Proc. Natl. Acad. Sci. USA*, 90:6909–6913, 1993.
10. Plunkett MJ, Ellman JA, Solid-Phase synthesis of structurally diverse 1,4-benzodiazepine derivatives using the Stille coupling reaction, *J. Am. Chem. Soc.*, 117:3306–3307, 1995.
11. Moroder L, Lutz J, Grams F, Rudolph-Bohner S, Osapay G, Goodman M, Kolbeck W, A new efficient method for the synthesis of 1,4-benzodiazepine-2, 5-dione diversomers, *Biopolymers*, 38:295–300, 1996.
12. Boojamara GC, Burow KM, Thompson LA, Ellman JA, Solid phase synthesis of 1,4-benzodiazepine-2,5-diones. Library preparation and demonstration of synthesis generality, *J. Org. Chem.*, 62:1240–1256, 1997.
13. Nicolaou KC, Pfefferkorn JA, Cao GQ, Selenium-based solid-phase synthesis of benzopyrans. I: Applications to combinatorial synthesis of natural products, *Angew. Chem., Int., Ed. Engl.*, 39:743–739, 2000.
14. Morton GC, Salvino JM, Labaudiniere RF, Herpin TF, Novee solid phase synthesis of 1,5-benzothiazepite-4-one derivatives, *Tetrahedron Lett.*, 41:3029–3033, 2000.
15. Ley SV, Mynett DM, Koot WJ, Solid phase synthesis of bicyclo[2,2,2]octane derivatives via tandem Michael addition reactions and subsequent reductive amination, *Synlett.*, 1017–1020, 1995.
16. Nefzi A, Ostresh JM, Meyer JP, Houghten RA, Solid phase synthesis of heterocyclic compounds from linear peptides: cyclic ureas and thioureas, *Tetrahedron Lett.*, 38:931–934, 1997.
17. Gordeev MF, Patel DV, Gordon EM, Approaches to combinatorial synthesis of heterocycles: a solid-phase synthesis of 1,4-dihydropyridines, *J. Org. Chem.*, 61:924–928, 1996.
18. Sternson SM, Wong JC, Grozinger CM, Schreiber SL, Synthesis of 7200 small molecules based on a substructural analysis of the histone deacetylase inhibitors trichostatin and trapoxin, *Org. Lett.*, 3:4239–4242, 2001.
19. Dressman BA, Spangle LA, Kaldor SW, Solid phase synthesis of hydantoins using a carbamate linker and a novel cyclization/cleavage step, *Tetrahedron Lett.*, 37:937–940, 1996.
20. Chen Y, Lam Y, Lai YH, Solid-phase synthesis of pyrazolines and isoxazolines with sodium benzenesulfinate as a traceless linker, *Org. Lett.*, 5:1067–69, 2003.

21. Ruhland B, Bhandari A, Gordon EM, Gallop MA, Solid-supported combinatorial synthesis of structurally diverse β-lactams, *J. Am. Chem. Soc.*, 118: 253–254, 1996.
22. Nefzi A, Ostresh JM, Houghten RA, Solid phase synthesis of 1,3,4, 7-tetrasubstituted perhydro-1,4-diazepine-2,5-diones, *Tetrahedron Lett.*, 38: 4943–4946, 1997.
23. Gordon DW, Steele J, Reductive alkylation on a solid phase: synthesis of a piperazinedione combinatorial library, *Bioorg. Med. Chem. Lett.*, 5:47–50, 1995.
24. Safar P, Stierandova A, Lebl M, Amino-acid like subunits based on iminodiacetic acid and their application in linear and DKP libraries, In HLS Maia, ed. *Peptides 94*, Proc. 23rd EPS, Leiden, ESCOM, 1995, pp. 471–472.
25. Marzinzik AL, Felder ER, Solid support synthesis of highly functionalized pyrazoles and isoxazoles; scaffolds for molecular diversity, *Tetrahedron Lett.*, 37:1003–1006, 1996.
26. Pei Y, Moos WH, Post-modification of peptoid side chains: (3 + 2) cycloaddition of nitrile oxides with alkenes and alkynes on the solid-phase, *Tetrahedron Lett.*, 35:5825–5828, 1994.
27. Beebe X, Schore NE, Kurth MJ, Polymer-supported synthesis of 2,5-disubstituted tetrahydrofurans, *J. Am. Chem. Soc.*, 114:10061–10062, 1992.
28. Murphy MM, Schullek JR, Gordon EM, Gallop MA, Combinatorial organic synthesis of highly functionalized pyrrolidines: identification of a potent angiotensin converting enzyme inhibitor from a mercaptoacyl proline library, *J. Am. Chem. Soc.*, 117:7029–7030, 1995.
29. Bhandari A, Li B, Gallop MA, Solid-phase synthesis of pyrrolo[3,4-*b*]pyridines and related pyridine-fused heterocycles, *Synthesis*, 11:1951–1960, 1999.
30. Gordeev MF, Hui HC, Gordon EM, Patel DV, A general and efficient solid phase synthesis of quinazoline-2,4-diones, *Tetrahedron Lett.*, 38:1729–1732, 1997.
31. Jagtap P, Baloglu E, Barron DM, Bane S, Kingston DG, Design and synthesis of a combinatorial chemistry library of 7-acyl, 10-acyl, and 7,10-diacyl analogues of paclitaxel (taxol) using solid phase synthesis, *J. Nat. Prod.*, 65:1136–1142, 2002.
32. Patek M, Drake B, Lebl M, Solid phase synthesis of "small" organic molecules based on thiazolidine scaffold, *Tetrahedron Lett.*, 36:2227–2230, 1995.
33. Bork J, Lee JW, Khersonsky SM, Moon HS, Chang YT, Novel orthogonal strategy toward solid-phase synthesis of 1,3,5-substituted triazines, *Org. Lett.*, 5:117–120, 2003.
34. Hoesl CE, Nefzi A, Houghten RA, Parallel solid-phase synthesis of 2-Imino-4-oxo-1,3,5-triazino[1,2-a]benzimidazoles *via* tandem Aza-Wittig/heterocumulene-mediated annulation reaction, *J. Comb. Chem.*, 5:155–160, 2003.
35. Souers AJ, Virgilio AA, Rosenquist Å, Fenuik W, Ellman JA, Identification of a potent heterocyclic ligand to somatostatin receptor subtype 5 by the synthesis and screening of β-turn mimetic libraries, *J. Am. Chem. Soc.*, 121:1817–1825, 1999.

6 Resins, Linkers and Reactions for Solid-Phase Synthesis of Organic Libraries

Eduard R. Felder and Daniela G. Berta

CONTENTS

In reviewing the chemical diversity systems outlined in the previous chapter, the utility of applying a variety of different reaction types productively on solid phase has become apparent. Combinatorial chemists must develop the skills to maintain a high level of flexibility, with respect to the pathways accessible in the solid-phase mode of operation. The diversity of building blocks, available in principle to everyone, needs to be complemented by a diversity of methods on how to link them. This skill base is available only to the advanced laboratories in the field. Furthermore, the know-how in the area of protecting group chemistry needs extension into the issues concerning solid phase linkers and the actual solid-phase matrices.

Here, we would like to briefly discuss basic peculiarities of the solid matrices, the various categories of linkers and some reaction types of established significance for combinatorial chemistry. For more comprehensive compilations of solid-phase reactions, the reader is referred to specific reviews [1–5] and commercial electronic databases (e.g. SPORE [6] or SPS [7]).

I. POLYMERIC SUPPORTS

Synthetic chemistry on solid or gel-like supports was initially developed for peptides and oligonucleotides by the pioneering contributions of Merrifield [8] and Letsinger [9], respectively. In the early days, cross-linked polystyrene was used almost exclusively, while, nowadays, a variety of other polymeric supports have been added to the choice. Polystyrene is normally cross-linked with 1% divinylbenzene and needs to swell in the reaction solvent, in order to allow reactants to reach all functional groups within the polymeric network. It is typical for all gelatinous supports that they are used in a beaded form, as spherical microparticles available in a choice of size distributions. The beads' diameters usually lie in the range of 80–200 μm on average. The majority of reactive sites are situated in the interior of the bead volume, with not more than 1% on the outer surface. The analogy with a sponge seems appropriate. The number of functional groups (available for derivatization) per mass of dry support is termed *loading capacity*. Commercial products are offered, with rough capacities between 0.2 and 1.5 mMol/g, depending on the type of chemical functionalization. This results in an order of magnitude of approximately 100 pMol/bead, but the exact number depends strongly on the actual volume of the particle. The swelling factors for each type of solid matrix vary considerably with the type of solvent used. Polystyrene swells up to 8 mL/g resin in dichloromethane, whereas lower alcohols hardly cause any swelling, and water fails to penetrate the bead interior adequately. Polystyrenes with a higher degree of cross-linking ($>2\%$) are mechanically more stable, but their degree of swelling and the loading capacity are reduced. In contrast to gelatinous resins, macroporous supports consist of a rigid, nonswellable matrix, like glass or highly cross-linked polystyrene, and they are not sensitive to the solvents used. The chemical accessibility is limited to the surface of the material, which, therefore, needs to be maximized without compromising the mechanical stability of the particles. High loading capacities are difficult to achieve.

Another very successful approach to circumvent the limitations on solvent types, consists of grafting polyethyleneglycol (PEG) on polystyrene [10]. Modern copolymers of this type contain about 60–70% PEG. Since the swelling properties of polystyrene no longer predominate, a broad range of solvents can be used, without drastic changes of bead volume, during an entire synthesis, and water may be added to the list of acceptable solvents. This has also particular significance upon cleavage of compounds into solution, in the last stage of a synthesis. Furthermore, the flexibility of the PEG tentacles is reflected in higher T1 values during NMR spectroscopy on beads, resulting in sharp resonance signals, a prerequisite for using this powerful analytical tool efficiently, with the analytes still grafted on the carrier matrix. Detailed spectroscopic data is often required in the important and time-consuming validation phase of a combinatorial chemistry reaction sequence.

Excellent compilations of commercially available solid supports have been published [11, 12].

II. LINKERS

For most purposes, combinatorial libraries need to be tested in solution at some stage. In the early days, it was not uncommon to screen a library by means of a simple binding assay [13]. This testing mode has increasingly lost significance and its role has, at times, shifted to a prescreening function [14]. Clearly, more meaningful characterization of compounds requires functional assays in solution and, therefore, the syntheses on solid phase need to make use of cleavable linker moieties, which anchor the reaction products to the carrier in the course of their assembly in a reversible manner. Moreover, in view of the possibility of simplifying postsynthetic work-up procedures, linkers should be cleavable without leaving traces of reagents in the sample. Some commonly used solid phase linkers and their cleavage conditions are listed in Figure 6.1.

Linker structure	Cleavage product/ reagent	References
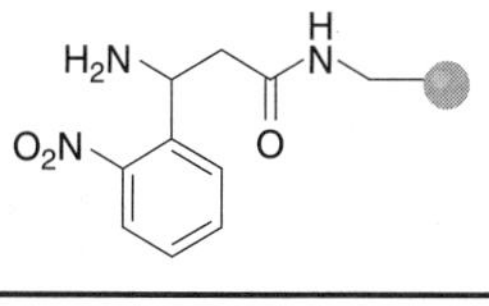 H_2N, O_2N, H N, O	R-$CONH_2$ photocleavage	Brown et al., 1995[35]
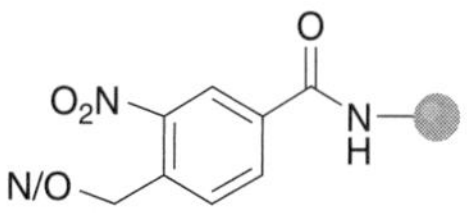 O_2N, N/O, O, N H	R-CONH R-$CONH_2$ R-OH photocleavage	Baldwin et al., 1995[36] Rich and Gurwara, 1975[37] Burbaum et al., 1995[38]
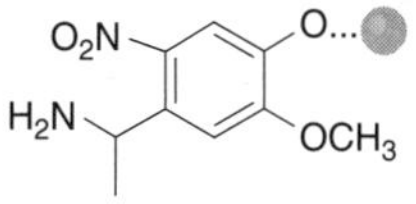 O_2N, H_2N, O..., OCH_3	R-$CONH_2$ photocleavage	Holmes and Jones, 1995[39]
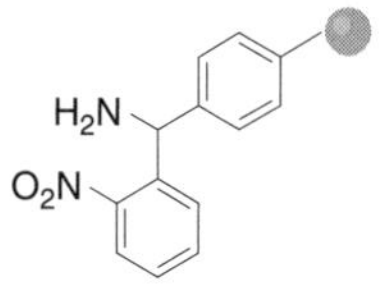 H_2N, O_2N	R-$CONH_2$ photocleavage	Ajayaghosh and Pillal, 1995[40]
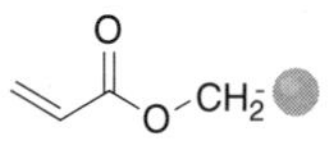 O, O, CH_2	tertiary amines $(R)_3N$ Hofmann elimination with mild base	Morphy et al., 1996[41]
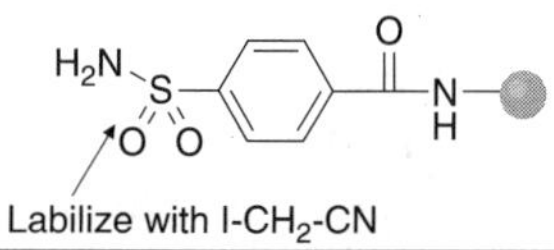 H_2N, S, O O, O, N H Labilize with I-CH_2-CN	R-$CONR_1R_2$ safety catch 1. iodoacetonitrile 2. R_1R_2NH	Ellman, 1996[42]

FIGURE 6.1 Some commonly used solid-phase linkers and their cleavage conditions.

Linker structure	Cleavage product/ reagent	References
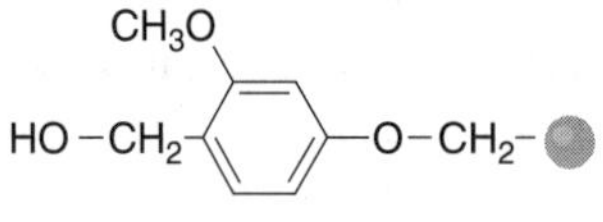 CH_3O; $HO-CH_2-C_6H_3-O-CH_2-$●	R-COOH 0.5% TFA	Sheppard and Williams, 1982[30]
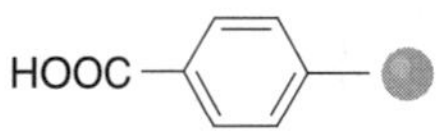 $HOOC-C_6H_4-$●	R-OH $NaOCH_3/MeOH$	Farall and Frechet, 1976[31] Meyers et al., 1995[32]
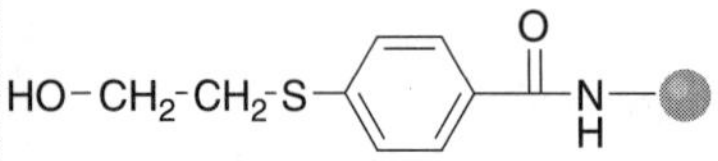 $HO-CH_2-CH_2-S-C_6H_4-C(=O)-NH-$●	R-COOH safety catch: 1. oxidize to sulfone 2. mild base (baryta)	Schwyzer et al., 1984[33]
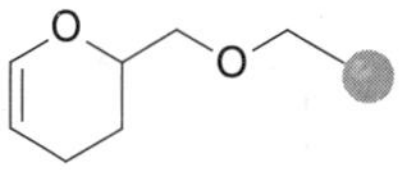 O; O; ●	R-OH Pyridine p-toluene-sulfonate (PPTS)	Thompson and Ellman, 1994[34]
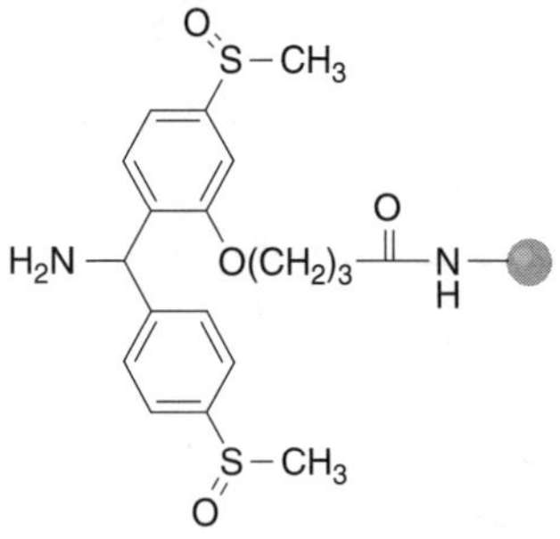 $O{=}S-CH_3$; H_2N; $O(CH_2)_3-C(=O)-NH-$●; $S(=O)-CH_3$	$R-CONH_2$ safety catch: 1. reduce to thioether 2. TFA	Patek and Lebl, 1991[20]

FIGURE 6.1 Continued.

A large choice of linkers already anchored to the solid support is commercially available. The selection of an appropriate linker for a particular synthesis always takes into consideration the various reaction conditions that must be applied in the course of the library assembly, and what type of functional group is an acceptable attachment point. Generally, the desired target

Linker structure	Cleavage product/ reagent	References
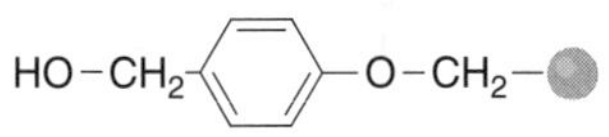 HO−CH2−C6H4−O−CH2−●	R-COOH R-OH Trifluoroacetic acid (TFA)	Wang, 1973[23] Sarshar et al., 1996[24]
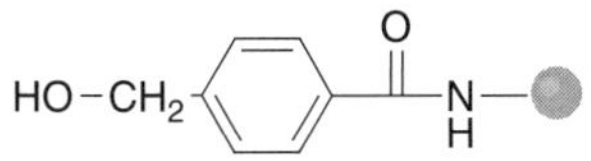 HO−CH2−C6H4−C(=O)−NH−●	R-CONH2 R-COOH Ammonia OH⁻	Atherton et al., 1981[25] Bray et al., 1994[26]
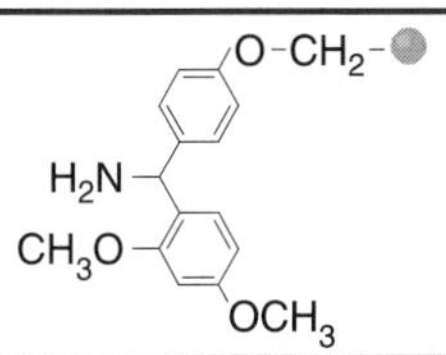 H2N−CH(C6H4−O−CH2−●)(C6H3(OCH3)2)	R-CONH2 TFA	Rink, 1987[27]
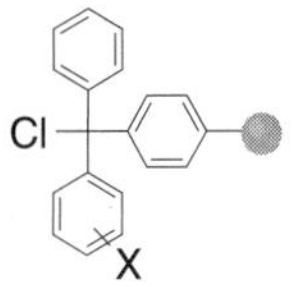 Cl−C(C6H5)(C6H4X)−C6H4−●	R-COOH R-NH2 R-OH various acid conditions depending upon X	Barlos et al., 1989[28]
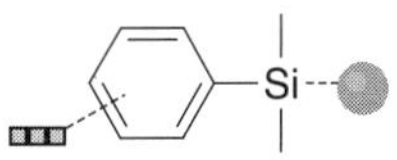 Aryl−Si−●	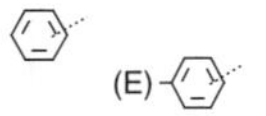 Ar-H (E)-Ar TFA HF E (electrophile)	Chenera et al., 1995[18] Plunkett and Ellman, 1995[17] Han et al., 1996[19]
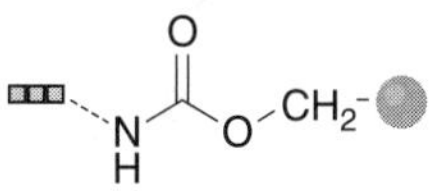 R−NH−C(=O)−O−CH2−●	R-NH2 TFA	Hauske and Dorff, 1995[29]

FIGURE 6.1 Continued.

structures already contain a functionality suitable to connect to the solid-phase linker. If this is not the case, an auxiliary functional group (e.g. a carboxylate, amide, amine, phenol, or alcohol) is introduced as a handle for linker attachment. The influence of such an additional group on biological activity is usually unpredictable. For this reason, there is an increasing interest for

Linker structure	Cleavage product/ reagent	References
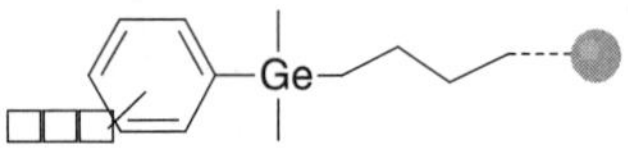 Aryl–Ge(CH3)2–(CH2)3–resin	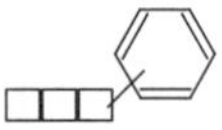 Aryl product / TFA	Plunkett and Ellman, 1997[43]
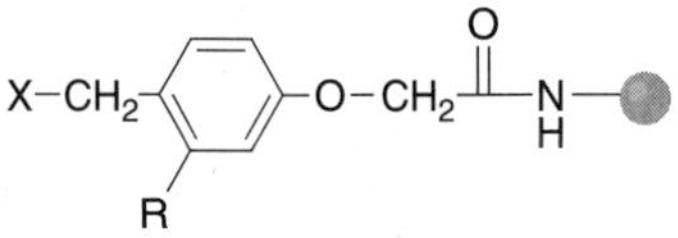 $X-CH_2-C_6H_3(R)-O-CH_2-C(=O)-NH$–resin; X = Cl or Br; R = H or OCH_3	sulfonamides carboxamides ureas / TFA	Raju and Kogan, 1997[44]
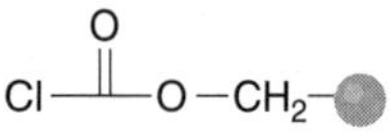 $Cl-C(=O)-O-CH_2$–resin	sulfonamides / $NaOCH_3/MeOH$	Raju and Kogan, 1997[45]
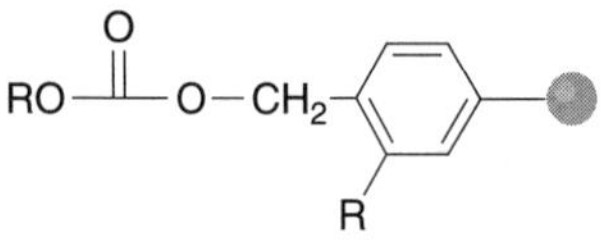 $RO-C(=O)-O-CH_2-C_6H_3(R)$–resin; R = H or NO_2	R-OH / HF, Photolysis (R = NO_2)	Alsina et al., 1997[46]
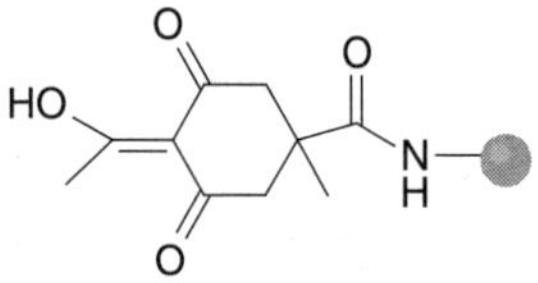 HO(CH3)C=dioxocyclohexyl(CH3)–C(=O)–NH–resin	R-NH_2 / Hydrazine 2%	Bannwarth et al., 1996[47]
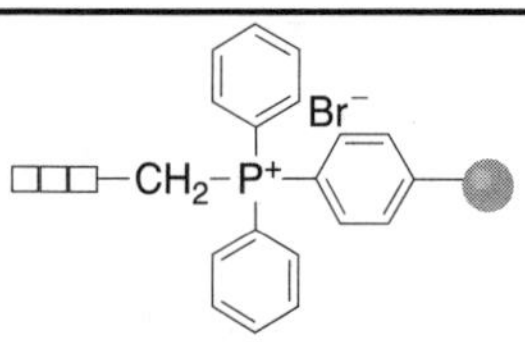 $-CH_2-P^+(Ph)_2-C_6H_4$–resin Br^-	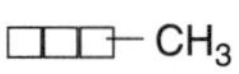 $-CH_3$ / $NaOCH_3/MeOH$	Hughes, 1996[48]

FIGURE 6.1 Continued.

"traceless" linkers, which do not impose a particular functional group on the target structures [15]. Examples include the preparation of *p*-tolyl derivatives by reductive cleavage of benzyl thioesters [16] and several silicon-based linkers, which release aromatic compounds by ipso-desilylation (proton or electrophile mediated) or *via* fluoride cleavage of a resin-bound arylsilane [17–19].

Linker structure	Cleavage product/ reagent	References
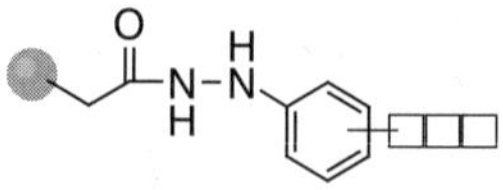	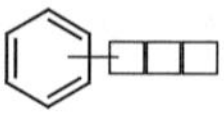	Millington et al., 1998[49]
	Cu(Oac)2, base, MeOH	
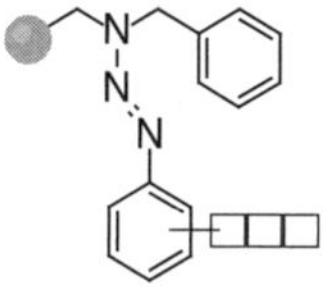	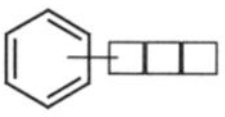	Braese et al., 1998[50]
	HCl/THF	
S, O, R	R–OH, R–C(O)–R1, R1(R)(R1)C–OH	May et al., 2000[51]
	$LiBH_4$ $R1_2CuLi$ R1MgBr	
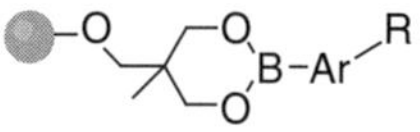 O, O, B–Ar–R	Ar–R	Pourbaix et al., 2000[52]
	$Ag(NH_3)_2NO_3$	
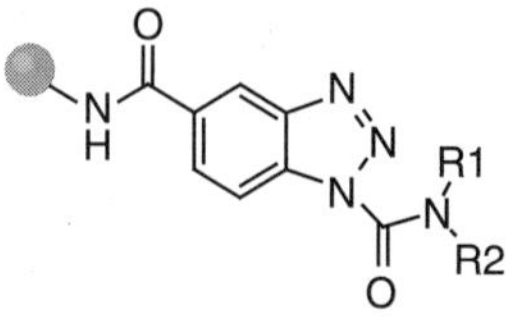 O, N, H, N, N, N, R1, R2, O	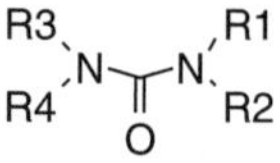 R3, R1, N, N, R4, R2, O	Paio et al., 2001[53]
	R3, NH, R4	

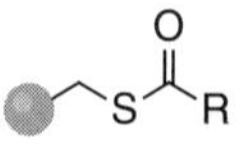

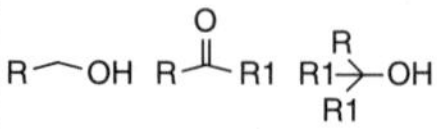

FIGURE 6.1 Continued.

A concept with attractive perspectives for increased flexibility in the choice of mutually compatible reaction conditions is termed *safety catch*. Here, particularly robust linkers are in their stable state during synthesis and are labilized by chemical transformation immediately prior to the final cleavage of the substrates. The SCAL linker [20] becomes acid labile only

Linker structure	Cleavage product/ reagent	References
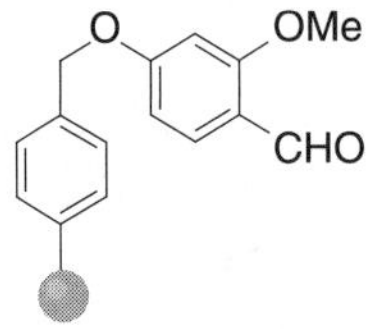 O, OMe, CHO	sulfonamides carboxamides ureas carbamates	Fivush and Wilson, 1997[54]
	TFA	
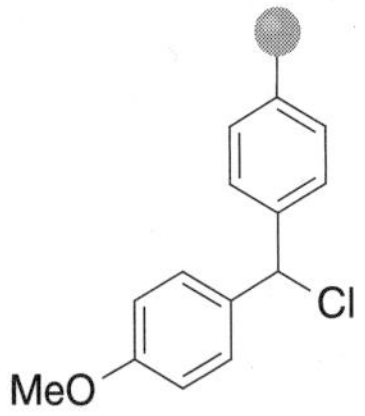 Cl, MeO	R-CONHR′	Brown et al., 1998[55]
	TFA	

FIGURE 6.1 Continued.

upon reduction of a sulfoxide to the thioether form. In another case, a photocleavable moiety of the 3''-methoxybenzoin type [21, 22] arises only after a dithiane protecting group is removed, thus avoiding the necessity to take precautions of light exclusion during synthesis.

III. SOLID-PHASE REACTIONS

Some significative examples of organic reactions are separated in Figure 6.2.

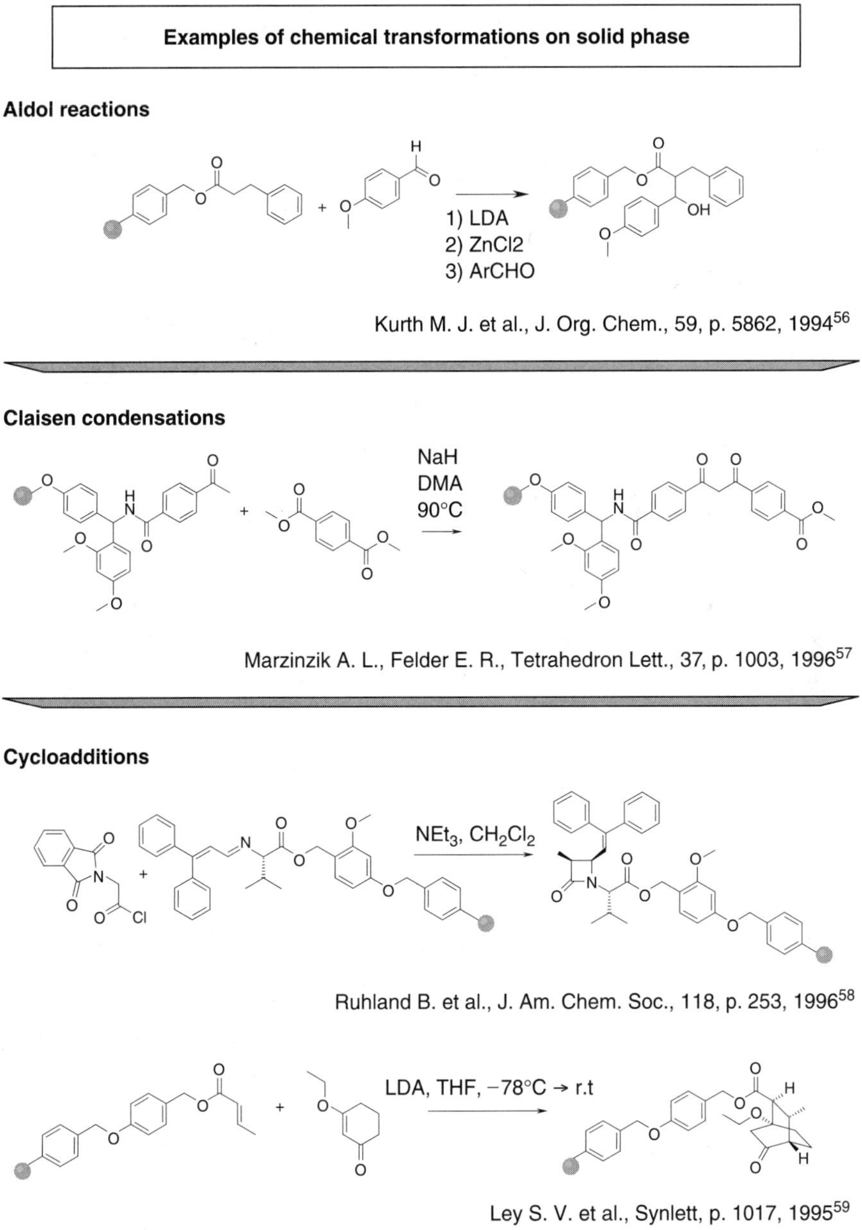

Examples of chemical transformations on solid phase

Aldol reactions

1) LDA
2) ZnCl2
3) ArCHO

Kurth M. J. et al., J. Org. Chem., 59, p. 5862, 1994[56]

Claisen condensations

NaH
DMA
90°C

Marzinzik A. L., Felder E. R., Tetrahedron Lett., 37, p. 1003, 1996[57]

Cycloadditions

NEt_3, CH_2Cl_2

Ruhland B. et al., J. Am. Chem. Soc., 118, p. 253, 1996[58]

LDA, THF, −78°C → r.t

Ley S. V. et al., Synlett, p. 1017, 1995[59]

FIGURE 6.2 Examples of chemical transformations on solid phase.

Cycloadditions (cont.)

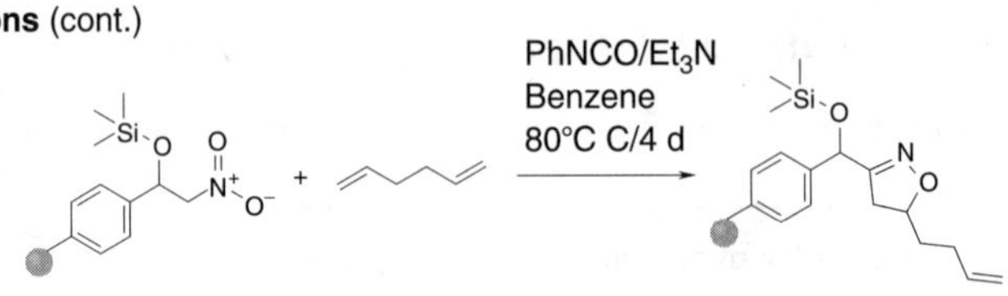

PhNCO/Et_3N
Benzene
80°C C/4 d

Beebe X. et al., J. Am. Chem. Soc., 114, p. 10061, 1992[60]

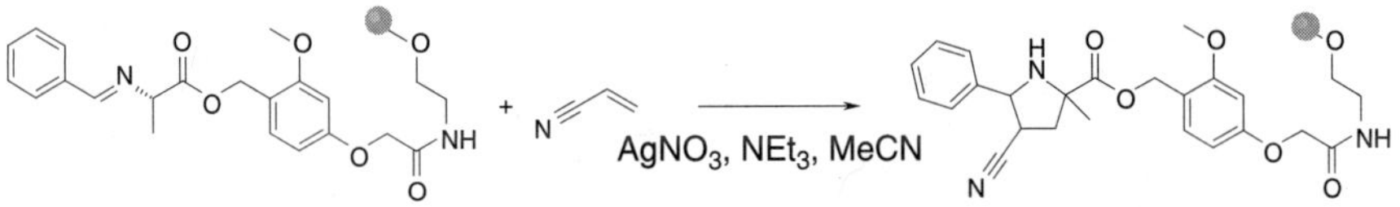

$AgNO_3$, NEt_3, MeCN

Murphy M. M. et al., J. Am. Chem. Soc., 117, p. 7029, 1995[61]

Heck reactions

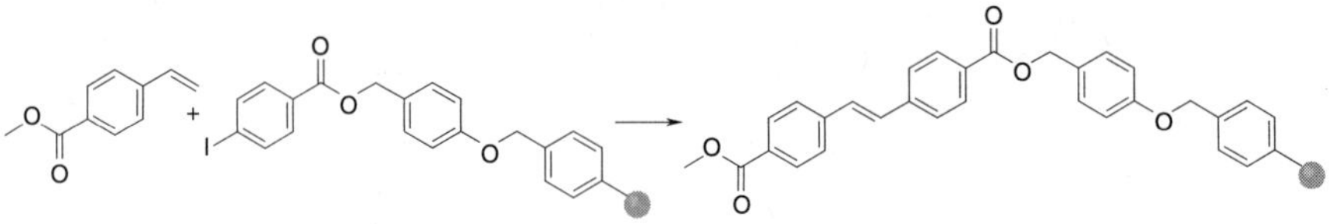

$Pd(OAc)_2$, $(n\text{-}Bu)_4NCl$, Et_3N, DMF, 90°C

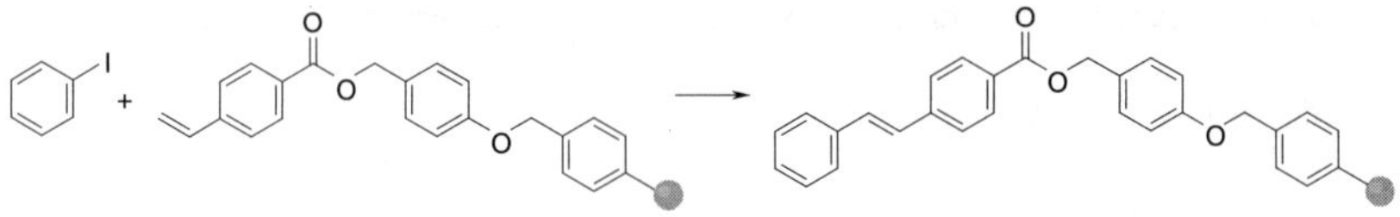

$Pd(OAc)_2$, $(n\text{-}Bu)_4NCl$, Et_3N, DMF, 90°C

Yu K.-L., Tetrahedron Lett., 35, p. 8919, 1994[62]

FIGURE 6.2 Continued.

Mitsunobu reactions

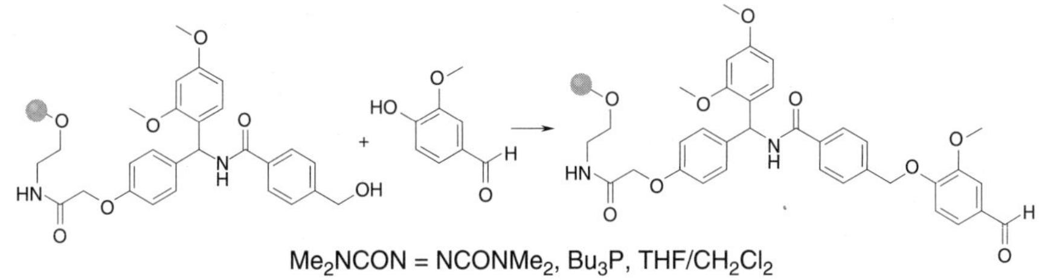

$Me_2NCON = NCONMe_2$, Bu_3P, THF/CH_2Cl_2

Rano T. A., Chapman K. T., Tetrahedron Lett., 36, p. 3789, 1995[63]

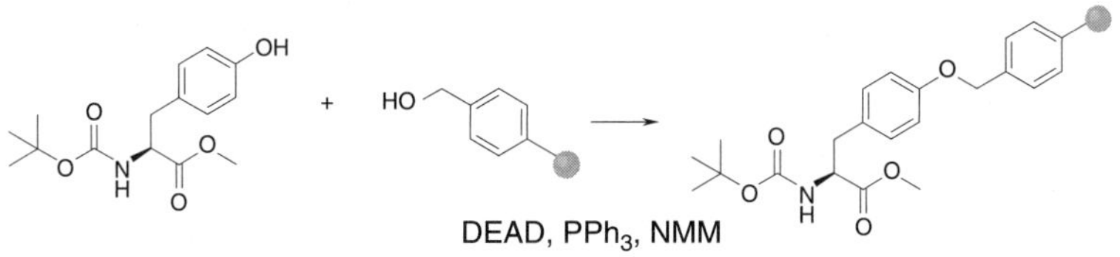

DEAD, PPh_3, NMM

Richter L. S., Gadek T. R., Tetrahedron Lett., 35, p. 4705, 1994[64]

Multiple-component condensations

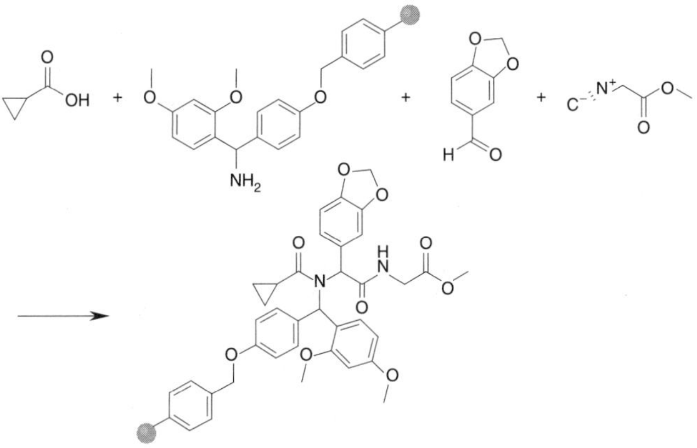

Armstrong R. W., Acc. Chem. Res., 29, p. 123, 1996[65]

FIGURE 6.2 Continued.

Oxidations

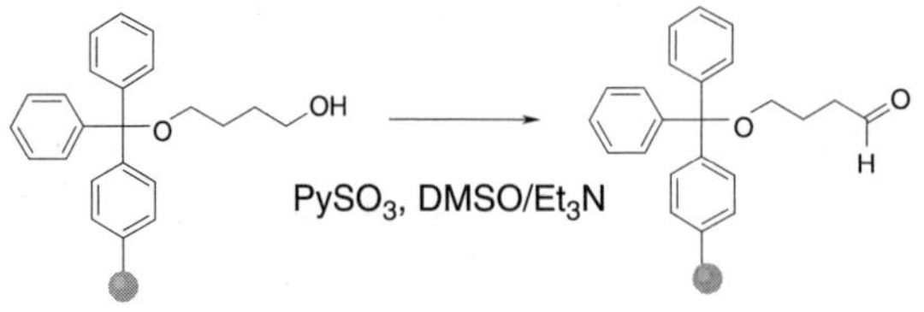

$PySO_3$, DMSO/Et_3N

Chen C. et al., J. Am. Chem. Soc., 116, p. 2661, 1994[66]

Reductions

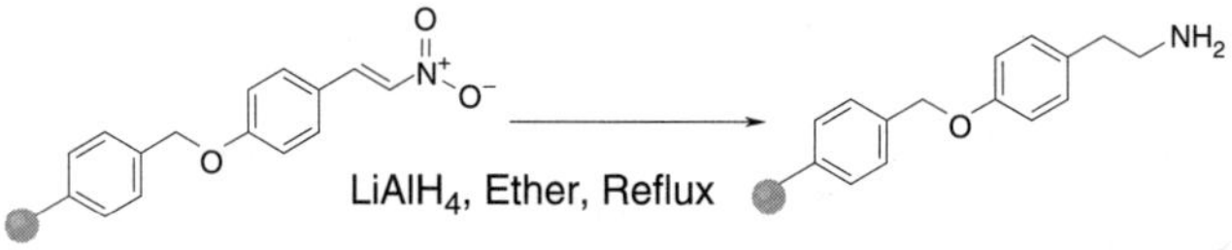

$LiAlH_4$, Ether, Reflux

Beebe X. et al, J. Org. Chem., 60, p. 4204, 1995[67]

Reactive Alkylations

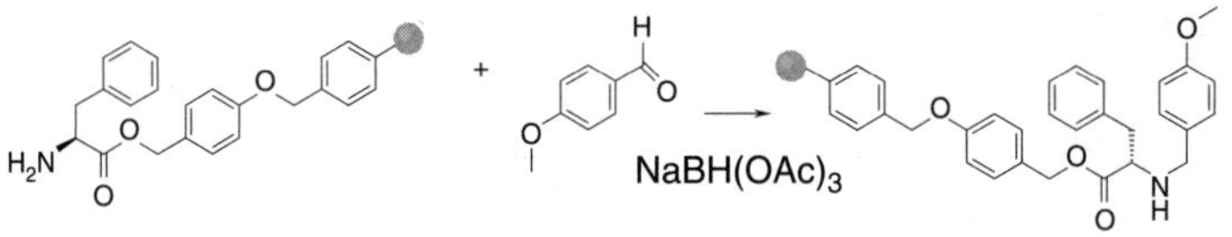

$NaBH(OAc)_3$

Gordon D. W., Steele J., Bioorg. Med. Chem. Lett., 5, p. 47, 1995[68]

FIGURE 6.2 Continued.

Stille couplings

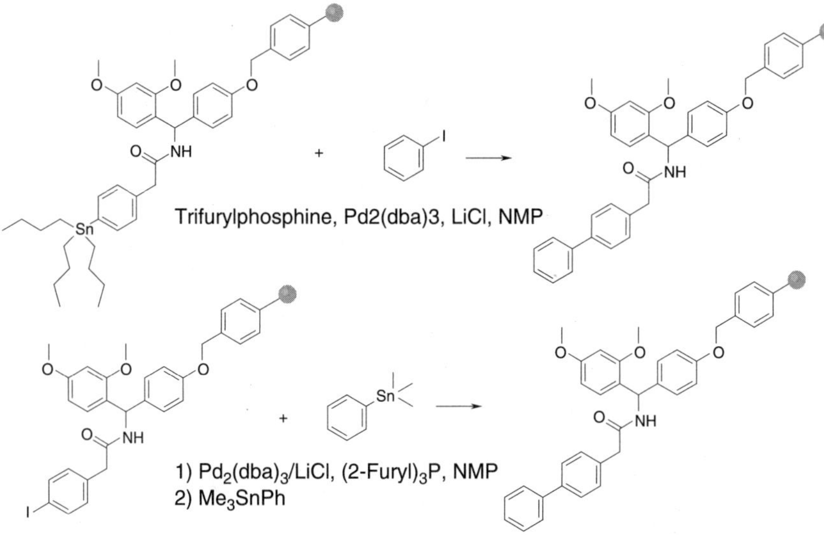

Trifurylphosphine, Pd2(dba)3, LiCl, NMP

1) $Pd_2(dba)_3$/LiCl, (2-Furyl)$_3$P, NMP
2) Me_3SnPh

Forman F. W., Sucholeiki I., J. Org. Chem., 60, p. 523, 1995[69]

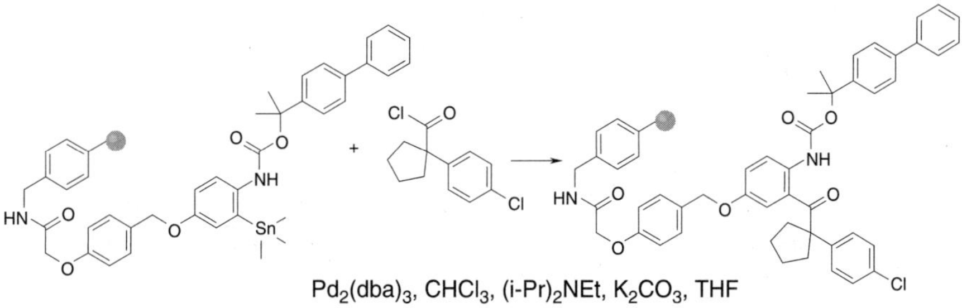

$Pd_2(dba)_3$, $CHCl_3$, (i-Pr)$_2$NEt, K_2CO_3, THF

Plunkett M. J., Ellman J. A., J. Am. Chem. Soc., 117, p. 3306, 1995[70]

FIGURE 6.2 Continued.

Suzuki couplings

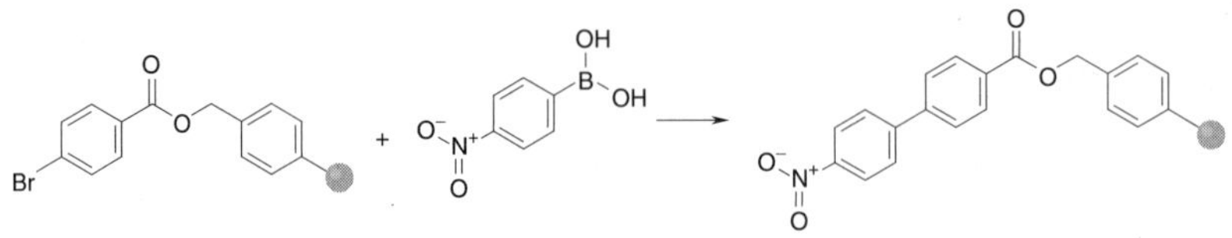

$Pd(PPh_3)_4$, Na_2CO_3, DME, Reflux

Frenette R., Friesen R. W., Tetrahedron Lett., 35, p. 9177, 1994[71]

Wittig reactions

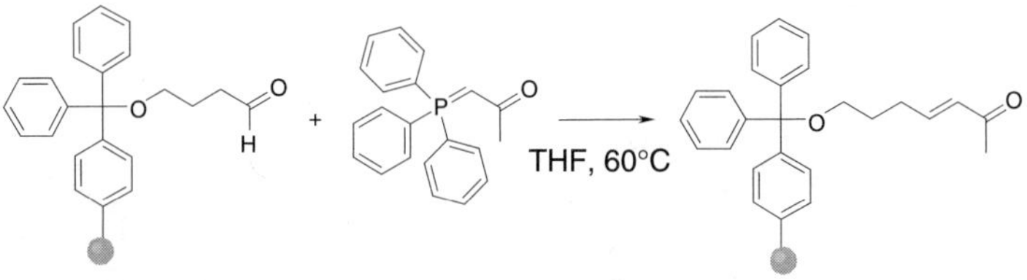

THF, 60°C

Chen C. et al., J. Am. Chem. Soc., 116, p. 2661, 1994[66]

FIGURE 6.2 Continued.

IV. CONCLUSIONS

As already stated, the spectrum of reaction types, which have been successfully adapted to the solid-phase format, has expanded noticeably over the last few years. The examples listed above illustrate the fact that a remarkable toolbox is now available to the organic chemist, whose challenge is to creatively combine reactions to powerful modular approaches with broad validity for whole sets of building blocks. Achievements in this area will pave the way toward molecular diversity for lead discovery, as well as efficient "analoging" of target structures.

REFERENCES

1. Früchtel JS, Jung G, Organic chemistry on solid supports, *Angew. Chem. Int., Ed. Engl.*, 35:17–42, 1996.
2. Hermkens PHH, Ottenheijm HCJ, Rees DC, Solid-phase organic reactions: review of the recent literature, *Tetrahedron*, 52:4527–4554, 1996.
3. Hermkens PHH, Ottenheijm HCJ, Rees DC, Solid-phase organic reaction II. A review of the literature, Nov 95–Nov 96, *Tetrahedron*, 53:5643–5678, 1997.
4. James I. *Solid phase chemistry publications*. Clayton, Australia, Chiron Mimotopes, 1995.

5. (a) White P, *Combinatorial chemistry catalog and solid phase organic chemistry (SPOC) handbook*, 2nd ed., Läufelfingen, Calbiochem-Novabiochem, 1997. (b) Scannell-Lansky A, Zechel C, In W Bannwarth, ER Felder, ed. *Combinatorial chemistry—A Practical Approach*, Weinheim, Wiley–VCH, pp. 329–421, 2000.
6. SPORE (Solid Phase Organic Reactions) database ver, 1, MDL Molecular Design Ltd., San Leandro. 2003.
7. SPS (Solid-Phase Synthesis) database ver 2003. 1, Accelrys Inc., San Diego, 2004.
8. Merrifield RB, Solid phase peptide synthesis, I. The synthesis of a tetrapeptide. *J. Am. Chem. Soc.*, 85:2149–2154, 1963.
9. Letsinger RL, Makadevan V, Oligonucleotide synthesis on a polymer support, *J. Am. Chem. Soc.*, 87:3526–3527, 1965.
10. Bayer E. Auf dem Weg zur chemischen Synthese von Proteinen, *Angew. Chem.*, 103:117–133, 1991.
11. Jung G, *Combinatorial Peptide and nonpeptide libraries: A handbook*, Weinheim: VCH Verlagsges, 2002.
12. Delgado M, Janda KD, Polymeric supports for solid phase organic synthesis, *Curr. Org. Chem.*, 6:1031–1043, 2002.
13. Lam KS, Salmon SE, Hersh EM, Hruby VJ, Kazmierski W, Knapp RJ, A new type of synthetic peptide library for identifying ligand-binding activity, *Nature*, 354:82–84, 1991.
14. Felder ER, Heizmann G, Matthews IT, Rink H, Spieser E, A new combination of protecting groups and links for encoded synthetic libraries suited for consecutive tests on the solid phase and in solution, *Mol. Diversity*, 1:109–112, 1995.
15. Blaney P, Grigg R, Sridharan V, Traceless solid-phase organic synthesis, *Chem. Rev.*, 202:2607–2624, 2002.
16. Sucholeiki I, Solid-phase photochemical C–S bond cleavage of thioethers—a new approach to the solid-phase production of nonpeptide molecules, *Tetrahedron Lett.*, 35:7307–7310, 1994.
17. Plunkett MJ, Ellman JA, A silicon-based linker for traceless solid-phase synthesis, *J. Org. Chem.*, 60:6006–6007, 1995.
18. Chenera B, Finkelstein JA, Veber DF, Protodetachable arylsilane polymer linkages for use in solid phase organic synthesis. *J. Am. Chem. Soc.*, 117:11999–12000, 1995.
19. Han Y, Walker SD, Young RN, Silicon directed ipso-substitution of polymer bound arylsilanes: preparation of biaryls *via* the Suzuki cross-coupling reaction, *Tetrahedron Lett.*, 37:2703–2706, 1996.
20. Patek M, Lebl M, Safety-catch anchoring linkage for synthesis of peptide amides by Boc/Fmoc strategy, *Tetrahedron Lett.*, 32:3891–3894, 1991.
21. Routledge AR, Abell C, Balasubramanian S, The use of a dithiane protected benzoin photolabile safety catch linker for solid-phase synthesis, *Tetrahedron Lett.*, 38:1227–1230, 1997.
22. Felder ER, Petriella P, Schneider P, Synthesis of a photolabile "safety catch" linker of the 3'-methoxybenzoin type, *Proc. of the First Int. Electr. Conf. on Synth. Org. Chem.* (ECSOC-1), www.mdpi.org/ecsoc/, September 1–30, 1997. Poster: http://www.unibas.ch/mdpi/ecsoc/b0003.htm1997.
23. Wang SS, *p*-Alkoxybenzyl alcohol resin and *p*-alkoxybenzyloxycarbonylhydrazide for solid phase synthesis of protected peptide fragments, *J. Am. Chem. Soc.*, 95:1328–1333, 1973.

24. Sarshar S, Siev D, Mjalli AMM, Imidazole libraries on solid support, *Tetrahedron Lett.*, 37:835–838, 1996.
25. Atherton E, Logan CJ, Sheppard RC, Peptide synthesis. Part 2. Procedure for solid-phase synthesis using *N*-fluorenylmethoxycarbonylamino-acids on polyamide supports. Synthesis of substance *P* and of acyl carrier protein 65–74 decapeptide, *J. Chem. Soc.*, Perkin I:538–546, 1981.
26. Bray AM, Jhingran AG, Valerio RM, Maeji NJ, Simultaneous multiple synthesis of peptide amides by multipin method. Application of vapor-phase ammonolysis, *J. Org. Chem.*, 59:2197–2203, 1994.
27. Rink H, Solid-phase synthesis of protected peptide fragments using a trialkoxy-diphenyl-methylester resin, *Tetrahedron Lett.*, 28:3787–3790, 1987.
28. Barlos K, Gatos D, Kallitsis J, Papaphotiu G, Sotiriu P, Wenqing Y, Schäfer W, Darstellung geschützter peptid-fragmente unter einsatz substituierter triphenylmethyl-harze, *Tetrahedron Lett.*, 30:3943–3946, 1989.
29. Hauske JR, Dorff P, A solid phase Cbz chloride equivalent—a new matrix specific linker, *Tetrahedron Lett.*, 36:1589–1592, 1995.
30. Sheppard RC, Williams BJ, A new protecting group combination for solid phase synthesis of protected peptides, *J. Chem. Soc. Chem. Commun.*, 587–589, 1982.
31. Farall MJ, Frechet JMJ, Bromination and lithiation: two important steps in the functionalization of polystyrene resins, *J. Org. Chem.*, 41:3877–3882, 1976.
32. Meyers HV, Dilley GJ, Durgin TL, Powers TS, Winssinger NA, Zhu H, Pavia MR, Multiple simultaneous synthesis of phenolic libraries. *Mol. Diversity*, 1:13–20, 1995.
33. Schwyzer R, Felder ER, Failli P, The CAMET and CASET links for the synthesis of protected oligopeptides and oligonucleotides on solid and soluble supports, *Helv. Chim. Acta.*, 67:1316–1327, 1984.
34. Thompson LA, Ellman JA, Straightforward and general method for coupling alcohols to solid supports, *Tetrahedron Lett.*, 35:9333–9336, 1994.
35. Brown BB, Wagner DS, Geysen HM, A single-bead decode strategy using electrospray ionization mass spectrometry and a new photolabile linker: 3-amino-3 (2-nitrophenyl) propionic acid, *Mol. Diversity*, 1:4–12, 1995.
36. Baldwin JJ, Burbaum JJ, Henderson I, Ohlmeyer MHJ, Synthesis of a small molecule combinatorial library encoded with molecular tags, *J. Am. Chem. Soc.*, 117:5588–5589, 1995.
37. Rich HD, Gurwara KS, Preparation of a new *o*-nitrobenzyl resin for solid-phase synthesis of *tert*-butyloxycarbonyl protected peptide acids, *J. Am. Chem. Soc.*, 97:1575–1578, 1975.
38. Burbaum JJ, Ohlmeyer MHJ, Reader JC, Henderson I, Dillard LW, Li G, Randle TL, Sigal NH, Chelsky D, Baldwin JJ, A paradigm for drug discovery employing encoded combinatorial libraries, *Proc. Natl. Acad. Sci. USA*, 92:6027–6031, 1995.
39. Holmes CP, Jones DJ, Reagents for combinatorial organic synthesis: development of a new *o*-nitrobenzyl photolabile linker for solid phase synthesis, *J. Org. Chem.*, 60:2318–2319, 1995.
40. Ajayaghosh A, Pillal VNR, Solid-Phase synthesis and C-terminal amidation of peptides using a photolabile *o*–Nitrobenzhydrylaminopolystyrene support, *Tetrahedron Lett.*, 36:777–780, 1995.

41. Morphy JR, Rankovic Z, Rees DC, A novel linker strategy for solid-phase synthesis, *Tetrahedron Lett.*, 37:3209–3212, 1996.
42. Ellman JA, Design, synthesis, and evaluation of small-molecule libraries, *Acc. Chem. Res.*, 29:132–143, 1996.
43. Plunkett MJ, Ellman JA, Germanium and silicon linking strategies for traceless solid-phase synthesis, *J. Org. Chem.*, 62:2885–2893, 1997.
44. Raju B, Kogan TP, Use of halomethyl resins to immobilize amines: an efficient method for synthesis of sulfonamides and amides on a solid support, *Tetrahedron Lett.*, 38:4965–4968, 1997.
45. Raju B, Kogan TP, Solid phase synthesis of sulfonamides using a carbamate linker, *Tetrahedron Lett.*, 38:3373–3376, 1997.
46. Alsina J, Chiva C, Ortiz M, Rabanal F, Giralt E, Albericio F, Active carbonate resins for solid-phase synthesis through the anchoring of a hydroxyl function, Synthesis of cyclic and alcohol peptides, *Tetrahedron Lett.*, 38:883–886, 1997.
47. Bannwarth W, Huebscher J, Barner R, A new linker for primary amines applicable to combinatorial approaches, *Bioorg. Med. Chem.*, 6:1525–1528, 1996.
48. Hughes I, Application of polymer-bound phosphonium salts as traceless supports for solid phase synthesis, *Tetrahedron Lett.*, 37:7595–7598, 1996.
49. Millington CR, Quarrell R, Lowe G, Aryl hydrazides as linkers for solid phase synthesis which are cleavable under mild oxidative conditions, *Tetrahedron Lett.*, 39:7201–7204, 1998
50. Braese S, Enders D, Koebberling J, Avemaria F, A surprising solid-phase effect: development of a recyclable "traceless" linker system for reactions on solid support, *Angew. Chem., Int. Ed.*, 37:3413–3415, 1998.
51. May P, Bradley M, Harrowven D, Pallin D, A new method of forming resin bound thioesters and their use as "traceless" linkers in solid phase synthesis, *Tetrahedron Lett.*, 41:1627–1630, 2000.
52. Pourbaix C, Carreaux F, Carboni B, Deleuze H, Boronate linker for "traceless" solid-phase synthesis, *Chem. Commun.*, 14:1275–1276, 2000.
53. Paio A, Crespo RF, Seneci P, Ciraco M, Solid-supported benzotriazoles. 2. Synthetic auxiliaries and traceless linkers for the combinatorial synthesis of unsymmetrical ureas, *J. Comb. Chem.*, 3:354–359, 2001.
54. Fivush MA, Willson TM, AMEBA: An acid sensitive aldehyde resin for solid phase synthesis, *Tetrahedron Lett.*, 38:7151–7154, 1997.
55. Brown DS, Revill JM, Shute RE, Merrifield, alpha-methoxyphenyl (MAMP) resin; a new versatile solid support for the synthesis of secondary amides, *Tetrahedron Lett.*, 39:8533–8536, 1998.
56. Kurth MJ, Ahlberg Randall LA, Chen C, Melander C, Miller R, Library-based lead compound discovery: antioxidants by an analogous synthesis deconvolutive assay strategy, *J. Org. Chem.*, 59:5862–5864, 1994.
57. Marzinzik AL, Felder ER, Solid support synthesis of highly functionalized pyrazoles and isoxazoles; scaffolds for molecular diversity, *Tetrahedron Lett.*, 37:1003–1006, 1996.
58. Ruhland B, Bhandari A, Gordon EM, Gallop MA, Solid-supported combinatorial synthesis of structurally diverse β-lactams, *J. Am. Chem. Soc.*, 118:253–254, 1996.
59. Ley SV, Mynett DM, Koot WJ, Solid phase synthesis of bicyclo[2.2.2]octane derivatives *via* tandem Michael addition reactions and subsequent reductive amination, *Synlett.*, 1017–1020, 1995.

60. Beebe X, Schore NE, Kurth MJ, Polymer-supported synthesis of 2,5-disubstituted tetrahydrofurans, *J. Am. Chem. Soc.*, 114:10061–10062, 1992.
61. Murphy MM, Schullek JR, Gordon EM, Gallop MA, Combinatorial organic synthesis of highly functionalized pyrrolidines: identification of a potent angiotensin converting enzyme inhibitor from a mercaptoacyl proline library, *J. Am. Chem. Soc.*, 117:7029–7030, 1995.
62. Yu K-L, Deshpande MS, Vyas DM, Heck reactions in solid phase synthesis, *Tetrahedron Lett.*, 35:8919–8922, 1994.
63. Rano TA, Chapman KT, Solid phase synthesis of aryl ethers *via* the Mitsunobu reaction, *Tetrahedron Lett.*, 36:3789–3792, 1995.
64. Richter LS, Gadek TR, A surprising observation about Mitsunobu reactions in solid phase synthesis, *Tetrahedron Lett.*, 35:4705–4706, 1994.
65. Armstrong RW, Combs AP, Tempest PA, Brown SD, Keating TA, Multiple-component condensation strategies for combinatorial library synthesis, *Acc. Chem. Res.*, 29:123–131, 1996.
66. Chen C, Ahlberg Randall LA, Miller RB, Jones AD, Kurth MJ, Analogous organic synthesis of small-compound libraries: Validation of combinatorial chemistry in small-molecule synthesis, *J. Am. Chem. Soc.*, 116:2661–2662, 1994.
67. Beebe X, Chiappari CL, Olmstead MM, Kurth MJ, Schore N, Polymer-supported synthesis of cyclic ethers: electrophilic cyclization of tetrahydrofuroisoxazolines, *J. Org. Chem.*, 60:4204–4212, 1995.
68. Gordon DW, Steele J, Reductive alkylation on a solid phase: synthesis of a piperazinedione combinatorial library, *Bioorg. Med. Chem. Lett.*, 5:47–50, 1995.
69. Forman FW, Sucholeiki I, Solid-phase synthesis of biaryls *via* the Stille reaction, *J. Org. Chem.*, 60:523–528, 1995.
70. Plunkett MJ, Ellman JA, Solid-phase synthesis of structurally diverse 1,4-benzodiazepine derivatives using the Stille coupling reaction, *J. Am. Chem. Soc.*, 117:3306–3307, 1995.
71. Frenette R, Friesen RW, Biaryl synthesis *via* coupling on a solid support, *Tetrahedron Lett.*, 35:9177–9180, 1994.

7 Solution Phase Combinatorial Libraries of Small Organic Molecules

Rafael Ferritto, Elisabetta de Magistris, Andrea Missio, Alfredo Paio and Pierfausto Seneci

CONTENTS

I. INTRODUCTION

The concept of combinatorial chemistry is often intended as closely related, or even coincident, with solid-phase chemistry [1–8]. This technique (thoroughly addressed by some chapters of this book [9–11]) has many advantages for an easy and reliable combinatorial synthesis, and, in fact, solid-phase combinatorial libraries with different formats and sizes have been dealt with in many excellent reviews [12–15]. Nevertheless, a significant amount of combinatorial efforts have been devoted to "solution" techniques. The term "solution" has to be intended in a broader sense, meaning that the chemical steps leading to library synthesis are performed in a homogeneous liquid medium, rather than at the interface between two phases, as in solid-phase combinatorial chemistry. This, as the reader may easily imagine, is a fundamental difference, which leads to completely different, and sometimes complementary, properties with respect to solid phase.

This review does not intend to provide absolute answers about the utility of solution-phase techniques versus solid phase, which are, anyway, strongly influenced by the specific requirements of each project. Rather, it will show the versatility of solution-phase techniques in generating combinatorial libraries through the detailed illustration of several different approaches. A list of the methodologies which will be covered in this review is shown in Figure 7.1.

We will start with classical solution phase combinatorial chemistry, using two examples, one of discretes and one of pools, describing the application of homogeneous phase library synthesis to multistep reaction sequences. The examples will be followed by a list of significant related references. Moving to new trends in the field of solution-phase chemistry, we will present multiple component condensations, which allow the simultaneous reaction of multiple monomer sets (typically three to four), thus increasing the size and the diversity of the library, combinatorial biocatalysis, where enzymatic catalysts are used to generate diverse compounds in solution via single biotransformations or sequential enzymatic reactions, and dynamic combinatorial libraries, where reversible reactions are used to build libraries containing larger quantities of the more biologically active compounds, if incubated in the presence of a specific biological target. Among the modifications aimed to improve the quality of homogeneous phase libraries and to automate the synthesis and purification procedures, we will illustrate solid support mediated solution synthesis, which

SOLUTION/HOMOGENEOUS LIBRARIES

Classical approaches: discrete libraries
pool libraries

Multiple component condensations

Combinatorial biocatalysis

Solid support mediated solution synthesis: reagents
catalysts/ligands

Fluorous multiphase combinatorial synthesis

Soluble supports: PEG
dendrimers

FIGURE 7.1 Solution phase techniques for the generation of combinational libraries.

uses solid supported reagents, catalysts or ligands, covalent or anionic reaction scavengers, or trapping reagents in homogeneous phase combinatorial chemistry, fluorous multiphase combinatorial synthesis, where perfluorinated molecules and a perfluorinated extraction layer allow purification of library components from reaction mixtures, and soluble supports, where the supports used can be dissolved in most solvents, but can eventually be separated from the solution phase at the end of the synthesis.

II. CLASSICAL SOLUTION-PHASE COMBINATORIAL CHEMISTRY

The vast majority of solution libraries synthesized as of today were small (< 1000) discrete focussed libraries, prepared via single reactions where high quality chemistries were used, and simple, if any, purification steps were needed. While a complete bibliography of these many examples would use too much space of this review, the reader may access some recent excellent reviews [16–18] which cover this topic extensively. We will rather move to more complex solution libraries, where multiple sequential chemical steps were performed. Two examples are presented here and, in the first [19], a medium size discrete library was prepared via a three-step sequence and purified by simple washings and extractions, obtaining pure compounds with high yields. In the second [20], a medium size pool library was prepared via the simultaneous addition of various monomers, providing liquid pools with high analytical quality. Several more recent reviews [21–24] cover the synthesis and the usefulness of more complex solution libraries.

A. MULTISTEP DISCRETE LIBRARIES

The selected example by Sim and Ganesan [19] reported the design of a 3078-member discrete thiohydantoin library, prepared without purification of intermediates, with a three-step synthesis, which is shown in Figure 7.2. Nine amino acid esters were reacted with eighteen aromatic aldehydes to produce imines, which were reduced by sodium triacetoxyborohydride. The resulting amines were treated with nineteen isothiocyanates in the presence of triethylamine (TEA), producing intermediate isoureas, which eventually cyclized to the final thiohydantoins.

The formation of the imines and their subsequent reduction was monitored by the disappearence of the starting amine (ninhydrin staining, TLC). Simple quenching with water and drying of the organic solutions with $MgSO_4$ constituted the reaction work-up. The condensation of the amines with isothiocyanates and TEA was monitored by TLC (blue bromophenol staining, disappearence of the amine) and treatment of the crudes with glycine; washing with water and concentration gave the final thiohydantoins. The yields (mass recovery) and the purities (calculated by HPLC) of the final library components were generally good to excellent for all the used building blocks.

The library synthetic steps were carefully adjusted to avoid complex purification procedures, but even the simple washing/drying/extraction

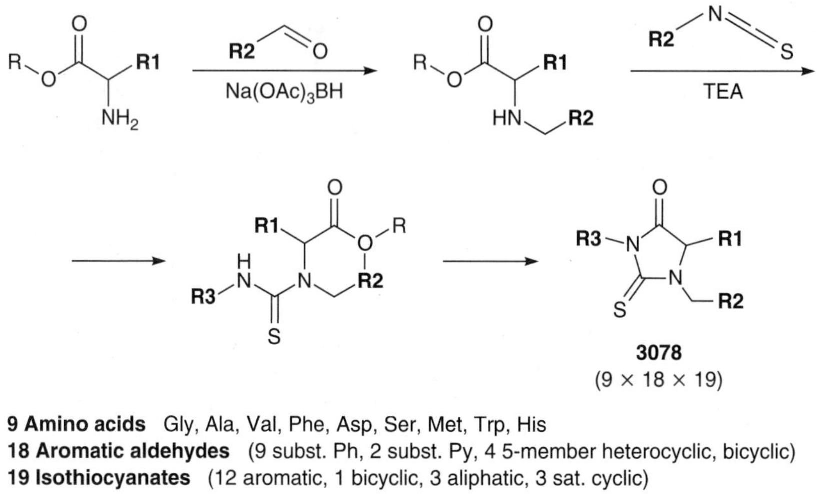

3078
(9 × 18 × 19)

9 Amino acids Gly, Ala, Val, Phe, Asp, Ser, Met, Trp, His
18 Aromatic aldehydes (9 subst. Ph, 2 subst. Py, 4 5-member heterocyclic, bicyclic)
19 Isothiocyanates (12 aromatic, 1 bicyclic, 3 aliphatic, 3 sat. cyclic)

FIGURE 7.2 Three step synthesis of 3078-membered discrete thiohydantoin library.

procedures made this library synthesis rather laborious and time consuming. The authors, in fact, synthesized only 600 discrete compounds out of the 3078 which were planned in the library design. This calls for either simpler and effective work-up procedures for multistep solution libraries, which are very difficult to identify, or new automated techniques to overcome the "purification/work-up" bottleneck (see some of the following paragraphs).

Among the multistep solution libraries, Cheng *et al.* [25, 26] reported a dipeptide mimetic template-based 78-member library, prepared via four synthetic steps and purified with an excellent protocol based on acid–base extractions. Thomas *et al.* [27] presented a >1000-member benzimidazole library, prepared via three steps, including a 2-ethoxy-1-ethoxycarbonyl-1,2-dihydroquinoline (EEDQ) assisted cyclization. Boger *et al.* reported two 600-member C_2-symmetrical [28, 29] and unsymmetrical [29] libraries, prepared via four to five synthetic steps, based on iminodiacetic acid as a scaffold and using the olefin metathesis reaction, and piperazinone libraries [30], derived from N-Boc-iminodiacetic acid for biological testing. Yang and coworkers [31] presented some dihydropyrimidinone arrays, prepared by using La-catalyzed Biginelli reactions.

B. Pool Libraries

The selected example, by An *et al.* [20], reported the use of novel polyazapyridinophanes scaffolds to prepare medium size solution pool libraries via solution-phase simultaneous addition of functionalities (SPSAF). The synthesis of the scaffolds is reported in Figure 7.3. The unsymmetrical 13-, 14- and 15-member scaffolds were prepared from the corresponding polyamines, bearing two free secondary amines, with either very similar **1**, **2** or completely

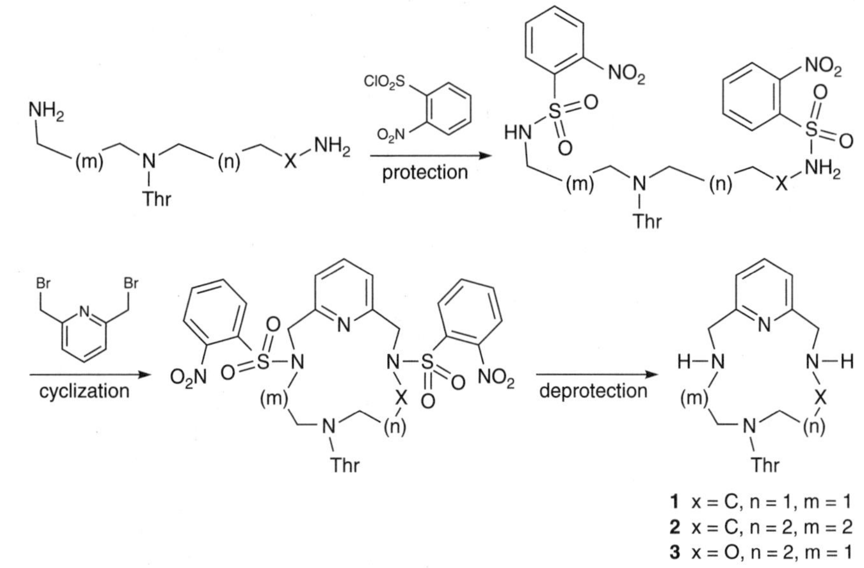

FIGURE 7.3 Synthesis of polyazapyridinophanes scaffolds.

different **3** reactivities, and another protected secondary amine. A monomer set of 10 substituted benzyl bromides was chosen for SPSAF, and the scaffold 2 was first used. This ensured an identical reactivity of the 2 free nitrogen atoms of the scaffold and of the 10 benzyl bromides, due to the limited effect of *meta* substituents on the total electrophilicity of the monomers. The homogeneity of the set of all monomers was confirmed by the synthesis and the full characterization of smaller model libraries, where sets of two to five monomers were simultaneously reacted with 2. Their capillary zone electrophoresis (CZE)-MS profile showed roughly equivalent quantities of each library component. The library synthesis was then performed, following the scheme depicted in Figure 7.4. The simultaneous alkylation with $R_{1,10}Br$ of the two free amines produced the 100-member library 4, then deprotected to give the library 5. This library was then reacted with 14 bromides (the 10 benzylic used for the first step, plus 4 substituted alkyl bromides), and the resulting fourteen 100-member pools 6–19 were tested with pools 4 and 5 for their antimicrobial activity. The quality of the libraries was good as judged from their CZE-MS profile, where all the expected peaks were present. Some individuals coming from the deconvolution of active pools produced interesting micromolar activities on various bacterial strains and also preliminary SAR data.

The use of the "fix last concept" [20] allowed the handling of a single reaction for each pool until the last step, where the addition of monomers in different reaction vessels produced the final pools. Each pool was prepared in large amounts (typically from 200 mg to 3 g of crude material), allowing

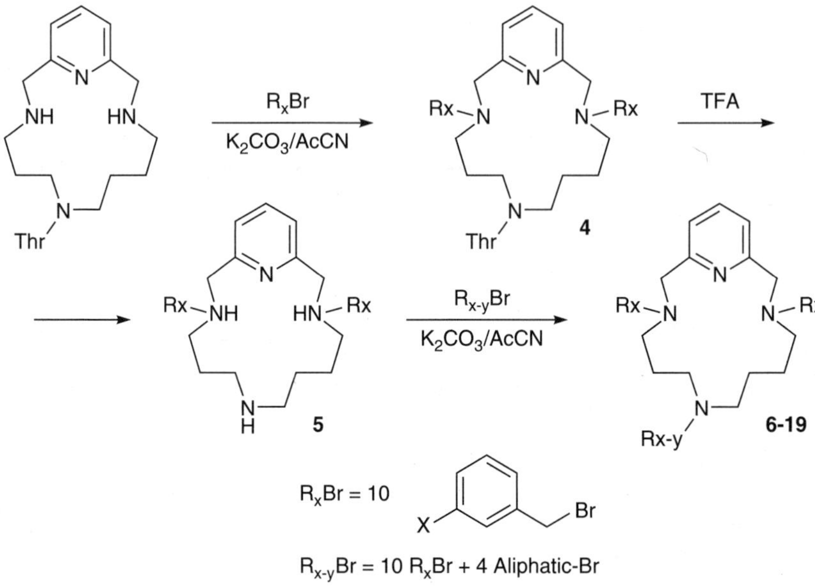

NH
HN
N
Thr
R$_x$Br
K$_2$CO$_3$/AcCN
Rx
4
TFA
5
R$_{x\text{-}y}$Br
Rx-y
6-19
R$_x$Br = 10
X
Br
R$_{x\text{-}y}$Br = 10 R$_x$Br + 4 Aliphatic-Br

FIGURE 7.4 Synthesis of polyazapyridinophanes-based library.

multiple testing of the library. The use of unsymmetrical scaffolds **1–3** and the rigorous reactivity assessment of both the scaffold nitrogens and of the 10 benzyl bromides, ensured the equimolar representation of each individual in a given pool. Equal representation of library components provided reliable biological activities which were confirmed after deconvolution. The concentration of pools for the biological assay was set to a total of 100 μM, thus having each compound present in a micromolar concentration. The "distortion" caused by multiple copies of an individual in a symmetrical library would rather bias toward the discovery of more represented individuals, as in a published example [32] of a four-position randomization with 19 building blocks on a symmetrical scaffold which produced a 65,321-member library out of the $19^4 = 130{,}321$ possible permutations.

Each of the library steps required a work-up followed, by chromatographic purification. While this was feasible for the 16 pools **4–19** prepared, larger monomer sets or intermediate splitting in pools with more reactions to be handled would definitely require automated purification/work-up procedures.

A comparison between solid-phase and solution-phase pool libraries can be made using this example. The detailed reactivity assessment produced good quality pools, but the almost identical reactivity for the functional groups of the scaffold and of the monomers allowed equimolar stoichiometric amounts of benzyl bromides to be reacted with the scaffold **2**. The use of a more "diverse" set of monomers, with significantly different reactivities, and/or the use of scaffolds containing functional groups with different reactivity, would require either the use of isokinetic mixtures [33] (large excess of monomers to be eliminated after the synthesis) or the sequential addition of substoichiometric equimolar mixtures of monomers [34] (significant complication of the library synthesis procedures). The decoration of scaffold positions with different reactivities, as the two free amines in **3**, would also "distort" the composition of the libraries and complicate the synthetic procedures, while the sequential decoration on the scaffold would require orthogonally protected functionalities and would increase the number of synthetic steps. In a word, as of today, the production of large and "diverse" pool libraries *via* solution-phase combinatorial chemistry cannot compare with the solid-phase "one bead—one compound" method [35] technique. Major improvements are needed for having similar performances from solution-phase synthesis of large pool libraries. The researchers of ISIS Pharmaceuticals are moving to more challenging libraries from scaffolds bearing functionalities with different reactivities, such as **3** (Figure 7.3), and to more complex monomer sets [36–43]. Their results could be helpful in opening new frontiers for solution-pool libraries.

Other groups reported efforts in pool solution libraries. Smith *et al.* [44] presented a 1600-member amide/ester library in solution (80 positional scanning pools of 40 compounds) which was tested on the NK_3 receptor and on the metalloproteinase MMP-1. Pirrung *et al.* [45, 46] deconvoluted the activity on acetylcholinesterase of two solution-phase indexed combinatorial libraries of 54 carbamates (15 orthogonal pools) and 72 tetrahydroacridines (18 orthogonal pools). Rebek and coworkers reported the decoration of a xanthene symmetrical scaffold, with amides [32] and ureas [47, 48], to give large solution libraries (single pool, several thousands of components) tested for trypsin inhibition [32] and transcription factors binding [47, 48]. Maehr and Yang [49] reported a 700-member pool library of functionalized thiazoles, tested as leukotriene D_4 antagonists *via* iterative deconvolution ($10+7+10$ pools). Chng and Ganesan [50] reported a 40,000-member β-aminoalcohol solution library (pools of four individuals), which was tested on a wide panel of biological assays. Boger *et al.* [51–53] reported the initial validation [51] and the full library synthesis [52] for a 64,980-member, solution-phase, pool library of biaryls, obtained via Pd-catalyzed iodoarene couplings, and the synthesis of a 2640-member pool library in solution, inspired by the structure of distamycin [53]. Olah and coworkers [54] validated the solution-phase pool library strategy, leading to 3,3-diaryloxyndoles via condensation of isatins and substituted benzenes.

III. MULTIPLE COMPONENT CONDENSATIONS

As early as 1961, Ugi [55] theorized the use of multicomponent reactions for the generation of large collections of compounds and, in the following years, other related works were reported [56–58]. It was only recently, though, that the full potential of multiple component condensations (MCCs) for the generation of combinatorial libraries became apparent, due to the efforts of major pharmaceutical industries and academic research groups.

MCCs are an easy and reliable way to obtain medium-large solution libraries *via* the simultaneous reaction of more than two reagents in solution. Even a single randomization step, in fact, may produce a 10000-member library, using 4 monomer sets of 10 individuals, and only a single purification/ work-up procedure is required.

The selected examples by Keating and Armstrong [59, 60] reported the synthesis of a small discrete solution library (eight compounds) of Ugi 4CC products and their further elaboration. The concerted mechanism of the Ugi 4CC is shown in Figure 7.5. The scarcity of commercial isocyanides encouraged the use of a "convertible isocyanide" [59, 60], which allowed its postcondensation transformations in other functional groups, which are shown in Figure 7.6.

The reaction of cyclohexenamides with nucleophiles such as water, alcohols, or thiols, produced carboxylic acid, esters, or thioesters. Reaction with acetylenic dipolarophiles in acidic conditions produced highly functionalized pyrroles *via* a complex mechanism, implying as intermediates 1,3-dipoles and bycyclic cycloaddition products. Reaction of cyclohexenamides containing protected hydroxylic functions with AcCl/MeOH produced δ-lactones, while cyclohexenamides, bearing in R_1 an *o*-aminophenyl group, easily cyclized to 1, 4-benzodiazepine-2, 5-diones.

The MCC reactions were exploited recently and, among the reported examples, Goebel and Ugi [61] prepared small discrete carbohydrate libraries, using 3CC reactions with sugars, trimethylsilylazide and dibenzyl disulfide. Weber *et al.* [62] employed a genetic algorithm [63, 64] for the selection of the best biological activities on a thrombin inhibition assay for a virtual 160,000-member library, prepared by Ugi 4CC reaction using aldehydes, amines, carboxylic acids, and isocyanides. Mukhopadhyay *et al.* [65] prepared a 20-member discrete library of β-acetamidocarbonyl compounds, using a $CoCl_2$-catalyzed 3CC reaction with ketones, aldehydes, and acetonitrile. Cavicchioli *et al.* [66] reported a Pd-catalyzed 3CC reaction between allylic

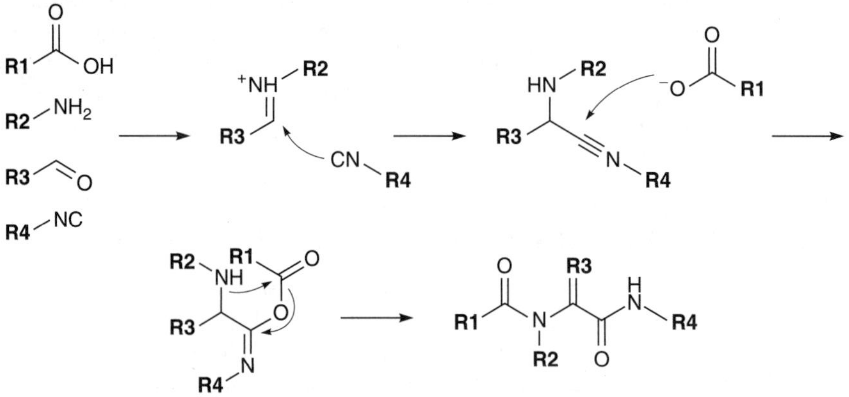

FIGURE 7.5 Concerted mechanism of Ugi 4CC reaction.

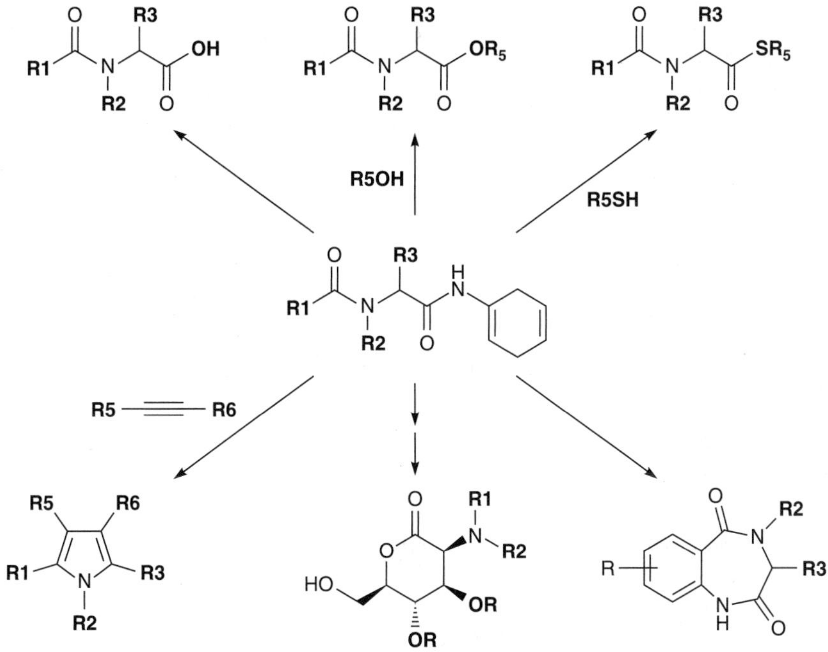

FIGURE 7.6 Postcondensation transformation of "convertible isocyanide".

alcohols, iodoaryls and Michael acceptors, to give a small discrete library (seven members). Kobayashi *et al.* [67] reported the synthesis of small discrete libraries of β-lactams, either via Lewis acid-catalyzed 3CC (silyl enolates, α,β-unsaturated thioesters, and imines) or *via* 4CC (silyl enolates, α,β-unsaturated thioesters, amines, and aldehydes) reactions. Nakamura *et al.* [68] reported the synthesis of discrete ketones via 3CC and *via* 4CC, using zincated hydrazones, vinyl Grignard reagents, and aldehydes or halides or disulfides or alcohols. Holmes *et al.* [69] prepared a small discrete library of 4-thiazolidinones and 4-metathiazanones, using 3CC reactions between amino acid esters, aldehydes, and mercaptoacids. The library was prepared both in solution and on solid phase, anchoring the amino acid *via* an ester bond to the resin.

This simple and versatile combinatorial one-pot method will surely provide, in future, many diverse libraries, and its use in combination with solution purification techniques (see the next sections) will help in automating the experimental procedures. A thorough search for new multicomponent condensations should even increase their applications in combinatorial library synthesis.

IV. COMBINATORIAL BIOCATALYSIS

Combinatorial chemistry has proven its usefulness for the synthesis of chemical libraries, with different degrees of complexity embedded in the scaffolds and/or the building blocks used. The synthesis of polyfunctionalized molecules in a combinatorial format, though, often requires the careful adjustment of experimental conditions to obtain and preserve the library intermediates from degradation during the synthesis, and the selection of orthogonal protecting groups to prevent side reactions or degradations, or regioselectivity problems. Moreover, the synthesis of chiral individuals has often been an unattainable target for combinatorial chemists.

Enzymes have been often used as reagents for classical organic chemistry reactions, and their application in many different examples has been reported [70–72]. Many of them are commercially available and inexpensive, and they perform a wide range of chemical transformations, including the introduction of new functional groups on a scaffold, the modification of existing functionalities, and the addition onto functional groups. Their appealing features as combinatorial reagents are the conversion of their substrates into reaction products under extremely mild conditions, the absence of reaction byproducts, their broad specificity in terms of substrates, which is common for many enzymes, and their complete regio- and stereoselectivity. Their use for combinatorial synthesis in solution, especially on complex substrates bearing different functional groups potentially susceptible to enzymatic reactions, could provide a library of complex modifications of the original scaffold, where also multiple sequential transformations could be performed.

The selected example by Khmelnitsky *et al.* [73] reported the synthesis of solution-phase libraries *via* combinatorial biocatalysis using different substrates. One of them, bicyclo[2, 2, 2]oct-5-ene-2,3-*trans* dimethanol (BOD), is shown in Figure 7.7, together with the biotransformation strategy that was applied. The selected reactions introduced new functional groups on BOD (halohydration), or selectively functionalized one of the existing alcoholic functions (glycosylation, acylation). The reactions were carried out either in organic or in aqueous solvents and the experimental conditions were adjusted to the enzyme stability and to the substrate solubility. A successful

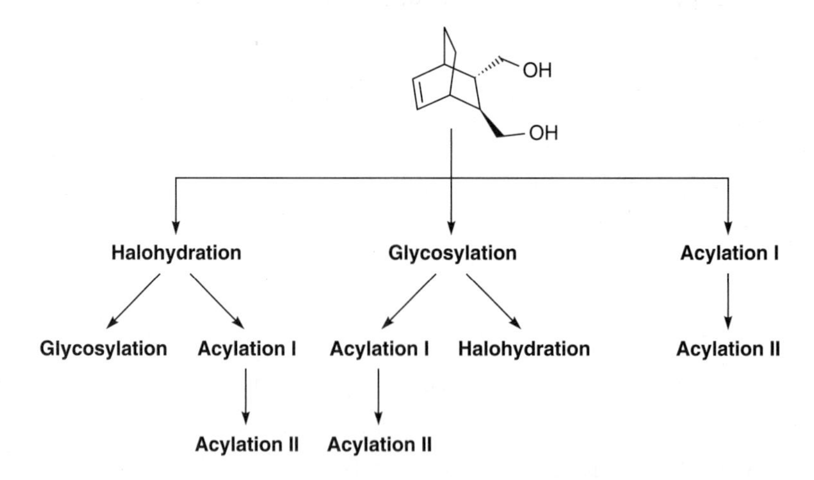

FIGURE 7.7 Biotransformations of BOD.

biotransformation was monitored with chromatographic methods, and the final products were characterized by MS and/or NMR. Selected structures of this biotransformation library (1222 derivatives in total) are shown in Figure 7.8. They illustrate the variety of functional groups, products and physico-chemical properties, which are obtained by such a "diverse" combinatorial library, where all the compounds are optically active.

Some other libraries were reported, using as substrates adenosine (ADS, 92 individuals) and 2, 3-(methylene dioxy)benzaldehyde (MDB, 457 individuals). Taxol was also used and submitted to various acylations, which produced

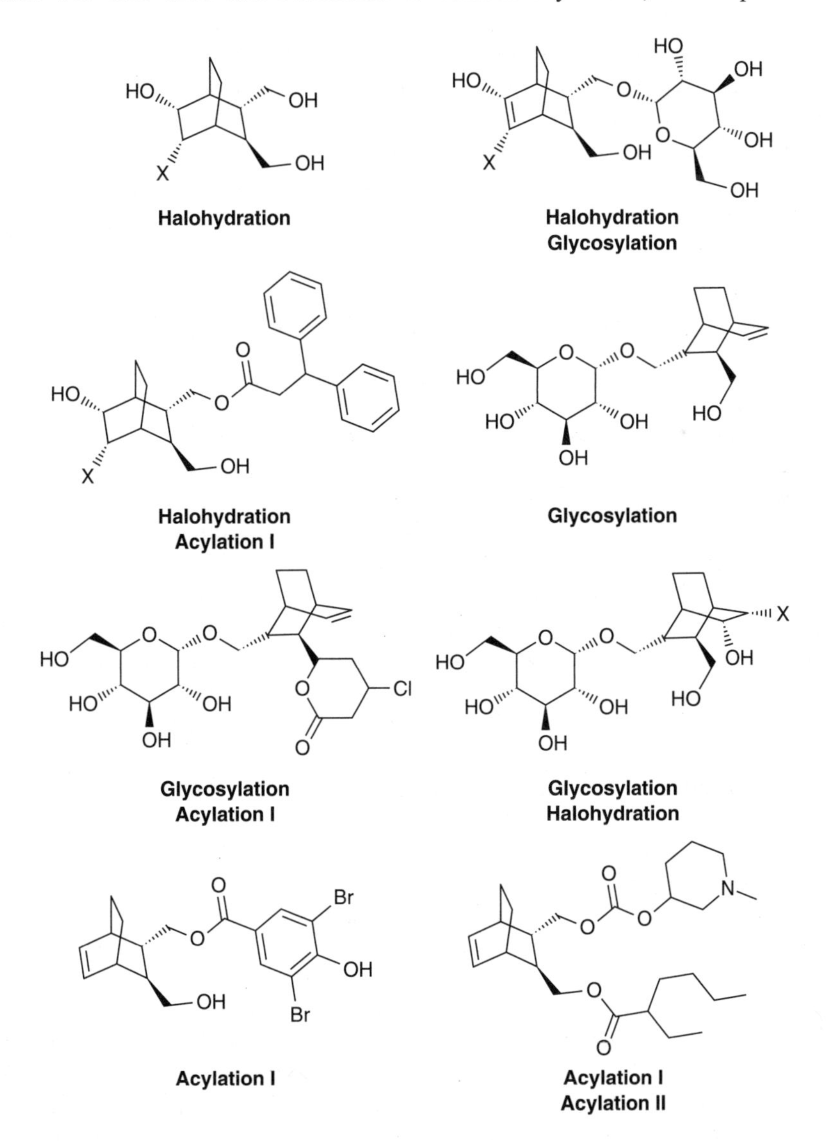

FIGURE 7.8 Selected structures of BOD's biotransformation library.

200 compounds, among which two showed a significant improvement in water solubility when compared to the parent compound. Khmelnitsky *et al.* [74] and Michels *et al.* [75] reported also further efforts in the derivatization of a diterpenoid [74] and of various natural products [75], including flavonoids and macrolides. Dordick and coworkers [76] prepared a library of α-acylaminoamides, using a combination of chemical synthesis and combinatorial biocatalysis, including a multicomponent reaction protocol.

A wider available panel of purified enzymes, or of characterized microorganisms performing a particular biotransformation, could increase the potential of this technique. Its main limitations are the stability of the enzymes, both to organic solvents and to different temperatures, which could be significantly improved by their immobilization on a solid support, and the solubility in aqueous media required in most cases for the biotransformation substrates. Anyway, the extreme usefulness of combinatorial biocatalysis for specific classes of products (complex natural products, polyfunctionalized chiral scaffolds) has already been assessed.

V. DYNAMIC COMBINATORIAL LIBRARIES

Synthetic organic libraries are made up of stable compounds, prepared through irreversible reactions, and these compounds are screened against one or more targets using different methodologies, depending on the library format, and the active molecules are identified and further profiled to assess their usefulness. The whole process is irreversible, because a library which shows no affinity for a specific target cannot adapt its structural features to interact better with the target structure.

Dynamic interactions between a target and a molecule, which lead to a reshaping of the molecule to better fit the target, have been known for many years [77] and represent the basis of so-called *template-directed synthesis* [78, 79]. The template acts on the macroscopic geometry of a reaction which could produce several products, by shifting the equilibrium toward a single product, but does not bind covalently to either the reagents or the reaction product. Templates may act *kinetically*, operating on irreversible reactions and accelerating the formation of a product *via* the stabilization of its transition state. In a hypothetical example (Figure 7.9), the reaction between A and B produces F, G, and H in different amounts through transition states C, D, and E (path A), while the template X binds non-covalently to the transition state E and leads only to the formation of H (path B). Other templates act *thermodynamically* when the reaction is reversible and the non-covalent binding of the template to a specific product shifts the equilibrium toward a single product. In path C, a reversible reaction between A and B produces an equilibrium mixture of monomer AB, dimer ABAB and cyclic dimer cABAB. When the template X is used, its affinity for the cyclic dimer cABAB shifts the equilibrium towards this compound, which is the sole reaction product (Figure 7.9, path D). If cABAB is submitted to the above reversible reaction

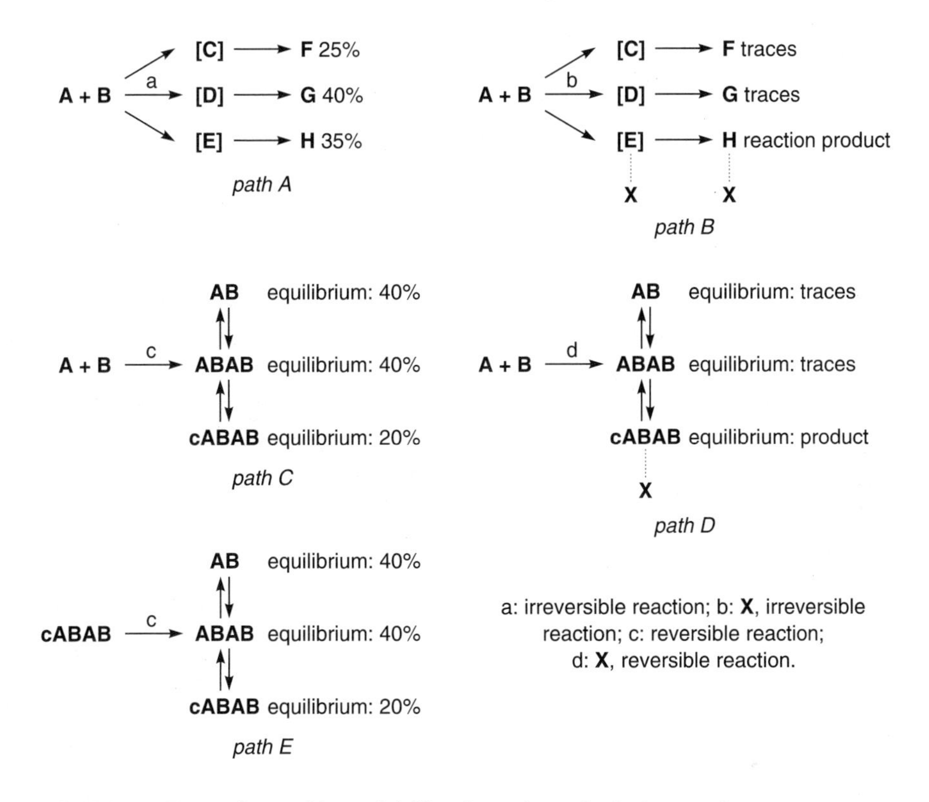

FIGURE 7.9 Dynamic combinatorial libraries: a hypothetical example.

conditions without the presence of X, it re-equilibrates, giving the same product mixture as in path C (Figure 7.9, path E).

Template-directed synthesis has also been exploited for combinatorial purposes in which a reversible reaction and the use of thermodynamic templates have been employed. Two different processes have been envisaged and validated, both of which consider the dynamic optimization of a receptor-ligand interaction where one of the partners is the template which drives the self-assembly of a reversible library of other partners, from which the best binder for the template is selected (Figure 7.10). If the receptor is a template, a library made using a reversible reaction is incubated with the receptor and the library component(s) giving the best fit with the template is selected and produced (A, Figure 7.10, top). Thus, the dynamic ligand virtual library consists of a mixture of all the potential reversible combinations of the library components (A–Z), but only the positives (A in the example) are isolated and their structure is determined. If the ligand is the template, a dynamic receptor library R_1–R_{20} is screened and the most active synthetic receptor (R_1 in the example) is identified, using the same approach (Figure 7.10, bottom).

The pioneering work in this area was reported by Lehn [80–83], both referring to ligand (imines) and to receptor libraries (barbiturate receptors and

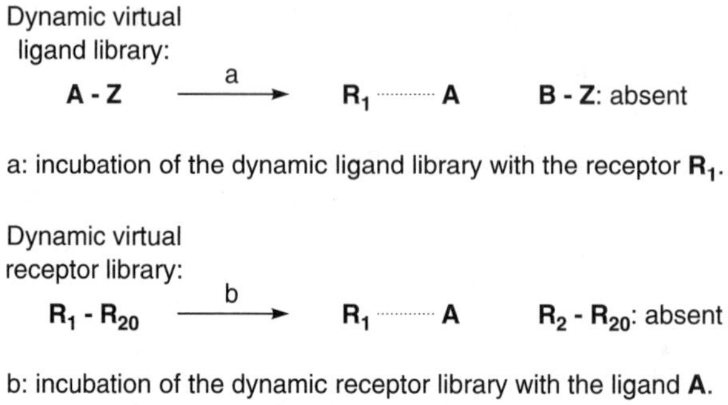

Dynamic virtual
ligand library:

A - Z $\xrightarrow{a}$ **R**$_1$ ········ **A** **B - Z**: absent

a: incubation of the dynamic ligand library with the receptor **R**$_1$.

Dynamic virtual
receptor library:

R$_1$ **- R**$_{20}$ $\xrightarrow{b}$ **R**$_1$ ········ **A** **R**$_2$ **- R**$_{20}$: absent

b: incubation of the dynamic receptor library with the ligand **A**.

FIGURE 7.10 Template-directed synthesis of dynamic combinatorial libraries.

bipyridine-metal complexes-selected receptors); the principle has been further applied by other workers, for example by Eliseev and Nelen [84], to select among isomeric forms of unsaturated dicarboxylates for their affinity toward arginine receptors [84] and to prepare a dynamic library of aryl and alkyl oximes [85]; by Sakai *et al.* [86] to discriminate among isomeric carbohydrates as lectin binders; by Miller to select among bis(salicylaldiminato) zinc complexes for their DNA-binding efficiency [87, 88]; by Benner and coworkers [89] to prepare dynamic libraries of alkenes; by Hinoka and Fujita [90] to select the best binders from a thermodynamic receptor library of Pd(II)-linked cages, driven by the presence of 1, 3, 5-benzenetricarboxylic acid as a ligand; by Rowan and Sanders and coworkers to make quinine macrocyclic libraries [91], to prepare reversible [2]-catenane libraries [92, 93], to produce reversible diversity by oligomerization of cinchona-based and xanthene-based building blocks [94], and to synthesize several cyclic peptidomimetic dynamic libraries [95]. Six reviews [96–101] have recently covered this subject.

While the applications of dynamic libraries have, so far, been limited to test cases, their potential to determine and influence the best molecular arrangement of ligands by using the relevant receptor/molecule as a template is significant, providing that reliable reversible chemical reactions are developed in the future, using a wide range of chemical diversity to generate large dynamic libraries.

In our selected example, Lehn and coworkers [80] reported the synthesis of a dynamic 12-member, template-directed imine library 1, obtained from the reversible condensation of three aldehydes (monomer set M_1, Figure 7.11), with four primary amines (monomer set M_2, Figure 7.11) in buffered aqueous conditions, followed by irreversible reduction to amines 2 with sodium cyanoborohydride. The library was prepared in the presence of a large excess of M_2, to prevent further condensation of an aldehyde onto the secondary amine product. A template-driven imine library 1 was prepared in the presence of the metalloenzyme carbonic anhydrase II (CAII). After the template-assisted, reversible dynamic reaction was complete, the reducing agent was added and the amine library **2** was produced (Figure 7.11). Without any

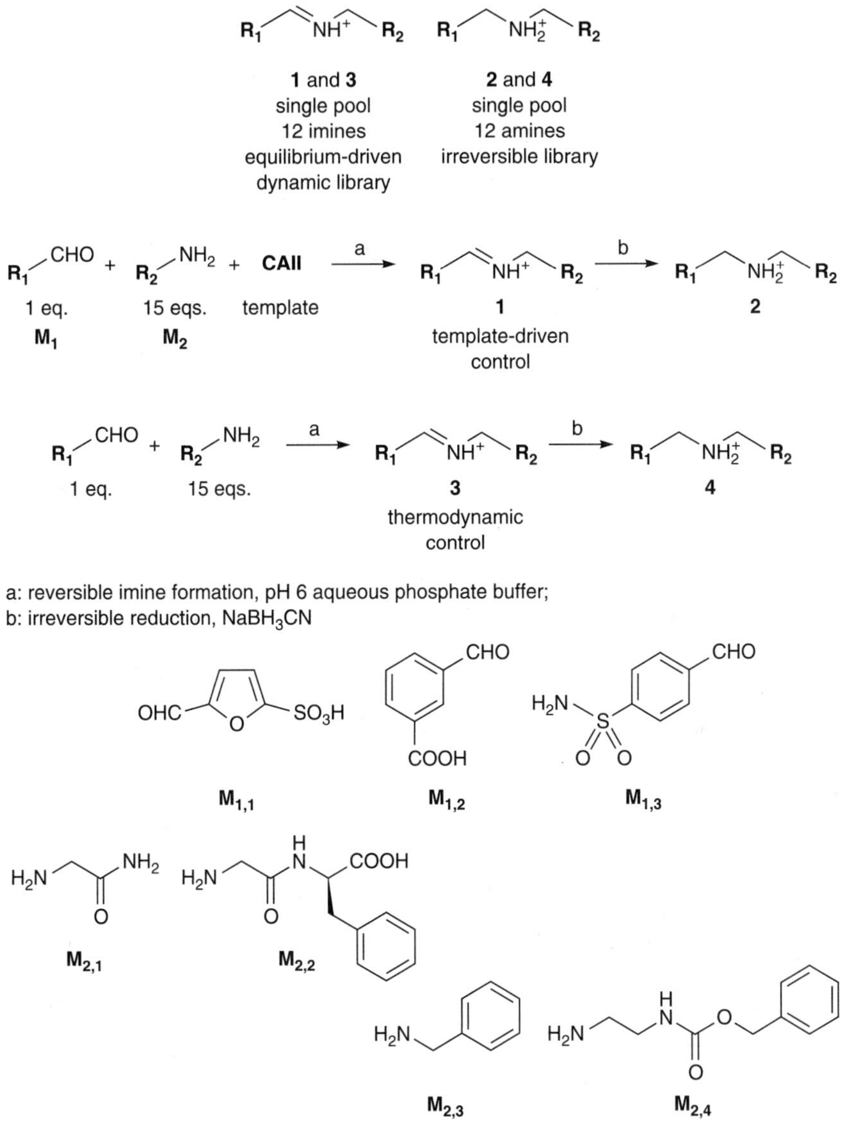

a: reversible imine formation, pH 6 aqueous phosphate buffer;
b: irreversible reduction, $NaBH_3CN$

FIGURE 7.11 Synthesis of a dynamic 12-member, template directed, imine library.

template the unbiased, equilibrated imine mixture **3** was then reduced to the mixture of amines **4** (Figure 7.11). The different abundance of library components **2** and **4**, reflecting the affinity of library components for CAII, was determined by comparing the HPLC traces of the stable amine mixtures.

The two aldehydes $M_{1,1}$ and $M_{1,2}$ produced the same relative amount of imines in the presence or absence of CAII, which implied that no interaction between these dynamic library members and the enzyme was observed. The relative abundance of imines from $M_{1,3}$ varied in the two libraries and the

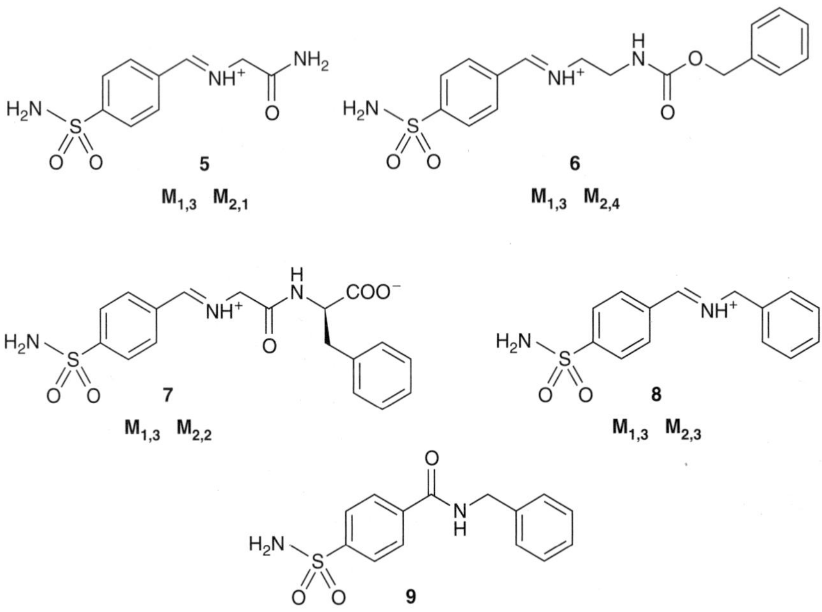

$M_{1,3}$ $M_{2,1}$ (5) $M_{1,3}$ $M_{2,4}$ (6) $M_{1,3}$ $M_{2,2}$ (7) $M_{1,3}$ $M_{2,3}$ (8) (9)

FIGURE 7.12 Library components which showed affinity for CAII.

two amines **5** and **6** almost disappeared from **2** when compared to **4**, while the formation of **7** and especially **8** was favored by the template (Figure 7.12). These results were confirmed in four validation experiments, in which the two amines were reacted with $M_{1,3}$ in the presence or absence of the enzyme. The results show how **8** was the favored library component in the template-assisted synthesis of the mixture. Further confirmation of the specificity of **8** toward the template was provided by adding the known CAII inhibitor **9** (Figure 7.12) to a binary mixture (entry *d*). The effect of CAII on the equilibrium of the dynamic mixture was significantly reduced in the presence of **9**, which bound the enzyme and reduced its template effect.

VI. SOLID SUPPORT MEDIATED SOLUTION SYNTHESIS

The advantages of heterogeneous solid phase in combinatorial chemistry, especially in terms of purification procedures, can be obtained also by solution-phase chemistry, using solid support assistance. The resin performs a specific function during the library synthesis and then is simply removed by filtration, leaving the pure library components in solution. A well-known adapted technique uses solid supported reagents or catalysts during a combinatorial synthesis, where the solid phase is filtered off and discarded after the reaction in which it was involved. A new application is represented by solid-phase purification, where one or more solid supports are added to trap the excess of reagents or the side products, and then filtered, leaving the pure reaction products in solution. Finally, an intriguing "hybrid" approach based on resin capture consists of performing the library synthesis in solution and then having a final step where the products are selectively captured by a complementary solid phase and thus purified from any other soluble impurities. Each of these techniques will be thoroughly illustrated in the next four sections. The reader should also access a monumental review [102], which covers anything in terms of solid support-mediated solution phase library synthesis until early 2000.

A. Solid Supported Reagents

The use of supported reagents has always been recognized as a powerful tool in classical organic chemistry for a large number of applications, and excellent reviews covered this topic in the past [103–105]. Recently, this technique has been applied to solution-phase combinatorial chemistry, where it looks really promising in terms of simpler work-up procedures, elimination of excess reagents, and isolation of pure reaction products.

The selected example by Pop *et al.* [106] reported a polymer-supported 1-hydroxybenzotriazole derivative as a reagent for acylations of *N*-nucleophiles in solution-phase combinatorial synthesis. The synthesis and the structure of the bound reagent are shown in Figure 7.13.

The resin bound electron-withdrawing substituted hydroxybenzotriazole was expected [107] to have enhanced reactivity in the carboxylic acid coupling and in the following substitution (Figure 7.14). The best acid activation conditions were found with PyBrOP in DMF for 3 h at room temperature, while the substitution was performed with stoichiometric amounts of *N*-nucleophiles in DMF for 20 h at room temperature. The work-up consisted of the washing of the resin after the activation step and the solid support filtration after the coupling with *N*-nucleophile obtaining pure amides in solution. Combinations of 700 carboxylic acids and 700 *N*-nucleophiles were tested, and among them all the acids without an α-hydrogen were compatible with the library chemistry, while a wide range of *N*-nucleophiles, including primary and secondary amines, aromatic amines, including moderately

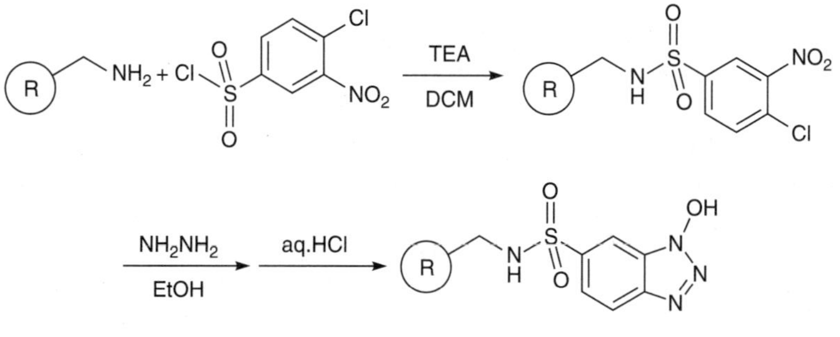

FIGURE 7.13 Synthesis of polymer-supported 1-hydroxybenzotriazole derivative reagent for acylation of N-nucleophiles.

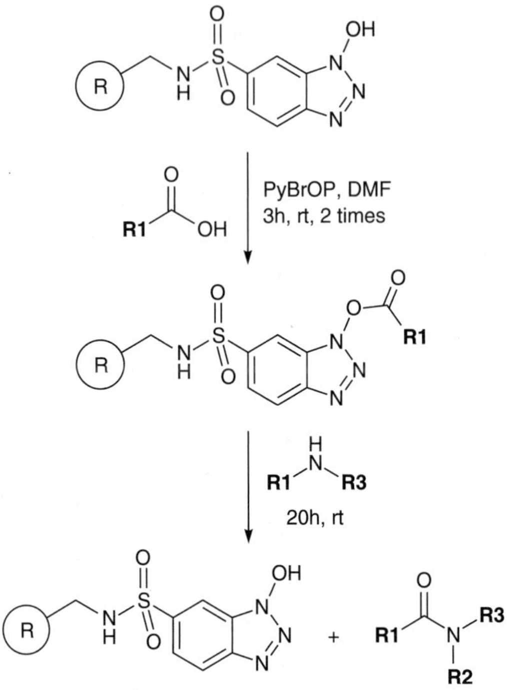

FIGURE 7.14 Acylation of N-nucleophiles using polymer supported 1-hydroxybenzotriazole derivative.

deactivated anilines, hydroxylamines, hydrazines, and hydrazones, was accepted. A small model library of 18 compounds was prepared with isolated yields >0% and with complete absence of any starting material or coupling reagent.

Among the other published works, Parlow [108] and Xu *et al.* [109] reported the use of anion exchange resins for the synthesis of solution-phase libraries of aryl and heteroaryl ethers *via* formation of a resin bound phenol salt and its subsequent alkylation [108] (two pools having 10 compounds each), or with a one-pot procedure [109] (13 discretes). Baxendale and Ley [110] reported the synthesis of a well-known drug, Sildenafil (Viagra™), employing a variety of supported reagents. Nicolaou *et al.* [111] used supported selenium-based reagent for the synthesis of solution-phase libraries, inspired by the natural product everninomycin. Organ and Dixon [112] used strong supported bases to prepare a library of biaryls in solution, and Dahmen and Braese [113] introduced the use of supported diazo derivatives in solution-phase combinatorial chemistry. An exhaustive review [114] regarding the use of supported reagents in classical and combinatorial synthesis in solution was also recently reported. Polymer supported reagents being established tools in classical organic chemistry, their use for the synthesis of high-quality solution-phase libraries, possibly in combination with other solid support mediated synthesis techniques reported in the following sections, has seen a steady increase, which will constantly continue in the near future.

B. Solid Supported Catalysts/Ligands

The use of solid supported, recyclable catalysts, is a well-assessed technique in classic organic chemistry, and many exhaustive reviews dealing with this subject are available [105, 115]. The use of solid supported catalysts for library synthesis in solution has also been reported. Among others, Kobayashi *et al.* presented the use of a new supported scandium catalyst for 3CC reactions leading to solution libraries of amino ketones, esters, and nitriles (24-member model discrete library) [116], or to quinolines (15-member model discrete library) [117], and Jang [118] presented a polymer bound Pd-catalyzed Suzuki coupling of organoboron compounds with halides and triflates. This area was also briefly reviewed recently [119].

The selected examples by Cole *et al.* [120] and Shimizu *et al.* [121] reported the parallel synthesis of a small library of solid supported dipeptide Schiff bases as ligands for the Ti-catalyzed enantioselective addition of trimethylsilyl cyanide to *meso* epoxides, and the determination of their catalytic activity on different substrates. The catalyzed addition reaction and the general structure of the dipeptide ligands are shown in Figure 7.15.

Using the well-known technique of positional scanning [122, 123], solid-phase libraries were inspired by a standard structure derived from preliminary experiments (AA1 = L-Phe, AA2 = L-Val, ArCHO = 2-hydroxy-1-naphthylaldehyde) and produced 45 different ligands in three sublibraries. AA1, AA2, or ArCHO were randomized, while keeping the other two building blocks as in

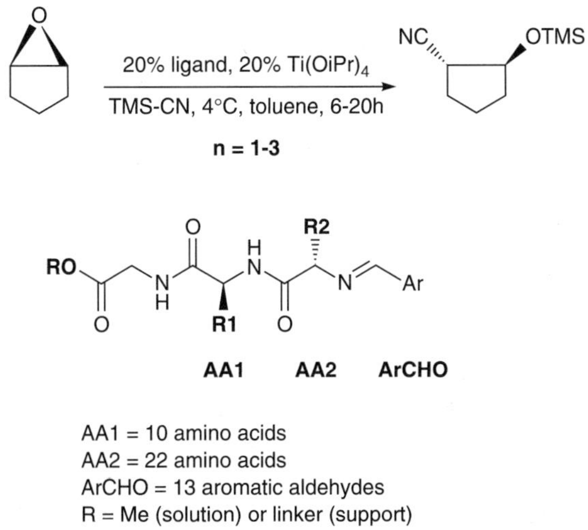

20% ligand, 20% $Ti(OiPr)_4$

TMS-CN, 4°C, toluene, 6-20h

n = 1-3

NC OTMS

RO N H R1 H N R2 N Ar

AA1 AA2 ArCHO

AA1 = 10 amino acids
AA2 = 22 amino acids
ArCHO = 13 aromatic aldehydes
R = Me (solution) or linker (support)

FIGURE 7.15 Ti-catalyzed enantioselective addition of trimethylsilyl cyanide to epoxides, and structure of dipeptide ligands.

the standard example. The compounds were hydrolyzed in solution (R = Me) and tested as catalysts for the addition of TMS-CN to cyclohexane oxide ($n = 2$), producing *ee* values for the protected β-cyanohydrins, which ranged from 0% to 84% and generally high substrate conversions. The same reaction was then performed with the resin bound ligands, and a correlation was found between the two sets of results for solution ligands and supported ligands. This allowed the preparation of relevant quantities of each ligand and its evaluation in the catalytized reaction on different substrates in a simple and rapid manner.

While 44 out of 45 ligands produced (S,R)-β-cyanohydrins, as depicted in Figure 7.15, only one of them (AA1 = L-tLeu, AA2 = L-Asn(Trt), ArCHO = 2-hydroxy-1-naphthaldehyde) gave (R,S)-β-cyanohydrins with $n = 1$ (cyclopentane oxide) and $n = 3$ (cycloheptane oxide). Moreover, for each of the three byciclo epoxides the best performing catalyst in terms of enantiomeric excess was different [121]. These results pointed to a general unpredictability of the stereochemical outcome for this reaction changing the substrate or the ligand, thus providing a general interest for support bound ligand libraries to be routinely tested on different valuable substrates for the identification of the best specific ligand.

Among other published reports, Gilbertson and Wang [124] prepared a 63-member supported phosphine containing oligopeptide library, whose members were complexed with rhodium and screened for their ability as chiral hydrogenation catalysts. Francis *et al.* [125] reported the solid-phase synthesis *via* "mix and split" [126, 127] of an encoded library of 12,000 peptides tested as novel coordination complexes for Fe^{III} and Ni^{II}. Shibata *et al.* [128] prepared a 7240-member solid-phase pool library of peptides showing specific and structure dependent strong binding to Co^{II}. An excellent review containing a critical evaluation of combinatorial "artificial" catalysis was also published recently [129].

C. Solid Support Purification

A different approach, recently introduced by Kaldor *et al.* [130, 131] and Virgilio [132], consists of using solid supported materials which function either as "scavengers," sequestering from the solution reaction mixture an excess of reagents or some reaction byproducts, or as "quenchers" which convert unstable library components or intermediates into stable compounds, thus making simpler the purification/work-up procedures.

The selected example by Flynn *et al.* [133] presented some purification strategies based on complementary molecular reactivity, molecular recognition, artificially imparted molecular recognition and solid support reaction quenching. All these concepts were exemplified by real applications and the first is reported in Figure 7.16.

Reactions between an excess of acylating agents (isocyanates, acyl chlorides, chloroformates, or sulfonyl chlorides) with secondary amines were performed in the presence of the resin bound tertiary amine **2** (see Figure 7.16). After reaction completion (16 h, DCM, room temperature) a resin bound

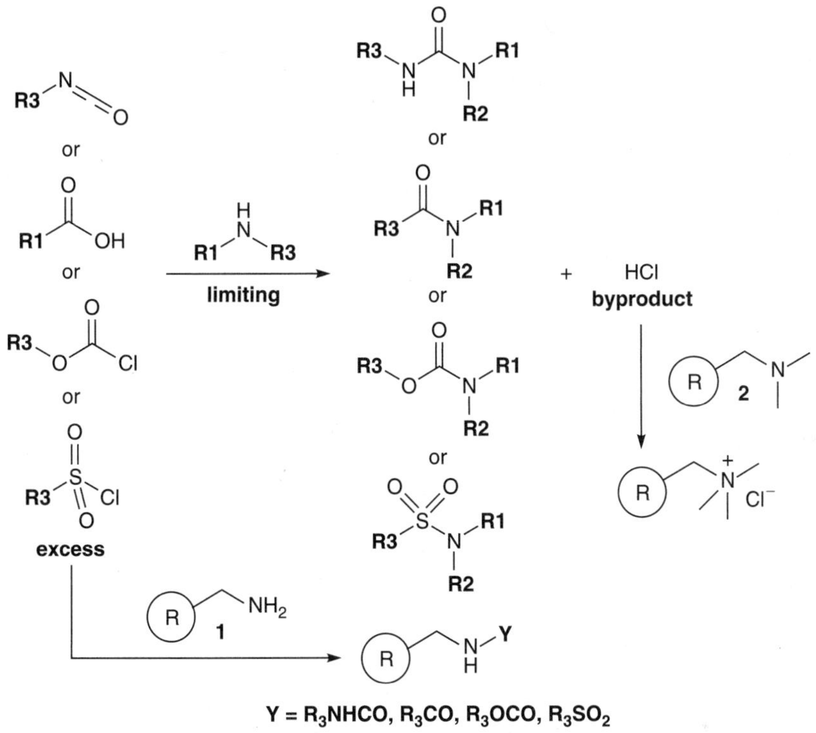

$Y = R_3NHCO, R_3CO, R_3OCO, R_3SO_2$

FIGURE 7.16 Acylation of secondary amines in presence of resin bound tertiary amines.

primary amine **1** was added and the reaction mixtures were stirred for 3 h. The solutions were collected and the reaction products (a model library of 12 discretes) were isolated after concentration with 80–90% yield, based on mass recovery, and with HPLC purity >95%. The tertiary amine resin **2** acted as a molecular recognition reagent, which sequestered a reaction byproduct (HCl) containing different chemical functionalities from the reaction products. The primary amine resin **1** acted as a molecular reactivity reagent which sequestered the excess of acylating agents containing different reactive functionalities from the reaction products.

An application of artificially imparted molecular recognition [133] is shown in Figure 7.17. A small model library (six discretes) of aminoketones was obtained by Moffatt oxidation of hydroxyethylamines, using dichloroacetic acid in DMSO and the tertiary amine-tagged EDC as a carbodiimide. After 24 h at room temperature the sulfonic acid resin **3** and the tertiary amine resin **2** were added, the mixtures were stirred for 20 h and the resulting solutions contained the pure library components (yields from 48% to 91%, HPLC purities around 90%) with no diimide or urea or dichloroacetic acid left. While resin **2** acted as a molecular recognition reagent, sequestering HCl and dichloroacetic acid, the sulfonic resin **3** recognized both EDC and its oxidation byproduct *via* the tertiary amine tag embedded in their structure.

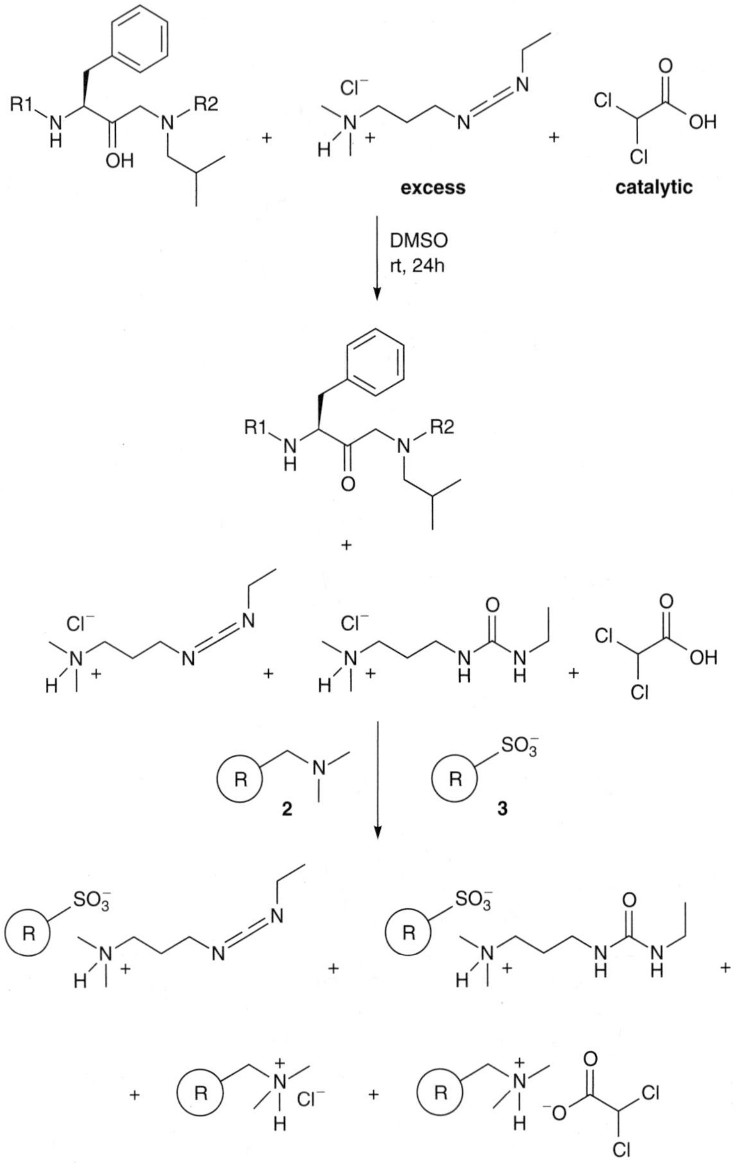

FIGURE 7.17 Model library of 6 aminoketones obtained by Moffatt oxidation.

It is remarkable that the simultaneous presence of cross-reactive groups, as in **3** and **2** was allowed by their site-isolation on different support beads.

Finally, an example of solid support reaction quenching [133] is reported in Figure 7.18. Six aldehydes were reacted with allylmagnesium chloride or with *n*-butyllithium to produce reactive alkoxides in solution. They were quenched by the carboxylic resin **4**, which also scavenged the excess of the

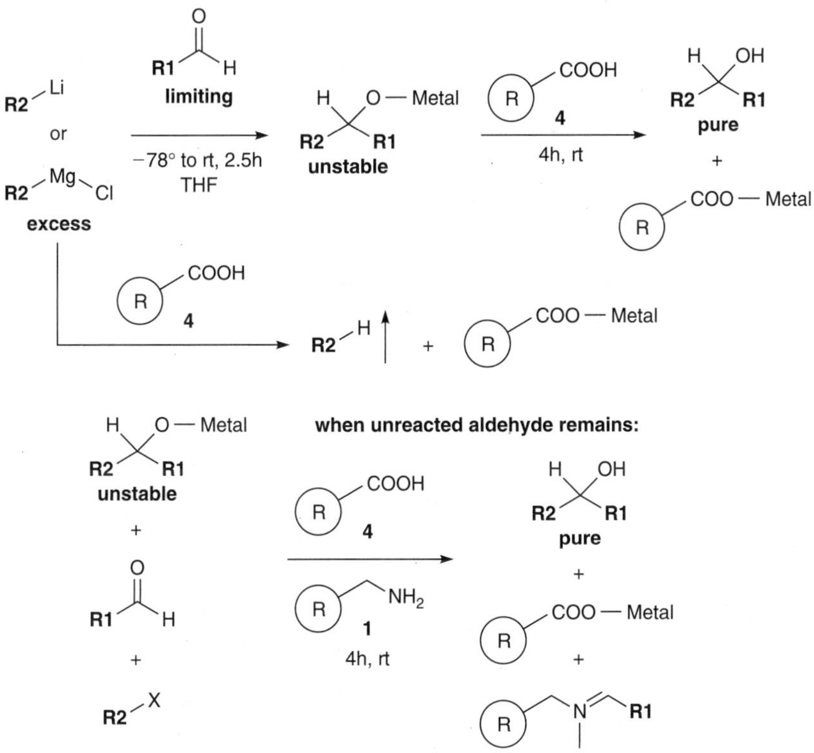

FIGURE 7.18 An example of solid support reaction quenching.

organometallic reagents, to give the corresponding alcohols (Figure 7.18, top). A single entry showed the presence of unreacted aldehyde after the reported procedure, so the reaction was repeated, adding the primary amine resin **1** together with the resin **4**. The former resin sequestered the starting aldehyde *via* imine formation (Figure 7.18, bottom). The model library (12 discretes) consisted of compounds obtained with yields >90% and HPLC purities >95%.

Among the other reports dealing with solid support purification, Gayo and Suto [134] studied the influence of nine anion exchange resins used as scavengers for the solution parallel synthesis of nine amides and esters. Siegel *et al.* [135] reported the use of cation exchange resins for the rapid purification of a parallel nine-member library of products prepared with different chemistries. Booth and Hodges [136] reported the synthesis and the use as solid support scavengers of three highly functionalized resins bearing an isocyanate (1.0 mmol/g), a tetraamine branched chain (4.3 mmol/g) and a morpholine moiety (3.5 mmol/g). Lawrence *et al.* [137] reported the automated synthesis and purification of a parallel library of 20 amides by ion exchange resins. Parlow *et al.* reported the parallel synthesis in solution of around 100 heterocyclic carboxamides using molecular reactivity and molecular recognition solid supported reagents [138], the use of sequestration enabling reagents for the purification of solution-phase combinatorial libraries [139] and the use of anthracene-derived tags for the purification of intermediates and library individuals in solution-phase library synthesis [140]. Stauffer and Katzenellenbogen [141] used a bicarbonate-supported scavenger for the synthesis of a 96-member pyrazole library. Frechet and coworkers [142] and Varady and coworkers [143] introduced novel scavenging-purification supports for semi-automated solution-phase combinatorial chemistry. These simple methods, which do not require expensive reagents and are easily automated, have already been, and will more and more be, widely used for library synthesis and will significantly contribute to the production of small-medium size solution libraries, especially when used in combination for simultaneous elimination of byproducts and excess of reagents.

D. Resin Capture

This approach, which was recently introduced by Keating and Armstrong [59], uses the solid support to trap the final reaction products of a solution-phase combinatorial library thus purifying them from the other components of the reaction mixture. The link between the support and the library components is then selectively cleaved to give the pure library in solution. The approach theoretically possesses the versatility of solution-phase chemistry, but it can also provide high-purity final products, or intermediates, as typical of solid-phase chemistry. A modification of this method implies the trapping of an intermediate during the solution library synthesis, which may be useful when early reactions in the library synthetic pathway are better performed in solution and the synthesis of this "hybrid" library is then continued on solid support, obtaining pure final compounds.

The selected example by Keating and Armstrong [59] reported the use of resin capture on a small solution model library (five individuals) produced by Ugi 4CC using a "convertible isocyanide" (details for this library synthesis were given in Section 3), as is described in Figure 7.19. A slight excess of the final cyclohexenamides was incubated with the support under anhydrous HCl/THF for 5 h at 55°C, then the resins were washed repeatedly and cleaved with 20% TFA in DCM for 20 min at room temperature. The cleavage solutions resulted to be the pure carboxylic acids (quantitative yields and >95% HPLC purities).

Other applications of resin capture were reported. Brown and Armstrong used resin capture *via* Suzuki condensation of vinylboronates with resin bound aryl iodides to give pure tetrasubstituted ethylenes in solution (discrete libraries of 10 [144] and 25 [145] individuals). Shuker *et al.* [146] used sulfonic acid ion exchange resin to capture and purify a discrete library of 48 ethanolamines, then simply released by methanolic ammonia (average yields > 70%, average HPLC purities > 90%). Kulkarni and Ganesan [147] used a trimethylammonium based ion exchange resin, both as a supported reagent to cyclize some α-amidoesters to 2, 4-pyrrolidinediones *via* Dieckmann condensation and as a

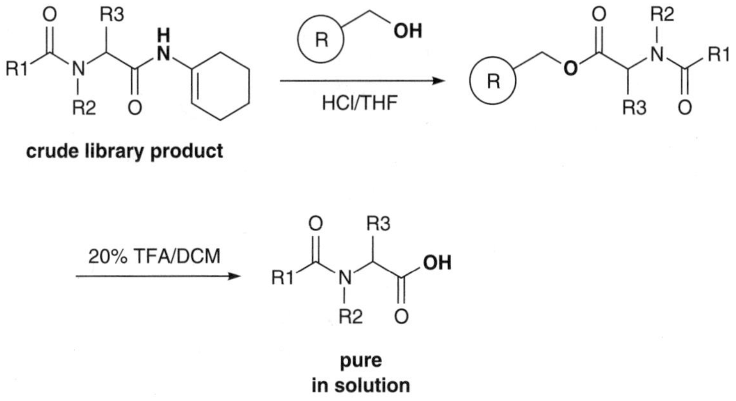

FIGURE 7.19 Use of resin capture on a solution library produced by Ugi 4CC.

resin capture of the cyclized products. The resin bound trimethylammonium salts (a model discrete library of 10 individuals) were released by 4% TFA/MeOH with yields >70% and HPLC purities around 90%. Chucholowski *et al.* [148] reported polymer bound triphenylphosphine as a resin capture reagent for the "hybrid" synthesis of benzodiazepines and benzodiazocines *via* Ugi 4CC in solution and cyclization *via* Aza-Wittig reaction on solid phase. Resin capture strategies were reported also by Hall and coworkers [149] (biaryls libraries *via* supported boronates and halides), Hamper *et al.* [150] (ketoimidazole libraries) and Rabinowitz *et al.* [151] (α-ketoamide libraries). The field was also thoroughly reviewed recently [152].

New applications of this technique, especially as a hybrid approach to perform each combinatorial step in the more convenient medium for its success, should become frequent in the near future. The use of capture resins bearing traceless linkers [153–155] could be particularly interesting, because the library components would then be released in solution without chemical modifications.

VII. FLUOROUS MULTIPHASE COMBINATORIAL SYNTHESIS

The use of fluorous biphase systems (FBS) received great attention, following the works by Zhu [156], which highlighted the usefulness of organic reactions performed in perfluorocarbon liquids and by Horvath and Rabai [157], which introduced the concept of fluorous biphasic catalysis. Curran [158] recently introduced the technique of fluorous multiphase combinatorial synthesis, where perfluorinated tagged reagents allowed the synthesis of discrete products in solution and their purification *via* simple liquid-liquid extractions by partitioning in a triphasic system (water, organic and perfluorinated liquids).

The selected examples by Studer *et al.* [159–161] reported the synthesis of two small discrete libraries of isoxazolines (eight compounds) and of amides derived from a Ugi 4CC reaction (ten compounds). The synthesis of the tagging silicon derivative and the scheme for isoxazoline synthesis are reported in Figure 7.20. A known silane [162] was coupled quantitatively with bromine in the fluorinated solvent FC-72 [163] and the tagging reagent was used to protect and label the allyl alcohols in THF. After the cycloaddition of the fluorinated dipolarophiles with the nitrile oxides, the protected isoxazolines were finally cleaved with HF/pyridine complex in ethyl ether. The scheme of the purification for each synthetic step is depicted in Figure 7.21.

The use of a triphasic extraction system, where an organic solvent, an aqueous phase and FC-72 [163] were used, allowed after any reaction step the isolation of the pure intermediates and eventually of the clean reaction products. The switch caused by the fluorous tag allowed the total partition of the library intermediates in the fluorous phase, where any other component of the reaction mixture was not dissolved, while after final deprotection the products were cleanly recovered from the organic phase and the tag moiety remained trapped by the fluorous phase. The eight isoxazoline alcohols were recovered with extremely high GC purities ($>91\%$, average $>95\%$) and with moderate to good yields (from 29% to 99%). The low yields were probably due to the volatility of some of the final products.

The same technique was also applied to the synthesis of a 10 member Ugi 4CC-derived solution library, where the fluorous tag was embedded into the carboxylic acid structure. The synthesis of the tag and the library synthetic scheme is shown in Figure 7.22. The tag was derived from a bromosilyl perfluorinated compound ($C_{10}F_{21}CH_2CH_2$, rather than $C_6F_{13}CH_2CH_2$, as for the isoxazolines) reacted with an orthothiobenzoate [164] and further elaborated with trivial chemistry. The library was produced using classical Ugi 4CC conditions and the tag detachment was obtained *via* treatment with

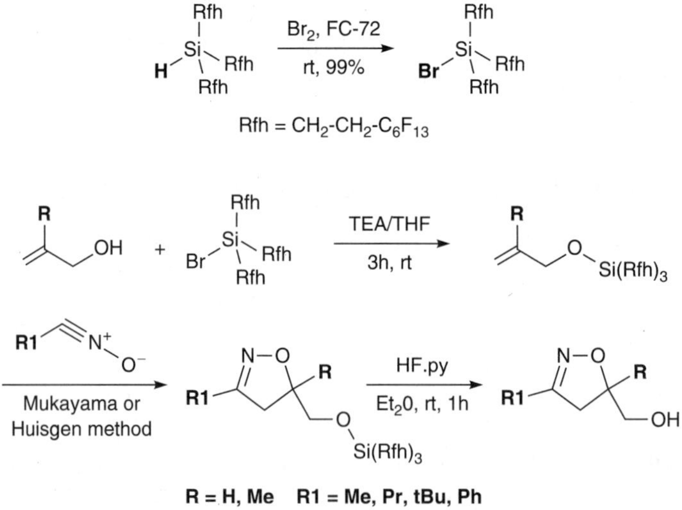

FIGURE 7.20 Synthesis of tagging silicon derivative and scheme for isoxazoline synthesis.

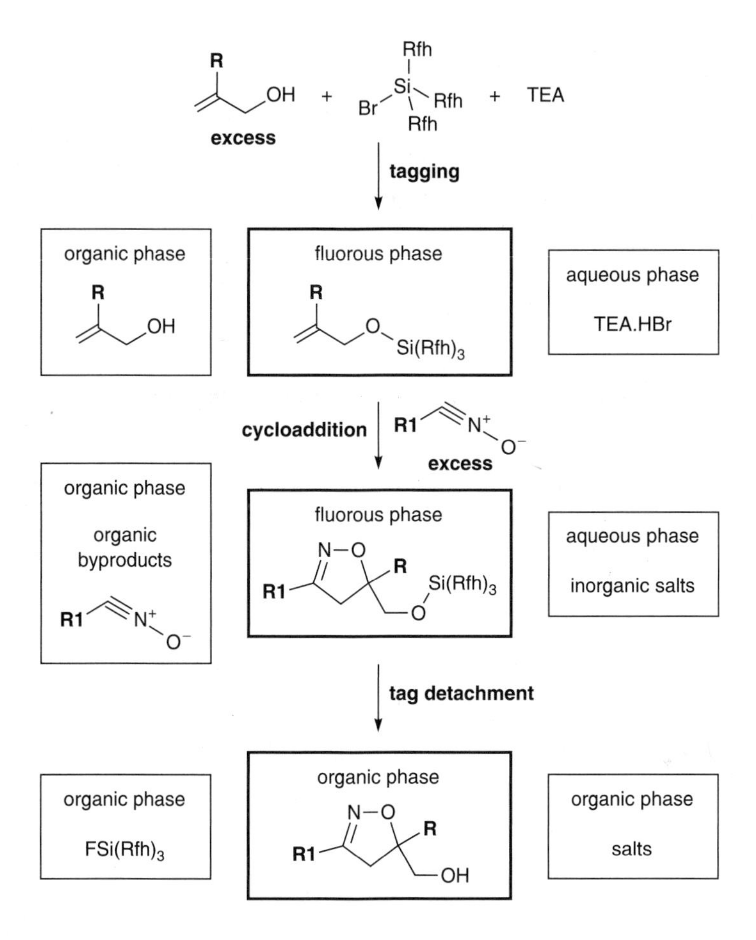

FIGURE 7.21 Purification of isoxazoline library.

TBAF. The partition process into the triphasic system allowed the isolation of pure compounds (>85% HPLC) in good yields (average 70%).

The same research group used this approach for Stille couplings in fluorous media [164–167] and for radical reactions amenable to the synthesis of solution libraries [168, 169]. Wipf *et al.* [170] reported the synthesis of a focussed library inspired by curacin A, an antimitotic natural product. The great attention which is currently devoted to FBS should provide in the future new tools (tagged reagents, solvents, experimental procedures and so on) to enlarge the set of reactions amenable to fluorous multiphase combinatorial synthesis and to thoroughly validate this procedure for automated purification of medium-large solution libraries.

VIII. SOLUBLE SUPPORTS

The use of solid supports which can be dissolved in many solvents allows to perform the reactions for library synthesis in a homogeneous medium, but also

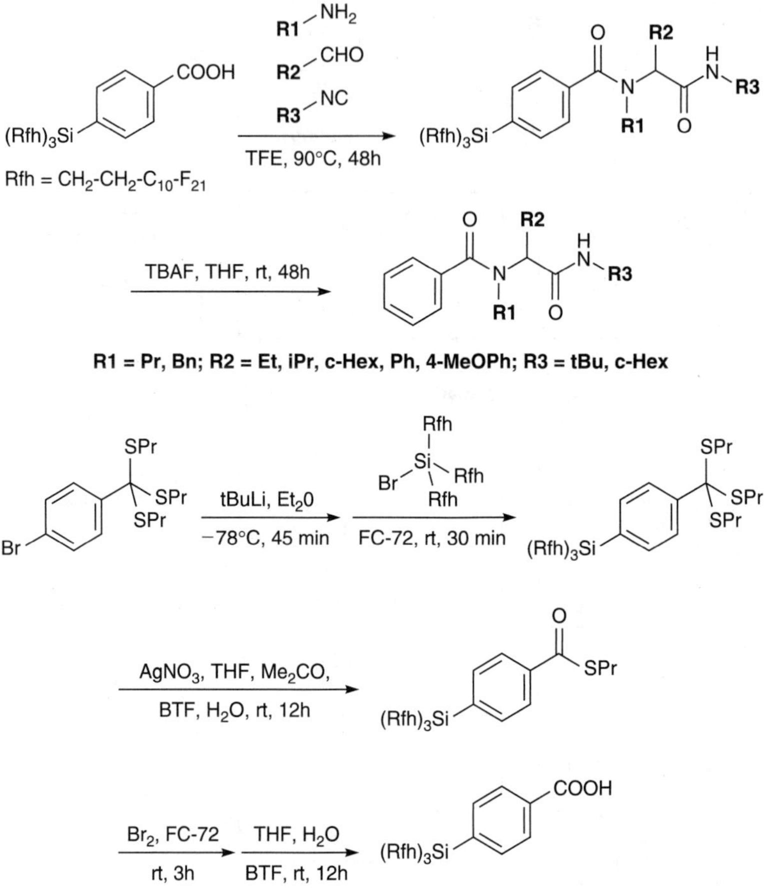

FIGURE 7.22 Synthesis of fluorous tag and Ugi 4CC derived solution library synthetic scheme.

allows to separate the intermediates from the reaction solution with common purification techniques (precipitation, crystallization, size exclusion chromatography and so on). We will present two examples related to liquid-phase combinatorial synthesis, where the soluble polymer is simply precipitated from the reaction mixture by addition of specific solvents, and to dendritic supports, where the intermediate dendrimer bound compounds are rapidly purified by analytical procedures.

A. Liquid-Phase Combinatorial Synthesis

The concept of liquid-phase combinatorial synthesis (LPCS) was introduced by Han *et al.* [171] in a paper where a peptide and a sulfonamide libraries were prepared using PEG (polyethylene glycol) as support. The support was chosen because of its successful application in peptide, oligosaccharide and oligonucleotide syntheses [172–174]. The main advantage of this and similar other "soluble supports" lies on performing the library synthesis in solvents where the PEG polymer is completely soluble (homogeneous reaction medium), while simply diluting the reaction mixtures with ether allows the precipitation of the polymer bound intermediates (heterogeneous purification).

The selected example by Chen and Janda [175] reported the validation of a method for the synthesis of prostanoid libraries on a soluble polymeric support made by non-crosslinked chloromethylated polystyrene (NCPS). The synthesis of supported PGE_2 methyl ester is shown in Figure 7.23.

The synthetic pathway required the use of a specific THP-based linker [176] and of different synthons 1 [177], 2 [178] and 3 [179], but most importantly the reaction conditions were highly demanding: metal hydrides, organometallic compounds, chlorosilanes and strong acids had to be used in various steps, and extreme reaction conditions (−78°C) were required. PEG was considered, due to its proven success in LPCS, but its solubility in THF at low temperatures was not satisfying and the removal of organometallic and inorganic reagents prior to ether precipitation was difficult for the high water solubility of polyethylene glycol. NCPS resulted to be somewhat complementary to PEG, being very soluble in any organic solvent, even at −78°C, but insoluble in water and methanol. The experimental procedure for the purification of intermediates in the synthesis of PGE_2 methyl ester simply consisted of diluting with DCM or AcOEt, the reaction mixture, extracting organometallics and inorganics with water, and finally precipitating the NCPS-bound intermediates by dilution with methanol. The final compound was obtained with an overall yield of 37%, thus showing the feasibility of a prostanoid library on NCPS *via* the use of different reagents 1, 2, and 3 and further elaboration of the protected functionalities of the polymer bound intermediates.

The final compound was obtained with an overall yield of 37%, thus showing the feasibility of a prostanoid library on NCPS *via* the use of different reagents 1, 2 and 3 and further elaboration of the protected functionalities of the polymer bound intermediates.

Different polyethylene glycol polymers were used in various papers and proved to be very reliable and useful for different classes of molecules: their use for the synthesis of peptides [180, 181], of peptidomimetics [182] and of oligosaccharide libraries [183] was reported as the development and the use of a new PEG-linked traceless linker [184, 185], the selection of ligands for asymmetric Sharpless dihydroxylation [186–188], the use of PEG-linked triarylphosphines for LPCS requiring Mitsunobu or Staudinger conditions [189], the use of PEG-based supports to prepare a library of [1,4]oxazepine-7-ones [190] and the use of PEG-supported Schiff bases for the synthesis of α-substituted amino acids [191]. Other examples of soluble polymers used for LPCS may include cellulose[192], polyacrylamide [193] polyvinyl alcohol [194, 195], various copolymers [196, 197] and NCPS [198–200]. Three excellent reviews [201–203] summarized the properties of PEG and other soluble polymers and their applications to the synthesis of peptides, oligonucleotides,

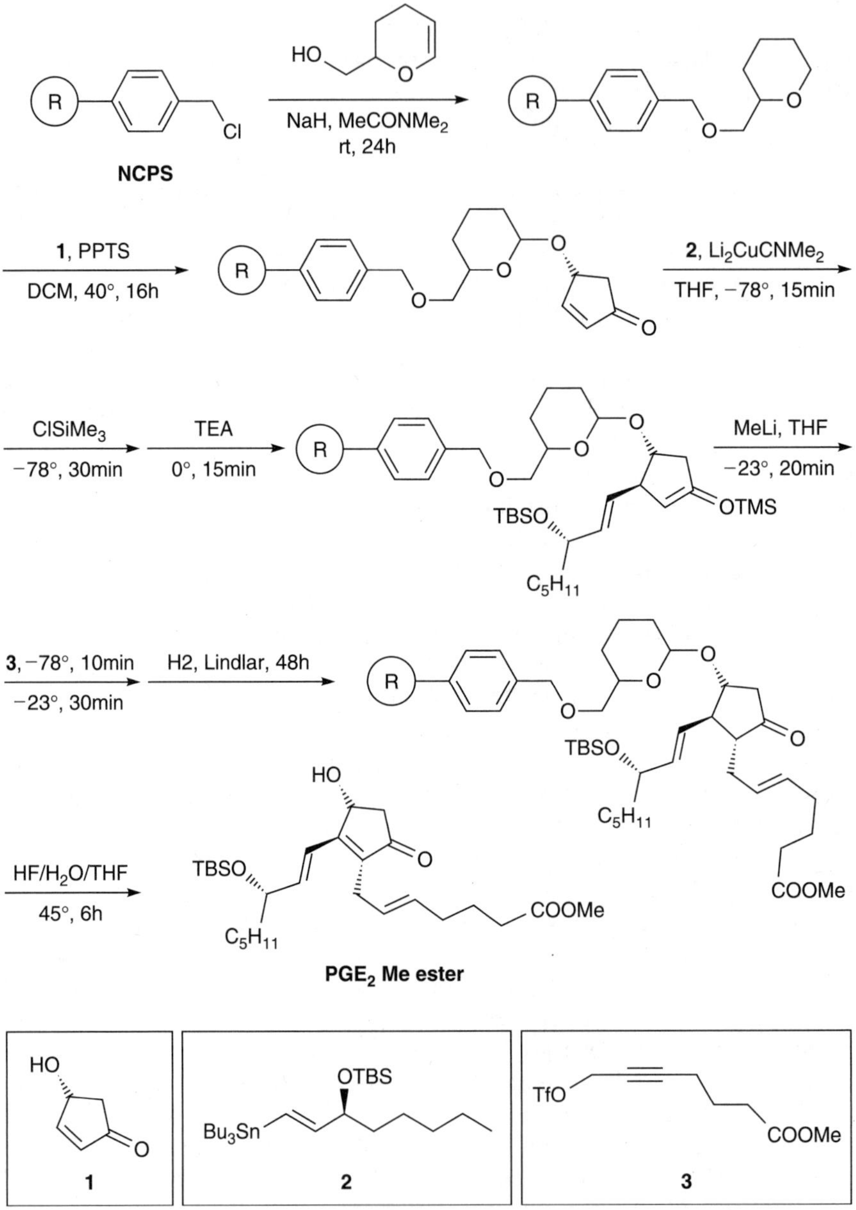

FIGURE 7.23 Synthesis of supported PGE_2 methyl ester.

oligosaccharides and small organic molecule libraries. The use of these soluble supports, and especially the application of new and more versatile homo- and copolymers not yet used for combinatorial libraries, should significantly increase in future, thus providing a panel of complementary choices for extremely different solution library chemistries.

B. Dendritic Supports

The use of dendrimers as soluble supports in combinatorial chemistry was recently introduced by Kim *et al.* [204] for the synthesis of a 27-member pool library of indoles (three pools by nine individuals). The structure of the dendritic support, which was prepared condensing the commercially available starburst polyamidoamine (PAMAM) dendrimer with the 4-hydroxymethyl benzoic acid (HMB) linker, is given in Figure 7.24.

Key features of such low generation dendrimers are their existence in an extended form, which causes a very high accessibility of the reaction sites in a single dendrimer molecule and inhibits cross-reactivity of different dendritic arms, and the extremely high loading of this support which allows synthesis of large quantities of library individuals. Considering a resin with a loading of 0.23 mmoles/g, 7 mg of this dendrimer has the same loading sites as 100 mg of the resin.

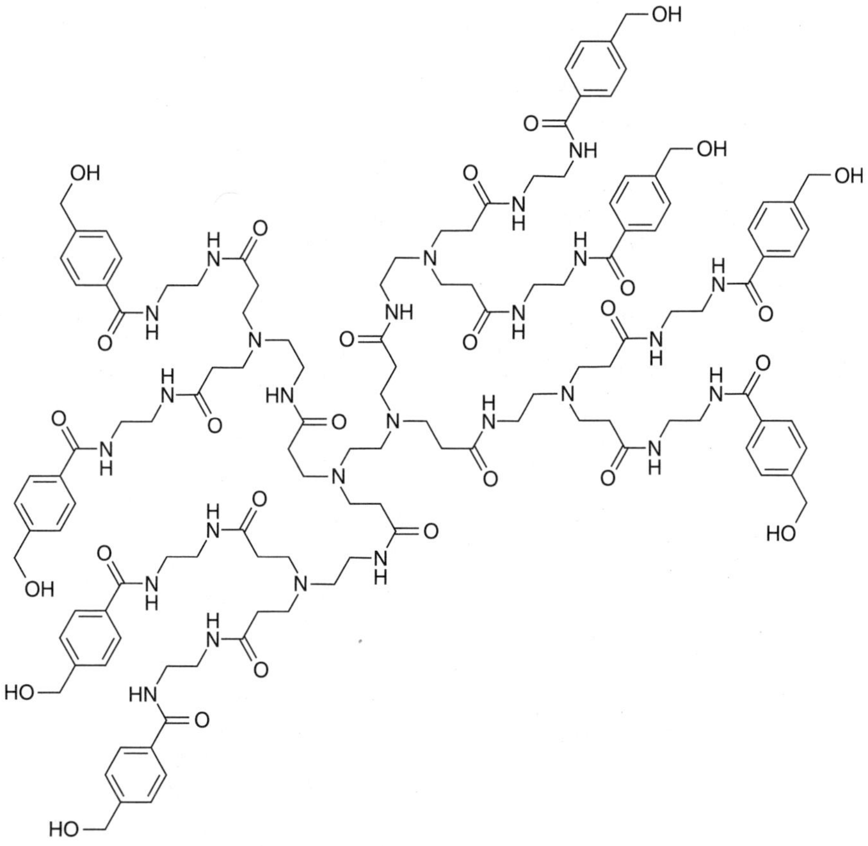

FIGURE 7.24 Structure of the dendritic support prepared condensing PAMAM and HMB.

The library synthetic pathway is depicted, where the dendritic structure shown in Figure 7.24 is simplified as a single functional site rather than the actual eight carried by the dendritic core. Coupling with a protected amino acid (A, I, or F, first monomer set) and deprotection produced a resin bound amine which was amidated with γ- or δ-ketoacids (second monomer set) and finally cyclized with substituted phenylhydrazines (third monomer set) to give the dendrimer bound indoles. These were cleaved by heating at 50°C with methanolic triethylamine, producing the pure library pools in solution (total yield >60%, HPLC purity >85%).

The purification of all the dendritic intermediates was obtained by SEC (size exclusion chromatography), by simply charging the raw reaction mixtures in DMF or DMA (dimethylacetamide) on a Sephadex LH-20 column and eluting the dendrimers with 15 mm runs. The intermediates were always obtained with HPLC purity >90% and could be characterized by various analytical methods such as NMR and HPLC/MS.

The research aimed at new dendritic structures received great attention recently [205–210], and many of the new constructs could be used in future for the generation of combinatorial libraries in a homogeneous reaction medium. The extreme flexibility of dendrimer chemistry, the high loading, and the reliability of their synthesis, should make these support one of the preferred options for combinatorial chemists in the near future. Among others, we can mention the works by Mulders *et al.* [211, 212] which reported the preparation of novel amino acid based dendrimers. Gudat *et al.* [213] reported the synthesis and the properties of silicon- and phosphorus-based dendrimers. Jayaraman and Stoddart [214] reported oligosaccharide-functionalized dendrimers. Gitsov *et al.* [215] reported the synthesis of polyether based dendrimers which can be viewed as tridimensional versions of PEG (see the previous section). Wyndham and Feher [216] presented new polyfunctional silsesquioxanes as supports for the synthesis of homogeneous phase combinatorial libraries. Swali *et al.* [217] reported the coupling of dendritic chains on resin beads to produce super high loading solid supports.

IX. SUMMARY AND OUTLOOK

The comparison of the main features of solution–phase versus solid-phase chemistry for combinatorial libraries clearly indicates the superiority of the former when synthesis of small discrete focused libraries is required. While such solution libraries are routinely prepared, many efforts which were briefly reported here are currently ongoing in order to fix the main limitations of solution-phase combinatorial chemistry and to broaden the panel of new homogeneous phase techniques for library generation. Major outcomes are to be expected, especially from the mutual use of solid supports and solution-phase techniques during the library synthetic scheme. For using the best method for a single step during library construction, the synthetic chemist needs to be aware of the features of each reported library synthesis technique and to be as imaginative as possible in considering and critically reviewing all the alternatives, to obtain the best result from the specific library synthesis. In this perspective, solution-phase techniques will eventually reinforce their presence in the scientific arena and will represent in future a useful tool for many, if not all, the combinatorial scientists facing a new library synthesis.

BIBLIOGRAPHY

1. Fruchtel JS, Jung G, Organic chemistry on solid supports, *Angew. Chem. Int. Ed. Eng.*, 35:17–42, 1996.
2. Hermkens PHH, Ottenheijm HCJ, Rees D, Solid-phase organic reactions: a review of the recent literarure, *Tetrahedron*, 52:4527–4554, 1996.
3. Hermkens PHH, Ottenheijm HCJ, Rees DC, Solid-phase organic reactions II: a review of the literature Nov 95–Nov 96, *Tetrahedron*, 53:5643–5678, 1997.
4. Sucholeiki I, Solid-phase methods in combinatorial chemistry, in *Combinatorial Chemistry: Synthesis and Application* (Eds. S.R. Wilson, A.W. Czarnik), pp. 119–133, 1997, Wiley, New York.
5. James IW, A compendium of solid-phase chemistry publications, in *Annual Reports in Combinatorial Chemistry and Molecular Diversity* (Eds. W.H. Moos, M.R. Pavia, B.K. Kay, A.D. Ellington), pp. 326–344, 1997, ESCOM Science Publishers, Leiden.
6. Merritt AT, Uptake of new technology in lead optimization for drug discovery, *Drug Disc. Today*, 3:505–510, 1998.
7. Calvert S, Stewart SP, Swarna, K, Wiseman, JS . *Curr. Opin. Drug Disc. Dev.*, 2:234–238, 1999.
8. Hird NW, Automated synthesis: New tools for the organic chemist, *Drug Disc. Today*, 4:265–274, 1999.
9. Furka A, Combinatorial chemistry: from split-mix to discrete, in *Combinatorial Chemistry and Combinatorial Technologies: Methods and Applications* (Eds. Fassina G, Miertus S), pp. 7–32, 2005, chapter 3, this volume.
10. Felder E, Solid phase synthesis of organic libraries, in *Combinatorial Chemistry and Combinatorial Technologies: Methods and Applications* (Eds. S. Miertus, G. Fassina), chapter 5 pp. 75–88, 2005, this volume.
11. Felder E, Resins, linkers and reactions for solid phase synthesis of organic libraries, in *Combinatorial Chemistry and Combinatorial Technologies: Methods and Applications* (Eds. G. Fassina, S. Miertus), chapter 6 pp. 89–106, 2005, this volume.
12. Balkenhohl F, von den Bussche-Hannefeld C, Lansky A, Zechel C, Combinatorial synthesis of small organic molecules, *Angew. Chem. Int. Ed. Eng.*, 35:2289–2337, 1996.
13. Thompson LA, Ellman JA, Synthesis and application of small molecule libraries, *Chem. Rev.*, 96:555–600, 1996.
14. Baldwin JJ, Henderson I, Recent advances in the generation of small-molecule combinatorial libraries: encoded split synthesis and solid-phase synthetic methodology, *Med. Res. Rev.*, 16:391–405, 1996.
15. Wendeborn S, De Mesmaeker A, Brill WKD, Berteina S, Synthesis of diverse and complex molecules on the solid phase, *Acc. Chem. Res.*, 33:215–224, 2000.

16. Storer R, Solution-phase synthesis in combinatorial chemistry: applications in drug discovery, *Drug Disc. Today*, 1:248–254, 1996.
17. Bailey N, Cooper AWJ, Deal MJ, Dean AW, Gore AL, Hawes MC, Judd DB, Merritt AT, Storer R, Travers S, Watson SP, Solution-phase combinatorial chemistry in lead discovery, *Chimia*, 51:832–837, 1997.
18. Coe DM, Storer R, Solution-phase combinatorial chemistry, in *Annual Reports in Combinatorial Chemistry and Molecular Diversity* (Eds. W.H. Moos, M.R. Pavia, B.K. Kay, A.D. Ellington), pp. 50–58, 1997, ESCOM Science Publishers, Leiden.
19. Sim MM, Ganesan A, Solution-phase synthesis of a combinatorial thiohydantoin library, *J. Org. Chem.*, 62:3230–3235, 1997.
20. An HY, Cummins LL, Griffey RH, Bharadwaj R, Haly BD, Fraser AS, Wilsonlingardo L, Risen LM, Wyatt JR, Cook PD, Solution phase combinatorial chemistry: Discovery of novel polyazapyridinophanes with potent antibacterial activity by a solution phase simultaneous addition of functionalities approach, *J. Am. Chem. Soc.*, 119:3696–3708, 1997.
21. Terrett N, *Combinatorial Chemistry*, pp. 55–76, 1998, Oxford University Press, Oxford, UK.
22. Suto MJ, *Curr. Opin. Drug Disc. Dev.*, 2:377–384, 1999.
23. Seneci P, *Solid-Phase Synthesis and Combinatorial Technologies*, pp. 339–421, 2000, Wiley Interscience, New York.
24. Collins I, Rapid analogue synthesis of heteroaromatic compounds, *J. Chem. Soc., Perkin Trans.*, 1:2845–2861, 2000.
25. Cheng S, Comer DD, Williams JP, Myers PL, Boger DL, Novel solution phase strategy for the synthesis of chemical libraries containing small organic molecules, *J. Am. Chem. Soc.*, 118:2567–2573, 1996.
26. Cheng S, Tarby CM, Comer DD, Williams JP, Caporale LH, Myers PL, Boger DL, A solution-phase strategy for the synthesis of chemical libraries containing small organic molecules: A universal and dipeptide mimetics template, *Bioorg. Med. Chem.*, 4:727–737, 1996.
27. Thomas JB, Fall MJ, Cooper JB, Burgess JP, Carroll FI, Rapid in-plate generation of benzimidazole libraries and amide formation using EEDQ, *Tetrahedron Lett.*, 38:5099–5102, 1997.
28. Boger DL, Ozer RS, Andersson CM, Generation of targeted C-2-symmetrical compound libraries by solution-phase combinatorial chemistry, *Bioorg. Med. Chem. Lett.*, 7:1903–1908, 1997.
29. Boger DL, Chai W, Ozer RS, Andersson CM, Solution-phase combinatorial chemistry *via* the olefin metathesis reaction, *Bioorg. Med. Chem. Lett.*, 7:463–468, 1997.
30. Boger DL, Goldberg J, Satoh S, Ambroise Y, Cohen SB, Vogt PK, Non-amide based combinatorial libraries derived from N-Boc iminodiacetic acid: Solution-phase synthesis of piperazinone libraries with activity against LEF-1/β-catenin-mediated transcription, *Helv. Chim. Acta*, 83:1825–1845, 2000.
31. Ma Y, Qian C, Wang L, Yang M, Lanthanide triflate catalyzed Biginelli reaction. One-pot synthesis of dihydropyrimidinones under solvent-free conditions, *J. Org. Chem.*, 65:3864–3868, 2000.
32. Wintner EA, Rebek J Jr, Combinatorial libraries in solution: polyfunctionalized core molecules, in *Combinatorial Chemistry: Synthesis and Application* (Eds. S.R. Wilson, A.W. Czarnik), pp. 95–118, 1997, John Wiley & Sons, Inc., New York.

33. Berk SC, Chapman KT, Spatially arrayed mixture (SPAM) technology: synthesis of two-dimensionally indexed orthogonal combinatorial libraries, *Bioorg. Med. Chem. Lett.*, 7:837–842, 1997.
34. Backes BJ, Virgilio AA, Ellman JA, Activation method to prepare a highly reactive acylsulfonamide "safety-catch" linker for solid-phase synthesis, *J. Am. Chem. Soc.*, 118:3055–3056, 1996.
35. Lam KS, Lebl M, Krchnak V, The "one-bead one-compound" combinatorial library method. *Chem. Rev.*, 97:411–448, 1997.
36. Wang T, An H, Vickers TA, Bharadwaj R, Cook PD, Synthesis of novel polyazadipyridinocyclophane scaffolds and their application for the generation of libraries, *Tetrahedron*, 54:7955–7976, 1998.
37. An H, Haly BD, Cook PD, New piperazinyl polyazacyclophane scaffolds, libraries and biological activities, *Bioorg. Med. Chem. Lett.*, 8:2345–2350, 1998.
38. An H, Haly BD, Fraser AS, Guinosso CJ, Cook PD, Solution phase combinatorial chemistry, Synthesis of novel linear pyridinopolyamine libraries with potent antibacterial activity, *J. Org. Chem.*, 62:5156–5164, 1997.
39. Kung PP, Bharadwaj R, Fraser AS, Cook DR, Kawasaki AM, Cook PD, Solution-phase synthesis of novel linear oxyamine combinatorial libraries with antibacterial activity, *J. Org. Chem.*, 63:1846–1852, 1998.
40. Kung PP, Cook PD, Solution-phase simultaneous addition of functionalities (SPSAF) and chemical transformations to prepare N,N′-disubstituted piperazine libraries, *Biotechnol. Bioeng.*, 61:119–125, 1998.
41. Kawasaki AM, Casper MD, Gaus HJ, Herrmann R, Griffey RH, Cook PD, *Nucleos. Nucleot.*, 18:659–660, 1999.
42. Gaus HJ, Kung PP, Brooks D, Cook PD, Cummins LL, Monitoring solution-phase combinatorial library synthesis by capillary electrophoresis, *Biotechnol. Bioeng*, 61:169–177, 1999.
43. Griffey RH, An H, Cummins LL, Gaus HJ, Haly B, Herrmann R and Cook PD, Rapid deconvolution of combinatorial libraries using HPLC fractionation, *Tetrahedron*, 54:4067–4076, 1998.
44. Smith PW, Lai JYQ, Whittington AR, Cox B, Houston JG, Synthesis and biological evaluation of a library containing potentially 1600 amides/esters, A strategy for rapid compound generation and screening, *Bioorg. Med. Chem. Lett.*, 4:2821–2824, 1994.
45. Pirrung MC, Chen J, Preparation and screening against acetylcholinesterase of a non-peptide "indexed" combinatorial library, *J. Am. Chem. Soc.*, 117:1240–1245, 1995.
46. Pirrung MC, Chau JHL, Chen J, Discovery of a novel tetrahydroacridine acetylcholinesterase inhibitor through an indexed combinatorial library, *Chemistry & Biology*, 2:621–626, 1995.
47. Shipps GW Jr, Spitz UP, Rebek J Jr Solution-phase generation of tetraurea libraries, *Bioorg. Med. Chem.*, 4:655–657, 1996.
48. Shipps GW Jr, Pryor KE, Xian J, Skyler DA, Davidson EH, Rebek J Jr, Synthesis and screening of small molecule libraries active in binding to DNA, *Proc. Natl. Acad. Sci. USA*, 94:11833–11838, 1997.
49. Maehr H, Yang R, Structure optimization of a leukotriene D-4 antagonist by combinatorial chemistry in solution, *Bioorg. Med. Chem.*, 5:493–496, 1997.
50. Chng BL, Ganesan A, Solution-phase synthesis of a beta-amino alcohol combinatorial library, *Bioorg. Med. Chem. Lett.*, 7:511–514, 1998.

51. Boger DL, Goldberg J, Andersson CM, Solution phase combinatorial synthesis of biaryl libraries employing heterogeneous conditions for catalysis and isolation with size exclusion chromatography for purification, *J. Org. Chem.*, 64:2422–2429, 1999.
52. Boger DL, Jiang W, Goldberg J, Convergent solution-phase combinatorial synthesis with multiplication of diversity through rigid biaryl and diarylacetylene couplings, *J. Org. Chem.*, 64:7094–7100, 1999.
53. Boger DL, Fink BE, Hedrick MP, Total synthesis of Distamycin A and 2640 analogues: a solution-phase combinatorial approach to the discovery of new, bioactive DNA binding agents and development of a rapid, high-throughput screen for determining relative binding affinity or DNA binding sequence selectivity, *J. Am. Chem. Soc.*, 122:6382–6394, 2000.
54. Klumpp DA, Yeung KY, Prakash GKS, Olah GA, Preparation of 3,3-diaryloxyndoles by superacid-induced condensations of isatins and aromatics with a combinatorial approach, *J. Org. Chem.*, 63:4481–4484, 1998.
55. Ugi I, Steinbruckner C, Isonitriles. II, Reaction of isonitriles with carbonyl compounds, amines, and hydrazoic acid, *Chem. Ber.*, 94:734–742, 1961.
56. Gokel G, Luedke G, Ugi I, Four-component condensations and related reactions, in *Isonitrile Chemistry* (Ed. I. Ugi), pp. 145–199, Academic Press, New York, 1971.
57. Failli A, Immer H, Gotz M, The synthesis of cyclic peptides by the four component condensation (4CC), *Can. J. Chem.*, 57:3257–3261, 1979.
58. Ugi I, From isocyanides by four-component condensation to antibiotic syntheses, *Angew. Chem.*, 94:826–835, 1982.
59. Keating TA, Armstrong RW, Molecular diversity via a convertible isocyanide in the Ugi four-component condensation, *J. Am. Chem. Soc.*, 117:7842–7843, 1995.
60. Keating TA, Armstrong RW, Postcondensation modifications of Ugi four-component condensation products: 1-isocyanocyclohexene as a convertible isocyanide. Mechanism of conversion, synthesis of diverse structures, and demonstration of resin capture, *J. Am. Chem. Soc.*, 118:2574–2583, 1996.
61. Goebel M, Ugi I, Beyond peptide and nucleic acid combinatorial libraries: applying unions of multicomponent reactions towards the generation of carbohydrate combinatorial libraries, *Tetrahedron Lett.*, 36:6043–6046, 1995.
62. Weber L, Wallbaum S, Broger C, Gubernator K, Optimization of the biological activity of combinatorial compound libraries by a genetic algorithm, *Angew. Chem. Int. Ed.*, 34:2280–2282, 1995.
63. Singh J, Ator MA, Jaeger EP, Allen MP, Whipple DA, Soloweij JE, Chowdhary S, Treasurywala AM, Application of genetic algorithms to combinatorial synthesis: A computational approach to lead identification and lead optimization, *J. Am. Chem. Soc.*, 118:1669–1676, 1996.
64. Brown RD, Martin YC, Designing combinatorial library mixtures using a genetic algorithm, *J. Med. Chem.*, 40:2304–2313, 1997.
65. Mukhopadhyay M, Bhatia B, Iqbal J, Cobalt catalyzed multiple component condensation route to β-acetamido carbonyl compound libraries, *Tetrahedron Lett.*, 38:1083–1086, 1997.
66. Cavicchioli M, Sixdenier E, Derrey A, Bouyssi D, Balme G, Three partners for a one pot palladium-mediated synthesis of various tetrahydrofurans, *Tetrahedron Lett.*, 38:1763–1766, 1997.

67. Kobayashi S, Akiyama R, Moriwaki M, Three-component or four-component coupling reactions leading to δ-lactams, Facile synthesis of γ-acyl-δ-lactams from silyl enolates, α, β-unsaturated thioesters, and imines or amines and aldehydes via tandem Michael-imino aldol reactions, *Tetrahedron Lett.*, 38:4819–4822, 1997.
68. Nakamura E, Kubota K, Sakata G, Addition of zincated hydrazone to vinyl grignard reagent, Ketone synthesis by one-pot assembly of four components, *J. Am. Chem. Soc.*, 119:5457–5458, 1997.
69. Holmes CP, Chinn JP, Look GC, Gordon EM, Gallop MA, Strategies for combinatorial organic synthesis: solution and polymer-supported synthesis of 4-thiazolidinones and 4-metathiazanones derived from amino acids, *J. Org. Chem.*, 60:7328–7333, 1995.
70. Dordick JS, Patil DR, Parida S, Ryu K, Rethwisch DG, Enzymic catalysis in organic media: prospects for the chemical industry, *Chem. Ind.*, 47:267–292, 1992.
71. Sutherland AG, Enzyme chemistry, *Annu. Rep. Prog. Chem., Sect. B*, 89:281–298, 1993.
72. Roberts SM, Preparative biotransformations: the employment of enzymes and whole-cells in synthetic organic chemistry, *J. Chem. Soc., Perkin Trans.*, 1:157–169, 1998.
73. Khmelnitsky YL, Michels PC, Dordick JS, Clark DS, Generation of solution-phase libraries of organic molecules by combinatorial biocatalysis, in *Molecular Diversity and Combinatorial Chemistry* (Eds. I.W. Chaiken, K.D. Janda), pp. 144–157, 1996, American Chemical Society, Washington.
74. Khmelnitsky YL, Budde C, Arnold JM, Usyatinsky A, Clark DS, Dordick JS, Synthesis of water soluble paclitaxel derivatives by enzymic acylation, *J. Am. Chem. Soc.*, 119:11554–11555, 1997.
75. Michels PC, Khmelnitsky YL, Dordick JS, Clark DS, Combinatorial biocatalysis for drug discovery and optimization, *Book of Abstracts, 213th ACS National Meeting*, San Francisco, April 13–17, 1997, American Chemical Society, Washington.
76. Liu X-C, Clark DS, Dordick JS, Chemoenzymatic construction of a four--component Ugi combinatorial library, *Biotechnol. Bioeng.*, 69:457–460, 2000.
77. Thompson MC, Busch DH, Reactions of coordinated ligands, VI. Metal ion control in the synthesis of planar nickel (II) complexes of α-diketo-bis-mercaptoimines, *J. Am. Chem. Soc.*, 86:213–217, 1964.
78. Anderson S, Anderson HL, Sanders JKM, Expanding roles for templates in synthesis, *Acc. Chem. Res.*, 26:469–475, 1993.
79. Hoss R, Vogtle F, Template synthesis, *Angew. Chem. Int. Ed. Engl.*, 33:375–384, 1994.
80. Hasenknopf B, Lehn JM, Kneisel BO, Baum G, Fenske D, Self assembly of a circular double helicate, *Angew. Chem. Int. Ed. Engl.*, 35:1838–1840, 1996.
81. Huc I, Lehn JM, Virtual combinatorial libraries: dynamic generation of molecular and supramolecular diversity by self-assembly, *Proc. Natl. Acad. Sci. USA*, 94:2106–2110, 1997.
82. Huc I, Krische MJ, Funeriu DP, Lehn JM, Dynamic combinatorial chemistry: Substrate H-bonding directed assembly of receptors based on bipyridine-metal complexes, *Eur. J. Inorg. Chem.*, 1415–1420, 1999.
83. Berl V, Huc I, Lehn JM, DeCian A, Fischer J, Induced fit selection of a barbiturate receptor from a dynamic structural and conformational/configurational library, *Eur. J. Org. Chem.*, 3089–3094, 1999.

84. Eliseev AV, Nelen MI, Use of molecular recognition to drive chemical evolution, 1. Controlling the composition of an equilibrating mixture of simple arginine receptors, *J. Am. Chem. Soc.*, 119:1147–1148.
85. Polyakov VA, Nelen MI, Nazarpack-Kandlousy N, Ryabov AD, Eliseev AV, Imine exchange in O-aryl and O-alkyl oximes as a base reaction for aqueous "dynamic" combinatorial libraries. A kinetic and thermodynamic study, *J. Phys. Org. Chem.*, 12:357–363, 1999.
86. Sakai S, Shigemasa Y, Sasaki T, *Tetrahedron Lett.*, 38:8145–8148, 1997.
87. Klekota B, Hammond MH, Miller BL, Generation of novel DNA-binding compounds by selection and amplification from self-assembled combinatorial libraries, *Tetrahedron Lett.*, 38:8639–8642, 1997.
88. Klekota B, Miller BL, Selection of DANN-binding compounds via multistage molecular evolution, *Tetrahedron*, 55:11687–11697, 1999.
89. Giger T, Wigger M, Audetat S, Benner SA, Libraries for receptor-assisted combinatorial chemistry (RACS). The olefin metathesis reaction, *Synlett*, 688–691, 1998.
90. Hiraoka S, Fujita M, Guest-selected formation of Pd(II)-linked cages from a prototypical dynamic library, *J. Am. Chem. Soc.*, 121:10239–10240, 1999.
91. Rowan SJ, Sanders JKM, Building thermodynamic combinatorial libraries of quinine macrocycles, *Chem. Commun.*, 1407–1408, 1997.
92. Try AC, Harding MM, Hamilton DG, Sanders JKM, Reversible five-components assembly of a [2]catenane from a chiral metallomacrocycle and a dinaphtho-crown ether, *Chem. Commun.*, 723–724, 1998.
93. Hamilton DG, Feeder N, Teat SJ, Sanders JKM, Reversible synthesis of *p*-associated [2]catenanes by ring-closing metathesis: Towards dynamic combinatorial libraries, *New J. Chem.*, 22:1019–1021, 1998.
94. Rowan SJ, Lukeman PS, Reynolds DJ, Sanders JKM, Engineering diversity into dynamic combinatorial libraries by use of a small flexible building block, *New J. Chem.*, 22:1015–1018, 1998.
95. Cousins GRL, Poulsen SA, Sanders JKM, Dynamic combinatorial libraries of pseudo-peptide hydrazone macrocycles, *Chem. Commun.*, 1575–1576, 1999.
96. Brady PA, Sanders JKM, Selection approaches to catalytic systems, *Chem. Soc. Rev.*, 327–336, 1997.
97. Ganesan A, Strategies for the dynamic integration of combinatorial synthesis and screening, *Angew. Chem. Int. Ed.*, 37:2828–2831, 1998.
98. Klekota B, Miller BL, Dynamic diversity and small-molecule evolution: a new paradigm for ligand identification, *Trends Biotechnol.*, 17:205–209, 1999.
99. Eliseev AV, Lehn JM, Dynamic combinatorial chemistry: Evolutionary formation and screening of molecular libraries, *Curr. Top. Microbiol. Immunol.*, 243:159–172, 1999.
100. Reinhoudt DN, Timmerman P, Cardullo F, Crego-Calama M, *NATO ASI Ser., Ser. C.*, 527:181–195, 1999.
101. Lehn JM, Dynamic combinatorial chemistry and virtual combinatorial libraries, *Chem.-Eur. J.*, 5:2455–2463, 1999.
102. Ley SV, Baxendale IR, Bream RN, Jackson PS, Leach AG, Longbottom DA, Nesi M, Scott JS, Storer RI, Taylor SJ, Multi-step organic synthesis using solid-supported reagents and scavengers: A new paradigm in chemical library generation, *J. Chem. Soc., Perkin Trans.*, 1:3815–4195, 2000.
103. Ando T, Inorganic solid supported reagents as acids and bases, *Stud. Surf. Sci. Catal.*, 90:9–20, 1994.

104. Smith K, Fry CV, Tzimas M, The use of solid supports and supported reagents in liquid phase organic reactions, in *Chem. Waste Minimization* (Ed. J.H. Clark), pp. 86–115, 1995, Blackie, Glasgow, UK.
105. Shuttleworth SJ, Allin SM, Sharma PK, Functionalized polymers. Recent developments and new applications in synthetic organic chemistry, *Synthesis*, 1217–1239, 1997.
106. Pop IE, Deprez BP, Tartar AL, Versatile acylation of N-nucleophiles using a new polymer-supported 1-hydroxybenzotriazole derivative, *J. Org. Chem.*, 62: 2594–2603, 1997.
107. Koenig W, Geiger R, New methods for the synthesis of peptides: Activation of the carboxyl group with dicyclohexylcarbodiimide by using 1-hydroxybenzotriazoles as additives, *Chem. Ber.*, 103:788–798, 1970.
108. Parlow JJ, The use of anion exchange resins for the synthesis of combinatorial libraries containing aryl and heteroaryl ethers, *Tetrahedron Lett.*, 37:5257–5260, 1996.
109. Xu W, Mohan R, Morrissey MM, Polymer supported bases in combinatorial chemistry: synthesis of aryl ethers from phenols and alkyl halides and aryl halides, *Tetrahedron Lett.*, 38:7337–7340, 1997.
110. Baxendale IR, Ley SV, Polymer-supported reagents for multi-step organic synthesis: application to the synthesis of sildenafil, *Bioorg. Med. Chem. Lett.*, 10:1983–1986, 2000.
111. Nicolaou KC, Mitchell HJ, Flyaktakidou KC, Suzuki H, Rodriguez RM, 1,2-Seleno migrations in carbohydrate chemistry: Solution and solid-phase synthesis of 2-deoxy glycosides, orthoesters, and allyl orthoesters, *Angew. Chem. Int. Ed.*, 39:1089–1093, 2000.
112. Organ MG, Dixon CE, The preparation of amino-substituted biaryl libraries: the application of solid-supported reagents to streamline solution-phase synthesis, *Biotechnol. Bioeng.*, 71:71–77, 2000.
113. Dahmen S, Braese S, The first stable diazonium ion on solid support—Investigations on stability and usage as linker and scavenger in solid-phase organic synthesis, *Angew. Chem. Int. Ed.*, 39:3681–3683, 2000.
114. Bhalay G, Dunstan A, Glen A, Supported reagents: Opportunities and limitations, *Synlett*, 1846–1859, 2000.
115. Reggelin M, Polymer catalysts, *Nachr. Chem., Tech. Lab.*, 45:1196–1201, 1997.
116. Kobayashi S, Nagayama S, Busujima T, Polymer scandium-catalyzed three-component reactions leading to diverse amino ketone, amino ester, and amino nitrile derivatives, *Tetrahedron Lett.*, 37:9221–9224, 1996.
117. Kobayashi S, Nagayama S, A new methodology for combinatorial synthesis. Preparation of diverse quinoline derivatives using a novel polymer-supported scandium catalyst, *J. Am. Chem. Soc.*, 118:8977–8978, 1996.
118. Jang SB, Polymer-bound palladium-catalyzed cross-coupling of organoboron compounds with organic halides and organic triflates, *Tetrahedron Lett.*, 38:1793–1796, 1997.
119. Kobayashi S, Immobilyzed catalysts in combinatorial chemistry, *Curr. Opin. Chem. Biol.*, 4:338–345, 2000.
120. Cole BM, Shimizu KD, Krueger CA, Harrity JPA, Snapper ML, Hoveyda AH, Discovery of chiral catalysts through ligand diversity: Ti-catalyzed enantioselective addition of TMSCN to meso epoxides, *Angew. Chem., Int. Ed. Engl.*, 35: 1668–1671, 1996.

121. Shimizu KD, Cole BM, Krueger CA, Kuntz KW, Snapper ML, Hoveyda AH, Search for chiral catalysts through ligand diversity: substrate-specific catalysts and ligand screening on solid phase *Angew. Chem., Int. Ed. Engl.*, 36:1704–1707, 1997.
122. Pinilla C, Appel JR, Blanc P, Houghten RA, Rapid identification of high affinity peptide ligands using positional scanning synthetic peptide combinatorial libraries, *BioTechniques*, 13:901–902, 904–905, 1992.
123. Dooley CT, Houghten RA, The use of positional scanning synthetic peptide combinatorial libraries for the rapid determination of opioid receptor ligands, *Life Sci.*, 52:1509–1517, 1993.
124. Gilbertson SR, Wang X, The combinatorial synthesis of chiral phosphine ligands, *Tetrahedron Lett.*, 37:6475–6478, 1996.
125. Francis MB, Finney NS, Jacobsen EN, Combinatorial approach to the discovery of novel coordination complexes, *J. Am. Chem. Soc.*, 118:8983–8984, 1996.
126. Furka A, Sebestyen F, Asgedom M, Dibo G, More peptides by less labour, *Abstract of Xth International Symposium on Medicinal Chemistry*, Budapest, p. 288, 1988.
127. Furka A, Sebestyen F, Asgedom M, Dibo G, General method for rapid synthesis of multicomponent peptide mixtures, *Int. J. Pept. Protein Res.*, 37:487–493, 1991.
128. Shibata N, Baldwin JE, Wood ME, Resin-bound peptide libraries showing specific metal ion binding, *Bioorg. Med. Chem. Lett.*, 7:413–416, 1997.
129. Borman S, Combinatorial chemists focus on small molecules, molecular recognition, and automation, *C&EN*, 29–54, 1996.
130. Kaldor SW, Siegel MG, Fritz JE, Dressman BA, Hahn PJ, Use of solid supported nucleophiles and electrophiles for the purification of non-peptide small molecule libraries, *Tetrahedron Lett.*, 37:7193–7196, 1996.
131. Kaldor SW, Fritz JE, Tang J, McKinney ER, Discovery of antirhinoviral leads by screening a combinatorial library of ureas prepared using covalent scavengers, *Bioorg. Med. Chem. Lett.*, 6:3041–3044, 1996.
132. Virgilio AA, Schurer SC, Ellman JA, Expedient solid-phase synthesis of putative β-turn mimetics incorporating the i+1, i+2 and i+3 sidechains, *Tetrahedron Lett.*, 37:6961–6964, 1996.
133. Flynn DL, Crich JZ, Devraj RV, Hockerman SL, Parlow JJ, South MS, Woodard S, Chemical library purification strategies based on principles of complementary molecular reactivity and molecular recognition, *J. Am. Chem. Soc.*, 119:4874–4881, 1997.
134. Gayo LM, Suto MJ, Ion-exchange resins for solution phase parallel synthesis of chemical libraries, *Tetrahedron Lett.*, 38:513–516, 1997.
135. Siegel MG, Hahn PJ, Dressman BA, Fritz JE, Grunwell JR, Kaldor SW, Rapid purification of small molecule libraries by ion exchange chromatography, *Tetrahedron Lett.*, 38:3357–3360, 1997.
136. Booth RJ, Hodges JC, Polymer-supported quenching reagent for parallel purification, *J. Am. Chem. Soc.*, 119:4882–4886, 1997.
137. Lawrence RM, Biller SA, Fryszman OM, Poss MA, Automated synthesis and purification of amides: exploitation of automated solid phase extraction in organic synthesis, *Synthesis*, 553–558, 1997.
138. Parlow JJ, Mischke DA, Woodard SS, Utility of complementary molecular reactivity and molecular recognition (CMR/R) technology and

polymer-supported reagents in the solution-phase synthesis of heterocyclic carboxamides, *J. Org. Chem.*, 62:5908–5919, 1997.
139. Parlow JJ, Naing W, South MS, Flynn DL, In situ chemical tagging: tetrafluorophthalic anhydride as a "sequestration enabling reagent" (SER) in the purification of solution-phase combinatorial libraries, *Tetrahedron Lett.*, 38:7959–7962, 1997.
140. Wang X, Parlow JJ, Porco JA, Parallel synthesis and purification using anthracene-tagged substrates, *Org. Lett.*, 2:3509–3512, 2000.
141. Stauffer SR, Katzenellenbogen JA, Solid-phase synthesis of tetrasubstituted pyrazoles, novel ligands for the estrogen receptor, *J. Comb. Chem.*, 2:318–329, 2000.
142. Tripp JA, Stein JA, Svec F, Frechet JMJ, "Reactive filtration": Use of functionalized porous polymer monoliths as scavengers in solution-phase synthesis, *Org. Lett.*, 2:195–198, 2000.
143. Nicewonger RB, Ditto L, Varady L, Alternative base matrices for solid phase quenching reagents, *Tetrahedron Lett.*, 41:2323–2326, 2000.
144. Brown SD, Armstrong RW, Synthesis of tetrasubstituted ethylenes on solid support via resin capture, *J. Am. Chem. Soc.*, 118:6331–6332, 1996.
145. Brown SD, Armstrong RW, Parallel synthesis of tamoxifen and derivatives on solid support via resin capture, *J. Org. Chem.*, 62:7076–7077, 1997.
146. Shuker AJ, Siegel MG, Matthews DP, Weigel LO, The application of high-throughput synthesis and purification to the preparation of ethanolamines, *Tetrahedron Lett.*, 38:6149–6152, 1997.
147. Kulkarni BA, Ganesan A, Ion-exchange resins for combinatorial synthesis: 2,4-pyrrolidinediones by Dieckmann condensation, *Angew. Chem., Int. Ed. Engl.*, 36:2454–2455, 1997.
148. Chucholowski A, Heinrich D, Mathys B, Muller C, Generation of benzodiazepine and benzodiazocine libraries through resin capture of Ugi-four component condensation reaction products, *Book of Abstracts, 214th ACS National Meeting*, Las Vegas, September 7–11, American Chemical Society, Washington, 1997.
149. Gravel M, Berubè CD, Hall DG, Resin-to-resin Suzuki coupling of solid supported arylboronic acids, *J. Comb. Chem.*, 2:228–231, 2000.
150. Hamper BC, Jerome KD, Yamalanchili G, Walzer DM, Chott RC, Mischie DA, Synthesis of highly substituted 5-(trifluoromethyl)ketoimidazoles using a mixed-solid/solution phase motif, *Biotechnol. Bioeng.*, 71:28–37, 2000.
151. Rabinowitz M, Seneci P, Rossi T, Dal Cin M, Deal M, Terstappen G, Solid phase/solution phase combinatorial synthesis of neuroimmunophilin ligands, *Bioorg. Med. Chem. Lett.*, 10:1007–1010, 2000.
152. Kirschning A, Monenshein H, Wittenberg R, The "resin capture-release" hybrid technique: A merger between solid and solution-phase synthesis, *Chem. Eur. J.*, 4445–4450, 2000.
153. Zhao XY, Jung KW, Janda KD, Soluble polymer synthesis: an improved traceless linker methodology for aliphatic C-H bond formation, *Tetrahedron Lett.*, 38:977–980, 1997.
154. Plunkett MJ, Ellman JA, Germanium and silicon linking strategies for traceless solid-phase synthesis, *J. Org. Chem.*, 62:2885–2893, 1997.
155. Woolard FX, Paetsch J, Ellman JA, A silicon linker for direct loading of aromatic compounds to supports. Traceless synthesis of pyridine-based tricyclics, *J. Org. Chem.*, 62:6102–6103, 1997.

156. Zhu DW, A novel reaction medium: Perfluorocarbon fluids, *Synthesis*, 953–954, 1993.
157. Horvath IT, Rabai J, Facile catalyst separation without water: fluorous biphase hydroformylation of olefins, *Science*, 266:72–75, 1994.
158. Curran DP, Combinatorial organic synthesis and phase separation: back to the future. *Chemtracts: Org. Chem.*, 9:75–87, 1996.
159. Studer A, Hadida S, Ferritto R, Kim SY, Jeger P, Wipf P, Curran DP, Fluorous synthesis: a fluorous-phase strategy for improving separation efficiency in organic synthesis, *Science*, 275:823–826, 1997.
160. Studer A, Jeger P, Wipf P, Curran DP, Fluorous synthesis: Fluorous protocols for the Ugi and Biginelli multicomponent condensations, *J. Org. Chem.*, 62:2917–2924, 1997.
161. Studer A, Curran DP, A strategic alternative to solid phase synthesis: preparation of a small isoxazoline library by "fluorous synthesis". *Tetrahedron*, 53:6681–6696, 1997.
162. Boutevin B, Guida-Pietrasanta F, Ratsimihety A, Caporiccio G, Gornowicz G, Study of the alkylation of chlorosilanes, Part I. Synthesis of tetra(1H,1H,2H,2H-polyfluoroalkyl)silanes, *J. Fluorine Chem.*, 60:211–223, 1993.
163. FC-72™ is a commercially available (3M) fluorocarbon liquid consisting mostly of isomers of C_6F_{14} (bp 56°C).
164. Breslow R, Pandey PS, A novel synthesis of aryl orthoesters: Trimethyl m-iodoorthobenzoate, *J. Org. Chem.*, 45:740–741, 1980.
165. Curran DP, Hoshino M, Stille couplings with fluorous tin reactants: Attractive features for preparative organic synthesis and liquid-phase combinatorial synthesis, *J. Org. Chem.*, 61:6480–6481, 1996.
166. Larhed M, Hoshino M, Hadida S, Curran DP, Hallberg A, Rapid fluorous Stille coupling reactions conducted under microwave irradiation, *J. Org. Chem.*, 62:5583–5587, 1997.
167. Hoshino M, Degenkolb P, Curran DP, Palladium catalyzed Stille couplings with fluorous tin reactants, *J. Org. Chem.*, 62:8341–8349, 1997.
168. Curran DP, Hadida S, Tris(2-(perfluorohexyl)ethyl)tin hydride: A new fluorous reagent for use in traditional organic synthesis and liquid phase combinatorial synthesis, *J. Am. Chem. Soc.*, 118:2531–2532, 1996.
169. Hadida S, Super MS, Beckman EJ, Curran DP, Radical reactions with alkyl and fluoroalkyl (fluorous) tin hydride reagents in supercritical CO_2, *J. Am. Chem. Soc.*, 119:7406–7407, 1997.
170. Wipf P, Reeves JT, Balachandran R, Giuliano KA, Hamel E, Day BW, Synthesis and biological evaluation of a focused mixture library of analogues of the antimitotic marine natural product curacin A, *J. Am. Chem. Soc.*, 122:9391–9395, 2000.
171. Han H, Wolfe MM, Brenner S, Janda KD, Liquid-phase combinatorial synthesis, *Proc. Natl. Acad. Sci. USA*, 92:6419–6423, 1995.
172. Bayer E, Mutter M, Liquid phase synthesis of peptides, *Nature*, 237:512–513, 1972.
173. Bonora GM, Scremin CL, Colonna FP, Garbesi A, HELP (high efficiency liquid phase) new oligonucleotide synthesis on soluble polymeric support, *Nucleic Acid Res.*, 18:3155–3159, 1990.
174. Douglas SP, Whitfield DM, Krepinsky JJ, Polymer-supported solution synthesis of oligosaccharides, *J. Am. Chem. Soc.*, 113:5095–5097, 1991.

175. Chen S, Janda KD, Synthesis of prostaglandin E2 methyl ester on a soluble-polymer support for the construction of prostanoid libraries, *J. Am. Chem. Soc.*, 119:8724–8725, 1997.
176. Thompson LA, Ellman JA, Straightforward and general method for coupling alcohols to solid supports, *Tetrahedron Lett.*, 35:9333–9336, 1994.
177. Paquette LA, Earle MJ, Smith GF, (4R)-(+)-tert-butyldimethylsiloxy-2-cyclopenten-1-one (2-cyclopenten-1-one, 4-[[(1,1-dimethylethyl)dimethylsilyl]oxy]-, (R)-), *Org. Synth.*, 73:36–43, 1996.
178. Corey EJ, Niimura K, Konishi Y, Hashimoto S, Hamada Y, A new synthetic route to prostaglandins, *Tetrahedron Lett.*, 27:2199–2202, 1986.
179. Gooding OW, An expedient triply convergent synthesis of prostaglandins, *J. Org. Chem.*, 55:4209–4211, 1990.
180. Erb E, Janda KD, Brenner S, Recursive deconvolution of combinatorial chemical libraries, *Proc. Natl. Acad. Sci. USA*, 91:11422–11426, 1994.
181. Vandersteen, AM, Han H, Janda KD, Liquid-phase combinatorial synthesis: In search of small-molecule enzyme mimics, *Mol. Diversity*, 2:89–96, 1996.
182. Han H, Janda KD, Azatides: solution and liquid phase syntheses of a new peptidomimetics, *J. Am. Chem. Soc.*, 118:2539–2544, 1996.
183. Park WKC, Auer M, Jaksche H, Wong CH, Rapid combinatorial synthesis of aminoglycoside antibiotic mimetics: use of a polyethylene glycol-linked amine and a neamine-derived aldehyde in multiple component condensation as a strategy for the discovery of new inhibitors of the HIV RNA Rev responsive elements, *J. Am. Chem. Soc.*, 118:10150–10155, 1996.
184. Jung KW, Zhou X, Janda KD, Development of new linkers for the formation of aliphatic C-H bonds on polymeric supports, *Tetrahedron*, 53:6645–6652, 1997.
185. Zhao XY, Janda KD, Synthesis of alkylated malonates on a traceless linker derived soluble polymer support, *Tetrahedron Lett.*, 31:5437–5440, 1997.
186. Han H, Janda KD, Soluble polymer-bound ligand-accelerated catalysis: Asymmetric dihydroxylation, *J. Am. Chem. Soc.*, 118:7632–7633, 1996.
187. Han H, Janda KD, A soluble polymer-bound approach to the Sharpless catalytic asymmetric dihydroxylation (AD) reaction: Preparation and application of a [(DHQD)2PHAL-PEG-OMe) ligand, *Tetrahedron Lett.*, 38:1527–1530, 1997.
188. Han H, Janda KD, Multipolymer-supported substrate and ligand approach to the Sharpless asymmetric dihydroxylation, *Angew. Chem., Int. Ed. Engl.*, 36:1731–1733, 1997.
189. Wentworth P Jr, Vandersteen AM, Janda KD, Poly(ethylene glycol) (PEG) as a reagent support: the preparation and utility of a PEG-triarylphosphine conjugate in liquid-phase organic synthesis (LPOS), *Chem. Commun.*, 759–760, 1997.
190. Raecker R, Doering K, Reiser O, Combinatorial liquid-phase synthesis of [1,4]oxazepine-7-ones via the Baylis-Hillman reaction, *J. Org. Chem.*, 65:6932–6939, 2000.
191. Sauvagnat B, Lamaty F, Lazaro R, Martinez J, Poly(ethylene glycol) as solvent and polymer support in the microwave assisted parallel synthesis of aminoacid derivatives, *Tetrahedron Lett.*, 41:6371–6375, 2000.
192. Kamaike K, Hasegawa Y, Ishido Y, Efficient synthesis of an oligonucleotide on a cellulose acetate derivative as a novel polymer-support using a phosphotriester approach, *Tetrahedron Lett.*, 29:647–650, 1988.

193. Wiemann T, Taubken N, Zehavi U, Thien J, Enzymic synthesis of N-acetyllactosamine on a soluble, light-sensitive polymer, *Carbohydr. Res.*, 257:C1–C6, 1994.
194. Brandstetter F, Schott H, Bayer E, Liquid phase synthesis of nucleotides, *Tetrahedron Lett.*, 2997–3000, 1973.
195. Zehavi U, Herchman M, Enzymic synthesis of oligosaccharides on a polymer support, light-sensitive, water-soluble substituted poly(vinyl alcohol), *Carbohydr. Res.*, 128:160–164, 1984.
196. Seliger H, Aumann G, Polymer support synthesis, 5, Oligonucleotide synthesis on a polymer support soluble in water and pyridine, *Tetrahedron Lett.*, 2911–2914, 1973.
197. Nishimura SI, Matsuoka K, Lee YC, Chemoenzymic oligosaccharide synthesis on a soluble polymeric carrier, *Tetrahedron Lett.*, 35:5657–60, 1994.
198. Enholm EJ, Gallagher ME, Jiang S, Batson WA, Free radical allyl transfers utilizing soluble non-cross-linked polystyrene and carbohydrate scaffold supports, *Org. Lett.*, 2:3355–3357, 2000.
199. Toy PH, Reger TS, Janda KD, Soluble polymer bound cleavage reagents: A multipolymer strategy for the cleavage of tertiary amines from REM resins, *Org. Lett.*, 2:2205–2207, 2000.
200. Lopez-Pelegrin JA, Janda KD, Solution- and soluble polymer-supported asymmetric synthesis of six-membered ring prostanoids, *Chem. Eur. J.*, 6:1917–1922, 2000.
201. Janda KD, Han H, Combinatorial chemistry: a liquid-phase approach, in *Methods in Enzymology: Combinatorial Chemistry* (Ed. J.N. Abelson), pp. 234–246, 1996, Academic Press, San Diego.
202. Gravert DJ, Janda KD, Organic synthesis on soluble polymer supports: Liquid-phase methodologies, *Chem. Rev.*, 97:489–510, 1997.
203. Toy PH, Janda KD, Soluble polymer-supported organic synthesis, *Acc. Chem. Res.*, 33:546–554, 2000.
204. Kim RM, Manna M, Hutchins SM, Griffin PR, Yates NA, Bernick AM, Chapman KT, Dendrimer-supported combinatorial chemistry *Proc. Natl. Acad. Sci. USA*, 93:10012–10017, 1996.
205. Frechet JMJ, Hawker CJ, Synthesis and properties of dendrimers and hyperbranched polymers, *Compr. Polym. Sci., 2nd Suppl.*, 71–132, 1996.
206. Moorefield CN, Newkome GR, A review of dendritic macromolecules, *Adv. Dendritic Macromol.*, 1:1–67, 1994.
207. Zeng F, Zimmerman SC, Dendrimers in supramolecular chemistry: From molecular recognition to self-assembly, *Chem. Rev.*, 97:1681–1712, 1997.
208. Reed NN, Janda KD, Stealth stars polymers: A new high loading scaffold for liquid-phase organic synthesis, *Org. Lett.*, 2:1311–1313, 2000.
209. Hovestad NJ, Ford A, Jastrzebski JTBH, van Koten G, Functionalized carbosilane dendritic species as soluble supports in organic synthesis, *J. Org. Chem.*, 65:6338–6344, 2000.
210. Freeman AW, Christoffels LAJ, Frechet JMJ, A simple method for controlling dendritic architecture and diversity: A parallel monomer combination approach, *J. Org. Chem.*, 65:7612–7617, 2000.
211. Mulders SJE, Brouwer AJ, van der Meer PGJ, Liskamp RMJ, Synthesis of a novel amino acid based dendrimer, *Tetrahedron Lett.*, 38:631–634, 1997.
212. Mulders SJE, Brouwer AJ, Liskamp RMJ, Molecular diversity of novel amino acid based dendrimers, *Tetrahedron Lett.*, 38:3085–3088, 1997.

213. Gudat D, Inorganic cauliflower: Functional main group element dendrimers constructed from phosphorusand silicon-based building blocks, *Angew. Chem. Int. Ed. Engl.*, 36:1951–1955, 1997.
214. Jayaraman N, Stoddart JF, Synthesis of carbohydrate-containing dendrimers. 5. Preparation of dendrimers using unprotected carbohydrates, *Tetrahedron Lett.*, 38:6767–6770, 1997.
215. Gitsov I, Wu S, Ivanova PT, Modular building blocks for combinatorial construction of polyether dendrimers and their hybrids, *Polym. Mater. Sci. Eng.*, 77:214–215, 1997.
216. Wyndham KD, Feher FJ, Polyfunctional silsesquioxanes as supports for liquid phase synthesis, *Book of Abstracts, 214th ACS National Meeting*, Las Vegas, September 7–11, 1997, American Chemical Society, Washington.
217. Swali V, Wells NJ, Langley GJ, Bradley M, Solid-phase dendrimer synthesis and the generation of super-high-loading resin beads for combinatorial chemistry, *J. Org. Chem.*, 62:4902–4903, 1997.

8 Direct Deconvolution Techniques for Pool Libraries of Small Organic Molecules

Pierfausto Seneci

CONTENTS

I. INTRODUCTION

The tremendous development of combinatorial chemistry during the last few years has contributed to increasing steadily the options for a chemist to synthesize a chemical library, and many excellent reviews have dealt extensively with different aspects of this exciting new discipline. Solid-phase [1] or solution-phase [2] libraries, parallel synthesis [3] or mixtures of compounds [4] and large or small libraries, are only a few of the alternatives, and each one of them may fit at best to some, but not to all the needs of the chemist. Each one of them will eventually be facing structure determination, at least for the "positives" showing up after the measurement of the activity of the library components. This process can be generically named "deconvolution" of the relevant structure/activity information embedded into any library, and its accuracy and reliability will strongly influence the quantity of the results obtained (number of significant positives/families of positives) and their quality (identification of the most active compounds/families).

The techniques for linking the structures of library components with specific parts of the library (wells, pools, single beads, and so on) will be reviewed in this and in the following contribution, with emphasis on small organic molecule libraries. This topic has been either part [5] or the main subject [6] of several recent reviews, and rather than duplicate any of these efforts, an example of each strategy, either from the literature or theoretical, will be thoroughly illustrated and discussed. More details for a specific reported example, especially referring to the experimental procedures, will be found in the original paper and references for other relevant applications of the same method will follow. This will help the reader in understanding both the potential of each deconvolution technique and its application to a specific problem by major scientists active in this field and will hopefully allow a critical evaluation of the different methodologies.

The most important distinction of a library for structure determination, and in general in combinatorial chemistry, is that of being a pool library or a library of discretes. The latter requires less efforts to match structures and activities, the compounds being prepared and tested as individuals, and will be briefly covered in the following paragraph.

Structure determination of components of pool libraries can be divided into two main classes. The first, which is the subject of this review, is called direct deconvolution. It does not require any additional construct in order to determine the structure of a library component and is normally applied to solid-phase libraries, although examples of solution libraries will also be presented. The second, which will be discussed in the following review, is called encoding because a unique code, either chemical or non-chemical, is linked to each library component and its reading allows the structure determination of the compound. It is applicable only to solid-phase libraries, and it requires additional constructs to be added to the library structure.

Pool libraries can be directly deconvoluted following many different approaches, the most relevant of which are shown in Figure 8.1.

These techniques can be broadly split into two groups, the first of which can be represented by pooling methods, where deconvolution is obtained via various chemical steps, run in parallel or after the library synthesis. Pooling methods normally require multiple synthesis of many library members, including inactive individuals, in different pool formats. They are not single bead methods, so they are independent from analytical methods for structure determination. This group includes iterative deconvolution, recursive deconvolution, subtractive deconvolution, positional scanning and mutational

Direct deconvolution

Libraries of discretes	Single bead methods
Pooling methods	Bioanalytical methods on-bead screening
Iterative deconvolution Recursive deconvolution Subtractive deconvolution Positional scanning/indexing Mutational surf other methods	**Theoretical and experimental comparisons** **Summary and outlook**

FIGURE 8.1 Direct deconvolution.

SURF, together with other related methods which have been presented in the last few years.

The second group can be represented by single bead methods, and relies on either bioanalytical methods to select the active compounds or on-bead screening to determine the beads carrying active compounds. It is limited to solid-phase chemistry and does not require chemical steps after library synthesis but does require sophisticated analytical methods to determine the structure of the active compounds. A recent hybrid deconvolution-single bead-decoding method named DRED (dual recursive deconvolution) requires both deconvolutive techniques and sophisticated analytical capacities.

A theoretical and experimental comparison of the efficacy of these methods when applied to different libraries will be presented, and a final summary and outlook of the trends for the future will conclude this review.

II. LIBRARIES OF DISCRETES

The attribution of the structure to a "positive" compound from a discrete library is always trivial, because the location of a particular well tells us which building blocks or reactants were added to this well during the library preparation. Scaling up discrete libraries from a few tens [7] to many thousands [8] of individuals only differs in the automation required for the preparation of large libraries, usually coupled with software recording the monomer's location in each one of the library wells. We will not deal further with this topic, but will move directly to pool libraries for their structure deconvolution.

III. POOLING METHODS

A. Iterative Deconvolution

This is, together with positional scanning, the most popular deconvolution technique. It was introduced by Geysen *et al.* [9] and exploited by Houghten *et al.* [10] and Ecker *et al.* [11] for the deconvolution of peptide and oligonucleotide libraries, but it has been applied to the deconvolution of many small molecule libraries.

Iterative deconvolution, as many other methods presented in the next paragraphs and chapters, requires the use of the "mix and split" [12] technique, which is briefly described in Figure 8.2. A hypothetical three-step library of $3 \times 3 \times 3 = 27$ members is prepared by first splitting the resin in three equivalent portions, then adding a different monomer of the subset **A** to each portion. The three portions are then mixed and split again in three portions, now containing similar amounts of the three **A** monomers but with a unique monomer on each resin bead. Repeating this procedure with monomers of the subset **B**, three pools where **B** is determined and **A** is randomized are obtained. Another "mix and split" step followed by addition of the subset **C**, produces three pools, each containing nine compounds, where the monomer **C** is determined while the other subsets are fully represented (Figure 8.2). This technique allows the preparation of mixtures of compounds

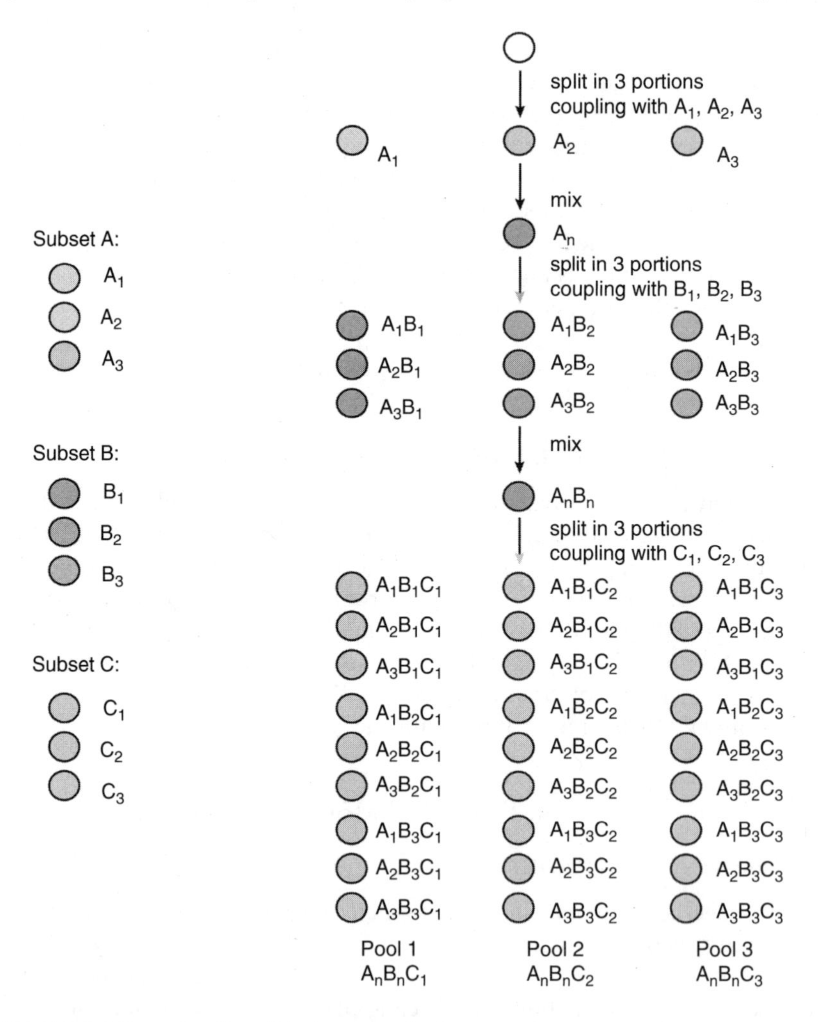

FIGURE 8.2 Mix and split solid phase synthesis.

with two important features: (1) *all the library components are represented in similar amounts* and (2) *each resin bead carries only one library component* because each coupling is performed separately on an equivalent amount of resin. Although this procedure has mostly been used for solid-phase libraries, it could also be used for solution libraries, providing that each reaction is very clean and there are no side products or coupling reagents to eliminate before the following step. An alternative is the use of automated or semi-automated purification procedures such as extractions [13], filtrations [14], resin capture [15], ion exchange resins [16], and so on.

The last library position remains physically determined by the last splitting, simplifying the deconvolution of the library. Biological screening of the final pools $A_nB_n\mathbf{C_1}$, $A_nB_n\mathbf{C_2}$ and $A_nB_n\mathbf{C_3}$ (Figure 8.3) identifies the best monomer of

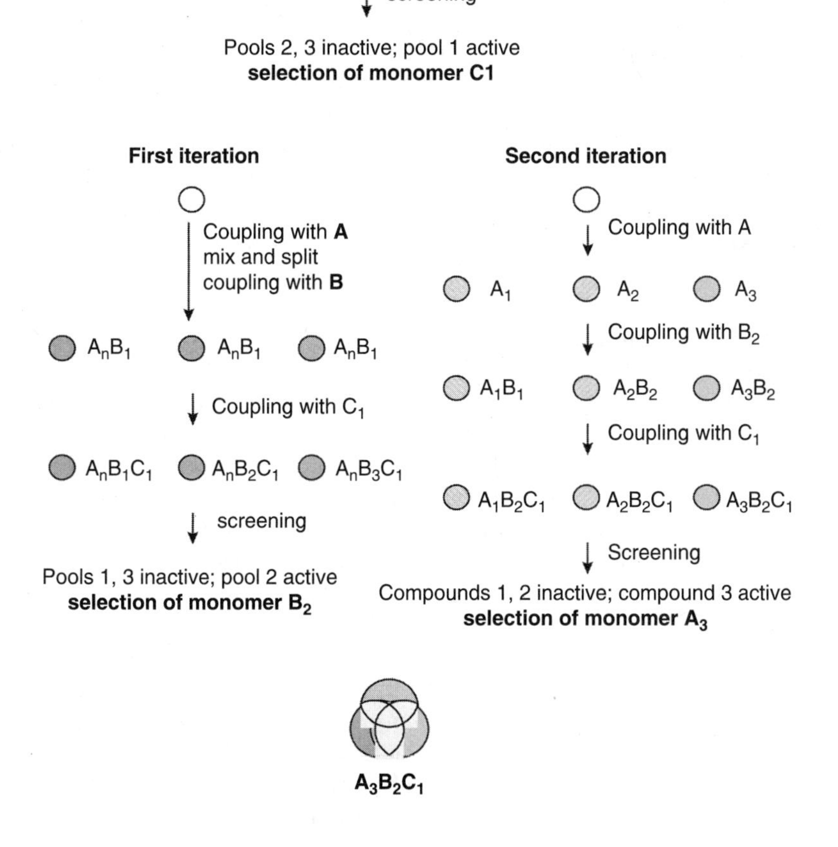

FIGURE 8.3 Iterative deconvolution of a pool library.

subset **C**, in our example $\mathbf{C_1}$. The first iteration is now performed as seen for library synthesis until the coupling of the subset **B**. Then, rather then mixing, the three pools are individually coupled with monomer $\mathbf{C_1}$ and the resulting pools $A_n\mathbf{B_1C_1}$, $A_n\mathbf{B_2C_1}$ and $A_n\mathbf{B_3C_1}$ are screened. In our example, pool $A_n\mathbf{B_2C_1}$ is the most active. The last iteration consists of coupling subset **A** on three portions of the resin, then coupling each portion with monomers $\mathbf{B_2}$ and $\mathbf{C_1}$ sequentially and without any mixing step. The final screening selects compound $\mathbf{A_3B_2C_1}$ as active *via* iterative deconvolution (Figure 8.3).

At any moment, the library synthesis and the iteration rounds handle a number of reactions, which is equal to the number of monomers which are coupled in the specific reaction. Whilst the example of Figures 2 and 3 is suitable for explaining both the "mix and split" technique and the iterative deconvolution in detail, a more complex library will highlight the effort- and cost-saving properties of iterative deconvolution of pools compared to synthesis of discretes. An example of a three-step library as in Figure 2 can be considered, but where subset **A** is constituted by 10 monomers, subset **B** by 20 and subset **C** by 15. During library synthesis the first coupling requires 10 reactions, the second 20 and the third 15. The first iteration requires a first coupling (10 reactions, then mix and split), a second (20, then no mixing) and a third (20, see Figure 8.3). The second iteration requires $10 + 10 + 10$ reactions (no mix and split steps). The complete scheme produces $15 + 20 = 35$ pools and 10 individuals to be screened, through $45 + 50 + 30 = 125$ chemical reactions. The same library, if prepared as a collection of discretes, produces $10 \times 20 \times 15 = 3000$ compounds to be screened and requires $10 + (10 \times 20) + (10 \times 20 \times 15) = 3210$ reactions to be handled. Moreover, if we assume a final quantity of 10 mg of resin for each discrete as the reasonable size for a single portion to be handled, the synthesis of discretes would require $0.01 \times 3000 = 30$ g of resin. The same final quantity of 10 mg for both pool library synthesis and iterative deconvolution requires $0.01 \times (15 + 20 + 10) = 0.45$ g of resin. A limitation of this technique is that the result of library deconvolution may not be the most active compound, especially when some pools have similar activities. In a hypothetical example of 10 pools with similar biological activity, Figure 8.4 (right) it is possible that pool **6** contains the population, but not the individual with the highest activity, which may be in another pool together with many other weaker compounds. This risk is less relevant when the most active pool is significantly better than all the others, as shown in Figure 8.4 (left). If pools with similar activities are generated during the first screening, it may well be that the deconvolution will select a final compound with a structure significantly different from the most active library component. Deconvolution of the library in Figure 8.4 (right) could produce compound $\mathbf{A_2B_3C_6}$, while the most active compound could well be $\mathbf{A_7B_3C_{10}}$, due to the original selection of monomer $\mathbf{C_6}$ rather than $\mathbf{C_{10}}$. The activity of each pool cannot be predicted *a priori*, but situations as in Figure 8.4 (right) can be solved performing the deconvolution of several pools rather than only of the best one. Unfortunately, the more pools that are deconvoluted, the more efforts are required, and the more we tend toward a parallel synthesis approach in terms

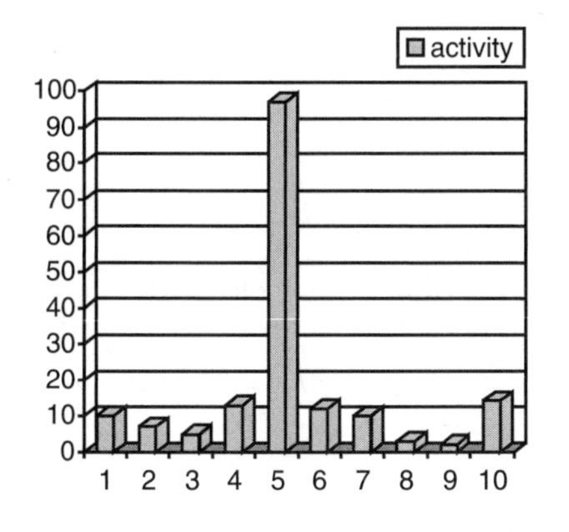

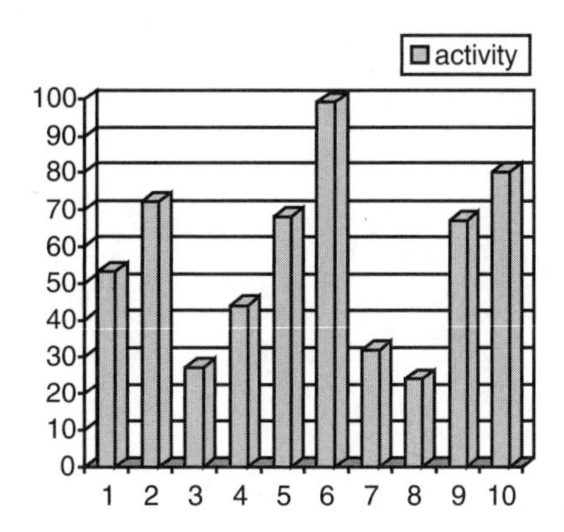

Pool 5: most active
low risk of a different monomer
in the most active compound

Pool 6: most active
high risk of a different monomer
in the most active compound

FIGURE 8.4 Activity profiles of pool libraries.

of time and effort. A modification [17] of iterative deconvolution where more than one position is determined in each cycle has been reported. Whilst this should better address the mutual influence of adjacent monomers on the activity of the final compound, significant additional efforts are required for the synthesis of the deconvolution pools. This technique will be compared to the classical one-position iterative deconvolution in the following pages.

The selected example was reported by Griffith *et al* [18], dealing with the preparation of a 43,472-member pool library of tetrahydroisoquinolinones. The reaction scheme is shown in Figure 8.5. Condensation of an amino acid (first monomer subset, $\mathbf{M_1}$) with MBHA resin gave a resin bound amine, which was reacted with an aromatic aldehyde (second subset, $\mathbf{M_2}$) to give a resin bound imine. The imine was cyclized with homophthalic anhydride and amidated with primary and secondary amines and hydrazines (third subset, $\mathbf{M_3}$). This reaction induced also isomerization of the *cis*/*trans* product mixtures to the more stable *trans* derivatives (>95%, see Figure 8.5). Cleavage *via* HF treatment produced the final library, derived from 11 amino acids, 38 aromatic aldehydes, and 52 nucleophiles, with a total number of $11 \times 38 \times 52 \times 2 = 43.472$ compounds (the two *trans* enantiomers were present in each final product). The library was prepared using "tea bags" [19] as solid supports and with the "mix and split" technique, so that 52 pools containing 836 products each were obtained.

Although the analytical control of each pool was impossible due to the large number of individuals, using a "tea bag" introduced especially for this purpose, a purity check for the reaction was performed on each pool after each step, and this was assumed as a quality control for the whole library. The library was tested in μ- and *k*-opioid receptor binding assays and in a σ-receptor binding assay. Some preliminary activities for one of the pools in

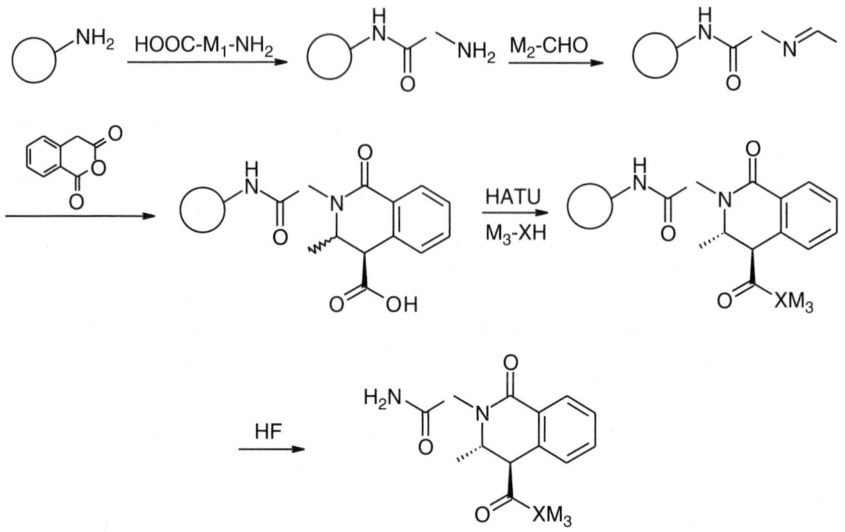

FIGURE 8.5 Synthesis of a tetrahydroisoquinolinone pool library.

the σ-receptor binding assay were reported, indicating that further deconvolution is in progress. Moreover, only a few structures of the monomers used for library synthesis were given in the paper. The lack of biological data in this and similar publications is common, and while it is perfectly understandable to protect interesting libraries and compounds, this reduces the possibility to show how pool libraries and iterative deconvolution allow the detection of active individuals in complex mixtures of compounds.

Among other applications of this method, the papers by Murphy *et al.* [20] on a 500-member solid-phase library of mercaptoacylprolines as ACE inhibitors, by Gordeev and coworkers [21] on a 100-member solid-phase library of dihydropyridines as calcium channel blockers, by Gordon and Steele [22] on a 1000-member solid-phase library of piperazinediones tested on the neurokinin-2 receptor, by Dankwardt *et al.* [23] on a 2000-member solid-phase library based on 3-amino-5-hydroxybenzoic acid, by Look *et al.* [24] on three 540-member solid-phase libraries of thiazolidinones as inhibitors of cyclooxygenase-1 inhibitors, by Chng and Ganesan [25] on a >40,000-member solution-phase library of β-aminoalcohols tested on many biological assays, by Maehr and Yang [26] on a 700-member solution-phase library of leukotriene D_4 antagonists, and by Tian and Coates [27] on a solution-phase pool library of Ti-salicylaldiminate complexes as catalytic systems for the synthesis of syndiotactic polypropylene, are worth mentioning.

B. Recursive Deconvolution

This modification of iterative deconvolution was reported by Janda and coworkers [28] and has since only been applied to the deconvolution of a

1024-member pentapeptide library, but the method is also suitable for small organic libraries. Considering the example of "mix and split" from Figure 8.2, recursive deconvolution requires saving and cataloging a portion of the resin after each coupling cycle, thus producing the final pools $\mathbf{A}_n\mathbf{B}_n\mathbf{C}_1$, $\mathbf{A}_n\mathbf{B}_n\mathbf{C}_2$ and $A_nB_n\mathbf{C}_3$, along with samples of the intermediate pools $\mathbf{A}_n\mathbf{B}_1$, $\mathbf{A}_n\mathbf{B}_2$, and $\mathbf{A}_n\mathbf{B}_3$ and subsets $\mathbf{A}_1$, $\mathbf{A}_2$, and $\mathbf{A}_3$ (Figure 8.6, top).

The three final pools are then tested, and the pool $A_nB_n\mathbf{C}_2$ shows biological activity. The intermediate pools where the subset **B** is defined are then coupled to the "active" monomer $\mathbf{C}_2$, producing the three pools $\mathbf{A}_n\mathbf{B}_1\mathbf{C}_2$, $\mathbf{A}_n\mathbf{B}_2\mathbf{C}_2$ and $\mathbf{A}_n\mathbf{B}_3\mathbf{C}_2$. Screening of these pools, retrieval of subset **A** and coupling with "active" monomers $\mathbf{B}_3$ and $\mathbf{C}_2$, produces the three pools $\mathbf{A}_1\mathbf{B}_1\mathbf{C}_2$, $\mathbf{A}_2\mathbf{B}_2\mathbf{C}_2$, and $\mathbf{A}_3\mathbf{B}_3\mathbf{C}_2$. Final screening shows the most active component of the library (in Figure 8.6, $\mathbf{A}_2\mathbf{B}_3\mathbf{C}_2$).

This method allows a faster deconvolution because some deconvolution steps are performed while preparing the library. When many subset

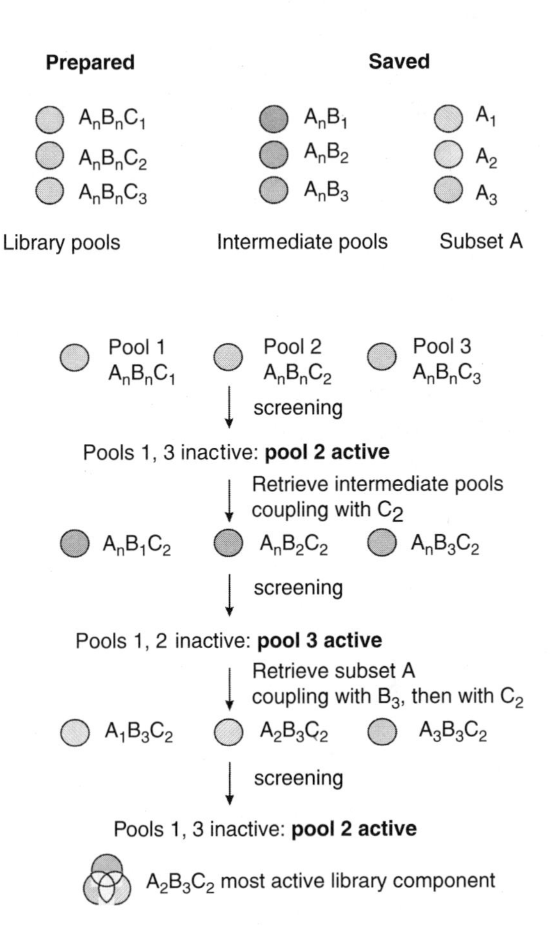

FIGURE 8.6 Recursive deconvolution of a pool library.

components and/or many reaction steps are required for library synthesis this may be an important issue. The "save and catalog" procedure for each coupling step can be adjusted depending on the predicted outcome of the screening. If several "positives" of significant potency and interest are expected, recursive deconvolution for more than one component will be required, thus larger amounts of resin must be saved after each coupling step. If timing is not crucial, iterative deconvolution allows better evaluation of the amounts of resin and monomers to be used in each deconvolutive cycle according to the screening results of the iteration rounds.

Recently, Sutherlin and Armstrong [29] published an interesting variation of this method, called recursive stereochemical deconvolution. A small array of oligosaccharides containing all the possible stereoisomers was made via non-stereoselective synthesis, and the "save and catalog" procedure for each chemical step was performed exactly as above. If any activity was found in the final library, each step could be repeated on the intermediate pools using a stereoselective procedure, which in the end allowed the determination of the active stereoisomer from the library mixture. While this example required substantial synthetic efforts and dealt with few compounds, libraries prepared by simpler chemistries and containing various chiral centers in their generic structures can be effectively deconvoluted if two parallel synthetic routes, one non-selective and another highly stereoselective, are available.

C. Subtractive Deconvolution

This approach was reported by Carell *et al.* [30] and, whilst it has often been cited with the name "subtractive deconvolution," it could be also named "tailored deconvolution." One of the libraries deconvoluted by this method, and targeted to trypsin inhibition, was prepared in solution adding equimolar amounts of 19 amino acids (Gln excluded) to a tetrafunctional xanthene scaffold, to produce a solution library composed of 65,341 individuals as a single pool (Figure 8.7). The extensive chemical and analytical assessment which assured the quality of the library included synthesis of six smaller model libraries (30–60 compounds), using simplified xanthene scaffolds, and their thorough MS characterization. Further details are not presented here, being out of the purpose of this review, but the conclusions about the library quality can be safely accepted.

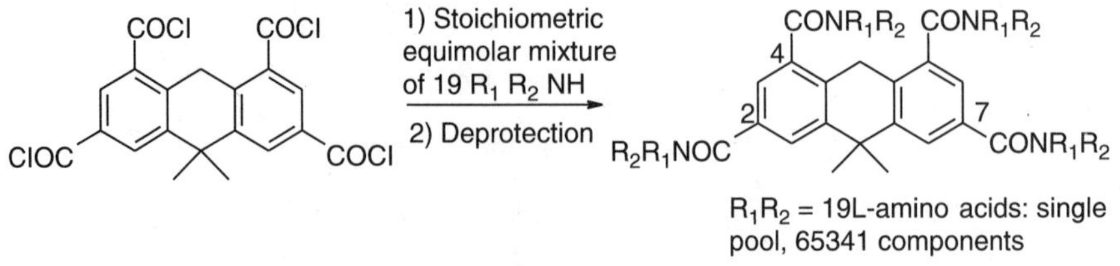

COCl COCl
ClOC COCl
1) Stoichiometric equimolar mixture of 19 $R_1 R_2$ NH
2) Deprotection
$CONR_1R_2$ $CONR_1R_2$
4 2 7
R_2R_1NOC $CONR_1R_2$
R_1R_2 = 19L-amino acids: single pool, 65341 components

FIGURE 8.7 Synthesis of a xanthene-based pool library.

The crude library, after the deprotection step, was washed with aqueous citric acid, aqueous sodium hydrogen carbonate and water, dried with magnesium sulfate and finally concentrated to a tan foam. The library was tested by adding 2.5 mg of this material to the trypsin assay and measuring the inhibition of the cleavage of a UV detectable *p*-nitroanilide substrate. Screening results for the library and all the deconvolutive efforts are reported in Figure 8.8, being expressed as percentages of trypsin activity (every screening round contained a blank well whose activity was set to 100%).

A substantial inhibition (>30%) of trypsin activity was measured for the whole library $\mathbf{A_1}$. A first iteration of 6 sublibraries, where 3 out of 18 monomers (Cys was removed to reduce artifacts) were removed from the set,

Screening:

A1 library 66

First iteration:

B1	-A,G,V	79
B2	-I,L,P	85
B3	-H,K,R	95
B4	-M,S,T	68
B5	-F,W,Y	74
B6	-D,E,N	54

Second iteration:

C1	A,G,H,I,K,L,P,R,V	31
C2	A,F,H,I,L,M,P,S,T,V,W,Y	90
C3	D,E,F,G,H,K,N,R,W,Y	85
C4	A,D,E,F,H,I,L,N,P,V,W	100
C5	F,G,H,K,M,R,S,T,W,Y	82
C6	A,D,E,F,H,I,L,N,P,V,W	90
C7	D,E,G,H,K,P,R,W	82

Third iteration:

D1	-R	28
D2	-K	93
D3	-H	21
D4	-L	46
D5	-I	50
D6	-P	80
D7	-G	32
D8	-A	27
D9	-V	59

Fourth iteration:

E1	I,K,P,V	20
E2	K,L,P,V	26
E3	G,K,P,V	45
E4	K,P,V	28
E5	I,P,V	100
E6	I,K,V	55
E7	I,K,P	100

Fifth iteration:

	4,5	2,7	
F1	K,V	I,P	37
F2	I,K	P,V	11
F3	K,P	I,V	100
F4	I,V	K,P	100
F5	P,V	I,K	100
F6	I,P	K,V	100

Sixth iteration:

	2	4	5	7	
G1	V	K	I	P	3
G2	P	K	I	V	20

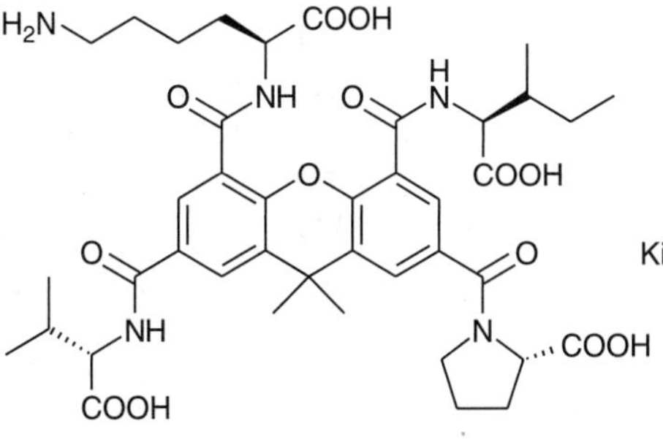

Ki = 9.4+/– 0.8 μM

FIGURE 8.8 Subtractive deconvolution of a xanthene-based pool library.

was prepared. Among these, $\mathbf{B_1}$, $\mathbf{B_2}$ and $\mathbf{B_3}$ were considered significant reversions to a higher trypsin activity, and the nine missing amino acids were selected for the second iteration. Their subset (A, G, H, I, K, L, P, R, V) represented the sublibrary $\mathbf{C_1}$, while $\mathbf{C_2}$–$\mathbf{C_7}$ were random selections of 9–12 amino acids from the initial 18. Increased inhibition for $\mathbf{C_1}$ and weak to no inhibition for $\mathbf{C_2}$–$\mathbf{C_7}$ confirmed the initial "subtractive" hypothesis. Removal of a single amino acid from the active subset of nine monomers (third iteration) showed significant reduction of trypsin inhibition for five out of the nine sublibraries, and hence these five monomers were progressed further. The fourth iteration consisted of seven sublibraries, where subsets were made of either three or four out of the five selected amino acids. $\mathbf{E_1}$ resulted the most active combination (four monomers I, K, P, V) and the reduction of activity for the three-member pools, where one of the four monomers was removed, showed the requirement of all these four monomers in the most active compound. The fifth iteration was made of six sublibraries where an orthogonal protection allowed the introduction, *via* a three-step synthesis, of two monomers in positions 2 and 7 and the other two in positions 4 and 5 (see the xanthene numbering in Figure 8.7). Sublibrary $\mathbf{F_2}$ was the most active, and the sixth iteration (three steps, one chromatography) selected the final compound out of the two isomeric candidates. Its structure, going from position 2 to 7, is V–K–I–P, and it exhibits a potency in the micromolar range (K_i of $9.4 \pm 0.8\,\mu M$, $72 \pm 7\,\mu M$ for the other isomer P–K–I–V).

This potency does not account for the inhibition observed for the library $\mathbf{A_1}$ (34%, 65,341 compounds), and examination of the various iterations clearly shows other "families" of active compounds. Nevertheless, a relatively short process (6 iterations, 51 reactions, 37 deprotections, 1 chromatography) detected a reasonably active lead compound as a starting point for chemical optimization. The sublibrary populations were reduced from 65341 to 12 in only four iterations (sublibrary $\mathbf{E_1}$, where only 12 permutations of the four monomers in the structure were possible), the second being a control of the validity of the first selection. Deconvolution of a few other "families of positives" (for example sublibraries $\mathbf{B_5}$, $\mathbf{E_2}$, $\mathbf{E_4}$ and $\mathbf{F_1}$) could have produced different lead structures while maintaining a relatively modest number of iterations and reactions.

The deconvolution process could have been realized differently (different sets or numbers of monomers to be substracted in the first or in the third iteration, different pools in the fourth iteration, and so on), but any process should have given inhibitors of similar potency in a similar time and number of iterations. A computer program for the theoretical analysis of this experiment was also designed, and its use showed that, while compounds with similar or even slightly better potency could have been missed, compounds showing significantly better activity (>1 order of magnitude) could not have been lost. For such a complex library, where mixtures of monomers are added to multiple sites at the same time, a more classical deconvolution method can not be used and this approach proves to be reasonably reliable in determining structures in a short time and with limited efforts.

While the presence of "families" of positives made the determination of the most active compounds more difficult, their activities contributed to the 34% inhibition of trypsin activity seen for the whole library. In other words, a moderately active single family of inhibitors could easily be lost due to the low activity of the whole library, thus this approach is only suitable for libraries where either very few highly potent or many weakly potent compounds are present, and for biological assays able to reliably detect activities in the medium micromolar range. The "redundancy" of the library (65,321 different individuals compared to the theoretical 130,321 permutations, producing multiple copies of some individuals) was due to the scaffold symmetry and could have caused anomalies in the assay results due to the different concentrations of the individuals.

Whilst the same researchers are using new scaffolds and non-peptide chemistries to generate other libraries [31], Kung *et al.* [32] used this deconvolutive technique on a SPSAF (solution-phase simultaneous Addition of Functionalities) generated library, and Davis *et al.* [33] reported the recursive deconvolution of a peptidomimetic library of potential artificial enzymes. A future application could be in the popular field of libraries from multicomponent reactions [34], either in solution or in solid phase, which are difficult to deconvolute with classical methods due to the mixtures of components reacting at the same time.

D. Positional Scanning/Indexing

Positional scanning, or 1-D indexing, was reported as a deconvolutive technique by Houghten and coworkers [35] and subsequently applied extensively to solid-phase peptide library synthesis. An example of solid-phase small organic molecule "2-D indexed" library synthesis and deconvolution was reported by Berk *et al.* [36]. Description of this strategy, called "spatially arrayed mixture" (SpAM), and a comparison with classical positional scanning, are shown in Figure 8.9.

In a four-dimensional library where each of the monomer subsets **A**, **B**, **C**, and **D** is composed of 10 individuals (Figure 8.9), the total number of compounds is 10,000. 1-D indexed positional scanning consists of preparing four copies of the library, each containing 10 pools made of 1000 compounds each, by "mix and split" synthesis. The first copy is made of 10 pools each containing a single monomer **A**, while **B**, **C**, and **D** are present as mixtures (1000 compounds). The others are identical but for **B**, **C**, and **D** being, respectively, present as unique monomers in each sublibrary (Figure 8.9, top right). The activity of each pool is measured, and the best monomer for each library position is selected according to the four most active pools in the library copies. While deconvolutive synthetic cycles after library synthesis are not required, thus speeding up the synthesis and biological screening of the library, four copies of the same library are synthesized, requiring larger amounts of monomers and scaffolds. The number of reactions is also increased. Iterative deconvolution requires 10 pools to be handled at every

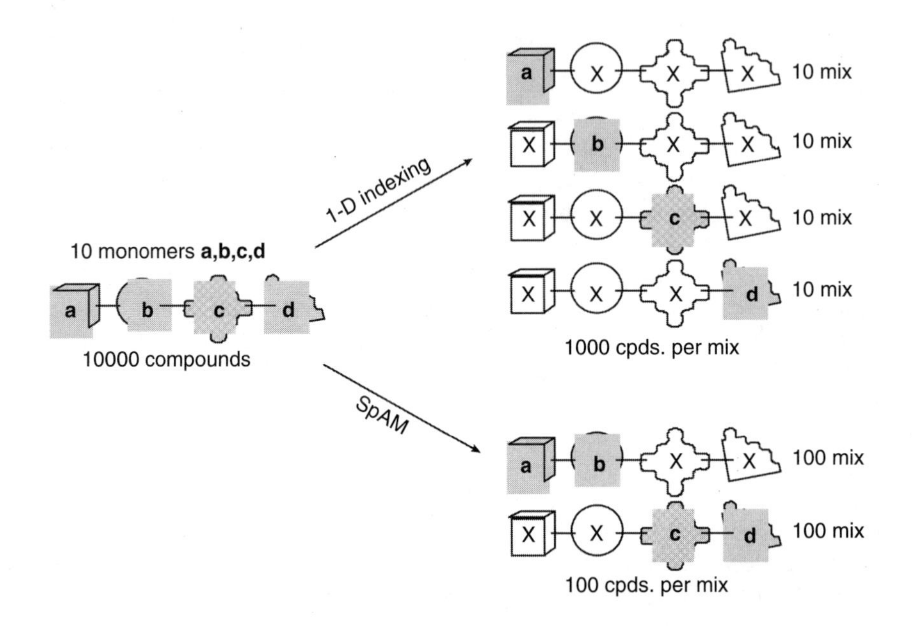

FIGURE 8.9 Positional scanning deconvolution.

stage during the synthesis for a total number of $(10+10+10+10)\times 4=$ 160 reactions. 1-D indexed positional scanning requires $(10+10+10+10)$ $+(10+10+10+100)+(10+10+100+100)+(10+100+100+100)=700$ reactions. Three of the library copies, in fact, require at different stages of the library synthesis the handling of 100 pools, which are eventually mixed to give the final 10 pools for each library copy. When subsets contain many monomers, this becomes a serious limitation.

The "spatially arrayed mixture", or SpAM [36] modification addresses pairs of neighboring monomer positions at the same time (Figure 8.9, bottom right), thus reducing the number of copies to be synthesized and partially solving the issue of mutual influence of the different monomeric units. A SpAM library was designed by Berk to mimic a known α_1-adrenergic receptor agonist (Figure 8.10, right), and its general structure is reported in Fig.10, left. It consisted of $8\times 12\times 8\times 12$ (9216) individuals, where the first monomer subset **A** was composed of bis electrophiles while **B**, **C**, and **D** were amine subsets and the linking units of the generic structure came from bromoacetic acid. The two library copies each consisted of 96 pools of 96 compounds, where the most active pool determined at the same time two of the library positions. Analysis of the pools showed that, while the **CD** position showed $\mathbf{C}_8\mathbf{D}_{12}$ as the most active pool with no other significant activities, for the **AB** position $\mathbf{A}_1\mathbf{B}_1$ resulted the most active pool, but a few others ($\mathbf{A}_1\mathbf{B}_8$, $\mathbf{A}_3\mathbf{B}_1$, $\mathbf{A}_3\mathbf{B}_8$, $\mathbf{A}_6\mathbf{B}_1$, $\mathbf{A}_8\mathbf{B}_1$) showed comparable activity. 1-D indexed positional scanning would have identified $\mathbf{C}_8$ and $\mathbf{D}_{12}$ as the best monomers in their subsets, but the **A** and **B** selection could have been more difficult, due to the comparable levels of

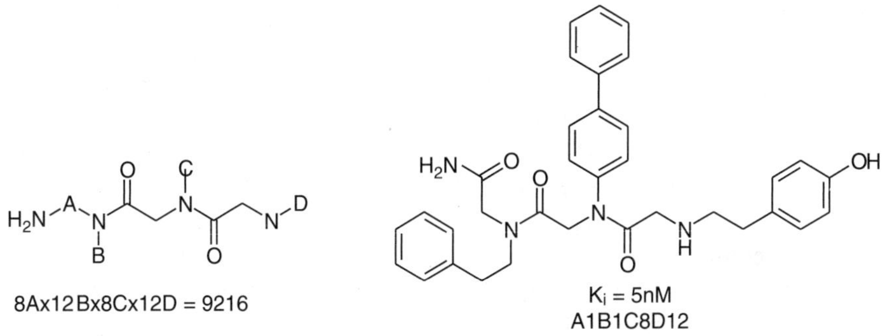

8Ax12Bx8Cx12D = 9216

K_i = 5nM
A1B1C8D12

FIGURE 8.10 Spatially arrayed mixture (SpAM) deconvolution.

activity produced by monomers $\mathbf{A}_3$, $\mathbf{A}_6$, $\mathbf{A}_8$, and $\mathbf{B}_8$. A major drawback of SpAM when compared to 1-D positional scanning is the higher number of reactions to be handled during the synthesis. In the example shown in Figure 8.9, 1-D indexing requires, as seen before, a maximum of 100 pools to be handled in each sublibrary to produce the four copies of the 1-D indexed library. SpAM synthesis of the first library copy proceeds as seen above until the coupling of the **B** monomers, but then the 100 pools must be kept divided to have a determined **AB** pool. A third "mix and split" cycle for **C** would require splitting to 1000 pools, which are extremely difficult to manage, even for automated instruments. Berk *et al.* [36] used both isokinetic mixtures of monomers and recursive additions of substoichiometric amounts of equimolar monomer mixtures [37] to avoid splitting, although neither solution completely guaranteed the same quantity of each individual in each pool. Moreover, kinetic studies for each monomer should prevent the use of compounds with significantly different reactivities in the same monomer subset, thus limiting the chemical diversity of the library.

Positional scanning was repeatedly used by Houghten and coworkers [38] in solid-phase peptide synthesis, but few examples of its application to small molecule libraries have been reported. Smith *et al.* [39] deconvoluted a solid-phase library of 1600 esters/amides, Andrus *et al.* [40] deconvoluted a solid-phase library of non-natural polyenes, Leone-Bay *et al.* [41] deconvoluted the activity of an indexed library in an *in vivo* assay, while Pirrung et al. reported both a solution phase 54-member carbamate library [42] and a solution-phase 72-member tetrahydroacridine library [43].

E. Mutational SURF

This method was presented by Freier *et al.* [44] and was intended as a refinement of other existing deconvolution techniques. While the only reported example deals with "virtual" oligonucleotide libraries, the technique looks promising and useful also for real small organic molecule libraries.

The choice of an oligonucleotide library came from the notion that oligonucleotide hybridization can be calculated accurately, even for large numbers of molecules, without the need of actually preparing a library. A hexameric (UGGGCA) and a nonameric target (GUGUGGGCA) were selected and the affinities for these targets of the individuals from a full 9-mer library, i.e. a 262,144-member virtual library, were calculated. This virtual library was deconvoluted both via iterative deconvolution and positional scanning, calculating each association constant for single components and then for all the library subsets, adding the affinity of each component of the pool. This comparison will be dealt with extensively in the following sections, but, for now, it is enough to say that in general the best binders were seldom the result of any deconvolutive computational cycle using either iterative deconvolution or positional scanning.

Freier *et al.* [44] showed that mutational synthetic unrandomization of randomized fragments (SURF) improves the quality of the selected structures obtained from any deconvolution technique. The structure of this approach, as applied to the above-mentioned virtual example, is depicted in Figure 8.11.

One of the suboptimal binders selected for its activity on the nonameric target was used as a starting point and mutational SURF was run. That is to say, all the oligonucleotidic permutations of a single base in the sequence (27, three bases × nine positions), plus two sequences where the last base was put at the beginning and another where the first was moved to the end were prepared. The most active compound became the starting point for a new cycle, and so on, until no better structures were found. The sequence of Figure 8.11 started from AGCCCGCAC, with an IC_{50} of 26.9 pM, and it was improved to 2.36 pM after selection of GCCCGCACA during the first mutational round. A second mutation produced GCCCACACA, which was one of the two known best binders with an IC_{50} of 1.05 pM. While the results of mutational SURF were not as good when applied to the 6-mer target, they significantly increased the number of good binders produced after a few mutation rounds. These computational experiments were repeated several times, and the results are reported in Figure 8.12. On 500 different runs for both targets the activity

starting sequence: AGCCCGCAC, IC_{50} = 26.9 pM

CGCCCGCAC **A**ACCCGCAC AG**A**CCGCAC AGC**A**CGCAC AGCC**A**GCAC
GGCCCGCAC A**C**CCCGCAC AG**G**CCGCAC AGC**G**CGCAC AGCC**G**GCAC
UGCCCGCAC A**U**CCCGCAC AG**U**CCGCAC AGC**U**CGCAC AGCC**U**GCAC

AGCCC**A**CAC AGCCCG**A**AC AGCCCGC**C**C AGCCCGCA**A** **C**AGCCCGCA
AGCCC**C**CAC AGCCCG**G**AC AGCCCGC**G**C AGCCCGCA**G**
AGCCC**U**CAC AGCCCG**U**AC AGCCCGC**U**C AGCCCGCA**U** *GCCCGCAC**A***

GCCCGCACA, IC_{50} = 2.36 pM

2nd mutation

GCCCCACACA, IC_{50} = 1.05 pM

FIGURE 8.11 Mutational SURF.

Target:	9-mer		6-mer	
Initial selection procedure:	*Iterative deconvolution*	*Positional scanning*	*Iterative deconvolution*	*Positional scanning*
No mutational SURF	0.6[a]	0.05	0.5	0.007
Mutational SURF	0.9/3.4[b]	0.9/3.7	0.7/14.2	0.7/13.9

[a]average activity (500 runs for each target) of selected molecules relative to the best binder in the library
[b]activity as above/average number of mutational SURF performed

FIGURE 8.12 Computational results using mutational SURF.

of the deconvoluted structure, coming from either positional scanning or iterative deconvolution, was consistently and significantly improved by using mutational SURF and the final results were independent from the technique that provided the starting structure.

Translating this method to small organic molecules could be difficult for the two "shuffled" sequences, where the last residue becomes the first and vice versa. While this is easily done for oligomeric synthesis (nucleotides, peptides, etc.) it could be either difficult or even impossible for libraries where the monomers do not belong to the same chemical class, or where a final cyclization step is planned. These two sequences were often the ones improving affinity [44], so the relevance of mutational SURF where the "head to tail" permutations are missing should be checked. This approach should be applied to some "real" libraries, possibly made of small organic molecules, in order to evaluate its real usefulness. The extreme simplicity and the limited number of operations required for a preliminary evaluation (the first cycle of mutational SURF for a 8000-member library, made from three sets of 20 monomers, would require 60 permutations plus two "head to tail" sequences, that is 62 compounds) make mutational SURF particularly appealing.

F. Other Deconvolutive Methods

Déprez *et al.* [45] presented the deconvolution of a 15,625-member tripeptide library using the "orthogonal library" approach, a technique which is similar to both positional scanning and SpAM methods. Its appealing features are the modulation of the number of components in each pool by dividing monomer sets into subsets without a proportional increase in the number of pools to be handled, the reduction of "redundancy" (each compound is synthesized twice for the orthogonal library deconvolution, regardless of the number of chemical steps, while positional scanning of an n-step library requires n copies of each compound to be prepared) and the grouping of significantly different monomers into a single subset (e.g. a five-member subset of amines containing a hydrophilic primary aliphatic, a bulky secondary aliphatic, an electron-poor primary aromatic, an electron rich secondary heteroaromatic, and a natural L-amino acid protected on the COOH) so as to reduce the probability of many active compounds in the same pool, their structures being more diverse that of a randomly generated pool. The main limitation is the rigidity of the method which requires splitting of monomers into equally populated subsets in any synthetic step, thus allowing only sets of 9 (3^2), 16 (4^2), 25 (5^2) monomers, and so on.

Blake and Litzi-Davis [46] deconvoluted a 50,625-member tetrapeptide library and a 16,777,216-member hexapeptide library by means of the "bogus coin" technique, which is conceptually similar to subtractive deconvolution and shares most of the features discussed earlier, but looks less "customizable" in the reported format and consequently less useful in identifying other positive structures in addition to the one resulting from the deconvolution process.

Boger *et al.* [47] introduced the so-called deletion synthesis deconvolution as a process typical for convergent dimerization libraries, based on the methodical elimination of one element from the variable library units and on the evaluation of loss vs. gain of activity. This method was compared with positional scanning in a following paper by the same group [48] and proved to be effective and useful.

IV. SINGLE BEAD METHODS

A. Bioanalytical Screening

When a library is deconvoluted *via* an iterative method, both time and monomers are used in excess for synthesis during iteration rounds. Moreover, the iterative process ends with a single structure which may not be the most active compound of the library (see previous chapters) and the deconvolution of additional positives, when it is possible, considerably lengthens the structure determination process.

Several analytical techniques, possibly in combination, allow structure determination for the compound anchored on a single bead. When this technology is coupled to the "mix and split" technique, which assures a single structure to be present on each bead according to the "one bead-one compound" concept [49], a library can be completely deconvoluted by means of MAS NMR [50], HPLC/MS [51], FTIR [52], and/or other analytical techniques, simply by analyzing each resin bead. Two major hurdles for an extensive application of such analytical techniques [53] to structure determination are evident. First, the compound often has to be cleaved off the bead for the analysis, which in turn commonly "destroys" the compound, thus preventing determination of activity in the following biological assay, and the same problem arises inverting the operations. Second, for a library of 10,000 compounds, sampling of 50,000 beads should be performed to represent

>99% of individuals, and a very optimistic average time of 30 minutes for full analytical characterization (NMR, FTIR, cleavage, HPLC/MS, etc.) adds three to four years to the time necessary to process the whole library.

When the analytical technique (usually MS or HPLC) is coupled with the biological assay (usually the binding to a receptor) the obstacles are removed because only a few positive compounds are selected, and the time required for their characterization becomes acceptable. A few examples dealing with small organic molecule libraries have been reported. Among them, a recent paper by van Breemen *et al.* [54] presented the characterization of a small library of adenosine analog for their affinity to adenosine deaminase using pulsed ultrafiltration coupled with electrospray MS spectrometry. An equimolar mixture of 20 adenosine analogs, including the known inhibitor EHNA (erhythro-9-(2-hydroxy-3-nonyl)adenine) as a positive control and the enzyme substrate (adenosine), was incubated with the enzyme for 15 min, then introduced into an ultrafiltration chamber (filter with a cutoff of 10,000 MW, enzyme MW = 41,250). Water was pulsed into the chamber for 8 min, removing the unbound and weakly bound compounds, then 30% MeOH was added to reversibly disrupt the receptor–ligand interaction. The methanolic solution contained only one compound, identified as EHNA through its mass spectrum (MW = 278). This result validated the method through the identification of the positive control and determined the lack of activity of the other adenosine analog. Another experiment performed with known inhibitors of human serum albumin (thyroxine, salicylate, furosemide, tryptophan, and warfarin) allowed the determination of the relative K_i of the inhibitors by comparison of their MS areas in the absence or in the presence of the enzyme in the assay mixture. Moreover, experiments with the poorly soluble thyroxine in a very diluted solution allowed the concentration of the inhibitor via receptor binding and subsequent methanolic elution. Testing dilute solutions of inhibitors (up to 10 nM) and recovering the receptor after the assay are the most valuable features of the biological assay, which may also be adapted to different purposes (different filter cutoffs, different binding disruption conditions, and so on). While this is the only reported application to a small pool library, the approach looks very promising, flexible and also unbiased toward oligomeric libraries. The presence of compounds with the same MW would require the use of more sophisticated MS techniques (MALDI-MS [55] or TOF-SIMS [56]). More data from different experimental sets and receptors are obviously needed to determine the drawbacks and the usefulness of the technique for the deconvolution of large pool combinatorial libraries.

Among others, Wieboldt *et al.* [57] used a similar technique, immunoaffinity ultrafiltration, in conjunction with ion spray HPLC/MS, on a small benzodiazepine library tested for binding to a polyclonal sheep IgG antibody. Dunayevskiy *et al.* [58] used gel filtration together with CE/MS and LC/MS on a small library of drugs tested for binding to human serum albumin, Chu and coworkers *et al.* [59–61] reported an application of affinity capillary electrophoresis (ACE)-MS to some tri- and tetrpeptide libraries (>1000 compounds) tested for affinity to vancomycin. McEwen *et al.* [62] and Blanchard *et al.* [63] reported the use of size exclusion chromatography in bioanalytical screening, while Karger *et al.* [64] used for this purpose capillary isoelectric focusing. Kelly *et al.* [65] presented library affinity selection (LAS)-MS applied to a 361-member peptide library tested for its affinity to the SH2 domain of PI-3 kinase, Bruce *et al.* [66] introduced bioAffinity characterization (BAC)-MS as a method to rapidly screen combinatorial libraries and, finally, Brummel *et al.* [67] evaluated sophisticated MS techniques, particularly imaging time of flight secondary ion MS (TOF-SIMS56), as a fast and automated method for quality control and deconvolution of pool libraries, "wasting" only small quantities of the bound compound for the MS analysis. Other more recent applications where MALDI-MS was involved, either alone [68] or coupled with size exclusion [69] or with immunoaffinity deletion [70], are also worth mentioning.

These techniques will likely find their applications in the near future for library deconvolution and structure determination of compounds on a single bead basis, and future advances in analytical techniques will make them more amenable to structure determination of large pool libraries. The application of MS techniques to target-assisted, off-bead structure determination of library actives has been recently reviewed [71, 72]. Two more general, screening-focused reviews, dealing with target-assisted methods using immobilized (supported) or confined (segregated in a compartment) targets for identification of positives from mixtures, have recently appeared [73, 74].

A major branch of bioanalytical screening is related to NMR spectroscopy. NMR has gained a great deal of importance as a screening technique in off-bead target assisted structure determination lately; several NMR-assisted screening methods providing either a qualitative or a quantitative estimation of the ligand–target binding strength, using simple ligand mixtures, will be briefly described here.

Fesik reported the so-called SAR by NMR^{TMM} method [75, 76] which monitors the chemical shift perturbation of ^{15}N–^{1}H heteronuclear single quantum correlation (HSQC) data for small proteins in the presence of small ligands. The method requires the use of uniformly labeled ^{15}N receptors, but can screen *via* sophisticated NMR CryoProbes [77] even 10,000 library components per day. Successful applications to identify inhibitors of human papillomavirus E2 protein [78], of stromelysin [79], of [80] and of Erm methyltransferases [81] have been reported by the same group. While the method is extremely reliable and accurate, often leading to relevant lead compounds, several bottlenecks must be highlighted: the need of significant quantities of uniformly ^{15}N labelled, purified receptors/proteins/enzymes; the restriction to low MW targets (only up to 30 kDa); the need of complete structure determination and NMR signal assignment for the target; the need of complex and expensive instrumentation which, as of today, confines this method to a few specialized laboratories. Its usefulness, thus, may be higher for later rather than early discovery phases *via* HTS.

A method named pulsed field gradient NMR (PFG-NMR) was recently reported by Shapiro [82–86] as being able to discriminate between bound and unbound library individuals, which had different diffusion coefficients in solution when complexed with large macromolecules. This property allowed the editing of the NMR spectrum to see only the signals of bound molecules. The authors applied this technique to detect interactions between small molecules [82–84], between small molecules and vancomycin [85] and between small molecules and medium length oligonucleotides [86]. A somewhat different application of PFG-NMR by Fesik and coworkers [87] obtained the spectra of bound library components by subtraction of the spectra of the mixtures in the presence and the absence of two different proteins, FKBP and stromelysin, and confirmed the detection of two known ligands in each NMR diffusion screening, thus validating this target-assisted screening protocol. The validation studies reported so far have only hinted toward the usefulness of PFG-NMR, and the publication of a "real" library screening would reinforce the confidence in this method; as of today its low throughput (several hours per sample) and the risk of experimental error (up to 100%) limit its scope.

Meyer *et al.* reported recently transfer NOE (trNOE) [88], which detects the strong negative trNOE effect of receptor-bound molecules compared to the weak, positive trNOE of unbound compounds without the need of receptor labelling; the method was used to identify an E-selectin antagonist from an artificially assembled 10-member library of saccharides [89]. The same approach has been cleverly exploited by Moore and coworkers [90] by using the so-called SHAPES strategy. The authors selected among the CDC database [91] the meaningful frameworks which are most frequent in the structures of known drugs, i.e. the cyclic arrays of atoms constituting several rings and the connecting atoms working as linkers. A careful computational search and selection [92] limited to 41 the number of frameworks representatives of a large part of CDC. These core scaffolds were used to screen the ACD database [93] and to select commercially available decorated/modified scaffolds without toxic or unstable moieties; with side-chains conferring aqueous solubility; with at least one N or O atom. Several novel compounds obtainable with simple routes were finally added, leading to a SHAPES library of 132 compounds with MWs from 68 to 341 Da and with log P from −2.2 to 5.5 (see Refs. [90, 92] for more details). This basic set was screened on several targets in a qualitative mode, that is without trying to quantify the trNOE phase switch of the ligand cross peaks but considering it a signal of ligand–target interactions; a reliable "yes-or-no" answer was unequivocably obtained for each SHAPES library individual. Several weak binding frameworks (micro- to low millimolar constants), which would have been missed by conventional biological screening protocols, were used to focus later larger screening efforts on biased collections or libraries containing these or similar frameworks; the frequency of positives from SHAPES-derived screening sets was significantly increased if compared with random screening [90], proving the validity of the approach. The method does not require labeled receptors and is ideally suited for large macromolecules (>60 kDa); the amount of macromolecular target needed for a SHAPES screen is relatively limited (tens of mgs of large proteins); a complete NMR assignation for the macromolecular target is not required; the library size allows the use of small mixtures of ligands (one to four ligands per sample), thus increasing reliability and simplicity; the SHAPES NMR screen takes only from several hours to a few days; a visual inspection of the spectra is enough to discriminate among binders and non-interacting compounds. Even though the information acquired is not quantitative (but can be quantified by progressing the positives through PFG-NMR experiments, as was done in Ref. [90]) and the binding specificity must be determined, this method will surely become a major asset for the early, information-poor phases of ligand fishing related to difficult targets, for which no ligands have been found using traditional HTS campaigns. The method could in future be tailored also for non-pharmaceutical applications.

NOE pumping [94] was recently reported by Shapiro. After the application of a diffusion filter suppressing all the ligand signals, the NOE experiment started to "pump" the magnetization from the target to the binders in the ligand mixture. By increasing the mixing time t_m of the target and the ligand mixture the magnetization transfer gradually decreased the target signals (horse serum albumin in Ref. [94]), and gradually increased the ligand signals (salicylic acid) without interferences from non-binders (glucose and ascorbic acid). Similar characteristics in respect with trNOE (no target labelling, no MW cut-offs for targets, no need of a complete NMR assignment for the target) also make this technique particularly appealing for NMR-assisted screening of mixtures.

Saturation transfer difference NMR (STD-NMR) was recently reported by Meyer and Mayer [95] as a fast and reliable screening method, and was used to spot the binding of saccharides to wheat germ agglutinin (WGA); the same assay was repeated successfully anchoring WGA on solid-phase and recording spectra from heterogeneous systems using magic angle spinning STD-NMR [96].

The application of NMR techniques to target-assisted, off-bead structure determination of library activities, has also been recently reviewed [97–100].

B. On-Bead Screening

Another technique of single bead direct deconvolution is based on on-bead screening. When the biological assay can be run in this format, each library component is tested as resin bound in an assay medium where a soluble receptor is present and the receptor/ligand, or inhibitor, interaction takes place at the surface of the bead. After the biological assay a few "positive" beads are identified, usually through colorimetric, fluorescent or radiolabeled detection. They are removed from the assay medium and analytically characterized after washing off the receptor. This technique was first described by Lam *et al.* [49], has since been used repeatedly by many groups and has been reviewed recently [4a]. Up to now it has always been used for peptide or peptidomimetic libraries, but nothing prevents its application to small organic molecule libraries.

In the selected example by Lam *et al.* [101] many peptide libraries were prepared using the "mix and split" technique and tested in different on-bead screens. "Incomplete libraries" were tested (the population of most of them was more than a million compounds), and the positive structures were exploited through focused libraries. Some libraries were screened against an anti-insulin monoclonal antibody tagged with alkaline phosphatase, which allowed an enzyme-linked colorimetric detection. Only the beads bound to the murine MAb showed a tourquoise color, while the vast majority remained colorless (details of the technical realization of the assay can be found elsewhere [101, 102]). The chemical structure linked to the positive beads was then easily determined via Edman degradation of the peptide sequences.

Initially, some linear libraries of penta-, hexa-, octa and decamers with structure $(X)_n$-β-ε-β-ε-TentaGelS (where $\beta = \beta$-alanine and $\varepsilon = \varepsilon$-aminocaproic acid) were tested. For each position 19 natural amino acids (cysteine was avoided) were used, and the active ligand structures are shown in Figure 8.13. Some recurrent motifs emerged, such as FDW and QDPR, especially for 5- and 6-mers, and WXXGF for 8- and 10-mers.

Some cyclic peptide libraries with the structure C-$(X)_n$-C-β-ε-β-ε-TentaGelS were also tested. The same set of 19 amino acids was used and cyclization was achieved by oxidation of the cysteines and formation of the disulfide bond. The active sequences are reported in Figure 8.14, top. Again, several motifs were detected but only WDXGF and FDW resembled motifs found from the linear libraries, while others, such as HGVQ and QDIXY, were peculiar to the cyclic libraries. Linear D-aminoacidic libraries were also tested and the results are given in Figure 8.14, bottom. A common motif, qxGsxG, was always observed.

5-MER	6-MER	8-MER	10-MER
FNWAI	K**W**GS**GF**	GSW**W**AQGF	TGQDF**W**AM**GF**
FNWKR	N**W**GH**GF**	EQA**W**HIGF	
FNWER	D**W**GY**GF**	DEL**W**GQGF	EQGAW**W**EV**GF**
Y**FNW**A	R**W**DL**GF**	FHK**W**ESGF	EDQLI**W**LT**GF**
	R**W**AH**GF**	HQN**W**GSGF	IIGHE**W**RY**GF**
FDWKR		MNY**W**KEGF	GDWEL**W**SK**GF**
FDWAH	**FDW**SQC	NLV**W**SMGF	
N**FDW**K	Q**FDW**YQ		PQQSNYYS**GF**
		GNGDDQ**GF**	GSAVQFTH**GF**
K**QNPR**	**FNW**AVG	FEGYQY**GF**	
F**QDPR**		QGQYTF**GF**	
R**QDPR**	KK**QDPR**		
P**QDPR**		**FDW**SNGGG	
	SQHGIW		
		KQTHLSWM	
		TNEWFAGK	
		RQDYARVM	
		NGQTWATG	

FIGURE 8.13 On-bead screened oligomeric peptide libraries.

Some compounds were prepared and tested, showing weak affinity (around 10 μM). One of the active families from the linear peptide libraries, containing the motif WXXGF, was optimized through a focused 8-mer library XXXWXXGF producing two families of active compounds, QXIWGXGF and XXXWKYGF. From these refined structures a second round of focused libraries, a 9-meric XXXXWKYFG and a 10-meric XXQXIWGXGF, was prepared and tested. The results of the whole process are shown in Figure 8.15. The binding affinities of the ligands from the second focused libraries were one to two orders of magnitude higher than the ligands from the original WXXGF sequence.

An on-bead screening often produces many active structures, sometimes significantly different, giving a preliminary SAR to orient further efforts. A panel of on-bead assays may allow a complete characterization on many different screens and a more detailed SAR for the molecules composing the library. Nevertheless, major concerns must be considered before thinking of

C-7-MER-C: WDKGFGY,WDMGFGI,WGQGFSA, QWGNHWN

C-8-MER-C: FLIMHGVQ, VANQHGVQ, GWARHGVQ, HENEHGVQ, KFDWMAGG, MYFFQRGE, TFQWGTGG, LYHRHGIQ, AQTERGNE

C-9-MER-C: QDISYQSNL, QDIAYENKM, VSQHDFPLQ, RQSTWGSTP, AHQQDYKYG, YDTGFGHHK

D-6-MER: pqrGst, pqiGst, pqGGst, yqeGst, vqeGst, vqmGst, vqGGst, rqlGst, rqlGsv, aqmGsi

D-8-MER: dqlGslGG, GwqpGsmG, asqaGsfG, fkiqGGsq, mniqfGss, psqqGGst, mqfmGss, tsyqaGst, rlfqmGtt

FIGURE 8.14 Active peptide sequences.

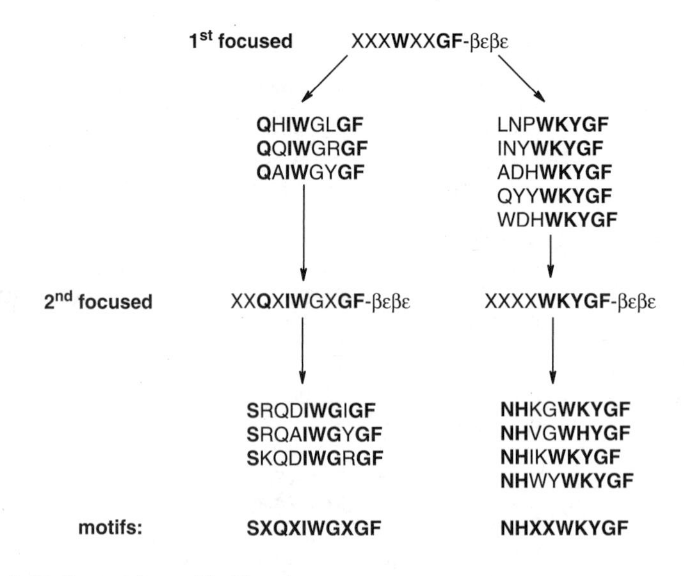

FIGURE 8.15 Focused peptide libraries.

such a technique for library screening. First, the interaction between ligand and receptor is influenced by the solid support hooked to the ligand. If the resin is bound to the compound in an essential area for the interaction, an on-bead assay will not show any activity while a solution assay would be successful. The steric influence of the support structure on the compound may also prevent the correct binding orientation of the hooked molecule (for example, lipophilic compounds tend to stick on lipophilic resins and to be shielded from any other interaction). The latter issue is solved using hydrophilic solid supports, such as TentaGel or ArgoGel beads, which allow a better swelling of the resin in aqueous media and consequently a greater accessibility of a bound ligand for the receptor. The former cannot be predicted, so that the presence of activities which are "hidden" by the solid support cannot be ruled out. If the library is designed around the structure of a known active compound, the standard can be provisionally hooked onto the resin and its on-bead activity compared with its solution activity. If the compound has more than one suitable chemical handle for solid-phase attachment, all the possible constructs should be prepared and tested to find the best match between on-bead and solution activities, and if none give the expected activity then on-bead screening has to be avoided. If a standard is not available, the on-bead screening should always be coupled with a corresponding solution assay, using a code to detect the structure of the active compound on the bead after cleavage (coding techniques related to on-bead screening will be presented in the following review). The two screening strategies, or a combination of the two [103], give the additional advantage of a direct comparison of activities in solution and on-bead for the whole library.

On-bead screening and structure determination of positives from pool libraries, using an enzyme-linked colorimetric assay, was used by Lam and coworkers [104–107] and by others [108–112], to probe the specificity of random peptide or peptidomimetic libraries. Fluorescence-labeled [107, 113, 114] and radionucleotide-labeled [115, 116] targets have also been reported. Aside from the above-mentioned binding assays, functional assays have also been performed on-bead [117, 120]. All these examples were able to detect one or more peptide sequences, which were reconfirmed after their identification as true binders/inhibitors interacting with the desired target. Recently an example of large oligocarbamate libraries (one cyclic trimer library, 19,863 individuals, and two tetramer libraries, 531,441 individuals) screened to search for human thrombin ligands [121] and another of glycopeptide libraries (around 300,000 individuals) screened for lectin binding [122] somewhat enlarged the applicability of on-bead screening to other library structures, both producing several binders to their respective targets.

V. THEORETICAL AND EXPERIMENTAL COMPARISONS

While many authors have used a particular deconvolution technique describing its usefulness for a specific application, few have analyzed different techniques in parallel for the same application. Four papers which have dealt with this topic are presented below.

Konings *et al.* [123] compared different deconvolution strategies on a computer-generated library of oligonucleotide 9-mers (262,144 individuals) tested for binding affinity on a 6-mer target (UGGGCA) and on a 9-mer target (GUGUGGGCA). After calculation of the binding affinities for each library component the virtual deconvolution of the library on the two targets was performed, splitting the population into different subsets and selecting pools after each iterative/scanning cycle. Standard Monte-Carlo techniques [124] were used to generate the activities of each pool. A two-fold error in subset activity was introduced in the simulations to account for the experimental errors encountered when measuring binding affinities (to a maximum of double or a minimum of half the calculated activity), making the selection process more "challenging". Molecules with a two-fold difference of affinity could easily be reversed in terms of selection by the computer program, slowing down or even irreversibly orienting the deconvolution process toward a "wrong" selection.

The same library, tested on the two targets, produced two significantly different patterns of activity. The calculated affinity for the 9-mer target gave two best binders and only twelve "good binders", defined as structures with less than 1 kcal/mol difference in binding affinity from the maximum (0.005% of the library), while 98.7% of the compounds had activities at least five orders of magnitude lower than that of the best binders. Affinity for the 6-mer target produced 16 best binders and 2414 molecules as "good binders" (0.9% of the library), while only 37% of the compounds had activities at least five orders of magnitude lower than that of the best binders. These patterns made the activity on the 9-mer easy to deconvolute, due to the extremely low number of significant actives, while activity on the 6-mer was extremely difficult to deconvolute to the structure of a best binder.

First, iterative deconvolution was examined, and 500 simulations of deconvolution were made for each target. For the 9-mer, roughly 50% of the simulations ended with one of the two best binders, and around 90% produced one of the twelve good binders. For the 6-mer, only about 1% of the simulations produced one of the 16 best binders and 97% produced one of the 2414 good binders. The influence of deconvolution order on the reliability of the deconvolution was also examined (all the permutations from 123456789 to 987654321), and, while for the 6-mer target any permutation produced the same result, slight differences were noted for the 9-mer target (from around 45%, giving a best binder from the sequence 957846321 to around 55% for 123456789). The conclusion was that the order of deconvolution did not play a major role, and that slight improvements were obtained when the most active individuals were grouped together in the early selection (as in 123456789 for the 9-mer) rather than being equally spread in different pools (as in 957846321).

Iterations where more than one position per round was fixed were also examined, using 1-, 3- and 9-fixed positions per round (that is, from usual iterative deconvolution to a parallel synthesis approach for 262,144 compounds). Results with 3-fixed positions per round were identical for the 9-mer compared to 1-fixed position, and slightly, but not significantly, better for the 6-mer target. Parallel synthesis performed better (80% of simulations gave a best binder for the 9-mer compared with 55% for 1- and 3-fixed positions per round), but introduction of the two-fold experimental error caused significant deviations from the best binders also for parallel synthesis (10% selection of one of the 16 best binders for the 6-mer, compared with 1% with 1- or 3- fixed positions per round).

This virtual library also allowed grouping of individuals in random pools, where all the positions are different, rather than having eight positions randomized and equally represented and one position fixed as in "mix and split" pools. The results of 500 simulations were slightly, but not significantly, worse than the results from "fixed" pools. Two extreme situations were also tested, namely "easy" and "hard" pooling. The former grouped the most active compounds in the same pool, the second most active group of compounds in a second pool, and so on, and this "easy pooling" was repeated after each iteration. Even when the best compounds are grouped together from the beginning, the results on the 9-mer and 6-mer were not significantly different from the "mix and split" pooling, showing the contribution of the two-fold error included in the calculation "equalized" the quality of the deconvoluted structures. Hard pooling was designed to make the identification of the best binder as difficult as possible, grouping all the best binders in different pools together with the less active individuals and grouping the other good (but not best) binders in the same pool, and repeating hard pooling for each iteration. The results were significantly worse than for any other pooling technique (for the 9-mer around 10% of best binder identification, compared with 45% to 60% for the other techniques), but this is a very unlikely situation to occur for real libraries because it needs repeated grouping of the most active compound with the least active compounds.

The conclusion for iterative techniques was that their quality was comparable, except for hard pooling, and their reliability was generally high for selecting good binders, as seen from the selection of good binders in 97% of the simulations for the difficult 6-mer target. Mutational SURF could be an easy method to further improve the quality of the selected structure(s).

Positional scanning was also examined, and results were much worse than for iterative deconvolution. For the 9-mer target, identification of the best binder only happened in 35% of the simulations, while 23% gave compounds at least three orders of magnitude less active than the best binders (a situation never observed either for iterative or for hard pooling). For the 6-mer target, a good binder was identified only in 20% of the experiments (97% for iterative deconvolution). This was due in part to the large number of subsets with similar activities, especially for the 6-mer target, but mainly to the two-fold experimental error introduced which gave equal, or reversed, measured affinities for most of the subsets.

Wilson-Lingardo *et al.* [125], in a related communication, applied this comparative approach to a real 729-member trimeric library of nucleic acid-like molecules. The monomers used are represented in Figure 8.16. The library was built coupling each monomer with the resin, then pooling together the subsets P, Q and R (see Figure 8.16). Each of the pools was reacted twice with the nine monomers using the "mix and split" technique, obtaining $81 \times 3\,s \times 3 = 729$ trimers, which were submitted to different direct deconvolution procedures. With the first position being partially fixed (P, Q and R), two different "mix and split" syntheses were performed where either the second or the third position was fixed and the other was randomized. This produced two sublibraries of 27 pools, each containing 27 compounds, which were tested in the PLA2 assay. Iterative deconvolution was then performed through a first (three compounds per well, nine pools) and a second iteration (three individuals). Both libraries produced GGG as the most active compound, confirming the independence of the starting fixed position for iterative deconvolution. The initial 27 pools were also further pooled to give nine pools of 81 compounds, and their iterative deconvolution (four steps rather than three) produced GGG again illustrating a relative insensitivity to pool size for this technique.

Random pooling was also examined: 243 pools of three compounds (P, Q, and R) were prepared and randomly mixed to give 27 pools of 27 compounds

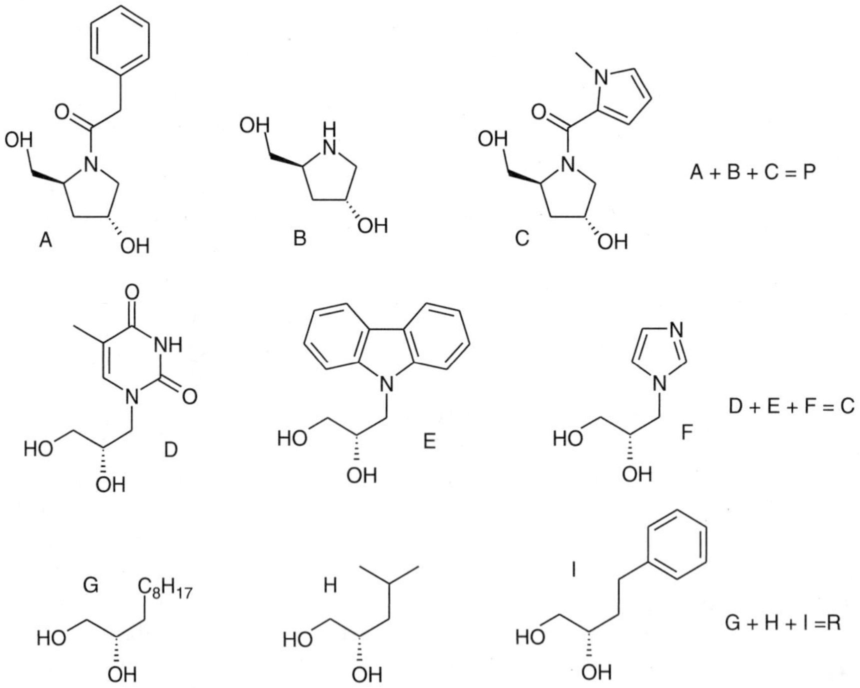

FIGURE 8.16 Monomers used for trimeric acid-like molecules.

each, and iterative deconvolution produced compound GGG. Finally, hard pooling was carried out, grouping the structure GGR (containing GGG) in a pool where no additional G-containing compounds were present, while GGQ, GGP and seven other G-containing compounds were grouped in another pool. Iterative deconvolution of this library produced individuals from the GGQ pool, thus confirming the worst possible outcome for the hard pooling technique. However, considering that "mix and split" could not produce such a "bad" pooling for GGG, the general extreme reliability of iteration techniques for deconvolution was proved (GGG resulted, in fact, to be the most active individual when the library was prepared as a discrete collection).

Positional scanning was also examined, where the first sublibrary included only three subsets (XXP, XXR, and XXQ) due to the library construction and the other two included nine subsets (XNX, and XXN respectively, with N = determined monomer). GGR was the result of the deconvolution, and the increased reliability compared to the theoretical experiment [123] was attributed to the simplicity of this "real" library, while for more complex situations iterative deconvolution should still perform better than positional scanning.

Very recently, Konings [126] presented a similar analysis to the previously reported [124], using the same library and oligonucleotidic targets. The authors also added to the compared iterative and non-iterative techniques procedures with overlapping subsets, namely orthogonal pooling [45], bogus coin pooling [46], and subtractive pooling [30]. While iterative deconvolution and positional scanning were comparable and performed well in different simulations, the less demanding overlapping strategies were generally producing worse results in the deconvolution process.

Stankova *et al.* [127] compared iterative deconvolution and on-bead screening for a tetrapeptide library of 927,360 compounds tested for inhibitor activity on factor Xa (384 pools of 2415 compounds each). This library contained a few known active structure in some pools, one of them being Ac-2Nal-Chg-Arg, with a $K_I = 270$ nM. Two pools of the library showed a significantly better activity than the pool containing the above-mentioned inhibitor. Deconvolution of one of them produced the compound 1–2–3–4 (structure not yet released for publication) with a $K_I = 650$ nM. The selection of compounds with activity lower than the known standard was also observed selecting two out of the four residues, both for the known and the unknown inhibitor, and keeping them fixed while randomizing the other two (sublibraries Ac-2Nal-X-X, X-X-Chg-Arg, 1-2-X-X, and X-X-3-4). For all of them the iterative deconvolution produced different compounds (only the % inhibition at 100 nM was given, and it varied from 11% for 1-2-X-X to 45% for X-X-Chg-Arg). On-bead screening obviously selected the best compounds out of thc library, being the assay related to single compounds rather than to partially defined mixtures. Unfortunately, the authors do not specify the number and the quality of the positives obtained from on-bead screening and do not compare the inhibitory potencies measured for a series of compounds as individuals in solution and bound on solid support in the on-bead screening.

VI. SUMMARY AND OUTLOOK

Direct deconvolution in general, and the techniques presented here in particular, have proved their usefulness in the fast structure determination of active compounds from pool libraries. Alternative strategies, such as parallel synthesis for small-to medium-size libraries and encoding for large bead-based pool libraries, may be competitive or even better choices for the chemist, but pooling methods are nevertheless very appealing for their independence from sophisticated analytical techniques and for the easy preparation and deconvolution of large libraries with little or no use of automation, which is often required for large numbers of discretes and for encoding techniques. Small laboratories with limited resources may especially benefit from these cost-effective approaches, once the library design and the direct deconvolution technique have been carefully selected.

Another interesting aspect is the small number of tests required for the analysis of a library. For example, a 10,000-member four-step library ($10 \times 10 \times 10 \times 10$ monomers) will require 10000 biological experiments when prepared as a library of discretes but only 40 tests when prepared and deconvoluted, either with iterative deconvolution or with positional scanning, with no additional decoding steps to be carried out. This may be of particular importance when the biological test is not suitable for automation (low throughput) and the evaluation of a large number of compounds is desirable to determine some preliminary SAR to be used for the design of focused libraries.

Despite all the *caveat* outlined in the previous chapter regarding on-bead screening, the enormous potential of libraries which could be tested on an indefinite number of on-bead assays, related to many receptors and to many therapeutic fields, makes this technique particularly appealing. More data for its application to small organic molecule libraries could be useful in evaluating the general applicability of this method. The fast progress of analytical techniques, of biological assay development and of polymer science applied to new solid supports, will probably drive researchers to consider this screening method more carefully and to validate its utility, either as such or coupled with encoding, in the near future for small organic molecule libraries. The scarcity of reported experimental applications is common for all the deconvolution techniques, for which mostly data regarding oligomeric libraries (peptides, oligonucleotides, peptidomimetics, etc.) are available. A larger panel of results using one or more or these techniques for the structure determination of active compounds from large pool combinatorial libraries will help in assessing their impact in future combinatorial chemistry efforts.

BIBLIOGRAPHY

1. (a) Fruchtel JS, Jung G, Organic chemistry on solid supports, *Angew. Chem. Int. Ed. Eng.*, 35:17–42, 1996. (b) Hermkens PHH, Ottenheijm HCJ, Rees D Solid-phase organic reactions: a review of the recent literarure, *Tetrahedron*,

52:4527–4554, 1996. (c) Hermkens PHH, Ottenheijm HCJ, Rees DC Solid-phase organic reactions II: a review of the literature Nov 95–Nov 96, *Tetrahedron*, 53:5643–5678, 1997. (d) Sucholeiki I, Solid-phase methods in combinatorial chemistry, in *Combinatorial Chemistry: Synthesis and Application* (Eds. Wilson SR, Czarnik AW), pp. 119–133, John Wiley & Sons, New York. (e) James IW, A compendium of solid-phase chemistry publications, in *Annual Reports in Combinatorial Chemistry and Molecular Diversity* (Eds. Moos WH, Pavia MR, Kay BK, Ellington AD), pp. 326–344, 1997, ESCOM Science publishers, Leiden.

2. (a) Storer R, Solution-phase synthesis in combinatorial chemistry: Applications in drug discovery, *Drug Disc. Today*, 1, 248–254, 1996. (b) Coe DM, Storer R, Solution-phase combinatorial chemistry, in *Annual Reports in Combinatorial Chemistry and Molecular Diversity* (Eds. Moos WH, Pavia MR, Kay BK, Ellington AD), pp. 50–58, 1997, ESCOM Science Publishers, Leiden. (c) Gravert, DJ, Janda KD, Organic synthesis on soluble polymer supports: Liquid-phase methodologies, *Chem. Rev.*, 97:489–510, 1997. (d) Kaldor SW, Siegel MG, Combinatorial chemistry using polymer-supported reagents, *Curr. Opin. Chem. Biol.*, 1:101–106, 1997.
3. (a) Selway CN, Terrett NK, Parallel-compound synthesis: methodology for accelerating drug discovery, *Bioorg. Med. Chem.*, 4:645–654, 1996. (b) Sashar S, Mjalli AMM, Techniques for single-compound synthesis, in *Annual Reports in Combinatorial Chemistry and Molecular Diversity* (Eds. Moos WH, Pavia MR, Kay BK, Ellington AD), pp. 19–29, 1997, ESCOM Science publishers, Leiden. (c) Cargill JF, Lebl M, New methods in combinatorial chemistry: Robotics and parallel synthesis, *Curr. Opin. Chem. Biol.*, 1:67–71, 1997.
4. (a) Lam KS, Lebl M, Krchnak V, The "one-bead one-compound" combinatorial library method, *Chem. Rev.*, 97:411–448, 1997. (b) Kiely JS, Techniques for mixture synthesis, in *Annual Reports in Combinatorial Chemistry and Molecular Diversity* (Eds. Moos WH, Pavia MR, Kay BK, Ellington AD), pp. 6–18, 1997, ESCOM Science publishers, Leiden.
5. (a) Balkenhohl F, von den Bussche-Hannefeld C, Lansky A, Zechel C, Combinatorial synthesis of small organic molecules, *Angew. Chem. Int. Ed. Eng.*, 35:2289–2337, 1996. (b) Thompson LA, Ellman JA, Synthesis and application of small molecule libraries, *Chem. Rev.*, 96:555–600, 1996. (c) Baldwin JJ, Henderson I, Recent advances in the generation of small-molecule combinatorial libraries: Encoded split synthesis and solid-phase synthetic methodology, *Med. Res. Rev.*, 16:391–405, 1996.
6. (a) Janda KD, Tagged versus untagged libraries: Methods for the generation and screening of combinatorial chemical libraries, *Proc. Natl. Acad. Sci. USA*, 91:10779–10785, 1994. (b) Baldwin JJ, Dolle RE, Deconvolution methods in solid-phase synthesis, in *Annual Reports in Combinatorial Chemistry and Molecular Diversity* (Eds. Moos WH, Pavia MR, Kay BK, Ellington AD), pp. 287–297, 1997, ESCOM Science publishers, Leiden. (c) Xiao X Y, Nova MP, Radiofrequency encoding and additional techniques for the structure elucidation of synthetic combinatorial libraries, in *Combinatorial Chemistry: Synthesis and Application* (Eds. Wilson SR, Czarnik AW), pp. 135–152, 1997, John Wiley & Sons, New York. (d) Czarnik AW, Encoding methods for combinatorial chemistry, *Curr. Opin. Chem. Biol.*, 1:60–66, 1997. (e) Maehr H, Combinatorial chemistry in drug research from a new vantage point, *Bioorg. Med. Chem.*, 5:473–491, 1997.

7. (a) Tarby CM, Cheng S, Boger DL, A solution-phase strategy for the parallel synthesis of chemical libraries containing small organic molecules: A general dipeptide mimetic and a flexible general template, in *Molecular Diversity and Combinatorial Chemistry* (Eds. Chaiken IW, Janda KD), pp. 81–98, 1996, American Chemical Society, Washington. (b) Hanessian S, Yang RY, Solution and solid phase synthesis of 5-alkoxyhydantoin libraries with a three-fold functional diversity, *Tetrahedron Lett.*, 37:5835–5838, 1996. (c) Shuker AJ, Siegel MG, Matthews DP, Weigel LO, The application of high-throughput synthesis and purification to the preparation of ethanolamines, *Tetrahedron Lett.*, 38:6149–6152, 1997.
8. (a) Baldino CM, Casebier DS, Caserta J, Slobodkin G, Tu C, Coffen DL, Convergent parallel synthesis, *Synlett*, 488–490, 1997. (b) Pop IE, Deprez BP, Tartar AL, Versatile acylation of N-nucleophiles using a new polymer-supported 1-hydroxybenzotriazole derivative, *J. Org. Chem.*, 62:2594–2603, 1997. (c) DeWitt SH, Czarnik AW, Parallel organic synthesis using Parke-Davis Diversomer method, in *High Throughput Screening: The Discovery of Bioactive Substances* (Ed. JP Devlin), pp. 191–208, 1997, Marcel Dekker Inc., New York.
9. (a) Geysen HM, Rodda SJ, Mason TJ, *A priori* delineation of a peptide which mimics a discontinuous antigenic determinant, *Mol. Immunol.*, 23:709–715, 1986. (b) Geysen HM, Rodda SJ, Mason TJ, Tribbick G, Schoofs PG, Strategies for epitope analysis using peptide synthesis, *J. Immunol. Methods*, 102:259–274, 1987. (c) Schoofs PG, Geysen HM, Jackson DC, Brown LE, Tang XL, White DO, Epitopes of an influenza viral peptide recognized by antibody at single amino acid resolution, *J. Immunol.*, 140:611–616, 1988.
10. (a) Houghten RA, Pinilla C, Blondelle SE, Appel JR, Dooley CT, Cuervo JH, Generation and use of synthetic peptide combinatorial libraries for basic research and drug discovery, *Nature* (London), 354:84–86, 1991. (b) Houghten RA, Appel JR, Blondelle SE, Cuervo JH, Dooley CT, Pinilla C, The use of synthetic peptide combinatorial libraries for the identification of bioactive peptides, *BioTechniques*, 13:412–421, 1992.
11. (a) Ecker DJ, Vickers TA, Hanecak R, Driver V, Anderson K, Rational screening of oligonucleotide combinatorial libraries for drug discovery, *Nucleic Acids Res.*, 21:1853–1856, 1993. (b) Wyatt JR, Vickers TA, Roberson JL, Buckheit RW, Klimkait T, DeBaets E, Davis PW, Rayner B, Imbach JL, Ecker DJ, Combinatorially selected guanosine-quartet structure is a potent inhibitor of human immunodeficiency virus envelope-mediated cell fusion, *Proc. Natl. Acad. Sci. USA*, 91:1356–1360, 1994.
12. (a) Furka A, Sebestyen F, Asgedom M, Dibo G, More peptides by less labour, *Abst. of* X^{th} *Int. Sym. on Med. Chem.*, Budapest, p. 288, 1988. (b) Furka A, Sebestyen F, Asgedom M, Dibo G, General method for rapid synthesis of multicomponent peptide mixtures, *Int. J. Pept. Protein Res.*, 37:487–493, 1991.
13. Studer A, Hadida S, Ferritto R, Kim SY, Jeger P, Wipf P, Curran DP, Fluorous synthesis: a fluorous-phase strategy for improving separation efficiency in organic synthesis, *Science* (Washington, DC), 275:823–826, 1997.
14. Kaldor SW, Siegel MG, Combinatorial chemistry using polymer-supported reagents, *Curr. Opin. Chem. Biol.*, 1:101–106, 1997.
15. Armstrong RW, Brown SD, Keating TA, Tempest PA, Combinatorial synthesis exploiting multiple-component condensations, microchip encoding and resin capture, in *Combinatorial Chemistry: Synthesis and Application* (Eds. Wilson SR, Czarnik AW), pp. 153–190, 1997, John Wiley & Sons, New York.

16. (a) Siegel MG, Hahn PJ, Dressman BA, Fritz JE, Grunwell JR, Kaldor SW, Rapid purification of small molecule libraries by ion exchange chromatography, *Tetrahedron Lett.*, 38:3357–3360, 1997. (b) Gayo LM, Suto MJ, Ion-exchange resins for solution phase parallel synthesis of chemical libraries, *Tetrahedron Lett.*, 38:513–516, 1997.
17. (a) Geysen HM, Meloen RH, Barteling SJ, Use of peptide synthesis to probe viral antigens for epitopes to a resolution of a single amino acid, *Proc. Natl. Acad. Sci. USA*, 81:3998–4002, 1984. (b) Houghten RA, Pinilla C, Blondelle SE, Appel JR, Dooley CT, Cuervo JH, Generation and use of synthetic peptide combinatorial libraries for basic research and drug discovery, *Nature* (London), 354:84–86, 1991.
18. Griffith MC, Dooley CT, Houghten RA, Kiely JS, Solid-phase synthesis, characterization, and screening of a 43,000-compound tetrahydroisoquinoline combinatorial library, in *Molecular Diversity and Combinatorial Chemistry* (Eds Chaiken IW, Janda KD), pp. 50–57, 1996, American Chemical Society, Washington.
19. Houghten RA, General method for the rapid solid-phase synthesis of large numbers of peptides: Specificity of antigen-antibody interaction at the level of individual amino acids, *Proc. Natl. Acad. Sci. USA.*, 82:5131–5135, 1985.
20. Murphy MM, Schullek JR, Gordon EM, Gallop MA, Combinatorial organic synthesis of highly functionalized pyrrolidines: Identification of a potent angiotensin converting enzyme inhibitor from a mercaptoacyl proline library, *J. Am. Chem. Soc.*, 117:7029–7030, 1995.
21. Patel DV, Gordeev MF, England BP, Gordon EM, Solid-phase and combinatorial synthesis of heterocyclic scaffolds: Dihydropyridines, pyridines, and pyrido[2,3-d]pyrimidines, in *Molecular Diversity and Combinatorial Chemistry* (Eds Chaiken IW, Janda KD), pp. 58–69, 1996, American Chemical Society, Washington.
22. Gordon DW, Steele J, Reductive alkylation on a solid phase: Synthesis of a piperazinedione combinatorial library, *Bioorg. Med. Chem. Lett.*, 5:47–50, 1995.
23. Dankwardt SM, Phan TM, Krstenansky JL, Combinatorial synthesis of small-molecule libraries using 3-amino-5-hydroxybenzoic acid, *Mol. Diversity*, 1:113–20, 1996.
24. Look GC, Schullek JR, Holmes CP, Chinn JP, Gordon EM, Gallop MA, The identification of cyclooxygenase-1 inhibitors from 4-thiazolidinone combinatorial libraries, *Bioorg. Med. Chem. Lett.*, 6:707–712, 1996.
25. Chng BL, Ganesan A, Solution-phase synthesis of a beta-amino alcohol combinatorial library, *Bioorg. Med. Chem. Lett.*, 7:1511–1514, 1997.
26. Maehr H, Yang R, Structure optimization of a leukotriene D-4 antagonist by combinatorial chemistry in solution, *Bioorg. Med. Chem.*, 5:493–496, 1997.
27. Tian J, Coates GW, Development of a diversity-based approach for the discovery of stereoselective polymerization catalysts: Identification of a catalyst for the synthesis of syndiotactic polypropylene, *Angew. Chem. Int. Ed. Eng.*, 39:3626–3629, 2000.
28. Erb E, Janda KD, Brenner S, Recursive deconvolution of combinatorial chemical libraries, *Proc. Natl. Acad. Sci. USA*, 91:11422–11426, 1994.
29. Sutherlin DP, Armstrong RW, Synthesis of 12 stereochemically and structurally diverse C-trisaccharides, *J. Org. Chem.*, 62:5267–5283, 1997.

30. (a) Carell T, Wintner EA, Bashir-Hashemi A, Rebek, J Jr, A novel procedure for the synthesis of libraries containing small organic molecules, *Angew. Chem. Int. Ed. Eng.*, 33:2059–2061, 1994. (b) Carell T, Wintner EA, Rebek J Jr, A solution-phase screening procedure for the isolation of active compounds from a library of molecules, *Angew. Chem. Int. Ed. Eng.*, 33:2061–2064, 1994. (c) Carell T, Wintner EA, Sutherland AJ, Rebek J Jr, Dunayevskiy YM, Vouros P, New promise in combinatorial chemistry: Synthesis, characterization, and screening of small-molecule libraries in solution, *Chemistry & Biology*, 2:171–183, 1995.
31. (a) Shipps GW Jr, Spitz UP, Rebek J Jr, Solution-phase generation of tetraurea libraries, *Bioorg. Med. Chem.*, 4:655–657, 1996. (b) Shipps GW Jr, Pryor KE, Xian J, Skyler DA, Davidson EH, Rebek JJr, Synthesis and screening of small molecule libraries active in binding to DNA, *Proc. Natl. Acad. Sci. USA*, 94:11833–11838, 1997. (c) Wintner EA, Rebek J Jr, Combinatorial libraries in solution: Polyfunctionalized core molecules, in *Combinatorial Chemistry: Synthesis and Application* (Eds. Wilson SR, Czarnik AW), pp. 95–117, 1997, John Wiley & Sons, New York.
32. Kung PP, Lingardo L, Greig M, Wyatt J, Cook DP, Antibacterial activity of tertiary nitrogen-based combinatorial libraries prepared by solution phase synthesis, *Book of Abstracts, 213th ACS National Meeting*, San Francisco, April 13–17, 1997, American Chemical Society, Washington.
33. De Muynck H, Madder A, Farcy N, De Clercq PJ, Perez-Payan MN, Ohberg LM, Davis AP, Application of combinatorial procedures in the search for serin-protease-like activity with focus on the acyl transfer step, *Angew. Chem. Int. Ed. Eng.*, 39:145–148, 2000.
34. (a) Armstrong RW, Combs AP, Tempest PA, Brown DS, Keating TA, Multiple-component condensation strategies for combinatorial library synthesis, *Acc. Chem. Res.*, 29:123–131, 1996. (b) Ugi I, Multi-component reactions (MCR), Part 1. Perspectives of multi-component reactions and their libraries, *J. Prakt. Chem./Chem.- Ztg.*, 339:499–516, 1997.
35. (a) Pinilla C, Appel JR, Blanc P, Houghten RA, Rapid identification of high affinity peptide ligands using positional scanning synthetic peptide combinatorial libraries, *BioTechniques*, 13:901–902, 904–905, 1992. (b) Dooley CT, Houghten RA, The use of positional scanning synthetic peptide combinatorial libraries for the rapid determination of opioid receptor ligands, *Life Sci.*, 52:1509–1517, 1993.
36. Berk SC, Chapman KT, Spatially arrayed mixture (SPAM) technology: Synthesis of two-dimensionally indexed orthogonal combinatorial libraries, *Bioorg. Med. Chem. Lett.*, 7:837–842, 1997.
37. Backes BJ, Virgilio AA, Ellman JA, Activation method to prepare a highly reactive acylsulfonamide "safety-catch" linker for solid-phase synthesis, *J. Am. Chem. Soc.*, 118:3055–3056, 1996.
38. (a) Pinilla C, Buencamino J, Appel JR, Hopp TP, Houghten RA, Mapping the detailed specificity of a calcium-dependent monoclonal antibody through the use of soluble positional scanning combinatorial libraries: Identification of potent calcium-independent antigens, *Mol. Diversity*, 1:21–28, 1995. (b) Pinilla C, Appel JR, Houghten RA, Tea bag synthesis of positional scanning synthetic combinatorial libraries and their use for mapping antigenic determinants, *Methods Mol. Biol.*, 66:171–179, 1996. (c) Houghten RA, Heterocyclic positional scanning combinatorial libraries, *Book of Abstracts, 213th ACS*

National Meeting, San Francisco, April 13–17, 1997, American Chemical Society, Washington.
39. Smith PW, Lai JYQ, Whittington AR, Cox B, Houston JG, Synthesis and biological evaluation of a library containing potentially 1600 amides/esters, A strategy for rapid compound generation and screening, *Bioorg. Med. Chem. Lett.*, 4:2821–2824, 1994.
40. Andrus MB, Turner TM, Sauna ZE, Ambudkar SV, The synthesis and evaluation of a solution phase indexed combinatorial library of non-natural polyenes for reversal of P-glycoprotein mediated multidrug resistance, *J. Org. Chem.*, 65:4973–4983, 2000.
41. Leone-Bay A, Freeman J, O'Toole D, Rosado-Gray C, Salo-Kostmayer S, Tai M, Mercogliano F, Baughman RA, Studies directed at the use of a parallel synthesis matrix to increase throughput in an *in vivo* assay, *J. Med. Chem.*, 43:3573–3576, 2000.
42. Pirrung MC, Chen J, Preparation and screening against acetylcholinesterase of a non-peptide "indexed" combinatorial library, *J. Am. Chem. Soc.*, 117:1240–1245, 1995.
43. Pirrung MC, Chau JHL, Chen J, Discovery of a novel tetrahydroacridine acetylcholinesterase inhibitor through an indexed combinatorial library, *Chemistry & Biology*, 2:621–626, 1995.
44. Freier SM, Konings DAM, Wyatt JR, Ecker DJ, "Mutational SURF": A strategy for improving lead compounds identified from combinatorial libraries, *Bioorg. Med. Chem.*, 4:717–725, 1996.
45. Déprez, B, Williard X, Bourel L, Coste H, Hyafil F, Tartar A, Orthogonal combinatorial chemical libraries, *J. Am. Chem. Soc.*, 117:5405–5406, 1995.
46. Blake J, Litzi-Davis L, Evaluation of peptide libraries: An iterative strategy to analyze the reactivity of peptide mixtures with antibodies, *Bioconjugate Chem.*, 3:510–513, 1992.
47. Boger DL, Chai W, Qing J, Multistep convergent solution-phase combinatorial synthesis and deletion synthesis deconvolution, *J. Am. Chem. Soc.*, 120:7220–7225, 1998.
48. Boger DL, Lee JK, Goldberg J, Jin Q, Two comparisons of the performance of positional scanning and deletion synthesis for the identification of active constituents in mixture combinatorial libraries, *J. Org. Chem.*, 65:1467–1474, 2000.
49. Lam KS, Salmon SE, Hersh EM, Hruby VJ, Kazmierski WM, Knapp RJ, A new type of synthetic peptide library for identifying ligand-binding activity, *Nature* (London), 354:82–84, 1991.
50. Dhalluin C, Boutillon C, Tartar A, Lippens G, Magic angle spinning nuclear magnetic resonance in solid-phase peptide synthesis, *J. Am. Chem. Soc.*, 119:10494–10500, 1997.
51. Brummel CL, Vickerman JC, Carr SA, Hemling ME, Roberts GD, Johnson W, Weinstock J, Gaitanopoulos D, Benkovic SJ, Winograd N, Evaluation of mass spectrometric methods applicable to the direct analysis of non-peptide bead-bound combinatorial libraries, *Anal. Chem.*, 68:237–242, 1996.
52. Gosselin F, Di Renzo M, Ellis TH, Lubell WD, Photoacoustic FTIR spectroscopy, a nondestructive method for sensitive analysis of solid-phase organic chemistry, *J. Org. Chem.*, 61:7980–7981, 1996.

53. Sepetov N, Issakova O, Analytical characterization of synthetic organic libraries, in *Combinatorial Chemistry and Combinatorial Technologies: Methods and Applications* (Eds. Fassina G, Miertus S), 237–268, 1998, 2005, Chapter 10, this volume.
54. van Breemen RB, Huang CR, Nikolic D, Woodbury CP, Zhao YZ, Venton DL, Pulsed ultrafiltration mass spectrometry: A new method for screening combinatorial libraries, *Anal. Chem.*, 69:2159–2164, 1997.
55. Youngquist RS, Fuentes GR, Lacey MP, Keough T, Matrix-assisted laser desorption ionization for rapid determination of the sequences of biologically active peptides isolated from support-bound combinatorial peptide libraries, *Rapid Commun. Mass Spectrom.*, 8:77–81, 1994.
56. Brummel CL, Lee INW, Zhou Y, Benkovic SJ, Winograd N, A mass spectrometric solution to the address problem of combinatorial libraries, *Science*, 264:399–402, 1994.
57. Wieboldt R, Zweigenbaum J, Henion J, Immunoaffinity ultrafiltration with ion spray HPLC/MS for screening small-molecule libraries, *Anal. Chem.*, 69:1683–1691, 1997.
58. Dunayevskiy YM, Lai JJ, Quinn C, Talley F, Vouros P, Mass spectrometric identification of ligands selected from combinatorial libraries using gel filtration, *Rapid Commun. Mass Spectrom.*, 11:1178–1184, 1997.
59. Chu YH, Dunayevskiy YM, Kirby DP, Vouros P, Karger BL, Affinity capillary electrophoresis-mass spectrometry for screening combinatorial libraries, *J. Am. Chem. Soc.*, 118:7827–7835, 1996.
60. Cheng CC, Chu YH, Affinity capillary electrophoresis-mass spectrometry in combinatorial library screening, *Am. Lab.*, 30:79–81, 1998.
61. Chu YH, Cheng CC, Affinity capillary electrophoresis in biomolecular recognition, *Cell. Mol. Life Sci.*, 54:663–683, 1998.
62. Blom KF, Larsen BS, McEwen CN, Determining affinity-selected ligands and estimating binding affinities by online size exclusion chromatography/liquid chromatography-mass spectrometry, *J. Comb. Chem.*, 1, 82–90, 1999.
63. Davis RG, Anderegg RJ, Blanchard SG, Iterative size exclusion chromatography coupled with liquid chromatographic mass spectrometry to enrich and identify tight binding ligands from complex mixtures, *Tetrahedron*, 55:11653–11667, 1999.
64. Lyubarskaya YY, Carr SA, Dunnington D, Prichett WP, Fisher SM, Appelbaum ER, Jones CS, Karger BL, Screening for high affinity ligands to the Src SH2 domain using capillary isoelectric focusing-electrospray ionization ion trap mass spectrometry, *Anal. Chem.*, 70:4761–4770, 1998.
65. Kelly MA, Liang H, Sytwu II, Vlattas I, Lyons NL, Bowen BR, Wennogle LP, Characterization of SH2-Ligand interactions via library affinity selection with mass spectrometric, detection *Biochemistry*, 35:11747–11755, 1996.
66. Bruce JE, Anderson GA, Chen R, Cheng X, Gale DC, Hofstadler SA, Schwartz BL, Smith RD, Bio-affinity characterization mass spectrometry, *Rapid Commun. Mass Spectrom.*, 9:644–650, 1995.
67. Brummel CL, Vickerman JC, Carr SA, Hemling ME, Roberts GD, Johnson W, Weinstock J, Gaitanopoulos D, Benkovic SJ, Winograd N, Evaluation of mass spectrometric methods applicable to the direct analysis of non-peptide bead-bound combinatorial libraries, *Anal. Chem.*, 68:237–242, 1996.

68. Hsieh F, Keshishian H, Muir C, Automated high throughput multiple target screening of molecular libraries by microfluidic MALDI-TOFMS. *J. Biomol. Screening*, 3:189–198, 1998.
69. Siegel MM, Tabei K, Bebernitz GA, Baum EZ, Rapid methods for screening low molecular mass compounds non covalently bound to proteins using size exclusion and mass spectrometry applied to inhibitors of human cytomegalovirus protease, *J. Mass Spectrom.*, 33:264–273, 1998.
70. Holtzapple CK, Stanker LH, Affinity selection of compounds in a fluoroquinone chemical library by on-line immunoaffinity deletion coupled to column HPLC, *Anal. Chem.*, 70:4817–4821, 1998.
71. Youngquist RS, Fuentes GR, Miller CM, Ridder GM, Lacey MP, Keough T, *Adv. Mass Spectrometry*, 14:423–448, 1998.
72. Sussmuth RD, Jung G, Impact of mass spectrometry on combinatorial chemistry, *J. Chromatogr., B: Biomed. Sci. Applic.*, 725:49–65, 1999.
73. Woodbury CP, Jr, Venton DL, Methods of screening combinatorial libraries using immobilized or restrained receptors, *J. Chromatogr., B: Biomed. Sci. Applic.*, 725:113–137, 1999.
74. Eliseev AV, Emerging approaches to target assisted screening of combinatorial mixtures, *Curr. Opin. Drug Discovery Dev.*, 1:106–115, 1998.
75. Shuker SB, Hajduk PJ, Meadows RP, Fesik SW, Discovering high-affinity ligands for proteins: SAR by NMR, *Science*, 274:1531–1534, 1996.
76. Hajduk PJ, Meadows RP, Fesik SW, Discovering high affinity ligands for proteins, *Science*, 278:498–499, 1997.
77. Marek D, Poster presented at *39th Experimental Nuclear Magnetic Resonance Conference,* Asilomar, CA, p. 227, 1998.
78. Hajduk PJ, Dinges J, Miknis GF, Merlock M, Middleton T, Kempf DJ, Egan DA, Walter KA, Robins TS, Shuker SB, Holzman TF, Fesik SW, NMR-based discovery of lead inhibitors that block DNA binding of the human papillomavirus E2 protein, *J. Med. Chem.*, 40:3144–3150, 1997.
79. Hajduk PJ, Sheppard G, Nettesheim DG, Olejniczak ET, Shuker SB, Meadows RP, Steinman DH, Carrera GM, Marcotte PA, Severin J, Walter K, Smith H, Gubbins E, Simmer R, Holzman TF, Morgan DW, Davidsen SK, Fesik SW, Discovery of potent nonpeptide inhibitors of stromelysin using SAR by NMR, *J. Am. Chem. Soc.*, 119:5818–5827, 1997.
80. Hajduk PJ, Gerfin T, Boehlen JM, Haeberli M, Marek D, Fesik SW, High-throughput nuclear magnetic resonance-based screening, *J. Med. Chem.*, 42:2315–2317, 1999.
81. Hajduk PJ, Dinges J, Schkeryantz JM, Janowick D, Kaminski M, Tufano M, Augeri DJ, Petros A, Nienaber V, Zhong P, Hammond R, Coen M, Beutel B, Katz L, Fesik SW, Novel inhibitors of Erm methyltransferases from NMR and parallel synthesis, *J. Med. Chem.*, 42:3852–3859, 1999.
82. Lin M, Shapiro MJ, Wareing JR, Diffusion-edited NMR affinity NMR for direct observation of molecular interactions, *J. Am. Chem. Soc.*, 119:5249–5250, 1997.
83. Lin M, Shapiro MJ, Wareing JR, Screening mixtures by affinity NMR, *J. Org. Chem.*, 62:8930–8931, 1997.
84. Shapiro MJ, Chin J, Chen A, Wareing JR, Tang Q, Tommasi RA, Marepalli HR, Covalent or trapped? PFG diffusion MAS NMR for combinatorial chemistry, *Tetrahedron Lett.*, 40:6141–6143, 1999.

85. Bleicher K, Lin M, Shapiro MJ, Wareing JR, Diffusion edited NMR: Screening compound mixtures by affinity NMR to detect binding ligands to vancomycin, *J. Org. Chem.*, 63:8486–8490, 1998.
86. Anderson RC, Lin M, Shapiro MJ, Affinity NMR: Decoding DNA binding, *J. Comb. Chem.*, 1:69–72, 1999.
87. Hajduk PJ, Olejniczak ET, Fesik SW, One-dimensional relaxation- and diffusion-edited NMR methods for screening compounds that bind to macromolecules, *J. Am. Chem. Soc.*, 119:12257–12261, 1997.
88. Meyer B, Weimar T, Peters T, Screening mixtures for biological activity by NMR, *Eur. J. Biochem.*, 246:705–709, 1997.
89. Henrichsen D, Ernst B, Magnani JL, Wang WT, Meyer B, Peters T, Bioaffinity NMR spectroscopy: Identification of an E-selectin antagonist in a substance mixture by transfer NOE, *Angew. Chem. Int. Ed. Eng.*, 38:98–102, 1999.
90. Fejzo J, Lepre CA, Peng JW, Bemis GW, Ajay Murcko MA, Moore JM, The SHAPES strategy: An NMR-based approach for lead generation in drug discovery, *Chem. Biol.*, 6:755–769, 1999.
91. Comprehensive Medicinal Chemistry (CDC) database, version CDC3D98.1, MDL Information Systems, San Leandro, CA, US.
92. Bemis GW, Murcko MA, The properties of known drugs. 1. Molecular frameworks, *J. Med. Chem.*, 39:2887–2893, 1996.
93. Available Chemical Directory (ADC) database, version ACD98.2, MDL Information Systems, San Leandro, CA, US.
94. Chen A, Shapiro MJ, NOE pumping: a novel NMR technique for identification of compounds with binding affinity to macromolecules, *J. Am. Chem. Soc.*, 120:10258–10259, 1998.
95. Mayer M, Meyer B, Characterization of ligand binding by saturation transfer difference NMR spectroscopy, *Angew. Chem. Int. Ed. Eng.*, 38:1784–1788, 1999.
96. Klein J, Meinecke R, Mayer M, Meyer B, Detecting binding affinity to immobilized receptor proteins in compound libraries by HR-MAS STD NMR, *J. Am. Chem. Soc.*, 121:5336–5337, 1999.
97. Keifer PA, NMR tools for biotechnology, *Curr. Opin. Biotechnol.*, 10:34–41, 1999.
98. Moore JM, NMR screening in drug discovery, *Curr. Opin. Biotechnol.*, 10:54–58, 1999.
99. Gounarides JS, Chen A, Shapiro MJ, Nuclear magnetic resonance chromatography: applications of pulse gradient diffusion NMR to mixture analysis and ligand-receptor interactions, *J. Chromatogr., B: Biomed. Sci. Applic.*, 725:79–90, 1999.
100. Chen A, Shapiro MJ, Affinity NMR, *Anal. Chem.*, 71:669A–675A, 1999.
101. Lam KS, Lake D, Salmon SE, Smith J, Chen ML, Wade S, Abdul-Latif F, Leblova Z, Ferguson RD, Krchnak V, Sepetov NF, Lebl M, A one-bead one-peptide combinatorial library method for B-cell epitope mapping, *Methods: Meth. Enzymol.*, 9:482–493, 1996.
102. Lam KS, Lebl, M, Selectide technology: Bead-binding screening, *Methods: Meth. Enzymol.*, 6:372–380, 1994.
103. Salmon SE, Lam KS, Lebl M, Kandola A, Khattri PS, Wade S, Patek M, Kocis P, Krchnak V, Thorpe D, Felder S, Discovery of biologically active peptides in random libraries: Solution-phase testing after staged orthogonal release from resin beads, *Proc. Natl. Acad. Sci. USA*, 90:11708–11712, 1993.

104. Lam KS, Wade S, Abdul-Latif F, Lebl M, Application of a dual color detection scheme in the screening of a random combinatorial peptide library, *J. Immunol. Methods*, 180:219–223, 1995.
105. Lam KS, Lebl M, From multiple peptide synthesis to peptide libraries, in *Peptide and Non-Peptide Libraries: A Handbook for the Search of Lead Structures* (Ed. Jung G), pp. 173–201, 1996, VCH, Weinheim, Germany.
106. Lam KS, Enzyme-linked colorimetric screening of a one-bead one-compound combinatorial library, *Methods Mol. Biol.*, 87:7–12, 1998.
107. Lebl M, Krchnak V, Sepetov NF, Seligmann B, Strop P, Felder S, Lam KS, One-bead-one-structure combinatorial libraries, *Biopolymers*, 37:177–198, 1995.
108. Ohlmeyer MHJ, Swanson RN, Dillard L, Reader JC, Asouline G, Kobayashi R, Wigler M, Still W Clark, Complex synthetic chemical libraries indexed with molecular tags, *Proc. Natl. Acad. Sci. USA*, 90:10922–10926, 1993.
109. Boyce R, Li G, Nestler HP, Suenaga T, Still W Clark, Peptidosteroidal receptors for opioid peptides. Sequence-selective binding using a synthetic receptor library, *J. Am. Chem. Soc.*, 116:7955–7956, 1994.
110. Nestler HP, Sequence-selective nonmacrocyclic two-armed receptors for peptides, *Mol. Diversity*, 2:35–40, 1996.
111. Yu Z, Chu YH, Combinatorial epitope search: Pitfalls of library design, *Bioorg. Med. Chem. Lett.*, 7:95–98, 1997.
112. Hwang S, Tamilarasu N, Ryan K, Huq I, Richter S, Still W, Clark Rana TM, Inhibition of gene expression in human cells through small molecule-RNA interactions, *Proc. Natl. Acad. Sci. USA*, 96:12997–13002, 1999.
113. Needels MC, Jones DG, Tate EH, Heinkel GL, Kochersperger LM, Dower WJ, Barrett RW, Gallop MA, Generation and screening of an oligonucleotide-encoded synthetic peptide library, *Proc. Natl. Acad. Sci. USA*, 90:10700–10704, 1993.
114. St. Hilaire PM, Willert M, Juliano MA, Juliano L, Meldal M, Fluorescence-quenched solid phase combinatorial libraries in the characterization of cysteine protease substrate specificity, *J. Comb. Chem.*, 1:509–523, 1999.
115. Kassarjian A, Schellenberger V, Turck CW, Screening of synthetic peptide libraries with radiolabeled acceptor molecules, *Pept. Res.*, 6:129–133, 1993.
116. Nestler HP, Wennemers H, Sherlock R, Dong DLY, Microautoradiographic identification of receptor-ligand interactions in bead-supported combinatorial libraries, *Bioorg. Med. Chem. Lett.*, 6:1327–1330, 1996.
117. Wu J, Ma QN, Lam KS, Identifying substrate motifs of protein kinases by a random library approach, *Biochemistry*, 33:14825–14833, 1994.
118. Meldal M, Svendsen I, Breddam K, Auzanneau FI, Portion-mixing peptide libraries of quenched fluorogenic substrates for complete subsite mapping of endoprotease specificity, *Proc. Natl. Acad. Sci. USA*, 91:3314–3318, 1994.
119. Meldal M, Svendsen I, *J. Chem. Soc.*, *Perkin Trans.*, 1:1591–1596, 1995.
120. Lou Q, Leftwich ME, Lam KS, Identification of GIYWHHY as a novel peptide substrate for human p60c-src protein tyrosine kinase, *Bioorg. Med. Chem.*, 4:677–682, 1996.
121. Cho CY, Liu CW, Wemmer DE, Schultz PG, Cyclic and linear oligocarbamate ligands for human thrombin, *Bioorg. Med. Chem.*, 7:1171–1179, 1999.
122. St. Hilaire PM, Lowary TL, Meldal M, Bock K, Oligosaccharide mimetics obtained by novel, rapid screening of carboxylic acid encoded glycopeptide libraries, *J. Am. Chem. Soc.*, 120:13312–13320, 1998.

123. Konings DAM, Wyatt JR, Ecker DJ, Freier SM, Deconvolution of combinatorial libraries for drug discovery: theoretical comparison of pooling strategies, *J. Med. Chem.*, 39:2710–2719, 1996.
124. Bevington PR, Robinson DK, Data reduction and error analysis for the physical sciences, pp. 75–95, 1992, McGraw-Hill, San Francisco.
125. Wilson-Lingardo L, Davis PW, Ecker DJ, Hubert N, Acevedo O, Sprankle K, Brennan T, Schwarcz L, Freier SM, Wyatt JR, Deconvolution of combinatorial libraries for drug discovery: Experimental comparison of pooling strategies, *J. Med. Chem.*, 39:2720–2726, 1996.
126. Konings DAM, Wyatt JR, Ecker DJ, Freier SM, Strategies for rapid deconvolution of combinatorial libraries: Comparative evaluation using a model system, *J. Med. Chem.*, 40:4386–4395, 1997.
127. Stankova M, Strop P, Chen C, Lebl M, Mixtures of molecules versus mixtures of pure compounds on polymeric beads, in *Molecular Diversity and Combinatorial Chemistry* (Eds Chaiken IW, Janda KD), pp. 136–141, 1996, American Chemical Society, Washington.

9 Encoding Techniques for Pool Libraries of Small Organic Molecules

Pierfausto Seneci

CONTENTS

I. INTRODUCTION

This review will cover encoding techniques for small organic molecule pool libraries, one of the most challenging and powerful techniques in combinatorial chemistry. A short introduction to illustrate briefly the advantages in library encoding and to compare the method with other structure determination techniques will be given, but for a more detailed introduction the reader should look elsewhere in this book [1].

Encoding

Chemical methods	**Non chemical methods**
Nucleotide tags Peptide tags Haloaromatic tags Secondary amine tags Isotopic tags Autotagging Non covalently bound tags Other chemical methods	Optical encoding Radiofrequency encoding **Experiment comparison with direct deconvolution** **Summary and outlook**

FIGURE 9.1 Encoding methods.

Parallel synthesis [2], or discrete library synthesis, allows the identification of each library component simply by its location (each well is treated with a single monomer in each reaction step); direct deconvolution by pooling methods [3] determines active structures by measuring the biological activity of complex pools and deconvoluting the chemical structure of active components through the selection of active pools from various synthetic cycles, run in either parallel or sequentially; direct deconvolution by bioanalytical methods [4] or by on-bead screening [5] determines the chemical structure of active library components (selected according to the biological assay results) by analytical methods. These techniques, which do not require any additional construct to be added to the library structure for the selection of active compounds, have been extensively covered elsewhere in this book [1, 6].

Encoding provides structure determination for libraries through the reading of a code which represents unambiguously a single component of the library. The coding entity may be chemical, where the tag structure is read using various analytical techniques, as for nucleotide tags, peptide tags, haloaromatic tags, secondary amine tags, isotopic tags, autotagging, noncovalently bound tags and other methods; it may be nonchemical, as for optical encoding and radiofrequency encoding, where the tag structure is encoded and decoded without chemical methods. Each technique will be presented and evaluated through the detailed illustration of an example and relevant references will be given at the end of the paragraph; for more details regarding any of the examples, the reader should refer to the original paper. A comparison among different encoding and direct deconvolution techniques will follow, based on experimental results, and a final summary and outlook toward future development in this area will conclude this review. The aim of this chapter is to provide the reader with some helpful information to decide if, when and how to apply any of the encoding techniques to the synthesis of libraries of any format; the complementarity, rather than the mutual exclusion, of the different methods will be often highlighted.

II. CHEMICAL METHODS

Chemical encoding aims to prepare a "one bead–one compound" [5, 7] solid-phase library. At first sight a comparison reveals apparent drawbacks with chemically encoded libraries: more chemical steps, either before or in parallel with the library synthesis process, the need for orthogonal protection of tags and library sites during the synthesis, the "waste" of some resin sites for the coding entities with the possibility of interferences, either during the synthesis or during the biological assay of the coding entity. Despite all of this, robust and simple tags which allow automated structure analysis of the positives after the library screening have become very popular; in particular small organic molecule libraries, where the direct structure determination of the library component from a single bead is often difficult and time consuming, receive significant benefits from encoding techniques.

All the techniques which are presented here either have clearly proven their utility through many applications, or were recently reported but nevertheless look promising for future applications.

A. Nucleotide Tags

The first published work on encoding by Brenner and Lerner [8] used nucleotide tags for structure determination of a peptide library. A formal example was presented where two amino acids (Gly and Met) were used and, using a six-base code for each amino acid, all the different trimers were prepared. The example is shown in Figure 9.2.

The use of nucleotides as tags allowed automated synthesis, cloning and amplification of the coding signal by means of PCR (polymerase chain reaction), and the whole decoding sequence became easy, extremely sensitive and automated. The use of PCR required two oligonucleotide sequences to be

Gly = **CACATG**
Met = **ACGGTA**

Step 1:

GGGCCCTATTCTTAG-LINK

Step 2:

CACATGGGGCCCTATTCTTAG-LINK-*Gly*
A**CGGTAG**GGCCCTATTCTTAG-LINK-*Met*

Step 3:

CACATGCACATGGGGCCCTATTCTTAG-LINK-*Gly-Gly*
ACGGTACACATGGGGCCCTATTCTTAG-LINK-*Gly-Met*
CACATGACGGTAGGGCCCTATTCTTAG-LINK-*Met-Gly*
ACGGTAACGGTAGGGCCCTATTCTTAG-LINK-*Met-Met*

Step 4 and 5:

AGCTACTTCCCAAGG**CACATGCACATGCACATG**GGGCCCTATTCTTAG-LINK-*Gly-Gly-Gly*
AGCTACTTCCCAAGG**ACGGTACACATGCACATG**GGGCCCTATTCTTAG-LINK-*Gly-Gly-Met*
AGCTACTTCCCAAGG**CACATGACGGTACACATG**GGGCCCTATTCTTAG-LINK-*Gly-Met-Gly*
AGCTACTTCCCAAGG**ACGGTAACGGTACACATG**GGGCCCTATTATTAG-LINK-*Gly-Met-Met*
AGCTACTTCCCAAGG**CACATGCACATGACGGTA**GGGCCCTATTCTTAG-LINK-*Gly-Met-Met*
AGCTACTTCCCAAGG**ACGGTACACATGACGGTA**GGGCCCTATTCTTAG-LINK-*Met-Gly-Met*
AGCTACTTCCCAAGG**CACATGACGGTAACGGTA**GGGCCCTATTCTTAG-LINK-*Met-Met-Gly*
AGCTACTTCCCAAGG**ACGGTAACGGTAACGGTA**GGGCCCTATTCTTAG-LINK-*Met-Met-Met*

FIGURE 9.2 Use of nucleotides for structure determination of a peptide library.

inserted before and after the coding strand; step 1 consisted in the anchoring (3′ to 5′ direction) of the sequence to the linker/support construct (here named LINK) on the coding sites. In step 2, the resin was split into two portions and the two amino acids were added on the library sites, then the two coding hexanucleotidic sequences were coupled on the coding sites. Step 3 produced four different dipeptides *via* the "one bead–one compound" method [7] and step 4 completed the synthesis of eight tripeptides each, encoded by eighteen nucleotides; finally, step 5 fused the second PCR oligonucleotide strand to the coding sites and gave the final library (Figure 9.2).

In recent examples [9] the protecting groups used were usually Fmoc for the peptide strand and dimethoxytrityl (DMT) for the coding strand. The two protecting groups were incorporated on a single serin linker, where the NH_2 function was used for peptide synthesis and the OH for nucleotide synthesis; the synthetic protocols for Fmoc–peptide chemistry and DMT–oligonucleotide chemistry were successfully adapted to the simultaneous synthesis of the two oligomeric chains.

A recent modification of this technique by Fenniri *et al.* [10] presented the encoded reaction cassette, where the detection of bond-breaking or bond-forming reactions *via* oligonucleotide codes and PCR signal amplification was realized. The general strategy is depicted in Figure 9.3.

The bond cleavage detection was validated using the enzymatic cleavage of peptides by α-chymotrypsin, and the important features for the success of the

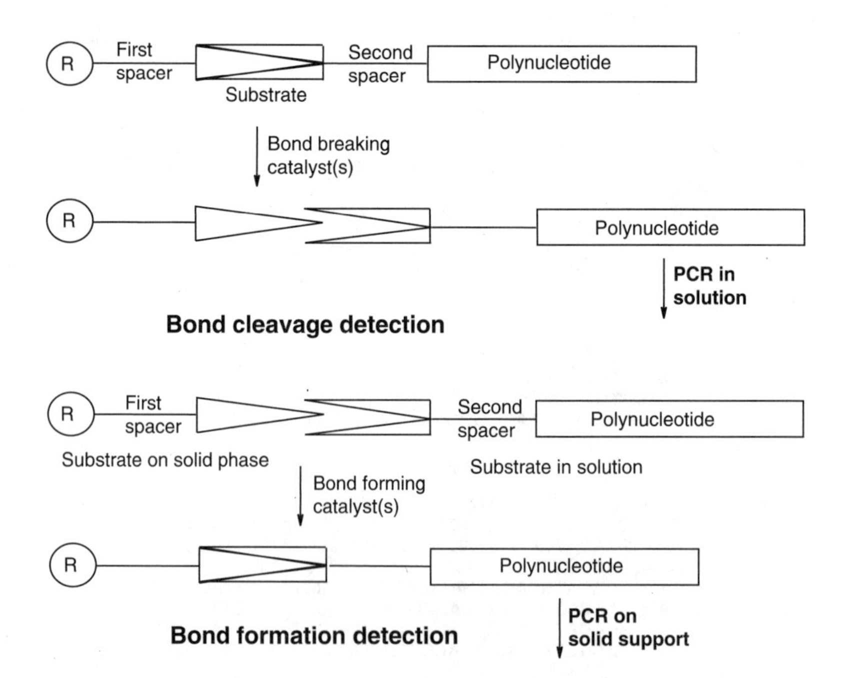

FIGURE 9.3 Encoded reaction on cassette: general strategy.

approach were reported. The resin used was TentaGel (good hydrophilicity and stability), so that its PEG chains functioned as a first spacer, separating the solid support from the substrate and allowing the enzyme–substrate interaction to take place. The second spacer divided the substrate from the tags, influencing the synthesis of the coding moieties and thus of PCR amplification, whose effectiveness was proportional to spacer length. The resin loading was also crucial, with loading higher than 85 μM showing poor nucleotide synthesis. The shortest nucleotide sequence that provided good and reliable PCR results (no mismatch, or wrong starting points for PCR) was 45 nucleotides in length, consisting of the two 15-mer PCR sequences and a 15-mer coding region. After careful optimization of the reaction and detection conditions, the authors detected PCR reaction in solution (see Figure 9.3, top) and also discriminated between different substrates in terms of affinity for the enzyme. Other proteases did not affect the reaction cassette, thus showing the specificity of the substrates for α-chymotrypsin. An interesting observation was the detection of PCR amplification in solution after 24 h in absence of enzyme. The phenomenon was attributed to the extremely slow uncatalyzed hydrolysis of the peptide ($t_{1/2}$ of few years), which was detectable due to the extreme sensitivity of the PCR technique. The enzymatic measurement, which took less than 30 min of incubation, was not affected.

The reverse reaction (bond formation, see Figure 9.3, bottom) did not work with α-chymotrypsin in a water/organic medium, probably due to extreme hindrance of the catalyst. The approach was successfully validated, detecting chemical bond formation by the use of a supported aldehyde, which was reacted with an amine and subsequently reduced with $NaBH_3CN$; the stable reaction product produced detectable resin bound PCR amplification (Figure 9.4).

Both results proved the validity of encoded reaction cassettes for enzymatic and classical organic applications; the complexity of the constructs for the encoded reaction cassette (substrate, two spacers and the nucleotidic code) can be outweighed by the preparation of large quantities of intermediates (second spacer plus different coding sequences, see Figure 9.3) to be used for any reaction cassette experiment, thus reducing the time needed for the synthesis of the encoded library. These experiments used a single substrate and a single enzyme, but a library of encoded substrates could be screened on a panel of catalysts for enzymatic reactions, or on a panel of synthetic conditions for classic organic reactions; the different coding strains, when amplified by PCR in the opportune medium, should allow the detection of the reaction substrate and also roughly determine the specificity of a catalyst, or an experimental condition, toward different substrates. Quantitation of any reaction for any encoded substrate should be obtained *via* detection with a fluorescent probe because the amount of the PCR product should be proportional to the efficiency of the chemical or enzymatic reaction.

The vast majority of reported applications of nucleotide encoding dealt with peptide libraries; among those Needels *et al.* [11] prepared an 823,543-member heptapeptide library on 10 μM beads with seven repeating building

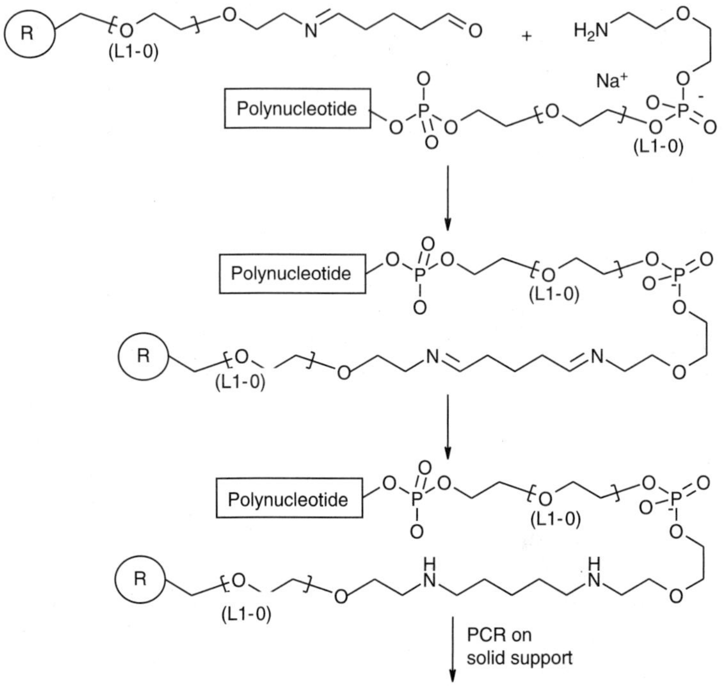

FIGURE 9.4 Encoded reaction on cassette: an example.

blocks encoded by a two-base sequence (TA, CT, TC, AT, TT, CA, and AC), tested their affinities for MAb 32.39 using on-bead screening and found many positives; Nielsen and Janda [12] reviewed the use of nucleotides as codes, with particular attention to their compatibility with classical peptide synthesis conditions and to their influence on the biological assay outcomes.

The most important drawback of oligonucleotide tags is their instability and incompatibility with many classical organic chemistry conditions. Even for peptide chemistry, the reaction conditions have to be carefully adapted to the presence of the nucleotide strands, and this limits significantly the use of this coding technique. Nevertheless, when the reaction conditions for library synthesis can be adapted to the presence and the stability of the nucleotide code, this technique should be considered for its many advantages (e.g. sensibility, automation of the synthetic and the decoding steps, assessed lack of interaction of the coding strand in the biological assay [12], and so on).

B. Peptide Tags

This subject will not be extensively covered in this review because peptide combinatorial chemistry is fully treated elsewhere in this book [13]. Only a few important contributions will be mentioned: Nikolaiev *et al.* [14] reported a general method for peptide encoding of nonsequenceable polymers using different sites of the solid support. Kerr *et al.* [15] presented the coding of non-natural peptide libraries (200 non-natural decapeptides) *via* L-amino acid peptide coding, using as orthogonal protecting groups the base-labile Fmoc and the acid-labile Ddz. Hornik and Hadas [16] presented some self-encoding peptide libraries, where an equimolar mixture of a natural and an unnatural amino acid was coupled in any synthetic step producing more than one peptidic sequence on a single bead, but having an internal sequenceable code on the same bead for the exotic building blocks. Felder *et al.* [17] presented a 100-member pentameric peptide library made by unnatural amino acids and other building blocks and encoded by sequenceable peptide tags, which was tested for thrombin inhibition with on-bead screening; Vagner *et al.* [18] presented the shaving technique, where a short polypeptide sequence was cut by proteolytic cleavage at the bead surface, leaving the inner sites available for peptide coding while the free amino group at the surface (orthogonally protected) could be used for library synthesis. A 100,000-member peptide encoded peptide library was tested and good ligands for three different receptors were identified *via* peptide decoding.

Advantages of peptide coding are that the chemistry is extremely well known, the availability of orthogonally protected amino acids, the possibility of automation for both the code synthesis and sequencing *via* Edman degradation. As for nucleotide tags, peptides are often unstable in classical organic reaction conditions and additionally they can interact with the biological target when an on-bead assay is performed, thus producing "false positives." Nevertheless, this coding technique could find use in small organic molecule libraries provided that the above limitations are not an issue for the library synthetic scheme.

C. Haloaromatic Tags

The first organic chemical tags, based on polyhalogenated phenols, were introduced by Ohlmeyer *et al.* [19] and subsequently modified by Nestler *et al.* [20]; an example for both strategies will be presented.

Yoon and Still [21] presented a 50,625-member *N*-acylated tripeptide library with the structure RESIN-$CH_2NHCO(CH_2)_5NH$-AA_1-AA_2-AA_3-COR; $AA_{1,2,3}$ were 15 L- or D-amino acids and RCOX were 15 acylating agents. The structure of the haloaromatic tags used, composed by an electrophoric tag and by a photosensitive linker, is shown in Figure 9.5, together with their synthetic pathway.

Eighteen tags $\mathbf{T_1}$–$\mathbf{T_{18}}$ were prepared [18] by reacting three halophenols with ten bromoalcohols, then acylating the resulting alcohols with phosgene and reacting them with the aromatic linker *t*-butyl ester. The final step was the hydrolysis of the ester to give the free tags as stable yellow solids (Figure 9.5). They differed in the number of carbon atoms in the alkyl chain and in the phenolic ring substitution and were ordered according to their elution ($\mathbf{T_1} = \mathbf{10A}$ was the most retained compound in the column, $\mathbf{T_{18}} = \mathbf{2C}$ the less

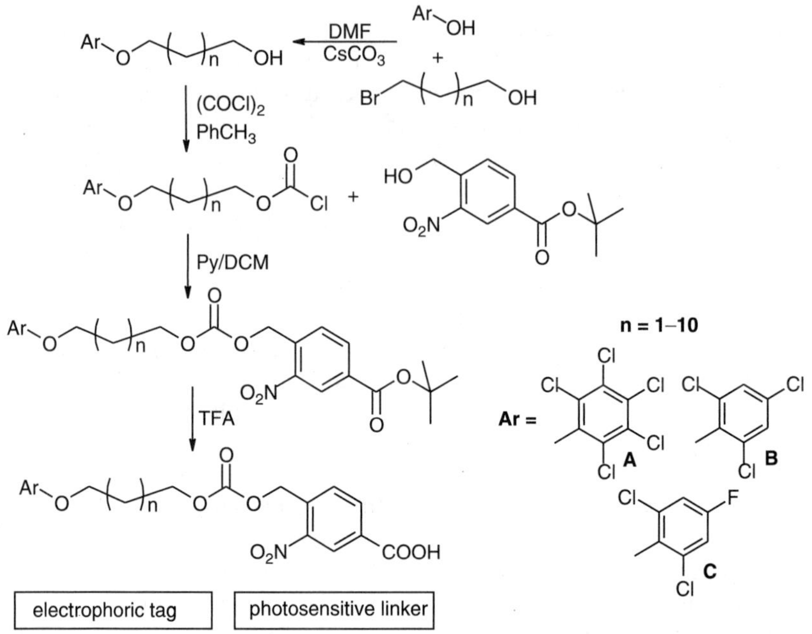

FIGURE 9.5 Chemical encoding: haloaromatic tags.

retained); their decoding was done by electron capture gas chromatography (EC-GC) [22].

In the 50,625-member *N*-acylated tripeptide library [21] each monomer position ($AA_{1,2,3}$ and RCOX) was coded by four tags using the binary code shown in Figure 9.6. The null code 0000 was excluded to have always at least one signal for each coding moiety. Each tag, or mixture of tags, coding for a monomer in a single position was coupled to the resin, so as to functionalize $\approx$0.5% of the resin loading prior to the monomer set coupling. The loading sites available for the library synthesis were not significantly reduced but this tag loading allowed its reading *via* EC-GC.

After library synthesis and solid-phase assay of $\approx$100,000 beads on a red-labeled synthetic receptor [23], 55 deep staining beads were selected and their code was released *via* photolysis at 350 nM; the released alcohols were silylated with *N*, *O*-bis(trimethylsilyl) acetamide and injected into a capillary GC with EC detection decoding 52 different structures, which are shown in Figure 9.7.

Screening the encoded library produced interesting structures to build an SAR and to design focused libraries of binders for the artificial receptor, but some limitations of these linkers for applications different from peptide chemistry are clearly evident. The carbonate bond connecting the electrophoric tag and the photocleavable linker (Figure 9.8) is sensitive to many organic reaction conditions, and the carboxylate requires an amino- or hydroxy function to be linked to the resin after each synthetic step to be encoded.

AA_1	T_1–T_4[a]		R(COX)	T_{13}–T_{16}
Gly	1000		Me	1000
D-Ala	0100		Et	0100
L-Ala	0010		iPr	0010
D-Ser	0001		tBu	0001
L-Ser	1100		tAmyl	1100
D-Val	0110	$AA_{2,3}$ as AA_1 using T_5–T_8 and T_9–T_{12} respectively	CF3	0110
L-Val	0011		iBu	0011
D-Pro	1001		MeOCH2	1001
L-Pro	1010		cyclopropyl	1010
D-Asn	0101		cyclobutyl	0101
L-Asn	1110		cyclopentyl	1110
D-Gln	0111		AcOCH2	0111
L-Gln	1011		Ph	1011
D-Lys	1101		Me2N	1101
L-Lys	1111		morpholino	1111

a: tagging molecules (0 denotes absence, 1 denotes presence of the tag)

FIGURE 9.6 Chemical encoding with haloaromatic tags: coding scheme.

	R	AA3	AA2	AA1		R	AA3	AA2	AA1
1	Me	V	n	G	27	MeOCH2	q	V	G
2	Ph	V	n	G	28	AcOCH2	q	V	G
3	cyclopentyl	V	n	K	29	Me	q	V	G
4	AcOCH2	V	n	Q	30	Et	q	V	q
5	Et	V	n	A	31	CF3	q	V	Q
6	MeOCH2	V	P	Q	32	Me2N	Q	V	N
7	AcOCH2	n	V	S	33	Me2N	G	V	S
8	Et	n	V	S	34	Ph	A	n	V
9	cyclopropyl	n	V	S	35	cyclopropyl	A	n	V
10	iBu	n	V	S	36	Et	A	n	V
11	Me	n	V	S	37	CF3	A	n	V
12	Ph	n	V	S	38	AcOCH2	a	n	V
13	Me2N	n	V	S	39	Me2N	S	n	V
14	iPr	n	V	S	40	AcOCH2	Q	n	V
15	IBu	n	V	G	41	morpholino	k	n	V
16	Ph	n	V	G	42	cyclopropyl	A	q	V
17	MeOCH2	n	V	G	43	Et	G	q	V
18	morpholino	n	V	K	44	Me	S	q	V
19	morpholino	n	V	Q	45	MeOCH2	S	q	V
20	Ph	n	V	n	46	cyclopentyl	p	q	V
21	AcOCH2	L	V	n	47	AcOCH2	s	G	N
22	cyclopentyl	L	V	n	48	MeOCH2	N	k	G
23	iPr	q	V	S	49	AcOCH2	G	K	S
24	iBu	q	V	S	50	tBu	q	G	G
25	CF3	q	V	S	51	iBu	q	G	S
26	MeOCH2	q	V	S	52	Ph	q	S	G

a: Capital and lower case stands for L- and D-amino acids respectively

FIGURE 9.7 Haloaromatic tags: decoded structures.

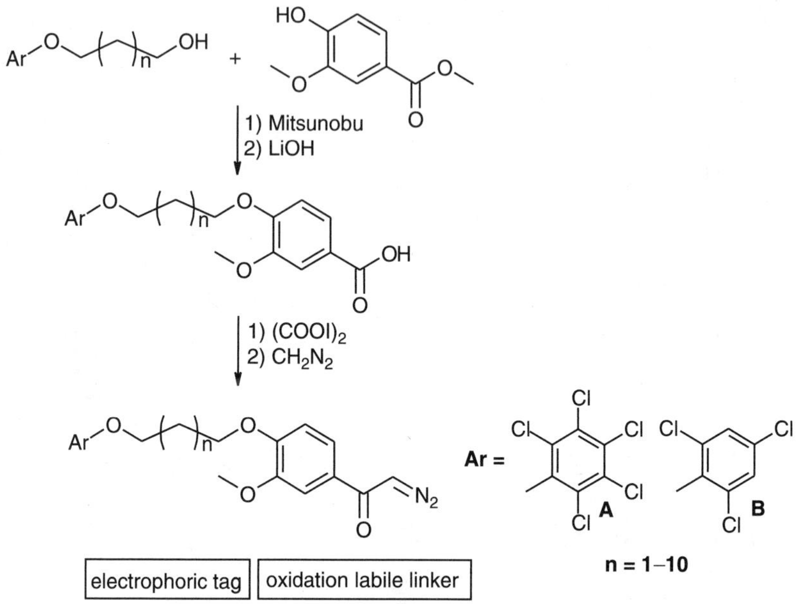

FIGURE 9.8 Chemical encoding: modified haloaromatic tags.

Nestler *et al.* [20] reported a modification of haloaromatic tags which solved the above mentioned limitations. The structure of the modified tags is shown in Figure 9.8, together with their synthesis.

Thirteen acylcarbene haloaromatic tags were prepared with trivial chemistry starting from the condensation of the known aryloxyalcohols (Figure 9.8) with methyl vanillate, saponification, formation of the acyl chloride and finally reaction with excess diazomethane to give the diazoketonic tags (Figure 9.8). Encoding to the solid support was obtained *via* rhodium tetratrifluoroacetate dimer catalyzed insertion of the acylcarbene on the arene structure of the resin, while decoding required an oxidative cleavage from the bead using cerium ammonium nitrate (CAN). Silylation with bis(trimethylsilyl)acetamide produced the Si-protected electrophoric tag decoded *via* EC-GC (Figure 9.9).

The experimental conditions for the attachment of the tags to the resin were carefully adjusted, and a procedure [20] for having roughly 1 pmol of tag per bead, coupled either as single compound or as mixture of tags according to the binary code presented in Figure 9.6, was obtained. With an average of 100 pmol per bead for the library members, even encoding multistep syntheses did not significantly affect the quality of the library. Moreover, a large part of the tag reacted with the arene moieties of the resin rather than with the library intermediates, further reducing the amount of "spoiled" library molecules. The insertion reaction did not require any handle on the solid support, thereby not limiting the library chemistry, and the stability of the electrophoric tag bound *via* an ether bond to the oxidatively labile linker was extremely good toward

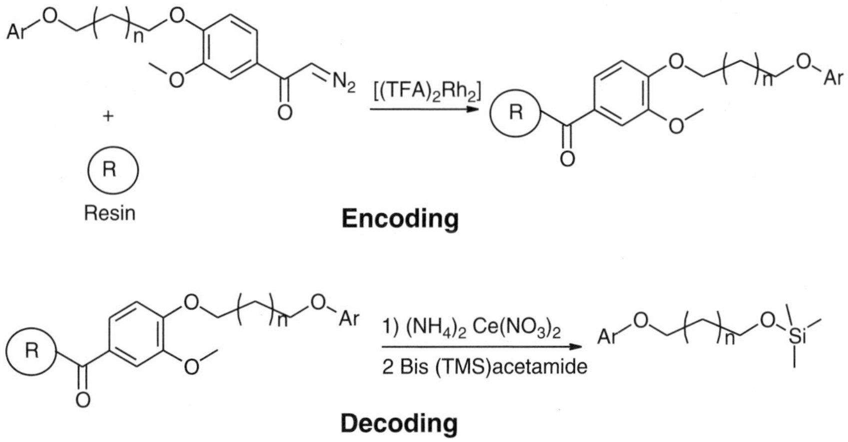

FIGURE 9.9 Modified haloaromatic tags: mechanism.

classical organic chemistry reaction conditions. The tags could be attached on the solid phase either before or after the library synthesis step which they were coding for, while the first version of electrophoric tags [19] allowed only their attachment prior to the library synthesis step. This allowed the coupling of a reactive library intermediate to the next monomer subset prior to its encoding, thus avoiding reaction with the substrate. The oxidative cleavage decoding, step was also carefully tuned to give general reliable conditions, and the lower limit of detection of EC-GC being below 0.5 pmol of compound, the structure determination of positive beads *via* haloaromatic encoding was successful in >95% of the attempts [20].

The selected example by Baldwin *et al.* [24] showed the application of this technique to a small molecule, nonoligomeric large organic library of dihydrobenzopyrans (six steps; over 85,000 members). The synthetic scheme for some among the prepared sublibraries is given in Figure 9.10. The first monomer set (seven amines) was condensed in solution to a bromophotolinker, then the construct was hooked onto TentaGel amino resin (step A). After TFA deprotection the resin bound secondary amines were coupled to the second monomer set (two subsets, six carboxylic acids, step B) prepared by condensation in solution of three dihydroxyacetophenones with either bromoacetic acid ($X = CH_2$, three compounds subset) or with *p*-carboxybenzyl bromide ($X = CH_2$-p-Ph, three compounds subset, Figure 9.10). The resin bound α-hydroxyacetophenones were then cyclized with the third monomer set (three *N*-protected cyclic aminoketones) to give the dihydrobenzopyran scaffold (step C). After TFA deprotection the free amine was coupled with the fourth monomer set (seven compounds including a sulfonyl chloride, a carboxylic acid, an acyl chloride, two aldehydes, a 2-chloropyrimidine and an isocyanate, step D). The resulting key functionalized scaffolds were split into four parts: one sublibrary was kept as such, another was reduced (step E), another was converted to a spirodithiolane scaffold (step F) and the fourth

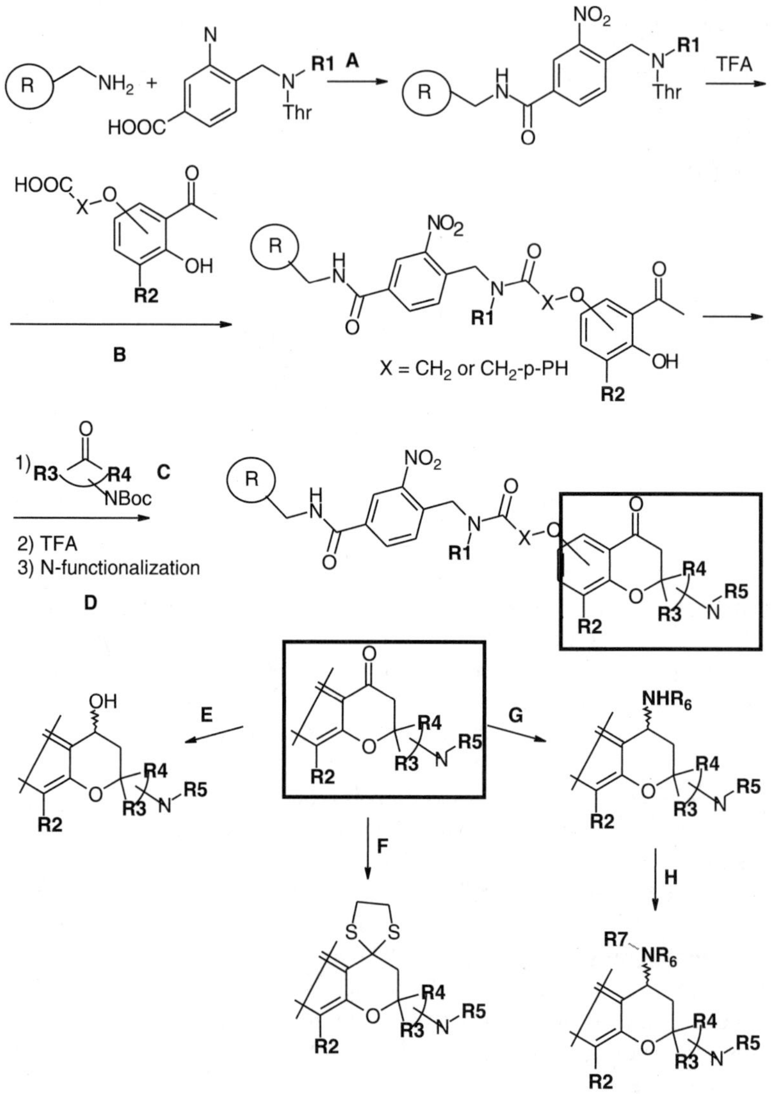

FIGURE 9.10 Encoded dihydrobenzopyran library.

was reductively aminated with the fifth monomer set (seven amines, step G). This last sublibrary was further split and reacted with the sixth monomer set (ten compounds, including two *N*-protected amino acids, two isocyanates, two isothiocyanates, a chloroformate, an acyl chloride, a carboxylic acid and a sulfonyl chloride, step H).

Baldwin [24] did not report details about the library encoding, but the number of necessary tags can be easily determined from the reaction scheme. The monomer set A required three tags for complete encoding of the monomers *via* the previously described binary code (Figure 9.6); set B required four tags in total, subdivided into two groups of two for the two monomer subsets B_1 and B_2; set C required two tags for encoding the three aminoketones, while set D required three tags for the seven *N*-functionalizing monomers used. To these 12 tags a single tag coding specifically for the unchanged sublibrary of 4-dihydropyranones could have been added, and the same could have been done for ketone reduction (step E) and its thioketalization (step F). More conveniently, the specific sublibraries E and F could have been kept divided, thus making their encoding unnecessary. Set G could have been encoded by three tags (seven monomers), while set H either could have been coded by four tags (ten monomers) or, more easily, the ten pools resulting from the last "mix and split" step could have been kept divided. This amounted to a minimum of 15 and a maximum of 22 tags. Additional encoding of steps E, F, and H for the library (the whole >85,000 individuals would have required additional three tags for the other sublibraries not reported here [24]); 40 electrophoric tags, individually detectable by EC-GC for this and for more complex libraries (more reaction steps or more monomers) were prepared [25], allowing the encoding of extremely large "one bead–one compound" [7] libraries (a five-step library where each monomer subset is composed by 31 individuals would contain $31^5 = 28{,}629{,}151$ individuals and would require $5 \times 5 = 25$ tags to be fully encoded).

A photocleavable linker was hooked onto the resin and the library compounds, allowing the on-bead screening of this library or its release in solution by UV irradiation after distribution of the beads in 96-well plates; the size of this large library and its structural features made it suitable for primary screening on different receptors. A smaller and slightly different version of this library [26] (1143 compounds) was tested in solution for carbonic anhydrase inhibition, producing a low nanomolar inhibitor after cleavage of over 2300 single beads and decoding of 33 positive structures.

Among other applications of electrophoric haloaromatic tags, Burger and Still [27] reported an encoded library of synthetic ionophores tested on solid-phase for its affinity for Cu^{2+} and Co^{2+}. Burbaum *et al.* [26] presented a 6727-member encoded acylpiperidine library tested to find isozyme-selective carbonic anhydrase inhibitors. Appell *et al.* [28] briefly described the biological screening of a 56,000-member encoded combinatorial library against two related G-protein coupled receptors. Nestler [29] reported two 10,000-member encoded peptidosteroid libraries tested for their affinity for enkhephalins. Liang *et al.* [30] reported a 1300-member di- and trisaccharide encoded library tested for its binding to lectin. Dolle *et al.* reported the identification of potent antimalarial compounds from an encoded statine-inspired library [31, 32], and the discovery of allosteric inhibitors of iNOS dimerization from an encoded library of pyrimidine imidazoles [33]. The same group has also reported a careful assessment of the quality of this encoding method [34]. The same method has also been repeatedly used, after some technical improvements impacting on the quality of the encoding-decoding process, by the group of Schreiber at Harvard for their "chemical genetics" approach [35–38].

This encoding method has proven to be very valuable and to have many useful properties, such as the use of inert tags linked in small amounts to the solid support with no need of particular functional groups present on the bead, the extreme sensitivity of the decoding method, and its general reliability. The only current limitation may be the time-consuming procedure for the code cleavage, silylation, and EC-GC analysis [19, 20] which is not an easily automated process. The application to small organic non-oligomeric large library synthesis [24, 26] has already validated this as a powerful and general encoding method, to be widely used for pool libraries to be screened on many different assays and formats.

D. Secondary Amine Tags

These chemical tags were first reported by Ni *et al.* [39] for the decoding of medium-size small organic molecule libraries. Their preparation and structure is given in Figure 9.11 (top). Eighteen different tags $\mathbf{T_1}$–$\mathbf{T_{18}}$ based on

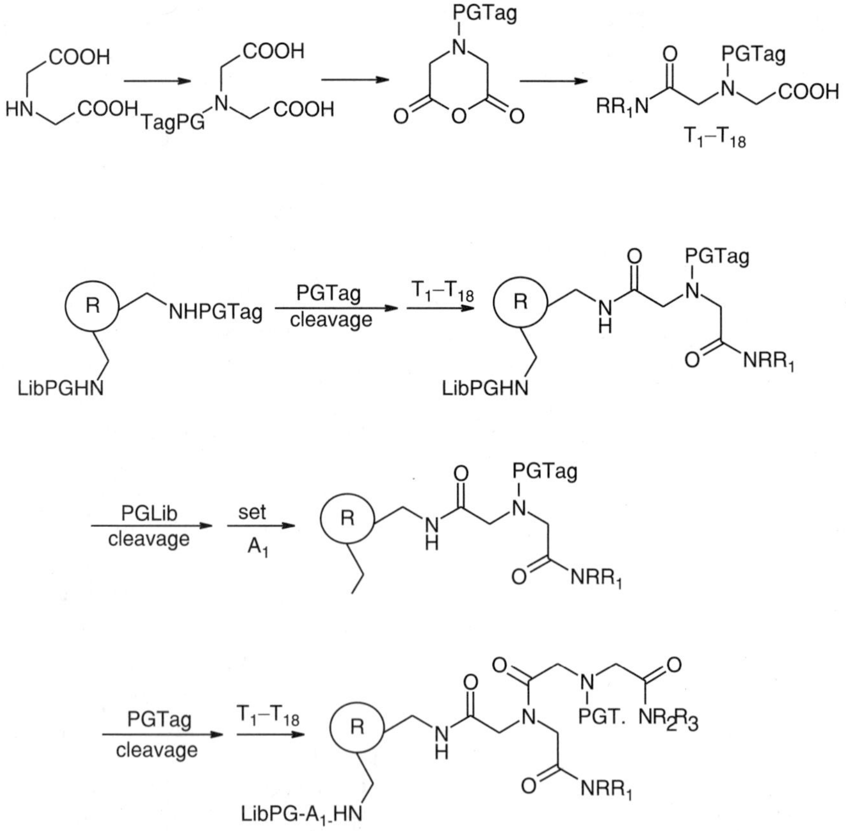

FIGURE 9.11 Chemical encoding: isotopic tags.

iminodiacetic acid were prepared and characterized as fluorescent-LC codes (see below). Their use as tags (Figure 9.11, bottom) required an amino or a hydroxy group on the resin, and the loading of the resin was partitioned between library sites and tagging sites by reacting 9/1 mixtures of *N*-orthogonally protected glycines (typically Fmoc for library sites and Alloc for tagging). Deprotection of PGTag (Figure 9.11) and coupling of the tag, or mixture of tags, on the free amino group produced the tagged resin. Deprotection of PGLib and coupling of the monomer set A_1 completed the first synthetic step, which was repeated with different tags and monomer sets until the completion of the library synthesis. The tags were used following a binary code as for the electrophoric tags (see Figure 9.6), and the decoding process consisted of hydrolyzing the tags in standard peptide acidic conditions, neutralizing with base, and dansylating the free amines. A small aliquot of this *N*-dansylated mixture was injected in HPLC and the corresponding structure was decoded.

The selected example by Maclean *et al.* [40] reported a 240-member encoded pyrrolidine library, whose synthetic scheme is reported in Figure 9.12. Coupling of the resin bound acid labile linker HMPB (4-hydroxymethyl-3-methoxyphenoxybutyric acid) with the first monomer set (four Fmoc amino acids, A) was followed by its coding (tags T_1–T_3), and by coding of the second monomer set B (tags T_4–T_6); then the second monomer set (four aldehydes, B) was added and the reactive imines were immediately reacted with the third

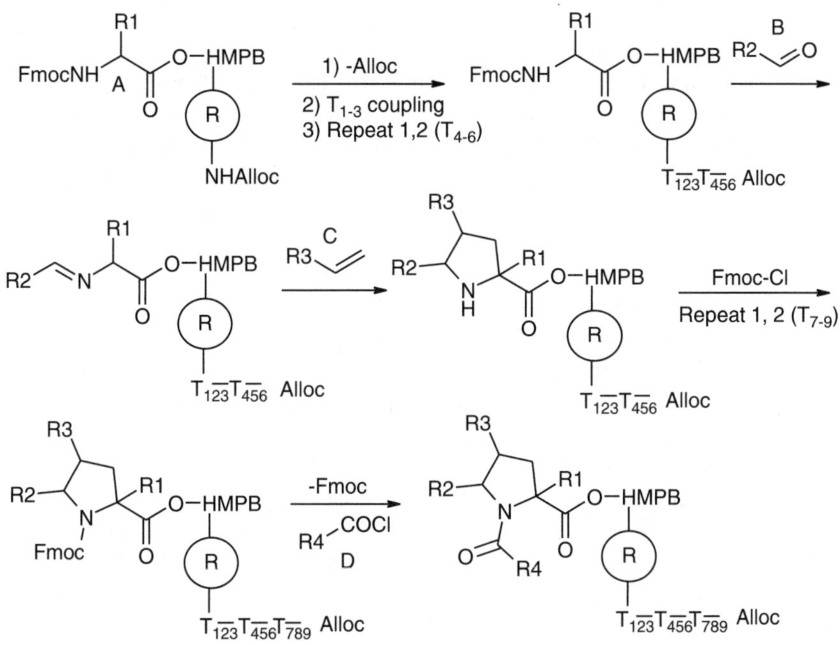

FIGURE 9.12

monomer set (five electron-poor alkenes, C) to give pyrrolidine scaffold *via* 2+3 cycloaddition. After protecting the free amine, the third monomer set was coded (tags T_7–T_9) and final coupling with the fourth monomer set (three mercaptoacyl chlorides, D) was done. These codes could be introduced either before or after the monomer set that they code. For example, the coding tags T_4–T_6 were added prior to the imine formation because of its reactivity, so as to avoid side reactions. The monomer set D was not coded because the three 80-compound sublibraries bearing the three different mercaptoacyl chlorides were kept divided.

As previously observed [41] this synthesis produced more than one isomer during the cycloaddition (three chiral centers are formed), so that the final library contained more than 500 individuals and the stereochemical outcome varied, depending on the nature of R_1, R_2 and R_3. Any chemical encoding could not account for stereochemical reaction outcomes, so that each stereoisomer from positive beads was prepared pure and its activity was determined after library synthesis, biological testing, and decoding.

The library was tested for ACE inhibitory activity through cleavage of single beads in 96-well plates, and approximately 1000 beads were cleaved. The most active compound identified was the same obtained with iterative deconvolution [41], but additional structures were obtained producing data for a preliminary SAR. A detailed comparison between the results obtained from this encoded library and from the same (non coded) library *via* iterative deconvolution will be presented in a following section (page 225).

The main advantages of this chemical encoding method are its robustness related to the well-known and reliable peptide bond chemistry, the automated encoding and decoding of the library structures, the stability of the tags, which are sensitive only to harsh organic conditions ($LiAlH_4$, strong acidic, and so on), the relatively low impact on the loading sites for library components on each bead (typically from 10 to 20% of sites used for coding), and the sensitivity of the fluorescence–LC decoding method (femtomolar range). The main drawbacks are the low throughput of the analytical procedure of decoding (hydrolysis, dansylation, and HPLC), the requirement of an amino- or hydroxy handle on the resin for the attachment of tags, and the requirement of at least three orthogonal protections on the same bead for the two syntheses—one for the codes (Alloc for the pyrrolidine library), one for the protection of functional groups to be reacted during the library synthesis (Fmoc in the example), and one for the protection of groups to be preserved during the library synthesis and made free before the biological assay (Ac in the example). These properties make the method both reliable and useful for the structure determination of some medium to large "one bead–one compound" [7] organic molecule libraries. Among reported examples, Ni *et al.* [39] briefly described β-lactam- and thiazolidinone-based encoded libraries. Atuegbu *et al.* [42] presented a 792-member encoded library, based on combinatorial modification of rauwolfscinic acid, a natural alkaloid, and Schullek *et al.* [43] reported a 324-member encoded dipeptide library which was tested with a miniaturized high-density screening for inhibitors of the matrix metalloproteinase matrilysin. Several recent reports have additionally provided optimized HPLC/fluorescence decoding protocols [44], which shortened the average decoding procedure from one hour to around six minutes, and also alternative analytical decoding techniques involving capillary electrochromatography (CEC) [45, 46], so as to produce coding patterns which can be attributed to all the positive library individuals. Finally, another recent report [47] introduced the use of so-called accurate isotopic difference analysis (AIDA) applied to secondary amine-tagged libraries.

E. Isotopic Tags

This procedure allows the identification of library compounds *via* reading a code made by isotopes of the most common elements (C, H, N, O) embedded into simple and commercial molecules used as tags and decoded *via* MS or NMR technique.

The selected example by Geysen and coworkers [48, 49] presented several coding strategies using 13_C or ^{15}N enriched glycines (G^0, G^1, G^2, G^3, G^4, G^5) and alanines (A^0, A^1, A^2, A^3, A^4) as tags and MS spectrometry for the decoding procedure. The coding entity (Figure 9.13) was realized by linking a first linker molecule to the resin, followed by the code represented by two sets of molecules. The first set was the bar code, made up of three isotopic glycines, G^0, G^1, and G^2, present in equimolar amounts in different combinations,

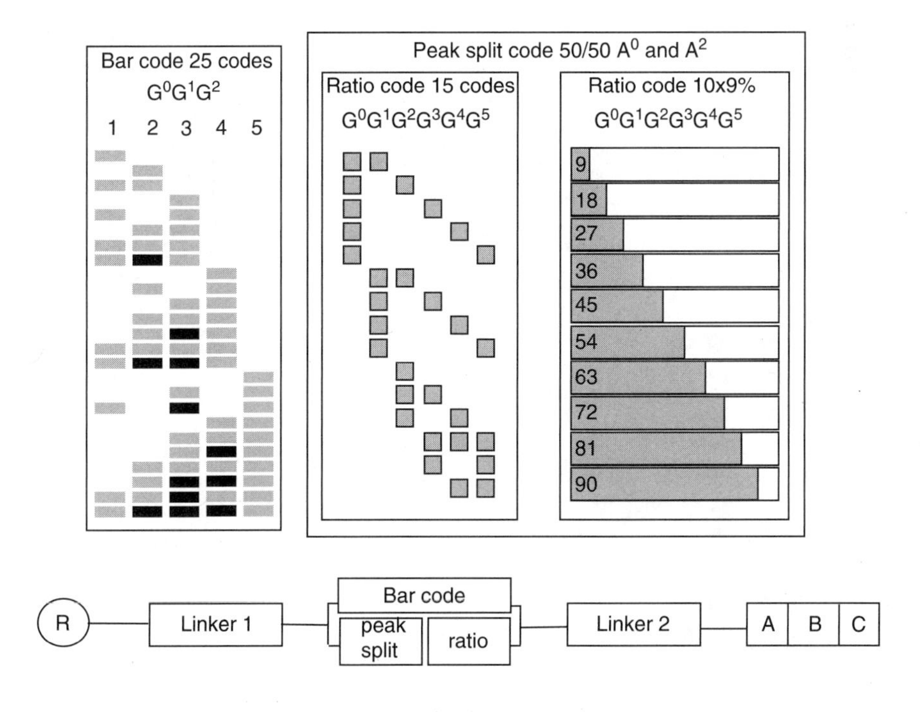

FIGURE 9.13 Chemical encoding: isotopic tags.

giving 25 different and distinguishable codes: a bar representation of this first set is shown in Figure 9.13 (left), where black bars stand for MS peaks taller than gray bars. The second set of molecules was the ratio code, made by binary mixtures of six isotopic glycines, G^0, G^1, G^2, G^3, G^4, and G^5: the resulting 15 combinations (Figure 9.13, middle) were further split by coupling each of the binary sets in 10 increments differing of 9% in the mixture composition (10 combinations, Figure 9.13, right). As an example, the binary mixture G^0G^1 could represent 10 codes *via* coupling in a 9 : 91 ratio, in a 18 : 82 ratio and so on until the 90 : 10 ratio. A peak splitter constituted by two isotopic alanines A^0 and A^2, which helped the reading of the coding moieties, was also inserted. The whole coding construct could thus encode for $25 \times 15 \times 10 = 3750$ different molecules. After the coding units a second linker, orthogonal to the first, was inserted followed by the library structure (A–B–C in our hypothetical example). Sequential release of the two linkers allowed first the biological assay, then the automated reading of the coding construct (typically around 1 min). The whole structure of the code-library construct is shown in Figure 9.13 (bottom).

As an example, the encoding of a hypothetical 8000-member three-step library, made by three sets of 20 monomers, is depicted in Figure 9.14. This example would require the preparation of 400 different coding constructs to be hooked with the first linker to the resin, then a second orthogonal linker would be added to start the library synthesis, as shown in Figure 9.13. The 400 coded resin portions would be divided into 20 subsets of 20 portions, and each subset

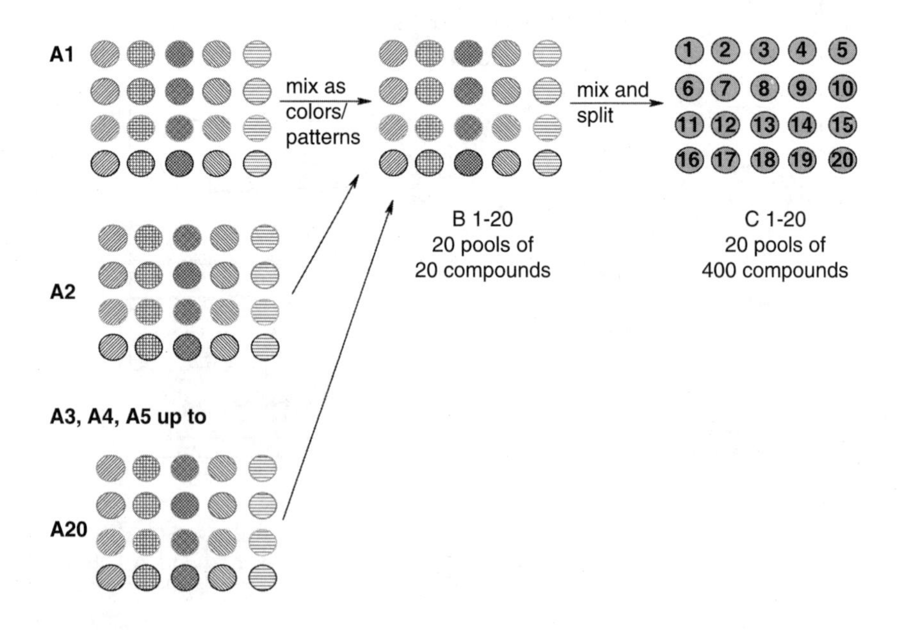

FIGURE 9.14 Isotopic tags: basic principles.

should be placed in a different reaction rack containing twenty reaction vessels. Each portion of a subset should be put in each vessel of a single rack, obtaining 20 racks labeled from A_1 to A_{20} as in Figure 9.14 (left). Each of the twenty monomers of the first set (A_1–A_{20}) would then be coupled to the 20 racks (monomer A_1 to the resin portions of rack A_1, and so on until monomer A_{20} to rack A_{20}). The resin portions would then be partially pooled, producing 20 pools containing 20 different codes and resin bound individuals; the portions in the same pool would be the portions in the same position of the 20 different racks A_1 to A_{20} (all the top left vessel portions from A_1 to A_{20} together to give the first pool B_1, and so on until the last pool B_{20}, made from all the bottom right vessel portions, Figure 9.14, middle). The second monomer set would be coupled to the resins (monomer B_1 to the pool B_1, and so on until monomer B_{20} to the pool B_{20}), and the pools would then be mixed and split to give twenty new pools; each pool would contain the same 400 different intermediates, and each individual would be coded by a different code (Figure 9.14, right). Finally, the third monomer set C_1–C_{20} would be coupled to the pools (monomer C_1 to the pool C_1, and so on until monomer C_{20} to the pool C_{20}), producing the 8000-member library as 20 pools of 400 compounds. Monomer sets A and B are encoded by the coding regions, while the set C is positionally encoded by the last coupling step (Figure 9.14).

By precoding the beads, as seen in the above hypothetical example, rather than running the code and the library syntheses in parallel, the large-scale preparation of coded samples of resin to be used for different libraries was realized. The insertion of a strongly ionizable group such as the free amine of a lysine as a "sensitivity maximizer" [48] could maximize the ionization response of the peaks related to the codes, suppressing the contamination from non-coding peaks resulting from any type of impurity, and increasing the sensitivity of the MS detection.

These coding strategies can be used with relatively large beads, where a few hundred picomoles are present so that single bead MS decoding with multiple coding moieties can be performed. If small beads (10 μm or similar) and/or low loadings are used only simpler coding constructs can be employed. Using 100 μm beads and several thousands of codes, large libraries can be decoded automatically. The decoding procedure is very reliable and significantly faster than for the previous approaches, allowing the decoding in a reasonable time of all molecules of interest coming from a large encoded library and the build-up of a more detailed SAR for the library components. The selection of different orthogonal linkers, and eventually of other isotopically enriched molecules as codes with different functional groups, could also allow to tailor the code chemistry to the library chemistry which has been designed.

The encoding constructs used require many additional molecules, such as glycine alanine, lysine and so on, which may prove sensitive to some reaction conditions for library synthesis. The physical bulk of the encoding moieties may also hinder the reactivity of some library intermediates, or produce side reactions which decrease the library quality and cannot be detected by the encoding method, or disturb on-bead screening, preventing or hindering the receptor–ligand interaction. Some applications of this technique to large organic molecule library encoding need to be reported to evaluate the flexibility and the reliability of the method, a method that looks extremely promising and useful also for other combinatorial applications [48, 49].

A recent patent by Garigipati and Sarkar [50] reported the use of isotopically enriched (^{13}C or ^{15}N) tags for decoding libraries with nano NMR probe, assessing the usefulness of this technique with few standard molecules. The nondestructive NMR detection method, which analyzes the resin-bound compounds without the need of cleavage, makes this technique extremely appealing provided that the technique can be validated for small organic molecule libraries.

F. Autotagging

This process refers to libraries where codes and library components are the same, but a clear distinction between coding molecules and library molecules can be made. Theoretically any "one bead–one compound" [7] library could be fully processed by bioanalytical methods [4] but the compounds must be cleaved off the beads, aliquoted and sent sequentially to the biological test and to structure determination for the test positives. Problems such as the stability of the components stored in solution, their concentration after prolonged storage, their solubility in the medium, and so on could arise from the total release in solution of the compounds. Partial controlled release of the library components in solution (library structures), their biological evaluation, and eventually their structure determination from the resin bound portion (coding structures) on "positive" beads makes a reliable process and has been the focus of published works, which will be presented in this section.

The selected example by Seneci *et al.* [51] reported the use of an IDA-DC [52] (iminoDiacetic acid based double cleavable) linker for structure determination and controlled release of pool libraries based on teicoplanin aglycone (TD) [53] and tested for their antimicrobial activities. The synthesis and the structure of the linker and the structure of teicoplanin aglycone are shown in Figure 9.15. Starting from iminodiacetic acid the linker was prepared in four trivial chemical steps (the cyclic anhydride used in step 4 was prepared in two steps from iminodiacetic acid); its structure was symmetrical with the amide arm bearing the carboxylic handle for the solid support attachment and the identical other arms bearing protected propylamine groups for the library synthesis.

The linker was hooked on AminoTentagel resin, then after Fmoc deprotection the teicoplanin aglycone derivative TD-Boc (see Figure 9.15) was coupled to the two linker arms to prepare the standard for chemistry and biology validation. The structure of this construct and its double cleavage mechanism are reported in Figure 9.16. The ammonium salt was stable after Boc deprotection, but dissolving it in an aqueous buffer at pH 7.5, an internal attack of the nitrogen on one of the two identical arms produced an intramolecular cyclization to a diketopiperazine, with simultaneous release in

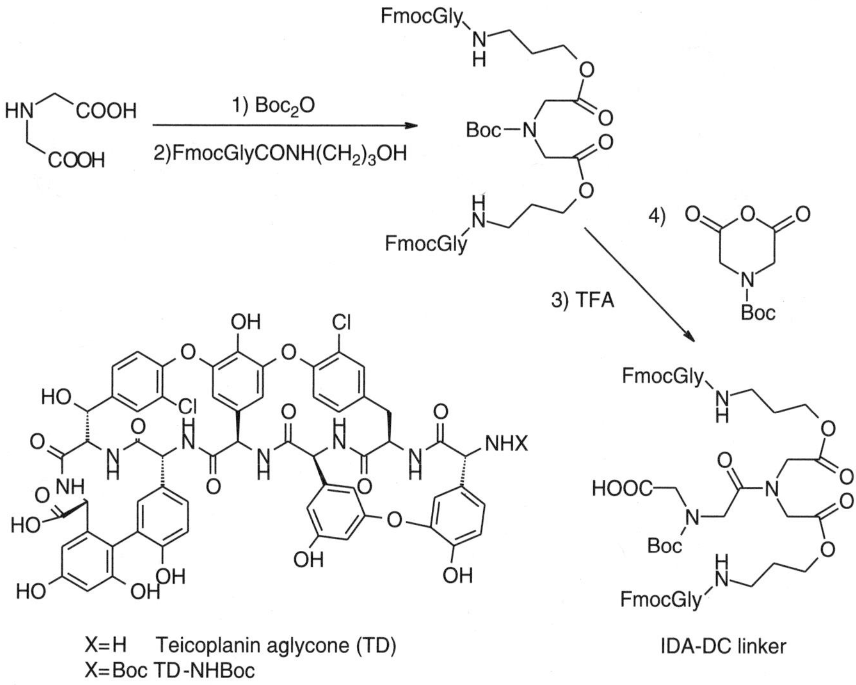

X=H Teicoplanin aglycone (TD)
X=Boc TD-NHBoc

IDA-DC linker

FIGURE 9.15 Chemical encoding: autotagging.

solution of TD-glyhydroxypropylamide, which was used for the biological assay. After treatment of the resin in mildly basic conditions the second molecule of TD-glyhydroxypropylamide was released for structure determination. The HPLC traces for both cleaved compounds were comparable, and preliminary attempts for library synthesis diversifying position N-15 (reductive amination and acylation) and O-56 (alkylation and acylation with protected N-15, see Figure 9.15 for TD numbering) were successful (for more details see the original paper) [51]. The release of the final compounds as free acids than as hydroxypropyl glycine amides has also been obtained introducing small modifications in the linker structure [51].

While this approach has limitations related to the peptidic nature of the linker such as stability to harsh reaction conditions and requirement of specific functional groups to be coupled with the linker arms, its application to particular libraries and chemistries may be useful. Its biological utility was assessed in a bead-based antimicrobial assay on bacterial cells [54], which produced good correlations between the MIC (minimal inhibitory concentration) values for the compounds released "in situ" in the culture medium from the beads or tested as standard solutions.

This linker was also adapted by Kocis and coworkers [53] and Lebl and coworkers [55] for peptide libraries having three different constructs for the same compounds on a bead, by simply inserting a lysine between the bead and

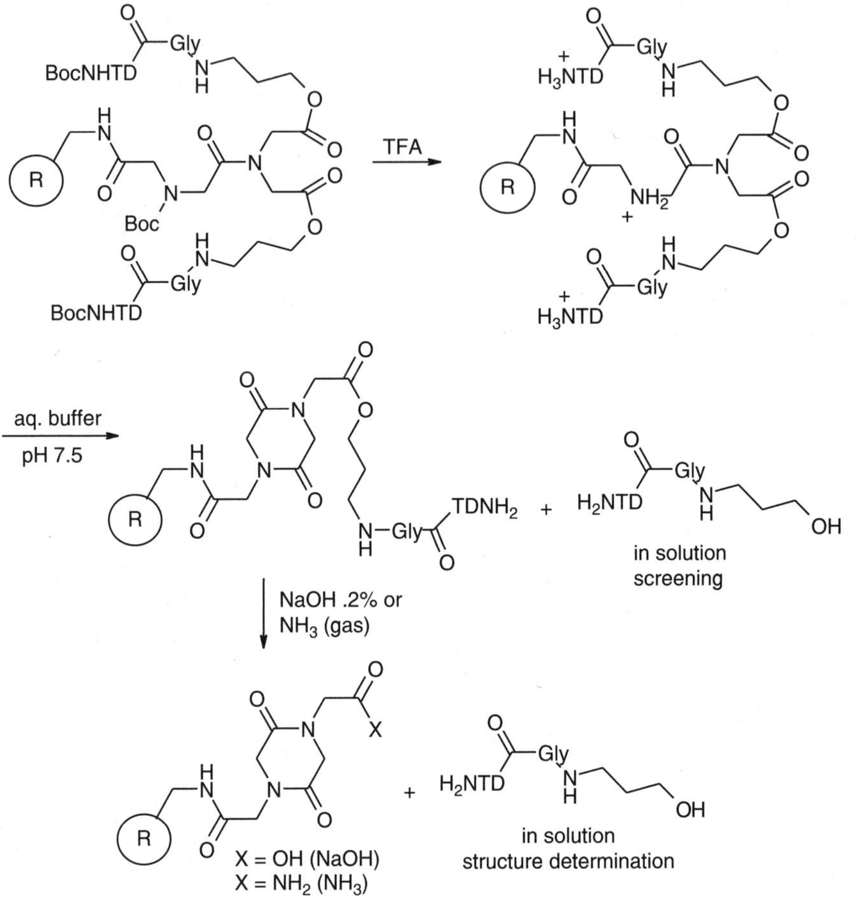

FIGURE 9.16 Autotagging: application to teicoplanin.

the IDA-DC linker. While the two linker arms could be released at neutral and basic pH, respectively, the third copy of the molecule, linked to the lysine α-amine (the ε-amine was coupled with the carboxyl function of the linker), was fixed permanently to the resin. A first release in pools (around 500 compounds per well) allowed rapid screening of extremely large libraries; the second release was performed on the redistributed single beads from active wells and the screening was repeated, while single bead structure determination was performed by Edman degradation, using the third resin bound copy of the library component corresponding to positive single beads. The approach should obviously be modified to allow structure determination for nonoligomeric organic molecule libraries.

Partial release has been attempted even using single release linkers, taking advantage of predictable reaction kinetics for the cleavage reaction; for example, the release of compounds from photocleavable linkers [56] was controlled by means of varying solvents and times of the cleavage reaction. Nevertheless, the extreme dependence of the success of this approach on the exact cleavage conditions (reaction time, concentration of reagents, washings, and so on) adds uncertainty to the results of the biological tests (amount of cleaved material, concentration, and so on).

Cardno and Brodley [57] recently reported a multiple release system where the solid support was treated with an equimolar mixture of three linkers with similar reactivity; after library synthesis (in the example a single standard tripeptide) one-third of the peptide was released by 1% TFA, another one-third TFA vapours, while the remaining one-third remained hooked onto the resin for sequencing. The analytical quantification of the tripeptide showed similar quantities for the three different aliquots. Similar methods using different families of linkers with similar coupling reactivities, or even using isokinetic mixtures of linkers with different reactivities, could easily produce interesting constructs for small molecule organic libraries autotagging.

G. Noncovalently Bound Tags

Rink *et al.* [58] recently reported an encoding procedure where heavy metal ions covalently bound to the resin beads can be used for encoding combinatorial libraries. An example of this approach is illustrated in Figure 9.17. The starting resin bound intermediate (prepared in four steps from AminoTentaGel resin with hydroxy methyl benzoic linker) was reacted with three monomer sets (respectively eight primary amines A, eight sulfonyl chlorides B, and 62 primary amines C) to produce a 3968-member pool library. After each monomer set coupling the resin portions were swollen into an aqueous solution of heavy metal salts (typically nitrates or chlorides with 99.99%+ purities, at 1 g/L concentration for 5–50 min) and surprisingly, even after thorough washings, some metal residues remained non- covalently bound to the resin. After library synthesis and cleavage of the library compounds in solution the tags could be detected *via* inorganic mass spectrometry (the only requirement was an MW of more than 100 to avoid technical problems). A "redundant" triplicate code was designed and it is shown in Figure 9.18 for the first and second monomer sets. The numeric code stands for the presence, 1, or absence, 0, of the corresponding element. As an example, MS detection of Cd, Dy, Gd, Ho, In, La, Lu, Os, and Pb decoded the amine 2 in the first monomer set and the sulfonyl chloride 4 in the second. The presence of three identical coding elements ensured a triple check of elements, presence, so as to rule out cross-contaminations of a single atom on a bead.

Typically, the resin (which had to be hydrophilic) was swollen and stirred in a solution containing the salt mixtures coding for a particular monomer of one of the sets. A thorough washing was done to eliminate the non-absorbed salts and to avoid cross-contamination of other beads during the "mix and split" operations. The mechanism by which the salts remained "bound" to the solid support was not determined; possibilities such as complex formation or inclusion effects were mentioned, and the use of resins containing heteroatomic

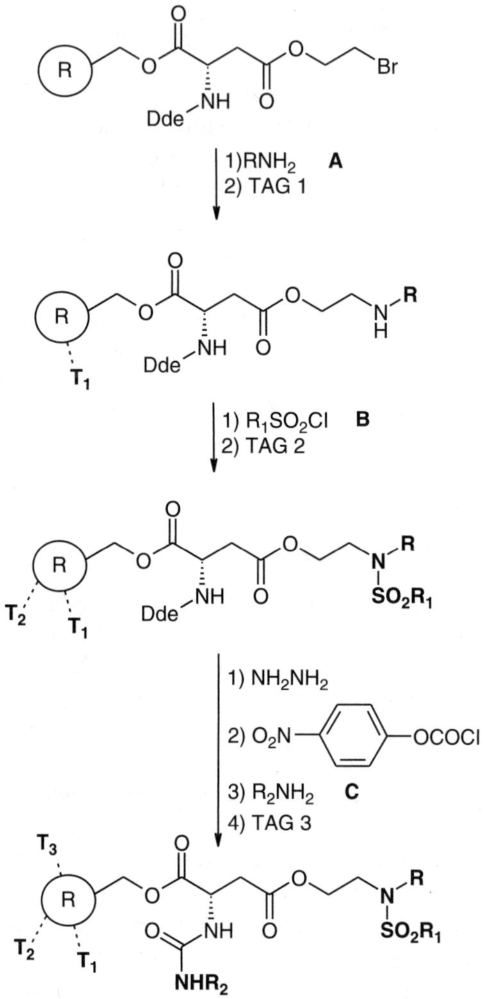

FIGURE 9.17 Chemical encoding: noncovalently bound tags.

chains (PEG, polyamide, and so on) was suggested in order to enhance this "absorption/complexation" of the metals onto the solid supports [58].

Appealing features of this method are the noncovalent interaction between the tag and the bead which does not require orthogonal chemistries and protecting groups, the easy and fast encoding procedure, the extremely sensitive and fast decoding technique; the possibility of preparing stock aqueous solution of tags to be used when necessary, and the "redundancy" which should prevent misreadings and technical problems. The presence of traces of toxic heavy metals, though, could seriously interfere in the screening, especially on bead, of these libraries.

Amines	Ba	Bi	Cd	Ce	Cs	Dy	Er	Eu	Gd	Hf	Ho	In	Ir	La	Lu	Nd	Os	Pb
1	0	0	0	0	0	0	0	0	0									
2	0	0	1	0	0	1	0	0	1									
3	0	1	0	0	1	0	0	1	0									
4	0	1	1	0	1	1	0	1	1									
5	1	0	0	1	0	0	1	0	0									
6	1	0	1	1	0	1	1	0	1									
7	1	1	0	1	1	0	1	1	0									
8	1	1	1	1	1	1	1	1	1									
Sulfonyl chlorides																		
1										0	0	0	0	0	0	0	0	0
2										0	0	1	0	0	1	0	0	1
3										0	1	0	0	1	0	0	1	0
4										0	1	1	0	1	1	0	1	1
5										1	0	0	1	0	0	1	0	0
6										1	0	1	1	0	1	1	0	1
7										1	1	0	1	1	0	1	1	0
8										1	1	1	1	1	1	1	1	1

FIGURE 9.18 Noncovalently bound tags: an example.

The encoding technique produced ligand structures for this and other two examples of peptide-like chemistry, validating the method also for large (>10,000 members) hydrophilic libraries. An extension of the technique, using more complex and harsh conditions for organic molecule libraries with different physico-chemical properties, various solid supports with different spacers/linkers, various loadings, and so on could really assess the general usefulness of this encoding technique.

H. ^{19}F ENCODING

The use of ^{19}F NMR for chemical encoding was recently reported by Pirrung *et al.* [59]. Nine fluoroaryls (**1a–i**, Figure 9.19) were prepared *via o*-metallation and further elaboration of the structures. These pro-codes were then used to alkylate the photosensitive reagent **2** to yield, after Boc deprotection, nine photolinkers **3a–i**, each carrying a different fluoro-based code. Each of them releases the corresponding alcohol **4a–i** under photolysis (Figure 9.19). The ^{19}F NMR signal for each alcohol is strong and well separated from each other due to the chemical environment surrounding the fluorine atom.

These tags were used to prepare the peptoid pool library L_1 (90 individuals, Figure 9.20). The aminoPS resin was acylated and split in nine aliquots (steps a, b), then a code from **3a–i** was added to each aliquot (step c). A new acylation step was followed by alkylation with the amine monomer set M_1 (steps a, d) to give **5a–i**. After pooling and splitting in aliquots (steps e and f), the cyclic anhydride monomer set M_2 was condensed to yield the library L_1 (step g), made by 90 individuals organized in 10 pools of 9 encoded individuals each.

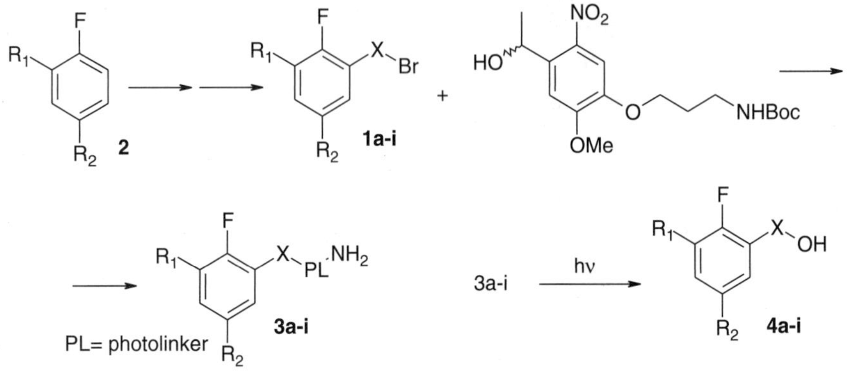

F
R_1
R_2 **2**
F
R_1
X
Br
R_2 **1a-i**
+
NO_2
HO
O
NHBoc
OMe

F
R_1
X
PL
NH_2
3a-i
PL= photolinker R_2
3a-i hν
F
R_1
X
OH
R_2 **4a-i**

3a: R_1=1, R_2=H, X=CH_2; **3b**: R_1=Br, R_2=H, X=CH_2;
3c:R_1=Br, R_2=Me, X=CH_2: **3d**: R_1=Me, R_2=H, X=CH_2;
3e: R_1=Me, R_2=OMe, X=CH_2: **3f**;R_1=OMe, R_2=H, X=CH_2;
3g: R_1=OMe, R_2=Me, X=CH_2, **3h**:R_1OMe, R_2=H, R_3=$O(CH_2)_4$–;
3i; R_1=Ome, R_2=H, X=OCH_2CH=$CHCH_2$–

FIGURE 9.19 ^{19}F encoding: fluoroaryls tags.

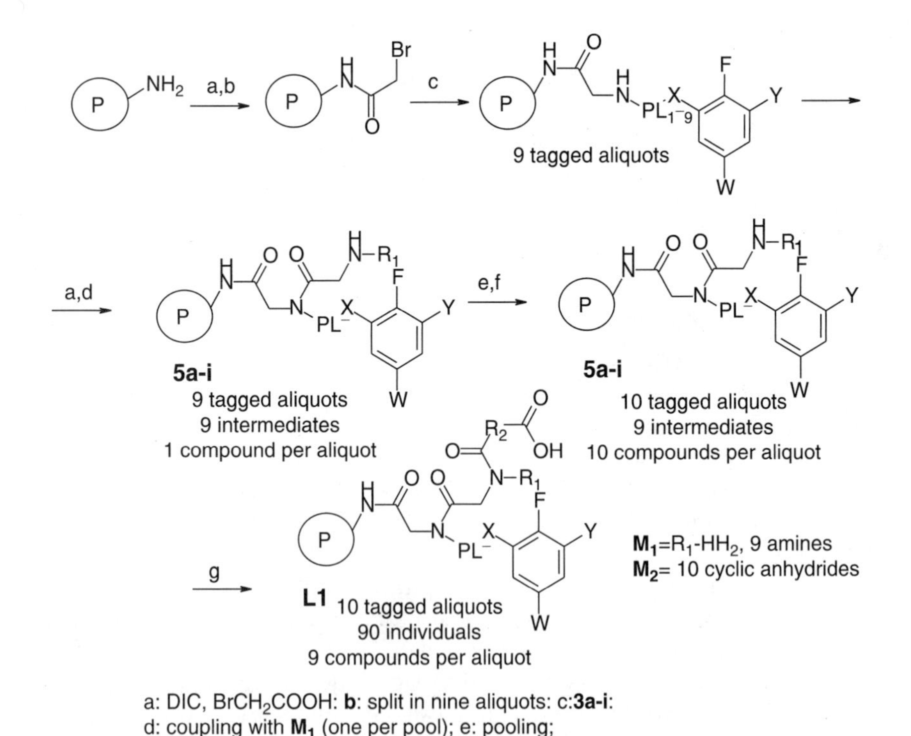

FIGURE 9.20 Fluoroaryls tags: application to a peptoid library.

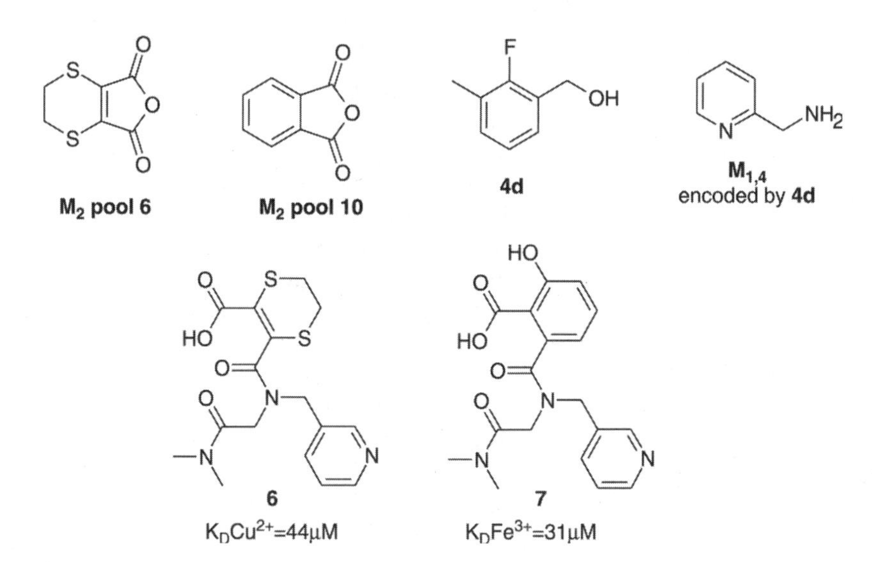

FIGURE 9.21 Fluoroaryls tags: confirmation of metal binding affinities of single entities.

L_1 was screened for its capacity to bind metal ions. An on-bead format was chosen, suspending some resin aliquots in presence of solutions of $Cu(OTf)_2$ in acetonitrile, or of Fe(2-ethylhexanoate)$_3$ in THF. Colored beads were picked from pools 6 and 10, corresponding to a specific anhydride (Figure 9.21). Irradiation and decoding determined the structure of alcohol 4d *via* ^{19}F NMR for beads from both pools. Compounds **6** and **7** were thus identified and reprepared as single entities to confirm their metal binding affinities (Figure 9.21).

Another group of researchers recently reported the validation of a similar ^{19}F NMR-based encoding method [60].

I. Other Methods

Garigipati and Adams [61] reported a chemical encoding procedure, based on novel photochemically cleavable aryl sulfonamides as inert tags with a robust and reliable chemistry. An example of a 1000-member encoded library was provided without experimental details, so additional applications of these chemical tags need to be disclosed in order to have a better understanding of their strengths and limitations.

Yamashita and Weinstock [62], Scott *et al.* [63, 64] and Egner *et al.* [65] reported the use of fluorophores in combinatorial chemistry and, in particular, as tags for pool library encoding. Multiple fluorophores were pre-encoded at a very low loading level ($>0.1\%$) and the codes were read *via* various fluorimetric detection techniques [63]. Small tripeptide libraries were tested and decoded with success [62], but many potential drawbacks were also highlighted [63] so that a careful assessment of all the relevant variables (solid support and loading influence, stability of the tags, quenching effects, detection methods, and so on) for some organic molecule libraries is necessary before considering this method to be reliable and easily applicable.

Edwards *et al.* [66] and Main *et al.* [67] reported the use of intrinsically labelled solid supports as pre-encoded beads for combinatorial libraries, encoding and decoding *via* MS spectroscopy of the single beads and examination of their decomposition patterns. Examples of primary tagging (functionalization of the solid support) and secondary tagging (elaboration of the primary tag structure) were given together with their decoding procedures. Applications of this technique to library decoding were not provided.

Rahman *et al.* [68] reported the use of infrared and Raman spectroscopy to decode libraries encoded with Raman and IR sensitive groups, such as 4-cyanobenzoylchloride and 3,5-di-tert-butyl-4-hydroxybenzoic acid. A proof-of-concept study involving seven individuals, all successfully identified with the two analytical techniques, was described. Applications of this technique to library decoding were not provided. Fenniri *et al.* [69, 70] recently introduced the use of dual recursive deconvolution (DRED) as a deconvolution-encoding hybrid method relying on bead self-encoding and multispectral bead imaging. IR and Raman spectroscopy were again used as analytical techniques.

III. NONCHEMICAL METHODS

The utility of encoding methods discussed above for chemical encoding is equally applicable to nonchemical encoding. In addition, nonchemical encoding entities avoid the need for two parallel chemistries, thus reducing the number of chemical steps, improving the overall quality of the synthesis and decreasing the risk of side reactions. At present, though, encoding of the solid support (small beads, typically from 10 to 130 μm) with nonchemical entities on single bead dimension has only been reported in a recent patent [71]. A nonchemical method which uses radio frequency encoding either on groups of beads separated by small reaction vessels or on new supports for solid-phase chemistry, has also been reported [72].

A. Optical Encoding

While considerable efforts have been spent in the past few years in the field of solid supports for combinatorial chemistry [73], most of them were devoted to modified polystyrenic beads with different sizes, loadings or swelling properties [74], or carrying different functionalities or linkers for library synthesis [75], or to solid supports different from resin beads (pins [76], cellulose [77], soluble supports [78], and so on). Few reports dealt with labelled solid supports prepared by chemical reactions (see the previous paragraphs) and significant efforts in the field of material sciences to obtain intrinsically labeled, nonchemically encoded, easily readable, combinatorial solid supports have not been reported.

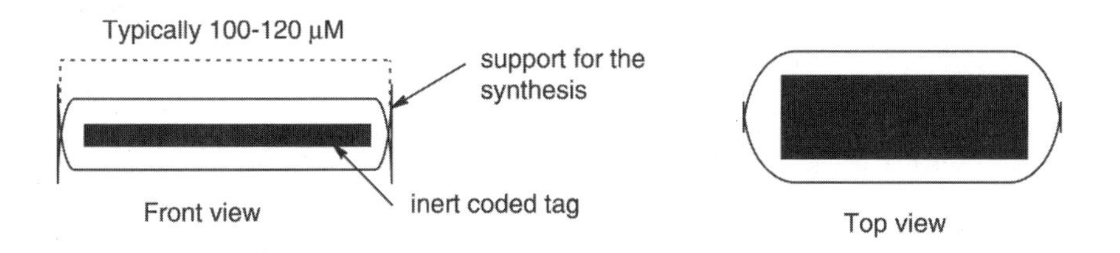

FIGURE 9.22 Nonchemical encoding: optical encoding.

A recent patent by Kaye and Tracey [71] reported intriguing applications of material sciences techniques for the preparation of optically encoded machine-readable solid supports. The most relevant features of some of the patent embodiments are presented and briefly discussed (see the original patent [53] for more detailed descriptions).

The typical reported constructs were made of two separate parts, one for the chemical synthesis and the other for encoding and reading the single particle. The latter was normally either totally or partially embedded into the former. The use of common, reliable, and automated techniques such as micromachining [79], which was developed for the microelectronic industry and worked with deposition and etching processes, allowed the fast preparation of millions of microparticles as coding units. Each one of them was "shape" encoded and optically readable as a combination of micropits, holes, hollows, grooves, or notches. The microparticles were made typically from inert materials such as silicon derivatives, which were then fully coated with the polymer selected for the solid-phase synthesis with such a thickness that allowed the optical reading of the internal code (typically 15 μm). A schematic representation of some particles is presented in Figure 9.22. Average total dimensions were 100–20 μm in length, 25 μm in height and 10 μm in width. These dimensions corresponded roughly to a resin bead of 130 μm in terms of loading sites. The elongated shape of the particle allowed its code reading during and after the library synthesis *via* optical instruments equipped with a capillary flow channel. This forced the microparticles to pass under the reading station one by one (typically 50–100 *per* second), allowing the reading of millions of particles during each synthesis step *via* conventional commercial image processing softwares and tracking of partial or total information carried by each particle during the whole synthesis.

Xiao *et al.* [80] recently reported LOSC (Laser Optical Synthesis Encoding), a technique where a polystyrene grafted synthetic support contained a two-dimensional laser-etched bar code on a ceramic substrate. Different chemical handles were introduced on the solid support and a small 27-member oligonucleotide library was encoded and subsequently decoded successfully by a small camera, linked to a combinatorial software for the automation of sorting/tracking procedures. More recently Nicolaou *et al.* [81] reported the synthesis of a $>$10,000-member library of benzopyrans using the same encoding technology.

The enormous potential of this technique is evident from the above brief descriptions, which highlighted the many "ideal" properties of such encoding in terms of ease, reliability, inertness and robustness, automation, complete compatibility with any biological assay, and so on. No apparent drawbacks can be found, except maybe for the lack of expertise in this field by combinatorial and organic chemists and their reluctance in getting involved in material science. Examples of applications of these, or of similar approaches where the solid support is intrinsically and nonchemically encoded, to real large encoded organic molecule libraries should appear in the near future, and will hopefully increase the interest of the combinatorial community toward these techniques.

B. Radiofrequency Encoding

Recently two different groups [82, 83] reported the use of radiofrequency tags for the encoding of chemical libraries. While the materials used were different, the coding strategy was rather similar and it is depicted schematically in Figure 9.20.

Nicolaou *et al.* [84] used microreactors composed by an inert porous wall, loaded with a SMART^{TM} memory chip, for radiofrequency signals and with small amounts of resin. While this is the only nonbead-based encoding method, the single entity (the microreactor) was reacted with the different monomer sets while being coded with the radiofrequency code. This allowed the sorting of microreactors at any moment simply by reading the stored RTF signal (for example, the partial pooling of an ABD triplet for each reaction vessel in the second coupling, Figure 9.23) and the complete pooling for the common reaction steps (protection or deprotection of common functional groups, introduction of nonvariable units in the library structure, and so on). Sorting of the microreactors could be done by automatic readout of the codes and appropriate separation of the different units (an automated microreactor autosorter is also commercially available).

The selected example by Xiao *et al.* [82] reported the synthesis of a 400-member encoded taxoid library whose synthetic scheme is reported in Figure 9.24. The precious starting material, baccatin III, was coupled to an excess of resin and TLC checked complete disappearance of the product in solution. The functionalized resin was distributed in 400 microreactors (20 mg of resin each) which were pooled together, then the Fmoc group was deprotected and the microreactors were split into twenty pools (20 reaction units); each pool being encoded by a different RTF signal. After coupling of the first monomer set (20 carboxylic acids A) under PyBOP/DIEA conditions, the microreactors were partially mixed and split into 20 new pools, each containing a microreactor from each of the previous pools (20 different RTF signals in the first coding position). The new pools were encoded with a second RTF signal, the second monomer set (20 carboxylic acids B) was added under DIC/DMAP coupling conditions and the final library of 400 taxoid compounds was prepared. The microreactors were then sorted by reading of

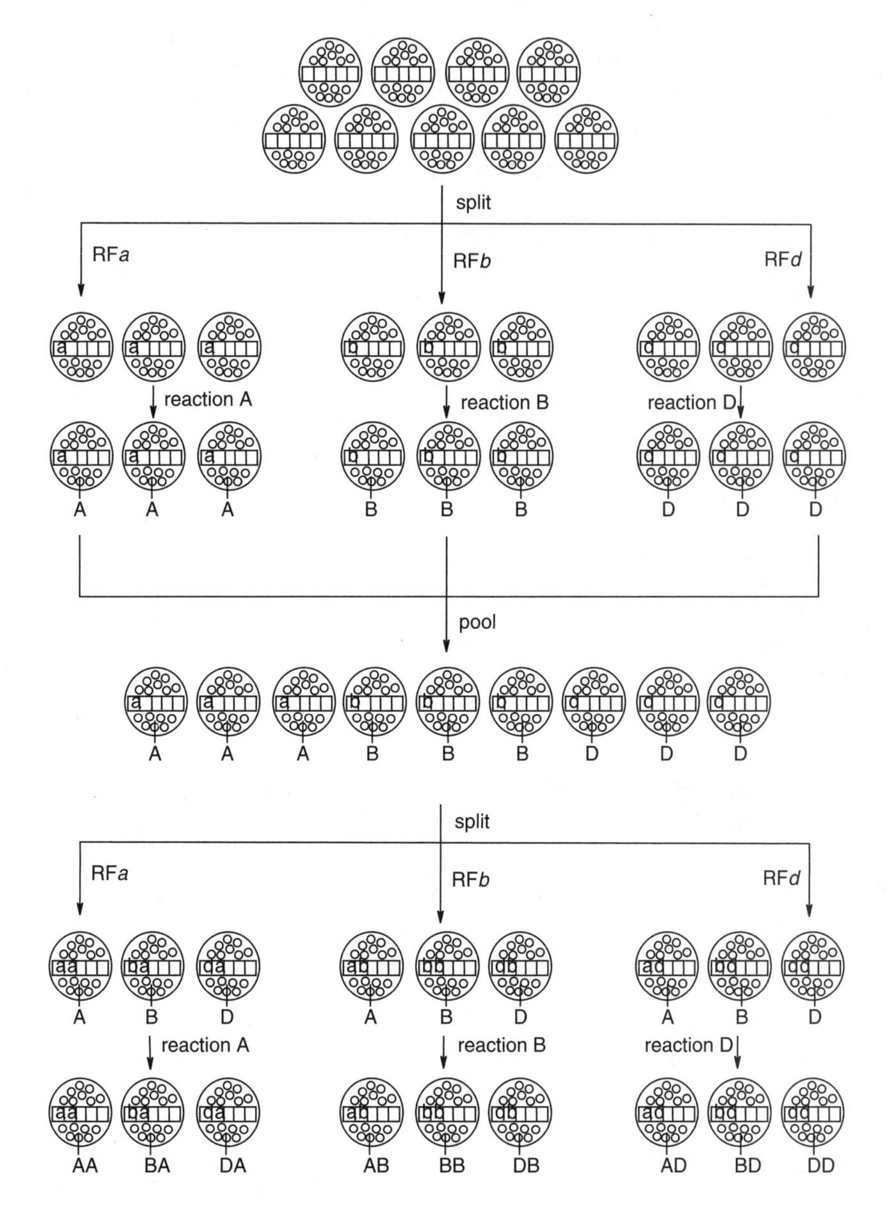

FIGURE 9.23 Radio frequency encoding: basic principles.

the codes and transferred in 400 vials, from which the pure individuals were cleaved as discretes in each microreactor, producing two 2–4 mg of each individual with purity between 50% and 100%, as judged by TLC and HPLC analyses.

The nonchemical, robust, and automated RTF tag both for the encoding and the decoding is very appealing, and while this is not a single bead method and does not allow complete "mix and split" procedures, the preparation of

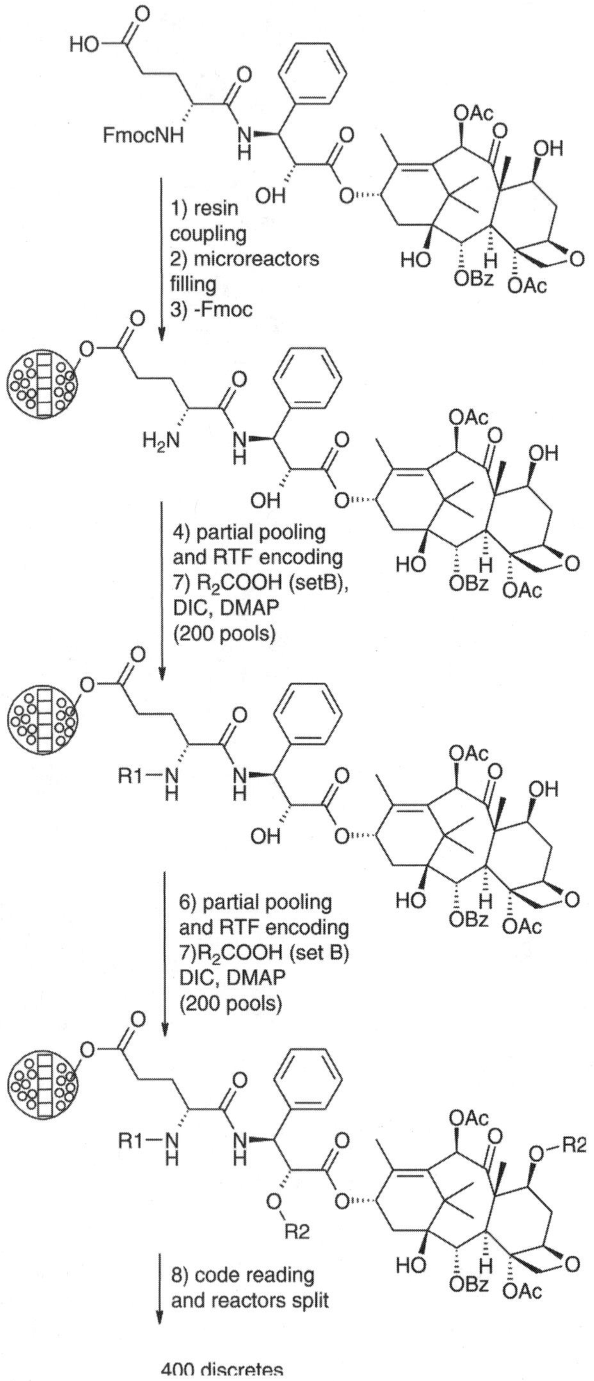

1) resin coupling
2) microreactors filling
3) -Fmoc

4) partial pooling and RTF encoding
7) R_2COOH (setB), DIC, DMAP
(200 pools)

6) partial pooling and RTF encoding
7)R_2COOH (set B)
DIC, DMAP
(200 pools)

8) code reading and reactors split

400 discretes

FIGURE 9.24 Radiofrequency encoding: an example.

multimilligram quantities of individuals for small/medium size libraries (up to few thousands) is largely simplified by the pooling for all the common steps. The intermediate sorting allows total flexibility in the partial pooling for intermediate randomization steps (see the second monomer addition in Figure 9.24).

Early reported applications of this technique were the preparation of a 24-member peptide library [83], of a 125-member tripeptide-substituted cinnamic acid library tested for inhibition of tyrosine phosphatase PTP1B [83], of a 64-member peptide-like library [83] and of libraries based on a natural product, epothilone, using also new polystyrene grafted solid supports [84]. Other applications, ranging from 1,5-benzodiazepin-2-one library synthesis [85] to chalcone library synthesis [86], were also recently reported. Commercialization of the basic components for this technique [87] (reaction supports and vessels, tags, software, sorters, reaction stations, and so on) will ensure its quick and effective use in combinatorial chemistry and also the implementation of new technical features and possibilities for more complex and demanding applications in future.

IV. EXPERIMENTAL COMPARISON WITH DIRECT DECONVOLUTION

A comparison between different encoding techniques was never attempted using the same library and building it with two, or more, different codes. While many authors reported or claimed specific advantages for a particular technique, a detailed examination of the available methods reveals a fundamental complementarity, rather than redundancy, of the approaches, such that different libraries may better benefit from different encoding methods. Moreover, variables such as the availability of analytical or synthetic instrumentation, resources, which may be devoted to the assessment for a chosen encoding technique and the expertise of the chemist(s) involved in the project, will also strongly influence the final choice of a particular method.

It is much easier, though, to compare encoding techniques with direct deconvolution (see elsewhere in this book) [1, 6]. The two main classes of structure determination methods for pool libraries are significantly different, and clear distinctions about their usefulness can be made. The example of a 240-member (without considering diastereoisomers) mercaptoacyl pyrrolidine library, which was prepared as an encoded (secondary amine tags) [40] or as a nonencoded library and then submitted to iterative deconvolution [88], will be used for this comparison. The following considerations were either reported by MacLean *et al.* [40] or derived from the critical analysis of the results obtained.

Any chemically encoded library requires a double orthogonal chemical strategy, or the construction of elaborate tags/linkers on the solid support prior to the synthesis. Even when these are easily prepared and inert, the synthetic scheme becomes more complicated than direct deconvolution methods, where only the library synthesis is required. Sometimes the tag chemistry and the library synthesis simply cannot be run in parallel, so a different approach is needed. Nonchemically encoded methods compare more favourably in this respect with direct deconvolution methods, but at present they require either sophisticated equipment and/or significant expertise.

While an encoded library, once prepared, requires only testing and decoding of positives, a nonencoded library almost always requires the synthesis of additional copies of the library in order to deconvolute the library structure. When some of or all the monomers used are not commercially available materials, the "waste" of valuable chemical intermediates may become an issue. An exception is the on-bead screening of nonencoded libraries, which anyway benefits of additional inserted coding molecules; both solution- and solid-phase-based biological assays can be performed so as to verify the correlation of the activities obtained from the two methods, thus reverting to solution assays when the on-bead interaction with the receptor/ligand does not take place (hindrance of the support, link of the ligand with the resin in an important area for the binding, and so on). The presence of additional coding molecules, though, may influence on-bead screening through interactions of the coding moiety with the receptor/enzyme tested. The results of various experiments tend to exclude this possibility (see the previous paragraphs).

Both classes suffer from issues related to the concentration of the material to be tested. Direct deconvolution controls less precisely the quantity of a single compound in a pool, so that its concentration can only be guessed, assuming ideal portioning of the resin at each step, similar reactivities of building blocks, and so on. Moreover, the simultaneous presence of other individuals in the cleaved pool solution may cause "false positive" (synergisms, side reactions, and so on) or "false negative" results (degradation of actives *via* side reactions, low concentration of an individual, and so on) or solubility problems (partial precipitation of material in a pool). Encoding and single bead screening, both in solution- and on solid-phase, do not have many of the above-mentioned problems because the bead contains only a single individual per well. Unfortunately, the used solid supports have typical loadings of 100–300 pmol, while cleavage solutions of 100–200 μL are the minimum volume acceptable; this makes the concentration of a high synthetic quality library individual from a bead not more than 1.5 μM. While this allowed the discovery of potent inhibitors [32], it could not find weak-medium inhibitors, which may have importance for uncharacterized targets or unexploited families of chemical compounds, or for more detailed SAR of a class of compounds.

Direct deconvolution needs to test pools of compounds which become large for large libraries, while encoding allows the "tailoring" of the screening. The decoding process for encoded libraries is generally the rate-limiting step; having a large number of beads show interesting activity in a primary screen requires that they all be decoded and for most of chemical encoding methods this process is in the range of hours for one single structure. Quite often a sort of statistical sampling is done, because a single bead-based assay for millions of compounds may become unpractical for the amount of effort required for separating, testing, and spotting single beads. Theoretical calculations [67] showed the necessity of redundancy of five for having the representation of 99% of the individuals in a library (for example, for a 100,000 member library, 500,000 beads must be screened). Screening less than five library equivalents may prevent the discovery of potent inhibitors/ligands which were not actually represented among the screened beads. Alternatively, random pooling can be done when necessary. Pools of 100–500 beads may be assembled and tested with partial release of the compound, then the beads from active pools may be redistributed and tested as single beads through a second release of the compound, and decoding is finally done on the positive beads.

Maybe the most important difference between the two classes is in the result of the structure determination process. Direct deconvolution methods usually produce a single structure which may or may not be the most active compound (see elsewhere in this book [1]), while the identification of additional structures for having a preliminary SAR strongly impacts on time, resources, and quantity of materials. Encoding produces results for the whole library, allowing the assessment of an SAR even when running a primary screening, and it often allows the test of the same library on many different biological assays. Moreover, direct deconvolution results are strongly dependent on the total activity of the library; if few compounds are significantly active, then the most active compound(s) will be easily identified, but when many compounds have weak to moderate activities the selection at each iteration step becomes trickier and less reliable. Encoding is independent from the overall activity of the library because the biological assay is performed on a single bead base.

Encoding methods provide generally a larger set of more detailed data but also require additional "complexity" to be added to the library structure. Direct deconvolution methods also have appealing features, such as being often independent from analytical techniques, easy to perform and often reliable. An "absolute" choice among the two main structure determination method classes, and among different methods in the same class, does not exist. The skilled chemist, by being aware of the many alternatives that exist, is more likely to pick the right method for his/her specific requirements.

V. SUMMARY AND OUTLOOK

The use of encoding techniques has already become a popular subject of research in the last few years in combinatorial chemistry. It is easy to think of a further, exponential increase in the application of some of the presented techniques and also of new ones yet to appear in the literature. The fields which will benefit more from further technological innovations are the decoding techniques, which are the rate-limiting steps especially for chemical coding approaches and need to be significantly shortened; the bead picking/handling methodologies, which could help in processing faster extremely large numbers of beads in many different assays; new high quality and equal size solid supports compatible with on-bead screening, allowing the interaction between the soluble receptor and the resin bound ligand on the bead surface; nondestructive and sensitive analytical techniques for fast on-bead structure determination, and powerful software tools for recording and tracking information about library compounds "on the fly" and at the end of the synthesis (in other words, automated instrumentation for the synthesis, encoding, and decoding of libraries in general).

Two subjects will probably be "hot topics" in the near future. The first is the further miniaturization of synthesis and biological assays, so as to increase possibly the concentration of cleaved compound from a single bead (more sensitive assays), to decrease the amount of monomers/scaffolds/reagents to be used for library synthesis (cost and time saving) and the number of "redundant" beads carrying the same compound. Theoretically, a single encoded bead for each compound to be reused for different on-bead screening and finally decoded is the "Holy Grail" for future efforts on large primary libraries. The second is the more thorough application of material science in particular, and physical disciplines in general, to combinatorial chemistry, which will easily allow nonchemical, fast and automated production of large number of encoded microparticles for solid-phase synthesis and decoding procedures using hitech instrumentation.

In conclusion, let me underscore once more the need for combinatorial chemists to be aware both of the current status of encoding/direct deconvolution techniques, of their main features and requirements, and of the new trends/methodologies/ideas emerging in literature. If the above is true, chemists will pick the best method to be applied to the combinatorial problem they are facing. It is also important to consider not only the synthesis of the library but also the screenings which will be used for testing it, the analytical and automation facilities available, and most of all the timelines of the project, so as to choose the best "compromise" which will give the best overall results for the project needs.

The field of chemical and nonchemical encoding has been in the past the subject of several reviews. A few of them [89–95] are provided here as references for the interested reader.

REFERENCES

1. Seneci P, Direct deconvolution techniques for pool libraries of small organic molecules, *Combinatorial Chemistry and Combinatorial Technologies: Methods and Applications* (Eds. Miertus S, Fassina G), 153–192, 2005, Chapter 8, this volume.
2. (a) Selway CN, Terrett NK, Parallel-compound synthesis: methodology for accelerating drug discovery, *Bioorg. Med. Chem.*, 4:645–654, 1996. (b) Sashar S, Mjalli AMM, Techniques for single-compound synthesis, *Annual Reports in Combinatorial Chemistry and Molecular Diversity* (Eds. Moos WH, Pavia MR, Kay BK, Ellington AD), ESCOM Science publishers, Leiden, pp. 19–29, 1997. (c) Cargill JF, Lebl M, New methods in combinatorial chemistry: robotics and parallel synthesis, *Curr. Opin. Chem. Biol.*, 1:67–71, 1997.

3. (a) Janda KD, Tagged versus untagged libraries: methods for the generation and screening of combinatorial chemical libraries, *Proc. Natl. Acad. Sci. USA*, 91:10779–10785, 1994. (b) Baldwin JJ, Dolle RE, Deconvolution methods in solid-phase synthesis, in *Annual Reports in Combinatorial Chemistry and Molecular Diversity* (Eds. Moos WH, Pavia MR, Kay BK, Ellington AD), ESCOM Science publishers, Leiden, pp. 287–297, 1997. (c) Xiao XY, Nova MP, Radiofrequency encoding and additional techniques for the structure elucidation of synthetic combinatorial libraries, in *Combinatorial Chemistry: Synthesis and application* (Eds. Wilson SR, Czarnik AW), John Wiley & Sons, New York. (d) Ref. 1 and references cited therein, pp. 135–152, 1997.
4. (a) van Breemen RB, Huang CR, Nikolic D, Woodbury CP, Zhao YZ, Venton DL, Pulsed ultrafiltration mass spectrometry: a new method for screening combinatorial libraries, *Anal. Chem.*, 69:2159–2164, 1997. (b) Dunayevskiy YM, Lai JJ, Quinn C, Talley F, Vouros P, Mass spectrometric identification of ligands selected from combinatorial libraries using gel filtration, *Rapid. Commun. Mass Spectrom.*, 11:1178–1184, 1997. (c) Chu YH, Dunayevskiy YM, Kirby DP, Vouros P, Karger BL, Affinity capillary electrophoresis-mass spectrometry for screening combinatorial libraries, *J. Am. Chem. Soc.*, 118:7827–7835, 1996. (d) Brummel CL, Vickerman JC, Carr, SA, Hemling ME, Roberts GD, Johnson W, Weinstock J, Gaitanopoulos D, Benkovic SJ, Winograd N, Evaluation of mass spectrometric methods applicable to the direct analysis of non-peptide bead-bound combinatorial libraries, *Anal. Chem.*, 68:237–242, 1996.
5. (a) Lam KS, Salmon SE, Hersh EM, Hruby VJ, Kazmierski WM, Knapp RJ, A new type of synthetic peptide library for identifying ligand-binding activity, *Nature*, 354:82–84, 1991. (b) Lam KS, Lebl M, Krchnak V, The one-bead one-compoundcombinatorial library method, *Chem. Rev.*, 97:411–448, 1997.
6. Furka, Ā, Combinatorial Chemistry: from mix-split to discrete in *Combinatorial Chemistry and Combinatorial Technologies: Methods and Applications* (Eds. Miertus S, Fassina G), 7–32, 2005, Chapter 2, this volume.
7. From here on the concept one bead–one compound will be used also in place of the more common synonym mix and split technique for library synthesis.
8. Brenner S, Lerner RA, Encoded combinatorial chemistry, *Proc. Natl. Acad. Sci. USA*, 89:5381–5383, 1992.
9. (a) Nielsen J, Brenner S, Janda KD, Synthetic methods for the implementation of encoded combinatorial chemistry, *J. Am. Chem. Soc.*, 115:9812–9813, 1993. (b) Jones DG, Applications of encoded synthetic libraries in ligand discovery, *Polym. Prepr.*, 35:981–982, 1994.
10. (a) Fenniri H, Janda KD, Lerner RA, Encoded reaction cassette for the highly sensitive detection of the making and breaking of chemical bonds, *Proc. Natl. Acad. Sci. USA*, 92:2278–2282, 1995. (b) Fenniri H, Rapid screening of biocatalysts, *CHEMTECH*, 26:15–25, 1996.
11. Needels MC, Jones DG, Tate EH, Heinkel GL, Kochersperger LM, Dower WJ, Barrett RW, Gallop MA, Generation and screening of an oligonucleotide-encoded synthetic peptide library, *Proc. Natl. Acad. Sci. USA*, 90:10700–10704, 1993.
12. Nielsen J, Janda KD, Toward chemical implementation of encoded combinatorial libraries, *Methods: Meth. Enzymol.*, 6:361–371, 1994.

13. (a) Dani M, Peptide display libraries: design and construction, in *Combinatorial Chemistry and Combinatorial Technologies: Methods and Applications* (Eds. Miertus S, Fassina G), 411–430, 2005, chapter 17, this volume.
14. Nikolaiev V, Stierandova A, Krchnak V, Seligmann B, Lam KS, Salmon SE, Lebl M, Peptide-encoding for structure determination of nonsequenceable polymers within libraries synthesized and tested on solid-phase supports, *Peptide Res.*, 6:161–170, 1993.
15. Kerr JM, Banville SC, Zuckermann RN, Encoded combinatorial peptide libraries containing non-natural amino acids, *J. Am. Chem. Soc.*, 115:2529–2531, 1993.
16. Hornik V, Hadas, E, Self-encoded, highly condensed solid phase-supported peptide library for identification of ligand-specific peptides, *React. Polym.*, 22:213–220, 1994.
17. Felder ER, Heizmann G, Matthews IT, Rink H, Spieser E, A new combination of protecting groups and links for encoded synthetic libraries suited for consecutive tests on the solid phase and in solution, *Molecular Diversity*, 1:109–112, 1996.
18. Vagner J, Barany G, Lam KS, Krchnak V, Sepetov NF, Ostrem JA, Strop P, Lebl M, Enzyme-mediated spatial segregation on individual polymeric support beads: application to generation and screening of encoded combinatorial libraries, *Proc. Nat. Acad. Sci. USA*, 93:8194–8199, 1996.
19. Ohlmeyer MHJ, Swanson RN, Dillard LW, Reader JC, Asouline G, Kobayashi R, Wigler M, Still WC, Complex synthetic chemical libraries indexed with molecular tags, *Proc. Natl. Acad. Sci. USA*, 90:10922–10926, 1993.
20. Nestler HP, Bartlett PA, Still WC, A general method for molecular tagging of encoded combinatorial chemistry libraries, *J. Org. Chem.*, 59:4723–4724, 1994.
21. Yoon SS, Still WC, Sequence-selective binding with a synthetic receptor, *Tetrahedron*, 51:567–578, 1995.
22. Hill HH, McMinn DG, Detectors for capillary chromatography, *Chemical Analysis*, 121:83–107, 1992.
23. (a) Yoon SS, Still WC, An exceptional synthetic receptor for peptides, *J. Am. Chem. Soc.*, 115:823–824, 1993. (b) Yoon SS, Still WC, Cyclooligomeric receptors for the sequence selective binding of peptides. A tetrahedral receptor from trimesic acid and 1, 2-diamines, *Tetrahedron Lett.*, 35:8557–8560, 1994.
24. Baldwin JJ, Design, synthesis and use of binary encoded synthetic chemical libraries, *Molecular Diversity*, 2:81–88, 1996.
25. Still WC, Discovery of sequence-selective peptide binding by synthetic receptors using encoded combinatorial libraries, *Acc. Chem. Res.*, 29:155–163, 1996.
26. Burbaum JJ, Ohlmeyer MHJ, Reader JC, Henderson I, Dillard LW, Li G, Randle TL, Sigal NH, Chelsky D, Baldwin JJ, A paradigm for drug discovery employing encoded combinatorial libraries, *Proc. Natl. Acad. Sci. USA*, 92:6027–6031, 1995.
27. Burger MT, Still WC, Synthetic ionophores. Encoded combinatorial libraries of cyclen-based receptors for Cu(2+) and Co(2+), *J. Org. Chem.*, 60:7382–7383, 1995.
28. Appell KC, Chung TDY, Ohlmeyer MJH, Sigal NH, Baldwin JJ, Chelsky D, Biological screening of a large combinatorial library, *J. Biomol. Screening*, 1:27–31, 1996.

29. Nestler HP, Sequence-selective nonmacrocyclic two-armed receptors for peptides, *Molecular Diversity*, 2:35–40, 1996.
30. Liang R, Yan L, Loebach J, Ge M, Uozumi Y, Sekanina K, Horan N, Gildersleeve J, Thompson C, Smith A, Biswas K, Still WC, Kahne D, Parallel synthesis and screening of a solid phase carbohydrate library, *Science*, 274:1520–1522, 1996.
31. Carroll CD, Patel H, Johnson TO, Guo T, Orlowski M, He ZM, Cavallaro CM, Guo J, Oskman A, Gluzman IY, Connelly J, Chelsky D, Goldberg DE, Dollé RE, Identification of potent inhibitors of Plasmodium falciparium plasmepsin II from an encoded statine combinatorial library, *Bioorg. Med. Chem. Lett.*, 8:2315–2320, 1998.
32. Carroll CD, Johnson TO, Tao S, Lauri G, Orlowski M, Gluzman IY, Goldnerg DE, Dolle RE, Evaluation of a structure-based statine cyclic diamino amide encoded combinatorial library against plasmepsin II and cathepsin D, *Bioorg. Med. Chem. Lett.*, 8:2315–2320, 1998.
33. McMillan K, Adler M, Auld DS, Baldwin JJ, Blasko E, Browne LJ, Chelsky D, Davey D, Dolle RE, Eagen KA, Erickson S, Feldman RI, Glaser CB, Mallari C, Morrissey MM, Ohlmeyer MHJ, Pan G, Parkinson JF, Phillips GB, Polokoff MA, Sigal NH, Vergona R, Whitlow M, Young TA, Devlin JJ, Allosteric inhibitors of inducible nitric oxide synthase dimerization discovered *via* combinatorial chemistry, *Proc. Natl. Acad. Sci. USA*, 97:1506–1511, 2000.
34. Dolle RE, Guo J, O'Brien L, Jin Y, Piznik M, Bowman KJ, Li W, Egan WJ, Cavallaio CL, Roughton AL, Zhao Q, Reader JC, Orlowski M, Jacob-Samuel, B, Carroll CD, A statistical-based approach to assessing the fidelity of combinatorial libraries encoded with electrophoric molecular tags, Development and application of tag decode-assisted single bead LC/MS analysis, *J. Comb. Chem.*, 2:716–731, 2000.
35. Tan DS, Foley MA, Shair MD, Schreiber SL, Stereoselective synthesis of over two million compounds having structural features both reminiscent of natural products and compatible with miniaturized cell-based assays, *J. Am. Chem. Soc.*, 120:8565–8566, 1998.
36. Tan DS, Foley MA, Stockwell BR, Shair MD, Schreiber SL, Synthesis and preliminary evaluation of a library of polycyclic small molecules for use in chemical genetic assays, *J. Am. Chem. Soc.*, 121:9073–9087, 1999.
37. Blackwell HE, Perez L, Stavenger RA, Tallarico JA, Cope Eatough E, Foley MA, Schreiber SL, A one-bead, one-stock solution approach to chemical genetics: Part 1, *Chem. Biol.*, 8:1167–1182, 2001.
38. Spring DR, Krishnan S, Blackwell HE, Schreiber SL, Diversity-oriented synthesis of biaryl-containing medium rings using a one bead/one stock solution platform, *J. Am. Chem. Soc.*, 124:1354–1363, 2002.
39. Ni ZJ, Maclean D, Holmes CP, Murphy MM, Ruhland B, Jacobs JW, Gordon EM, Gallop MA, Versatile approach to encoding combinatorial organic syntheses using chemically robust secondary amine tags, *J. Med. Chem.*, 39:1601–1608, 1996.
40. Maclean D, Schullek JR, Murphy MM, Ni ZJ, Gordon EM, Gallop MA, Encoded combinatorial chemistry: synthesis and screening of a library of highly functionalized pyrrolidines, *Proc. Natl. Acad. Sci. USA*, 94:2805–2810, 1997.
41. Murphy MM, Schullek JR, Gordon EM, Gallop MA, Combinatorial organic synthesis of highly functionalized pyrrolidines: identification of a potent

angiotensin converting enzyme inhibitor from a mercaptoacyl proline library, *J. Am. Chem. Soc.*, 117:7029–7030, 1995.

42. Atuegbu A, Maclean D, Nguyen C, Gordon EM, Jacobs JW, Combinatorial modification of natural products: preparation of unencoded and encoded libraries of Rauwolfia alkaloids, *Bioorg. Med. Chem.*, 4:1097–1106, 1996.
43. Schullek JR, Butler JH, Ni ZJ, Chen D, Yuan ZY, A high-density screening format for encoded combinatorial libraries: assay miniaturization and its application to enzymatic reactions, *Anal. Biochem.*, 246:20–29, 1997.
44. Fitch WL, Baer TA, Chen W, Holden F, Holmes CP, Maclean D, Shah N, Sullivan E, Tang M, Waybourn P, Fischer SM, Miller CA, Snyder LR, Improved methods for encoding and decoding dialkylamine-encoded combinatorial libraries, *J. Comb. Chem.*, 1:188–194, 1999.
45. Lane SJ, Pipe A, Unambiguous bead decoding by microelectrospray capillary electrochromatography tandem mass spectrometry of dansylated secondary amine tags from encoded combinatorial organic synthesis, *Rapid Commun. Mass Spectrom.*, 12:667–674, 1998.
46. Lane SJ, Pipe A, A single generic microbore liquid chromatography/time-of-flight mass spectrometry solution for the simultaneous accurate mass determination of compounds on single beads, the decoding of dansylated orthogonal tags pertaining to compounds and accurate isotopic difference target analysis, *Rapid Commun. Mass Spectrom.*, 13:798–814, 1999.
47. Lane SJ, Pipe A, Single bead and hard tag decoding using accurate isotopic difference target analysis-encoded combinatorial libraries, *Rapid Commun. Mass Spectrom.*, 14:782–793, 2000.
48. Geysen HM, Wagner CD, Bodnar WM, Markworth CJ, Parke GJ, Schoenen FJ, Wagner DS, Kinder DS, Isotope or mass encoding of combinatorial libraries, *Chemistry & Biology*, 3:679–688, 1996.
49. Wagner DS, Markworth CJ, Wagner CD, Schoenen FJ, Rewerts CE, Kay BK, Geysen HM, Ratio encoding combinatorial libraries with stable isotopes and their utility in pharmaceutical research, *Combi. Chem. High Throughput Screen*, 1:143–153, 1998.
50. Garigipati RS, Sarkar SK, A binary coding method for use in combinatorial chemistry, *WO 9714814 A1*, 970424, p. 13, 1997.
51. Seneci P, Sizemore C, Islam K, Kocis P, Combinatorial chemistry and natural products, Teicoplanin aglycone as a molecular scaffold for solid phase synthesis of combinatorial libraries, *Tetrahedron Lett.*, 37:6319–6322, 1996.
52. Kocis P, Krchnak V, Lebl M, Symmetrical structure allowing the selective multiple release of a defined quantity of peptide from a single bead of polymeric support, *Tetrahedron Lett.*, 34:7251–7252, 1993.
53. Sizemore CF, Seneci P, Kocis P, Wertman KF, Islam K, Combinatorial chemistry and natural products: determination of the biological activity of on-bead, double cleavable teicoplanin aglycon (TD), *Protein Pept. Lett.*, 3:253–260, 1996.
54. Lebl M, Krchnak V, Salmon SE, Lam KS, Screening of completely random one-bead one-peptide libraries for activities in solution, *Methods: Meth. Enzymol.*, 6:381–387, 1994.
55. Salmon SE, Lam KS, Lebl M, Kandola A, Khattri PS, Wade S, Patek M, Kocis P, Krchnak V, Thorpe D, Felder S, Discovery of biologically active peptides in random libraries: solution-phase testing after staged orthogonal release from resin beads, *Proc. Natl. Acad. Sci. USA*, 90:11708–11712, 1993.

56. Brown BB, Wagner DS, Geysen HM, A single-bead decode strategy using electrospray ionization mass spectrometry and a new photolabile linker: 3-amino-3-(2-nitrophenyl)-propionic acid, *Molecular Diversity*, 1:4–12, 1995.
57. Cardno M, Bradley M, A simple multiple release system for combinatorial library and peptide analysis, *Tetrahedron Lett.*, 37:135–138, 1996.
58. Rink H, Vetter D, Gercken B, Felder E, Preparation of combinatorial compound libraries coded with element atom tags, *WO 9630392 A1*, 961003, p. 49, 1996.
59. Pirrung MC, Park K, Discovery of selective metal-binding peptoids using ^{19}F encoded combinatorial libraries, *Bioorg. Med. Chem. Lett.*, 10:2115–2118, 2000.
60. Hochlowski JE, Whittern DN, Sowin TJ, Encoding of combinatorial chemistry libraries by fluorine-19 NMR. *J. Comb. Chem.*, 1:291–293, 1999.
61. Garigipati RS, Adams JL, Photolytically cleavable arylsulfonamide encoding and linking agents for use in combinatorial chemistry, *WO 9630337 A1*, 961003, p. 32, 1996.
62. Yamashita DS, Weinstock J, Encoded combinatorial libraries, *WO 9532425 A1*, 951130, p. 42, 1995.
63. Scott RH, Balasubramanian S, Properties of fluorophores on solid phase resins: implications for screening, encoding and reaction monitoring, *Bioorg. Med. Chem. Lett.*, 7:1567–1572, 1997.
64. Scott RH, Barnes C, Gerhard U, Balasubramanian S, Exploring a chemical encoding strategy for combinatorial synthesis using Friedel-Crafts alkylation, *Chem. Commun.*, pp. 1331–1332, 1999.
65. Egner BJ, Rana S, Smith H, Bouloc N, Frey J, Brocklesby WS, Bradley M, Tagging in combinatorial chemistry: the use of colored and fluorescent beads, *Chem. Commun.*, pp. 735–736, 1997.
66. Edwards PN, Main BG, Shute RE, Chemical libraries, labelling and deconvolution thereof, *WO 9623749 A1*, 960808, p. 13, 1996.
67. Main BG, Shute RE, Intrinsically labelled solid support, *WO 9703931 A1*, 970602, p. 10, 1997.
68. Rahman SS, Busby DJ, Lee DC, Infrared and Raman spectra of a single resin bead for analysis of solid-phase reactions and use in encoding combinatorial libraries, *J. Org. Chem.*, 63:6196–6199, 1998.
69. Fenniri H, Hedderich HG, Haber KS, Achkar J, Taylor B, Ben-Amotz D, Towards the DRED of resin-supported combinatorial libraries: a noninvasive methodology based on bead self-encoding and multispectral imaging, *Angew. Chem. Int. Ed.*, 39:4483–4485, 2000.
70. Fenniri H, Ding L, Ribbe AE, Zyrianov Y, Barcoded resins: a new concept for polymer-supported combinatorial library self-deconvolution, *J. Am. Chem. Soc.*, 123:8151–8152, 2001.
71. Kaye PH, Tracey MC, Coded particles for process sequence tracking in combinatorial compound library preparation, *WO 9715390 A1*, 970501, p. 36, 1997.
72. (a) Nicolaou KC, Xiao XY, Parandoosh Z, Senyei A, Nova MP, Radio-frequency encoded combinatorial chemistry, *Angew. Chem. Int. Ed.*, 34:2289–2291, 1995. (b) Moran EJ, Sarshar S, Cargill JF, Shahbaz MM, Lio A, Mjalli AMM, Armstrong RW, Radio frequency tag encoded combinatorial library method for the discovery of tripeptide-substituted cinnamic acid inhibitors of the protein tyrosine phosphatase PTP1B, *J. Am. Chem. Soc.*, 117:10787–10788, 1995.

73. For a recent review see Felder E, Resins, linker and reactions for solid phase synthesis of organic libraries, in *Combinatorial Chemistry and Combinatorial Technologies: Methods and Applications* (Eds. Miertus S, Fassina G), 89–106, 2005, chapter 6, this volume.
74. (a) Rapp W, PEG grafted polystyrene tentacle polymers: physico-chemical properties and application in chemical synthesis, in *Molecular Diversity and Combinatorial Chemistry* (Eds Chaiken IW, Janda KD), American Chemical Society, Washington, pp. 425–464, 1996. (b) Porco JA Jr, Deegan TL, Gooding OW, Heisler K, Labadie JW, Newcomb WS, Nguyen C, Tran TH, Van Eikeren P, New poly(styrene-oxyethylene) graft copolymers for solid-phase organic synthesis: linkers and applications, *Book of Abstracts, 212th ACS National Meeting*, Orlando, FL, August 25–29, American Chemical Society, Washington, 1996. (c) Kempe M, Barany G, CLEAR: a novel family of highly cross-linked polymeric supports for solid-phase peptide synthesis, *J. Am. Chem. Soc.*, 118:7083–7093, 1996.
75. (a) Backes BJ, Ellman JA, Solid support linker strategies, *Curr. Opin. Chem. Biol.*, 1:86–93, 1997. (b) Blackburn C, Albericio F, Kates SA, Functionalized resins and linkers for solid-phase synthesis of small molecules, *Drugs Future*, 22: 1007–1025, 1997.
76. (a) Geysen HM, Meloen RH, Barteling SJ, Use of peptide synthesis to probe viral antigens for epitopes to a resolution of a single amino acid, *Proc. Natl. Acad. Sci. USA*, 81:3998–4002, 1984. (b) Valerio RM, Bray AM, Campbell RA, Dipasquale A, Margellis C, Rodda SJ, Geysen HM, Maeji NJ, Multipin peptide synthesis at the micromole scale using 2-hydroxyethyl methacrylate grafted polyethylene supports, *Int. J. Pept. Protein Res.*, 42:1–9. (c) Chin J, Fell B, Shapiro MJ, Tomesch J, Wareing JR, Bray AM, Magic angle spinning NMR for reaction monitoring and structure determination of molecules attached to multipin crowns, *J. Org. Chem.*, 62:538–539, 1997.
77. (a) Frank R, Spot synthesis: an easy technique for the positionally addressable, parallel chemical synthesis on a membrane support, *Tetrahedron*, 48:9217–9232. (b) Frank R, Strategies and techniques in simultaneous multiple peptide synthesis based on the segmentation of membrane type supports, *Bioorg. Med. Chem. Lett.*, 3:425–430, 1993.
78. (a) Gravert DJ, Janda KD, Synthesis on soluble polymers: new reactions and the construction of small molecules, *Curr. Opin. Chem. Biol.* 1:107–113, 1997. (b) Gravert DJ, Janda KD, Organic synthesis on soluble polymer supports: liquid-phase methodologies, *Chem. Rev.* 97:489–509, 1997. (c) Kim RM, Manna M, Hutchins SM, Griffin PR, Yates NA, Bernick AM, Champman KT, Dendrimer-supported combinatorial chemistry, *Proc. Natl. Acad. Sci. USA*, 93:10012–10017, 1996.
79. (a) French PJ, Gennissen PTJ, Sarro PM, New silicon micromachining techniques for microsystems, *Sens. Actuators A*, A62, 652–662. (b) Bao M, Li X, Shen S, Chen H, A novel micromachining technology for multilevel structures of silicon, *Sens. Actuators A*, A63:217–221, 1997. (c) Navarro M, Lopez-Villegas JM, Samitier J, Morante JR, Bausells J, Merlos A, Electrochemical etching of porous silicon sacrificial layers for micromachining applications, *J. Micromech. Microeng.*, 7:131–132, 1997.
80. Xiao X, Zhao C, Potash H, Nova MP, Combinatorial chemistry with laser optical encoding, *Angew. Chem. Int. Ed.*, 36:780–782, 1997.

81. Nicolaou KC, Pfefferkorn JA, Mitchell HJ, Roecker AJ, Barluenga S, Cao GQ, Affleck RL, Lillig JE, Natural product-like combinatorial libraries based on privileged structures, 2, Construction of a 10,000-membered benzopyran library by directed split-and-pool chemistry using Nanokans and optical encoding, *J. Am. Chem. Soc.*, 122:9954–9967, 2000.
82. Xiao X, Parandoosh Z, Nova MP, Design and synthesis of a taxoid library using radiofrequency encoded combinatorial chemistry, *J. Org. Chem.* 62:6029–6033, 1997.
83. Armstrong RW, Tempest PA, Cargill JF, Microchip encoded combinatorial libraries, Generation of a spatially encoded library from a pool synthesis. *Chimia*, 50:258–260, 1996.
84. Nicolaou KC, Vourloumis D, Li T, Pastor J, Winssinger N, He Y, Ninkovic S, Sarabia F, Vallberg H, Roschangar F, King NP, Finlay MRV, Giannakakou P, Verdier-Pinard P, Hamel E, Designed epothilones: combinatorial synthesis, tubulin assembly properties, and cytotoxic action against taxol-resistant tumor cells, *Angew. Chem., Int. Ed.*, 36:2097–2103, 1997.
85. Herpin TF, Van Kirk KG, Salvino JM, Yu ST, Labaudiniere RF, Synthesis of a 10,000 member 1,5-benzodiazepine-2-one library by the directed sorting method, *J. Comb. Chem.*, 2:513–521, 2000.
86. Nicolaou KC, Pfefferkorn JA, Cao GQ, Selenium-based solid-phase synthesis of benzopyrans I: Applications to combinatorial synthesis of natural products, *Angew. Chem. Int. Ed.*, 39:734–739, 2000.
87. Radiofrequency-based parallel synthesis tools are commercialized by IRORI Quantum Microchemistry, 11025 North Torrey Pines Road, La Jolla, CA 92037, USA.
88. Burgess K, Liaw AI, Wang N, Combinatorial technologies involving reiterative division/coupling/recombination: statistical considerations, *J. Med. Chem.*, 37:2985–2987, 1994.
89. Jacobs JW, Ni ZJ, Encoded combinatorial chemistry, in *Combinatorial Chemistry and Molecular Diversity in Drug Discovery*, (Eds. Gordon EM, Kerwin J), Wiley-Liss, New York, pp. 271–290, 1998.
90. Seneci P, *Solid-Phase Synthesis and Combinatorial Technologies*, Wiley Interscience, New York, pp. 264–338, 2000.
91. Terrett N, *Combinatorial Chemistry*, Oxford University Press, Oxford, UK, pp. 77–94, 1999.
92. Baldwin JJ, Dolle R, Deconvolution tools for solid-phase synthesis, *A Practical Guide to Combinatorial Chemistry*, (Eds. Czarnik AW, DeWitt SH), ACS, Washington, 153–174, 1997.
93. Barnes C, Scott RH, Balasubramanian S, Progress in chemical encoding for combinatorial synthesis, *Recent Res. Dev. Org. Chem.*, 2:367–379, 1998.
94. Barnes C, Balasubramanian S, Recent developments in the encoding and deconvolution of combinatorial libraries, *Curr. Opin. Chem. Biol.*, 4:346–350, 2000.
95. Xiao XY, Li R, Zhuang H, Ewing B, Karunaratne K, Lillig J, Brown R, Nicolaou KC, Solid-phase combinatorial synthesis using MicroKan reactors, Rf tagging, and directed sorting, *Biotechnol. Bioeng.*, 71:44–50, 2000.

10 Analytical Characterization of Synthetic Organic Libraries

Nikolai Sepetov and Olga Issakova

CONTENTS

I. INTRODUCTION

Combinatorial chemistry was developed in the 1990s as a tool for rapid identification of new drug candidates, and was used at the beginning by only a few companies. Today, it is one of the fastest-growing and fastest-changing research fields and is well incorporated into the pharmaceutical, biotechnological and agrochemical industries [1–4]. The most dramatic changes in the field of combinatorial technology within the last several years include a shift from production of peptide libraries to combinatorial synthesis of small organic molecules and the expansion of automation in both the synthesis process and screening technologies [2]. These changes, especially with the development of automation in synthesis, have shifted the bottleneck from production of new compounds to analytical characterization of small-molecule combinatorial libraries.

In this chapter we are intentionally using the term "analytical characterization" instead of "quality control (QC)" to stress that analytical chemistry is not just a "gatekeeper," passing libraries from production to screening. While methods for the generation of peptide libraries are now well established, preparation of small-organic-molecule libraries remains a very challenging and rapidly evolving area of research. To maintain the speed of development and production of small-organic-molecule libraries, analytical chemistry should not only provide chemists information about the final quality of libraries (quality control and quality assurance) but help them during development of libraries with selection of building blocks, optimization of reaction conditions, etc.

It is worth mentioning that a subject for analytical characterization is a *library*, but technically *compounds* are analyzed. What is the difference between analysis of a combinatorial library and analysis of a collection of many compounds? There are at least two major differences related to quantitative and qualitative aspects of combinatorial libraries. The first is that combinatorial synthesis is a practically unlimited source of compounds. The number of compounds in a library may very easily reach the point at which none of the existing analytical methods will allow measurement of all compounds in a library. The second major difference between a combinatorial library and a collection of compounds is that all components of a combinatorial library share something common (design, synthetic history, etc.), and successful analysis of combinatorial libraries is not possible without utilization of this built-in information.

In this chapter we will discuss not only the methods and techniques used for analysis of small-organic-molecule libraries but also the approaches for generation of these libraries, with the understanding that synthesis, analysis, and screening of combinatorial libraries are the integrated parts of combinatorial technology.

II. HOW TO CHARACTERIZE A COMBINATORIAL LIBRARY

A. General Strategy of Analytical Characterization of Combinatorial Libraries

When analyzing any compound obtained as a result of subsequent chemical reactions, chemists always address three major issues: identity (Did we synthesize what we intended?), quality (How pure is our compound? What side product(s) do we have in our sample?), and quantity (How much did we synthesize? What is the yield of the reactions?). In the process of analysis of organic compounds, the focus has always been on obtaining comprehensive information with a variety of analytical methods. The traditional scheme of analysis includes structure identification by NMR (^{1}H and ^{13}C) and MS (including high-resolution mass spectrometry), with additional confirmation of structure provided by IR spectroscopy, X-ray, etc. [5].

Estimation of purity was performed by elemental analysis, TLC, HPLC with UV/VIS detection, NMR, and often many other methods, assuming that impurities are detectable by those methods [6].

1. Determination of Amount of Synthesized Compound, Usually by Weighing

This scheme worked well for many years, being enforced by the "unwritten" but strict rule of medicinal chemistry laboratories that every compound must be thoroughly characterized prior to biological testing. With the appearance of combinatorial technology enabling the production of thousands and even hundreds of thousands of compounds in a very short period of time, it became clear that the approach of detailed characterization of each compound in a combinatorial library is not feasible for several reasons. In many cases, the amount of each compound synthesized in a combinatorial library is not enough for complete characterization with all of these methods. Most of the above-mentioned analytical methods do not allow fast analysis of samples, and their application would create an obvious bottleneck. For libraries with medium and high complexity (more than a few thousand compounds), analysis of all components of the library is obviously impossible, even by utilizing a single high-speed analytical technique, in a reasonable time frame. This means that many uncharacterized compounds are subjected to biological screening. A new focus was placed on providing useful and practical information about a combinatorial library without analyzing all compounds and using a limited number of analytical techniques. This required development of novel approaches to analytical chemistry and even new ways of thinking by analytical chemists.

In this chapter we will discuss current approaches for analytical characterization of combinatorial libraries in a pharmaceutical industry environment. Recently, several analytical groups have presented very similar strategies for analysis of libraries [7–9]. As will be shown later, the key to successful analytical characterization of a combinatorial library is to perform analytical and chemical work in parallel with the library development. The accumulation of data and "analytical experience" during this process results in an assessment of library quality with a high level of confidence, even if as little as 5–10% of the library components are analyzed. Utilization of the strategy will be demonstrated using two examples: analysis of a library synthesized on a robotic station in spatially addressed format and analysis of a library synthesized in accordance with "split-and-mix" technology.

An example of a scheme for analysis of a library synthesized on a robotic station in 96-well microtiter plates is shown in Figure 10.1. The milestones that need to be reached during the library development process are presented on the left side of the figure, and the goals to be achieved in order to surpass the corresponding milestone are on the right.

Any library begins with an idea, and a chemist usually spends a lot of time and effort until the idea is transformed into the concept of a particular library. However, the first stage at which analytical chemistry becomes heavily involved is synthesis and analysis of "standard compounds." Standard compounds are synthesized after chemical reactions to be used during library

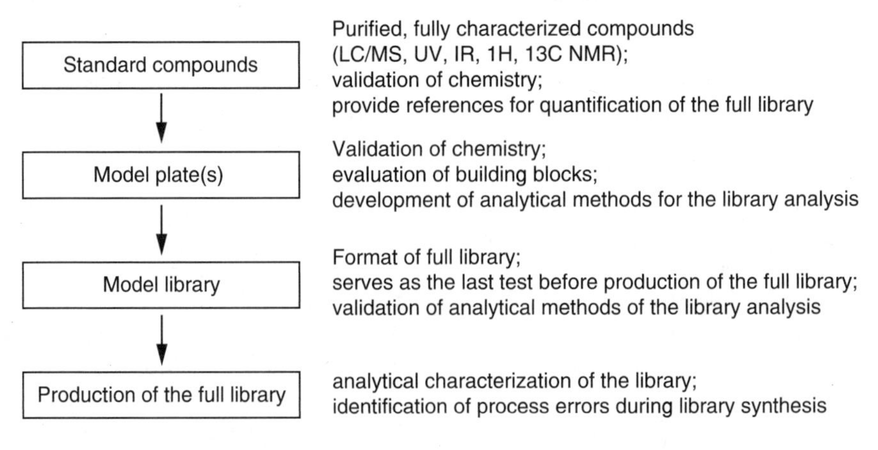

FIGURE 10.1 The scheme of analysis of a library synthesized in spatially addressed format (96-well plate).

synthesis have been developed. Synthesis of standard compounds targets two major goals: validation of chemistry and providing a set of reference compounds for the following stages of library development. Usually the standard compounds are a set of 10–15 compounds with very diverse properties synthesized in amounts that are enough to be used for full characterization. These compounds will be included in the model plates, the model library, and the full library and will serve as a basis for quantification.

At the stage of model plate(s) synthesis, the chemical objectives are the validation and/or optimization of chemistry in high-throughput mode (96-well plate format) as well as the evaluation of building blocks that will be used in the future library. Reaching this goal generally requires the analysis of several hundred samples, providing an opportunity to the analytical chemist for the development and/or optimization of analytical methods for the particular library. At the same time the analytical chemist takes a substantial part of workload from the chemist by performing analyses of relatively big arrays of homologous compounds.

The model library (a few 96-well plates) serves as the last test for synthetic and analytical chemistry before production of the full library. It is synthesized using the optimized conditions developed during the model plate stage, which are as close as possible to the conditions which are expected to be used in the synthesis of the production-scale library. All components of the model library are analyzed, and the results provide a solid basis for assessing the quality of the production library. Surpassing this milestone means that we know what we can expect from the full library in terms of purity and quantity of its components. If the results of analysis of the full library fit the conclusions made after the analysis of the model library, the quality of the library may be assessed with a high level of confidence, even if only a small portion of the full library has been analyzed. With the premise that analysis was performed correctly, disagreement between the analytical results for the full library and those of the model library usually implies that a process error such as misfunction of instrumentation occurred during the library synthesis. In this case the objective of analysis is to identify the possible source(s) of error.

B. Requirements for Analytical Techniques Used in Combinatorial Chemistry

The key questions of compound characterization concerning purity, quantity, and identity are the same whether compounds are synthesized as components of combinatorial libraries or according to traditional medicinal chemistry schemes. However, as already pointed out, analysis of a combinatorial library is quite challenging. Actually, the major analytical challenges in the characterization of a combinatorial library originate in the definition of a good combinatorial library. Usually a good combinatorial library is defined as a large set of very diverse, relatively pure compounds, synthesized in relatively low quantities, sometimes intentionally as mixtures of several components. This means that an analytical chemist should be able to deal with a huge number of samples which have very different properties, develop methods which are applicable to the entire library and not biased to a particular component, work with small quantities of material, and provide results fast and cost effectively. The challenges of analytical characterization of combinatorial libraries put significant limitations on the analytical methods that can be used in combinatorial chemistry. The analytical methods should be

informative,
fast, robust, and easy to automate (requirements for high-throughput analysis, in order to handle large numbers of samples),
sensitive (to analyze small amounts of material),
not based on very specific properties of analytes (to be able to analyze most of the compounds in the library),
easily combined with separation techniques (to handle mixtures of compounds).

There is no ideal method which would satisfy all these requirements. For example, NMR is a very informative method but it is time consuming, optical spectroscopy is sensitive and fast but not very informative, capillary electrophoresis is biased toward charged molecules, etc. Thus it is important to develop an "analytical toolkit" for combinatorial chemistry.

Demands for analytical techniques are different at different stages of library development. The information content of analytical methods is crucial in the early stages of library development, when it is necessary to elucidate the structure of intended compounds and any side products. During the optimization of reaction conditions and the evaluation of building blocks, the ability to combine analytical methods with separation techniques plays an essential role. At the later stages, when hundreds and even thousands of compounds need to be measured, high throughput becomes one of the primary requirements.

C. Analytical Techniques Used in the Early Stages of Combinatorial Library Development

In the early stages of combinatorial library development an analytical chemist deals with a small number of samples synthesized in large quantities, so high throughput of analysis is not an issue. Therefore, the analysis of these samples should not be any different from the analysis of compounds synthesized by traditional medicinal chemistry approaches. Methods of analysis of organic molecules in solution, e.g. NMR, IR, and MS, are well established. However, most of the synthetic strategies used in combinatorial chemistry rely on solid-base techniques. It is generally accepted that optimization of solid-phase reactions is one of the most time-consuming steps of library development. One of the reasons for this is the lack of analytical methods to establish the degree to which the expected chemistry has proceeded on polymeric support. The apparent way to evaluate a reaction on solid support is to cleave the products from the support and characterize them, but this is not always efficient. In certain cases cleavage from the support is not possible, because intermediates in a multistep synthesis may be unstable to cleavage conditions. In this section we will discuss analytical methods used for analysis of compounds attached to a solid phase and the methods of monitoring solid-phase reactions.

1. Nuclear Magnetic Resonance (NMR)

Undoubtedly, NMR is the most informative method for characterization of organic compounds. However, it has limited application in combinatorial chemistry due to several factors. NMR is a relatively insensitive and slow method, requires homogeneous samples, and consumes quite expensive deuterated solvents. Here we will discuss the most recent developments of this method that overcome the major limitations and make NMR one of the promising techniques in combinatorial chemistry. It relates to the application of NMR, not only for analyzing compounds attached to polymer support and for monitoring reactions on a solid phase, but also as a detector for liquid chromatography (LC/NMR). For the most recent review, see [10].

The NMR analysis of organic compounds is effective only if the NMR data have good spectral resolution. Typically, good resolution can be achieved for freely moving (rotating) molecules in a homogeneous magnetic field. Organic molecules attached to solid support give broad resonances in NMR spectra due to restricted molecular motion and inhomogeneities in the magnetic field [11]. The latter arises from the difference in the magnetic susceptibility of beads and surrounding solvent. There are a few ways to overcome these problems, and only those most frequently used will be discussed in this section. One of the ways to narrow NMR resonances from molecules attached to solid support is to swell resin in solvent, thus giving the molecules a significant amount of motional freedom compared to nonswollen resin. The technique of acquiring NMR data on these samples is called "gel-phase" NMR [12–15]. However, NMR resonances are still quite wide, and application of this technique is practical only for nuclei with a large chemical-shift range, primarily for ^{13}C NMR. Since natural abundances of ^{13}C NMR spectra exhibit poor sensitivity, ^{13}C-labeled compounds have to be used to monitor the progress of solid-phase reactions [16, 17]. A very powerful approach to narrow signals in NMR spectra is magic angle spinning (MAS) NMR. NMR theory predicts that spinning of sample at the "magic angle" (54.7° relative to the axis of the magnetic field) at high speed (a few kilohertz) eliminates magnetic susceptibility mismatch, which causes line broadening. Obviously, MAS NMR can be done using solvent-swollen resin, thus combining the advantages of both approaches [18]. Numerous publications describe one- and two-dimensional NMR experiments utilizing MAS for combinatorial chemistry applications [19–23].

As was mentioned above, NMR is a relatively insensitive method. Quite often combinatorial synthesis results in small amounts of material.

Improvements in sensitivity have been achieved by using specially designed NMR probes [24, 25]. The combination of improved sensitivity and the ability to suppress background signals arising from nondeuterated solvents opened a new avenue in the application of NMR in combinatorial chemistry: high-performance liquid chromatography (HPLC) NMR. This method combines the high information content of NMR technique with the ability to analyze complicated mixtures produced by combinatorial synthesis. HPLC/NMR can be used as a complementary technique to HPLC/mass spectrometry and would be especially useful for isomeric compounds [26]. The need to analyze a large number of samples generated by combinatorial methods led to the development of "columnless" LC/NMR, which allowed for high-throughput analysis in automated mode (~3 min per sample) [10]. Thus the new developments in NMR spectroscopy offer substantial advantages and make NMR truly available as an analytical tool in combinatorial chemistry.

2. Infrared (IR) Spectroscopy

Contrary to NMR, which allows obtaining information about each nuclei in a molecule, IR spectroscopy provides information only about some functional groups. Consequently, the major applications of this technique include determining the extent of conversion from starting material to product and monitoring the time course of a reaction by observing the changes of intensity of bands corresponding to the functional group of interest. Two major techniques allow one to obtain IR spectra from solid-phase support: the measurement of a KBr pellet made from several milligrams of solid support and Fourier transform (FT) IR microspectroscopy of single beads. Since KBr pellet FT-IR requires laborious sample preparation and consumes substantial amounts of resin, most of the recent publications describe the application of FT-IR to single-bead analysis [27–29]. Overall, IR spectroscopy is a valuable technique employed by many researchers for evaluating the solid-phase transformation of functional groups.

3. Mass Spectrometry (MS)

Most of the mass spectrometry applications for combinatorial chemistry will be described in the following sections of this chapter. Here we will give a short overview of MS techniques utilized for the characterization of resin-bound molecules. The majority of publications in this field describe applications of matrix-assisted laser desorption ionization (MALDI), combined with time-of-flight (TOF) detection. The major difference of MS application for analysis of resin-bound molecules from the above-described NMR and IR applications is that analyte should not be covalently bound to solid support prior to mass measurement. Detachment of compound molecules from resin can be done chemically (for example, by bead exposure to TFA vapors) [30,31] or photochemically, such that cleavage, desorption, and ionization of molecules occur simultaneously upon stimulation by laser radiation [32]. Since the application of MS to real-time solid-phase reaction monitoring is limited by the need for cleavage of the molecules from resin, or the inefficient ionization of the analytes, a method which relies on the use of a derivatized resin was developed to overcome these shortcomings [33]. The combination of a photocleavable linker and the ionization tag on the resin enables direct analysis by MS regardless of whether the attached compound contains a protonation site for MALDI.

D. High-Throughput Analytical Techniques Used for Characterization of Combinatorial Libraries

After the library development process passes the stage of development and optimization of chemical reactions, the requirement of high throughput becomes more and more important. Usually, hundreds of samples need to be analyzed for evaluation of building blocks and the characterization of the full production library requires analysis of thousands of samples. At this moment detailed structural analysis of synthesized compounds is substituted by *confirmation* of the fact that intended compound has been synthesized. *Purity assessment* and *quantification* of intended compounds are the other two goals of analysis.

1. Confirmation of Structure

Among numerous analytical techniques, mass spectrometry has been identified as the method of choice for the high-throughput structure confirmation of compounds in combinatorial libraries. There are a few reasons for this choice. MS is based on the measurement of a very fundamental parameter of a compound: mass-to-charge ratio of molecules. Thus the method does not depend on the presence of chromophores or any functional group in a molecule. High sensitivity is another advantage of MS; as little as femtomoles of a compound can be easily measured. Mass spectrometry is a fast method, with the measurement time approximately several seconds, and it can be easily automated. It is worth mentioning that measurement of molecular weight, especially by low-resolution MS, which is typically used for high-throughput applications, is not secure confirmation of structure, since many compounds with different structures may have identical molecular weights. However, utilization of mass spectrometry as the only method of structure confirmation of compounds in the late stages of library development is justified by the high level of confidence in chemistry which was developed in the previous stages.

Any MS experiment begins with ionization of molecules of analyte. Numerous ionization techniques (electron ionization, fast atom bombardment, plasma desorption, electrospray ionization, etc.) allow MS analysis of a wide range of organic molecules. In most cases the characterization of combinatorial libraries means analysis of crude compounds; i.e. one can expect not only the intended compound to be present in the analyte, but also products of side reactions. Thus the ability to ionize molecules without substantial fragmentation is essential to avoid the confusion between ions generated from side products and fragments of intended compounds. The most commonly used ionization techniques for analysis of combinatorial libraries include electrospray ionization (ESI), chemical ionization (CI), and matrix-assisted laser desorption/ionization (MALDI). However, application of soft ionization techniques has a negative side. One can expect that different components of a library will have different abilities to be ionized by soft ionization techniques. Thus compounds with the same concentration might give substantially different response in mass spectra, which makes quantification of compounds by mass spectrometry problematic. In the extreme case a compound will not be detected by mass spectrometry if it is not ionizable by the particular technique.

2. Purity Assessment

Just as mass spectrometry dominates among other analytical techniques for structure confirmation, gradient high-performance liquid chromatography (HPLC) is the method of choice for purity assessment. This method is easily conjugated with many different types of detectors, some of which will be discussed later. Practically all HPLC systems are equipped with an ultraviolet (UV) detector. HPLC possesses many features essential for high-throughput analysis. It is robust, easy to automate, quite "universal" (not biased to specific features of a molecule such as, for example, charge), and has high resolving power. However, along with numerous advantages, gradient HPLC also has certain limitations. The major one is that it is a relatively slow method due to gradient time, required column cleaning, and re-equilibration time. A typical HPLC experiment for analysis of organic compounds takes dozens of minutes. To address the necessity of high throughput, fast gradient methods have been developed. As a rule, an HPLC experiment for analysis of combinatorial libraries does not exceed 15 min, although even shorter run times have been reported [33, 34].

3. Quantification

The major limitation of both UV and MS detectors is that neither can provide quantitative or even semiquantitative information without reference standards. Ultraviolet response depends on the presence of a chromophore in a molecule and evidently might vary from one molecular species to another in a library. Although successful application of electrospray mass spectrometry for quantitative analysis of peptides has been reported [35], one should always keep in mind that signal intensity in a mass spectrum depends on the ability of a molecule to ionize. The ability to produce ions, especially with soft ionization techniques, might be very different for different molecules within one library, and the difference might be even bigger from one library to another.

Recently, new analytical techniques incorporating instruments such as the evaporative light-scattering detector (ELSD) [36] and the chemiluminescent

$$R_3N + O_2 \xrightarrow{> 1000°C} H_2O + NO + \text{other oxides}$$

$$\bullet NO + O_3 \text{ —— } {}^*NO_2 \text{ —— } NO_2 + h\nu$$

FIGURE 10.2 Mechanism for the chemiluminescent nitrogen detector. R = any group attached to a nitrogen.

nitrogen detector (CLND) [37] have begun to be used to address the issue of quantitative analysis. Due to the relative novelty of these methods in their application to combinatorial chemistry, we will spend some time describing them in more detail.

(a) Chemiluminescent Nitrogen Detector (CLND)

The response of a chemiluminescent nitrogen detector is linearly proportional to nitrogen content in a sample [37]. Application of the detector for compound quantification is based on the very simple idea that, by measuring nitrogen content of a sample and knowing the number of nitrogen atoms in a molecule of analyte, one can determine the sample quantity. CLND is a universal method for nitrogen-containing compounds in the sense that it does not require any reference standard for quantification. Although it is limited to the analysis of nitrogen-containing compounds, its application for quantitative analysis of combinatorial libraries is very promising. Most developmental and marketed drugs contain nitrogen [38], as do most scaffolds described for use in combinatorial libraries. One of the limitations of the method, although not a very important one, is that it does not allow the use of any nitrogen-containing solvent during analysis. For example, acetonitrile, the most popular organic modifier in the HPLC mobile phase, cannot be used.

The mechanism of chemiluminescent nitrogen detection is presented in Figure 10.2. Each compound eluting from the HPLC column undergoes high-temperature oxidation, resulting in the formation of nitric oxide from all chemically bond nitrogens. The gases are then mixed with ozone, thereby forming nitrogen dioxide in the excited state, which, upon relaxation to the ground state, emits light, which is detected by a photomultiplier tube. Several applications of this method to peptide mixture analysis [39, 40] and to the direct measurement of yield and purity of solid-phase reactions [38] have been described in the literature. From our point of view, CLND will gain more popularity in combinatorial chemistry analytical laboratories as its robustness increases.

(b) Evaporative Light-Scattering Detector (ELSD)

The idea of light-scattering detection is very attractive due to its simplicity. When an aerosol is created from a sample and light is shone through the cloud formed, the amount of diffused light depends on the density of the cloud (size and number of aerosol particles). By measuring the amount of diffused light, one can estimate the amount of sample used to create aerosol. The relationship between the amount of diffused light and the amount of analyte can be precisely described by the mathematical model, and the size of the particles is a very important parameter of the model. Since the size of the particles depends on the nature of the analyte as well as on the conditions at which the aerosol was formed (temperature, solvent, etc.), one can use only an empirical relationship rather than a mathematical model. The amount of diffused light A as a function of mass m obeys the relationship $A = am^b$, where a and b are empirical parameters which depend on the nature of the analyte, the temperature, solvent, etc. Thus, ELSD can be used mostly as a semiquantitative method.

ELSD is commonly used in combination with separation techniques such as HPLC. Each compound eluting from a chromatographic column is nebulized to form a homogenous mist of droplets, which enters a heated tube where the mobile phase is evaporated. Then the nonvolatile analyte goes through a light beam, and light scattering caused by nonvolatile analyte(s) is collected by a photomultiplier. Signal from the multiplier is recorded, integrated, and the amount of analyte is determined based on the above-mentioned empirical equation, $A = am^b$. Quantitative determination of the amount of analyte is possible only if the calibration curve was constructed using compounds with properties similar to those of the analyte. The mobile-phase evaporation step affects the detector applications. Generally, a greater response is obtained at lower temperature, and for high-sensitivity applications it is important to have the evaporation temperature significantly below the boiling point of the solvent used for the mobile phase. The evaporation step limits the detector's application to nonvolatile molecules only. As a rule, the molecular weight of organic molecules should be greater than 200 Da in order to be detected by ELSD. In addition, detection of some compound classes, e.g. amines, carboxylic acids, by ELS is pH dependent [7]. Two major types of light sources are currently used in ELSD instruments: laser sources and polychromatic light (tungsten-halide) sources. ELS detectors with laser sources are generally more sensitive, while the ones with tungsten-halide sources are more "universal" (less dependent on the nature of the analyte, solvent, etc.).

Independence of the detector's response from a compound chromophore or ionizability makes it more universal for quantification applications than UV or MS. Recently, the evaporative light-scattering detector has been introduced as a useful tool for quantification and purity assessment of compounds in combinatorial libraries [36, 41].

The limitations of MS, UV detection, ELSD, and CLND in terms of compound quantification, especially when combined with the requirement of high throughput, make the task of quantitative analysis of libraries very difficult. From our point of view the key to quantitative analysis of libraries is in using reference standard compounds and in making conclusions based on a combination of results from several techniques whenever possible.

III. ANALYTICAL CHARACTERIZATION OF VARIOUS TYPES OF LIBRARIES

A. Strategies of Combinatorial Libraries Synthesis and Analytical Challenges Associated with Them

A short overview of synthetic strategies that are currently used in combinatorial chemistry is important to get a better understanding of how to characterize a combinatorial library in the most efficient way. Figure 10.3 presents different strategies for synthesis of combinatorial libraries on an example of a library with three points of randomization (X, Y, Z), three building blocks in each randomization (XI, X2, X3, Y1, Y2, etc., correspondingly), and total complexity of 27 compounds.

In the parallel synthesis strategy (Figure 10.3a), all 27 compounds are synthesized in parallel in separate reaction vessels, and this is not any different from traditional organic chemistry synthesis of compounds "one at a time," except that in most cases synthesis is automated. This strategy requires performing $27 \times 3 = 81$ reaction steps in order to obtain 27 compounds. Compounds are synthesized by either solution-phase or solid-phase chemistry in relatively large quantities (milligram scale). Since the compounds are synthesized independently of each other, analytical data about one

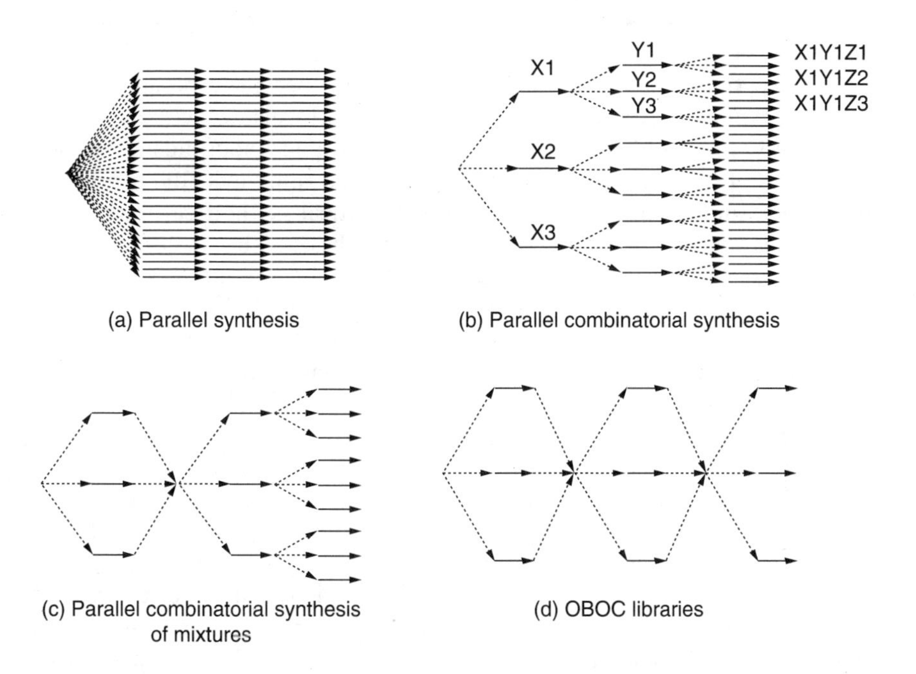

FIGURE 10.3 Different strategies for synthesis of combinatorial libraries. Dotted arrows indicate the step of resin split or combination; solid arrows indicate a chemical reaction step.

of the synthesized compounds cannot be applied to other compounds in the array.

Another strategy for producing libraries of individual compounds by solid-phase chemistry is presented in Figure 10.3b. According to this strategy, resin is first split into three batches, and building blocks X1, X2, X3 are correspondingly coupled to each batch of resin. Then each of the batches is split into three portions, coupling of building blocks Yl, Y2, Y3 is performed, each of those nine batches is again split into three batches, and coupling of Z building blocks is performed. As a result, 27 individual compounds are synthesized using $3+9+27=39$ reaction steps instead of 81 in the strategy described in Figure 10.3a. The important difference between this strategy and parallel synthesis is that the former produces arrays of related compounds rather than independently synthesized ones. For example, compounds X1Y1ZI, X1Y1Z2, and X1Y1Z3 have the common precursor X1 Y1. Utilization of knowledge about relationships between compounds allows reducing the amount of analytical work while maintaining the quality of analysis. If analysis of intermediates is performed during synthesis of a library, analytical data obtained for one intermediate (X1YI) will be meaningful for three library compounds (X1Y1Z1, XIYIZ2, and X1YIZ3). This approach may be applied to analysis of libraries retrospectively. As an example, if compound XlYIZ1 is of a good quality, this means that the first and the second steps of synthesis were successful. Since compounds XIY1Z2 and X1Y1Z3 have the same steps in their synthetic history, low quality of these compounds should be associated only with the last step of synthesis.

The two synthetic approaches described above became especially popular with the expansion of automated synthetic platforms, which provide the opportunity for parallel synthesis of hundreds and thousands of organic compounds per day. Due to relatively large amounts of synthesized compounds (milligram scale), almost any analytical method can be used for characterization of products of parallel synthesis. The major analytical challenge associated with the libraries of individual compounds is to keep up with the high productivity of the automated synthetic platforms.

The strategies presented in Figures 10.3c and 10.3d are variations of the quite popular "split and mix" approach that is based mostly on solid-phase techniques. The "split and mix" strategy was historically the first one to produce large combinatorial libraries of peptides [42–44] and, later, libraries of small organic molecules [45]. In the minds of many scientists this approach is associated with the generation of very large mixtures and is not considered a vital method for the synthesis of libraries of small organic molecules. However, two variations of this method, allowing for synthesis of small mixtures (4–12 compounds per well) and "one bead, one compound" technology, have been successfully applied for organic libraries of small molecules [46, 47]. The strategy presented in Figure 10.3c describes parallel combinatorial synthesis of mixtures of compounds. The difference from combinatorial synthesis of individual compounds is that the resin is mixed after the first coupling. With the following steps being the same as described in

Figure 10.3b, this strategy leads in the described example to nine mixtures with three components in each mixture. Synthetically, this strategy is quite attractive, allowing cost-efficient production of large numbers of compounds. However, analytical characterization of even small mixtures (4–10 components) is not a trivial task. One of the additional issues in the analysis of mixtures, compared to that of individual compounds, is the ambiguity in determining the identity of each compound in the mixture by mass spectrometry. The reason for ambiguity is that the appearance of the expected molecular ion in mass spectrum does not necessarily mean that this compound is present in the mixture, because possible side products of another component of the mixture may have a coinciding molecular weight. In most cases this problem could be resolved with tandem mass spectrometry techniques (MS/MS). However, the requirement for high throughput does not allow detailed analysis of each component of a mixture. Thus, the major challenge of analytical characterization of mixtures of compounds is to perform high-throughput interpretation of huge volumes of information with some uncertainty for each point of data.

The strategy of synthesis of "one bead, one compound" (OBOC) libraries is presented in Figure 10.3d. This approach relies exclusively on solid-phase synthesis techniques. The mixing of resin occurs after each randomization, enabling the production of libraries in which single beads contain individual compounds. This technique allows the screening of single compounds that are cleaved from individual beads or mixtures of compounds cleaved from pools of beads. Complexity of these mixtures may vary from several to thousands of compounds per mixture. The essential difference of this strategy from the combinatorial chemistry techniques described above is that it does not track the chemical history of library compounds during synthesis, and therefore structure elucidation is necessary once activity has been identified in biological screening. The two ways which are currently used to address this problem are the synthesis of encoded libraries [48, 49] and the use of direct structure elucidation by analytical methods [50, 51]. Quality assurance of OBOC libraries is analytically very challenging. The additional step of structure elucidation is required, in comparison with analysis of spatially addressed libraries, and it has to be performed using very limited amounts of material from the individual bead (picomoles to nanomoles). In addition, OBOC libraries are usually, although not necessarily, of high complexity (up to hundreds of thousands of components and even more).

Thus, there are several strategies for the synthesis of combinatorial libraries, and many pharmaceutical companies use them concurrently to address the need for different types of libraries (small directed libraries of individual compounds for SAR studies, big libraries synthesized in "split-and-mix" format, or big arrays of individual compounds synthesized in spatially addressed format for use in initial screening against multiple targets, etc.). The variety of synthesis strategies provides an additional challenge for analytical laboratories in these companies, in terms of finding the optimal strategy to support all those activities.

B. Analysis of Single-Component, Small-Size Libraries

Single-component libraries of small complexity (up to 1000 components) that are synthesized in a spatially addressed format can be considered as a transition from traditional medicinal chemistry to combinatorial chemistry. These libraries are usually target specific and/or are built around a known hit/lead. They serve the same goals as arrays of compounds synthesized "one by one" in medicinal chemistry laboratories: generation of structure activity data, optimization activity and determination of the therapeutic characteristics of a lead, etc. Actually, their major difference from arrays of compounds synthesized according to the traditional medicinal chemistry approach is the extent of automation used during the synthesis, allowing dramatic increase of productivity of a medicinal chemist. In addition to being a source of potential leads with improved characteristics, low-complexity, single-component libraries play a very important role during the development of a big combinatorial library. In this case they serve as the goals of evaluating building blocks, testing cleavage conditions, validating the performance of automation, etc. Usually, comprehensive analytical information about each compound in a library (purity, quantity, structures of side products, etc.) is required in order to make a decision about the next step of a combinatorial library development and/or to generate data about structure-activity relationship. The relatively small size of the libraries (100–1000 compounds) allows for the characterization of 100% of their components. Actually, the possibility of prescreening characterization of 100% of the compounds is another similarity of these libraries, with the arrays of compounds made by traditional medicinal chemistry and, possibly, one of the major reasons of their popularity in the drug discovery process. The quantity of synthesized compounds (low milligram or submilligram amounts) is sufficient to perform analysis by several analytical techniques, and "there is no excuse for not fully characterizing compounds made by parallel synthesis" [52].

Analysis of single-compound libraries by HPLC/UV/MS and, more recently, by HPLC/UV/ELSD/MS techniques, is generally accepted as the most appropriate means of characterization. The major steps of the analysis are the following:

1. acquisition of data using HPLC with on-line UV/MS/ELSD detection;
2. data processing in which peak areas are integrated, expected molecular ions are extracted, etc;
3. reduction of processed data to a form which can be easily used for quality assessment of the library.

To avoid bottlenecks in the analysis, each step should be automated. Automation of data acquisition is achieved by automated HPLC/UV/MS/ELSD instruments, equipped with autosamplers that allow compounds to be batch analyzed in unattended fashion. Recently, fast generic HPLC methods have been implemented for analysis of combinatorial libraries [33–53].

Short columns (5 cm) and fast gradients (typically 3–10 min per sample) allow reaching throughput of 100–200 samples per day per instrument.

Utilization of several analytical techniques (UV, MS, ELSD) during one HPLC run requires coordination of the output from different detectors. In most cases modern mass spectrometers come with software packages that allow a wide range of manipulations with mass spectroscopic data (extraction of ion current corresponding to ion of interest, integration of peak area, background subtraction, etc.) as well as inclusion of data from external detectors such as UV, ELSD, CLND [54, 55]. As a rule, output of the software can be relatively easily customized to fit particular needs. For example, users of the quite popular mass spectrometers from PE Sciex may customize data processing with the help of a macro language for Macintosh computers called Apple Script. Figure 10.4 demonstrates an example of automated data processing by Apple Script. This script, using as input a list of expected masses, automatically processes the data for each sample and automatically prints the results in the output format shown in the figure. The graphical output consists of the UV and ELSD traces (Figures 10.4A and 10.4B), the total ion chromatogram (TIC) (Figure 4C), the extracted ion-chromatogram (XIC) for

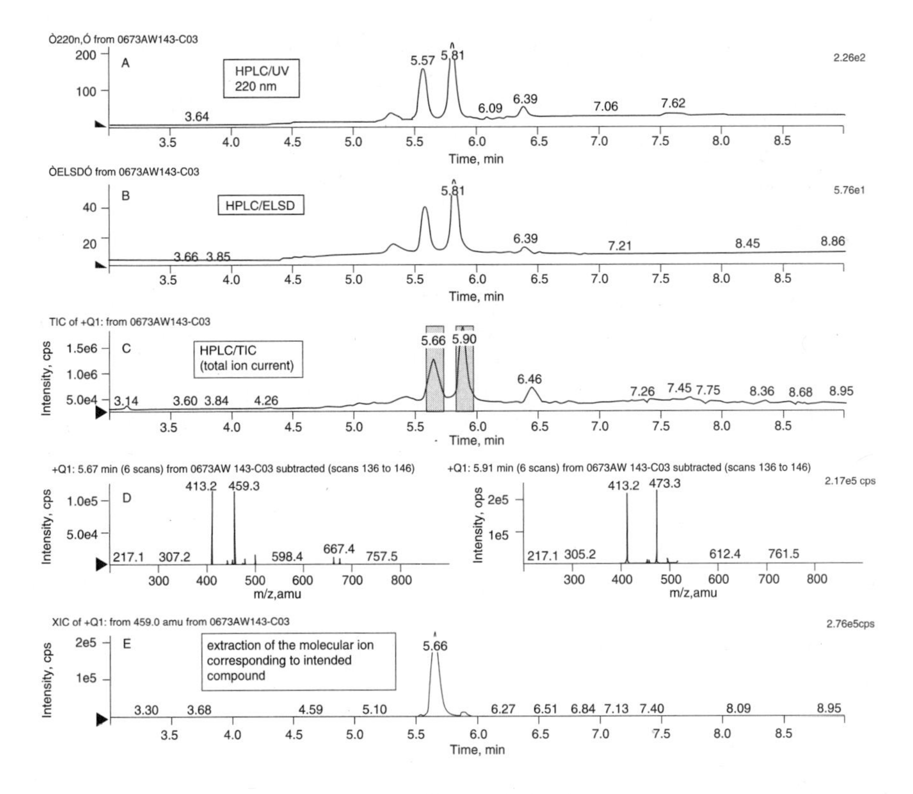

FIGURE 10.4 Graphical output of LC/MS analysis of a compound in the combinatorial library.

the expected mass (Figure 10.4E), and the mass spectra corresponding to each HPLC peak (Figure 10.4D).

Another example of software for postacquisition data processing is the OpenLynx-Diversity system utilized on the Platform mass spectrometer (Micromass, UK) running within the Windows NT operating system. The software is specially designed for single-component libraries and requires information about the mass of the expected component in each well prior to data processing [56].

Thus, in the result of automated data processing one can confirm the identity of library components, perform purity assessment based on several independent techniques (UV, MS, ELSD), estimate quantities of compounds, obtain information about side products, etc. The results may be presented using tabular output, which has certain strengths and weaknesses. On one side, a lot of different information about compounds can be captured in one table, such as sample ID, purity, estimated amount of material, etc.; the data can be easily sorted and transferred to various databases. An example of tabular output where compounds were sorted based on their average purity is presented in [34]. On the other side, the amount of information usually captured in a table is much more than what is necessary for quick assessment of library quality. For instance, in many cases a user of analytical information wants to know if the compound is pure enough but is not interested in the exact percentage of purity. To meet this request, a few laboratories have developed graphical output of data, typically presenting the HPLC/MS data about compounds synthesized in microtiter plates as a color-coded image of the plate. If an ion that is representative of the expected product is above the threshold defined by the user, then the well is colored green. If the expected ion is below the threshold, the well is colored red [57]. In our laboratory we are using a similar color-coding scheme to show quality of compounds. We use the colors green, yellow, and red to enable a user to set two thresholds.

Within the last couple of years, the focus of effort in characterization of combinatorial libraries has started to shift from the development of methods for analysis of crude compounds to the development of methods for parallel purification and characterization of purified compounds [34, 41]. Although many synthetic reactions can be conducted with high yields providing reasonable purity, it is unrealistic to expect a high quality of synthesis for all components of a combinatorial library. Hence, high-throughput purification approaches need to be developed in order to increase the overall purity of combinatorial libraries. Also, with a system for purification and characterization of all components of a library in place, the results of biological screening become more valuable. One can be sure that the negative result from biological screening of a compound from the library is truly due to this particular component of library not being active, and not to the fact that it was not successfully synthesized.

One of the most interesting achievements in the area of high-throughput analytical characterization and purification is the development of an integrated, fully automated system for analysis and purification [34]. The system incorporates fast, reverse-phase HPLC (5–10 min per run) and electrospray mass spectrometry analyses, with the purity of each compound in the library automatically assessed using Apple Script. The compounds which are expected to fall below a threshold level of purity (< 90%) are subjected to automatic on-line preparative HPLC-MS purification. An essential part of the system is fraction collection triggered by the mass spectrometer, when the ion signal corresponding to the intended compound reaches a set threshold. Thus only one fraction is collected per sample, and no post-purification analysis is required to identify the fraction containing the desired compound. Also, there is no need for large fraction collection beds when big arrays of compounds need to be purified. Overall, the "one sample, one fraction" paradigm leads to considerable saving of time and allows the throughput required for purification and analysis of small-sized combinatorial libraries.

C. Analysis of Single-Component, Large-Size Libraries

In the previous section we mentioned that requirements for reasonable separation of HPLC peaks in single-component libraries limit the throughput of HPLC/UV/MS analysis to about 100–200 samples per day per instrument. Thus, if the synthesis rate exceeds this level, either the number of instruments in the analytical laboratory has to be increased or methods of analysis should be altered. When the rate of synthesis becomes as high as 400–1000 individual compounds per day (which is not unusual today), most combinatorial chemistry companies change their analytical strategy. Basically, two approaches are used to cope with such a high synthesis rate: limited analysis of all library components and detailed analysis of a certain percentage of the representative samples from a library.

1. Limited Analysis of All Components in a Library

In this strategy, dramatic increase of analytical throughput is achieved at the expense of completeness of information about each component of a library. In a recently published example [58], only one analytical technique, flow injection mass spectrometry, is used for characterization of libraries. Compounds are automatically sampled from a 96-well microtiter plate, injected, and mass analyzed. The parent ion, adduct ion, and isotopic distribution of these ions are the criteria used to determine whether the intended compound was successfully synthesized. With several autosamplers put on one mass spectrometer, the throughput can be less than 1 min per sample. Obviously limitations of this approach include absence of purity information and the danger of getting misleading results about compounds, which are not ionizable or are suppressed by more easily ionizable impurities. On the other hand, its attractiveness is its ability to provide information, although limited, about each component of a library. This approach is driven by the idea that high speed is one of the major advantages in the application of combinatorial chemistry to the drug discovery process. Consequently, the goal of prescreening analysis is to obtain only important information about a library, and that is the information about the identity of its components, thereby confirming that the desired diversity is achieved. Thorough analytical characterization of a library's components is postponed until later stages and performed only on those compounds that have shown biological activity.

The above described approach can be modified by adding off-line HPLC analysis [59]. In this case the time-consuming HPLC/UV analysis is performed in parallel on several HPLC systems, while MS characterization is performed in high-throughput flow injection mode. This is not equal substitution of LC/MS analysis, since, when HPLC and MS analyses are handled separately, the signals in mass spectrum cannot be assigned to specific HPLC peaks, which might be a source of misinterpretation of results. Another problem is how to combine the results from independent analyses (HPLC/UV and MS) to make a conclusion about the quality of a library. One of the ways to address this problem is to keep the results of HPLC/UV/ELSD and MS analyses in separate tables, assigning number flags to each compound depending on its quality, determined by each technique [59]. For example, compounds which show purity over 75% in HPLC/UV/ELSD analysis get flag "2"; those between 25% and 75% get flag "1," and those below 25% get flag "0." 1 or 0 flags are a result of MS analysis when the expected mass was observed or not observed, respectively. In the output table the corresponding columns are multiplied. This very simple approach to data interpretation allows for a fast look at the quality of a library by counting corresponding flags.

2. Detailed Analysis of a Certain Percentage of Representative Samples from a Library

This approach is based on limited sampling of a library with subsequent detailed analysis of the samples, which is similar to the analysis of small-sized libraries described above. The first step in this approach is the selection of components of the library, such that their analysis will allow for extrapolation of results to the whole library. Two sampling strategies can be used. The first one is based on utilizing statistics for library analysis. Compounds for analysis are chosen randomly, and a "statistically significant" number of library components are analyzed to provide the basis for objective quality assessment. The "statistically significant" number is calculated based on the formulas of mathematical statistics and depends on the size of the library, confidence interval, desired accuracy, etc. The second sampling strategy is based on the fact that a combinatorial library is not an ensemble of random chemical entities. For example, if all compounds in one plate have the same building block in the first randomization, and the coupling of building blocks in the second and third randomizations was performed by columns and rows, respectively, selecting a diagonal pattern of compounds for QC permits evaluation by analysis of all building blocks used for synthesis [59]. In our laboratory we utilize information about the reactivity of different building blocks in addition to information about the array layout. Instead of the diagonal pattern we use "cross sections" of the array in two orthogonal directions, where the position of each "cross section" is defined by an "easily reactive" building block. This approach will be discussed in detail in the next section.

D. Analysis of Libraries Synthesized in the Format of Small Mixtures of 4–12 Compounds per Mixture

Combinatorial libraries are synthesized in the format of mixtures when high complexity (over 10,000 compounds) has to be achieved. Two types of analytical challenges are associated with libraries of this kind: the uncertainty of each data point (see section "Strategies of Combinatorial Libraries Synthesis and Analytical Challenges Associated with Them") and the problems related to large-sized libraries, which were discussed above. To decrease the uncertainty of data points, detailed analysis (LC/MS, LC/MS/MS) must be performed for each mixture chosen for analysis. Due to the in-depth analysis required for mixtures, only a limited number of them can be measured. The sampling strategy in the case of libraries synthesized as mixtures is very important, because we wish to limit the vast amount of data while maximizing the information gained as a result of analysis. Below we will describe an approach developed in our laboratory for analysis of these libraries, which not only assesses the quality of the library, but also provides information about the reactivity of all building blocks used for the library synthesis.

We illustrate our scheme of analysis, using as an example a small organic molecule library with three points of randomization and a complexity of 7488 compounds (Figure 10.5). The number of building blocks in the first, second, and third positions are 12, 26, and 24 respectively. After the first coupling the resin was mixed and split into 26 batches, and after the second coupling each of the 26 batches was further split into 24 batches, resulting in $26 \times 24 = 624$ batches with 12 compounds in each. Synthesis, cleavage, and extraction of compounds were performed in 96 deep-well plates.

Since each well contains 12 compounds, analysis of even one well is not an easy task. Thus, in developing our strategy, we tried to minimize the number of wells required for analysis. If we randomly choose a limited number of wells (as an example, 50 wells), some of the building blocks used for synthesis of the library will not be evaluated (for example, z7, zll, z16, y3, y22, etc., in Figure 10.6). Our scheme is based on rational rather than random selection of wells for analysis. Since our library can be considered as a two-dimensional array of wells, we decided to select wells for analysis so as to sample "cross sections" of this array in two orthogonal directions (Figure 10.7). Using this rational approach we ensure that each of the building blocks used for synthesis of the library has been met at least once. The choice of the position of each cross section is based on the information obtained during the library development process. During the building-block evaluation step, we identify which building blocks perform best (with high yield, without substantial amount of side products, etc.) in each randomization and assume that they behave the same way

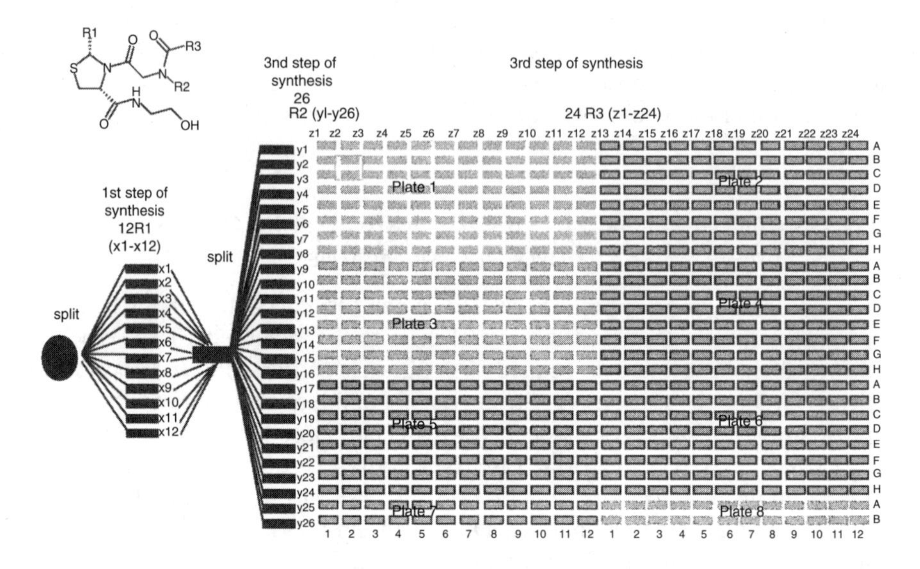

FIGURE 10.5 The scheme of library synthesis ($12 \times 26 \times 24 = 7488$ compounds; 12 compounds per well).

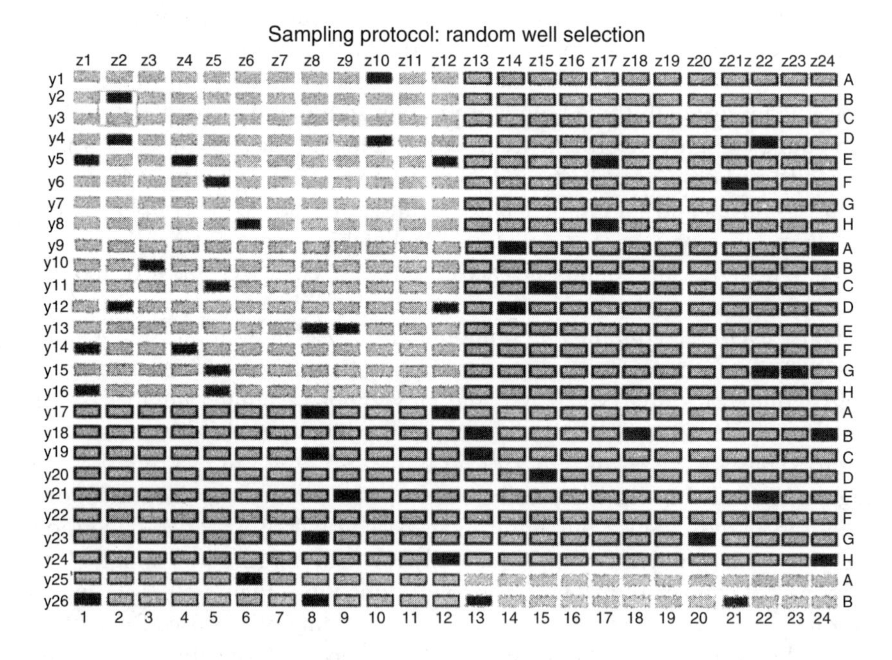

FIGURE 10.6 Layout of 96-well plates with the library ($12 \times 26 \times 24$). Each well contains 12 compounds; compounds within each well contain the same building blocks in the second (y) and third positions (z) but differ by the building blocks in the first position (x).

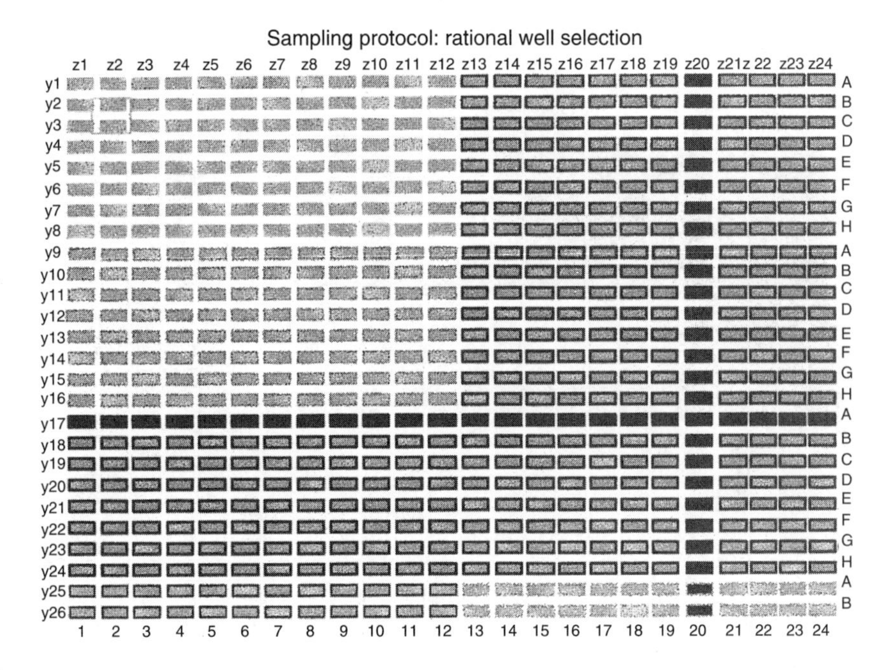

FIGURE 10.7 Layout of 96-well plates with the library xyz (12 × 26 × 24). The wells chosen for analysis according to the scheme of "cross sections" are shaded.

during the library synthesis. We suggested using these building blocks to define position of "cross section" during analysis. The logic behind this suggestion is the following: if, for example, a compound fails QC and we know that the building block used in the second randomization was usually coupled successfully, then there is most probably a problem with building blocks used in the first and/or third randomization.

Mixture of compounds from each well, which was chosen according to the scheme of "cross sections," was analyzed by electrospray LC/MS. Ion currents corresponding to all 12 expected compounds were extracted from the total ion current, and the areas under each peak in the ion-current chromatogram were determined. Analysis of each well resulted in 12 numbers reflecting the amount of each component in the mixture. We used color coding for visualization of data output. If the area under a peak is below the set threshold, the color white is assigned, indicating the absence of the corresponding component in the mixture. If the area is within the expected range, gray is used, and if it is larger than expected, black is used. The data obtained about all the wells analyzed within one "cross section" are then combined in one color table. As an example, output of analysis of "cross section" of z20 is presented in Figure 10.8. Each column in this table represents the results of the analysis of an individual well. White columns in Figure 10.8 indicate problems with certain building blocks used in the third randomization (y21, y22, y23, y24), and white rows show a problem with building blocks in the first randomization (x7). Thus, the

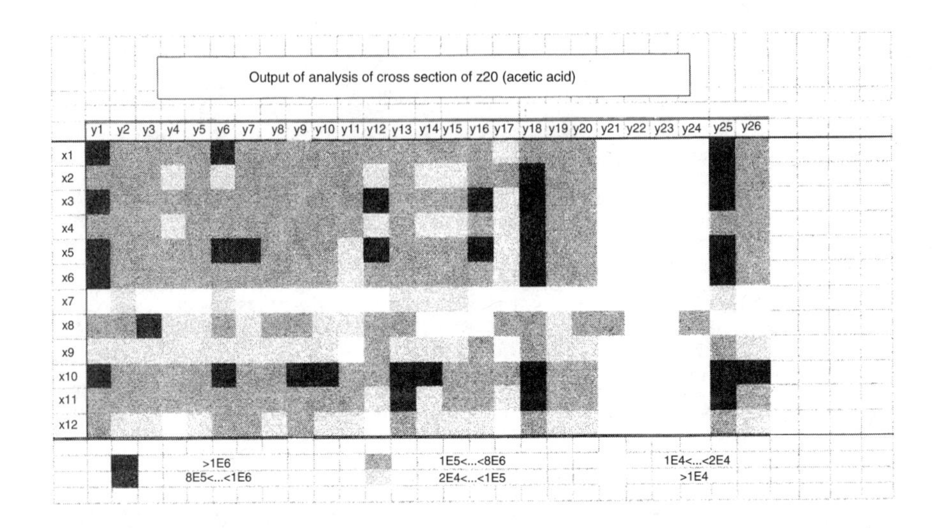

FIGURE 10.8 Output of LC/MS analysis of "cross section" of z20 (26 wells) analyzed. White columns in the table reveal problematic building blocks in the second randomization (y21, y22, y23, y24); white rows indicate problematic building blocks in the first randomization (x7).

suggested approach of rationally choosing wells for analysis in combination with a color-coding scheme allows for easy identification of building blocks, which did not work in each randomization. A set of Apple Script was written, which automatically performs the data-processing steps described above. By automating this analysis we have dramatically decreased our time for library analysis and building-blocks evaluation and substantially increased throughput.

E. Analysis of "Split-and-Mix" Libraries

Although a variety of automated parallel synthesis strategies have recently been described and have gained a lot of popularity, "split-and-mix" technology remains a powerful method in the drug discovery process. This technology allows access to very large numbers of compounds suitable for high-throughput screening, making it attractive for the task of identifying new leads.

As was mentioned above (see section "Strategies of Combinatorial Libraries Synthesis and Analytical Challenges Associated with Them"), the OBOC technique does not track library compounds during its synthesis. As a consequence, structure elucidation of a compound from an individual bead is required during analytical evaluation of the library and after identification of active bead(s) from biological screening. Structure elucidation can be performed directly with analytical methods or indirectly by using tagging techniques. In this chapter we will not discuss encoded libraries; some information can be found in other chapters of this book or elsewhere [48, 49]. It should be mentioned that once the issue of structure elucidation of a compound from an individual bead is addressed, the task of analytical characterization of OBOC libraries is generally not any different from analytical characterization of large-sized libraries with random selection of compounds for analysis (see section "Analysis of Large-Size Libraries"). However, analysis of OBOC libraries is the most difficult case of large-sized libraries, due to the limited amount of material on individual beads and the high complexity of these libraries.

There are two approaches to analytical evaluation of OBOC libraries: analysis of compounds from individuals bead [51, 60] and analysis of mixtures of compounds detached from multiple beads [61, 62]. It should be mentioned that analysis of mixtures from OBOC libraries is quite different from that described in section "Analysis of Libraries Synthesized in the Format of Small Mixtures of 4–12 Compounds per Mixture." In the mixture of compounds cleaved from a few beads, which are randomly picked for analysis from the OBOC library, the structures of components of the mixture are unknown. Consequently, it is not possible to apply the strategy developed for the analysis of mixtures synthesized in spatially addressed format, which is based on information about the structures of components of each mixture. In the case of OBOC libraries, analysis of only very large mixtures is meaningful when the theoretical distribution of molecular weights within the whole library is compared with that experimentally observed [61, 62]. The focus of this section is on characterization of OBOC libraries through analysis of individual beads.

Analytical characterization of OBOC library will be discussed for an example of a small-molecule combinatorial library with a complexity of 54,150 ($57 \times 25 \times 38$) compounds (Figure 10.9) [63]. In the result of an OBOC library analysis, identity, purity, and quantity of library components should be evaluated, but it is unknown which one out of thousands of compounds in a library is synthesized on the particular bead picked for analysis (Figure 10.10). General schemes of analytical characterization of combinatorial libraries (analysis of a small number of standard compounds, analysis of the model library, and analysis of the production-scale library) can be applied to analysis of OBOC libraries, but the issue of structure elucidation for OBOC libraries should be addressed through the scheme.

The analysis of standard compounds for OBOC libraries is very similar to characterization of compounds for other types of libraries discussed above. Keeping in mind the necessity for structure elucidation of compounds for OBOC libraries, standard compounds have to be used as a starting point for development of methods for structure elucidation. Due to high sensitivity and high information content, tandem mass spectrometry is generally applied for direct structure elucidation of components of OBOC libraries. Utilization of tandem mass spectrometry requires knowledge of fragmentation patterns for the particular library, and initial study of fragmentation is performed using a set of standard compounds.

The next step, which is the analysis of the model library, serves two goals in the case of OBOC libraries: development of methods for structure elucidation and validation of chemistry used for library synthesis. Synthesis

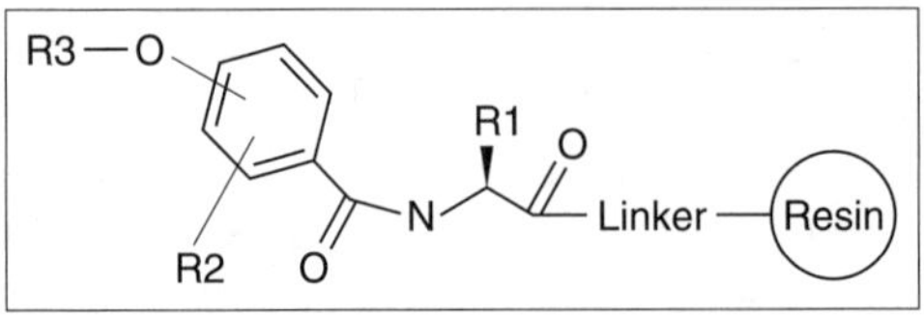

R3—O, R2, R1, O, N, O, Linker—Resin

Complexity	57 × 25 × 38 = 54,150 compounds
	R1 - 57 amino acids R2 - 25 aromatic hydroxy acids R3 - 38 alcohols
Resin	PEG/PS beads, 220 μm Belford, MA
Linker	Iminodiacetic-acid-based double cleavable

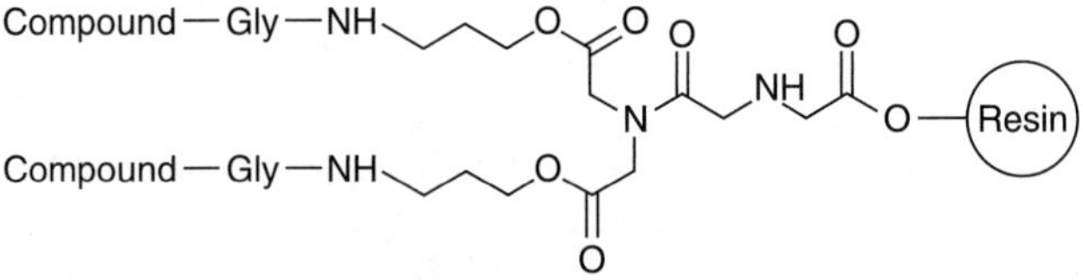

Compound—Gly—NH … O … O O O NH N O—Resin

Compound—Gly—NH … O … O

FIGURE 10.9 An OBOC library synthesized according to the "split-and-mix."

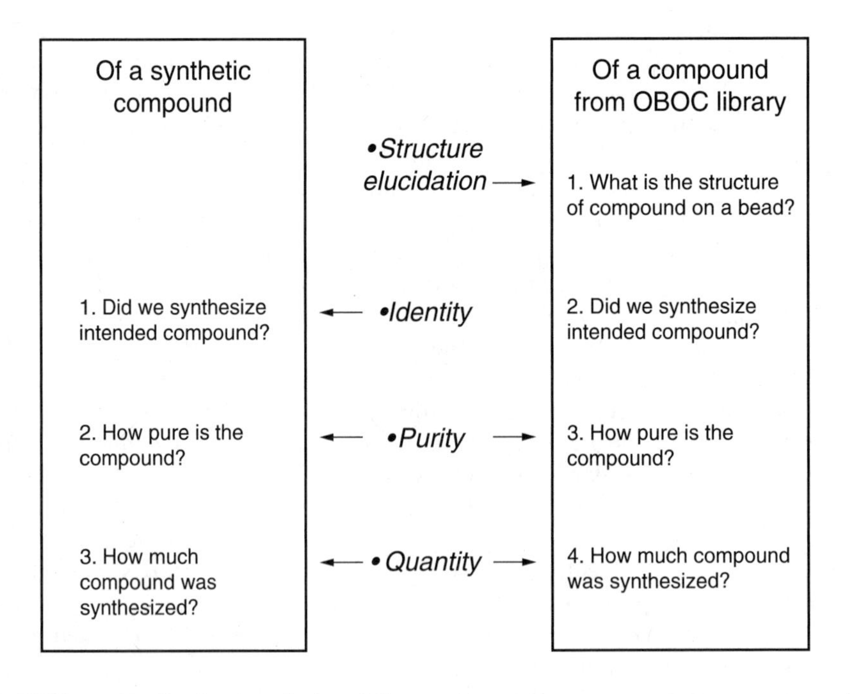

FIGURE 10.10 Checkpoints during QC.

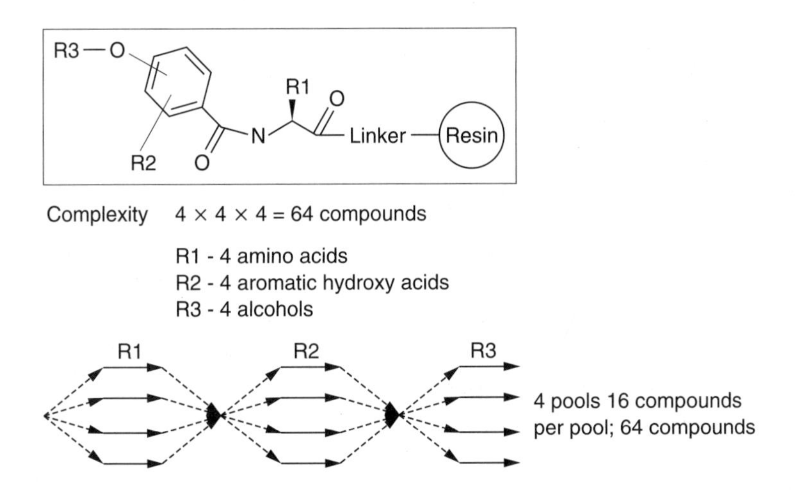

FIGURE 10.11 Scheme of synthesis of the model library.

of a model library is the perfect way to generate rapidly many diverse compounds, which can then be used to perform an intensive fragmentation study. In the reported example, a model library was synthesized using four diverse building blocks in each of three randomizations, consisting of 64 compounds (Figure 10.11). The high diversity of building blocks in the model library (Figure 10.12) is necessary to account for the difference in ionizability of compounds detected by electrospray MS. LC/MS and LC/MS/MS analysis is done for compounds released from individual beads. As a result of the analysis of a model library, the chemistry is validated (the purity of compounds from individual beads is estimated, the structure(s) of side product(s) are identified), and rules of fragmentation for the compounds are investigated and incorporated into software written "in house" for automated structure elucidation of compounds from individual beads.

For analysis of the production-scale library we use an orthogonal approach that is similar to the one discussed in the previous section describing analysis of small mixtures. First, we thoroughly analyze many beads picked randomly from a sublibrary where we expect successful coupling of the building block in the third randomization. Automated LC/MS analysis of beads from this sublibrary enables us to estimate successful coupling of building blocks in the first and second randomizations. Finally, the analysis of a few beads from each of the remaining sublibraries gives us an idea about how successful the coupling of building blocks in the third randomization was.

Concluding this chapter, we would like to stress that the analytical characterization of combinatorial libraries is a very rapidly developing and changing area of analytical chemistry. The variety of synthetic strategies used in combinatorial chemistry continues to grow and requires development of new analytical methods. The authors would not be surprised if new ones replaced many of the approaches described in this chapter or radically

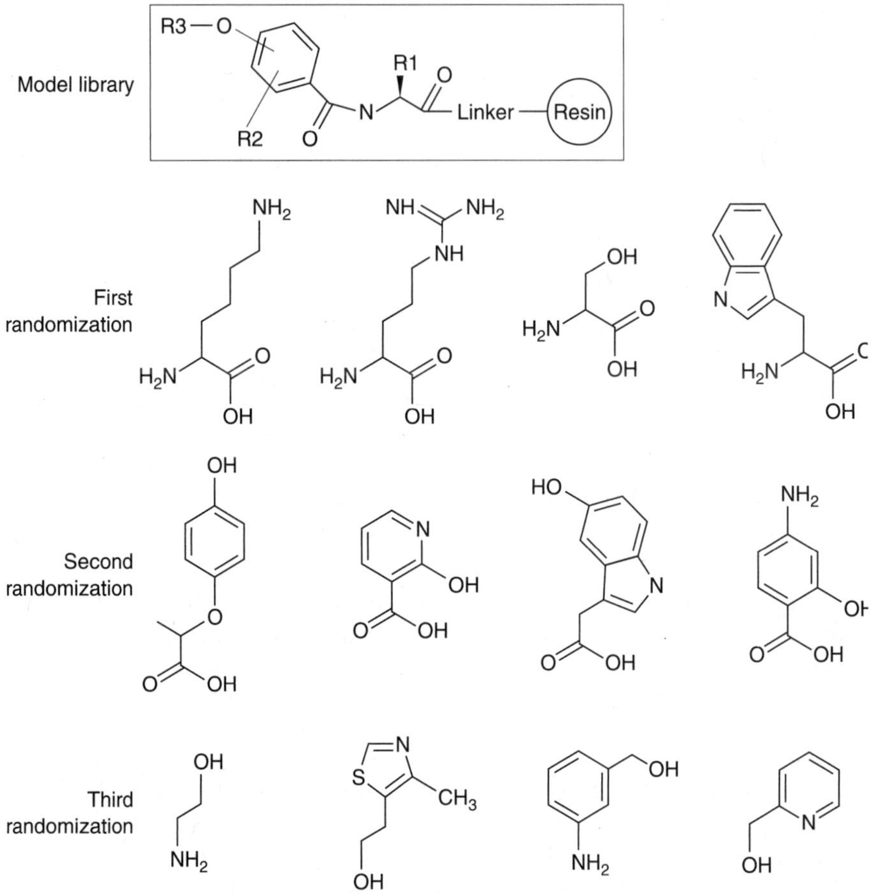

FIGURE 10.12 Building blocks used for synthesis of the model library.

modified in the nearest future. However, we believe that the basic ideas and general strategies of characterization of libraries will remain unchanged and, most important, will provide the solid basis for development of new applications of analytical chemistry in combinatorial technology.

ACKNOWLEDGMENTS

The authors gratefully acknowledge their colleagues in the Analytical Department of Selectide, Dr. Victor Nikolaev, Dr. John Isbell, Nina Ma, and Shelly Wade, for their contributions in creating and developing analytical concepts in combinatorial chemistry. Writing this manuscript would not have been possible without numerous fruitful discussions with our colleagues. Special thanks to Dr. John Isbell for his time and effort for editing the chapter.

REFERENCES

1. Gordon EM, Barrett RW, Dower WJ, Fodor SPA, Gallop MA, Applications of combinatorial technologies to drug discovery, 2. Combinatorial organic synthesis, library screening strategies, and future directions, *J. Med. Chem.*, 37:1385–1401, 1994.
2. Thompson LA, Ellman, Synthesis and applications of small molecule libraries, *Chem. Rev.*, 96:555–600, 1996.
3. Jung G, ed., Combinatorial Peptide and Non-peptide Libraries, Weinheim, Germany, *Verlag Chemie*, 1996.
4. Desai MC, Zuckermann RN, Moos WH, Recent advances in the generation of chemical diversity libraries, *Drug. Dev. Res.*, 33:174–188, 1994.
5. (a) Wender A, Introduction: Frontiers in Organic Synthesis, *J. Org. Chem.* 61:71–2, 1996, (b) Resnati G, Soloshonok VA Eds, Fluoroorganic Chemistry: Synthetic challenges and biomedicinal rewards, *Tetrahedron*, 52:1–330, 1996.
6. Stavely LAK, *The Characterization of Chemical Purity: Organic Compounds*, Butterworths, London, 1971, p. 173.
7. Morand K, Burt TM, Dobson RLM, Wilson LJ, Development of a mass spectrometry laboratory for support of pharmaceutical combinatorial chemistry programs, *Proc. 45th ASMS Conf.*, p. 1261, 1997.
8. Fang L, Dreyer M, Chen Y, Plunkett M, Hacker M, Analysis of small organic combinatorial libraries by LC-MS, *Proc. 45th ASMS. Conf.*, p. 401, 1997.
9. Issakova O, Nikolaev V, Ma N, Wade S, Sepetov N, Analytical characterization of one-bead-one-compound combinatorial libraries by LC/MS and LC/MS/MS, *Proc. 45th ASMS. Conf.*, p. 403, 1997.
10. Keifer P, High-resolution NMR techniques for solid-phase synthesis and combinatorial chemistry, *Drug Discovery Today*, 2(11): 468–478, 1997.
11. Keifer PA, Baltusis L, Rice DM, Tymiak AA, Shoolery JN, A comparison of NMR spectra obtained for solid-phase-synthesis resins using conventional high-resolution, magic-angle-spinning, and high-resolution magic-angle-spinning probes, *J. Magn. Res*, 119:65–75, 1996.
12. Look GC, Holmes CP, Chinn JP, Gallop MA, Methods for combinatorial organic synthesis: The use of fast ^{13}C NMR analysis for gel-phase reaction monitoring, *J. Org. Chem.*, 59:7588–7590, 1994.
13. Svensson A, Fex T, Kihlberg G, Use of ^{19}F NMR spectroscopy to evaluate reactions in solid phase organic synthesis, *Tetrahedron Lett.*, 37(42):7649–7652, 1996.
14. Quarrel R, Claridge TDW, Weaver GW, Lowe G, Structure and properties of tentagel resin beads: Implications for combinatorial library chemistry, *Mol. Diversity* 1:223–232, 1995.
15. MacDonald AA, Dewitt SH, Ghosh S, Hogan EM, Kieras L, Czarnik AW, Ramage R, The impact of polystyrene resins in solid-phase organic synthesis, *Mol. Diversity*, 1:183–186, 1996.
16. Gordeev M, Patel DV, Wu J, Gordon EM, Approaches to combinatorial synthesis of heterocycles: Solid-Phase synthesis of pyridines and pyrido [2, 3-d] pyrimidines, *Tetrahedron Lett.*, 37:4643–4646, 1996.
17. Barn DR, Morphy JR, Rees DC, Synthesis of an array of amides by aluminum chloride assisted cleavage of resin-bound esters, *Tetrahedron Lett.*, 37:3213–3216, 1996.

18. Stoever HDH, Frechet JMJ, Direct polarization ^{13}C and ^{1}H magic angle spinning NMR in the characterization of solvent-swollen gels, *Macromolecules*, 22:1574–1576, 1989.
19. Anderson RC, Jarema MA, Shapiro MJ, Stokes JP, Ziliox M, Analytical techniques in combinatorial chemistry: MAS CH correlation in solvent-swollen resin, *J. Org. Chem.*, 60:2650–2651, 1995.
20. Wehler T, Westman I, Magic angle spinning NMR: A valuable tool for monitoring the progress of reactions in solid phase synthesis, *Tetrahedron Lett.*, 37:4771–4774, 1996.
21. Sarkar SK, Garigipati RS, Adams JL, Keifer PA, An NMR method to identify nondistructively chemical compounds bound to a single solid-phase-synthesis bead for combinatorial chemistry applications, *J. Am. Chem. Soc.*, 118:2305–2306, 1996.
22. Shapiro MJ, Chin J, Marti RE, Jarosinski MA, Enhanced resolution in MAS NMR for combinatorial chemistry, *Tetrahedron Lett.*, 38(8):1333–1336, 1997.
23. Chain J, Fell B, Pochapsky S, Shapiro MJ, Wareing JR, 2-D SECSY NMR for combinatorial chemistry—High resolution MAS spectra for resin-bound molecules, 63:1309–1311, 1998.
24. Gordon EM, Galop MA, Patel DV, Strategy and tactics in combinatorial organic synthesis, Application to drug discovery, *Accounts. Chem. Res.*, 29:144–154, 1996.
25. Olson DL, Peck TL, Webb AG, Magin RL, Sweedler JV, High resolution macrocoil ^{1}H NMR for mass-limited, nanoliter samples, *Science*, 270:1967–1970, 1995.
26. Chin J, Fell JB, Jarosinski M, Shapiro M, Wareing JR, HPLC/NMR in combinatorial chemistry, *J. Org. Chem.*, 63:386–390, 1998.
27. Yan B, Kumaravel HA, Wu A, Petter RC, Jewell C, Wareing J, Infrared spectrum of a single resin bead for real time monitoring of solid phase reactions, *J. Org. Chem.*, 60:5736–5738, 1995.
28. Yan B, Gstach H, An indazole synthesis on solid phase support monitored by single bead FT-IR microspectroscopy, 37(46):8325–8328, 1996.
29. Pivonka DE, Simpson TR, Tools for combinatorial chemistry: real time single-bead infrared analysis of a resin-bound photocleavage reaction, *Anal. Chem.*, 69:3851–3853,1997.
30. Haskins NJ, Humter DJ, Organ AJ, Rahman SS, Thom C, Combinatorial chemistry: Direct analysis of bead surface associated materials, *Rapid. Common. Mass. Spect.*, 9:1437–1440, 1995.
31. Egner BJ, Langly GJ, Bradly M, Solid phase chemistry: Direct monitoring by matrix-assisted laser desorption/ionization time of flight mass spectrometry, A tool for combinatorial chemistry, *J. Org. Chem.*, 60:2652–2653, 1995.
32. Fitzgerald MC, Harris K, Shevlin CG, Siuzdak G, Direct characterization of solid phase resin-bound molecules by mass spectrometry, *Bioorg. Med. Chem. Lett.*, 6(8):979–982, 1996.
33. Li LYT, Kyranos JN, Automated multi-dimensional HPLC/UV/MS for qualitative method development, *Proc. 44th ASMS. Conf.*, p. 1041, 1996.
34. Zeng L, Burton L, Yung K, Shushan B, Kassel DB, Automated analytical/preparative high-performance liquid chromatography-mass spectrometry system for the rapid characterization and purification of compound libraries, *J. Chromatogr.*, 794:3–13, 1998.

35. Smart SS, Mason TJ, Bennell PS, Maeji NJ, Geysen HM, High-throughput purity estimation and characterization of synthetic peptides by electrospray mass spectrometry, *Int. J. Peptide. Protein. Res.*, 47:47–55, 1996.
36. Kibbey CE, Quantitation of combinatorial libraries of small organic molecules by normal-phase HPLC with evaporative light-scattering detector, *Molec. Diversity.*, 1:247–258, 1995.
37. Fujinary E, Manes JD, Nitrogen-specific detection of peptides in liquid chromatography with a chemiluminescent nitrogen detector, *J. Chromatogr A*, 676:113–120, 1994.
38. Fitch W, Szardenings K, Chemiluminescent nitrogen detection for HPLC: An important new tool in organic analytical chemistry, *Tetrahedron Lett.*, 38(10): 1689–1692, 1997.
39. Fujinary E, Manes JD, Bizanek R, Peptide content determination of crude synthetic peptides by reversed-phase liquid chromatography and nitrogen-specific detection with a chemiluminescent nitrogen detector, *J. Chromatogr A*, 743:85–89, 1996.
40. Bizanek R, Manes JD, Fujinari E, Chemiluminescent nitrogen detection as a new technique for purity assessment of synthetic peptides separated by reversed-phase HPLC, *Peptide. Res.*, 9(l):40–44, 1996.
41. Garr C, The use of evaporative light scattering in quality control of combinatorial libraries, *Proceedings of the Materials of Solid and Solution Phase Combinatorial Synthesis Meeting*, Princeton, NJ, 1997.
42. Furka A, Sebestyen F, Asgedom M, Dibo G, A general method for the rapid synthesis of multicomponent peptide mixtures, *Int. J. Peptide. Protein. Res.*, 37:487–493, 1991.
43. Lam KS, Salmon SE, Hersh EM, Hruby VJ, Kazmierski WM, Knapp RJ, A new type of synthetic peptide library for identifying ligand-binding activity, *Nature*, 345:81–82, 1991.
44. Lam K, Lebl M, Krchnak V, The "one-bead-one-compound" combinatorial library method, *Chem. Rev.*, 97:411–448, 1997.
45. Houghten RA, Pinilla C, Blondelle SE, Appel JR, Dooley CT, Cuevro JH, Generation and use of synthetic peptide combinatorial libraries for basic research and drug discovery, *Nature*, 345:84–86, 1991.
46. Appell KC, Chung TDY, Ohmeyer MHJ, Sigal NH, Baldwin JJ, Chelsky D, Biological screening of a large combinatorial library, *I. Biomol. Screen.*, 1:27–31, 1996.
47. Krchnak V, Weichseil AS, Issakova O, Lam KS, Lebl M, Bifunctional scaffolds as templates for synthetic combinatorial libraries, *Molec. Diversity*, 1:177–182, 1995.
48. Ohlmeyer MJ, Swanson RN, Dillard LW, Reader JC, Asouline G, Kobayashi R, Wigler M, Still WC, Complex synthetic chemical libraries indexed with molecular tags, *Proc. Natl. Acad. Sci. USA.*, 90:10922–10926, 1993.
49. Chabala JC, Baldwin JJ, Barbaum JJ, Chelsky D, Dillard LW, Henderson I, Li G, Ohlmeyer MHJ, Randle TL, Reader JC, Rokosz L, Sigal NH, Binary encoded small-molecule libraries in drug discovery and optimization, *Perspect. Drug Discovery Des.*, 2:305–318, 1995.
50. Kocis P, Issakova O, Sepetov N, Lebl M, Kemp's triacid scaffolding for synthesis of combinatorial nonpeptide uncoded libraries, *Tetrahedron Lett.*, 36:6623, 1996.

51. Issakova O, Ma N, Kocis P, Sepetov N, Non-peptide libraries: Structure determination by LC/MSIMS, *Proc. 43th ASMS Conf.*, p. 483, 1995.
52. Fitch W, Analytical methods for the quality control of combinatorial libraries, *Annu. Rep. Combinatorial Chem. Molec. Diversity*, 1:59–68, 1997.
53. Weller HN, Young MG, Michalczyk SJ, Reitnauer GH, Cooley RS, Rahn PC, Loyd DJ, Fiore D, Fischman SJ, High throughput analysis and purification in support of automated parallel synthesis, *Molec. Diversity.*, 3:61–70, 1997.
54. Bonner R, Burton L, Mann M, Wilm M, Automating mass spectrometry with Apple Script, *Proc. 43th ASMS Conf.*, p. 739, 1995.
55. Bonner R, Burton LL, A new mass spectrometry data processing program, *Proc. 43th ASMS Conf.*, p. 740, 1995.
56. Batt J, McDowall MA, Preece SW, Rontree JA, Automated LC/MS for the characterization of combinatorial libraries, *Proc. 44th ASMS Conf.*, p. 1033, 1996.
57. Vestal C, Li LYT, Towle MR, Kyranos IN, Evaluation of an automated high throughput LC/MS system for the analysis of combinatorial libraries, *Proc. 45th ASMS Conf.*, p. 1249, 1997.
58. Fang L, Automated high-speed mass spectrometric analysis of large combinatorial libraries, Materials of the Third *Annual CHI Meeting for High Throughput Screening for Drug Discovery and Rapid Compound Characterization*, San Diego, CA, 1996.
59. Kyranos, Analytical characterization of combinatorial libraries: the short and long of it, Materials of the Third *Annual CHI Meeting for High Throughput Screening for Drug Discovery and Rapid Compound Characterization*, San Diego, CA, 1996.
60. Lewis K, Fitch W, Maclean D, Characterization of split/pool combinatorial libraries, *Proc. 45th ASMS. Conf.*, p. 1265, 1997.
61. Steinbeck C, Berlin K, Richert C, MASP—a program predicting mass spectra of combinatorial libraries, *J. Chem. Information Comput. Sci.*, 37:449–457, 1997.
62. Boutin JA, Hennig P, Lambert PH, Bertin S, Petit L, Mahiev JP, Serkiz B, Volland JP, Fauchere JL, Combinatorial peptide libraries: Robotic synthesis and analysis by nuclear magnetic resonance, mass spectrometry, tandem mass spectrometry, and high-performance capillary electophoresis techniques, *Anal. Biochem.*, 234:126–141, 1996.
63. Issakova O, Nikolaev V, Ma N, Wade S, Sepetov N, Analytical characterization of one-bead one-compound combinatorial libraries by LC/MS and LC/MS/MS, *Proc. 45th ASMS Conf.*, p. 403, 1997.

11 Key Ingredients for Efficient High-Throughput Screening

Lucia Carrano and Stefano Donadio

CONTENTS

I. INTRODUCTION

High-Throughput Screening (HTS) is the process by which a large number of compounds can be tested, in an automated fashion, for activity as inhibitors (antagonists) or activators (agonists) of a certain biological function. The purpose of HTS is to identify, amid a large collection of samples, a few that exert the desired activity. The rate (the efficiency of the process) is given by the number of compounds that can be tested per unit of time, typically one year. The origin of this technology lies in the screening of microbial natural products for the discovery of new antibiotics in the 1950s. Millions of samples from natural sources (microbes, plants, animals) have been screened since then for potentially useful medicinal properties. The screening of natural products declined in the 1970s as a result of the decreasing number of new chemical entities discovered and the strong development of medicinal chemistry, which allowed successful chemical modification of existing drugs or active principles.

In the 1990s, advances in disparate disciplines such as laboratory automation and robotics, which allow the rapid testing of enormous numbers of samples while minimizing human labor, transformed screening into HTS. These major strides in automation have been paralleled by the availability of large collections of natural and synthetic compounds accumulated in the store through the years by major pharmaceutical companies; and by the development of combinatorial chemistry, which can generate large chemical libraries in a very short time. In addition, major advances in molecular biology and the entrance into the "genomic era" have opened the possibility of identifying every target involved in almost any pathology. These technological advances came into a period when competitive pressure forced first pharmaceutical and then agrochemical companies to revolutionize their research process, with the goal of reaching the market in a shorter time. HTS seems to offer the "magic" solution of filling the "pipeline" of developmental products that pharmaceutical and agrochemical companies need to maintain to be commercially successful.

Nowadays, HTS has the potential to revolutionize not only compound discovery but also the development phase. By automating routine processes and miniaturizing assay format, HTS facilities can screen as many as 10,000 samples a day, in comparison with the hundreds per week in the 1980s. This factor alone greatly accelerates the discovery process.

II. THE DRUG DISCOVERY PROCESS

Pharmaceutical and agrochemical companies have recently undergone enormous changes both in the way of doing business and in the discovery of new drugs. The drug discovery and development process (from idea to market) is risk taking and time consuming, a big economic effort involving several disciplines. This process can be divided (see Figure 11.1) in the search for new compounds that are active in model systems, in their optimization and/or characterization into developmental compounds, and in the "trial" phases, when these biologically active compounds are evaluated for efficacy in the clinic (pharmaceuticals) or in the field (agrochemicals). For simplicity, we will consider in this chapter the pharmaceutical field. Nonetheless, many concepts expressed here apply, *mutatis mutandis*, to the agrochemical industry as well. The clinical phases transform a "drug candidate" into a drug (Figure 11.1).

HTS is a multidisciplinary process directed at the identification of leads (see Figure 11.2). Samples are tested against a series of assays in order to identify *hits* samples that exhibit desirable properties such as potency, specificity, selectivity, etc. Hits are submitted to further tests for a deeper evaluation of their biological properties. Those compounds that exhibit the desirable biological properties and represent patentable structure are usually defined as *leads*. Lead compounds may require improvements of their properties. HTS can also be employed during lead optimization, when thousands of compounds can be generated by combinatorial methods. HTS can therefore be performed

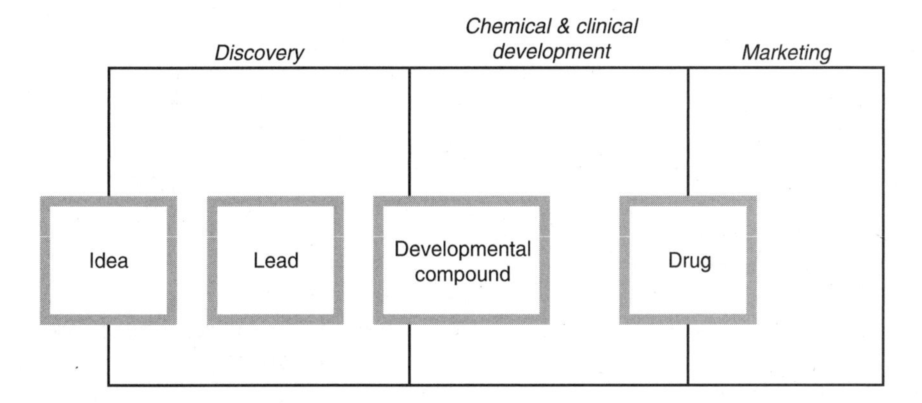

FIGURE 11.1 The drug discovery process.

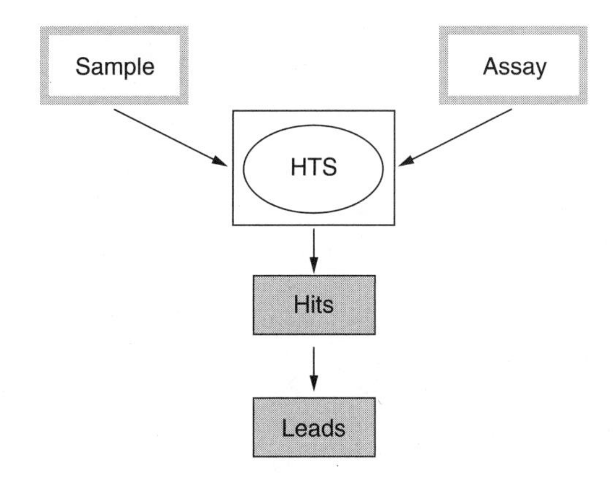

FIGURE 11.2 The HTS process. Samples and assays are the inputs and hits and leads the outputs.

at the very beginning of the discovery cycle, allowing the selection of hits amid large collections of compounds; or during the evaluation of hits and the optimization of leads. In addition, the data obtained during screening operations may give valuable insight into structure–activity relationships, which in turn may direct improved approaches to drug discovery.

As an example, we can consider the discovery and development of an antiviral drug. Inhibitors of the HIV protease, an enzyme essential for the maturation of the virus, can potentially cure HIV-infected people. The discovery process may consist of finding samples able to inhibit the viral aspartic protease over a certain threshold, while having little or no effect on another protease of the same class, such as pepsin. The hits will be submitted to further biological tests to identify leads, patentable compounds that are capable, for example, of inhibiting viral replication in cellular models.

The lead optimization process introduces structural variations in the molecule in order to identify the best drug candidate. After a thorough evaluation of its toxicity and pharmacological activity in animal models, the molecule enters the clinical phases, when it is evaluated for tolerability, efficacy, and potential side effects on human subjects. After submittal of all the required documentation and registration to the appropriate authority (e.g. the U.S. Food and Drug Administration), the HIV protease inhibitor is ready for launching on the market.

The drug discovery process involves several stages of compound advancement. Few leads will become drug candidates; still fewer will reach the market. It is estimated that approximately 100 leads must be generated to obtain a compound that eventually reaches the market. This is due to the high attrition rate between one stage and the next. Since the number of leads depends on the number of hits, HTS bears promise as an important answer to the economic pressures that have forced industry to speed up the discovery process while reducing costs. Indeed, if one considers hit generation as a purely stochastic process, the number of hits will be proportional to

$$\text{(Number of samples)} \times \text{(number of assays)}$$

However, HTS is not an entirely stochastic process. In fact, the attrition rate between hits and leads will depend greatly on the quality of the hits, which in turn stems from the appropriate combination of samples, targets, and data analysis. Therefore, quality as well as quantity of hits are extremely important. It should be kept in mind that the progression of a hit into a lead, and of a lead into a drug candidate, involves strong commitment of multidisciplinary resources. To avoid costly investment in too many leads or drug candidates, a judicious selection of the hits, the immediate output of HTS, is extremely important. In the remainder of this chapter we will briefly describe the key ingredients for HTS and the impact they may have on drug discovery.

III. THE COMPONENTS OF HTS

The key components of HTS are samples, assays, automation, and data management. Without any one of these ingredients, HTS would not be possible. Each plays a crucial role in the process: samples are the source of chemical diversity that will generate leads; assays are the baits to fish out bioactive compounds; automation allows testing large numbers of compounds in a relatively short time; and data management enables the retrieval of meaningful information from the flood of data generated. In these years, we are witnessing a rapid evolution of these HTS ingredients—for example, new sample collections become available; improvements in detection technology make new types of assays suitable for HTS; tests are continuously miniaturized; operations once performed manually are automated; data can be accessed directly from a desktop computer. We will try to give the general concepts behind each HTS ingredient. This chapter is not an attempt to cover the latest technological advances in HTS, but rather a global view of the framework in which it operates.

A. Automation

Automation eliminates or dramatically reduces several error sources associated with manual work, and enables continuous execution of routine tasks. Automation can be applied to several facets of HTS. For example, sample preparation can be automated by executing the same treatment on all samples, e.g. dilution or extraction with organic solvent and allocation in test plates suited for the assays. Large libraries of compounds can thus be built in a ready-to-test format through a safe and flexible process, which enables several dispositions of samples. Next, the automated execution of the assays ensures high throughput (number of samples tested per week), quality, and repeatability. The assays are executed by robotic systems, which commonly (see Figure 11.3) have the following elements in common: (1) plate storage arrays: some systems hold plates in a stack or in a carousel of stacks, from which they can be placed on trays and moved from one station to another; (2) robotic arm(s), with fittings that can handle plates, trays or individual vials

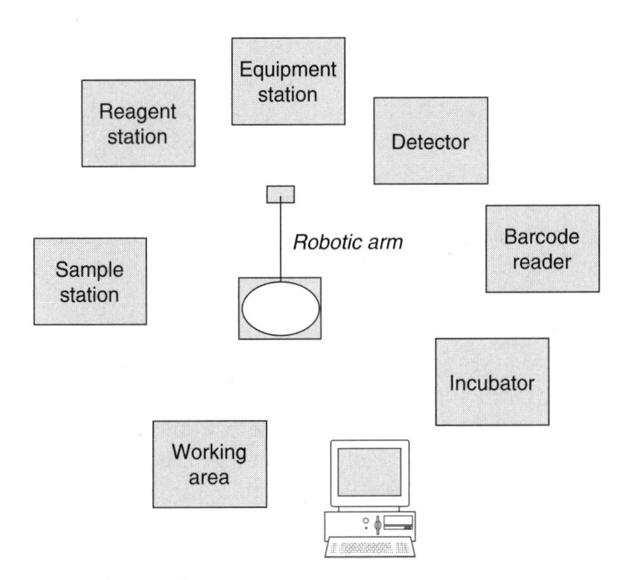

FIGURE 11.3 Robot layout. A typical operating scheme may be as follows. The robotic arm picks up a plate from the sample station, brings it to the barcode reader and puts it on the working area. The robotic arm then grabs the 96-well multichannel pipettor from the equipment station, pipette reagents from the reagent station, and dispenses them into the microtiter plate in the working area. The plate is agitated and placed in the incubator. The same sequence of steps is then performed on another plate. After the required incubation time, the robotic arm transports the microtiter plate to the barcode reader and then to the detector.

and move them to the appropriate location; (3) liquid-handling equipment, pipettors in different configurations that can range from single-channel to parallel pipettors such as the 96- or 384-well multichannel pipettors; (4) incubators that can monitor temperature and humidity; (5) readers that detect the results of the assays, usually by measuring fluorescence, radioactivity, or color, and that produce files listing numeric results for each well in a plate; (6) barcode readers, integrated systems to track plates; (7) integration systems that allow the automatic acquisition of raw data and their transformation in results.

The activity of the robots must be scheduled for best performance. Robots can be programmed to work 24 h a day. For example, different assays can be performed during different days of the week by the same robot, while the necessary cleaning of each robotic component is executed overnight and routine maintenance is done on weekends. The robots perform analytical measurements such as withdrawing and dispensing fixed volumes, and registering absorbance values at different wavelengths. Thus, their proper functioning requires Standard Operating Procedures (SOPs), including periodical assessment of the correct functioning of each robotic component. These procedures ensure the quality and consistency of the results.

B. Samples

Samples are the second key component for the success of HTS—they are the source of new leads. The number and diversity of the chemical entities present in a sample library are crucial for providing a reasonable chance of success. Sample libraries can be collections of natural compounds, combinatorial products, or previously synthesized molecules. As a first approximation, the probability of finding a hit is correlated directly to the number of different chemical entities present in the screened library. However, this is true only if the compounds present are indeed unique and unrelated. This is seldom the case.

Synthetic compounds accumulated through decades by large corporations represent a source of chemical diversity. A synthetic library has the advantage of consisting of pure compounds with known physicochemical properties. Furthermore, the concentration of the compounds can be adjusted, and a defined synthetic scheme is usually available for future scale-up. The big disadvantage of these libraries is that they are biased: the compounds have been synthesized in the past for specific projects and are thus nonrandom. This often results in limited chemical diversity and low affinity for the target. In addition, hits identified in these libraries may not be patentable because of previous disclosure of their structure.

Combinatorial libraries are one significant source of compounds. Their preparation is described in great detail elsewhere in this volume. Millions of compounds can easily be generated by combinatorial techniques using relatively simple chemistry. This approach is very useful in the lead optimization phase, when subtle variations on a lead structure can be introduced that may lead to quick identification of the best compound. However, the initial bias introduced by the synthetic scheme severely limits the chemical diversity present in combinatorial libraries for their use in lead identification.

Natural products, usually secondary metabolites, are a notable source of novel molecules with diverse and unpredictable structures. Natural sources may include plants, insects, and microbes. The last class of producers represents the easiest scalable and renewable source of compounds. Microbial secondary metabolites have been the source of many commercial products. These molecules may have high affinity for their cognate target, as they may interact with it directly or with an analogous target in the ecosystem of the producer organism. Natural-product libraries are usually made up of samples consisting of many different molecules prepared as extracts from the producer organism. Disadvantages of natural-compound libraries are the cost per sample, and the quite complicated deconvolution procedures necessary to identify and characterize the active principle from a complex mixture of unknown and unrelated molecules.

In summary, the ideal library should be large, chemically diverse, and unexploited. Each sample in it must be fully trackable to the appropriate source (e.g. structure, synthetic scheme, deconvolution code, producing organism). The amount of each sample in the library should permit all the tests necessary to obtain the relevant information about its bioactivity and for proceeding to the next stage, since sample preparation is expensive. Unrelated assays can be used on previously screened libraries—however, previous uses must be carefully evaluated. For example, if a chemical library has already been used to identify inhibitors of the metallo protease MMP3 (involved in rheumatoid arthritis), there will be little chance of finding new hits by screening the same library against another zinc protease (unless the new assay is more sensitive than the older one). It may make more sense to test only the samples that were positive on the MMP3 assay. It should be noted that already-described molecules may exhibit unexpected properties on unreported assays. However, a patent for their novel use can only protect them. New, undescribed chemical derivatives are usually synthesized when the identified hit is a described chemical entity.

C. Targets and Assays

The assays used to identify hits are derived from a target, related to a disease that we need to combat (modulate). A target is the biological system on which the desired compounds acts. For example, a certain microorganism is responsible for an infection that causes a disease. We may want to find a molecule that is able to kill that microorganism without affecting a human being. The medical target is represented by the infection, but it is necessary to design a test to identify anti-infective molecules. We could use the microorganism directly in the assay and follow its ability to duplicate after exposure to different samples. Most microorganisms contain multiple molecular targets essential for viability, and we may want to hit a specific essential function of that microorganism, for example, DNA replication. Or we may want to look for inhibitors of a particular enzyme involved in DNA replication, for example, DNA gyrase; or for noncompetitive inhibitors of this enzyme. Usually, there is a medical target, the disease, and a molecular target which is the basis of the assay. In the above example, the molecular target could be an entire organism, an essential function, an enzyme, or a specific site of a protein. Thus, while killing a particular pathogen is the final aim of an anti-infective drug discovery process, considerations about hit rate, specificity, efficiency of the process, and, above all, the perceived chance of success, may lead to choosing one molecular target or another. The molecular target needs to be reduced to an assay—therefore, it must be amenable to miniaturization and high throughput. In a HTS program, a series of tests is usually introduced in parallel or serial mode to determine accurately the specificity or selectivity of samples.

Ideally, the role of the target in the pathology should be demonstrated. However, such a target has a high chance of having been utilized by a competitor. Consequently, a certain risk may be taken in target selection. Selecting a target for a particular pathology is an extremely critical step that affects the quality of the hits and leads that will be generated. It requires deep knowledge of the biological system, and its description clearly goes beyond the scope of this chapter. Once the target has been identified, it is possible to design an assay. An assay suited for HTS has to obey some rules: it must run in a standard microplate format; produce a simple readout; provide consistent results; and operate using stable and available reagents. Deciding which assay to run and then creating the robotic infrastructure to support it is an important part of implementing a successful HTS program. Each assay is designed for a particular purpose. For example, when analyzing a large library of mixtures of compounds, the assay must be sensitive enough to detect traces of active compounds or weak activities, while at the same time discriminating the signal from background noise efficiently. On the other hand, when screening a combinatorial library during lead optimization, the assay must detect those molecules that act selectively over a certain threshold. About a half dozen basic assay types are suited for most HTS programs. The most common are cell-free assays, such as enzyme inhibition or receptor binding, and whole-cell assays. Cell-based assays closely mimic the inner workings of a living cell and present a number of implementation problems. For example, cells must remain viable for the duration of the assay, without adverse effects from the delivery of reagents or the sample solvent. In addition, whole-cell assays based on lack of or reduced signal must take into account that any toxic compound will give the same signal as a specific inhibitor of the target. On the other hand, cell-free assays are more apt to identify molecules that act specifically on the target. They do not address the ability of the sample to access a cell's interior, however. The type of assay chosen has important consequences on the type of hits that will be identified.

D. Assay Validation

The assay is the only HTS ingredient whose output is a measurement. Consequently, it is an analytical method that needs validation. This is achieved by measuring some parameters such as reproducibility, variation in measurements of replicas during the same experiment, and repeatability, changes in measurement of replicas on different days. The precision of the test is established by measuring the ratio between repeatability and reproducibility—the closer this value is to 1, the more precise is the test. Other important parameters to consider are: the confidence range (the accuracy of the mean value), the prediction range (the accuracy of a single value), and the coefficient of variation (the deviation of single determinations from the mean value). The importance of assay validation becomes evident if we consider that the current format for HTS assays is the 96-well plate. The trend is toward miniaturization, so 384-well plates, together with the corresponding robots, will rapidly replace existing systems. Miniaturization reduces reagent costs and the required amount of sample, so more tests can be performed per unit of time, more data can be generated on a single assay. As data increase, so does the possibility of generating false results due to statistical fluctuations, instrumental answer, or background noise. A thorough evaluation of the quality of the data and continuous monitoring of the execution of the assay are necessary to detect any malfunctioning or false results.

Let us consider as an example one of the most-used HTS tests: a transcription assay (see Figure 11.4). We are looking for a modulator of the expression of a certain gene (for example, an inhibitor of the expression of a growth factor). The cells have been engineered by placing a reporter gene under the control of the promoter we want to modulate. The reporter gene is expressed if it is appropriately stimulated and codes for an enzyme (e.g. luciferase), whose activity can be easily detected by measuring the light emitted by a chemiluminescent substrate. When the promoter is down-regulated, the reporter gene is expressed to a lesser extent, producing a smaller amount of luciferase and consequently less signal (light). The light emitted is recorded by a robotic reader. The microwells, each containing the same amount of engineered cells, are exposed to the samples for the appropriate time; then the signal is recorded. An inhibitor of gene expression gives no or little signal. The robotic software converts the signal by calculating the percentage of inhibition. This is done by applying a simple calculation:

$$\%\ \text{Inhibition} = 100 - \left(\frac{\Delta S_{\text{sample}}}{\Delta S_{\text{control}}}\right) \times 100$$

where ΔS_{sampie} and $\Delta S_{\text{control}}$ are the optical densities (the differences between the values measured at time zero and at the time required to develop light) for sample and control, respectively. In HTS jargon this operation is called "applying a macro." The macro is a mathematical calculation that is applied directly to the output of the reader to transform raw data into results.

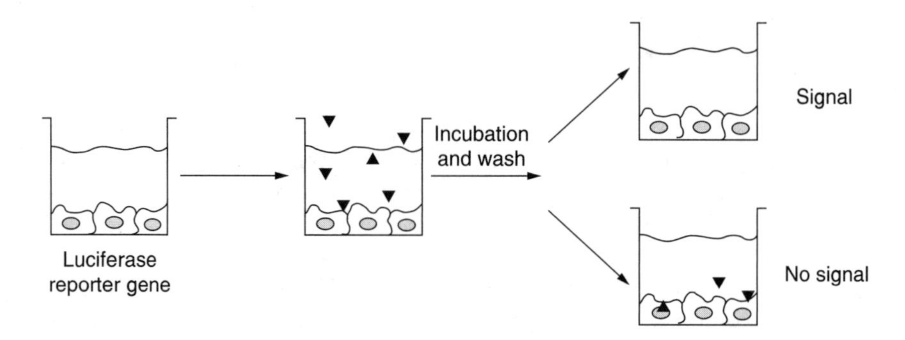

FIGURE 11.4 Schematic representation of a transcription assay. Cells are engineered by placing the luciferase-encoding gene under the control of a promoter of interest. Each microwell contains the same amount of cells, which are exposed to different samples and incubated for the required time. After the appropriate incubation, the sample is washed out and the light emitted by luciferase is measured.

Obviously, it will be necessary to eliminate those samples that generally inhibit transcription or luciferase activity (for example, by running a parallel test where the luciferase encoding gene is placed under the control of a "housekeeping" promoter).

The first step in a validation procedure is to evaluate the instrumental answer, which needs to be clear, meaningful, and independent of position in the microplate. To verify this, the same samples are distributed to different positions within the same plate. If a modulator of the activity is known, a dose-response curve can be built. Alternatively, an inhibitor of the reporter enzyme can be used to evaluate the instrumental answer and to define the parameters of the assay. The same experiment will be executed on different days and the values obtained plotted and compared. In the best case the dose-response curves will coincide; more often they establish the error affecting the measured values, hence the confidence interval. This type of analysis identifies the cutoff for the assay—the minimum percent inhibition for a meaningful activity. Possible artifacts, such as the "frame effect" (higher evaporation at the microtiter frame results in slightly different signals between central and border wells) can also be detected, and appropriate controls introduced. Figure 11.5 shows an example of how the signal can be evaluated.

E. Generation and Management

HTS accelerates the identification of drug leads through the analysis of enormous numbers of samples in a series of automated assays (primary assays) run repetitively by robots. To decide whether to accept or reject the assay results, scientists validate the assay, analyze the data collected, and monitor whether the assay runs according to protocol. Several quality control procedures have been developed to evaluate the screening results and to

	Signal														
	1	2	3	4	5	6	7	8	9	10	11	12	AVG	STDEV	CV%
A	1.103	1.205	1.120	1.131	1.140	1.190	1.121	1.130	1.140	1.190	1.190	1.150	1.16	0.03	2.51
B	1.150	1.209	1.208	1.190	1.110	1.230	1.204	1.190	1.120	1.230	1.230	1.240	1.19	0.05	4.32
C	1.200	1.150	1.140	1.230	1.210	1.106	1.142	1.230	1.210	1.109	1.230	1.110	1.17	0.06	4.86
D	1.190	1.121	1.190	1.260	1.210	1.110	1.190	1.260	1.210	1.110	1.110	1.210	1.18	0.06	4.96
E	1.222	1.210	1.209	1.106	1.230	1.233	1.201	1.109	1.230	1.230	1.210	1.130	1.20	0.05	4.12
F	1.110	1.115	1.208	1.231	1.271	1.283	1.210	1.230	1.270	1.303	1.120	1.240	1.24	0.06	4.63
G	1.120	1.150	1.283	1.150	1.290	1.293	1.290	1.150	1.259	1.288	1.115	1.121	1.23	0.08	6.67
H	1.150	1.120	1.197	1.190	1.230	1.190	1.198	1.195	1.230	1.190	1.210	1.120	1.20	0.03	2.89
AVG	1.16	1.17	1.21	1.19	1.22	1.21	1.21	1.19	1.22	1.21	1.18	1.17			
STDEV	0.04	0.06	0.05	0.05	0.06	0.07	0.05	0.05	0.05	0.07	0.05	0.06			
CV%	3.81	5.23	4.08	4.50	4.97	5.82	4.17	4.45	4.36	5.97	4.36	4.77			

FIGURE 11.5 A microplate map. Rows are indicated by letters A–H and columns by numbers 1–12. Each well contains the same sample, in this case a blank consisting of 2% sample solvent. The measurements and the calculations performed indicate variations in signal. The average value obtained along each column and row is calculated with standard deviation (StDev), and coefficient of variation (CV, equal to StDev/average). The overall values for the plate are also indicated. Meaningful results are usually obtained when CV < 5%, but assays performing with CVs up to 10% are also accepted. In the example shown, the border columns perform better than average. This effect is normally taken into account with appropriate controls. The experiment relates to an ATPase reaction measured with a purified enzyme. The amount of ADP released is measured through a coupled enzymatic reaction by determining the optical density at 340 nm.

identify biologically active compounds, few among thousands of samples examined. The early identification of hits is an important factor for success. Another is the assessment of structure–activity relationships, and HTS can be applied to further profiling hits or leads. In general, HTS is first performed at a single concentration (one data point per sample for each assay), possibly in triplicate (three data points), to find samples that display some activity above a certain threshold. These samples are then examined in the same assay(s) using multiple sample concentrations. The data are fitted in a dose-response curve to obtain an evaluation of the sample's potency or selectivity. Those samples that exhibit satisfactory properties undergo further testing in secondary assays for additional features.

HTS is a substantial economic effort, so it is necessary to understand its intricacies for a true payoff. Efficient data analysis ensures high quality of the results. Usually, the same sample library is run through more than one assay simultaneously (Figure 11.6). A huge amount of data is generated, which needs to be accessed, analyzed, and correlated among the different assays, for the identification of hits.

In order to describe the characteristics of an efficient data management system, let us consider in detail a typical HTS program. A library of 100,000 samples is used against 10 different primary assays, all performed within a one-year period. Each assay generates an average of three data points per sample; thus the throughput is 3,000,000 data points per year, or about 15,000 data points per day. All these data are acquired automatically, processed, and transformed into results through a simple algorithm, the macro. Results are stored in appropriate computer space. If the positive rate is 0.5% on average, secondary assays will be performed on 5000 samples for each of the 10 assays. Usually, secondary assays are more elaborate and may require dose-response curves. Assuming that each sample generates, during secondary assays, an average of 10 data points per assay, the output is an additional 500,000 data points. In this example, 3,500,000 data points are generated per year. In order to identify hits, it is thus necessary to manipulate the 3,500,000 data points; to access primary and secondary assay results for each sample, and to retrieve the sample track for evaluation of the meaningfulness of the hits. The data management system has to make easily accessible—(1) all data for a given assay to identify positives; (2) all data for a time period or plate position to identify artifacts due to malfunctioning of the robots (for example, a broken pin not delivering reagents can cause the same signal as a true inhibitor; a lesser amount of reagents delivered can affect the results for all samples tested in that time period; (3) all data of a single assay for appropriate statistical analysis; (4) all samples that are positive in more than one assay, to identify false positives; (5) all data necessary to track the sample and allow its deconvolution (i.e. sources, structure, dilution, freezing and thawing cycles, etc.).

A data management system handles all the data generated during different HTS projects. This system needs to be centralized but user friendly and flexible to allow query execution and satisfy the users' needs, while keeping

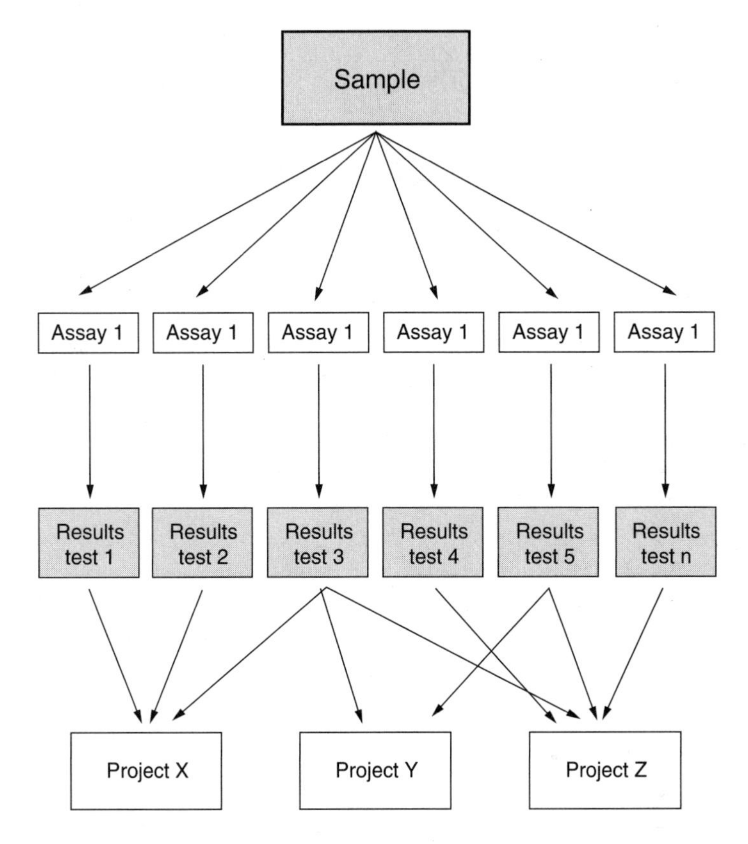

FIGURE 11.6 Scheme of data management and integration. Samples undergo several simultaneous assays, generating data that are transformed into results. Each project consists of a combination of results from different tests. The data management system correlates all data relative to each box (assays, results, projects). In this example, samples are declared positives to project X when the proper combination of results from assays 1, 2, and 3 is encountered.

all the HTS data. The identification of the correct signal and the evaluation of the confidence limit are critical for the success of a HTS program. Bad data cause a great deal of useless work and increase the cost of HTS. The data can be visualized in one of several representations; a useful one is illustrated in Figure 11.7. The majority of the values should fall around the control; only a few should deviate substantially from the average. Another way to visualize data is illustrated in Figure 11.8, where artifacts such as the plate effect (increased number of positives in a certain row or column, due generally to the plate manufacturing) are clearly visible. An efficient system has to satisfy some requisites: automatic data entry and capture, easy access to data, quality control and uniformity of results. It has to allow the execution of personalized queries, for example, results from all samples with certain characteristics or for all cell-based assays. The data may be viewed transversally (the same

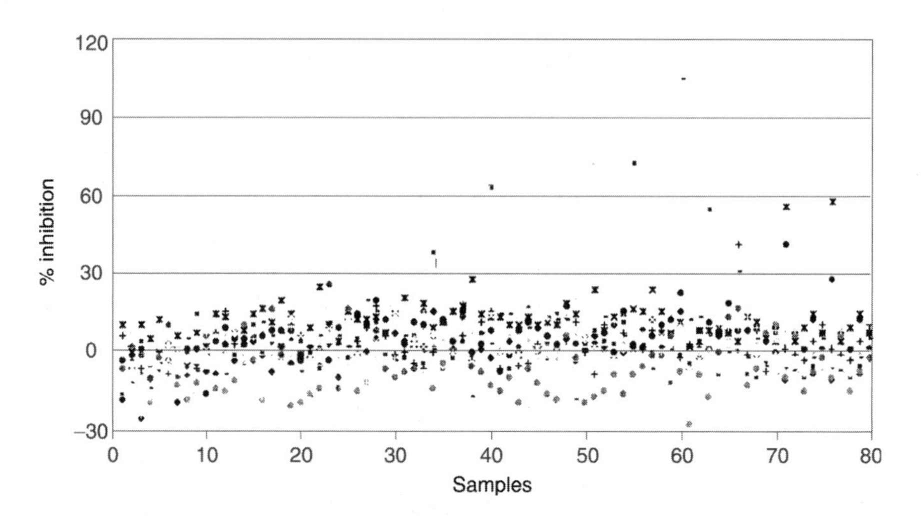

FIGURE 11.7 Sample distribution. The well number and the percent inhibition are reported on the *x*- and *y*-axes, respectively. Different symbols denote values from different microplates. The results from a total of 200 plates, equivalent to 16,000 samples, are included in the graph. Horizontal lines mark 30%, 60%, and 90% inhibition. The vast majority of samples give an inhibition oscillating +15% of the controls. In this case the threshold was set at 40% inhibition and only 8 of 16,000 (0.05%) samples performed above this value. The overall positive rate for the entire project was 0.3%. The data refer to the same HTS project as Figure 11.5.

samples on different primary tests), or sequentially (primary and secondary tests performed with the same sample). The first data management systems were custom built, nowadays several systems are commercially available, and they only need to be slightly modified and adapted to individual uses.

IV. HTS AS AN INTEGRATED PROCESS

The power of HTS lies in the large amount of data that can be generated in a short time with a small amount of sample. Technological developments will allow further miniaturization and throughput of the process, increasing HTS potential to revolutionizing compound discovery and speeding up the developmental phases. Under these conditions, factors affecting performance become important determinants for the efficiency of the process. Optimization of robot capacity obviously becomes a critical step for shortening the time and increasing the efficiency of the process. This involves scheduling the use of robots, organizing work in shifts, executing accurate hardware maintenance, customizing robots for the HTS process, etc. However, like all big improvements in science and technology, HTS is not the magic bullet that will identify new drugs. It is a powerful tool that makes use of rational and empirical approaches for drug discovery. The ingenuousness of

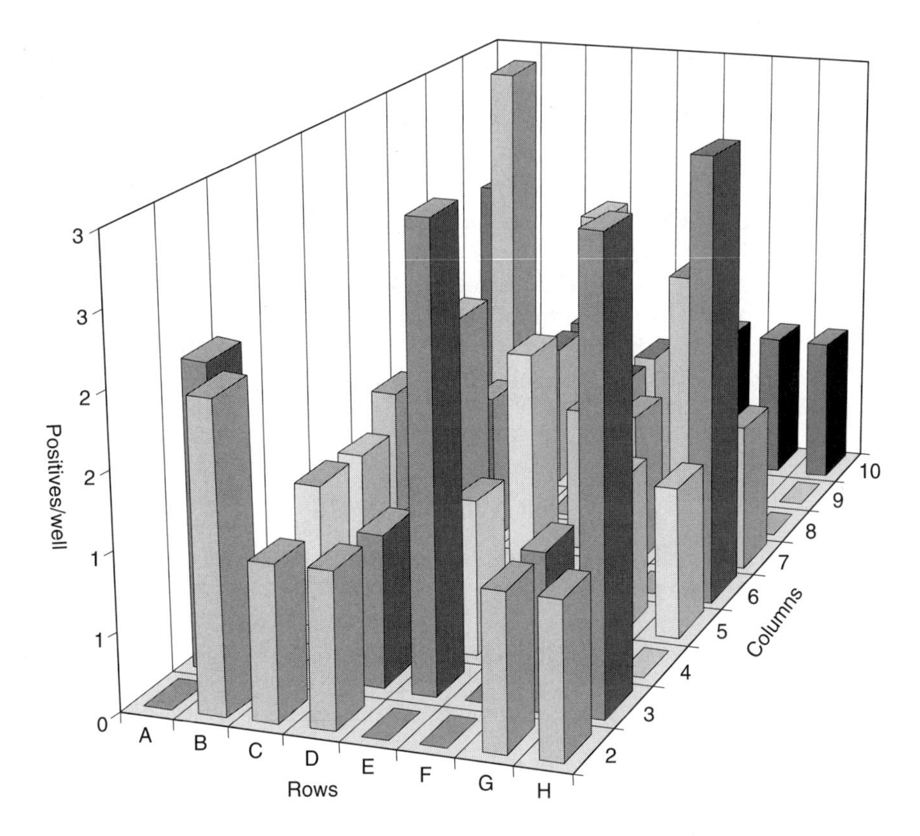

FIGURE 11.8 Topographical distribution of samples performing above a defined threshold. Each plate position is marked by a letter (rows) and a number (columns). The vertical axis reports the number of positives found in each position. Data refer to the same HTS project as Figures 11.5 and 11.7 for a total of 32,000 samples.

the tests employed and the characteristics of the sample library will continue to play key roles in the quality of the hits that will be generated. It is important to bear in mind that only rarely is a single assay sufficient to arrive to a hit. Using as an example the luciferase reporter system of Figure 11.4, inhibitors of the enzyme, general inhibitors of transcription, cytotoxic substances, just to name a few, will all give the same answer as true, specific inhibitors of the desired promoter. Therefore, it is extremely important that appropriate control tests are introduced in parallel or serial mode to identify true hits. The more the different tests synergize with each other, the more meaningful the results will be (see Figure 11.6), with a smaller number of data points.

HTS is an integrated process, where a large amount of input samples are distilled into a few valuable ones through the use of appropriate biological (the assays), mechanical (the robots), and statistical (data analysis) tools. The desired outcome of HTS is usually a molecule endowed with a desired bioactivity whose structure could not be predicted a priori. The success of

HTS as a drug discovery tool lies in the key ingredients described in this chapter, No single ingredient can ensure success. Unexpected leads that will become future drugs can be found with HTS by a careful balance of these various components. For example, if the samples are truly novel and unexploited, well-established targets can be employed. On the other hand, previously screened libraries necessitate innovative targets or assays. An increased risk is associated with unexploited samples and with innovative targets, but so is a higher chance of payoff for the HTS program. In our opinion the proper combination of samples and assays, the chemical entities and the bait, ensures the greatest chance of success in a drug effort. However, only a rigorous analysis of the data allows the identification of the high-quality hit that has the highest chance of becoming a drug candidate.

ACKNOWLEDGMENTS

We are grateful to all our colleagues at Biosearch Italia for advice, comments, and valuable insights.

SUGGESTED READING

1. Broach JP, Thornsen J, High throughput screening for drug discovery, *Nature*, 384:14–16, 1996.
2. Emory K, Schledgely J, Eds, Cost-effective Strategies for Automated and Accelerated High-throughput Screening, IBC Biomedical Library Series, Southborough, 1996.
3. Bevan P, Hamish R, Shaw I, Identifying small molecules lead compounds: the screening approach to drug discovery, *Tibtech*, 13:115–118, 1995.
4. Harding D, Banks M, Fogarty M, Binnie A, Development of an automated high-throughput screening: a case history, *Drug Discovery Today*, 2:385–388, 1997.
5. Rishton GM, Reactive compounds and in vitro false positives in HTS, *Drug Discovery Today*, 2:382–384, 1997.
6. All the topics covered in this chapter are treated in *Journal of Biomolecular Screening*, Mary Ann Liebert Inc., Publishers.

12 NMR in Combinatorial Chemistry

Edith S. Monteagudo and Daniel O. Cicero

CONTENTS

I. INTRODUCTION

Combinatorial chemistry was developed with the clear aim of generating large collections of diverse molecular entities in a very short period of time. In this context, the need for efficient sample analysis remains a challenge, either if the structural characterization is performed in solution or in the solid phase synthesis context. In the first case, the main difficulty is associated with the large number of compounds to be analyzed, and in the second case with the physical heterogeneity associated with the use of solid supports. The first part of the present chapter will focus on recent developments introduced in NMR to tackle the structural characterization of combinatorial libraries (single compounds and mixture of compounds).

There is a second area of combichem where NMR is finding an increasing popularity—the screening of libraries of compounds. This exciting stage of NMR development is currently opening the possibility of the use of this spectroscopy, especially in the early phases of the drug-discovery process, where lead generation remains the major effort. An overview of these methods will be presented in the second part of the chapter.

II. ANALYTICAL NMR AND COMBINATORIAL CHEMISTRY

Some particular features of the analysis of products obtained by combinatorial methods have impaired the use of NMR spectroscopy in the initial phase of the development of this technique. Combinatorial chemistry produces large number of compounds in a very short period of time, in small quantities and instead of using traditional glassware for synthesis employs 96-well microtiter plates to store, transport and sometimes even to synthesize the compounds of interest. Another issue is the need to characterize solution and solid samples, since solid phase synthesis is extensively used in combichem. In this context, the need of an efficient and "universal" sample analysis remains a challenge. Actually, most combichem programs obtain mass spectrometry and UV (photodiode-array detection) data on their samples but clearly the use of NMR spectroscopy provides a structural characterization unparalleled by the aforementioned techniques. In the last years an increasing number of new NMR methods opened the possibility for the utilization of this analytical technique for monitoring combinatorial chemistry reactions. The first part of this chapter will focus on the recent developments introduced in NMR spectroscopy to overcome these difficulties.

A. Liquid Samples

Combinatorial chemistry can be performed by producing any given library either as a collection of pure compounds or as a mixture of compounds. In the first case, the major challenge for the application of NMR spectroscopy remains the repetitive analysis of hundreds or thousands of compounds. The use of the so-called flow NMR, together with automation of both the acquisition and the analysis of the data, can be of great help in reducing the total measuring time. Mixture analysis requires some means of resolving the individual components. This can be done with some form of chromatography or using spectroscopic methods. Chromatographic separations can be accomplished using either traditional off-line fractionation prior to NMR analysis or with online fractionation as is used in LC-NMR. The last possibility is without doubt the most appropriate in combichem strategies.

1. Flow NMR

Homogeneous liquid samples (samples which have a uniform magnetic susceptibility) can be analyzed not only by conventional 5 mm precision glass tubes but also using the so-called "on-flow" probes, in which samples flow through narrow-bore tubing directly into the NMR coil [1] (Figure 12.1). In flow probes the NMR sensitivity is relatively high due to their increased filling factor that makes it possible to detect samples in the range from 1–100 nanomoles. Since the fixed flow cell does not allow sample spinning, the RF coil can be mounted closer to the sample. This type of probe is particularly appropriate for the analysis of large numbers of samples, since the fixed sample geometry reduces the need to re-shim every time the sample is changed. Moreover, the possibility of using an automatic liquid handler allows to take the sample stored in a microtiter plate and inject it directly into the probe. The primary goal is to eliminate the need to transfer the samples into the precision glass tubes that are typically used for high-resolution NMR [2].

Different methodologies of flow analysis were developed—FIA-NMR (Flow Injection-NMR), DI-NMR (Direct Injection-NMR) and other hybrid techniques [3]. Each technique obviously has its own set of advantages and disadvantages that need to be evaluated considering the specific problem to solve (Figure 12.2). The analysis of combichem libraries is facilitated if the sample can be employed without any pretreatment (drying to remove protonated solvents and reconstitution in deuterated solvents). This possibility not only reduces the time, cost and effort of sample handling but allows the use

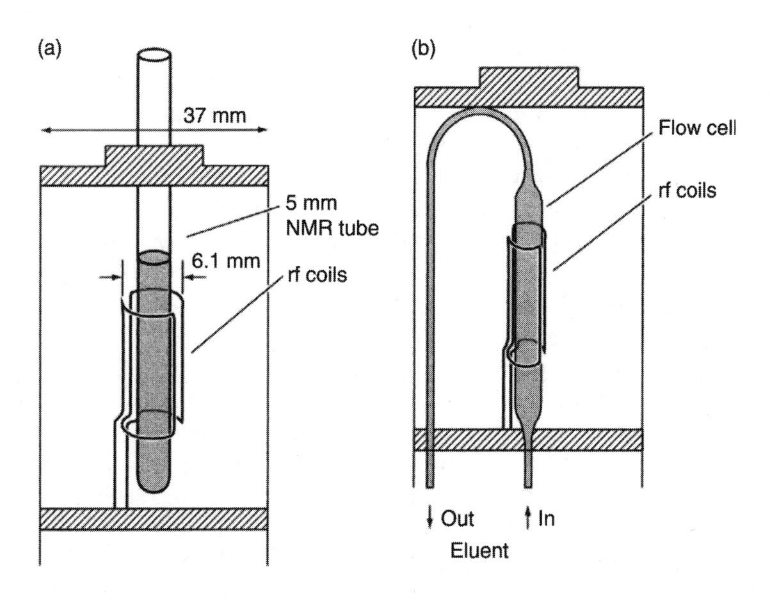

FIGURE 12.1 (a) Conventional 5 mm probe and (b) flow probe.

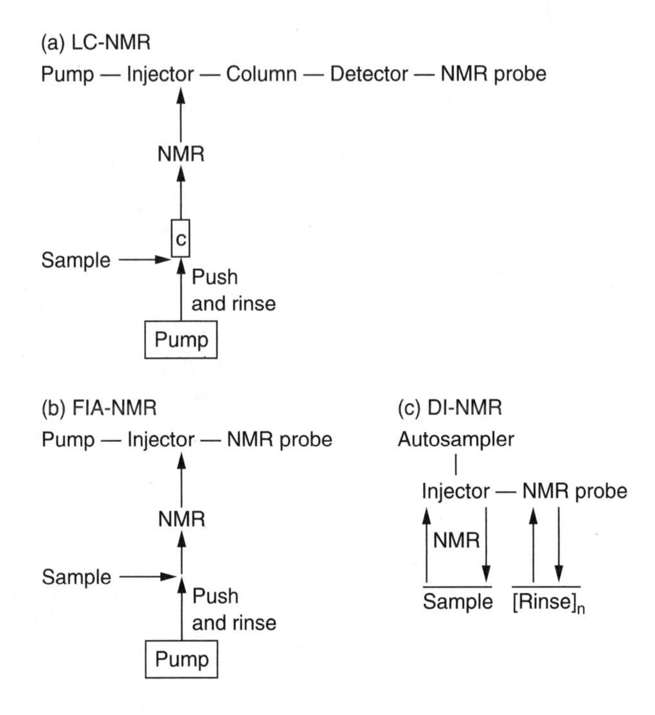

FIGURE 12.2 Schematic block diagrams of the (a) LC-NMR, (b) FIA-NMR and (c) DI-NMR techniques. Reprinted with permission from P.A. Keifer, *Magn. Reson. Chem.*, 2003, 41, 509–516. Copyright © 2003 John Wiley & Sons, Ltd.

of the same sample for NMR and mass spectrometry analysis. A typical NMR solvent like D_2O can complicate the mass spectral data if variable amounts of ^2H exchange occur within the sample. So NMR solvent suppression plays an important role in this kind of analysis. It is necessary to detect signals from low concentrations of analytes in the presence of large ^1H NMR signals from solvents. The first attempts to overcome this difficulty, including solvent saturation [4], were neither very effective nor very general. Particularly important was the contribution given by Smallcombe and coworkers [5]; developing the "WET" solvent-suppression pulse sequence (Figure 12.3). WET is composed of a series of four frequency-selective shaped RF pulses, each followed by a Pulsed Field Gradient (PFG). ^{13}C decoupling is applied during the RF pulses to eliminate ^{13}C satellites signals since they are often much larger than the signals of the analytes. Broad signals and multiplets are effectively suppressed with no polarization transfer to exchangeable protons (OH,NH). New techniques for solvent suppression have been developed [6, 7] that are able to cope with mixed solvents, such as methanol-water, acetonitrile-water and even more complex solvent combinations. Shift laminar pulses [8], magic angle pulsed field gradients [9] and excitation sculpting [10] are suitable methods for multiple-resonance solvents or mixture of solvents suppression.

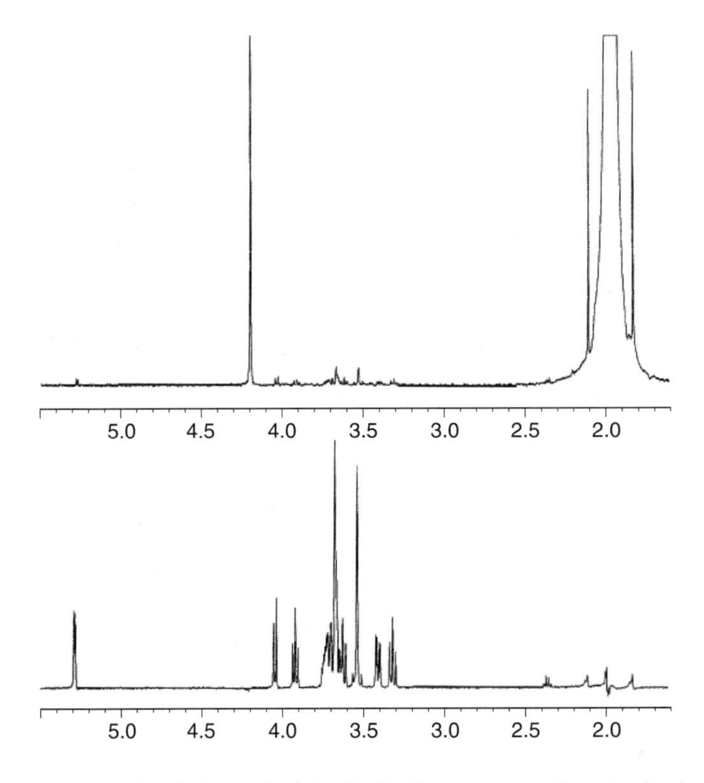

FIGURE 12.3 Example of the typical level of solvent suppression obtained using the WET sequence. Reprinted with permission from S.H. Smallcombe, S.L. Patt and P.A. Keifer, *J. Magn. Reson.*, 1995, 117, 295–303. Copyright © 1995 Academic Press Inc.

The introduction of gradient shimming (that makes this procedure very easy to automate) together with the possibility of performing automatic tuning and matching (*via* the use of motors mounted on the probe body) has further improved the whole procedure.

Another important issue has been the development of robust protocols to automate data acquisition and analysis. Clearly, data analysis is the more complicated stage of the process to be automated. Depending on the problem, particular displays can be adopted that quickly and simultaneously scan all the spectra of a library for characteristic patterns and diagnostic signals that permit the structural characterization. These software tools avoid manual spectral interpretation by automatically analyzing selected regions of each spectrum and translating the results into parameters such as the extent of conversion of starting materials or the amount of product.

An example of the strategy described above was employed to characterize the 88-member (one-compound-per-well) test plate used for the synthesis of imides [2]. In the plate, eight different anhydrides were reacted with eleven different primary amines to afford intermediate amide-acids that were dehydrated at elevated temperatures to afford 88 different imides. Aliquots of the crude solution-state samples were directly injected by using a robotic liquid handler into a flow probe of 60 μl active volume and a total flow cell volume (minimal sample volume) of 115 μl. To fill the system (including the transfer tubing connecting the injector port to the probe) a total of 350 μL of sample was required. Much smaller sample volumes can be analyzed if push solvents are used. The composition of this push solvent should usually be matched as closely as possible to the composition of the sample solution to avoid magnetic susceptibility line broadenings. NMR software was used to automatically find and suppress the intense NMR signals from any nondeuterated solvent used. The flow NMR analysis described before can be satisfactorily employed in most rapid analogue synthesis approaches.

Within the past few years, various analytical techniques have been applied to the molecular characterization of single Solid Phase Synthesis (SPS) beads. Approaches to screening analytes from single SPS are particularly useful in the application of split and pool synthesis methodology to generate "one bead-one compound libraries" [11–13]. A capillary NMR flow probe was designed to generate high-resolution ^{1}H NMR spectra at 600 MHz from the cleaved product of individual 160 μm Tentagel combinatorial chemistry beads [14]. By injecting a dissolved sample sandwiched between an immiscible "push," perfluorinated organic liquid directly to the probe, NMR spectra of the cleaved product was acquired in just 1 hour of spectrometer time without diffusional dilution.

2. Spectroscopic Editing of Mixtures

All the spectroscopic approaches applied for structural characterization of mixtures derive from methods originally developed for screening libraries for their biological activities. They include diffusion-ordered spectroscopy [15–18], relaxation-edited spectroscopy [19], isotope-filtered affinity NMR [20] and SAR-by-NMR [21]. These applications will be discussed in the last part of this chapter. As usually most of the components show very similar molecular weight, their spectroscopic parameters, such as relaxation rates or self-diffusion coefficients, are not very different and application of these methodologies for chemical characterization is not straightforward. An exception is diffusion-edited spectroscopy, which can be a feasible way to analyze the structure of compounds within a mixture without the need of prior separation. This was the case for the analysis of a mixture of five esters (propyl acetate, butyl acetate, ethyl butyrate, isopropyl butyrate and butyl levulinate) [18]. By the combined use of diffusion-edited NMR and 2-D NMR methods such as Total Correlation Spectroscopy (TOCSY), it was possible to elucidate the structure of the components of this mixture. This strategy was called diffusion encoded spectroscopy "DECODES." Another example of combination between diffusion-edited spectroscopy and traditional 2-D NMR experiment is the DOSY-NOESY experiment [22]. The use of these experiments have proven to be useful in the identification of compounds from small split and mix synthetic pools.

3. HPLC-NMR

The development of HPLC-NMR represents a major contribution to the characterization of individual components in complex mixtures [6]. Although the sensitivity of HPLC-NMR cannot be favourably compared with MS techniques, it can nevertheless play a role in the identification of iso-molecular weight compounds (stereo and regioisomers) exploiting the possibility of applying the complete set of modern NMR experiments and also in the absence of "complete" chromatographic resolution.

Direct online coupling of an NMR spectrometer as a detector for chromatographic separation has required the development of technical features already discussed for flow analysis of combichem libraries. Strictly speaking, most of these technological improvements were originally developed for HPLC-NMR during the last 20 years. Also in this case, it is necessary to detect signals from low concentrations of analytes in the presence of large ^{1}H NMR signals from the HPLC solvents, requiring solvent suppression, small volume probes, automatic tuning and matching, etc., with the extra challenge of interfacing the separation procedure with the spectroscopy. The design of flow probes in the microliter range pushed the detection limits to circa 50 ng working in LC-NMR and to 5 ng using capillary LC-NMR. A further improvement in sensitivity that will greatly benefit high-throughput NMR methods will be achieved using "cryoflowprobes." The cryoflowprobes combine the advantages of flow probes with the superb sensitivity of cryoprobes, with an enhancement of up to 400% in sensitivity over conventional flow NMR probes. Cryoprobes provide improved Signal/Noise (S/N) ratios by reducing the operating temperature of the coil and the pre-amplifier.

There are some examples of the use of HPLC-NMR in combinatorial chemistry. A synthetic mixture of 27 closely related tripeptides [23] formed from all the combinations of alanine (A), methionine (M) and tyrosine (Y), as the C-terminal amide could be resolved and characterized with the help of this methodology (REF is given in 24 which is listed at the end). Based on chemical shifts and peak multiplicities, the on-flow HPLC-NMR characterization of the majority of the components of the mixture was achieved. This application demonstrated that this approach is likely to be an effective method for compounds mixtures.

A similar application has been the investigation of a model mixture of four regioisomeric (dimethoxybenzoyl)glycines [24]. Both continuous-flow and stop-flow ^{1}H NMR spectroscopies at 500 MHz were employed to characterize the four molecules, using a gradient elution of acetonitrile/D_2O containing phosphoric acid. The determination of the structure for each compound was straightforward by consideration of the chemical shifts patterns, even though two components were overlapped in the chromatography. Figure 12.4 shows the 500 MHz HPLC-NMR stopped-flow stacked spectra and Figure 12.5 shows the on-flow 2-D plot for this mixture. The last figure is a contourplot with ^{1}H NMR chemical shifts on the horizontal axis and retention times on the vertical axis.

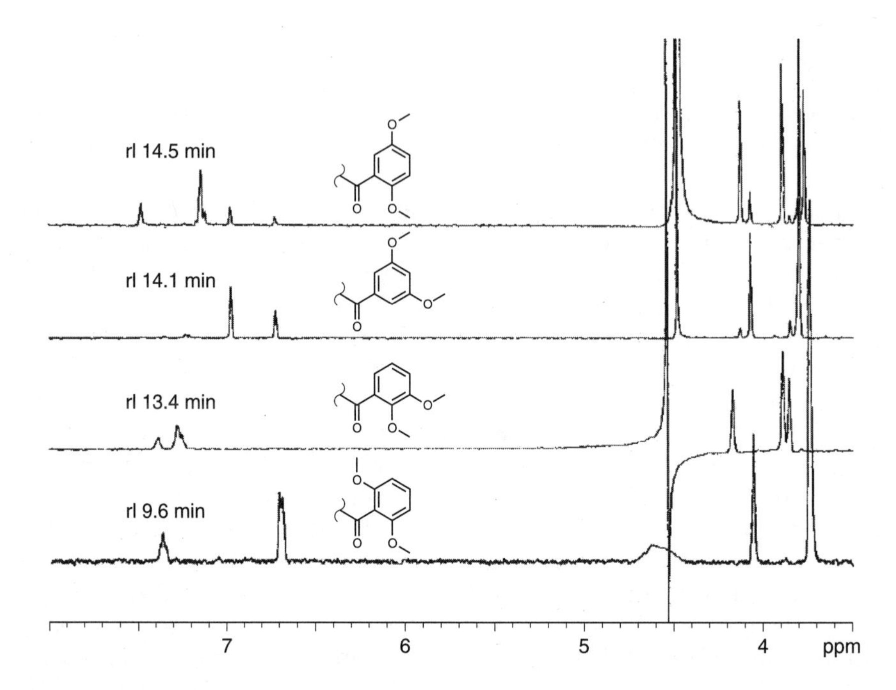

FIGURE 12.4 500 MHz HPLC/NMR stopped-flow stacked NMR plot for compounds 1–4. Reprinted with permission from J. Chin, J.B. Fell, M. Jarosinski *et al.*, *J. Org. Chem.*, 1998, 63, 386–390. Copyright © 1998 American Chemical Society.

4. Hyphenation HPLC-MS-NMR

HPLC-MS has been employed for many years since the introduction of electrospray ionization as a characterization tool for compounds synthesized using combinatorial technologies. The hyphenation of HPLC-MS-NMR has been extensively described by different authors [25, 26]. The large sensitivity difference between the NMR and MS, of at least 10^3 to 10^6, the higher speed for collecting data of the MS unit, the solvent selection, are all important issues that have to be compromised between the ideal requirement of each instrument. In this experimental set-up, the NMR and mass spectrometers are most frequently linked in parallel with the LC (Figure 12.6).

The flow from the LC is split approximately 95:5 between the NMR and MS spectrometer with the possibility of further dilution of the portion to be sent to the MS to prevent saturation of this detector. The main advantages of using the simultaneous NMR-MS approach are: the possibility of collecting all the data with limited quantities of sample, avoiding degradation that might occur while waiting for separate NMR and MS experiments to be run and analyzing the "exact" sample with both spectroscopical techniques. In this high throughput analysis the mass data can be used as a filter so that NMR data are only collected of samples that fulfil a molecular weight and fragmentation pattern.

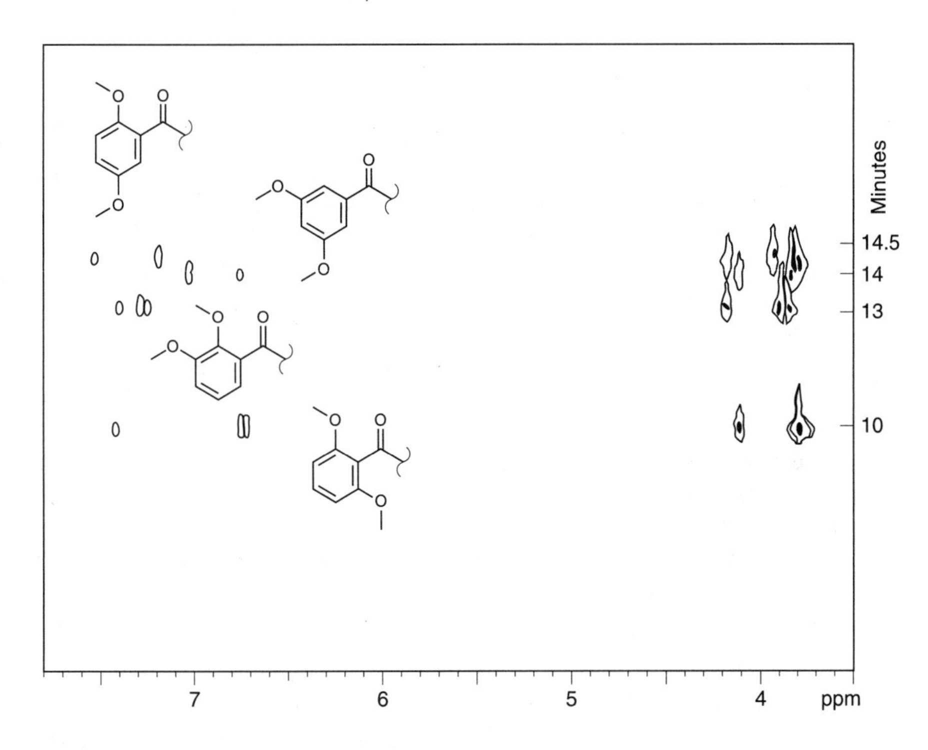

FIGURE 12.5 500 MHz HPLC/NMR on-flow 2-D plot for compounds 1–4. Reprinted with permission from J. Chin, J.B. Fell, M. Jarosinski *et al.*, *J. Org. Chem.*, 1998, 63, 386–390. Copyright © 1998 American Chemical Society.

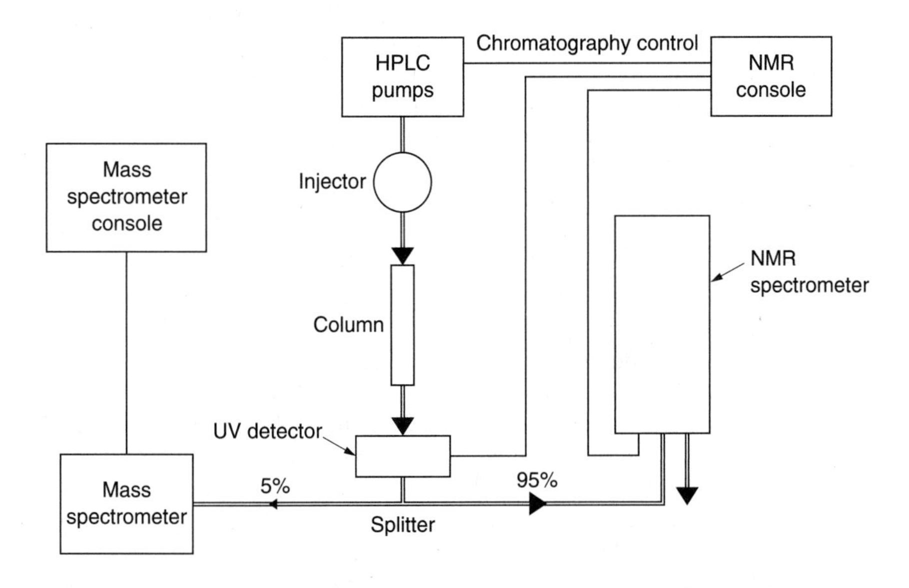

FIGURE 12.6 Schematic representation of HPLC-NMR-MS system.

III. MONITORING COMBINATORIAL CHEMISTRY REACTIONS

A. Solid Phase Analysis

Solid-phase synthesis can be considered as the prototypical technique of combinatorial chemistry [27, 28]. Since its introduction by Merrifield [29], it has been used for the synthesis of a wide variety of peptides [30], nucleotides [31], organic chemicals [32] and small molecules [33, 34]. An important limitation in the use of the solid-phase synthesis remains the difficulty of characterizing the synthesized products while they are still bound to the resin. The more direct solution to this problem is the cleavage of the product from the resin in order to characterize it using common solution-phase spectroscopical tools, in particular MS. The obvious drawback of this approach is that it is time-consuming, can alter the product and lowers the yield. NMR spectroscopy is currently the only analytical technique available for obtaining detailed structure information on a compound covalently bound to the resin, using different methodologies that will be discussed in this chapter.

B. GelPhase NMR

In this method, resin samples are swollen with solvent and placed into a standard NMR tube. Spectral data are then collected using conventional sequences and hardware of solution-state NMR. The quality of the NMR data depends on the mobility of the individual nuclei in the sample. This molecular motion can be sufficiently high in order to produce narrow linewidths. The study of these solvent-swollen slurries of the SPS resins is known as gel-phase.

Since its first description in 1971 [35], gel-phase NMR was applied to peptide chemistry by Manatt and coworkers [36, 37]. These authors used ^{13}C NMR to determine the extent of chloromethylation of crosslinked polymers and ^{19}F NMR to monitor protection-deprotection reactions. These two nuclei are the most commonly used in these types of studies, mainly because of their significant chemical shift dispersion, which can alleviate in part the resolution loss due to the non ideal linewidth obtained in the gel state. Apart from restricted molecular motion, that shortens T2 because of an efficient transverse relaxation, other sources of line-broadening derive from magnetic susceptibility variations within the sample (due to the physical heterogeneity of the system) and residual dipolar couplings.

Since the line-broadening effects of the magnetic susceptibility variations are directly proportional to the NMR frequency, lower frequency nuclei (like ^{13}C) may produce narrower linewidths. In fact, this technique has been mostly used for this nucleus [37–43]. In some cases, the poor sensitivity associated with ^{13}C detection was compensated by incorporating ^{13}C labels near the reaction site of interest. This approach was called "fast ^{13}C NMR" [44–48] and allowed to obtain data in a few minutes for each sample.

^{19}F NMR is also very popular, mainly because of the fact that structural modifications quite remote from the fluorine can give rise to useful shift changes [49]. Gel-phase NMR was also employed associated to the detection of ^{31} P [50, 51] and ^{2}H [52]. This approach is inaccessible for ^{1}H-NMR because of the broadness of the peaks in the order of 100 to 300 Hz.

C. Magic Angle Spinning NMR

As stated before, differences in magnetic susceptibility arising from heterogeneity within the sample and residual dipolar coupling are the main factors contributing to the broadening of signals of solvent-swollen slurries of the SPS resins. A large number of studies, including powdered solids [53], heterogeneous solid-liquid mixtures [54], compartmentalized liquid samples [55, 56], membranes [57] and seeds [58] have shown that the broadening because of both these factors disappears when the sample is spun about the magic angle (54.7° relative to the static magnetic field). These results suggested that Magic Angle Spinning (MAS) NMR would be useful also for obtaining high quality proton and carbon NMR data on solid support. That this is the case is demonstrated by the large number of recent publications showing proton NMR on swollen resins with linewidths in the range of 6–10 Hz [59]. This residual line broadening comes mainly from the anisotropic bulk magnetic susceptibility differences that are not averaged out by MAS and most likely derives from the aromatic rings from the polystyrene resin [60, 61]. As a consequence of this, a new resin that does not contain aromatic rings has been developed [36]. These PEG-based resins give higher resolution spectra than the PEG-grafted resins Tentagel and Argogel and all polystyrene-based resins. The work by Keifer [62] has extensively analyzed the different factors that play a role to obtain high quality MAS spectra, in particular the influence of the solvent in the line-width. The choice of solvent for MAS-NMR studies is dependent on the solvation properties of a solvent for a given polymer, the mobility of the compound on the resin, and on the solvation of the compound [62, 63]. Apart from PEG resins, where the compounds have an increased mobility due to the PEG spacer, the line-width in proton MAS-NMR is not sufficiently reduced to allow direct measurement of coupling constants. In these cases spectral analysis can be done using chemical shift criteria only. Several strategies were described [64, 65] to obtain coupling constants information for structure determination. Recently, 2-D Spin-Echo spectroscopy (SECSY) [66, 67] was used to obtain high-resolution MAS spectra of resin-bound molecules, with a resolution similar to that observed in solution NMR. This 2-D J-resolved experiment can be collected in less than 10 minutes. Another useful experiment to measure coupling constants or to determine multiplicity and correlate coupled spins is MAS E.COSY [64]. The mobility of the compound on the resin is also a useful information to evaluate different reaction conditions during the solid phase synthesis [68, 69]. Because linewidths are directly related to chain mobility, the lack of mobility due to aggregation is a potential source of difficulties in this synthetic approach.

The interpretation of these resin spectra is sometimes complicated by the presence of broad signals arising from the polymer support itself. To alleviate this problem both spin echoes techniques [71, 72] and selective presaturation can be used [59]. In the latter case, it was observed that it is sufficient to saturate an isolated signal of the resin to reduce the signal intensities of other polymer resonances. The signal saturation is presumably spread by a spin-diffusion process.

New probes for high resolution MAS NMR include pulse field gradients so it is possible to improve both quality and throughput [72], especially in ^{1}H-^{13}C correlations at natural abundance. Field gradients are very important in order to attenuate artefacts arising from non-ideal suppression of protons bound to ^{12}C, thus making possible to acquire high quality artefact-free spectra such as MAS-HMBC. As it is well known, this long range correlation experiment represents one of the most powerful structure determination techniques in organic chemistry. An example that illustrates well the application of MAS-NMR for monitoring reaction sequences on resin is the synthesis of a trisubstituted amine on hydroxymethyl polystyrene resin [73]. The ^{1}H and ^{13}C spectra collected are shown in Figures 12.7 and 12.8. In this case the intermediates formed could not easily be cleaved from the resin. Both ^{1}H and ^{13}C spectra of the polystyrene resin were acquired to determine the starting background. The introduction of the acrylate moiety is shown by the proton resonances at around 6 ppm and the carbon resonances at 166.4, 131 and 129 ppm. Complete disappearance of the vinyl protons and the appearance of six new aliphatic protons peaks characterized the addition of the butylamine. In the ^{13}C spectrum the vinyl carbons' resonances have been replaced by six distinct aliphatic resonances in the 14–50 ppm range and the carbonyl peak shifted downfield to 173.1 ppm. The progress of all five reactions in this synthesis can be monitored in this simple way. Only in cases where spectral overlap makes assignment difficult can more sophisticated experiments be used (e.g. MAS-COSY).

Examples for determining enantiomeric excess on a resin has been reported. ^{13}C MAS-NMR was used to monitor the asymmetric dihydroxylation of 10-undecenoic acid supported on Wang-resin and to determine the ee of the dihydroxylation product derivatized with R-(+)-Mosher's acid [74]. The MAS-NMR results agreed to better than 1% with those obtained from HPLC performed on the freed material following cleavage from the resin.

Another cutting edge application, made possible because of the increased sensitivity of the new generation of probes, is the structural characterization on a single bead [75, 76]. This kind of analysis is particularly important for the analysis of "one bead-one compound" libraries [60, 77].

Two-dimensional MAS NOESY spectroscopy of peptides on beads was used to determine the conformation of an immunogenic peptide while bound to polystyrene-based tentagel resin [78]. These results suggested the existence of a role for three-dimensional structure determination to obtain a conformation-based understanding of chemical reactivity and the influence of the resin environment on molecular conformation [79]. This information would be extremely valuable during the design phase of a combinatorial library or when assaying biological activity of a compound still attached to the resin [56].

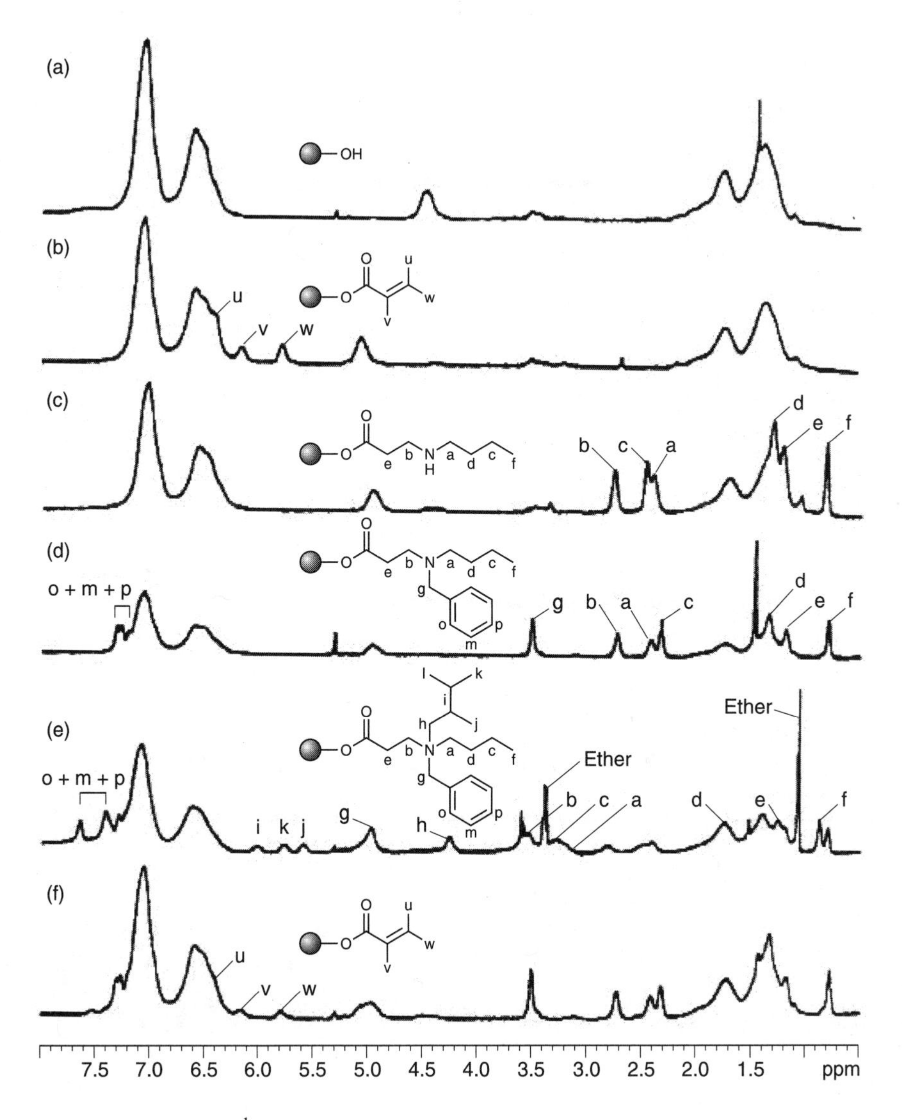

FIGURE 12.7 MAS ^{1}H-NMR spectra of the solid-phase synthesis of a trisubstituted amine on the hydroxymethyl polystyrene resin. Reprinted with permission from Y. Luo, X. Ouyang, R.W. Armstrong and M.M. Murphy, *J. Org. Chem.*, 1998, 63, 8719. Copyright © 1998 American Chemical Society.

Although the vast majority of published resin data consist of ^{1}H and ^{13}C NMR, multinuclear NMR is also possible. Different examples of the use of ^{19}F MAS spectrometry were described in the literature [36, 80]. This nucleus was used to monitor different reactions on solid support, as coupling of fluorobenzyloxycarbonyl with aminoacids and S_NAr nucleophilic displacements [80].

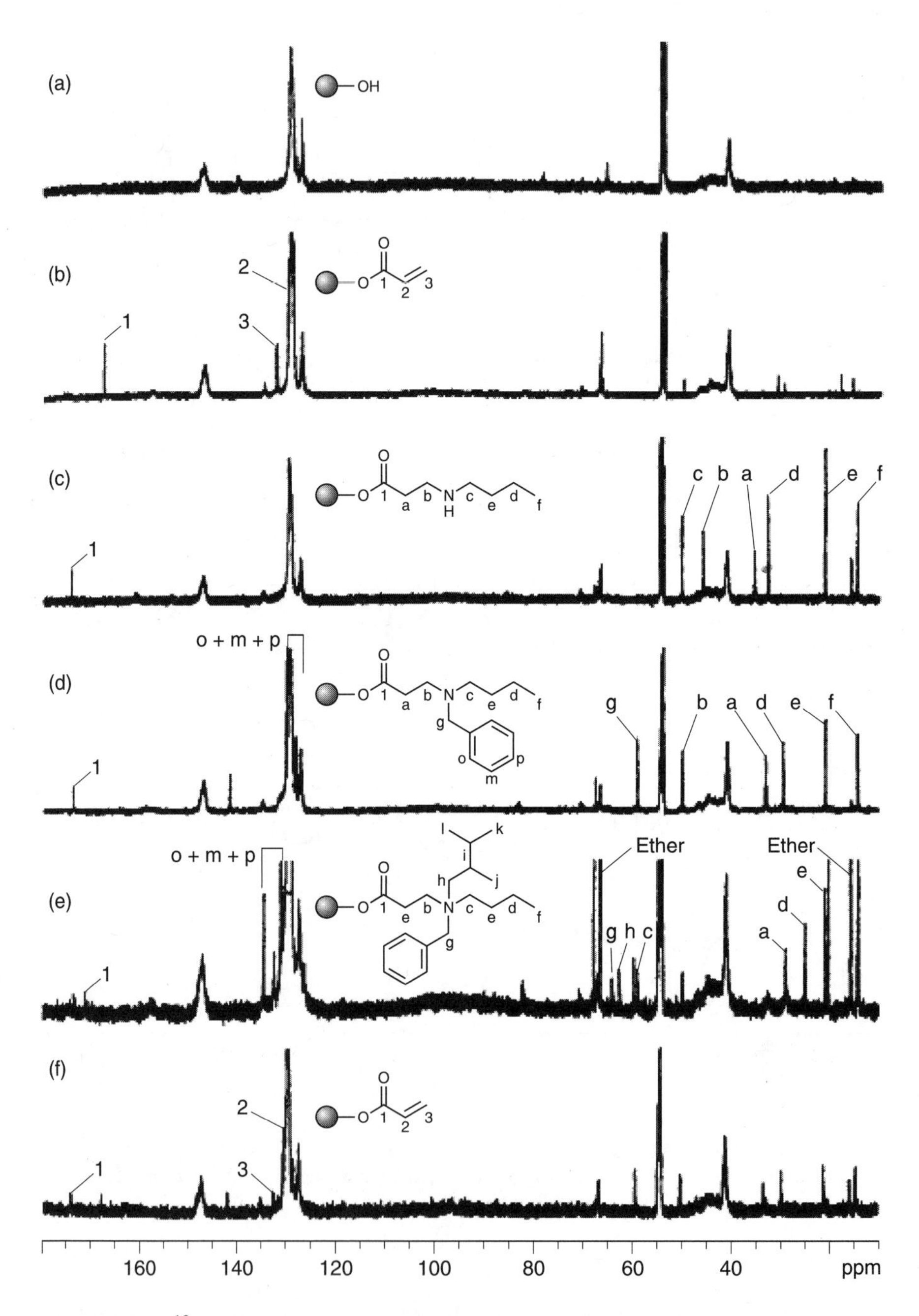

FIGURE 12.8 ^{13}C MAS spectra of the solid-phase synthesis of a trisubstituted amine on the hydroxymethyl polystyrene resin. Reprinted with permission from Y. Luo, X. Ouyang, R.W. Armstrong and M.M. Murphy, *J. Org. Chem.*, 1998, 63, 8719. Copyright © 1998 American Chemical Society.

IV. NMR AS A SCREENING TOOL

One of the most important activities within the drug discovery process is the detection and identification of lead molecules. This process is often carried out by high-throughput screening of large chemical libraries. There is a need though for deconvolution to identify the active entity, reducing as much as possible the risk of "false positives" during the analysis. In what follows we will see how NMR can be used in certain cases to screen mixtures for their biological activity and to identify the active component of the mixture [81–85].

The main purpose when using NMR as a screening tool is to identify weak binding small molecule scaffolds, and subsequently use this information to identify/create a tight binding drug. There are three main advantages inherent to NMR. First, NMR is a very sensitive methodology to identify binding, as interaction in the range of millimolar can be clearly detected. There are a number of NMR parameters that change upon intermolecular interaction—chemical shifts, diffusion rates, relaxation rates (like T2 and NOE). Second, NMR detects the binding itself and not a signal coming from a related process. This fact translates into the use of an essentially undisturbed assay system, compared with the need for immobilization or attachment of chemical labels, as in other screening techniques such as RIA, EIA or fluorescence based assays. Third, it can yield details of the interaction at an atomic level, which can be used to derive bound conformations of ligands, topologies of receptor sites or binding epitopes of ligand molecules.

Among the possible limitations for the use of NMR as a screening tool, the relative insensitivity of NMR as compared with other spectroscopical methods is the more important. This fact imposes the use of high total compound concentration for the ligand molecules and limits the mixture size that can be tested because of compound solubility, protein precipitation and non-specific binding. We shall see in several examples that this limitation turns out to be inherently less important as the use of new probes and higher fields become available, lowering the threshold of sample concentration to be used in these studies.

There are several articles highlighting diverse experimental approaches [86–92], but generally speaking we can divide the different methods into two different categories—those that are based on the observation of the protein signals and those based on the observation of the ligand signals. The main example of the first group is the so-called SAR by NMR, whereas the Saturation Transfer Difference (STD) Experiments, the WaterLOGSY (Water-Ligand Observation with Gradient Spectroscopy), the Transfer-NOE, and experiments based on relaxation or diffusion filters belong to the second class. The decision on what kind of approach is to be used depends on a number of factors and cannot be easily generalized. Methods based on the observation of protein signals offer the unique opportunity to not only detect binding but also to define where binding occurs. In this respect, it can be seen as the combination of both a screening assay and a competition experiment. It has the main drawback that it requires isotopic labelling of the target and that it is confined to a certain size of proteins that are amenable to be assigned by multidimensional NMR experiments. On the other hand, those experiments based on the observation of ligand signals are not limited by the size of the protein and there is no requirement of isotope labeling. They can define the binding epitope of the ligand, which is a valuable information for the directed development of drugs. A clear drawback of this approach is that binding can occur also in regions not related to the active site of the target without changing the output of the experiment. This problem can be faced performing competition experiments. They remain, however, the only choice for protein targets that cannot be labeled or that are too large to be assigned. Whenever it is possible, the combination of the two strategies gives the best result.

Whatever the NMR approach chosen on the basis of the case under study, there is the need to combine the technical aspect with the design of intelligent compound libraries and especially with the design of compound libraries that are tailor made for NMR screening protocols [93]. One such approach is based on the SHAPES strategy, that recognized that there are a limited number of scaffolds efficiently describing a significant fraction of all known bioactive molecules [94–96]. These basic molecular shapes constitute the base for a library which can be modified after detection of binding activity to yield high affinity ligands.

A. Structure-Activity Relationship (SAR) by NMR

The possible use of NMR as a screening tool was first developed by Stephen Fesik and coworkers [97]. In this approach, the ^{15}N chemical shifts of the backbone amides of the protein that act as biological target are monitored. Binding with a small molecule will cause a change in the chemical shifts of the NH's that are proximal to the interaction site. If the assignment of at least the backbone of the protein is available it is possible, by studying these induced changes, to identify the subsite of interaction of this small scaffold. The main idea behind this method is that if two binders are identified that interact very close one to the other, a linked fragment containing both subunits will present a much higher affinity for the target.

The first application of this approach was related to the FK506-binding protein FKBP [21]. FK506 is an important drug that shows significant immuno suppression by mediating the binding of FKBP to calcineurin. One of its uses is to repress rejection reactions after organ transplants, but has a strong associated toxicity. Therefore, new drugs are needed that maintain the FKBP binding activity showing less toxicity. The SAR by NMR was applied to this system, using uniformly ^{15}N,^{13}C-labeled FKBP at 2 mM concentration and a library of around 1000 substances. In a first round, compound 2 (Figure 12.9) was identified that binds FKBP with a K_D of 2 μM. The second step consisted in the search for a second binder that interacts in a second site not competing with the binding of 2 to the first site. To this purpose, the system for the second screening was composed by both FKBP and compound 2 at a concentration of 2 mM. The benzanilide derivative 3 was so identified and a synthetic strategy

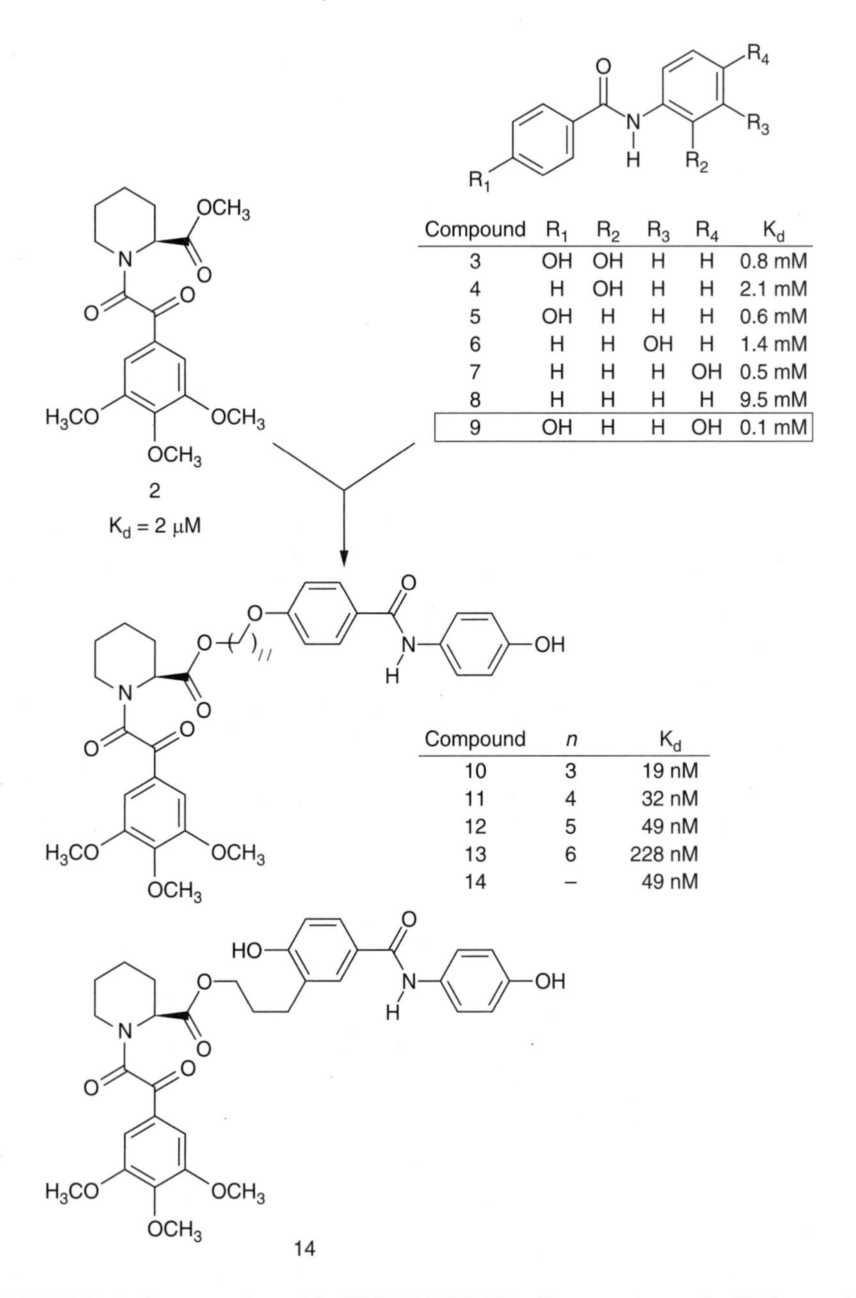

Compound	R_1	R_2	R_3	R_4	K_d
3	OH	OH	H	H	0.8 mM
4	H	OH	H	H	2.1 mM
5	OH	H	H	H	0.6 mM
6	H	H	OH	H	1.4 mM
7	H	H	H	OH	0.5 mM
8	H	H	H	H	9.5 mM
9	OH	H	H	OH	0.1 mM

Compound	*n*	K_d
10	3	19 nM
11	4	32 nM
12	5	49 nM
13	6	228 nM
14	–	49 nM

FIGURE 12.9 Compounds used in 1H/15N-HSQC binding experiments by Shuker *et al.*, 1996. Reprinted with permission from S.B. Shuker, P.J. Hajduk, R.P. Meadows and S.W. Fesik, *Science*, 1996, 274, 1531–1534. Copyright © 1996 American Association for the Advancement of Science.

was applied to enhance its low affinity (K_D 0.8 mM). The optimized compound 9 (K_D 0.1 mM) was found to be a good candidate, and different linkers were tested that combined both fragments (compounds 10 to 14). In this way, a series of binders with nM affinities could be identified.

1. The former example illustrates the five main steps of SAR by NMR—1. Identification of ligands with high affinity from a library of compounds utilizing $^{1}H/^{15}N$-HSQC experiments
2. Optimization of ligands by chemical modification
3. Identification of ligands binding in the presence of saturating amounts of optimized ligands from step 2 utilizing $^{1}H/^{15}N$-HSQC experiments
4. Optimization of ligands for the second site
5. Linking the two ligands for the primary and the secondary binding sites. Since its application to FKBP, SAR by NMR has been successfully used for the design of ligands for several other proteins, like stromelysin [98, 99], the E2 protein of the human papilloma virus [100], FKBP12 [101], ErmAM and ErmC' [102], the Lck SH2 domain [103], urokinase [104], adenosine kinase [105] and antagonists of the leukocyte function-associated antigen-1/intracellular adhesion molecule-1 interaction [106].

SAR by NMR is a totally NMR based strategy, and being that the macromolecule is the source of NMR information, the target was originally limited to 20 kDa of molecular weight. This threshold is changing continuously because of the introduction of higher field magnets, more sensitive probes and new experiments, like Transverse-Relaxation Optimized Spectroscopy [TROSY, 105]. The resolution and sensitivity enhancement associated with this experiment, which at very high fields allows to obtain sharper signals for

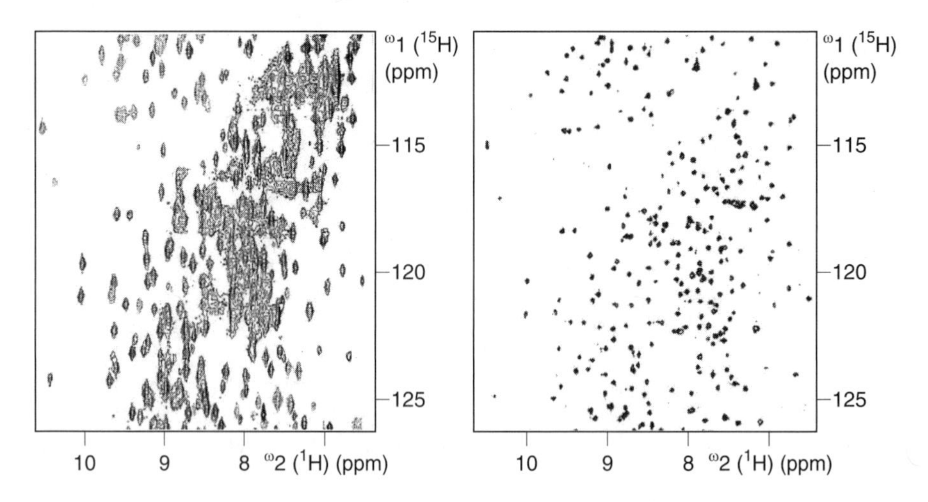

FIGURE 12.10 Resolution and sensitivity enhancement associated with 1H/15N TROSY. Reprinted with permission from P. Andersson, A. Annila and G. Otting, *J. Magn. Reson.* 1998, 133, 364–367. Copyright © 1998 Elsevier Inc.

the backbone amide HN group, is illustrated in Figure 12.10 for the 45 kDa fragment of 15N/2H labelled Staphylococcus aureus gyrase B [108]. Although the potential use of high molecular weight protein targets is clearly unlocked by these type of experiments, the problem of the assignment of an increasing number of nuclei still remains. The resulting overcrowded spectra can be simplified by selective labelling of a given type of aminoacid [109] or segmental isotope labeling using peptide splicing [110–112], as an alternative for uniform labeling. Another strategy involves the selective detection, using the TROSY experiment, of exposed amides, which are often those more readily available for the interaction [113].

If large libraries have to be screened, hundreds of mgs of ^{15}N labeled protein will be required. However, by using cryo-probe technologies, the ^{15}N labeled protein concentration can be reduced to 50 μM [114]. This protein concentration not only lowers the quantity of protein sample needed, but also allows an increase in the number of potential ligands that can be tested simultaneously. In these conditions, it was demonstrated that it is possible to test mixtures of 100 compounds in one sample at a 5 mM overall ligand concentration, leading to the possibility of testing 200 Ca compounds per month, including deconvolution of actives libraries [114].

A second strategy was recently proposed that both increases the sensitivity and allows in certain cases more direct information about the binding site of the interacting compounds [115]. It is based on ^{1}H-^{13}C correlation experiments of δ1 protons of the methyl groups of valine, leucine and isoleucine selectively ^{13}C labeled proteins. The use of protein targets labeled in this way presents several advantages—first, it reduces the complexity of the ^{1}H-^{13}C HSQC spectrum and, second, the resulting sensitivity is enhanced with respect to the ^{1}H-^{15}N HSQC experiments because of the presence of three protons vs. a single proton in an NH group, and because of the favourable relaxation properties of methyl groups. For FKBP (12 kDa) and Bcl-xL (19 kDa) it was found that the use of ^{1}H-^{13}C HSQC experiments results in a nearly 3-fold higher sensitivity [115], even when compared to the ^{1}H-^{15}N-TROSY experiment. Unfortunately, this sensitivity increment cannot be further enhanced by the application of the TROSY approach, because aliphatic carbons show, even at ultra high fields, very small chemical shift anisotropy, one of the two interfering relaxation mechanisms together with dipole-dipole coupling that give rise to the TROSY effect.

The main advantage of using the methyl groups as reporters for ligand binding is that changes in chemical shifts are more directly related to an interaction occurring in the proximity of the involved methyl group. On the contrary, the HN chemical shift is extremely sensitive and conformational changes occurring far away from the binding site can be erroneously assigned as a binding perturbation. Given the more uniform distribution of HN groups in a protein, a first analysis could indicate that the amide groups would result in a better reporter group. However, it was shown that methyl groups of valine, leucine or isoleucine are statistically always present within 6 Å of a heavy atom of ligands, by analyzing 191 proteins bound to ligand molecules [115].

Another variant of this methodology involves the analysis of differential chemical shift perturbations of a series of closely related ligands to rapidly determine the precise location of the ligand binding site and the orientation of the ligand in the binding pocket, which provides critical structural information for lead development and optimization [116, 117].

B. Saturation Transfer Difference (STD) Experiments

These experiments are based on the fact that if nuclei in the protein are irradiated, saturation will propagate to protons of an interacting ligand, resulting in a decrease of their intensities. This effect will be monitored by making a difference between an experiment irradiating a region where only protein signals are present (e.g. around 7 ppm, where applicable or at -2 ppm if there are methyl groups of the protein), and another experiment irradiating at a value far from any signal (e.g. 30 ppm). Irradiation of a single proton of the protein will spread out the saturation on the entire proton network through spin diffusion and will reach protons of the bound ligand. In the difference spectrum, all noninteracting ligands will cancel out, and only signals from the bound molecules will show up.

The STD principle may be combined with other NMR experiments, yielding 2-D experiments like STD HSQC or STD TOCSY [118, 119]. These kind of experiments have proven to be very effective in the detection of binding epitope of ligands [120, 122] and was used to determine that the L-fucose residue was the major binding epitope of the Sialyl Lewis^x tetrasaccharide when bound to the L-fucose recognizing lectin *Aleuria Aurantia* Agglutinin (AAA) [122]. Another impressive example of the power of this approach was the deconvolution of a library of 20 carbohydrate derivatives, which was tested for binding towards a lectin from elderberry, *Sambucus Nigra* Agglutinin (SNA) [119]. This library consisted of a mixture of monomethylated derivatives of the methyl α- and β-D-galacto- gluco- and mannopyranosides. It was particularly challenging because all the components presented the same molecular weight and almost identical polarity; the only difference among them was the stereochemistry. Such a library was synthesized by a onestep random methylation of the three methyl glycopyranosides and a classical deconvolution would have been difficult, because of the need for a clean separation of all different compounds. Using STD TOCSY two components of the library, the 2-O-methyl- and the 6-O-methyl derivatives of methyl-β-D-galacto-pyranoside were clearly identified as the only two compounds that bind to the protein.

The sensitivity of this kind of experiment depends on the degree of saturation that reaches the protons of the ligand. The main factors that influence this value are the residence time of the ligand in the protein binding pocket, the irradiation time and the excess of ligand molecules used. The residence time of the ligand depends on the affinity. Only those binders that ensure efficient exchange between bound and free ligands will be suitable for this technique. Tight binders, in the low nanomolar range, will not be detected using STD. The most suitable irradiation time will be a function of the effectiveness of the spin diffusion to bring about the effect on the entire spin network and of the longitudinal relaxation rates of the protons, which in turn depend on the molecular tumbling of the target protein. The higher the molecular weight of the target, the more efficient the spin diffusion process will be. This explains why this approach was used also to search for binders to membrane bound proteins, even within the natural membrane environment [123]. Finally, using irradiation times of the order of 2 s, it was shown that a 100-fold excess of ligand gives good results. The typical protein amount used in this approach is close to 1 to 10 nmol of protein, underlining one of the main strengths of the method. This fact, combined with a high degree of automation, consents to screen compound libraries at HTS rates (50,000 compounds per day).

C. Water-Ligand Observation with Gradient Spectroscopy (WaterLOGSY)

Water protons constitute a huge reservoir of magnetization that can be used instead of protein protons as the starting point to reach, *via* NOE or saturation transfer, the nuclei of a bound ligand. In this respect, the WaterLOGSY strategy [124] constitutes a variant of STD NMR. The principle behind this approach is that bound water located at the protein-ligand interfaces will mediate efficiently the transfer of magnetization because of the slow tumbling that those molecules acquire when associated to the protein. As in the case of STD NMR, only signals belonging to a bound compound will experience the negative NOE or the saturation transfer, and this allows to identify the binder in a way that resembles the experiments discussed in the point above.

The origin of these experiments can be traced to the observation of negative intermolecular water-ligand NOEs, which was explained either by bound water squeezed in between ligand and protein or by the water shell surrounding the ligand [125]. Based on this observation, it was suggested that the bulk water can be used to detect the binding of ligands to proteins [126]. Two variants of the experiment can be used. In the first, the water resonance is selectively inverted, constituting the so-called NOE-ePHOGSY NOE (Nuclear Overhauser Effect) ePHOGSY (enhanced protein hydration observed through gradient spectroscopy) scheme [127]. The second uses on-resonance saturation at the water chemical shift. To avoid problems associated with radiation damping and to attenuate artefacts, pulsed field gradients were employed. This technique was used to study the binding of ten low molecular weight ligands (100 μM each) to cyclin-dependent kinase 2 (34 kDa, 10 μM) [126]. As Figure 12.11 shows, only signals arising from the interacting molecule, an indole derivative, show a positive sign, whereas signals from noninteracting molecules are negative. This difference in sign associated to the NOE regimes of the small and large molecules constitutes the basis of other popular methods for NOE screening, based on the Transfer NOE experiment.

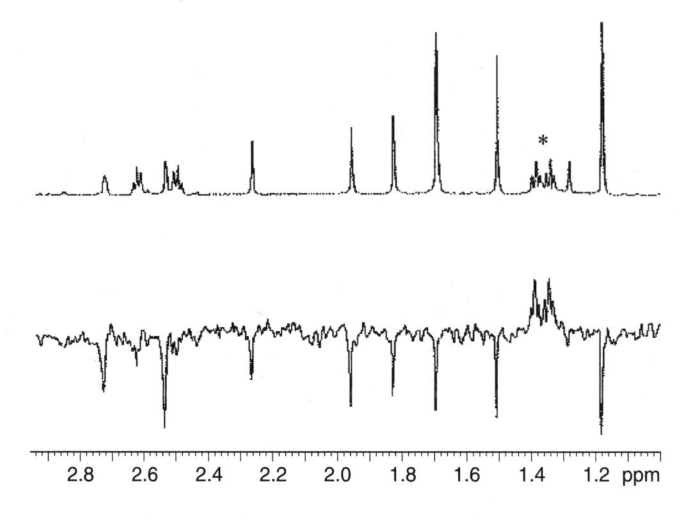

FIGURE 12.11 WaterLOGSY spectrum with NOE-ePHOGSY of a library of ten compounds in the Presence of cdk2 (lower) and reference spectrum (upper). The signals belonging to the binding ligand are marked with an asterisk. Reprinted with permission from C. Dalvit, P. Pevarello, M. Tatò *et al.*, *J. Biomol. NMR*, 2000, 18, 65–68. Copyright © 2000 Kluwer/Escom.

When compared with STD experiments, the general sensitivity and the possibility of application as a high-throughput screening approach for the two methods are similar. However, it requires the presence of water in the interface, which makes it particularly suited when dealing with polar interfaces like the ones normally present in nucleic acids-protein complexes.

D. Using NOE as a Screening Tool

The well known Transferred NOE Effect (TrNOE) [128–132] can be very efficiently used for screening libraries of compounds for binding to macromolecules. The efficiency of this method reflects the great increase of the build-up rate of NOEs between ligand protons which occurs upon complex formation. The sample conditions are those typical of the TrNOE experiment—low concentration of macromolecule ($< 50\,\mu M$), higher concentration of potential ligands ($> 500\,\mu M$) and conditions of relatively fast exchange between free and bound forms. The NOEs between ligand protons in the absence of the macromolecule can be conveniently observed in the small molecule regime (sign of cross-peaks opposite to diagonal peaks). In presence of the macromolecule, compounds that are involved with complex formation, even for a small fraction of time, will undergo a drastic change in their cross relaxation behaviour (sign of cross-peaks same as diagonal peaks). This technique was used to identify oligosaccharides that bind to *Aleuria Aurantia* agglutinin [133]. Two libraries, consisting of six and fifteen individual carbohydrate derivatives, respectively, were selected to test their binding affinity and the "active" molecule was readily identified without the need for deconvolution. Other examples have been published more recently, underlying other practical aspects of the method [134–137].

The main advantage of this technique, as compared with other ligand resonances techniques like diffusion- or relaxation-edited NMR, is the sensitivity to detect the binding signal, which makes it the method of choice when dealing with very weak interactions. On the other hand, it uses a smaller amount of ligand excess compared to STD experiments. For the observation of TrNOEs ligand to protein ratios of about 10 : 1 to 30 : 1 are usually optimal, which implicates the need for using an order of magnitude more protein than in the case of STD. Both types of experiments are complementary, as STD may reveal the binding epitope and trNOE experiments can be used for the analysis of the bound conformation of the ligand.

Another approach based on the transfer of magnetization via NOE is the so-called "NOE pumping" technique, which makes use of the fact that protons of a ligand bound to a protein contribute to the relaxation of protons of the protein and vice versa. This effect was exploited in two different ways to detect binding. In the first approach, ligand signals are suppressed using a diffusion filter, followed by a NOESY mixing time in order to make relaxation pathways from the protein to the ligand visible by way of subtraction from a reference spectrum [138]. A second and more recent version of the experiment uses the reversed setup [139]. In a first experiment, a T2 filter is applied after excitation, followed by a NOESY, in order to suppress protein signals and to allow the "NOE pumping" to occur from the ligand to the protein only. In a second reference spectrum the T2 filter is applied after the mixing period, so that during the mixing time both processes of NOE pumping from the ligand to the protein and vice versa cancel each other. The difference of the two experiments yields signals only from those compounds that were in contact with the protein binding site.

E. Diffusion-Edited NMR

Self-diffusion is one parameter that readily differentiates large and small molecules in solution. The key idea behind the use of diffusion as a screening tool is that the diffusion coefficient of a small molecule binding to a "receptor" in solution will be altered with respect to that of a small compound that does not interact. This observation prompted the utilization of pulse field gradient technology to detect changes in the diffusion coefficient induced by intermolecular interactions. Using this principle, it was demonstrated for a mixture of nine compounds and stromelysin, that it is possible to "fish out" the active compound without the need of separation [19]. Moreover, it was also demonstrated that, given a mixture of compounds with different affinities for a biological target, it is possible to tune the system to a desired sensitivity level, just by adjusting the gradient strength [15, 18, 140–142].

F. Relaxation-Edited NMR

As a consequence of binding, the long T2 normally observed for small molecules is reduced by the complexation with a macromolecule. The observed effect depends both on the affinity and the kinetic off-rate of the binding process. In fact, the first evidence of binding is usually the broadening of the signal of the small molecule in presence of its biological target. This effect, in combination with a subtraction scheme, was used to identify a 200 μM inhibitor of FK506 binding protein (FKBP), within a mixture of eight non-interacting compounds [19]. As in the case of diffusion-edited NMR, this methodology can also be optimized for detecting ligand binding within a range of target affinities, just by adjusting the length of the relaxation filter. An optimal screening strategy would employ multiple relaxation filter times for each sample to allow the detection of both high and low affinity ligands. In the case of using compound mixtures, there is the need for a careful selection of the compounds in order to observe well separated signals [136].

A recently proposed variant to this method is the use of a spin labelled first ligand, that binds to the main protein binding site, and in this way the procedure is applied only for the screening of a second ligand [143, 144]. The presence of a spin label induces a very fast T2 relaxation on ligands that bind nearby the first fragment, and the effect on the line shape is therefore enhanced.

REFERENCES

1. Keifer PA, New methods for obtaining high-resolution NMR spectra of solid-phase synthesis resins, natural products, and solution-state combinatorial chemistry libraries, *Drugs of the Future*, 23:301–317, 1998.
2. Keifer PA, Smallcombe SH, Williams EH, *et al.*, DirectInjection-NMR (DI-NMR): a flow NMR technique for the analysis of combinatorial chemistry libraries, *J. Comb. Chem.*, 2:151–171, 2000.
3. Keifer PA, Flow injection analysis NMR (FIA-NMR): a novel flow NMR technique that complements LC-NMR and direct injection NMR (DI-NMR), *Magn. Reson. Chem.*, 41:509–516, 2003.
4. Sidelmann UG, Gavaghan C, Carless HAI, *et al.*, 750-MHz directly coupled HPLC-NMR: Application for the sequential characterization of the positional isomers and anomers of 2-, 3-, and 4-fluorobenzoic acid glucuronides in equilibrium mixtures, *Anal. Chem.*, 67:4441–4445, 1995.
5. Smallcombe SH, Patt SL, Keifer PA, WET Solvent Suppression and Its Applications to LC NMR and High-Resolution NMR Spectroscopy, *J. Magn. Reson.*, 117:295–303, 1995.
6. Lindon JC, Nicholson JK, Wilson ID, Direct coupling of chromatographic separations to NMR spectroscopy, *Prog. NMR Spectrosc.*, 29:1–49 1996.
7. Korhammer SA, Bernruether A, Hyphenation of high-performance liquid chromatography (HPLC) and other chromatographic techniques (SFC, GPC, GC, CE) with nuclear magnetic resonance (NMR): A review, *J. Anal. Chem.*, 354:131–135, 1996.

8. Patt SL, Single and multiple-frequency-shifted laminar pulses, *J. Magn. Reson.*, 96:94–102, 1992.
9. Dalvit C, Bohlen IM, Multiple-Solvent Suppression in Double-Quantum NMR Experiments with Magic Angle Pulsed Field Gradients, *Magn. Reson. Chem.*, 34:829–833, 1996.
10. Hwang TL, Shaka AJ, Water suppression that works. Excitation sculpting using arbitrary wave-forms and pulsed-field gradients, *J. Magn. Reson.*, 112:275–279, 1995.
11. Burbaum JJ, Ohlmeyer MHJ, Reader JC, *et al.*, A paradigm for drug discovery employing encoded combinatorial libraries, *Proc. Natl. Acad. Sci. USA*, 92:6027–6031, 1995.
12. Salmon SE, Lam KS, Lebl M, Kandda A, Khattri PS, Wades, Patek M, Kocis P, Krchnak V, Thorpe D, Felder S, Discovery of biologically active peptides in random libraries: solution-phase testing after staged orthogonal release from resin beads, *Proc. Natl. Acad. Sci. USA*, 90 :11708–11712, 1993.
13. Lam KS, Salmon SE, Hersh EH, Hruby VJ, Kaamierski WH, Knapp RJ, A new type of synthetic peptide library for identifying ligand-binding activity, *Nature*, 354:82–84, 1991.
14. Lacey ME, Sweedler JV, Larive CK, Pipe AJ, Farrant RD, 1H NMR Characterization of the product from single solid-phase resin beads using capillary NMR flow probes, *J. Magn. Reson.*, 153:215–222, 2001.
15. Lin M, Shapiro MJ, Wareing JR, Diffusion-edited NMR-affinity NMR for direct observation of molecular interactions, *J. Am. Chem. Soc.*, 119:5249–5350, 1997.
16. Morris KF, Johnson CS Jr., Resolution of discrete and continuous molecular size distributions by means of diffusion-ordered 2-D NMR spectroscopy, *J. Am. Chem. Soc.*, 115:4291–4299, 1993.
17. Barjat H, Morris GA, Smart S, Swanson AG, Williams SCR, High-resolution diffusion-ordered 2-D spectroscopy (HR-DOSY)—A new tool for the analysis of complex mixtures, *J. Magn. Reson. B*, 108:170–172, 1995.
18. Lin M, Shapiro MJ, Mixture analysis in combinatorial chemistry. Application of diffusion-resolved NMR spectroscopy, *J. Org. Chem.*, 61:7617–7619, 1996.
19. Hajduk PJ, Olejniczak ET, Fesik SW, One-dimensional relaxation- and diffusion-edited NMR methods for screening compounds that bind to macromolecules, *J. Am. Chem. Soc.*, 119:12257–12261, 1997c.
20. Gonnella N, Lin M, Shapiro MJ, Wareing JR, Zhang X, Isotope-Filtered affinity NMR, *J. Magn. Reson.*, 131:336–338, 1998.
21. Shuker SB, Hajduk PJ, Meadows RP, Fesik SW, Discovering high-affinity ligands for proteins: SAR by NMR, *Science*, 274:1531–1534, 1996.
22. Gozansky EK, Gorenstein DG, DOSY-NOESY: Diffusion-Ordered NOESY, *J. Magn. Reson. Series B*, 111:94–96, 1996.
23. Lindon JC, Farrant RD, Sanderson PN, Doyle PM, Gough SL, Spraul M, Hofmann M, Nicholson JK, Separation and characterization of components of peptide libraries using on-flow coupled HPLC-NMR spectroscopy, *Magn. Reson. Chem.*, 33:857–863, 1995.
24. Chin J, Fell JB, Jarosinski M, Shapiro MJ, Wareing JR, HPLC/NMR in combinatorial chemistry, *J. Org. Chem.*, 63:386–390, 1998.
25. Lindon JC, Nicholson JK, Wilson ID, Directly coupled HPLC-NMR and HPLC-NMR-MS in pharmaceutical research and development, *J. Chromat. B*, 748:233–258, 2000.

26. Taylor SD, Wright B, Clayton E, Wilson ID, Practical aspects of the use of high performance liquid chromatography combined with simultaneous Nuclear Magnetic Resonance and Mass Spectrometry, *Rapid Commun. Mass Spectrom.*, 12:1732–1736, 1998.
27. Jung G, Beck-Sickinger AG, Multiple peptide synthesis methods and their applications, *Angew. Chem. Int. Ed. Eng.*, 31:367–383, 1992.
28. Gallop MA, Barret RW, Dower WI, Fodor SPA, Gerdon EM, Applications of combinatorial technologies to drug discovery. 1. Background and peptide combinatorial libraries, *J. Med. Chem.*, 37:1233–1251, 1994.
29. Epton R. (Ed.), *Innovation and Perspectives in Solid Phase Synthesis-Peptides, Proteins and Nucleic Acids*, Mayflower Worldwide, Kingswinford, UK, 1994.
30. Lloyd-Williams P, Albericio R, Giralt E, *Convergent Solid-Phase Peptide Synthesis*, 49:11065–11133, *Tetrahedron*, 1993.
31. Beaucage SL, Iyer RP, Advances in the synthesis of oligonucleotides by the phosphoramidite approach, *Tetrahedron*, 48:2223–2311, 1992.
32. Bunin BA, Plunkett MJ, Ellman JA, The combinatorial synthesis and chemical and biological evaluation of a 1, 4-benzodiazepine library, *Proc. Natl. Acad. Sci. USA*, 91:4708–4712, 1994.
33. Randolph JT, McClure KF, Danishefsky SJ, Major simplifications in oligosaccharide syntheses arising from a solid-phase based method: An application to the synthesis of the Lewis b Antigen, *J. Am. Chem. Soc.*, 117:5712–5719, 1995.
34. Wenschuh H, Beyennann M, Haber H, Seydel JK, Krause E, Bienert M, Carpino LA, El-Fahan A, Albericio F, Stepwise automated solid phase synthesis of naturally occurring peptaibols using FMOC amino acid fluorides, *J. Org. Chem.*, 60:405–410, 1995.
35. Schaefer J, High-resolution pulsed carbon-13 nuclear magnetic resonance analysis of some crosslinked polymers, *Macromolecules*, 4:110–112, 1971.
36. Manatt SL, Horowitz D, Horowitz R, Pimnell RP, Solvent swelling for enhancement of carbon-13 nuclear magnetic resonance spectral information from insoluble polymers: chloromethylation levels in crosslinked, *Anal. Chem.*, 52:1529–1532, 1980.
37. Blossey EC, Cannon RG, Ford WT, Periyasam YM, Mohanraj S, Synthesis, reactions and carbon-13 FT NMR spectroscopy of polymer-bound steroids, *J. Org. Chem.*, 55:4664–4668, 1990.
38. Auzanneau FI, Melda M, Bock K, Synthesis, characterization and biocompatibility of PEGA resins, *J. Pept. Sci.*, 1:31–44, 1995.
39. Forman FW, Sucholeiki I, Solid-phase synthesis of biaryls via the Stille reaction, *J. Org. Chem.*, 60:523–528, 1995.
40. Albericio F, Pons M, Pedroso E, Giralt E, Comparative study of supports for solid-phase coupling of protected-peptide segments, *J. Org. Chem.*, 54, 360–366, 1989.
41. Giralt E, Rizo J, Pedroso E, Application of gel-phase carbon-13 NMR to monitor solid phase peptide synthesis, *Tetrahedron*, 40:4141–4152, 1984.
42. Epton R, Goddard P, Ivin KJ, Gel phase (13)C NMR spectroscopy as an analytical method in solid (gel) phase peptide synthesis, *Polymer*, 21:1367–1371, 1980.
43. Sternlicht H, Kenyon GL, Packer EL, Sinclair J, Carbon-13 nuclear magnetic resonance studies of heterogeneous systems. Amino acids bound to cationic exchange resins, *J. Am. Chem. Soc.*, 93:199–209, 1971.

44. Look GC, Holmes CP, Chinn JP, Gallop MA, Methods for Combinatorial Organic Synthesis: The Use of Fast ^{13}C NMR Analysis for Gel Phase Reaction Monitoring, *J. Org. Chem.*, 59:7588–7590, 1994.
45. Barn DR, Morphy JR, Rees DC, Synthesis of an array of amides by aluminium chloride assisted cleavage of resin-bound esters, *Tetrahedron Lett.*, 37:3213–3216, 1996.
46. Holmes CP, Jones DG, Reagents for combinatorial organic synthesis: development of a new o-nitrobenzyl photolabile linker for solid phase synthesis, *J. Org. Chem.*, 60:2318–2319, 1995.
47. Gordeev MF, Patel DV, Gordon EM, Approaches to combinatorial synthesis of heterocycles: A solid-phase synthesis of 1, 4-Dihydropyridines, *J. Org. Chem.*, 61:924–928, 1996.
48. Gordeev MF, Patel DV, Wu J, Gordon EM, Approaches to combinatorial synthesis of heterocycles: Solid-phase synthesis of pyridines and pyrido[2,3-d] pyrimidines, *Tetrahedron Lett.*, 37:4643–4646, 1996.
49. Svensson A, Fex T, Kihlberg J, Use of 19F NMR spectroscopy to evaluate reactions in solid phase organic synthesis, *Tetrahedron Lett.*, 37:7649–7652 1996.
50. Bardella F, Eritja R, Pedroso E, Giralt E, Gel-phase ^{31}P-NMR. A new analytical tool to evaluate solid phase oligonucleoside synthesis, *Bioorg. Med. Chem. Lett.*, 3:2793–2796, 1993.
51. Johnson CR, Zhang B, Solid phase synthesis of alkenes using the Horner-Wadsworth-Emmons reaction and monitoring by gel phase (31)P NMR, *Tetrahedron Lett.*, 36:9253–9256, 1995.
52. Ludwick AG, Jelinski LW, Live D, Kintanar A, Dumais JJ, Association of peptide chains during Merrifield solid-phase peptide synthesis. A deuterium NMR study, *J. Am. Chem. Soc.*, 108:6493–6496, 1986.
53. Stoll ME, Majors TJ, Elimination of magnetic-susceptibility broadening in NMR using magic-angle sample spinning to measure chemical shifts in niobium hydride (NbHx), *J. Phys. Rev. B*, 24:2859–2862, 1981.
54. Doskocilova D, Tao DD, Schneider B, Effects of macroscopic spinning upon line width of NMR signals of liquid in magnetically inhomogeneous systems, *Czech. J. Phys. B*, 25:202–209, 1975.
55. Garroway AN, Magic-angle sample spinning of liquids, *J. Magn. Reson.*, 49: 168–171, 1982.
56. Edmonds DT, Wormald MR, Theory of resonance in magnetically inhomogeneous specimens and some useful calculations, *J. Magn. Reson.*, 77:223–232, 1988.
57. Forbes J, Bowers J, Shan X, Moran L, Oldfield E, Moscarello MA, Some new developments in solid-state nuclear magnetic resonance spectroscopic studies of lipids and biological membranes, including the effects of cholesterol in model and natural systems, *J. Chem. Soc. Faraday Trans.*, 84:3821–3849, 1988.
58. Rutar V, Kovac M, Lahajnar G, Improved NMR spectra of liquid components in heterogenous samples, *J. Magn. Reson.*, 80:133–138, 1988.
59. Keifer PA, Baltusis L, Ricer DM, Tymiak A, Shoolery JN, A comparison of NMR spectra obtained for solid-phase-synthesis resins using conventional high-resolution, magic-angle-spinning, and high-resolution magic-angle-spinning probes, *J. Magn. Reson. Series A*, 119:65–75, 1996.
60. Grøtli M, Grotfredsen CH, Radermann J, Buchardt J, Clark AJ, Duus J, Meldav M, Physical properties of poly(ethylene glycol) (PEG)-based resins for

combinatorial solid phase organic chemistry: A comparison of PEG-cross-linked and PEG-grafted resins, *J. Comb. Chem.*, 2:108–119, 2000.

61. Elbayed K, Bourdonneau M, Furrer J, Richert T, Raya J, Hirschinger J, Pioho M, Origin of the residual NMR linewidth of a peptide bound to a resin under magic angle spinning, *J. Magn. Reson.*, 136:127–129, 1999.
62. Keifer P, Influence of resin structure, tether length, and solvent upon the high-resolution ^{1}H NMR spectra of solid-phase-synthesis resins, *J. Org. Chem.*, 61:1558–1559, 1996.
63. Chili E, Oliveira E, Marchetto R, Nakaie CR, Correlation between solvation of peptide-resins and solvent properties, *J. Org. Chem.*, 61:8992–9000, 1996.
64. Meissner A, Bloch P, Humpfer E, Spraul M, Sørensen OW, Reduction of inhomogeneous line broadening in two-dimensional high-resolution MAS NMR spectra of molecules attached to swelled resins in solid-phase synthesis, *J. Am. Chem. Soc.*, 119:1787–1788, 1997.
65. Shapiro MJ, Chin J, Marti RE, Jarosinski MA, Enhanced resolution in MAS NMR for combinatorial chemistry, *Tetrahedron Lett.*, 38:1333–1336, 1997.
66. Nagayama K, Wüthrich K, Ernst RR, Two-dimensional spin echo correlated spectroscopy (SECSY) for 1H NMR studies of biological macromolecules, *Biochem. Biophys. Res. Commun.*, 90:395, 1979.
67. Chin J, Fell B, Pochapsky S, Shapiro MJ, Wareing JR, 2-D SECSY NMR for combinatorial chemistry, high-resolution MAS spectra for resin-bound molecules, *J. Org. Chem.*, 63:1309–1311, 1998.
68. Dhalluin CF, Boutillon C, Tartar AL, Lippens G, Magic angle spinning nuclear magnetic resonance in solid-phase peptide synthesis, *J. Am. Chem. Soc.*, 119:10494–10500, 1997.
69. Pop IE, Dhalluin CF, Deprez BP, Melnyk PC, Lippens GH, Tartar AL, Monitoring of a three-step solid phase synthesis involving a Heck reaction using magic angle spinning NMR spectroscopy, *Tetrahedron*, 52:12209–12222, 1996.
70. Ganapathy S, Badiger MV, Rajamohana PR, Mashelkar RA, High-resolution solid-state proton MASS NMR of superabsorbing polymeric gels, *Macromolecules*, 22:2023–2025, 1989.
71. Garigipati RS, Adams B, Adams JL, Sarkar SK, Use of spin echo magic angle spinning ^{1}H NMR in reaction monitoring in combinatorial organic synthesis, *J. Org. Chem.*, 61:2911–2914, 1996.
72. Mass WE, Laukien FH, Cory DG, Gradient, High resolution, magic angle sample spinning NMR, *J. Am. Chem. Soc.*, 118:13085–13086, 1996.
73. Luo Y, Ouyang X, Armstrong RW, Murphy MM, A case study of employing spectroscopic tools for monitoring reactions in the developmental stage of a combinatorial chemistry library, *J. Org. Chem.*, 63:8719, 1998.
74. Rieddl R, Tappe R, Berkessel AJ, Probing the scope of the asymmetric dihydroxylation of polymer-bound olefins. Monitoring by HRMAS NMR allows for reaction control and on-bead measurement of enantiomeric excess, *J. Am. Chem. Soc.*, 120:8994, 1998.
75. Pursch M, Schlotterbeck G, Tseng L, Rapp W, Monitoring the reaction progress in combinatorial chemistry: ^{1}H MAS NMR investigations on single macro beads in the suspended state, *Angew. Chem. Int. Ed. Engl.*, 35:2867–2869, 1996.
76. Sarkar S, Garigipati RS, Adams JL, Keifer PA, An NMR method to identify nondestructively chemical compounds bound to a single solid-phase-synthesis

bead for combinatorial chemistry applications, *J. Am. Chem. Soc.*, 118:2305–2306, 1996.
77. Lebl M, Krchnak V, Sepetov F, Seligmann B, Strop P, Felder S, Lan KS, One-bead-one-structure combinatorial libraries, *Biopolymers*, 37:177–198, 1995.
78. Jelinkek R, Valente AP, Valentine KG, Opella SJ, Two-dimensional NMR spectroscopy of peptides on beads, *J. Magn. Reson.*, 125:185–187, 1997.
79. Lippens G, Bourdonneau M, Dhalluin C, Warras R, Richter T, Seetharaman C, Boutillen C, Pioho M, Study of compounds attached to solid supports using high resolution magic angle spinning NMR, *Curr. Organ. Chem.*, 3 :147–169, 1999.
80. Shapiro MJ, Kumaravel G, Petter RC, Beveridge R, ^{19}F NMR monitoring of a S(N)Ar reaction on solid support, *Tetrahedron Lett.*, 37:4671–4674, 1996.
81. Sem, Pellechia M, NMR in the acceleration of drug discovery, *Curr. Opin. Drug Discovery Dev.*, 4:479–492, 2001.
82. Pochasky SS, Pochasky TC, Nuclear magnetic resonance as a tool in drug discovery, metabolism and disposition, *Curr. Top. Med. Chem.*, 1:427–441, 2001.
83. Pellecchia M, Sern DS, Wüthrich K, NMR in drug discovery, *Nat. Rev. Drug Disc.*, 1:211–219, 2002.
84. Peng JW, Lepre CA, Fejzo J, Abdul-Manan N, Moore JH, Nuclear magnetic resonance-based approaches for lead generation in drug discovery, *Methods Enzymol*, 338:202–230, 2001.
85. Ross A, Schlotterbeck G, Kalus W, Senn H, Automation of NMR measurements and data evaluation for systematically screening interactions of small molecules with target proteins, *J. Biomol. NMR*, 16:139–146, 2000.
86. Stockman BJ, NMR spectroscopy as a tool for structure-based drug design, *Progr. NMR Spectrosc.*, 33:109–151, 1998.
87. Chen A, Shapiro MJ, NOE pumping 2. A high-throughput method to determine compounds with binding affinity to macromolecules, *J. Am. Chem. Soc.*, 122:414–415, 2000.
88. Moore JM, NMR screening in drug discovery, *Curr. Opin. Biotechnol.*, 10:54–58, 1999.
89. Roberts GCK, NMR spectroscopy in structure-based drug design, *Curr. Opin. Biotechnol*, 10:42–47, 1999.
90. Roberts GCK, Applications of NMR in drug discovery, *Drug Disc. Today*, 5:230–240, 2000.
91. Lepre CA, Library design for NMR-based screening, *Drug Disc. Today*, 6:133–140, 2001.
92. Wyss DF, McCoy MA, Senior MM, NMR-based approaches for lead discovery, *Curr. Opin. Drug Disc. & Devel.*, 5:630–647, 2002.
93. Jack RM, Smith JS, Villar HO, Sem DS, Couhs SM, Triad therapeutics: Integration of NMR structural determinations and smart chemistry to speed drug discovery, *Drug Disc. Today*, 7:S35–S38, 2002.
94. Bemis GW, Murko MA, The properties of known drugs. 1. Molecular frameworks, *J. Med. Chem.*, 39:2887–2893, 1996.
95. Bemis GW, Murko MA, Properties of known drugs. 2. Side chains, *J. Med. Chem.*, 42:5095–5099, 1999.
96. Fejzo J, Lepre CA, Peng JW, Bemis GW, Muroko MA, Moore JM, The SHAPES strategy: An NMR-based approach for lead generation in drug discovery, *Chem. Biol.*, 6:755–769, 1999.

97. Hajduk PJ, Meadows RP, Fesik SW, NMR-based screening in drug discovery, *Q. Rev. Biophys*, 32:211–240, 1999.
98. Hajduk PJ, Sheppard G, Nettesheim DG, Olejniczak ET, Shuker SB, Meadows RP, Steinman DH, Carrera GM, Marcotte PA, Severin J, Walter K, Smith M, Gubbins E, Simmer R, Holzman TF, Morgan DW, Davidsen SK, Summers JB, Fesik SW, Discovery of potent nonpeptide inhibitors of stromelysin using SAR by NMR, *J. Am. Chem. Soc.*, 119:5818–5827, 1997.
99. Olejniczak ET, Hajduk PJ, Marcotte PA, Nettesheim DG, Meadows RP, Edalji R, Holzman TF, Fesik SW, Stromelysin inhibitors designed from weakly bound fragments: Effects of linking and cooperativity, *J. Am. Chem. Soc.*, 119: 5828–5832, 1997.
100. Hajduk PJ, Dinges J, Miknis GF, Merlock M, Middlton T, Kemph DJ, Egan DA, Walter KA, Robins TS, Shuker SB, Holzman TF, Fesik SW, NMR-Based discovery of lead inhibitors that block DNA binding of the human papillomavirus E2 protein, *J. Med. Chem.*, 40:3144–3150, 1997.
101. Wiedermann PE, Fesik SW, Petros SW, Nettesheim DG, Mollison KW, Lane BC, Or SY, Luly JR, Retention of immunosuppressant activity in an ascomycin analogue lacking a hydrogen-bonding interaction with FKBP12, *J. Med. Chem.*, 42:4456–4461, 1999.
102. Hajduk PJ, Dinges J, Schkeryantz JM, *et al.*, Novel inhibitors of erm methyltransferases from NMR and parallel synthesis, *J. Med. Chem.*, 42:3852–3859, 1999.
103. Hajduk PJ, Zhou MM, Fesik SW, NMR-based discovery of phosphotyrosine mimetics that bind to the Lck SH2 domain, *Bioorg. Med. Chem. Lett.*, 9:2403–2406, 1999.
104. Hajduk PJ, Boyd S, Nettesheim D, Nienaber V, Severin J, Smith R, Davidson D, Rockway T, Fesik SW, Identification of novel inhibitors of urokinase via NMR-based screening, *J. Med. Chem.*, 43:3862–3866, 2000.
105. Hajduk PJ, Gomtsyan A, Didomenico S, Cowart M, Bayburt EK, Solomon L, Severin J, Smith R, Walter K, Holzman TF, Stewart A, McGaraughty S, Yarvis M, Kavaluk E, Fesik SW, Design of adenosine kinase inhibitors from the NMR-based screening of fragments, *J. Med. Chem.*, 43:4781–4786, 2000b.
106. Liu G, Huth JR, Olejniczak ET, Herdoza R, DeVries P, Lietza S, Reilly EB, Okosinski GF, Fesik SW, vonGeldern TW, Novel p-arylthio cinnamides as antagonists of leukocyte function-associated antigen-1/intracellular adhesion molecule-1 interaction. 2. Mechanism of inhibition and structure-based improvement of pharmaceutical properties, *J. Med. Chem.*, 44:1202–1210, 2001.
107. Pervushin K, Riek R, Wider G, Wüthrich K, Attenuated T2 relaxation by mutual cancellation of dipole-dipole coupling and chemical shift anisotropy indicates an avenue to NMR structures of very large biological macromolecules in solution, *Proc. Natl. Acad. Sci. USA*, 23:12366–12371, 1997.
108. Andersson P, Annila A, Otting G, An alpha/beta-HSQC-alpha/beta experiment for spin-state selective editing of IS cross peaks, *J. Magn. Reson.*, 133:364–367, 1998.
109. Kainosho M, Tsuji T, Assignment of the three methionyl carbonyl carbon resonances in Streptomyces subtilisin inhibitor by a carbon-13 and nitrogen-15 double-labeling technique. A new strategy for structural studies of proteins in solution, *Biochemistry*, 21:6273–6279, 1982.

110. Yamazaki T, Otomo T, Oda N, Kyogoku Y, Uegaki K, Ito N, Ishino Y, Nakamura H, Segmental isotope labelling for protein NMR using peptide splicing, *J. Am. Chem. Soc.*, 120:5591–5592, 1998.
111. Otomo T, Teruya K, Uegaki K, Yamazaki T, Kyogoku Y, Improved segmental isotope labeling of proteins and application to a larger protein, *J. Biomol. NMR*, 14:105–114, 1999.
112. Xu R, Ayers B, Cowbum D, Muir TW, Chemical ligation of folded recombinant proteins: Segmental isotopic labelling of domains for NMR studies, *Proc. Natl. Acad. Sci. USA*, 96:388–393, 1999.
113. Pellechia M, Meininger D, Shen AL, Jack R, Kasper CB, Sern DS, SEA-TROSY (solvent exposed amides with TROSY): A method to resolve the problem of spectral overlap in very large proteins, *J. Am. Chem. Soc.*, 123: 4633–4634, 2001.
114. Hajduk PJ, Gerfin T, Boehlen JM, Mäberli M, Marek D, Fesik SW, High-throughput nuclear magnetic resonance-based screening, *J. Med. Chem.*, 42:2315–2317, 1999.
115. Hajduk PJ, Augeri DJ, Mack J, Mendoza R, Yang J, Betz SF, Fesik SW, NMR-based screening of proteins containing ^{13}C-labeled methyl groups, *J. Am. Chem. Soc.*, 122:7898–7904, 2000.
116. Cicero DO, Barbato G, Koch U, Ingallinella P, Bianchi E, Nandi MC, Steinkuhler C, Cortese R, Matassa V, De Francesco R, Pessi A, Bazzo R, Structural characterization of the interactions of optimized product inhibitors with the N-terminal proteinase domain of the Hepatitis C Virus (HCV) NS3 protein by NMR and modelling studies, *J. Mol. Biol.*, 289:385–396, 1999.
117. Medek A, Hajduk PJ, Mack J, Fesik SW, The use of differential chemical shifts for determining the binding site location and orientation of protein-bound ligands, *J. Am. Chem. Soc.*, 122:1241–1242, 2000.
118. Mayer M, Meyer B, Characterization of ligand binding by saturation transfer difference NMR spectroscopy, *Angew. Chem. Int. Ed. Engl.*, 38:1784–1788, 1999.
119. Vogtherr M, Peters T, Application of NMR based binding assays to identify key hydroxyl groups for intermolecular recognition, *J. Am. Chem. Soc.*, 122: 6093–6099, 2000.
120. Mayer M, Meyer B, Group epitope mapping by saturation transfer difference NMR to identify segments of a ligand in direct contact with a protein receptor, *J. Am. Chem. Soc.*, 123:6108–6117, 2001.
121. Maaheimo H, Kosma P, Brade L, Brade H, Peter T, Mapping the binding of synthetic disaccharides representing epitopes of chlamydial lipopolysaccharide to antibodies with NMR, *Biochemistry*, 39:12778–12788, 2000.
122. Haselhorst T, Weimar T, Peters T, Molecular Recognition of Sialyl Lewis^{x} and Related Saccharides by Two Lectins, *J. Am. Chem. Soc.*, 123:10705–10714, 2001.
123. Klein J, Meinecke R, Mayer M, Meyer B, Detecting binding affinity to immobilized receptor proteins in compound libraries by HR-MAS STD NMR, *J. Am. Chem. Soc.*, 121:5336–5337, 1999.
124. Dalvit C, Fogliatto G, Stewart A, Veronesi M, Stockman B, WaterLOGSY as a method for primary NMR screening: Practical aspects and range of applicability, *J. Biomol. NMR*, 21:349–359, 2001.
125. Dalvit C, Cottens S, Ramage P, Hommel U, Half-filter experiments for assignment, structure determination and hydration analysis of unlabelled ligands bound to ^{13}C/^{15}N labelled proteins, *J. Biomol. NMR*, 13:43–50, 1999.

126. Dalvit C, Pevarello P, Tatò M, *et al.*, Identification of compounds with binding affinity to proteins *via* magnetization transfer from bulk water, *J. Biomol. NMR*, 18:65–68, 2000.
127. Dalvit C, Hommel U, Sensitivity-improved detection of protein hydration and its extension to the assignment of fast-exchanging resonances, *J. Magn. Reson. Series B*, 109: 334–338, 1995.
128. Balaram P, Bothner-By AA, Dadok J, Negative nuclear Overhuaser effects as probes of macromolecular structure, *J. Am. Chem. Soc.*, 94:4015–4017, 1972.
129. Balaram P, Bothner-By AA, Breslow E, Localization of tyrosine at the binding site of neurophysin II by negative nuclear overhouser effects, *J. Am. Chem. Soc.*, 94:4017–4018, 1972b.
130. Clore GM, Gronenborn AM, Theory and applications of the transferred nuclear Overhauser effect to the study of the conformations of small ligands bound to proteins, *J. Magn. Reson.*, 48:402–417, 1982.
131. Clore GM, Gronenborn AM, Theory of the time dependent transferred nuclear Overhauser effect: application to the structural analysis of ligand-protein complexes in solution, *J. Magn. Reson.*, 53:423–442, 1983.
132. Ni F, Recent developments in transferred NOE methods, *Prog. NMR Spectr.*, 26:517–606, 1994.
133. Meyer B, Weimar T, Peters T, Screening mixtures for biological activity by NMR, *Eur. J. Biochem.*, 246:705–709, 1997.
134. Henrichsen D, Ernst B, Magnani JL, Wang WT, Meyer B, Peter ST, Bioaffinity NMR spectroscopy: identification of an E-selectin antagonist in a substance mixture by transfer NOE, *Angew. Chem. Int. Ed.*, 38:98–102, 1999.
135. Mayer M, Meyer B, Mapping the active site of angiotensin-converting enzyme by transferred NOE spectroscopy, *J. Med. Chem.*, 43:2093–2099, 2000.
136. Moore JM, NMR techniques for characterization of ligand binding: utility for lead generation and optimization in drug discovery, *Biopolymers*, 51:221–243, 1999.
137. Herfurth L, Weimar T, Peters T, Application of 3D-TOCSY-tr-NOESY for the assignment of bioactive ligands from mixtures, *Angew. Chem. Int. Ed. Engl.*, 39:2097–2099, 2000.
138. Chen A, Shapiro MJ, NOE pumping—A novel NMR technique for identification of compounds with binding affinity to macromolecules, *J. Am. Chem. Soc.*, 120:10258–10259, 1998.
139. Chen A, Shapiro MJ, Affinity NMR, *Anal. Chem.*, 71:669A–675A, 1999.
140. Lin M, Shaprio MJ, Wareing JR, Screening mixtures by affinity NMR, *J. Org. Chem.*, 62:8930–8931, 1997b.
141. Bleicher K, Lin M, Shapiro MJ, Wareing JR, Diffusion edited NMR: screening compound mixtures by affinity NMR to detect binding ligands to vancomycin, *J. Org. Chem.*, 63:8486–8490, 1998.
142. Anderson RC, Lin M, Shapiro MJ, Affinity NMR: Decoding DNA Binding, *J. Comb. Chem.*, 1:69–72, 1999.
143. Jahnke W, Perez LB, Paris CG, Strauss A, Ferdrich G, Nalin CM, Second-site NMR screening with a spin-labeled first ligand, *J. Am. Chem. Soc.*, 122: 7394–7395, 2000.
144. Jahnke W, Spin labels as a tool to identify and characterize protein-ligand interactions by NMR spectroscopy, *ChemBioChem.*, 3:167–173, 2002.

13 Automation in Combinatorial Chemistry

Valery V. Antonenko

CONTENTS

I. INTRODUCTION

An eternal dream of many organic and medicinal chemists is to automate the tedious process of synthesis and liberate themselves for more creative work. In combinatorial chemistry, due to the number of chemical manipulations necessary to synthesize a library of compounds, automation is an unavoidable exigency. The field is too young to benefit widely from commercially available equipment that would meet all the diverse needs. Major instrumentation companies are only beginning to pay attention to this market. Most of the progress in automation was originally achieved by companies that pioneered the field and in many cases is not well publicized. Later, many commercial products became available from both well established and newly founded corporations. To escape the burden of equipment production and technical support, some drug discovery and even technology-oriented companies are forming alliances with instrumentation firms. As a result, a lot of options are available for scientists interested in automation of combinatorial chemistry. This review will attempt to classify and describe major directions in the field, with focus on commercially available products.

II. GENERAL CONSIDERATIONS

Combinatorial chemistry can be carried out in solution or on solid support. Most solution combinatorial chemistries are typically limited to one-step reactions, whereas solid-phase chemistries often involve multistep processes that include resin manipulation, washing, drying, cleavage of the products from the resin, etc.

These and other specifics of solution and solid-phase chemistries impose different demands on the instrumentation design for the two approaches. Depending on the goal (synthesis of primary libraries for lead discovery or synthesis of individual compounds for lead optimization, for example), drastically different numbers and amounts of compounds are sought. Most of the time, when the size of a library is larger than 10^4, the split-and-pull method [1, 2] is applied. Each bead of polymer support contains a single compound, which can be released from the polymer and screened for biological activity individually or as part of a larger pool of compounds. Some companies actively pursue various encoding strategies to determine the structure of an active hit in a single bead approach [3–11]. Therefore, automation should be compatible with these encoding technologies. Lead optimization, on the other hand, is more efficient via parallel synthesis of individual compounds, when knowledge of the quality and quantity of each compound submitted for bioassay is critical for interpreting the data. Nearly all the instrumentation for parallel synthesis can be used in the split-and-pull approach with manual intervention. Instrumentation for automated split-and-pull solid-phase synthesis in a closed system is protected by a patent [12] issued to the Affimax Research Institute/GlaxoWellcome, and is not commercially available.

Another important aspect of combinatorial drug discovery is chemistry development. It is crucial to discover experimental conditions that ensure high purity and yield of each member of the desired library. Combination of different solid supports, linkers, solvents, temperatures, reagents, etc. can easily lead to multidimensional experiments with hundreds of data points to be collected. Automation is as important for this stage of combinatorial drug discovery as it is for all other aspects.

It is important to keep in mind that not all stages of combinatorial synthesis can be equally easily automated. Gathering, weighing out, and making solutions of dozens of building blocks for a combinatorial synthesis, arranging them on a synthesizer, entering locations of each into a software, and some other tasks remain largely manual and time-consuming operations. Some instruments do not offer automatic cleavage of synthesized compounds from polymer support. Conducting the cleavage manually, evaporating cleavage solution, distributing compounds for analysis, and bioassay can be more time and labor consuming than the synthesis itself. Keeping track of all synthesized compounds, their structures, molecular weights, analytical and biological data is another challenging part of the process. Unfortunately, there is no product on the market which allows automation of all the above operations.

At the beginning, combinatorial chemistry was a somewhat separate field, with its own views on automation, dedicated staff for library synthesis, operation, and maintenance of the synthesis equipment, and, sometimes, high tolerance to the price, reliability, usefulness, and sophistication of the instrumentation. Typically, a dedicated user was not concerned very much with all of these issues. Recently, however, combinatorial chemistry has become more and more an integral part of the drug discovery process in the pharmaceutical industry, with its high standards for instrumentation. Traditional medicinal chemists are in need of practical and easy-to-understand and easy-to-operate devices for combinatorial synthesis, especially for parallel synthesis. They would like to be able to use the instrumentation as freely and easily as they use chemical glassware, to apply all of the traditional techniques or organic chemistry to combinatorial synthesis. Moreover, it is not possible to automate everything (for example, it is difficult to automate a Dean-Stark distillation or the addition of powdered reagents in a parallel synthesis mode) at the present level of technological development and resources allocated to the field. Accordingly, there has been a noticeable recent shift in commercial instrumentation toward meeting the most practical and concurrent needs. Instrumentation is becoming less expensive, not overengineered, more reliable, and more practical.

III. AUTOMATION FOR SOLUTION COMBINATORIAL CHEMISTRY

Most instrumentation described in the following sections is designed for and is more suitable for solution-phase chemistry. However, some users, who are willing to carry out a lot of manual operations, find them useful for solid-phase experiments as well. It is also possible to conduct some chemistry optimization experiments on these instruments.

A. HP 7686 Solution-Phase Synthesizer by Hewlett Packard

The first published report on the utilization of the HP 7686 Solution-Phase Synthesizer (Figure 13.1) for lead optimization of corticotropin-releasing factor receptor antagonists appeared in 1996 [13]. Scientists from Neurocrine

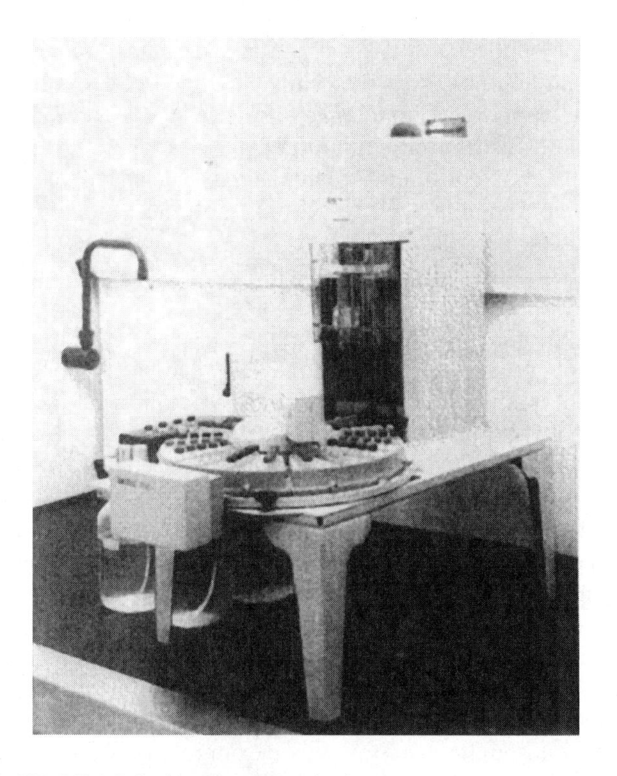

FIGURE 13.1 HP 7686 Solution-Phase Synthesizer.

Biosciences were able to use the synthesizer to prepare 350 individual analogs of triazine **1** in about 4 weeks, using the reaction shown in Scheme 1 [13].

Almost two-thirds of the synthesized analogs of triazine **1** were isolated in 70–95% purity and in milligram quantities. Preparation of amines **2** in an additional reaction sequence on the HP 7686 synthesizer demonstrates its application for automating chemical reactions requiring inert atmosphere and dry solvents.

The synthesizer fits easily into a standard chemical hood, leaving plenty of free space for other regular chemical activities. Its carousel hosts up to 100 vials, which can hold reagents, reaction intermediates, and/or HP solid-phase extraction cartridges (100- or 300-mg size) for purification of synthesized compounds. It has a heating station capable of maintaining temperatures in the range 30–120°C (in 1°C increments) for a time specified by the operator. The content of the vials can be easily transferred in a wide dispensing-volume range (0.001–1.6 mL) to any location on the synthesizer with high precision (±3% for 0.01 mL, ±0.5% for 0.1 mL, and ±0.2% for 1 mL of delivered volume).

Reagent transfers are carried out by syringes (0.5, 1.0, 2.5, 10.0, or 25.0 mL, 2.5 mL standard; reagent carryover on the instrument is minimized by an internal/external needle wash system) with no tubing lines to fill out, and therefore no wasted material. This feature is especially useful when expensive and/or rare reagents are used in the synthesis. Many other commercially available synthesizers are almost useless in a fully automated mode, when it comes to reagent consumption issues, and call for manual addition of valuable building blocks as the only alternative. The HP 7686 Solution-Phase Synthesizer is capable of carrying out both liquid–liquid and solid-phase extractions. It is equipped with a bar code reader/mixer, which offers three preset speeds for automatically mixing vial contents. Very few instruments offer automatic evaporation of solvents and solvent exchange. The HP 7686 offers these features. Evaporation is achieved by blowing a

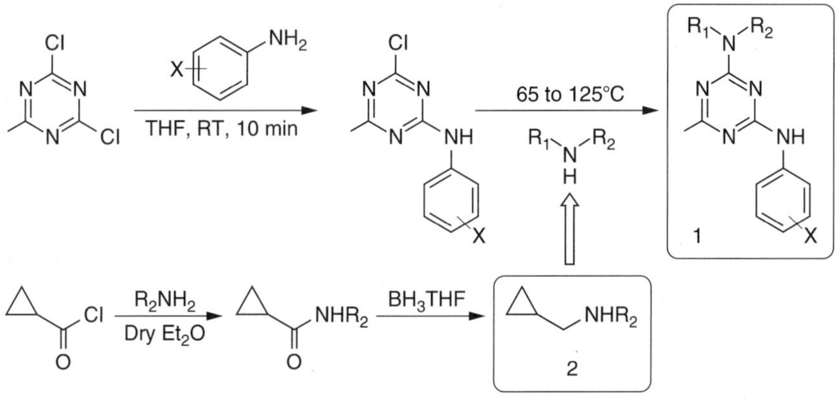

SCHEME 1 Synthesis of triazine analogs on the HP 7686 Solution-Phase Synthesizer.

stream of an inert gas into a vial while the needle tracks the surface of the liquid as it descends. Evaporation can be carried out with or without heating.

The synthesizer can produce as many as 200 compounds a week through single-step automated reactions that require minor cleanup routines; or about 60 compounds a week through multistep reaction sequences that include labor-intensive liquid–liquid extractions or solid-phase cleanup.

IV. MANUAL AND SEMIAUTOMATED DEVICES FOR SOLUTION-PHASE CHEMISTRY

A. MultiReactor™ by RoboSynthon Inc.

RoboSynthon Inc. manufactures a manual synthesizer, MultiReactor (Figure 13.2). The synthesizer consists of a reactor block accommodating up to 24 reaction vessels, glassware, control unit, and software to control simultaneous reactions. The reaction block is relatively small (582 mm × 160 mm × 132 mm) and has 24 locations for glass reaction tubes arranged in two parallel rows. The maximum volume of each reaction tube is 20 mL, allowing for the synthesis of compounds in gram quantities. The reaction block maintains identical temperatures in all reaction vessels (−60 to 200°C). Unfortunately, temperatures below ambient can only be achieved using an external liquid cooling bath, which can be bulky and expensive. Each tube is stirred by means of individual motors situated beneath each of the reaction vessel chambers. A magnetic stir bar is placed in each reaction vessel. The MultiReactor control module allows controlling

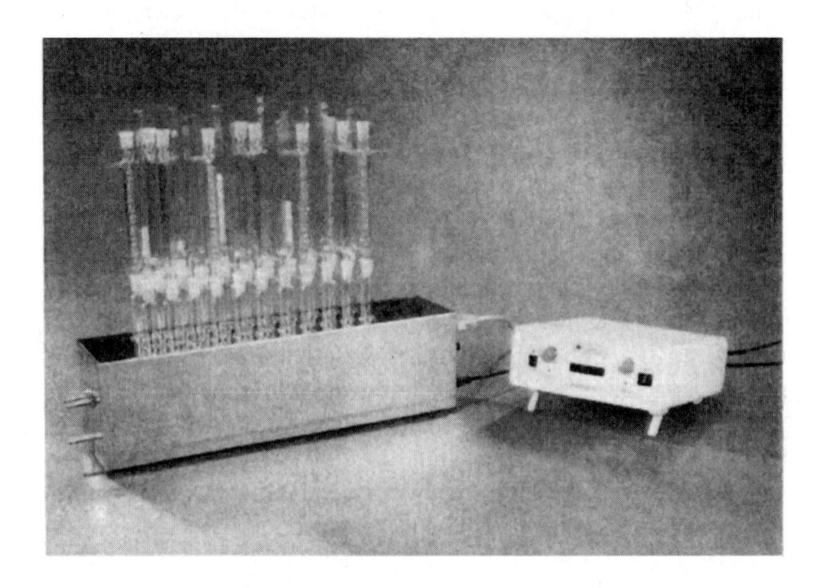

FIGURE 13.2 MultiReactor™ by RoboSynthon Inc.

reaction temperatures and mixing rates by simply dialing chosen values on the instrument or via a Windows^{TM}-based program. In PC control mode, temperatures can be adjusted according to a custom profile. This allows one to perform special stirring programs for solid-phase reactions or viscous suspensions. Each reaction vessel is supplied with an air condenser. Water-cooled glass condensers are supplied as an option. The MultiReactor can be used to produce more than 200 compounds per month in 100- to 200-mg quantities using solution-phase chemistry [14].

B. STEM Reacto-Stations™ by STEM Corporation

STEM Corporation developed its Reaction-Stations, presented in Figure 13.3 (heated: RS 1000, RS 2000, RS 25000, RS 5000; chilled: RS 1050, RS 2450, RS 2550, RS 5050) "in conjunction with two leading pharmaceutical companies." Reaction-Stations are designed for use with automated sampler systems and robots as well as in benchtop applications. Reaction blocks have, depending on the specific model, 10, 25, or 50 vessel shafts for the accommodation of reaction tubes. Temperature stability across all positions is better than ±0.5°C. A magnetic stirrer is installed under each tube position to ensure that the samples are thoroughly mixed in a closed tube if required. Purposely designed stir magnets ensure the maintenance of maximum coupling between the stir bar in the sample and the powerful drive motors. Even very viscous samples can be mixed efficiently using this technology. The powerful magnetic stirrer also warrants a uniform

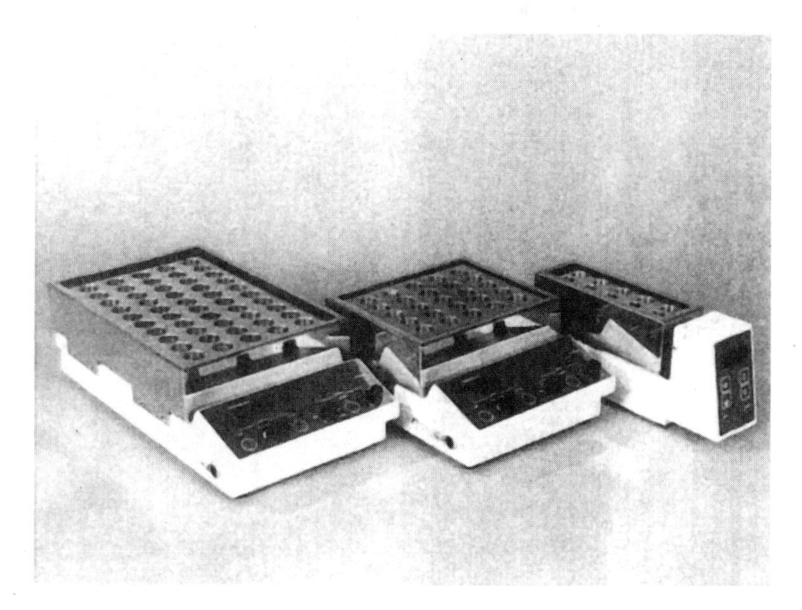

FIGURE 13.3 STEM Reacto-Stations^{TM} by STEM Corporation.

temperature throughout the sample in tubes up to 150 mm high. Moreover, all models have a "soft start" or ramping feature to allow slow buildup to the set speed. Selection and adjustment of temperature (ambient +5 to 150°C), stirring speed (400–2000 rpm), and soft start rate can be done on a keypad.

These parameters can also be controlled by external software as part of a fully automated system. The standard hole diameter of 25 mm can be reduced using adapter sleeves to accommodate 16- or 20-mm tube sizes. The company accepts requests for custom sizes as well. STEM Corporation supplies flat bottomed tubes to ensure efficient stirring. These tubes can be fitted with pierceable caps for easy access by autosamplers. Reflux action is accommodated either by simply optimizing hole depth to use the exposed length of tube for natural reflux action or by using forced air and water reflux modules. The latter mount on top of the heated reaction stations (Figure 13.4) and form a cooled environment for the unheated section of the tubes. To minimize exposure of samples to moisture or oxygen, both forced-air and water-cooled modules are fitted with a removable lid and an inlet port for purging with inert gas. The lid also contains holes, corresponding to the position of the tubes underneath, which are normally covered on the inside by replaceable silicone gaskets. These can be pierced to introduce or withdraw chemicals under inert atmosphere. Chilled reaction station units have similar

FIGURE 13.4 RS 1000 with air-cooled reflux module.

features, with one difference: the temperature range is −30 to 70°C (with a supplied chiller unit).

V. AUTOMATION FOR SOLID-PHASE COMBINATORIAL CHEMISTRY

A. AUTOMATED RAM™ SYNTHESIZER BY BOHDAN AUTOMATION INC.

The Bohdan Automation Inc. RAM synthesizer (Figure 13.5) is designed to fully automate the entire process of solution- and solid-phase syntheses: preparation of reagents (weighing, adding solvents, and mixing), transfer of chemicals to reaction vessels, quenching solution-phase reactions, cleavage of products from solid supports, liquid–liquid work-up, distribution of synthesized products to a variety of vials or microtiter plates for further analytical characterization, biological screening, and storage [15, 16].

The system has a modular design. The Automatic Weigh Station allows an operator to place reagents into preweighed vials. The station then automatically records the weight of the reagents and adds solvent to achieve specified molar concentration. The specific amount of solvent is calculated by the software. The Vortex Mixer is used automatically by the system to solubilize dry reagents. The mixer is also used for liquid–liquid extractions.

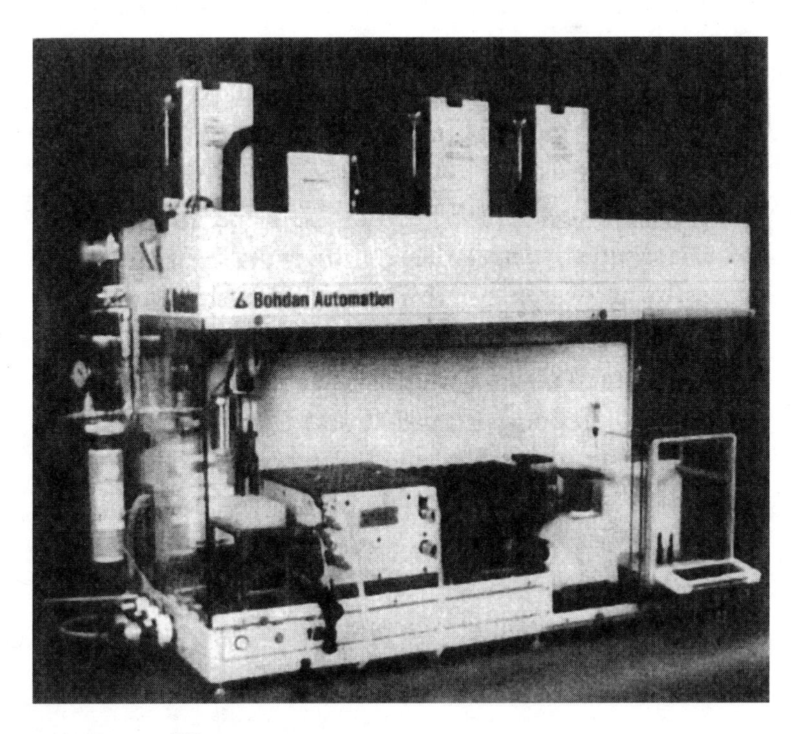

FIGURE 13.5 RAM™ Synthesizer.

A multifunction robotic arm automatically moves reagent vials to software-specified locations, significantly diminishing the possibility of human error. Liquid delivery is accomplished by two cannulas; one is used for aqueous solutions and the other is for organic solvents. Each cannula consists of three concentric tubes. The inner tube is designed to pierce the septum and to add or remove liquid. The middle tube delivers an inert gas into the vials or reaction vessels. The outer tube vents the contents of the vials or reaction vessels to allow for displacement of atmospheric gases. The Reaction Block contains 48 glass reaction vessels in a 6×8 array and accommodates chemistries requiring temperature control (−40 to 150°C), refluxing, elevated pressure (up to 2 atm), and inert atmosphere. After the addition of reagents, the Reaction Block can be transferred to the separate heat/cool/shake station, freeing the workstation for a new application.

Productivity of the instrument is 100–250 compounds per week for solution-phase chemistries and over 50 compounds for solid-phase syntheses.

B. Nautilus™ 2400 Synthesizer by Argonaut Technologies Inc.

The Nautilus 2400 Synthesizer [16, 17] (Figure 13.6) is engineered as a closed-to-the-outside-atmosphere fluid delivery system. It is the first commercially available synthesizer of organic compounds that is not based on a robotic system.

The synthesizer consists of six major modules: the Reagent Solvent Enclosure, the Reaction Vessel Enclosure, an autosampler, a fraction collector, a gas chiller, and a computer. It fits into an 8-ft chemical fume hood.

The Reaction Vessel Enclosure stores solvents and common reagents in 250-mL to 4-L bottles. Reagent delivery is mediated by a valve system. The accuracy of delivery is under the control of optical sensors. The auto-sampler handles up to 176 different reagents. The Reaction Vessel Enclosure contains 24 reaction vessels, which are organized in three banks of eight.

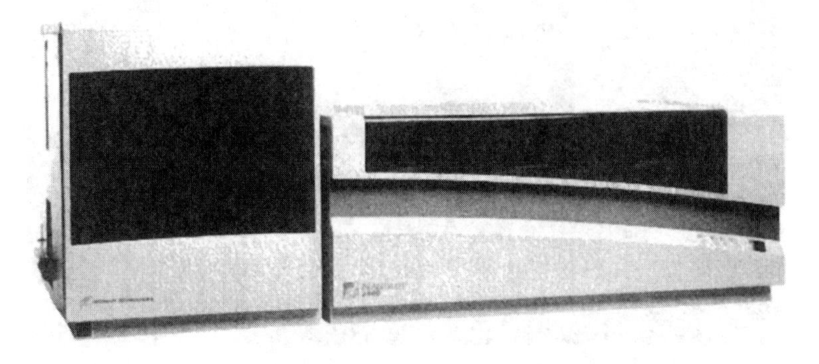

FIGURE 13.6 Nautilus™ 2400 Synthesizer.

Reaction vessels come in three different sizes: 8, 15, and 23 mL. The latter can be used for carrying out a synthesis with up to 1.2 g of resin. The minimum amount of a reagent that can be added automatically to each reaction vessel is 200 μL. The temperature of each reaction vessel can be individually controlled in the range of −40 to 150°C by a combination of chilled inert gas and a computer-controlled heating mantle. The content of all the reaction vessels is visible at all times (it is often important to observe the behavior of a specific reaction visually). Efficient mixing is achieved by a rocking agitator that inverts the reaction tubes by 217°. Reagents and solvents are delivered to and withdrawn from the reaction vessels through a Teflon tube reaching the bottom of the vessels. The tube is equipped with a disposable 30-μm frit.

Synthesized compounds can be cleaved from the resin and delivered to any of the positions of the integrated fraction collector. Development of different challenging chemistries on the Nautilus 2400 Synthesizer has been recently reviewed [18].

C. Trident 4192 Library Synthesizer by Argonaut Technologies Inc.

Argonaut Technologies developed a system capable of simultaneous synthesis of 192 compounds. All 192 glass reaction vessels (4 mL each) are organized into four modular reactions cassettes. Each cassette accommodates 48 reaction vessels in a 6×8 array. The cap of each reaction vessel is made of Teflon and serves as part of a uniquely designed valve system. The synthesizer is capable of delivering eight different reagents serially to eight reaction vessels using one delivery manifold. Reagents can be delivered in 10-μL increments, with a minimum deliverable volume 50 μL. The temperature of each cassette can be controlled individually on the synthesizer (−40 to 150°C). Efficient mixing is achieved through a single-axis, variable-speed oscillation. The cassettes can also be easily removed from the synthesizer and placed on a separate agitation-thermal unit, thus freeing the synthesizer for delivering reagents to another set of cassettes. After the completion of the reaction, cassettes can be returned to the synthesizer for conducting washing cycles and further chemical transformations. The system is equipped with a device designed for manual delivery of the reagents to the cassettes under inert atmosphere. The synthesizer is useful for both solution and solid-phase chemistries. The cleavage of synthesized compounds from the solid support is carried out automatically. The final products are distributed by a fraction collector to a specified vessel by an operator. The instrument will cost $250,000.00 in the United States.

D. SYRO II Synthesizer by MultiSynTech GmbH

MultiSynTech GmbH, a German-based company, introduced the multiple organic synthesizer SYRO a few years ago. It evolved, as have some other commercial instruments for synthesis of organic compounds, from a peptide synthesizer [19]. The SYRO II is based on a robotic liquid dispenser system with two dispensing arms. All solvents and reagents used for synthesis are stored under an inert atmosphere. A standard monomer rack contains 32 vessels (50 mL each). Three different reaction blocks are available. They contain 40 removable 10-mL glass or polypropylene reaction vessels (syringes) with a glass or a PTFE frit, 60 removable 5-mL reaction vessels, or 96 removable 2-mL reaction vessels. A stirring bar is placed into each vessel. It floats above the bottom of the syringes, preserving the resin from grinding during the mixing, which takes place simultaneously in all vessels of the block. Reactions can be carried out at a wide temperature range (−60 to 150°C).

E. ADVANCED CHEMTECH SYNTHESIZERS

Advanced ChemTech was the first company to release a combinatorial library synthesizer to the market. Currently the company offers several instruments for combinatorial chemistry. All fully automated models are based on a robotic system.

1. Model 496 Multiple Organic Synthesizer

The model 496 Multiple Organic Synthesizer (Figure 13.7) has two robotic arms to deliver reagents and solvents to the Teflon reaction block. Standard monomer racks accommodate 10, 36, or 128 vessels (custom configurations are easy to adapt). The reagents in the racks are kept under an inert atmosphere. To pick up a reagent, the robotic arms pierce the septum at the top of the rack. In addition, six 100-mL reservoirs are available to store common reagents.

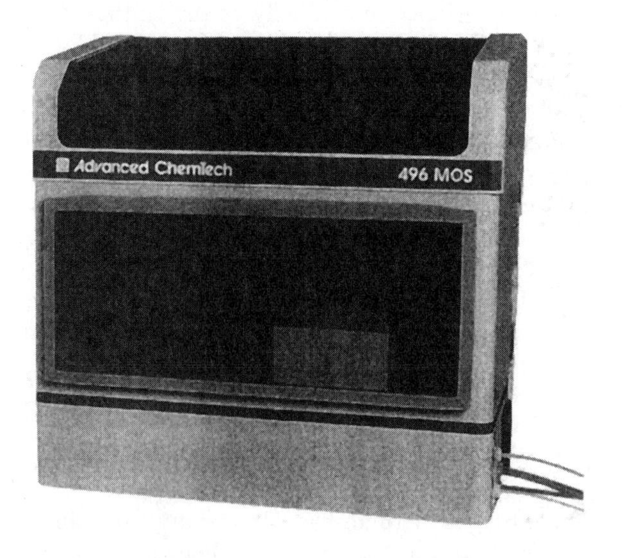

FIGURE 13.7 Model 496 Multiple Organic Synthesizer.

The content of the reservoirs is protected from the atmosphere by a specially designed double septum flushed with an inert gas. The needles of the robotic arms are washed inside and outside in a needle wash station located on the instrument. The tip of the needle is designed to spray liquids evenly and efficiently in all directions. This allows for efficient washing of the walls of each reaction well on the reaction block. The block has 96 reaction vessels with Teflon frits at the bottom of each. The vessels are organized in an 8 × 12 array. The top of the block has a gasket which separates the contents from the atmosphere. Reaction mixtures in the block are agitated by a vortex mixer under an inert gas. Reaction vessels can be filtered simultaneously in groups of 24. The filtration process is assisted by an inert gas. Published temperature range for the synthesizer is −70 to 150°C. Cleavage from the solid support can be carried out automatically. A collection tray is placed under the reaction block. After the cleavage, the reaction vessels are cleaved into individual beakers.

2. Model 440 Multiple Organic Synthesizer

The model 440 Multiple Organic Synthesizer is smaller than the model 496 (24 in. wide × 25 in. deep × 36 in. high). It is otherwise very similar to the model 496. The major difference is the reaction block, which has only 40 reaction vessels. However, the volume of each is 8 mL.

3. Model 384 High Throughput Synthesizer

The model 384 High Throughput Synthesizer is essentially a large version of the model 496. It is designed for parallel synthesis of 384 compounds. It has four reaction blocks with 96 reaction vessels in each. Most of the other parameters are identical to those of the 496.

F. AccuTag™ 100 Combinatorial Chemistry System by IRORI

The AccuTag 100 system is based on radiofrequency (RF) tags. The RF tag is a small (8 mm × 1 mm) chip. It consists of a memory with a unique 40-bit alphanumeric code (2^{40} unique codes are available), a circuit with which radiofrequency energy is converted into electrical energy, and an antenna that sends and receives radiofrequency signals. These signals can be initiated and read by a transceiver, which is connected to a computer. Any information about structure or about a sequence of chemical manipulations necessary to obtain a specific compound can be easily associated by appropriate software with the unique signal of each tag [20, 21]. Each RF tag is encapsulated in a thick-walled glass shell and placed into a synthesis unit called a microreactor. IRORI offers several miniature reactors, including MicroKans® and Micro-Tubes® (Figure 13.8). The MicroKan is a small cylindrical container with mesh walls. The container has a 330-μL internal volume. It holds a RF tag and up to 30 mg of most commercially available resins. MicroTube reactors are polypropylene- or fluoropolymer-based tubes with a functionalized polystyrene-grafted surface and a completely enclosed RF tag. Any number

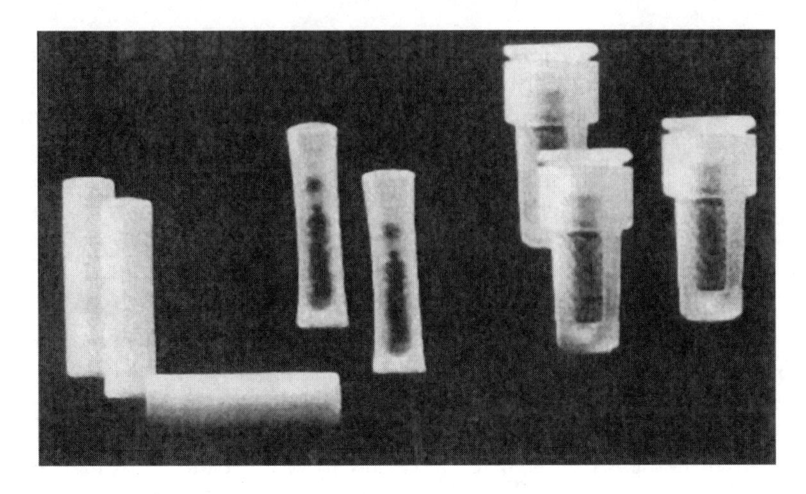

FIGURE 13.8 MicroKan® and MicroTube®.

of microreactors can be combined in one reaction and then redistributed with microreactors from other reactions for further chemical manipulations. By simply placing a microreactor above the reader, an operator allows software to recognize the code and receive instructions as to the location of the microreactor for the next chemical transformation. One of the major advantages of this technology is the ability to use conventional glassware while being more comprehensive than any of the other contemporary automated synthesizers.

Manual sorting of microreactors can become tedious when a large number (more than 1000–2000) of them are used. IRORI's AutoSort™-10K Microreactor Sorting System (Figure 13.9) [22, 23]) is designed to solve this

FIGURE 13.9 AutoSort™-10K Microreactor Sorting System.

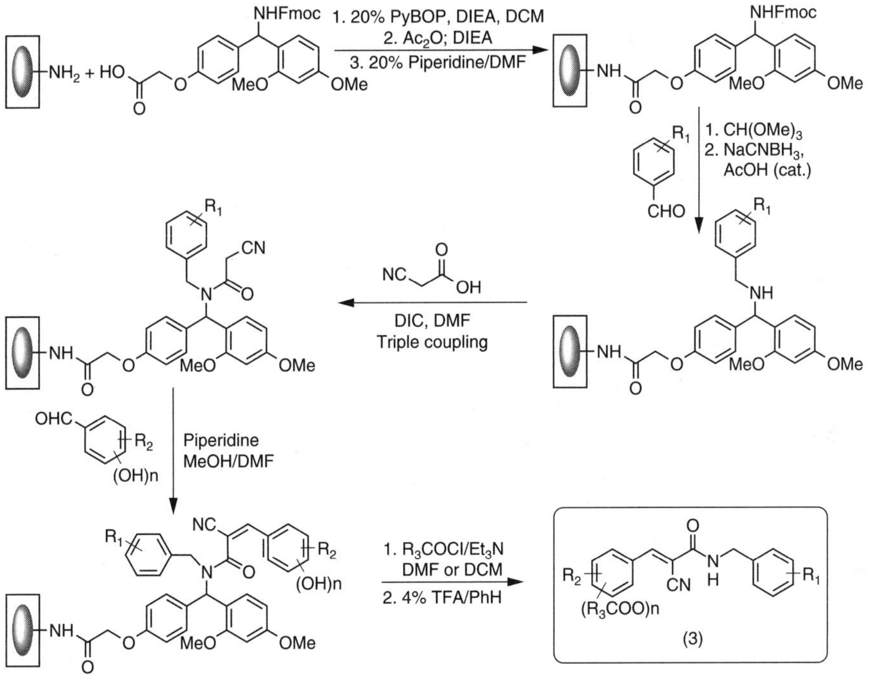

SCHEME 2 A combinatorial synthesis of 432 tyrphostins.

problem by fully automating the sorting process. The system accommodates up to 10,000 microreactors, which can be distributed between chemical steps into 48 different containers. The rate of distribution is 1000 per hour. The device is also very useful for sorting MicroKans and MicroTubes into microreactor carriers for placement into the AccuCleave-96 cleavage station. Twelve 8×12 microreactor carriers can be used simultaneously on the AutoSort-10K system.

Synthesis of a 432-member library ($18 \times 8 \times 3$ array) or tyrphostins (3) (Scheme 2), using IRORI's technology, has been reported [23].

G. SPOC Synthesizer by ABO-TECAN

TECAN introduced the CombiTech Synthesizer in 1996. Recently it was announced [24] that support for this product, including sales, service, and distribution, will be integrated into Perkin-Elmer's global structure. The instrument is based on TECAN's Genesis liquid-handler platform. One of the major advantages of this instrument over other robotic devices is its ability to use, simultaneously, four robotic arms for dispensing. Another four robotic arms are used for aspiration and removal of the reactants and wash solvents to waste. The Teflon reaction block has 48 threaded glass round-bottom reaction vessels (10 mL each). They are screwed onto the block.

The neck of each reaction vessel is connected to two channels, which are sealed with two dual septa. A flow of an inert gas between the septa creates an inert atmosphere over each reaction chamber. A glass tube is inserted through one of the channels. The tube is fritted at the end and reaches the bottom of the flask. Reagents and solvents are removed from the reaction vessels through the tube by the robotic arms. The other channel is used for delivering reagents to the flasks. The Teflon part of the block has jackets for recirculating coolant and serves as an efficient condenser. It is capable of creating an 80–90°C temperature drop within the reflux zone compared to the reaction chamber temperature. The reaction block can be agitated on an on-board shaking platform. It can also be removed from the synthesizer to an off-line heating/cooling/refluxing mixing station, an orbital shaker coupled to a heating bath (up to 150°C), and a recirculating chiller (−30°C). After the synthesis, a cleaving agent can be added to the reaction vessels in the same manner as other compounds. The cleaved products are collected by the robotic arms and transferred to an appropriate set of collection vessels.

H. SOPHAS M Solid-Phase Synthesizer by Zinsser Analytic

SOPHAS M (Figure 13.10) is an automatic modular solid-phase synthesizer based on a robotic system. Synthesis can be carried out in a variety of reaction vessels, such as 96-well microtiter plates, tubes, or vials. The vessels are mowed on the 1- or 1.2-m-length workbench in aluminum carriers (12 mm × 86 mm) by a robotic arm. The content of the reaction vessels is isolated from the atmosphere by a pierceable double seal. There are four independent pipetting probes on the synthesizer. Each probe has three independent channels. The channels allow the synthesizer to simultaneously aspirate and add washing solvents and nitrogen.

FIGURE 13.10 SOPHAS M Solid-Phase Synthesizer.

The excess reagents and washing solvents are removed from the reaction vessels by filtration from the top by the pipetting probes without any loss of resin. The synthesizer can deliver 6 or 11 different system liquids. Chemical reactions can be carried out within the temperature range of −60 to 150°C (one heating station for two carriers and a cooling station for one carrier). Mechanical agitation is available (integrated vortex mixers in three positions) at all times during the synthesis at any position on the workbench.

It is difficult to find an instrumentation company that is not involved in the development of automation for combinatorial chemistry. Zymark Corporation, Zeneca, and SCITEC Inc. offer comprehensive solutions for automation of solution- and solid-phase chemistries [25, 26]. Ontogen, Selectide Corporation, and others have published about their efforts in automation [27–30].

VI. MANUAL AND SEMIAUTOMATED DEVICES FOR SOLID-PHASE CHEMISTRY

Recently there has been an obvious trend toward simplifying instruments for combinatorial synthesis. This is driven by several factors. Most of the fully automated synthesizers are expensive. Complexity of instrumentation sometimes has a negative effect on reliability, producing a need for dedicated operators and labor-intensive maintenance. As a reflection of the complexity of the instruments, the software is also complicated and is not intuitive in most cases. Chemists who do not use instrumentation on an everyday basis tend to have difficulties operating synthesizers. Often it is difficult to find space in laboratories to accommodate rather large instruments. Most medicinal and organic chemists would like to use devices that are simple to operate, can improve their productivity, and make their work less tedious at the same time. Some of the instrumentation firms (e.g. Advanced ChemTech, MultiSynTech, and Argonaut Technologies) have responded to the need by offering simpler, easier to use, more practical, and less expensive versions of the fully automated instruments.

A. ReacTech Synthesizer by Advanced ChemTech

The ReacTech synthesizer has a Teflon reaction block containing 40 separate reaction vessels. The volume of each of the reaction vessels is 8 mL. The reactors are sealed by a Teflon-coated septum and are kept under inert atmosphere during the synthesis. They are organized into four individually controlled reactor chambers. This provides independent control of reaction strategies for every 10-reaction vessels. Notably, there is no robotic handler for reagent delivery. Reagents can be delivered manually with a syringe. The synthesizer also has a 50-mL measuring vessel with a photocell. The machine allows users to specify the amount of up to three common reagents or solvents to be delivered to all wells within a chamber simultaneously. The heating and cooling systems allow a wide temperature range (−70 to 150°C). Variable-speed vortex mixing, like other synthesis parameters, is controlled by a computer.

B. SYRO II O. S. Manual Block by MultiSynTech GmbH

MultiSynTech produces a manual version of its SYRO II synthesizer, SYRO II O. S. Manual Block. It does not have a robotic delivery system or a computer. Reaction blocks can be used just like with the SYRO II synthesizer. Reagent delivery is manual. All reaction parameters can be set up by a separate control unit. The Manual Block is substantially less expensive and less space consuming than its robotic equivalent.

C. Quest 210 by Argonaut Technologies

The Quest 210 synthesizer (Figure 13.11) fits into a regular fume hood. It has a small footprint of 18 in. and can be placed on a rotating table, providing easy access to both sides of the instrument. Each side has a set of 10 reaction vessels arranged in a parallel manner. The vessels are made from a transparent fluoropolymer and are disposable. They are provided in 5- and 10-mL sizes. Liquid and solid reagents are added manually through the tops of the vessels, which have a frit at the bottom. The temperature of each group of 10 reaction vessels can be individually controlled between −40 and 150°C. A magnetic bar is placed in each vessel. Agitation is achieved by vertical

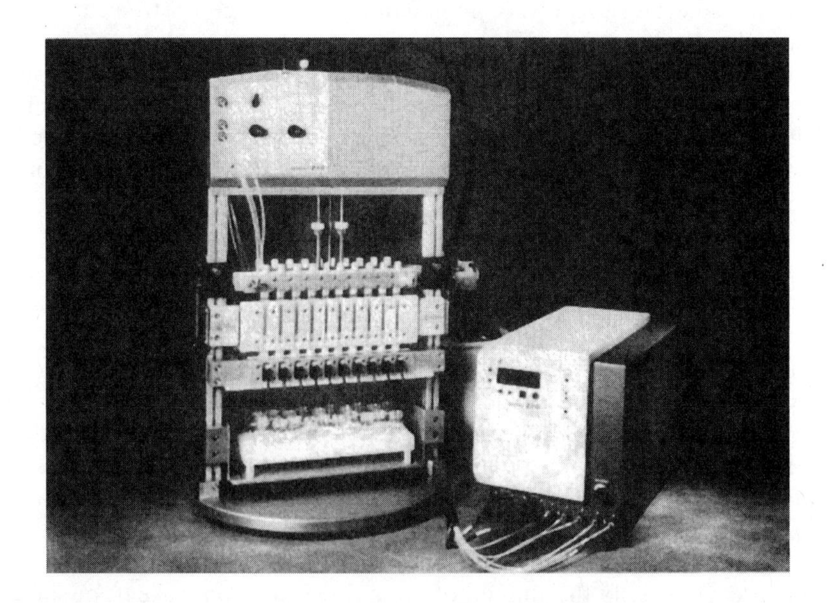

FIGURE 13.11 Quest 210 Synthesizer.

oscillation (moving the bar up and down the vessel). Reactions are carried out under the positive pressure of an inert gas. The synthesizer can be used for both solution- and solid-phase reactions. Liquid–liquid extractions can also be carried out. The content of each vessel can be collected or sent to waste individually or simultaneously with the others. Reaction parameters are controlled by a simple controller, not a computer with a synthesis software. Multiple wash processes are automated with an optional automated solvent wash accessory. After cleavage from solid support or after a solution synthesis or liquid–liquid extraction, compounds are collected into a collection rack, which accommodates 20- and 40-mL scintillation vials.

D. APOS 1200 Synthesizer by Rapp Polymere GmbH

The APOS 1200 is a small synthesizer (Figure 13.12) designed for the synthesis of 12 individual compounds. The synthesizer is small and can be used as a module for building a larger system. The reaction vessels are glass columns with glass frits at the bottom and a ball joint at the top. Each vessel can hold up to 3 mL of solvent and can be used with 200 mg of resin. The reactors are organized in a block in two rows of six (Figure 13.13). The temperature of this block can be controlled between −60 and 150°C. Reagents are delivered manually through the tops of the reactors or by a robotic system. The top joints of all reactors can be connected simultaneously to 12 condensers organized in a cooling block.

The block with the reaction vessels can align automatically with any two positions on a base unit (Figure 13.14). This alignment is carried out automatically by pneumatic actuators on a transfer unit. One position of the base unit is a gas-flow position for the bubbling of an inert gas (a gas can be used as

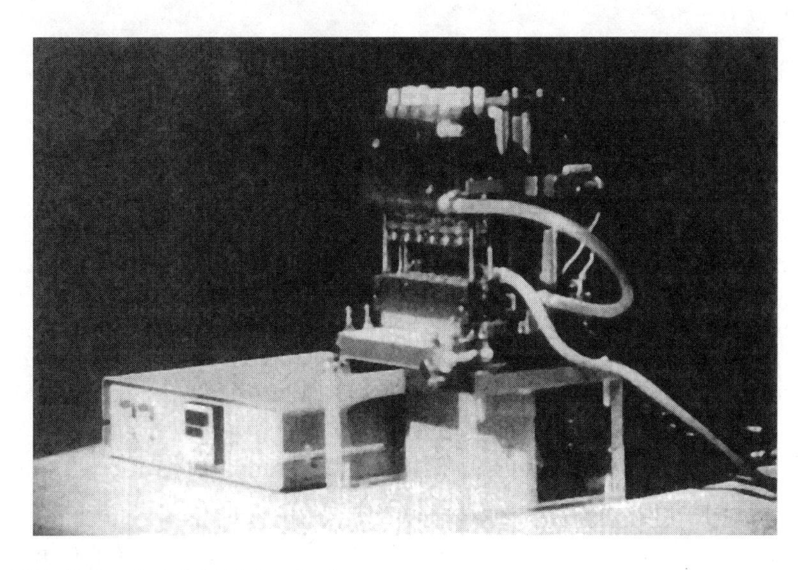

FIGURE 13.12 APOS 1200 Synthesizer.

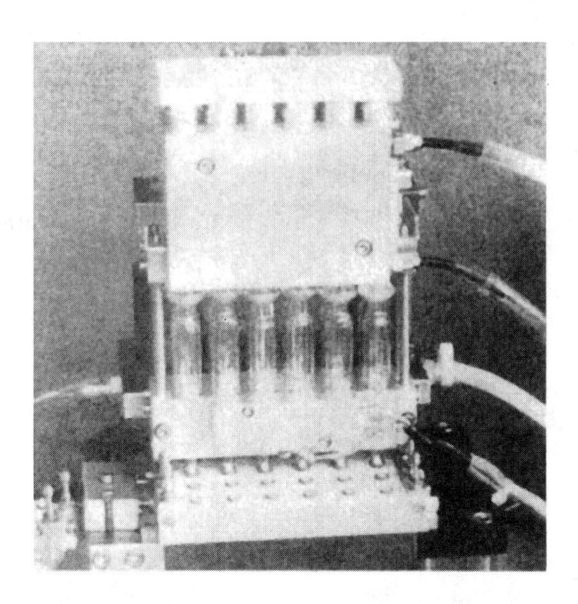

FIGURE 13.13 APOS 1200 Reaction Block.

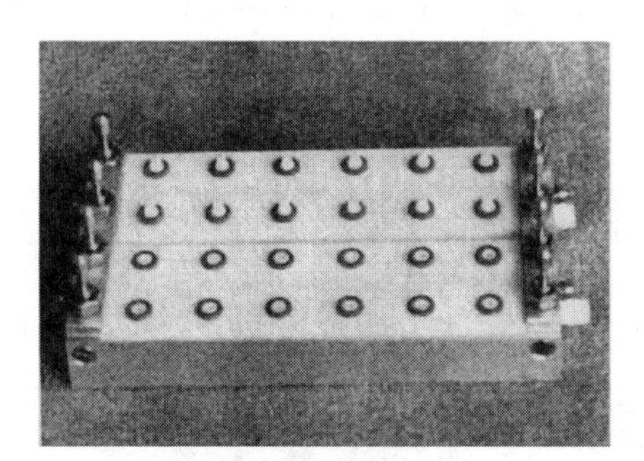

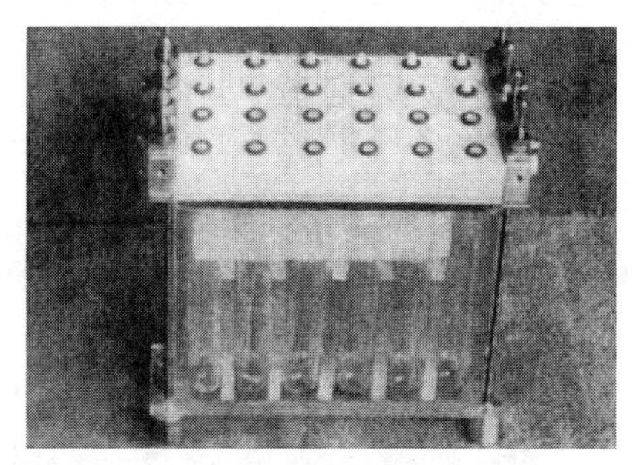

FIGURE 13.14 APOS 1200 Base Unit.

a reactant as well) through the reactors in the course of the synthesis, or during the delivery of the reagents. The other position is for filtering and washing. There are also two types of base units: the reaction base and the collection base. The latter is used to collect synthesized products in twelve 10- to 15-mL tubes or vials. The temperature of the reactors and alternation between the two positions on the base units is determined by a controller, which can be operated manually or by downloading parameters from a computer.

E. Diversomer Technology by Parke-Davis Pharmaceutical Research Division

The Diversomer 8-pin Synthesizer is one of the first examples of a technology developed entirely by researchers in a pharmaceutical company and successfully commercialized [31–35]. The major component of the synthesizer is a PIN, which is simply a piece of a glass tubing, ending with a glass frit. Two reactor sizes, for 100 mg and for 800 mg of solid support, are available. Eight PINs are arranged in a block in a 4×2 format. The frit contains all resin and allows reactants and washing solvents to penetrate in and out of the PIN. The glass construction of the reactors allows for an unlimited choice of reagents and a wide range of temperatures (−78 to 200°C) for the syntheses. The top portions of the PINs are enclosed in a manifold, which separates the openings of the PINs from the atmosphere by a gasket. The top portions also serve as condensers when a chilled gas is circulated through the manifold. Conventional laboratory equipment is sufficient for operating the synthesizer. Several units can be easily combined together on a platform of a robotic liquid handler to automate the synthesis.

VII. SYSTEMS BASED ON 96-WELL MICROTITER FORMAT

The 96-well microtiter format has been a common platform for high-throughput screening for years. It also offers many advantages as a platform for parallel synthesis. There are multiple liquid-handling devices (multichannel pipettes, robotic systems, etc.), which significantly simplify and accelerate the process of reagent delivery to the wells on the plate. Moreover, all 96 compounds can be handled as one synthesis entity. Their structures are spatially encoded by the location on the plate. Contemporary analytical instrumentation, such as mass spectrometers and HPLC systems, can also easily handle compounds in the 96-well microtiter format. Combined with the commonly used 96-well-based systems for screening, they eliminate the need for time-consuming redistributing and relabeling of a large number of synthesized compounds for analytical and biological characterization.

A. 96-Deep-Well-Based Synthesis Device by Sphinx Pharmaceuticals

Researchers from Sphinx Pharmaceuticals were the first to report [36, 37] an apparatus based on the 96-well microtiter format for parallel solid-phase synthesis on resin. This apparatus utilizes a polypropylene 96-deep-well plate. Each well has a polyethylene frit and a small hole in the bottom. The plate is placed in a clamping device, which presses the bottoms of the wells against a gasket sealing the holes. The reagents and resin can be loaded to the plate, and the wells can be sealed with eight-well strip caps. The entire plate can be agitated manually or on an orbital shaker. Upon completion of a reaction step, the plate is removed from the clamping device and transferred to a receiving vessel. The caps are removed, allowing one to filter off the liquid reagents either by gravity or by using vacuum. At this stage the resin can be readily washed. The final products are collected by filtering into a rack of 96 microdilution tubes or into another 96-deep-well plate.

B. Calypso System™ by Charybdis Technologies

Charybdis Technologies undertook a similar approach and advanced it further in the development of the Calypso System [38]. A variety of disposable deep-well filtration plates, manufactured by Polyfiltronics, or reusable Multi-Well Reaction Arrays, supplied by Charybdis Technologies, can be used on the system. The clamping mechanism consists of a frame into which the plate can be placed, leaving the top and the bottom fully accessible. Top and bottom cover plates can be individually attached to the frame. When the bottom cover is in place, a gasket (which is part of the cover) seals the bottoms of all wells, allowing delivery of the reagents to the plate. Then the top cover can be connected to the frame, sealing the plate. The top cover has holes which allow reagent delivery into the sealed wells by piercing the gasket. In completely assembled form, the Calypso System can be heated, cooled, and agitated. Charybdis Technologies offers 12 specialized Multi-Well Reaction Arrays (PTFE and Multi-Temp) for solution- and solid-phase synthesis applications. The PTFE Arrays are available in three different formats: 96 (8×12, well volume is 2.0 mL), 48 (8×6, well volume is 5.0 mL), and 24 (4×6, well volume is 10.0 mL). These arrays have round-bottom wells for solution applications or bottom-filtration wells with removable frits for solid-phase applications. They can be used to carry out reactions from ambient temperature up to 150°C. The Multi-Temp arrays are also available in three formats (like for the PTFE arrays). They feature all-glass reaction wells assembled within a glass-filled PTFE shell. The shell has an internal cavity through which a fluid can be circulated for precise temperature control. The Multi-Temp arrays withstand temperatures from −80 to 150°C and internal pressures of 30 psi per well. For each of the three array formats, the company offers gas manifold systems, which can be used in place of the top cover. The manifolds can be used to charge the wells with an inert gas or a gaseous reagent. A vacuum manifold system permits draining of the Calypso Reaction Block™ assembly. The content of the wells can be either sent to waste or individually collected.

Charybdis Technologies offers a line of supplementary equipment, such as the Calypso 4X™ Shaker Station and ILIAD PSZ™ Personal Synthesis System. The shaker has a reduced orbital radius and faster rotational speed to ensure efficient mixing. The ILIAD PS^2 is a two-armed liquid-handling robot, produced by Cavro Scientific Inc., which can support up to four Calypso Reaction Blocks. Each arm is individually controlled and can deliver liquids and gases. The robot has six system fluids along with 18 large and 80 small reagent vessels.

C. FlexChemp™ Solid-Phase Chemistry System by Robbins Scientific

Robbins Scientific offers its own version of the original "clamping" approach. The FlexChem System [39, 40] utilizes a proprietary polypropylene 96-well filtration plate specifically designed for synthetic applications. The plate is manufactured by the company. It has channels molded between the well openings to prevent cross-contamination, which can result from solvent "creeping." The bottom of each 2.0-mL well is shaped to direct cleavage products into 96-well plates. The clamping principle is the same as for the Calypso System. The top and bottom sealing covers can be attached individually directly to the plate by spring clamps. There is a choice of three different materials for the sealing gaskets: rubber, Viton™, or ChemTuf™ (a laminate of Viton with a thin fluoropolymer layer designed by Robbins Scientific). Their stability toward a wide range of solvents at different temperatures has been reported by the company [39]. Robbins Scientific recommends using the FlexChem Rotating Incubator to agitate reaction mixtures and to carry out reactions at low or elevated temperatures (up to 100°C). The incubator can accommodate up to four FlexChem plates. The Hydra-96 Microdispenser is another Robbins Scientific product, which can be used for delivering resin or reagents simultaneously into 96 wells. The addition of air-sensitive reagents can be achieved by using the Gilson 215 Personal Synthesizer System. Its robotic arm can easily pierce the top gasket of the fully assembled FlexChem Reaction Assembly. No loss of methylene chloride was observed during the incubation of the plate assembly with the pierced seal at 35°C [39]. The process of cleaving compounds from resin and collecting them is also similar to the Charybdis Technologies approach.

D. HiTOPS System by Affymax Research Institute Polyfiltronics/Whatman

Affymax Research Institute (a member of the Glaxo Smithkline group) has developed a different approach for the synthesis of organic compounds in 96-deep-well plates. The HiTOPS (High Throughput Organic Parallel Synthesis) System [41, 42] (Figure 13.15) utilizes a variety of deep-well filtration plates available from Polyfiltronics/Whatman. They are made of polypropylene and some other polymers and are available with a selection of different filters. Contrary to the clamping method, reactants are retained in wells by the positive pressure of an inert gas (Figure 13.16). At the beginning of the synthesis, a filtration plate is placed into a synthesis device. The bottom portion of the device has an opening in the center connected to an internal channel, which leads to two valves trough a T-connector. One valve regulates the delivery of an inert gas to the bottom of the plate. The other valve links the bottom with a waste container connected to vacuum. When the inert gas

FIGURE 13.15 HiTOPS System: synthesis device (left) and cleavage device (right).

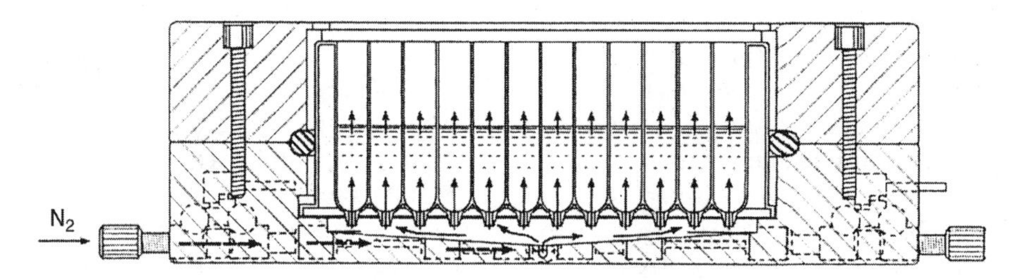

FIGURE 13.16 Reactants are retained by positive pressure.

valve is open, a positive pressure is created at the bottom. Resins and reagents can be delivered to the plate and kept in place by the positive pressure for the duration of the synthesis. The pressure can be increased to allow the inert gas to bubble through the wells, providing for the mixing and inert atmosphere. Mixing can also be achieved either by placing the device on an orbital shaker or by using magnetic stirring bars in each well. At the end of a chemical step, liquid reagents or washing solvents can be removed into the waste container by opening the vacuum valve. At all times the plate remains in the synthesis device. There is no need to disassemble the device in the course of the synthesis, making the whole process very ergonomic. One person can easily operate four devices simultaneously, synthesizing 384 compounds. The synthesis device is also equipped with clamping inserts, which allow the user to carry out synthesis in a closed system, similarly to the Sphinx Pharmaceuticals' approach. For the final cleavage the plate is transferred to the cleavage device (Figure 13.17), which is made of Teflon. It is deeper (to accommodate a collection plate) than the synthesis device and has the same design. After creating positive pressure at the bottom, a cleavage reagent is delivered to the top plate. Upon completion of the cleavage, vacuum is applied and filtrates are collected in the bottom 96-deep-well collection plate.

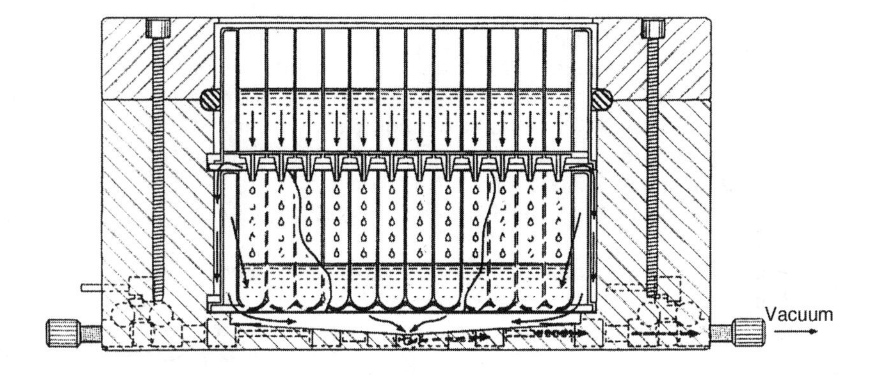

FIGURE 13.17 Collection of cleaved products in cleavage device.

REFERENCES

1. Furka A, Sebestyen F, Asgedom M, Dibo G, General method for rapid synthesis of multicomponent peptide mixtures, *Int. J. Peptide Protein Res.*, 37:487–493, 1991.
2. Lam KS, Salmon SE, Hersh EM, Hruby VJ, Kazmierski WM, Knapp RJ, A new type of synthetic peptide library for identifying ligand-binding activity. *Nature.*, (London) 354:82–84, 1991.
3. Brenner S, Lerner RA, Encoded combinatorial chemistry, *Proc. Natl. Acad. Sci. USA*, 89:5381–5383, 1992.
4. Nielsen J, Brenner S, Janda KD, Synthetic methods for the implementation of encoded combinatorial chemistry, *J. Am. Chem. Soc.*, 115:9812–9813, 1993.
5. Nikolaiev V, Stierandova A, Krchnak V, Seligmann B, Lam KS, Salmon SE, Lebl M, Peptide-encoding for structure determination of nonsequenceable polymers within libraries synthesized and tested on solid-phase supports, *Peptide. Res.*, 6:161–170, 1993.
6. Salmon SE, Lam KS, Lebl M, Kandola A, Khattri PS, Wades, Patek M, Kocis P, Krchnák V, Thorpe D, Felder S, Discovery of biologically active peptides in random libraries: solution-phase testing after staged orthogonal release from resin beads, *Proc. Natl. Acad. Sci. USA*, 90:11708–11712, 1993.
7. Kerr JM, Banville SC, Zuckerman RN, Encoded combinatorial peptide libraries containing non-natural amino acids, *J. Am. Chem. Soc.*, 115:2529–2531, 1993.
8. Ohlmeyer MHJ, Swanson RN, Dillard L, *et al.*, Complex synthetic chemical libraries indexed with molecular tags, *Proc. Natl. Acad. Sci. USA*, 90:10922–10926, 1993.
9. Nestler HP, Bartlett RA, Sill WC, A general method for molecular tagging of encoded combinatorial chemistry libraries, *J. Org. Chem.*, 59:4723–4724, 1994.
10. Baldwin JJ, Burbaum JJ, Henderson I, Ohimeyer MHJ, Synthesis of a small molecule library encoded with molecular tags, *J. Am. Chem.* 117:5588–5589, 1995.
11. Ni Z-J, Maclean D, Holmes CP, Murphy MM, Ruhland B, Jacobs JW, Gordon EM, Gallop MA, Versatile approach to encoding combinatorial organic syntheses using chemically robust secondary amine tags, *J. Med. Chem.*, 39:1601–1608, 1996.

12. *U.S. Patent* 5, 503,805, April 2, 1996.
13. Whitten JP, Yon Feng Xie, Erickson PE, Webb TR, De Souza EB, Grigariadis DE, McCarthy JR, Rapid microscale synthesis, a new method for lead optimization using robotics and solution phase chemistry: application to the synthesis and optimization of corticotropin-releasing factor, receptor antagonists, *J. Med. Chem.*, 39:4354–4357, 1996.
14. Powers JL, Scott W, Personal synthesizer for HTS, *Genet. Eng. New.*, 18(3):14, 1998.
15. Harness JR, Automation of high-throughput synthesis, Automated laboratory workstations designed to perform and support combinatorial chemistry, In Chaiken IM, Janda KD, eds., Molecular Diversity and Combinatorial Chemistry: Libraries and Drug Discovery, ACS Conference Proceedings Series, Washington, DC, *American Chemical Society*, pp. 188–198, 1996.
16. Rivero RA, Greco MN, Maryanoff BE, Equipment for the high-throughput organic synthesis of chemical libraries, In Czarnik AW, DeWitt SH, eds., A Practical Guide to Combinatorial Chemistry, Washington, DC, *American Chemical Society*, pp. 281–307, 1997.
17. Gooding O, Hoeprich PD, Labadie JW, Porco JA, Paul van Eikeren, Wright P, Boosting the productivity of medicinal chemistry through automation tools, Novel technological developments enable a wide range of automated synthetic procedures, In Chaiken IM, Janda KD, eds., Molecular Diversity and Combinatorial Chemistry: Libraries and Drug Discovery, ACS Conference Proceedings Series. Washington, DC, *American Chemical Society*, pp. 199–206, 1996.
18. Porco JA, Deegan TL, Devonport W, Gooding OW, Labadie JW, Mac Donald AA, Newcomb WS, Van Eikeren P, Automated chemical synthesis: chemistry development on the Nautilus 2400™, *Drugs of the Future*, 23:71–78, 1998.
19. Floyd CH, Lewis CH, Patel S, Whittaker M, The automated synthesis of organic compounds—some newcomers have some success, *In Proceedings of the International Symposium on Laboratory Automation and Robotics* 1996, Zymark Corporation, Zymark Center, Hopkinton MA, pp. 51–76, 1997.
20. Nicolau KC, Man XY, Parandoosh Z, Senyei A, Nova MS, Radiofrequency encoded combinatorial chemistry, *Angew. Chem. Int. Ed.*, 34:2289–2291, 1995.
21. Moran EJ, Sarshar S, Cargill JF, *et al.*, *J. Am. Chem. Soc.*, 117:10787–10788, 1995.
22. Czarnik AW, No static at all: using radiofrequency memory tubes without (human) interference, Abstract at The Association for Laboratory Automation Labautomation' 97 Conference, 1997 January 18–22, San Diego. URL: http:/wwwlabautomation.org.
23. Czarnik T, Nova M. No static at all, *Chem. Brit.*, 39–41, October, 1997.
24. Bergot JB, Combinatorial chemistry workstation to facilitate pharmaceutical development, Presented at Cambridge Healthtech Institute's Third Annual Symposium, *High-Throughput Organic Synthesis*, *Coronado*, *CA*, March 5–6, 1998.
25. Lightbody B, Automated technologies for the handling and preparation of reagents and formulation reactants for combinatorial synthesis, Presented at Strategic Research Institute's Symposium, Solid and Solution Phase Combinatorial Synthesis, New Orleans, LA, April 28–29, 1997.
26. Campbell J, Automating solid-phase synthesis without compromise, Presented at Cambridge Healthtech Institute's Third Annual Symposium, High-Throughput Organic Synthesis, Coronado CA, March 5–6, 1998.

27. Cargill IF, Maiefski RR, Toyonaga BE, Automated combinatorial chemistry on solid phase, In Proceedings of the International Symposium on Laboratory Automation and Robotics, 1995, Zymark Corporation, Zymark Center, Hopkinton, MA, pp. 221–234, 1995.
28. Cargill JF, Maiefski RR, Automated combinatorial chemistry on solid phase, *Lab Robotics Automation*, 8:139–148, 1996.
29. Mjalli AMM, Application of automated parallel synthesis, In Czarnik AW, DeWitt SH, eds., A Practical Guide to Combinatorial Chemistry, Washington, DC, *American Chemical Society*, pp. 327–354, 1997.
30. Lebl M, Krchnak V, Sepetov NF, Seligmann B, Strop P, Felder S, Lam KS, One-bead-one-structure combinatorial libraries, *Biopolymers*, 37:177–198, 1995.
31. DeWitt SH, Kiely JS, Stankovic CJ, Schroeder MC, Cody DMR, Pavia MR, Diversomers, an approach to nonpeptide, nonoligomeric chemical diversity, *Proc. Natl. Acad. Sci. USA*, 90:6906–6913, 1993.
32. DeWitt SH, Kiely JS, Pavia MR, Schroeder MC, Stankovic CJ, *U.S. Patent*, 5:324,483, 1994.
33. DeWitt SH, Czarnik AW, Combinatorial organic synthesis using Park-Davis' Diversomer method, *Accounts Chem. Res.*, 29:114–122, 1996.
34. DeWitt SH, Schroeder MC, Stankovic CJ, Czarnik AW, Diversomer technology: solid phase synthesis, automation, and integration for the generation of chemical diversity, *Drug. Dev. Res.*, 33:116–124, 1994.
35. DeWitt SH, Bear BR, Brussolo JS, Duffield MJ, Hogan EM, Kibbey CE, Mac Donald AA, Nickell DG, Rhoton RL, Robertsen GA, A modular system for combinatorial and automated synthesis. In Chaiken IM, Janda KD, eds., Molecular Diversity and Combinatorial Chemistry: Libraries and Drug Discovery, ACS Conference Proceedings Series, Washington DC, *American Chemical Society*, pp. 188–198, 1996.
36. Meyers HV, Dilley GJ, Durgin TL, Powers TS, Wissinger NA, Zhu M, Pavia HR, Multiple simultaneous synthesis of phenolic libraries, *Molec. Diversity.*, 1:13–20, 1995.
37. Meyers HV, Dilley GJ, Powers TS, Winssinger NA, Pavia MR, Versatile method for parallel synthesis, *Methods Molec. Cell. Biol.*, 6:1–7, 1996.
38. Baiga TJ, Integrated instrumentation for high-throughput organic synthesis. Presented at Cambridge Healthtech Institute's Third Annual Symposium, *HighThroughput Organic Synthesis*, Coronado, CA, March 5–6, 1998.
39. Stanchfield J, FlexChem^{TM}: a modular system for high throughput synthesis of small molecules, *Robbins Innovations*, 5(3):1–6, 1997.
40. Stanchfield JE, A flexible, modular system for performing high-throughput synthesis of small molecules, Presented at Cambridge Healthtech Institute's Third Annual Symposium, *High-Throughput Organic Synthesis*, Coronado, CA, March 5–6, 1998.
41. Antonenko VV, Manual and automated parallel synthesis: its development and application for drug discovery at Affymax-GlaxoWellcome, Presented at the First European Conference on Combinatorial Technologies, ComTech97, Freiburg, Germany, February 20–21, 1997.
42. Antonenko VV, Manual and automated parallel synthesis of individual compounds. Presented at Strategic Research Institute's Symposium, *Solid and Solution Phase Combinatorial Synthesis*, New Orleans, LA, April 28–29, 1997.

14 Computational Aspects of Combinatorial Chemistry

Valerie J. Gillet

CONTENTS

I. INTRODUCTION

Combinatorial chemistry is the process whereby large numbers of compounds can be synthesized simultaneously. The technology has developed as a means of providing large numbers of compounds for high throughput screening in the search for new lead compounds. The different technologies and strategies used in the actual syntheses of combinatorial libraries are described elsewhere in this book. These techniques are now sufficiently well developed that it is easy to plan synthetic schemes that could potentially generate massive numbers of compounds. The bottleneck in the use of combinatorial chemistry for the discovery of new leads is the rate at which compounds can be screened. There is, therefore, a great need to be selective about the compounds which are synthesized. Computational methods are increasingly being used to assist in the design of combinatorial libraries. This chapter discusses various computational approaches that are applied in library design, including molecular descriptors, compound selection strategies, combinatorial library design and the integration of structure-based design methods with combinatorial library design.

II. LIBRARY DESIGN STRATEGIES

The earliest combinatorial libraries were peptide libraries, which are linear sequences of amino acids. There are 20 possible amino acids available at each position in the sequence; thus, there are 20^2 (400) dipeptides; 20^3 (8000) tripeptides; 20^4 (160 000) tetrapeptides and so on. For libraries of small organic molecules, typically, there are many more building blocks available at each position of variability and hence the numbers increase much more rapidly. For example, N-Substituted Glycine (NSG) "peptoids" [1] are built from primary amines and carboxylic acids of which there are thousands of each available in commercial databases. Combining all of these combinatorially would give rise to millions of products. The current capacity of high throughput screening and combinatorial synthesis programs is not sufficient to handle libraries of this size. Hence, virtual screening techniques must be employed to select compounds for synthesis and screening, where virtual screening is the in silico equivalent of high-throughput screening [2–4].

In the early days of combinatorial chemistry the emphasis was on the synthesis of large numbers of diverse compounds, in the expectation that simply increasing the number of molecules that are screened would inevitably lead to more hits. The rationale for diversity lies in the similar property principle, which states that structurally similar compounds tend to have similar properties [5]. If structurally similar compounds are likely to exhibit similar activity then maximum coverage of activity space should be achieved by selecting a structurally diverse set of compounds. Thus, a diverse combinatorial library should increase the chances of finding compounds with different activities, while minimizing the number of "redundant" compounds that share the same activity. Results from early diverse libraries were disappointing, however, with lower hit rates than expected and the hits that were found often had properties that made them unsuitable as lead compounds (e.g. they were too large, too insoluble, or contained inappropriate functional groups).

The recent emphasis in library design is directed towards the design of smaller, more focused libraries that incorporate as much information about the therapeutic target as possible. The amount of knowledge available will vary from one project to another. In some cases, the 3-D structure of the protein may be known in which case library design can be coupled with structure-based drug design techniques. More often the 3-D structure of the target is unknown; however, one or more active compounds may have been identified. In this case, ligand-based library design strategies can be used whereby the library is focused on properties of the known active ligands. Thus, libraries can be designed to be enriched in compounds that have similar physicochemical properties or high 2-D or 3-D similarity to the known actives. A recent trend is to focus libraries on a family of targets, such as proteases or kinases, in which case the design should incorporate properties that are known to be important for all targets in the family. When little is known about the target then more diverse libraries are relevant. Such libraries are sometimes referred to as primary screening libraries and are designed to give a broad coverage of chemistry space, so that they can be screened against a range of structural targets. In general, a balance between diversity and focus is needed, with the amount of diversity required being inversely related to the amount of information available about the target [6]. It is also important to try to ensure that the compounds contained within libraries have "drug-like" physicochemical properties, so that they constitute good start points for further optimization.

A variety of different compound selection methods have been developed. These techniques are dependent on the use of molecular descriptors which are numerical values that characterize the properties of molecules. In addition, many compound selection methods are based on quantifying the degree of similarity or dissimilarity of compounds based on molecular descriptors. This requires the use of similarity or distance coefficients.

III. STRUCTURAL DESCRIPTORS

To be effective in library design, molecular descriptors must be able to distinguish between molecules that exhibit different biological activities. If a structural descriptor is a good indicator of biological activity, then molecules with similar activities should occupy similar regions in the descriptor space whereas molecules that are separated in the descriptor space should cover a range of different activities. Another factor affecting the choice of structural descriptors is the speed with which they can be calculated; this should be sufficiently rapid to allow the analysis of the huge numbers of compounds (potentially millions) that characterize combinatorial libraries.

Biological activity is determined by a range of different properties. Receptor binding is clearly important and is determined by physical properties such as hydrophobicity, electrostatic interactions, the ability to form hydrogen bonds between the receptor and a ligand, and 3-D shape. The variety of descriptors used in library design have been reviewed recently [7–9]. They include 2-D and 3-D structural descriptors, topological indices and a range of different physicochemical properties. A summary of these various types of descriptors is given below.

A. 2-D Descriptors

2-D fragment-based descriptors were originally developed for substructure searching [10]. These systems are based on a predefined dictionary of fragments where the presence or absence of each fragment in a molecule is recorded in a binary bitstring. To be effective for substructure searching, fragments should be statistically independent of each other and should be equifrequent. Hence, the optimum choice of fragment dictionary is dataset dependent. Although they were developed for substructure searching, 2-D fragment descriptors have been used successfully in similarity and diversity studies. The similarity between two structures is determined from the number of fragments they have in common and is typically calculated using the Tanimoto coefficient [11], as described later. Examples of fragment-based descriptors include the MACCS structural keys [12], which include atom counts, ring types and counts, augmented atoms and short linear sequences, and BCI fingerprints [13].

An alternative approach to the fragment-based bitstring is the hashed fingerprint which was designed to overcome the dataset dependency associated with the use of fragment dictionaries. In Daylight fingerprints [14], all the paths of predefined length in a molecule are generated exhaustively and hashed to several bit positions in a bitstring. Unity 2-D fingerprints [15] are also based on paths and additionally denote the presence of specific functional groups, rings or atoms encoded in 60 of the total 988 bits.

Several groups have developed atom–typing methods, whereby atoms are represented by their physicochemical properties rather than by element types. Kearsley *et al.* [16] identified atom types as belonging to seven binding property classes—cations, anions, neutral hydrogen bond donors and acceptors, atoms which are both donor and acceptor, hydrophobic atoms and all others. They used two structural descriptors called atom-pairs and topological torsions that are based on these atom types in structural activity relationship studies. Their results showed that the new descriptors based on binding classes are complementary to the original descriptors based on element types.

Pearlman [17] has developed novel molecular descriptors that are called BCUT values. These can be used to define a low-dimensional chemistry space. BCUT values are derived by first creating a $n \times n$ matrix representation for a molecule consisting of n atoms. The rows and columns in the matrix correspond to the atoms in the structure; atomic properties are recorded on the diagonal of the matrix and the off-diagonals record the connectivity of the molecule, for example, the off-diagonals are assigned the value 0.1 times the bond type if the atoms are bonded and 0.001 otherwise. The highest and lowest eigenvalues are then extracted and used as descriptors. Three different matrices are typically used—one with atomic-charge related values on the diagonal; a second with atomic polarizabilities on the diagonal and a third with atomic hydrogen bonding ability on the diagonal. The six descriptors can then be used to represent chemistry space. BCUT values have also been developed that encode 3-D properties. The same atomic properties are encoded on the diagonals of the matrices and the off-diagonals encode interatomic distances.

B. 3-D Descriptors

The fact that receptor binding is a 3-D event would suggest that 3-D descriptors are best able to model biological activity; however, there are significant problems inherent in the use of such descriptors. These problems relate to representation of the conformational space available to molecules—for example, the method used to generate the 3-D structures and the way in which conformational flexibility is handled. However, despite these difficulties 3-D descriptors have found widespread use.

3-D screens were originally designed for 3-D substructure searching [18] (*cf.* 2-D fragment-based descriptors). The screens encode spatial relationships such as distances and angles between the features in a molecule which can be atoms, ring centroids and planes. For example, distance and angle ranges are specified for each pair of features and each range is then divided into a series of bins by specifying a bin width. A distance range of 0..20 Å between two nitrogen atoms might be represented by ten bins each of width 2 Å. The 3-D features are then represented by a bitstring where the number of bits is equal to the total number of bins for all feature pairs. The presence or absence of feature pairs at certain distance ranges is recorded as for 2-D fragments.

Unity 3-D rigid screens [15] are based on a single conformation of a molecule, usually that conformation that is generated by CONCORD [19]. Unity 3-D flexible screens record all possible distances between the same types of features (atom types, rings and planes) based on the incremental rotation of all the rotatable bonds between the two features.

Pharmacophore keys are 3-D structural keys that are based on features of a molecule that are thought to have relevance for receptor binding. The features typically include hydrogen bond donors, hydrogen bond acceptors, charged centers, aromatic ring centers and hydrophobic centers. Pharmacophore keys are usually based on combinations of three or four pharmacophoric features and their associated distances. Flexibility can be taken into account by combining the keys for all distinct conformations of a molecule.

In the ChemDiverse software [20], developed by Chemical Design Ltd., the pharmacophore key is based on 3-point pharmacophores generated for seven features over 32 distances. This gives over 2 million theoretical combinations; however, this number reduces to 890,000 by geometric and symmetry considerations. A single molecule can be represented as a pharmacophore key by recording the presence of each 3-point pharmacophore it contains; however, due to its large size the key is normally used to represent a whole library of compounds. This is achieved by taking the union of keys of the individual molecules. In the Pharmacophore Derived Queries (PDQ) method [21] a much smaller key is used, based on six features and six distance bins, giving a total of 5916 potential pharmacophores. Thus, these pharmacophore keys can be used to represent individual molecules and can be used to measure the similarity and dissimilarity between pairs of molecules. More recently, 4-point pharmacophore keys have been used. These descriptors contain more information since they can distinguish between stereoisomers; however, there are many more potential 4-point pharmacophores leading to much longer bitstring representations [22].

C. Topological Indices

Topological indices [23] are single valued integers or real numbers that characterize the bonding patterns in molecules and that can be calculated from 2-D representation of molecules. They describe structures according to their size, degree of branching and overall shape. Many hundreds of different indices have been developed; examples include the molecular connectivity indices and the kappa shape indices. In diversity studies it is usual to use a large number of different indices in order to attempt to fully to describe a structure. As many of the indices are correlated, normally some data reduction technique, such as principal component analysis, is used to obtain a smaller set of uncorrelated variables.

D. Physicochemical Properties

Physicochemical properties and other whole molecule properties, such as molecular weight, logP and molar refractivity, have been used as molecular descriptors in diversity and clustering studies [9]. For example, Downs *et al.* [24] describe the use of 13 physical properties in clustering experiments. Rose *et al.* [25] have shown that additive bulk properties such as logP and molar refractivity, 2-D structural properties, including fingerprints and connectivity indices and 3-D parameters such as dipole moments and moments of inertia, each describe different aspects of the chemical properties of molecules and hence they are complementary to one another. Hudson *et al.* [26] have used three calculated properties (molecular weight, calculated logP and intrinsic binding energies) in their compound selection method.

E. Quantifying Similarity and Dissimilarity

The similarity, or dissimilarity, of a pair of compounds is quantified using a similarity or distance coefficient that is applied to the descriptor representation of the molecules. As mentioned, the Tanimoto coefficient is commonly used when the molecules are represented by binary bitstrings. The Tanimoto similarity between compounds A and B, S_{AB} is:

$$S_{AB} = \frac{c}{a + b - c}$$

where there are a bit set to "1" in molecule A, b bit set to "1" in molecule B and c "1" bits that are common to A and B. The value of the Tanimoto coefficient varies from zero to one, where one indicates that the bitstrings are identical (note this does not necessarily mean that the molecules are identical) and zero indicates that there are no bits in common between the two bitstrings, i.e. that the molecules are maximally dissimilar. The Tanimoto coefficient can also be used with continuous data such as topological indices and physicochemical properties. Many different similarity coefficients have been developed; some measure similarity directly, such as the Tanimoto coefficient, others measure the distance (or dissimilarity) between compounds, such as the Hamming and Euclidean distance coefficients. When similarity coefficients give a value in the range zero to one, such as the Tanimoto coefficient, then taking the complement provides a way to interconvert between similarity and distance e.g. the Soergel distance is the complement of the Tanimoto coefficient. Willett *et al.* provide a recent review of similarity coefficients [11]. Many of the subset selection methods described in the next section involve calculating pairwise similarities or dissimilarities.

IV. SELECTION METHODS

Many different methods have been developed for compound selection. They include: clustering, dissimilarity-based compound selection, partitioning a collection of compounds into a low-dimensional space and the use of optimization methods such as simulated annealing and genetic algorithms. Filtering techniques are often employed prior to compound selection to remove undesirable compounds.

A. Filtering

The simplest way to reduce the number of compounds to be considered in library design is to apply computational filters to remove those that are known to have undesirable properties. Thus filtering techniques are typically applied early in the library design process. Filters can be based on substructure search techniques to remove compounds that contain functional groups that will interfere with the biological assay, or which are known to be inappropriate for drug discovery [27]. Other filters are used to ensure that the compounds in the resulting library are "drug-like." For example, Lipinski's "rule of five" is commonly applied [28]. The "rule of five" predicts that oral absorption of a molecule is unlikely when two or more of the following conditions are met—molecular weight >500, $\log P > 5$, >5 hydrogen bond donors, >10 hydrogen bond acceptors. Typically molecules that violate two or more of the rules will be eliminated.

B. Cluster-Based Compound Selection

Clustering is the process of dividing a collection of objects into clusters in such a way that there is high intracluster similarity and high intercluster dissimilarity. Once a compound collection has been clustered, a representative subset of compounds can be chosen by taking one or more compounds from each cluster. The basic clustering algorithms all involve choosing structural descriptors for the molecules; measuring the pairwise structural similarities between the compounds using a similarity or distance coefficient and then using a clustering method to group similar compounds together. Since cluster-based selection involves the calculation of very large numbers of intermolecular similarities it can be a computationally expensive process.

Clustering methods can be divided into hierarchical and nonhierarchical methods [29–32]. In hierarchical clustering, smaller clusters of very similar molecules are nested within larger clusters. At one extreme, all individual compounds form a cluster and at the other extreme all compounds are contained in a single cluster. Hierarchical agglomerative methods are bottom-up in that they start with the individual compounds as clusters and progressively combine clusters together until a single cluster is left. Divisive hierarchical clustering is top-down—it starts with a single cluster and recursively divides it into smaller and smaller clusters until single compounds are reached.

Generally, nonhierarchical methods produce a single set of nonoverlapping clusters. For example, Jarvis-Patrick clustering [33] operates by forming nearest neighbour lists for each compound in the set. Compounds are then clustered together if they have a minimum number of near neighbours in common. Although this method has been found to be less effective than hierarchical methods at producing useful clusters, it has the advantage of being fast, hence it has been applied to the clustering of large datasets. For example, McGregor and Pallai [34] have used Jarvis-Patrick clustering and the MACCS keys as structural descriptors to select representative compounds from several commercially available libraries in compound acquisition studies.

Brown and Martin [35] have compared various clustering methods and a range of structural descriptors on their ability to cluster compounds together with similar biological activity. The clustering methods were the Jarvis-Patrick (including an improved version), Ward's hierarchical agglomerative clustering [36] and Guenoche [37], or minimum-diameter, clustering which is a hierarchical divisive method. Their criterion for judging the effectiveness of the clustering methods was to examine their ability to separate active and inactive compounds. A good clustering method will cluster compounds together sharing the same activity and also separate active and inactive compounds into different clusters. Choosing a representative compound from each cluster will ensure that every activity class is represented, that there is no redundancy and that there is a minimal chance of selecting an inactive compound as a representative of cluster that also contains active compounds. They found that Ward's hierarchical agglomerative clustering was the most effective of the methods tested. More recently, Bayada *et al.* [38] have also found Ward's clustering to be effective.

C. Dissimilarity-Based Compound Selection

Dissimilarity-Based Compound Selection (DBCS) involves identifying directly the subset comprising the n most dissimilar compounds in a database containing N compounds, where typically $n \ll N$. Identification of the most dissimilar subset is not computationally feasible since it requires consideration of all possible n-member subsets of the database and, therefore, approximations have been developed that are not guaranteed to be optimal. Maximum dissimilarity methods operate by maximizing the diversity of a subset with respect to a set of descriptors and some associated dissimilarity measure. The algorithms operate by selecting the first compound, for example either at random or as the one that is most dissimilar to all the rest. The subset is then built up by selecting one compound at a time. The next compound to be chosen is the one which is most dissimilar to the compounds already in the subset and so on. There are several ways in which the most dissimilar compound to a set of compounds can be determined [39].

In the Maximum Dissimilarity (MD) selection method described by Lajiness [40] the first compound is selected at random and subsequent compounds are then chosen iteratively, such that the distance to the nearest of the compounds already chosen is a maximum. This method is known as MaxMin. In this study, the compounds were represented by COUSIN 2-D fragment-based bitstrings. Polinsky *et al.* [41] use a similar algorithm in the LiBrain system. In this case, the molecules are represented by a feature vector that contains information about the following *affinity* types—aliphatic hydrophobic, aromatic hydrophobic, basic, acidic, hydrogen bond donor, hydrogen bond acceptor and polarizable heteroatom.

An alternative way of measuring the dissimilarity of one compound to a set of compounds is to sum the pairwise dissimilarities between the compound and all compounds in the set, a method known as MaxSum. The most dissimilar compound to a set of compounds is the compound which has the maximum sum of pairwise dissimilarities. Holliday *et al.* [42] have implemented an efficient version of MaxSum that uses the cosine coefficient as the (dis)similarity coefficient. Their algorithm operates in $O(nN)$ time complexity and can thus be applied to very large datasets. However, as Snarey *et al.* [43] have pointed out, there is a tendency for the algorithm to focus on outliers.

Chapman [44] describes a method for selecting a diverse set of compounds that is based on 3-D similarity. The diversity of a set of compounds is computed from the similarities between all conformers in the dataset, where multiple conformers are generated for each structure. The similarity between two conformers is determined by aligning them and measuring how well they can be superimposed in terms of steric bulk and polar functionalities. A diverse subset is built by adding one compound at a time and the compound that would contribute the most diversity to the subset is chosen in each step. The high computational cost of this method restricts its use to small datasets.

Hudson *et al.* [26] describes a method called the "Most Descriptive Compound" (MDC) method for selecting representative subsets and a "sphere exclusion" method for selecting sets of compounds that cover the available property space. The MDC method aims to select subsets that most effectively represent the compounds in the original collection. It operates by calculating a vector I of N elements where there are N compounds. For each compound, the other compounds are ranked in order of distance to it. The reciprocal of the rank of each compound n is then stored in vector position I_n. The process of ranking and adding the reciprocal of the ranked positions to vector *I* is repeated for each compound. Once this is complete the MDC is the compound with the highest score in *I*. The next compound is selected by removing the contributions of the MDC from *I* and then taking the compound with the highest value in the modified vector.

Sphere exclusion algorithms are closely related to DBCS methods. The algorithms use a dissimilarity threshold as a minimum exclusion radius for a hyper-sphere in multidimensional descriptor space. The basic algorithm operates by selecting a compound and then excluding from consideration all the compounds within a sphere centered on that compound. Hudson *et al.* [26] choose the first compound as the one that has the smallest sum of dissimilarities to all other compounds in the database. In subsequent iterations the next compound chosen is that which is least dissimilar to the compounds already chosen. The algorithm continues until all compounds are either selected or excluded and hence it is not possible to specify the final size of the subset. Pearlman *et al.* [45] have also implemented a sphere exclusion algorithm in the Diverse Solutions package. In this algorithm, the compound to be added to the subset in each iteration is chosen at random. This means that different subsets will result from different runs of the algorithm.

Clark [46] has recently described a subset selection algorithm called OptiSim which includes maximum and minimum dissimilarity based selection as special cases. A parameter is used to adjust the balance between representativeness and diversity in the compounds that are selected.

Snarey *et al.* [43] compared a number of DBCS and sphere exclusion algorithms for compound selection. The effectiveness of the methods was measured by examining the range of biological activities that result from selecting diverse subsets from the World Drugs Index (WDI) [47] (based on the assumption that a diverse set of compounds in structural space corresponds to a range of biological activities). Their results suggest that the maximum dissimilarity algorithm described by Lajiness [40] is most effective at selecting compounds associated with a range of bioactivities. The performance is sometimes exceeded by the sphere exclusion method of Pearlman, but the random element in the latter means that its effectiveness varies substantially from one run to the next.

D. Partitioning

Partitioning or cell-based methods are based on the definition of a low-dimensional property space. The range of values for each property is divided into a set of bins and the combinatorial product of all bins then defines the set of cells that make up the space. Each molecule in a library is assigned to a cell according to its physical properties. A diverse subset of compounds can be selected by taking one or more molecules from each cell, whereas a focused set of compounds can be selected by choosing compounds from a limited number of cells, for example from those cells that surround the cell occupied by a known active compound.

Mason *et al.* [48–50] first described partitioning methods based on global physicochemical properties. In the Diverse Property-Derived (DPD) method, six descriptors, that are thought to describe important features in drug/receptor interactions, were chosen by analyzing a larger set of 49 calculated molecular descriptors, including: atoms and group counts, electronic and flexibility indices and calculated logP and molar refractivity. The six descriptors resulting from the analysis measure hydrophobicity, flexibility, shape, hydrogen bonding properties and aromatic interactions. Each descriptor was then split into two to four partitions to give 576 theoretical combinations or bins. When the Rhône-Poulenc Rorer (RPR) corporate collection was mapped onto the space it was found to occupy 86% of the bins. A diverse screening set was chosen by selecting three compounds from each bin.

Pearlman [17] has developed a partitioning method based on the BCUT values described earlier. Different methods are available for calculating the atomic properties (charge, polarizability and hydrogen bonding ability) encoded in the BCUT values and so a chi-squared approach is included to enable descriptors to be chosen that give an approximately uniform distribution of compounds. This allows the coverage of chemical space to be tailored for particular libraries. Sampling from the partitions can be un-biased or biased by criteria such as cost of reactant or physicochemical property.

The PDQ method [21] is a partitioning scheme that is based on 3-point pharmacophores. In this scheme, each potential 3-point pharmacophore in a pharmacophore key is considered as a single cell. A molecule is mapped onto the key by identifying the 3-point pharmacophores that it contains. Thus, a molecule will typically occupy more than one cell; this is in contrast to partitioning schemes based on physicochemical properties in which each molecule occupies a single cell. The pharmacophore keys for a set of molecules can be combined into an ensemble pharmacophore which is the union of the individual keys. The resulting ensemble key can be used to measure total pharmacophore coverage, to identify pharmacophores that are not represented in the set of molecules and to compare different sets of molecules.

The PDQ approach was used to calculate the total number of possible pharmacophores within each of three different libraries at RPR and also to find the number of pharmacophores that are in common to any two of the three libraries [49]. Library comparisons can be used to decide which library to synthesis; next in an iterative screening programmee for example, the next library to be screened could be the one that covers the greatest range of new pharmacophores or the one that covers a more focused region of space around the areas already covered in earlier iterations.

E. Optimization Methods

Optimization techniques provide effective ways of sampling large search spaces and hence several methods have been applied to compound selection.

Martin *et al.* [1] have described a selection technique that is based on statistical experimental design. A wide range of molecular descriptors was used including: calculated log P, 81 topological indices, 70 connectivity indices and seven shape descriptors. Principal components analysis was used to reduce the dimensionality to five latent variables. Five chemical functionality descriptors were calculated by performing multidimensional scaling on the distance matrix, calculated using Daylight fingerprints and the Tanimoto coefficient. Five receptor recognition descriptors were calculated by performing multidimensional scaling on atom layer descriptors. Compound selection was carried out using the statistical technique of D-optimal design, with the objective of selecting compounds that are evenly spread in property space and that are also orthogonal in their properties. The method was applied to select representative sets of amines and carboxylic acids in the design of peptoid combinatorial libraries. A potential disadvantage of D-optimal design, however, is that it has a tendency to select molecules that are at the edges of descriptor space. The selection can be made to be more "space-filling" by including new descriptors in the design that are powers and cross-terms of the original descriptors and that have the effect of moving points from the edges into the interior of the descriptor space.

Higgs *et al.* [51] described two experimental design algorithms for selecting molecules from large databases that approximate spread and cover which are linear with respect to time complexity. Spread designs are used to identify a subset of molecules that are maximally dissimilar with respect to each other. Coverage designs are used to identify subsets of molecules that are maximally similar to the candidate set of molecules.

Monte Carlo search methods have been applied to compound selection, often combined with simulated annealing [52, 53]. An initial subset is chosen at random and its diversity is calculated. A new subset is then generated from the first by replacing some of the compounds with others chosen at random from the data set. The diversity of the new subset is measured—if it is more diverse than the previous subset it is accepted for use in the next iteration; if it is less diverse, then the probability that it is accepted depends on the Boltzmann factor. The process continues for a fixed number of iterations or until no further improvement is observed in the diversity function. In the simulated annealing variant the temperature of the system is gradually reduced, so increasing the chance of finding the globally optimal solution.

Typical diversity functions used in optimization methods include MaxMin and MaxSum. In recent work, Waldman *et al.* have identified a new diversity function that is based on computing the *minimum spanning tree* for the set of molecules [54]. A spanning tree is a set of edges that connect a set of objects without forming any cycles. The objects in this method are the molecules in the subset and each edge is labeled by the dissimilarity between the two molecules it connects. A minimum spanning tree is the spanning tree that connects all molecules in the subset with the minimum sum of pairwise dissimilarities. The diversity then equals the sum of the intermolecular similarities along the edges in the minimum spanning tree.

V. VALIDATION OF METHODS

Given the variety of different descriptors and subset selection methods that are available, several studies have been carried out in an attempt to validate both the compound selection methods and the various descriptors. To some extent the choice of descriptors and subset selection methods are interlinked. For example, partitioning schemes are restricted to low-dimensional descriptors such as physicochemical descriptors, whereas clustering and dissimilarity-based methods can be used with high dimensional descriptors such as fingerprints.

The relative efficiencies of the various compound selection methods requires consideration. The relative efficiencies of the various clustering algorithms varies enormously. Hierarchical clustering methods typically require storage or memory space proportional to N^2 (written $O(N^2)$) and the time required to perform the clustering is $O(N^3)$, where there are N compounds. Thus, hierarchical clustering has traditionally been limited to relatively small data sets, although fast implementation methods and the introduction of parallel clustering algorithms means that this is becoming less of a restriction [32].

The basic DBCS algorithm has time complexity $O(n^2N)$, where n compounds are selected from N. Since n is generally a small fraction of N, the time is thus cubic in N. DBCS can also be very computational demanding; however, fast implementations have been developed, for example the MaxSum method described by Holliday *et al.* [42] and a MaxMin method described by Agrafiotis and Lobanov that can be used with low-dimensional descriptors [55].

Partitioning methods based on physicochemical properties are fast, hence they can be applied to very large datasets.The computational requirements of the optimization methods are dependent on the diversity function that is used since this is applied very frequently during the optimization process.

A number of studies have compared the effectiveness of different descriptors for compound selection. For example, Brown and Martin [35, 56] compared a range of 2-D and 3-D descriptors and clustering methods and assessed their effectiveness according to how well they were able to distinguish between active and inactive compounds. They found that the 2-D descriptors were more effective than the 3-D descriptors and concluded that they are sufficient to characterize much of the 3-D structure of molecules, even though they were originally developed for optimum screenout during substructure search. However, the poor performance of the 3-D descriptors may be due to the incomplete handling of conformational flexibility when generating the 3-D descriptors.

Brown and Martin [57] also investigated the extent to which a number of different descriptors (2-D and 3-D) are able to encode information that relates to the interaction forces which must exist if a ligand is to bind to a receptor. Thus, the information content of each descriptor was assessed by its ability to accurately predict values for the physical properties of a structure from the known values of other structures which are shown to be structurally similar to it, using the descriptor in question. The predicted properties included measured logP values and calculated properties that explore the shape and flexibility of the molecules, including the number of hydrogen bond donors and acceptors within a molecule. Two methods of predicting properties were used—a similarity-based prediction where the predicted value for a molecule was taken as the mean value of all molecules within a given similarity threshold to it; and cluster-based prediction where the predicted value was calculated as the mean values of all molecules in the same cluster. Their results confirmed their previous findings.

Matter [58] has also validated a range of 2-D and 3-D structural descriptors on their ability to predict biological activity and on their ability to sample structurally and biologically diverse datasets effectively. The compound selection techniques used were maximum dissimilarity and clustering. Their results also showed the 2-D fingerprint-based descriptors to be the most effective in selecting representative subsets of bioactive compounds.

Briem and Kuntz [59] compared similarity searching using Daylight fingerprints [14] with fingerprints generated using the DOCK program. The DOCK fingerprints are based on shape and electrostatic properties. The Daylight 2-D descriptors performed better than the DOCK 3-D descriptors at identifying known active compounds, thus providing more evidence in support of the use of 2-D descriptors in (dis)similarity studies. The DOCK descriptors were, however, found to be complementary to the 2-D descriptors.

VI. DESIGNING COMBINATORIAL LIBRARIES

The compound selection methods described thus far can be used to select compounds for screening from an in-house collection, or to select which compounds to purchase from an external supplier. In combinatorial library design, however, it is necessary to select subsets of reactants for actual synthesis. The two main strategies for combinatorial library design are *reactant-based selection* and *product-based selection*. In reactant-based selection, optimized subsets of reactants are selected without consideration of the products that will result and any of the compound selection methods already identified can be used. An early example of reactant-based design is that already described by Martin and colleagues which is based on experimental design and where diverse subsets of reactants were selected for the synthesis of peptoid libraries [1].

In product-based selection, the properties of the resulting product molecules are taken into account when selecting the reactants. Typically this is done by enumerating the entire virtual library that could potentially be made. Any of the subset selection methods described previously could be used to select a diverse subset of products, however the resulting subset is very unlikely to represent a combinatorial subset. This process is known as cherry-picking and is synthetically inefficient as far as combinatorial synthesis is concerned. Synthetic efficiency is maximized by taking the *combinatorial constraint* into account whereby a combinatorial subset is selected directly with every reactant selected at each position of variation appearing in a product with every reactant selected at the other positions.

Product-based selection is much more computationally demanding than reactant-based selection, however it has been shown that better optimized libraries can result [60, 61], especially when the aim is to optimize the properties of a library as a whole, such as diversity or the distribution of physicochemical properties. In addition, product-based selection is usually more appropriate for focused libraries which require consideration of the properties of the resulting products.

Enumerating the product space of combinatorial libraries and calculating the descriptors can be expensive computationally. Recently, Downs and Barnard [62] have described an efficient method of generating fingerprint descriptors of the molecules in a combinatorial library without the need for enumeration of the products. This method is based on earlier technology for handling Markush structures. An alternative approach is to use random sampling techniques to derive a statistical model of the property under consideration [63].

A. Product-Based Library Design

Product-based selection is typically implemented using optimization techniques such as Genetic Algorithms (GAs) or simulated annealing [64–67]. For example, the SELECT program [65] is based on a GA in which each chromosome encodes one possible combinatorial subset. Thus, for a two component combinatorial synthesis, in which n_A of a possible N_A first reactants are to be combined with n_B of a possible N_B second reactants, a chromosome contains $n_A + n_B$ elements, with each element specifying one possible reactant. The fitness function quantifies the "goodness" of the combinatorial subset encoded in the chromosome and the GA evolves new potential subsets in an attempt to maximize this quantity. The fitness function could be designed to maximize the diversity of the subset, using MaxSum, MaxMin or via a partitioning scheme, or it could be designed to focus the subset around a known target compound; for example, by maximizing the sum of similarities to the target compound.

Alternative product-based approaches to library design have been developed that do not require enumeration of the full virtual library. These methods have been termed molecule-based methods to distinguish them from library-based methods [68]. They are based on identifying a set of product molecules with desired properties. The product molecules selected are then examined to identify reactants that occur frequently within them with the frequently occurring reactants being used to define a combinatorial library. An early example of this approach is the genetic algorithm developed by Kearsley *et al.* [69]. The chromosome of the GA represents a single product molecule constructed from reactants extracted from reactant pools, rather than a combinatorial library as in SELECT. The fitness is based on similarity to a known active molecule calculated using 2-D fingerprints. The final population of the GA represents a set of highly scoring product molecules and the reactants that occur frequently in these products can be identified and used in a subsequent combinatorial synthesis.

Weber *et al.* [70] have developed a GA that optimizes actual biological response of compounds within a virtual combinatorial library. Each chromosome in the GA represents a single product compound of the reaction. The fitness function involves performing the corresponding reaction and actually testing the product for activity. Crossover and mutation are performed to generate new chromosomes in an iterative manner. The method was able to find compounds with micromolar activity after synthesizing and testing a small fraction of the virtual library.

B. Multiobjective Library Design

A recent trend in library design is to optimize libraries over a number of properties simultaneously; for example, whether a library is designed to be diverse, focused or some combination of the two, it is desirable that the library is cheap to synthesis and that the compounds contained within the library have "drug-like" physicochemical properties. Most approaches to multiobjective library design combine the different properties *via* a weighted-sum fitness function. For example, in the SELECT program the fitness function can have the following form:

$$f = w_1 \text{diversity} + w_2 \text{cost} + w_3 \Delta\text{MW}$$

Diversity is typically measured using a distance-based or cell-based method; cost is typically given as reactant cost/gm and physiochemical properties such as ΔMW are typically measured as the difference in the distribution of the property in the library compared to the distribution of the same property in a collection of known drugs. The weights, w_1, w_2, w_3, are user-defined and are typically set so that diversity is maximized, while the cost and physicochemical properties are minimized. This weighted-sum approach leads to a single solution that represents one particular compromise in the objectives. Several other groups have also adopted this approach [67, 71–73].

Recently the limitations of this approach have been recognized [74]. For example, it can be difficult to choose appropriate weights, especially when the objectives are in competition, which is usually the case and a single somewhat arbitrary compromise solution is produced when in fact usually a family of different compromise solutions exists. These limitations have been addressed through the application of a Multiobjective Genetic Algorithm (MOGA) in the MoSELECT program [74, 75]. In MoSELECT, multiple objectives are handled independently and a family of equivalent solutions is found, where each solution represents a different compromise in the objectives. This approach allows the library designer to investigate the relationships between the various objectives and to make an informed choice on what represents an appropriate compromise solution.

VII. LIBRARY DESIGN AND STRUCTURE-BASED DRUG DESIGN

When the biological target is known, the active site can be used to assist in the design of targeted combinatorial libraries [76]. The chemical properties of an active site, such as its size, hydrophobicity and charge, can be used to suggest an appropriate template for a library and complementary substituents. Once initial library members have been synthesized and found to be active, their binding modes can be studied and used to guide the design of subsequent libraries. For example, Graybill *et al.* [77] have used structural information about the active site of thrombin in order to identify nonpeptide scaffolds that were complementary to the S2 binding site, but that also contained sufficient substitution points to be functionalized using combinatorial techniques.

The PRO_SELECT program [78] uses elements of structure-based drug design to assist in the design of combinatorial libraries for known targets. A synthetically accessible template, or scaffold, is positioned within the active site of the target. 3-D database searching is then used to generate lists of potential substituents for each substitution position on the template. The substituents are selected on the basis of their being able to couple to the template using known synthetic routes and on their possessing functionality that will enable them to interact with residues in the active site. Additional substituents can be generated by performing molecular transformations on the substituents found in the database search. A number of filters are then applied to the substituent lists, starting with molecular property screens such as log P and molecular weight, followed by more elaborate methods involving positioning the substituents in the active site with a substituent that results in either intramolecular or intermolecular steric clashes being eliminated. The lists can be further reduced using diversity analysis based on clustering. The final substituents are scored and ranked according to their predicted binding affinities.

Jones *et al.* [79] have reported the application of the ligand docking program GOLD to the screening of combinatorial libraries. GOLD was used to screen the reactant lists of a library rather than the product molecules themselves. 406 carboxylic acids and 105 amines were used as ligand fragments. Three-dimensional structures of the acids and amines were generated and each fragment was docked into the active site of lipase. Each docked acid was then compared with each docked amine to see if the reactants could be joined to form a product molecule, by forming an amide bond. The energy of the product molecule was calculated and if its energy was below a given threshold it was considered to be a good prediction of the binding mode of the product formed from the two reactants, otherwise it was rejected. The fully enumerated library consisted of 44,730 products. The docking studies resulted in the binding modes for 129 products with reasonable predicted energies which were generated from 34 unique acids and 49 unique carboxylic acids. This represents a significant reduction in the number of reactants used and hence in the combinatorial library that would be produced.

One of the earliest successes of a combined approach to library design has been presented by Kick *et al.* [80]. Libraries were designed against Cathepsin *D* which is an aspartyl protease. Two library design strategies were compared. In the first, libraries were designed within the context of the active site of the target protein. Molecular modeling techniques were used to position a scaffold within the active site and then docking techniques were used to score reactants at each position of variability according to their ability to give favourable interactions with the receptor. A subset of high scoring reactants was chosen and the library was synthesized and tested for activity. In the second library, reactants were selected based on diversity alone. Again the library was synthesized and tested. The targeted library resulted in three times as many compounds with micromolar activity as compared with the diverse library. This study clearly showed the power of integrating combinatorial synthesis methods with structure-based drug design. Since this early work, several other successes have been reported [81–83].

Recently, there has been a great deal of interest in docking large databases of molecules to protein active sites; for example, these could be in-house databases, databases provided by external suppliers or virtual libraries such as those that could be synthesized using combinatorial chemistry [84–86]. Although the docking process itself is computationally expensive, it is very straightforward to implement in a parallel computing environment, thus enabling very large databases to be processed.

VIII. CONCLUSIONS

The vast number of compounds that could be made using combinatorial chemistry has resulted in the development of a variety of computational techniques to assist in the design of combinatorial libraries. Many of these techniques are based on established methods such as substructure searching and similarity methods; however, an increasing number of new methods is also emerging. These methods include new subset selection methods, novel molecular descriptors and the application of data analysis methods that were originally developed in the data mining community [87]. Furthermore, the requirement to take into account multiple properties when designing libraries has led to the introduction of new approaches to multiobjective library design.

Since the introduction of combinatorial chemistry techniques, the emphasis has shifted from the design of large diverse libraries towards smaller more focused libraries. Indeed, as the number of protein crystal structures that are available has increased, so too has the interest in using this structural knowledge for combinatorial library design, compound acquisition programmes and virtual screening. This trend is likely to continue in the near future.

The emphasis in this chapter has been on designing libraries based on diversity, similarity to known actives, or complementarity to a protein binding site. It is also important, however, that a drug molecule is able to reach its site of action within the body. For example, it should be able to pass through one or more physiological barrier such as a cell membrane or the blood brain barrier, it should remain within the body for an appropriate period of time and it should not be toxic. The properties of molecules that govern these effects are often referred to as ADMET (Absorption, Distribution, Metabolism, Excretion and Toxicity) properties and there is increasing interest in taking these properties into account as early in the drug discovery process as possible. Thus, considerable effort is now being directed towards the development of models for the accurate prediction of ADMET properties [88].

BIBLIOGRAPHY

1. Martin EJ, Blaney JM, Siani MS, Spellmeyer DC, Wong AK, Moos WH, Measuring diversity—experimental design of combinatorial libraries for drug discovery, *J. Med. Chem.*, 38:1431–1436, 1995.
2. Walters WP, Stahl MT, Murcko MA, Virtual screening—an overview, *Drug Discovery Today*, 3:160–178, 1998.
3. Böhm HJ, Schneider G, eds., *Virtual Screening for Bioactive Molecules*, Weinheim: Wiley-VCH, 2000.
4. Bajorath J, Integration of virtual and high-throughput screening. *Nature* Reviews, *Drug Discovery*, 1:882–894, 2002.
5. Johnson MA, Maggiora GM, *Concepts and Applications of Molecular Similarity*, New York, Wiley, 1990.
6. Valler MJ, Green DVS, Diversity screening versus focused screening in drug discovery, *Drug Discovery Today*, 5:286–293, 2000.
7. Brown RD, Descriptors for diversity analysis, *Perspect. Drug. Discov. Design*, 7/8:31–49, 1997.
8. Downs GM, Molecular descriptors, In Bultinck P, Winter H De, Langenaeker W, Tollenaere JP, eds., *Computational Medicinal Chemistry and Drug Discovery*, New York, Dekker Inc., 2003.
9. Livingstone D, The characterization of chemical structures using molecular properties—A survey, *J. Chem. Inf. Comput. Sci.*, 40:195–209, 2000.
10. Barnard JM, Substructure searching methods—old and new, *J. Chem. Inf. Comput. Sci.*, 33:532–538, 1993.
11. Willett P, Barnard JM, Downs GM, Chemical similarity searching, *J. Chem. Inf. Comput. Sci.*, 38:983–996, 1998.
12. MACCS II, MDL Information Systems, Inc., 14600 Catalinsa Street, San Leandro, CA 94577, http://www.mdli.com.
13. Barnard Chemical Information Ltd., 46 Uppergate Road, Stannington, Sheffield S6 6BX, UK, http://www.bci.gb.com.
14. Daylight Chemical Information Systems, Inc., Mission Viejo, CA, USA, www.daylight.com at http://www.daylight.com.
15. Unity, Tripos Inc., 1699 South Hanley Road, St. Louis, MO 63144-2913, USA, http://www.tripos.com.

16. Kearsley SK, Sallamack S, Fluder EM, Andose JD, Mosley RT, Sheridan RP, Chemical similarity using physiochemical property descriptors, *J. Chem. Inf. Comput. Sci.*, 36:118–127, 1996.
17. Pearlman RS, Smith KM, Novel software tools for chemical diversity, *Perspect. Drug. Discov. Design*, 9/10/11:339–353, 1998.
18. Sheridan RP, Nilikantan R, Rusinko A, Bauman N, Haraki K, Venkataraghavan R, 3-D SEARCH: A system for three-dimensional substructure searching, *J. Chem. Inf. Comput. Sci.*, 29:255–260, 1989.
19. Rusinko A III, Skell JM, Balducci R, McGarity CM, Pearlman RS, CONCORD: A program for the rapid generation of high quality 3-D molecular structures, The University of Texas at Austin and Tripos Associates, St Louis MO, 1988.
20. ChemDiverse, Chemical Design Ltd., Roundway House, Cromwell Park, Chipping Norton, Oxfordshire, OX7 5SR, UK.
21. Pickett SD, Mason JS, McLay IM, Diversity profiling and design using 3-D pharmacophores: Pharmacophore-Derived Queries (PDQ), *J. Chem. Inf. Comput. Sci.*, 36:1214–1223, 1996.
22. Mason JS, Morize I, Menard PR, Cheney DL, Hulme C, Labaudiniere RF, New 4-point pharmacophore method for molecular similarity and diversity applications: Overview of the method and applications, including a novel approach to the design of combinatorial libraries containing privileged substructures, *J. Med. Chem.*, 42:3251–3264, 1999.
23. Randić M, The connectivity index 25 years after, *J. Mol. Graphics Modelling*, 20:19–35, 2001.
24. Downs GM, Willett P, Fisanick W, Similarity searching and clustering of chemical-structure databases using molecular property data, *J. Chem. Inf. Comput. Sci.*, 34:1094–1102, 1994.
25. Rose VS, Rahr E, Hudson BD, The use of Procrustes analysis to compare different property sets for the characterisation of a diverse set of compounds, *Quant. Struct-Act. Relat.*, 13:152–158, 1994.
26. Hudson BD, Hyde RM, Rahr E, Wood J, Parameter based methods for compound selection from chemical databases, *Quant. Struct-Act. Relat.*, 15:285–289, 1996.
27. Roche O, Schneider P, Zuegge J, Guba W, Kansy M, Alanine A, Bleicher K, Danel F, Gutknecht EM, Rogers-Evans M, Neidhart W, Stalder H, Dillon M, Sjögren E, Fotouhi N, Gillespie P, Goodnow R, Harris W, Jones P, Taniguchi M, Tsujii V, von der Saal W, Zimmermann G, Schneider G, Development of a virtual screening method for identification of "Frequent Hitters" in compound libraries, *J. Med. Chem.*, 45:137–142, 2002.
28. Lipinski CA, Lombardo F, Dominy BW, Feeney PJ, Experimental and computational approaches to estimate solubility and permeability in drug discovery and development settings, *Advanced Drug Delivery Reviews*, 23:3–25, 1997.
29. Willett P, Similarity and Clustering in Chemical Information Systems, Letchworth, Research Studies Press, 1987.
30. Downs GM, Willett P, Clustering of chemical structure databases for compound selection, In van de Waterbeemd H, ed., *Advanced Computer-Assisted Techniques in Drug Discovery*, Vol. 3, Weinheim, VCH, 1994.
31. Dunbar JB Jr, Cluster-based selection, *Perspect. Drug Discov. Design*, 7/8: 51–63, 1997.

32. Downs GM, Barnard JM, Clustering methods and their uses in computational chemistry, In Lipkowitz KB, Boyd DB, eds., *Reviews in Computational Chemistry*, Vol. 18, New York, VCH Publishers, pp. 1–40, 2002.
33. Jarvis RA, Patrick EA, *IEEE Trans. Comput.* C22:1025–1034, 1973.
34. McGregor MJ, Pallai PV, Clustering of large databases of compounds: Using the MDL keys as structural descriptors, *J. Chem. Inf. Comput. Sci.*, 37:443–448, 1997.
35. Brown RD, Martin YC, Use of structure-activity data to compare structure-based clustering methods and descriptors for use in compound selection, *J. Chem. Inf. Comput. Sci.*, 36:572–584, 1996.
36. Ward JH, Hierarchical grouping to optimise an objective function, *J. Am. Statist. Assoc.*, 58:236–244, 1963.
37. Guenoche A, Hansen P, Jaumard B, Efficient algorithms for divisive hierarchical clustering with diameter criterion, *J. Classif.*, 8:5–30, 1991.
38. Bayada DM, Hamersma H, van Geerestein VJ, Molecular diversity and representativity in chemical databases, *J. Chem. Inf. Comput. Sci.*, 39:1–10, 1999.
39. Holliday JD, Willet P, Definitions of "dissimilarity" for dissimilarity-based compound selection, *J. Biomolec. Screening*, 1:145–151, 1996.
40. Lajiness MS, Dissimilarity-based compound selection techniques, *Perspect. Drug. Discov. Design*, 7/8:65–84, 1997.
41. Polinsky A, Feinstein RD, Shi S, Kuki A, (1996) LiBrain: software for the automated design of exploratory and targeted combinatorial libraries, In Chaiken IM, Janda KD, eds., *Molecular Diversity and Combinatorial Chemistry*, Washington DC, American Chemical Society, pp. 219–232, 1996.
42. Holliday JD, Ranade SS, Willett P, A fast algorithm for selecting sets of dissimilar structures from large chemical databases, *Quant. Struct-Act. Relat.*, 15:285–289, 1996.
43. Snarey M, Terrett NK, Willett P, Wilton DJ, Comparison of algorithms for dissimilarity-based compound selection, *J. Mol. Graphics.*, 15:372–385, 1997.
44. Chapman DJ, The measurement of molecular diversity: A three-dimensional approach, *J. Comput-Aided. Mol. Design.*, 10:501–512, 1996.
45. *Diverse Solutions User's Manual*, St Louis, MO, Tripos Inc, 1996.
46. Clark R, OptiSim: An extended dissimilarity selection method for finding diverse representative subsets, *J. Chem. Inf. Comput. Sci.*, 37:1181–1188, 1997.
47. World Drug Index, Thomson Derwent, 14 Great Queen St., London W2 5DF, UK, http://www.derwent.com.
48. Mason JS, McLay IM, Lewis RA, Applications of computer-aided drug design techniques to lead generation, In Dean PM, Jolles G, Newton CG, eds., *New Perspectives in Drug Design*, London, Academic Press, pp. 225–253, 1994.
49. Mason JS, Pickett SD, Partition-based selection, *Perspect. Drug. Discov. Design*, 7/8:85–114, 1997.
50. Lewis RA, Mason JS, McLay IM, Similarity measures for rational set selection and analysis of combinatorial libraries: The diverse property-derived (DPD) approach, *J. Chem. Inf. Comput. Sci.*, 37:599–614, 1997.
51. Higgs RE, Bemis KG, Watson IA, Wikel JH, Experimental designs for selecting molecules from large chemical databases, *J. Chem. Inf. Comput. Sci.*, 37:861–870, 1997.

52. Hassan M, Bielawski JP, Hempel JC, Waldman M, Optimization and visualization of molecular diversity of combinatorial libraries, *Molec. Diversity*, 2:64–74, 1996.
53. Agrafiotis DK, Stochastic algorithms for molecular diversity, *J. Chem. Inf. Comput. Sci.*, 37:841–851, 1997.
54. Waldman M, Li H, Hassan M, Novel algorithms for the optimization of molecular diversity of combinatorial libraries, *J. Mol. Graphics Modelling*, 18:412–426, 2000.
55. Agrafiotis DK, Lobanov VS, An efficient implementation of distance-based diversity measures based on k-d trees, *J. Chem. Inf. Comput. Sci.*, 39:51–58, 1999.
56. Martin YC, Kofron JL, Traphagen LM, Do structurally similar molecules have similar biological activity? *J. Med. Chem.*, 45:4350–4358, 2002.
57. Brown RD, Martin YC, The information content of 2-D and 3-D structural descriptors relevant to ligand binding, *J. Chem. Inf. Comput. Sci.*, 37:1–9, 1997.
58. Matter H, Selecting optimally diverse compounds from structural databases: A validation study of two-dimensional and three-dimensional molecular descriptors, *J. Med. Chem.*, 40:1219–1229, 1997.
59. Briem H, Kuntz ID, Molecular similarity based on dock-generated fingerprints, *J. Med. Chem.*, 39:3401–3408, 1996.
60. Gillet VJ, Willett P, Bradshaw J, The effectiveness of reactant pools for generating structurally-diverse combinatorial libraries, *J. Chem. Inf. Comput. Sci.*, 37:731–740, 1997.
61. Jamois EA, Hassan M, Waldman M, Evaluation of reagent-based and product-based strategies in the design of combinatorial library subsets, *J. Chem. Inf. Comput. Sci.*, 40:63–70, 2000.
62. Downs GM, Barnard JM, Techniques for generating descriptive fingerprints in combinatorial libraries, *J. Chem. Inf. Comput. Sci.*, 37:59–61, 1997.
63. Beroza P, Bradley EK, Eksterowicz JE, Feinstein R, Greene J, Grootenhuis PDJ, Henne RM, Mount J, Shirley WA, Smellie A, Stanton RV and Spellmeyer DC, Applications of random sampling to virtual screening of combinatorial libraries, *J. Mol. Graphics Modelling*, 18:335–342, 2000.
64. Brown RD, Martin YC, Designing combinatorial library mixtures using a genetic algorithm, *J. Med. Chem.*, 40:2304–2313, 1997.
65. Gillet VJ, Willett P, Bradshaw J, Green DVS, Selecting combinatorial libraries to optimize diversity and physical properties, *J. Chem. Inf. Comput. Sci.*, 39:169–177, 1999.
66. Zheng W, Cho SJ, Tropsha A, Rational combinatorial library design, 1. Focus-2-D: A new approach to the design of targeted combinatorial chemical libraries, *J. Chem. Inf. Comput. Sci.*, 38:251–258, 1998.
67. Zheng W, Hung ST, Saunders JT, Seibel GL, PICCOLO: A tool for combinatorial library design via multicriterion optimization, In Atlman RB, Dunkar AK, Hunter L, Lauderdale K, Klein TE, eds., *Pacific Symposium on Biocomputing 2000*, Singapore, World Scientific, pp. 588–599, 2000.
68. Sheridan RP, SanFeliciano SG, Kearsley SK, Designing targeted libraries with genetic algorithms, *J. Mol. Graphics Modelling*, 18:320–334, 2000.
69. Sheridan RP, Kearsley SK, Using a genetic algorithm to suggest combinatorial libraries, *J. Chem. Inf. Comput. Sci.*, 35:310–320, 1995.

70. Weber L, Wallbaum S, Broger C, Gubernator K, Optimization of the biological activity of combinatorial libraries by a genetic algorithm, *Angew. Chem. Int. Ed. Engl.*, 34:2280–2282, 1995.
71. Good A, Lewis RA, New methodology for profiling combinatorial libraries and screening sets: Cleaning up the design with HARPick, *J. Med. Chem.*, 40:3926–3936, 1997.
72. Agrafiotis DK, Multiobjective optimization of combinatorial libraries, *J. Comput-Aided. Mol. Design.*, 5/6:335–356, 2002.
73. Brown JD, Hassan M, Waldman M, Combinatorial library design for diversity, cost efficiency, and drug-like character, *J. Mol. Graphics Modelling*, 18:427–437, 2000.
74. Gillet VJ, Khatib W, Willett P, Fleming P, Green DVS, Combinatorial library design using a multiobjective genetic algorithm, *J. Chem. Inf. Comput. Sci.*, 42:375–385, 2002.
75. Gillet VJ, Willett P, Fleming P, Green DVS, Designing focused libraries using MoSELECT, *J. Mol. Graphics Modelling*, 20:491–498, 2002.
76. Beavers MP, Chen X, Structure-based combinatorial library-design: Methodologies and applications, *J. Mol. Graphics Modelling*, 20:463–468, 2002.
77. Graybill TL, Agrafiotis DK, Bone R, Illig CR, Jaeger EP, Locke KT, Lu T, Salvino JM, Soll RM, Spurlino JC, Subasinghe N, Tomczuk BE, Salemme FR, Enhancing the drug discovery process by integration of high-throughput chemistry and structure-based drug design, In Chaiken IM, Janda KD, eds., *Molecular Diversity and Combinatorial Chemistry*, Washington DC, American Chemical Society, 1996.
78. Murray CW, Clark DE, Auton TR, Firth MA, Li J, Sykes RA, Waszkowycz B, Westhead DR, Young SC, PRO_SELECT: Combining structure-based drug design and combinatorial chemistry for rapid lead discovery. 1. Technology, *J. Comput-Aided. Mol. Design.*, 11:193–207, 1997.
79. Jones G, Willet P, Glen RC, Leach AR, Taylor R, Further development of a genetic algorithm for ligand docking and its application to screening combinatorial libraries, ACS Symposium Series, *Rational Drug Design: Novel Methodology and Practical Applications*, 719:271–291, 1999.
80. Kick EK, Roe DC, Skillman AG, Liu G, Ewing TJA, Sun Y, Kuntz ID, Ellman JA, Structure-based design and combinatorial chemistry yield low nanomolar inhibitors of Cathepsin D, *Chemistry and Biology*, 4:297–307, 1997.
81. Kubinyi H, Combinatorial and computational approaches in structure-based drug design, *Current Opinion in Drug Discovery and Development*, 1:16–27, 1998.
82. Bohm H-J, Stahl M, Structure-based library design: Molecular modelling merges with combinatorial chemistry, *Current Opinion in Chemical Biology*, 4:283–286, 2000.
83. Stahl M, Structure based library design, *Methods and Principles in Medicinal Chemistry*, 10:229–264, 2000.
84. Klebe G, ed., Virtual screening: An alternative or complement to high throughput screening? *Perspect. Drug. Discov. Design*, 20:2000.
85. Muegge I, Rarey M, Small molecule docking and scoring, In Lipkowitz KB, Boyd DB, eds., *Reviews in Computational Chemistry*, Vol. 17, New York, VCH Publishers, pp. 1–60, 2001.

86. Lyne PD, Structure-based virtual screening: An overview, *Drug Discovery Today*, 7:1047–1055, 2002.
87. Leach AR, Gillet VJ, *An Introduction to Chemoinformatics*, Amsterdam, Kluwer, 2003.
88. van de Waterbeemd H, Gifford E, ADMET, In silico modelling: Towards prediction paradise? *Nature Reviews Drug Discovery*, 2:192–204, 2003.

15 High Throughput Combinatorial Methods for Heterogeneous Catalysts Design and Development

J.M. Domínguez

CONTENTS

I. INTRODUCTION

The heterogeneous catalysts have a profound impact on the chemical industry in general; for example 60% of all chemical processes, 75% of oil refining processes, nearly 100% of polymers and about one hundred petrochemicals depend on the action of catalysts, as well as a significant part of environmental technologies (VOCs, automotive emissions control, stationary sources, etc.) and fine chemical production. Actually, the worldwide catalysts market is worth about 10 billion USD, (i.e. 10×10^9 USD) a year and, according to some reports, the catalysts cost represents only about 0.2% of the value of products they generate, which is about 5×10^{12} USD [1–7].

Nowadays, the main driving forces of the catalysts markets are the upgrading of heavier crude fractions and the strict environmental regulations [6, 7]. Also, the competitive edge in the catalysts industry seems to depend more on a clear differentiation of products in the marketplace. For these reasons, innovative research plays an essential role to keep ahead of the competence. However, the traditional route for catalysts innovation is a long pathway from the basic R&D laboratory, through the product engineering and the scaling up process, before use in the chemical plant for five to seven years later [8]. All along this pathway industrial interaction is necessary, especially in the early stages of process analysis and the scaling-up of leads for implementing industrial products [8].

Industrially-oriented R&D on catalysts generally involves interdisciplinary work at different levels, for example the synthesis, characterization and catalytic testing of the more promising materials. The detection of some hits out of dozens of candidates at the laboratory scale may eventually end up with one or two leads of industrial interest for specific applications. More often than not, the research work focuses on the screening of diverse catalytic materials, with the aim of understanding their intrinsic activity patterns and to determine the reaction kinetics without the influence of other parameters like heat and mass transfer [9, 10]. Searching for a correlation between the intrinsic catalytic properties of the solids and their composition, structural and textural properties [11], is of prime importance. However, this pathway is often limited because it involves a great number of experiments, before having a complete picture of the activity pattern and the influence of all the possible variables. In fact, traditional R&D is rather empirical and based on trial and error, which is often uneconomical and labor-intensive, as well as dependent on the knowledge available [12]. However, this lengthy and costly process has proven some effectiveness to develop most of the industrial catalysts used nowadays [13]. In contrast to this, a rational approach tends to understand first the fundamental steps in catalysts using theoretical and simulation methods, as well as for example single crystal, clean surfaces and *in situ* studies [14, 15], but the experimental conditions used in these methods are rather different with respect to the industrial practice, i.e. low temperatures and high vacuum. In spite of such differences, at least in a few cases, the results of the rational approach have been translated successfully to industrial catalysts design [16, 17]. Although the prediction of reactivity between specific surface configurations and organic molecules is possible using advanced methods [18, 19], still the more significant hypothesis need validation by a number of experiments, which is often limited because they are too costly or time consuming.

In any of the above mentioned pathways, the advent of High Throughput (HT) and combinatorial methods may have a real significance for accelerating the pace of discovery, through speeding up the process of preparing and testing diverse molecules and materials. In particular, the pharmaceutical industry have used combinatorial chemistry techniques for years and this has dramatically increased the speed of drug discovery [20, 21]. Also, in the field of materials science these methods have demonstrated their potential for discovery, for example in the fields of new luminescent devices [22], superconducting [23], magnetoresistance [24] and dielectric/ferroelectric materials [25]. The term Combinatorial Catalysis refers to methods for creating chemical libraries, i.e. vast collections of catalytic compounds of varying composition or other properties (i.e. textural, structural) that are tested in order to identify a subset of promising compounds. Also, the term "High Throughput" (HT) refers to methods for obtaining data collections resulting from the preparation and testing of large numbers of catalytic materials in a short time scale, often a few minutes [26]. One of the advantages of these methods has the possibility to study the chemical diversity using a small sample size. For example, in the first screening mode some individual library samples may contain from about a tenth of a milligram to about 1 mg, while the secondary screening mode employs typically sample sizes between 10 and 100 mg. In contrast to this, conventional fixed bed reactors at the bench scale typically requires 10 g or more for each candidate compound, thus limiting the use of extended libraries for upgrading the tests of experimental materials.

Therefore, heterogeneous catalysts present a greater potential for the application of HT and Combinatorial methods, because they involve diverse compositional phases that are usually formed by interfacial reactions during their synthesis, which in turn produce a variety of structural and textural properties, often too vast to prepare and test by traditional methods. In this respect the HT and Combinatorial methods extend the capabilities of the R&D cycle, which comprises the synthesis, the characterization of physico-chemical properties and the evaluation of catalytic properties. The primary screening HT method gives the possibility of performing a rapid test of hundreds or thousands of compounds using infrared detection methods [27–29]. Alternatively, a detection method called REMPI (Resonance Enhanced Multi Photon Ionization) has been used, which consists of the *in situ* ionization of reaction products by UV lasers, followed by the detection of the photoions or electrons by spatially addressable microelectrodes placed in the vicinity of the laser beam [30, 31].

In addition, a secondary screening HT method is well suited to test the catalytic properties of materials using a parallel flow-through reactor system (i.e. the Multi Fixed Bed Reactor System—MFBR) fitted with parallel detection and quantification techniques by MS and GC [32–35].

Nowadays, the High Throughput concept is being extended further to the characterization of physical and chemical properties, with the aim of searching out potential correlations of the library members (6×8 or 12×16) with their catalytic behavior. The HT characterization techniques are complementary of the primary and secondary screening facilities, thus closing up the HT R&D cycle for discovery and optimization of catalytic materials. The HT methods increase the rate of preparing and testing catalytic species by a factor of 10^4 to 10^5 with respect to conventional R&D methods, using primary screening methods, or by a factor of 10^2 to 10^3 using secondary screening methods [34, 36]. Also, new challenges arise, for example (1) the development of data managing systems to keep abreast of the high volume of data, (2) the incorporation of the HT concept in most of the characterization techniques and (3) the development of potential systems or algorithms for establishing a virtual correlation between the data provided by the HT methods and the more traditional systems of industrial significance.

II. THE CONCEPTUAL INPUT

One hundred years ago A. Mittasch [37, 38] realized the massive search of catalytic materials for ammonia synthesis. This effort is recognized nowadays as an early combinatorial work leading to the industrial catalysts, with little modifications. This was a trial and error method focused on the discovery and optimization of a few leads out of four thousand compositions, which were based on combinations of transition metal oxides and alkaline promoters that led to perform about 20,000 tests. In contrast with this pioneering work, HT Combinatorial methods take advantage of computer controlled automates for parallel preparation and testing of hundreds of samples per day, thus multiplying the throughput by several orders of magnitude with respect to conventional R&D work. These novel techniques give the possibility of exploring the multiple parameter space (i.e., chemical concentration, phase dispersion, pH, time, etc.) in a very short time, thus increasing the capacity of probing basic hypothesis and trends, as well as improving the accuracy of liquids handling and dispensing of solutions, thus improving the reproducibility of the physical and chemical properties of the solids.

In spite of the high number of potentially active compounds and the considerable number of diverse catalytic systems studied so far [7], the knowledge available on catalytic materials with more than two components is rather limited. To illustrate this point, one can take 75 of the more stable elements of the periodic table, then divide them in subgroups of *n* members each, which gives 75^n possible compounds, that is about 5,600 binary compounds, or 4.2×10^5 ternary, 3.1×10^7 quaternary and about 5.6×10^{18} decanary compositions. In addition, the relative concentrations between two or more elements and the modifications driven by special treatments contribute to increase those numbers [39]. The main point here is that, using traditional R&D methods, the possibilities of scanning a substantial portion of the multiparameter space are rather limited, that is why HT and Combinatorial methods become more important.

On the other hand, significant developments in catalysis arise from the knowledge of reaction kinetics and engineering, which are crucial for designing industrial catalysts and processes [11]. Of course, the actual HT and Combinatorial catalysis methods do not override traditional semi-industrial or Pilot Plant tests, but they are complementary and powerful techniques for exploring the multiparameter space under variable conditions and kinetic

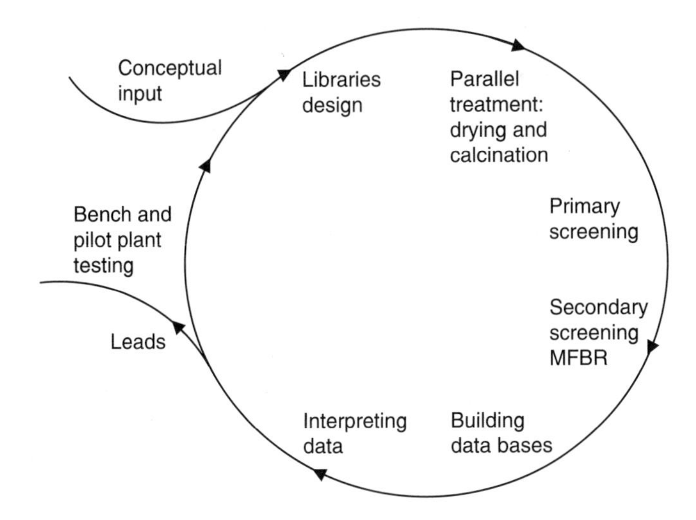

FIGURE 15.1 Rational & Complementary pathway in catalysis R&D—combinatorial + conventional.

regime, which can provide useful data for designing new catalysts and reaction engineering altogether (Figure 15.2). An advantage of these methods is the use of small amounts of samples, which allows to extend the compositional libraries beyond the limits imposed by traditional methods. The use of such tiny amounts of samples in the synthesis of compounds is comparable with respect to bulk synthesis, as demonstrated recently by the hydrothermal synthesis of zeolites, i.e. pentasil Ti-silicalite at the one-microliter scale [39], which showed these solids to present full structural characteristics as the materials prepared by conventional bulk synthesis. Other recent works reported mesoporous aluminosilicates synthesized at the milliliter scale using HT and Combinatorial methods [40]. These works demonstrate that small scale synthesis of micro and mesoporous aluminosilicates is reproducible and susceptible to scaling up at higher scale.

The master concept lying behind the libraries design in HT and Combinatorial methods is essential for carrying out the experiments and the choice of the appropriate screening level. The rational pathway followed in catalysts R&D using HT and Combinatorial methods is illustrated in Figure 15.1. Here, the primary and secondary screening methods are clearly indicated, as well as the amount of catalyst and the instrumentation used. In the former case, i.e. primary screening, tiny amounts of samples equivalent to 0.1 to 1 mg are distributed on silicon wafers from liquid solutions or other sources, using thin film vapor deposition, radiofrequency sputtering, capillary ink-jet technology, solution or electrochemical methods [41–45]. In contrast to this approach, the secondary screening method deals with samples of about 10 to 100 mg each. With respect to the evaluation of catalytic properties, the primary screening approach uses IR technology for detecting the heat released

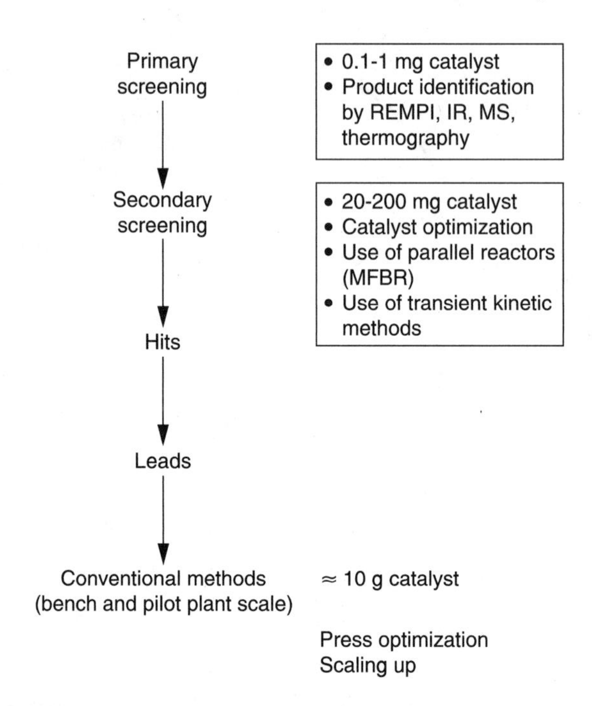

FIGURE 15.2 Complementary High Throughput Synthesis and Combinatorial Strategies for Catalysts Design.

by the reaction (ΔH), thus giving a 2-D map of the activity pattern across the library members [27–29, 46, 47]. Alternatively, the product analysis can be made individually by Mass Spectrometry, using small tubes that are mounted on moving plates, which carry out the product gases from each composition spot to the analysis chamber [42, 48]. A 16×12 or 8×6 library is prepared in a few minutes and then can be treated under specific conditions for drying, calcination or reduction, which may take some minutes or hours. The relative activity of the library members is usually made using the IR maps, REMPI or MS. In contrast to this, the secondary screening approach makes use of conventional gas chromatography or MS methods, which are well suited for analysis of the products coming out of the microreactors system. Unlike the primary screening mode, the Multi Fixed Bed Reactors (MFBR) used in the secondary screening are comparable with the plug flow type reactor working under differential regime, with the gas reactants flowing through the catalytic bed [49], which improves the contact between the solid particles and the gas phase. Therefore, the secondary screening unit may be used for studying the kinetics and the catalysts behavior along a period of time, i.e. the deactivation profile and the effects of some variables like space velocity (VHSV), reaction temperature (T) and reactants flow. The typical VHSV interval in this system vary from about 5×10^4 to 7×10^4 Vol(gas)/Vol(cat)/h, while the typical mass velocity values are between 2×10^3 and 3×10^3 Kg/m^2/h.

Therefore, the secondary screening method is a powerful tool for determining the catalytic behavior of materials under kinetic regime, but further developments of industrial catalysts require the use of conventional pilot plant scale units (Figure 15.2), which are well suited for testing the catalysts under the influence of additional parameters like mass and heat transfer [8, 9].

In summary, the catalysts innovation process, a master concept is necessary before doing the experimental design, which in turn leads to the libraries design. This implies a solid foundation on chemistry and materials science, as well as a clear hypothesis in mind, with enough potential for advancing the knowledge in catalysts or to generate the kinetic data needed in reaction engineering and industrial catalysts development. The operational part is not the main focus of Combinatorial Catalysts, because the automated computer controlled robots are used to carry out the accurate liquid handling and data gathering operations, thus leaving room for scientists for better planning and data interpretation. This approach closes the entire HT R&D cycle, thus multiplying the potential of discovery of new catalysts by several orders of magnitude.

III. THE TOOLS

The common practice followed in HT and Combinatorial Catalysts makes use of two robots, one for handling the synthesis of the catalytic materials and the second one for performing the evaluation of their catalytic properties. In the former case, the robot is fitted with liquid handling pipettes for picking up chemical components which are distributed across the members of the 2-D library (6×8 or 12×16). This system is able to deliver liquids in a range varying from the microliter scale (10^{-6} L) to several milliliters (10^{-3} L), with a precision better than 2 μL, using mother solutions previously prepared. This system has been used for implementing some common methods of catalysts preparation, as for example (1) solid supports impregnation, (2) precipitation of metal salts from solution, (3) sol gel and (4) surfactants-assisted self-assembling synthesis, etc. In a typical run, a 6×8 library containing 48 vials of 100 ml each can be prepared in a few minutes. Also, thermal treatments can be performed in separate units to induce specific structural and textural characteristics to the solids, for example drying, reduction and calcining under special environments. Thus, different thermal treatments can be performed to identical libraries or a specific treatment can be applied to a series of different libraries, which may induce diverse structural and textural features to the solids.

One of the golden rules for a successful work in HT experiments is the automation of the entire workflow, i.e. one must be able to produce what one is able to analyze, because a variety of solid phases and distinct structural features can be obtained across a typical 6×8 matrix. Therefore, the characterization of structural properties of catalytic materials, as for example crystallinity degree, requires the application of X-ray microdiffractometers [50] fitted with GADDS (Brucker). The collimated beams of about 500 μ diameter can be focused by Göbel mirrors to scan areas of about 50 μ size of individual members of a given library. As the beam has a high intensity on the spot, the recording of X-ray diffraction patterns may take a very short time, usually a couple of hours for a 12 × 16 library of solid samples. The GADDS system gives a rapid evaluation of the crystalline solid phases as well as structural variations induced during the synthesis procedures. Other techniques for characterizing the physical properties of catalytic materials are being adapted to the HT concept, for example NMR(MAS)-Solid State (Brucker), which uses a special sample holder for obtaining the series of NMR spectra of different nuclei. Also, the textural properties of heterogeneous catalysts, such as surface area (S, m^2/g), pore volume (V_p, cc/g) and mean pore diameter (D_p, Å) can be determined by means of automated sorptometers that are able to run several samples at once.

The chemical analysis of the fluids coming in and out of each individual reactor can be performed in at least two modes. The first screening of a 12 × 16 library usually takes a couple of minutes. The secondary screening of the catalytic properties of at least six library members usually takes about three minutes. Therefore, it takes about 24 to 36 minutes to complete the secondary screening and detection of a 48 library members [42, 49, 51]. Thus, the average time for evaluating a library member is about 36/48 or 0.75 minutes. Figure 15.3 shows the basic reactor system used for the rapid screening of libraries containing catalytic materials of potential interest in the secondary screening mode. A single microreactor can hold from about 10 to 100 mg of

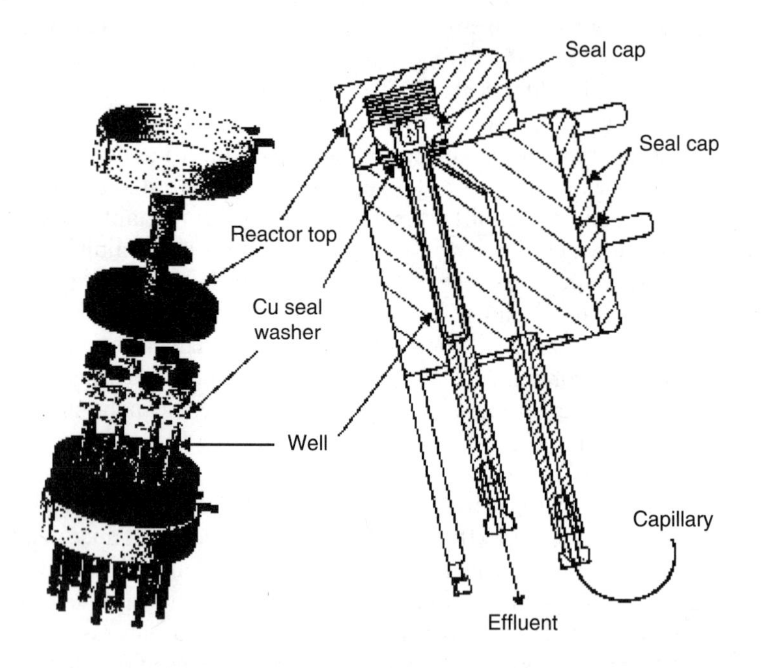

FIGURE 15.3 Typical Microreactor used in MFBR parallel arrays (49).

material, depending on the density. The vessels are made of stainless steel and they are hollow right circular cylinders, each having a fluid permeable upper end and lower end. A quartz paper frit in the lower end of each of the vessels holds the sample in place, but allows fluid to pass through. Any multiple of six library members up to forty eight members can be screened at a time [49, 51].

The detectors used for the secondary screening of catalytic properties are usually Gas Chromatographs (GC) or Mass Spectrometers (MS), or both coupled together, which can measure the concentration of the reaction products in a parallel mode.

IV. REPRODUCIBILITY AND ACCURACY

A key concept of the synthesis and screening of heterogeneous catalysts is the reproducibility of their catalytic properties, which implies the control of their structural and textural features during the synthesis process. As the solid phases evolve through complex interfacial interactions and ever changing conditions such as pH, volume of liquid, temperature, aging time, etc. [52], the degree of confidence in the preparation procedures is a key aspect for obtaining specific active phases by HT and Combinatorial methods. Also, solid phases might evolve when they enter in contact with reactive gases during the distinct stages of synthesis and testing [53] but, in many other cases, the reproducibility level reached might be high enough in such a way that HT and Combinatorial methods can be representative, as illustrated in the following section by the study of the variations of structural and textural properties of silica mesoporous sieves (MSS).

A. Synthesis of Mesoporous Silica Sieves (MSS)

A series of MSS materials were synthesized from nanostructured liquids based on CTAB (CetylTrimethylAmmonium Bromide)/H_2O/NH_4OH, which are usually constituted by molecular self-assemblies in solution, or liquid crystals at higher concentration, i.e. 25% wt. The self-assembling properties of the organic surfactants like CTAB, together with the cooperative interaction of silicon oxide precursors in water solution (i.e. $[Si_3O_6(OH)_3]^{3-}$ and $[SiO_2(OH)_2]^{2-}$) lead to materials with pore array symmetries such as the hexagonal, cubic or 1-D (i.e. the MCM-41, MCM-48 and MCM-50 type structures), respectively [42]. Both textural and structural variations may occur when a series of solvents (co-surfactants) are added to the CTAB/H_2O/NH_4OH system, i.e. acetone (CH_3COCH_3) and the series of low linear chain alcohols CH_3OH, C_2H_5OH and C_3H_7OH, because this causes the systematic variation of the surface tension (γ), the dielectrical constant (ε) and dipolar moment (μ) from pure water ($\gamma = 72\,mJ\,m^{-2}$, $\varepsilon = 78.3$, $\mu = 1.76$ Debye) to the pure co-surfactant ($\gamma = 23\,mJ\,m^{-2}$, $\varepsilon = 32.66$, $\mu = 1.7$ Debye for MeOH, $\gamma = 22.8\,mJ\,m^{-2}$, $\varepsilon = 24.59$, $\mu = 1.6$, $\mu = 1.73$ Debye for EtOH, $\gamma = 22.5\,mJ\,m^{-2}$,

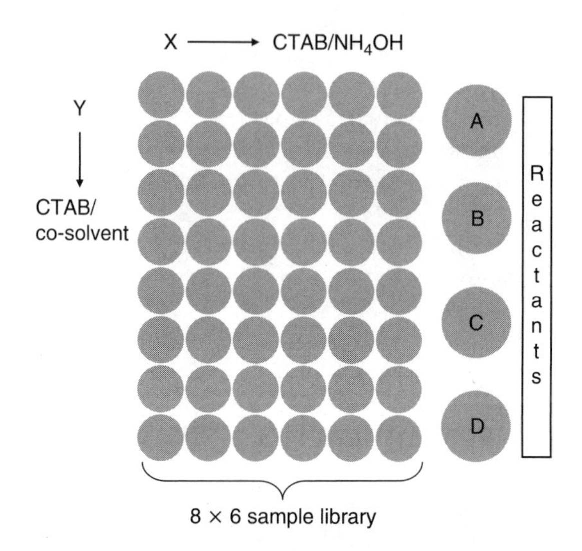

FIGURE 15.4 Building combinatorial libraries.

$\varepsilon = 20.45$, $\mu = 1.64$ Debye for PrOH and $\gamma = 23.7\,\mathrm{mJ\,m^{-2}}$, $\varepsilon = 20.56$, $\mu = 2.69$ Debye for Acetone).

Then, the silica mesoporous sieves were prepared at room temperature using a Cavro [34] liquids handling robot (i.e. Symyx Inc). A compositional library was designed in function of two main variables, the CTAB/NH_4OH and CTAB/Co-surfactant molar ratios. The aqueous solutions of the main surfactant were prepared using demineralized water and TEOS (i.e. $SiO[C_2H_5OH]_4$) as silica source, the ammonium hydroxide (NH_4OH) being used for controlling the pH. The synthesis was performed at room temperature with a constant CTAB/TEOS (silica source) ratio, i.e. 4.5 wt.% CTAB, CTAB/TEOS = 0.4 and pH equal to 11.5. The experimental settings consisted of four 6 × 8 Combinatorial libraries designed for synthesizing 48 silica structures per unit, where the main variables were CTAB/NH_4OH (x axis) and (y axis) ratios (Figure 15.4). After dispensing the basic reagents about 50 to 100 mg of gel was obtained within a few minutes, then a subsequent drying at 40°C and calcining in air at 550°C led to the final solids.

The main properties of these materials were characterized by means of x-ray diffraction (Siemens D-500 with λ_{Cu} radiation of 1.54 Å), Transmission Electron Microscopy (Phillips-CM200) and N_2 adsorption (Micromeritics ASAP-2000), ^{29}Si-NMR(MAS). As the textural properties of the catalytic materials, for example the inner pore structure, is a key parameter for their performance, in the present work the N_2 adsorption isotherms of the calcined mesoporous SiO_2-based solids were determined. The solids were prepared using different CTAB surfactant and some co-surfactants based in the light alcohols, i.e. MeOH, EtOH and PrOH. Thus, Figure 15.5 shows the isotherms of the mesoporous solids prepared with MeOH (co-surfactant). In all

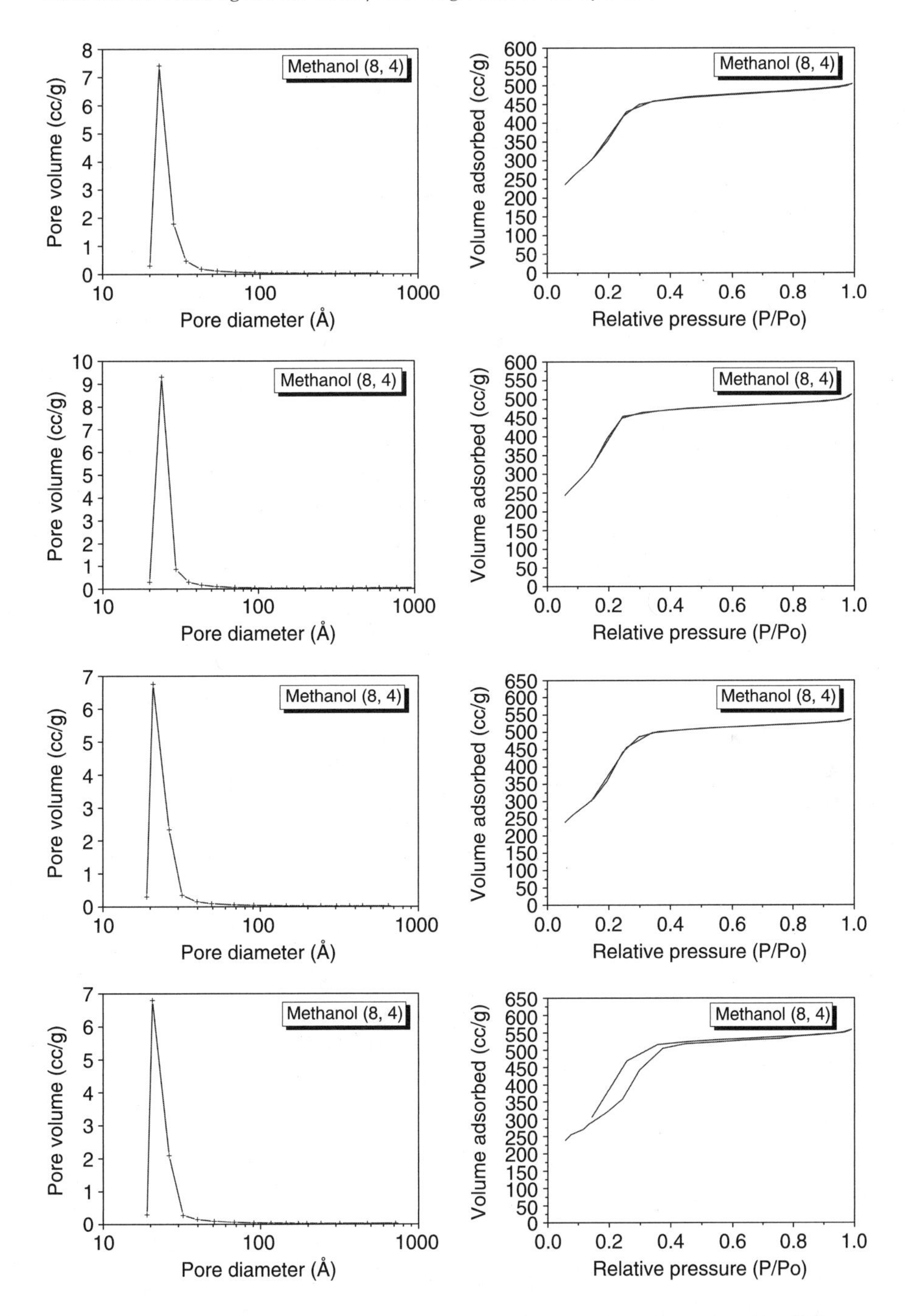

FIGURE 15.5 Textural properties of mesoporous solids prepared with different concentration of co-surfactant (Methanol), along 4th column of a 6×8 library.

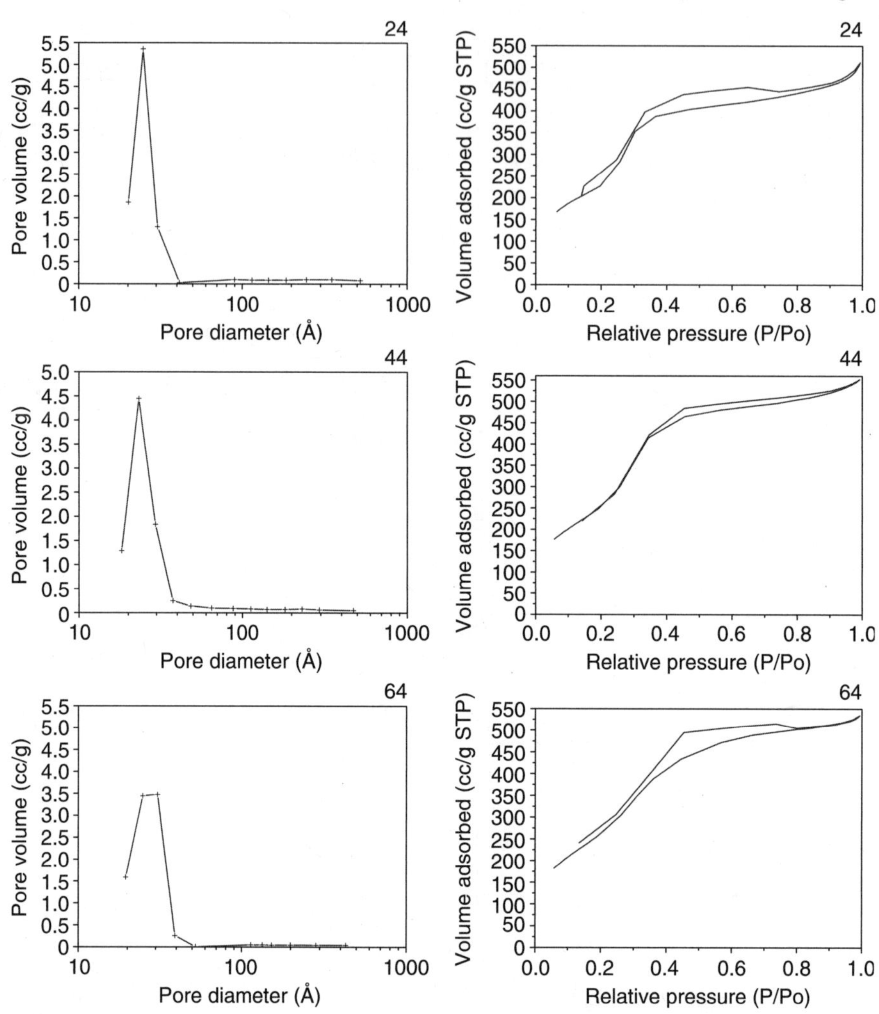

FIGURE 15.6 Textural properties of mesoporous solids prepared with different concentration of co-surfactant (Ethanol), along 4th column of a 6×8 library.

these cases the isotherms profile looks about the same, except perhaps at the lower end, where a small hysteresis loop is more apparent. The maxima of the pore diameters distribution are in all cases between 2.2 and 2.5 nm, which is a remarkable uniformity of the textural properties of this series. Furthermore, Figures 15.6 and 15.7 illustrate the textural properties of another series prepared with distinct co-surfactants, i.e. ethanol and prophanol, respectively, which confirm the previous trend, that is the similarity of the isotherms profile and pore size distribution for each series. Again, the isotherms profile is about the same, except at the lower end of each series, but the maxima of the pore diameter distribution remains around 2.7 nm.

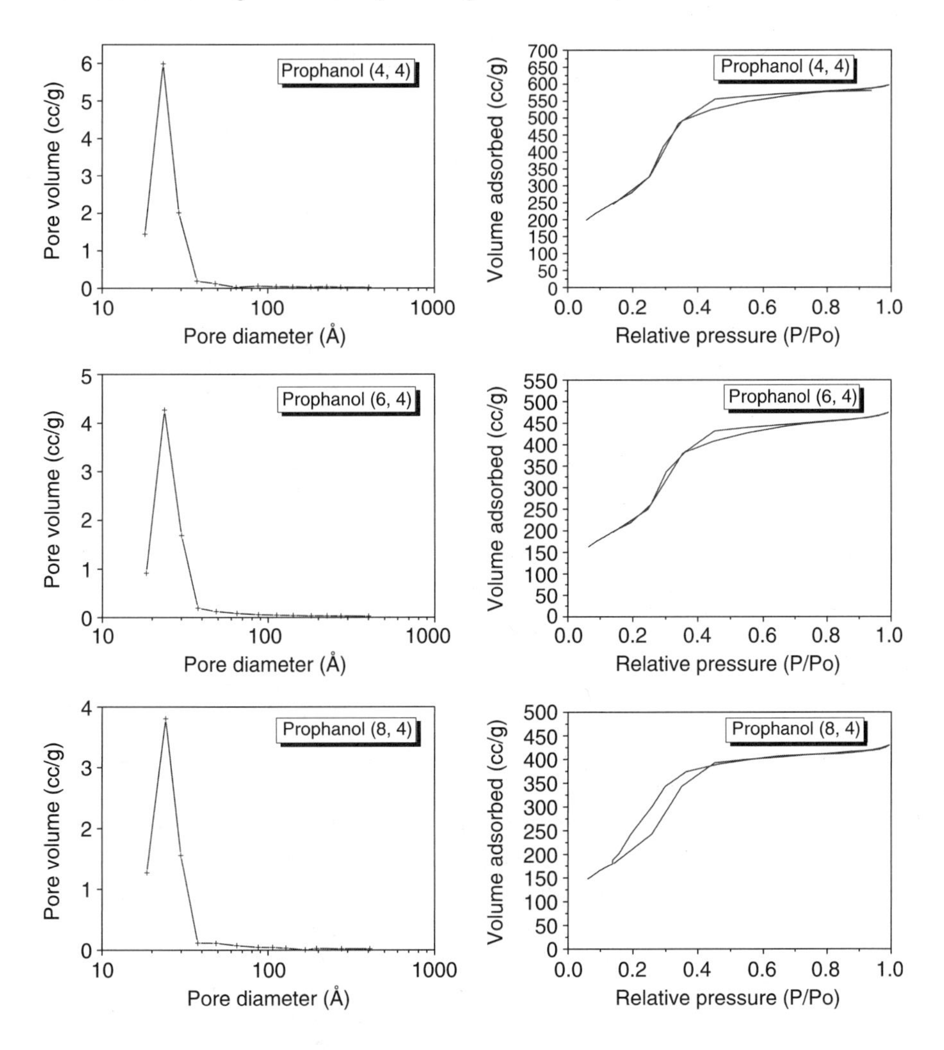

FIGURE 15.7 Textural properties of mesoporous solids prepared with different concentration of co-surfactant (Prophanol), along the 4th column of a 6×8 library.

In summary, textural parameters that are essential for the catalysts performance were prepared from variable combinations of CTAB/NH_4OH/H_2O in the presence of co-surfactants, i.e. acetone and the light alcohols (MeOH, EtOH, PrOH). The resuts indicate that the porous structure of the materials thus obtained are maintained along the distinct TEOS and CTAB concentration ratios, even with the influence of diverse co-surfactants. Then, the textural properties of the mesoporosus MSS, measured by N_2 adsorption, indicate a reproducibility of the textural properties, i.e. pore volume, mean pore size distribution and total surface area (Figure 15.8).

On the other hand, the MSS materials are not crystalline, but their pore ordering is prone to be followed by x-ray diffraction (XRD) patterns, as the

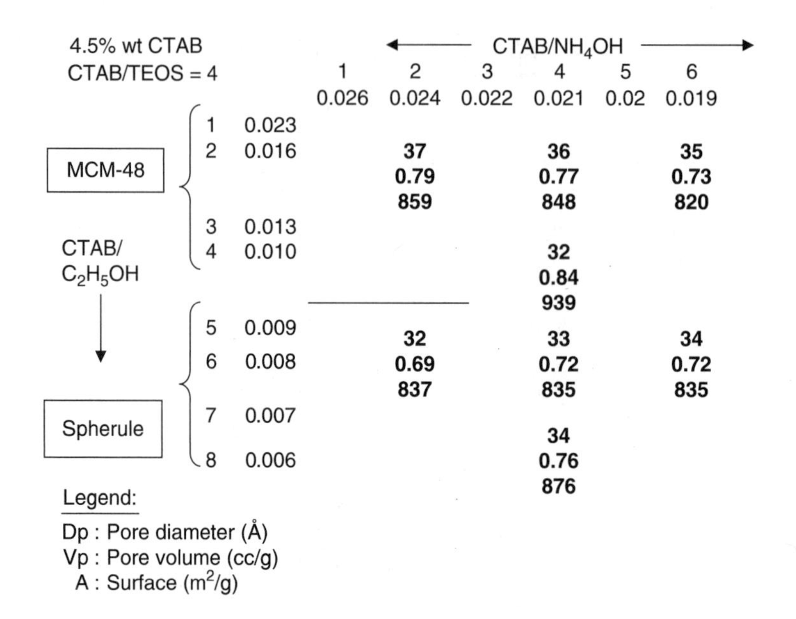

FIGURE 15.8 Textural properties of EtOH-MSS.

ones shown in Figure 15.9. This illustrates the main peaks of the MSS materials prepared with ethanol (Figure 15.9a) and other co-surfactants (Figure 15.9b), like Di-MethylHydrazine (DMH), Di-MethylFormamide (DMF) and Di-MethylSulfoxide (DMS). As it is observed, the XRD profiles of the EtOH-MSS materials vary along the compositional axis of column four, that is, from the second row at the bottom to the eighth row at the top of Figure 15.9a. The symmetry of the XRD profile decreases along with increasing the CTAB/EtOH ratio, thus indicating the loss of pore ordering with respect to the materials having lower CTAB/EtOH ratios. In contrast to this case, Figure 15.9b shows the XRD profiles of the MSS materials prepared with DMH, DMF and DMS, which are very similar to each other, for example the d-values are equal to 35.1, 33.6 and 34.6 Å. Furthermore, the XRD peaks maintain about the same width. This indicates that no significant variations of the pore ordering occur with the addition of the co-surfactants. Also, this indicates a high reproducibility achieved when using the HT and Combinatorial methods for synthesizing MSS materials.

B. Evaluation of the Catalytic Properties with the MFBR System

The Multi Fixed Bed Reactors system (MFBR) is the heart of the secondary screening unit used in this work [49]. Figure 15.3 shows a single reactor, which allows the introduction of gas or vapor reactants by the top, following either the capillary exhaust or passing through the catalyst bed. The latter is

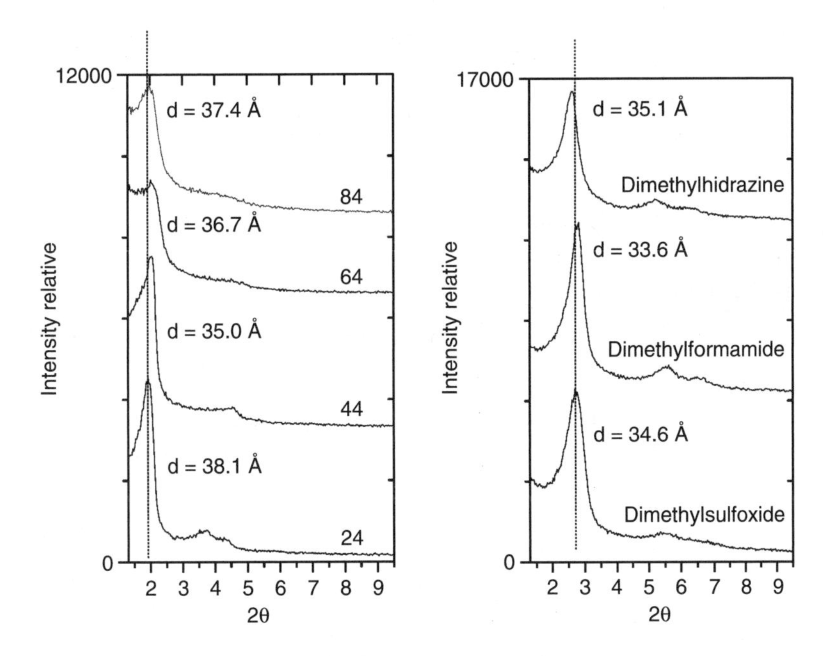

FIGURE 15.9 Structural properties of MSS materials prepared by HT and Combinatorial methods at T = 25°C, by X-ray Diffraction.

usually composed by a powder packed softly inside the microtube, where a microporous quartz paper holds the powder in place. The system consists of six heads with eight microreactors each, which are connected to the detection unit by a series of revolving valves. The gas exhaust or the product gases follow their way towards the battery of six GC, which can analyze the products in a conventional way.

For a correct catalysts testing a quality guide must be followed; one having more acceptance is the Total Quality Management for Catalysts Testing [54]. At this point, the effective experimental strategies apply for enhancing both effectiveness and efficiency. A series of methods available for experimental planning and optimization were reported previously [55, 56], which may lead to obtaining more information per experiment, significant time savings and organized collections of results.

Following the guidelines reported elsewhere [57] the relative activities of two catalysts can be determined using the following equation:

$$a/WHSV = \tilde{\int} dX_B/R_BM_B/X_Bf(X_B,C_{Bo},T)$$

where function $f(X_B, C_{Bo}, T)$ depends on the conversion of reactant B (X_B), the feed concentration of reactant B (C_{Bo}) and temperature (T). This relationship assumes implicitly the validity of the same rate law for all the catalysts under evaluation. In a practical case, a series of catalysts are tested and compared at the same temperature, with the same feed composition and if the experiment is performed up to the same final conversion, then the activity ratio of two catalysts, numbers 1 and 2, is a function of the space velocities:

$$a_2/(WHSV)_2 = a_1/(WHSV)_1$$

According to this definition, the relative activities of two catalysts can be obtained without knowing function f, but they may be readily compared by fixing the temperature and varying the Weight Hourly Space Velocity (WHSV), to obtain a chosen degree of conversion [54]. This can be done with the MFBR system [34, 49], where space velocity can be varied individually for each reactor across the 48 library members.

It is possible to compare the activity of two catalysts by measuring the temperatures at which the catalysts give the same conversion under the same feed concentration, then:

$$a_2/a_1 = f(X_B, C_{Bo}, T_2)/f(X_B, C_{Bo}, T_1)$$

In this way the catalyst presenting more activity will achieve the same conversion at lower temperature, but in this case there is not a quantitative evaluation, unless the function f (X_B, C_{Bo}, T) is known. Therefore, similar catalysts having a similar kinetics will be more readily compared by fixing the temperature and varying only the space velocity to obtain a specific degree of conversion.

One of the experimental tests recommended for diagnosing the undesirable gradients present in fixed bed reactors during catalysts evaluation, i.e. intrareactor, interphase or intraparticle gradients, consists of measuring the reaction conversion while varying the flow rate, without varying the space velocity. Thus, the region where conversion remains constant is free of the influence of interphase and intrareactor effects, which in turn ensures that the activity measurements are not disguised by heat or mass transfer effects. In the MFBR system these effects are minimized because of the small diameter of the microreactors and the small particle size that are typical of the materials tested in those units.

Another feature of importance in measuring and comparing catalytic activities is the influence of deactivation rate, due to the overall loss of activity with reaction time. Thus, a continuous evaluation under stream is important and it can be performed using the MFBR system.

A library consisting of distinct composition matters may generate an activity plot of the 48-reactor MFBR system, as illustrated in Figure 15.10. The bars correspond to the total conversion of n-hexane across the library members, under the same temperature and reactant flow concentration. In this particular array only S2 and S6 rows correspond to the same set of materials, those having identical chemical and physical properties. Thus, a similar conversion is verified along those rows, while the other rows have much

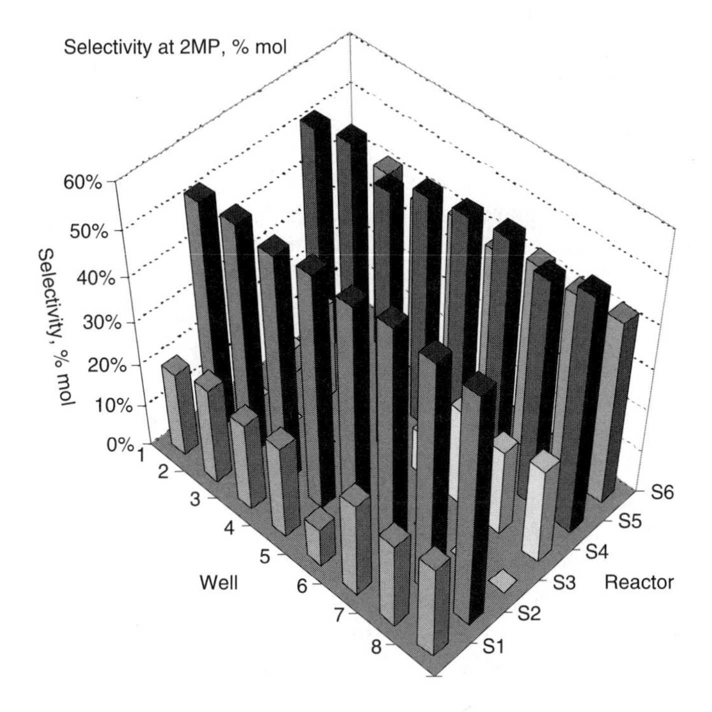

FIGURE 15.10 HT and Combinatorial evaluation of catalytic properties by means of the Multi Fixed Bed Reactor System (MFBR).

different values. Cross measurements of catalytic conversion in the 48 library members have given a figure of merit of less than 1% mean fluctuation, thus having an acceptable confidence level. Nevertheless, the best practice recommends taking reference wells at random across the 48 library members, i.e. standard catalytic materials are tested first in the single-composition mode across the 48 library members and, once the confidence level is validated, they are distributed randomly across the library, placed between the other materials to evaluate, in order to verify their activity pattern and, in this way, to have a representative activity picture.

V. DISCUSSION AND MAIN CONCLUSIONS

In the search for new catalytic materials the High Throughput and Combinatorial Catalysts methods are powerful techniques for exploring the multiparameter space. The primary synthesis and screening methods use tiny amounts of samples, usually in the range between 0.1 to 1 mg each, thus allowing the rapid preparation and testing of thousands of compounds. In this case, infrared detection methods (nonquantitative), REMPI (Resonance Enhanced Multi Photon Ionization) or Mass Spectrometry are commonly used. The secondary synthesis and screening methods deal with samples of about 10–100 mg each and the catalytic tests are performed in a parallel flow-through reactor system, fitted with parallel detection and quantification techniques like MS or GC. The MFBR systems are able perform the secondary screening of catalytic materials, including the quantitative product analysis (MS and GC), thus making it possible to study the selectivity, as well as kinetic and deactivation studies. This system allows varying the feed composition (gas or vapor), the space velocity and temperature, which are common variables that influence the catalytic activity of materials having similar kinetic behavior. A study of the intrinsic properties of catalytic materials without the influence of heat and mass transfer is of interest for reaction engineering and kinetics, because these are two basic ingredients for catalysts and reactor design. In the industrially-oriented R&D of new catalysts the HT and Combinatorial methods are visualized as complementary rather than competitors of conventional pilot plant and semi-industrial methods. In fact, the HT and Combinatorial methods extend the capabilities of searching out the multi-parameter space in a rapid way, thus leading the way for discovery of new catalysts and their optimization. Also, new challenges arise with the advent of these novel methods, as for example the definition of the best strategies for data mining and data managing for dealing with massive amounts of information [58–60]. Also, some fields of future work are scaling up and reproducibility of the general properties of materials synthesized at the microscale level. In the present work and other recent publications the synthesis of porous aluminosilicate materials at the microscale level may produce structural and textural properties that are similar to bulk materials [34, 50]. Other complex systems like the mixed oxides will require further attention in order to look for the appropriate routes for the synthesis and evaluation under controlled conditions.

REFERENCES

1. Chemical & Petroleum Catalysts, *Industrial Study 1178*, Freedonia Group Inc., Cleveland OH, USA, *Freedonia Report*, 1999.
2. Advanced Catalysts, Spurring Markets in Petroleum Refining, Chemicals, Polymers and the Environment, *Report D222*, Technical Insights, Frost & Sullivan, NY, USA 2001.
3. Courty PR, Chauvel A, Catalysis the turntable for a clean future, *Catal, Today*, 29:3–5, 1996.
4. Roth JF, Future Catalysis for the Production of Chemicals, in Catalysis 1987, *Studies on Surface Science and Catalysis*, No. 38, Elsevier, Ward JW, Ed. 925–943, 1988.
5. Roth JF, *Chemtech*, 357, 1991.
6. Kulakowsky M, Reformulated Gasoline: Defining the Challenge, ACS Symposium on the Impact of Reformulated Fuels, *ACS Division Petr. Chem. Prepr.*, 39:494, 1994.
7. Martino G, Catalysts for Oil Refining and Petrochemistry, Recent Developments and Future Trends, Proc. 12th International Congress on Catalysis, Granada (Spain), *Studies in Surface Science and Catalysis*, 130:83, 2000.

8. Derouane EG, Lemos F, Corma A, Ramoa F, Eds. Rostrup-Nielsen JR, Scale-up of catalytic processes, *Combinatorial Catalysis and High Throughput Catalyst Design and Testing*, 337–361, Ed. Kluwer Academic Publishers, The Netherlands, 2000.
9. Fogler HS, *Elements of Chemical Reaction Engineering*, 3rd Ed., Prentice Hall International Series, Inc., 1999.
10. Veser G, Friedrich G, Freygang M, Zengerle R, A Simple and flexible microreactor for investigation of heterogeneous catalytic gas phase reactions, in Reaction kinetics and development of catalytic processes, Froment GF, Waugh KC, Eds., *Studies in Surface Science and Catalysis*, 122:237–246, Elsevier, Amsterdam, 1999.
11. Kaptejin F, Marin GB, Moulijn JA, *Catalysis: An integrated Aprroach to Homogeneous, Heterogeneous and Industrial Catalysis*, pp. 251–306, Elsevier, Amsterdam, 1993.
12. Schlögl R, *Angew. Chem. Int., Ed.*, 37(17):2333–2336, 1998.
13. Le Page JF, *Applied Heterogeneous Catalysis*, Ed. Technip, Paris, 1986.
14. Thomas JM, Somorjai GA, Catalyst characterization under reaction conditions, *Topics in Catal.*, 8:1, 1999.
15. Hutchings GJ, Desmartin-Chomel A, Ollier R, Volta JC, *Nature*, 368:41, 1994.
16. Undergaard NR, Bak Hansen JH, Hanson DC, Stal JA, *Oil & Gas J.* 90:62, 1992.
17. Rostrup-Nielsen JR, *Chem. Eng. Sci.*, 50:4061, 1995.
18. Kramer CJ, Van Santen RA, Emeis CA, Nowak AK, *Nature*, 363:529, 1993.
19. Gay DH, Rohl AL, Chem J, *Soc. Faraday Trans.*, 91(5):925, 1994.
20. Gallop MA, Barret RW, Dower WJ, Fodor SPA, Gordon EM, Applications of combinatorial technologies for drug discovery, 1.—Background and peptide combinatorial libraries, *J. Med. Chem.*, 37:1233–1251, 1994.
21. Thompson LA, Ellman JA, Synthesis and applications of small molecule libraries, *Chem. Rev.*, 96:555–600, 1996.
22. Danielson E, Golden J, Mc Farland E, Reaves C, Weinberg WH, Wu XD, A combinatorial approach to the discovery of new luminescent materials, *Nature*, 389:944–948, 1997.
23. Xiang XD, Sun X, Briceno G, Lou Y, Wang KA, Chang H, Wallace-Freedman WG, Chen SY, Schultz PG, *Science*, 268:1738, 1995.
24. Briceno G, Chang H, Sun X, Schultz PG, Xiang XD, *Science*, 270:273, 1995.
25. Francis MB, Finney NS, Jacobsen EN, *Am. J. Chem. Soc.*, 118:8983, 1996.
26. Jandeleit B, Schaefer DJ, Powers TS, Turner HW, Weinberg WH, Combinatorial Materials Science and Catalysis, *Angew. Chem. Int. Ed.*, 38:2494–2532, 1999.
27. Moates FC, Somani M, Annamalai M, Richardson JT, Luss D, Wilson RC, *Ind. Eng. Chem. Res.*, 34:4801, 1996.
28. Holzwarth A, Schmidt PW, Maier WE, *Angew. Chem. Int. Ed.*, 37:2644, 1998.
29. Reetz MT, Becker MH, Kuhling KM, Holzwarth A, *Angew. Chem.*, 110:2792, 1998.
30. Senkan S, *Nature*, 394:350, 1998.
31. Senkan S, *Angew. Chem. Int. Ed.*, 40:312–329, 2001.
32. Hoffman C, Wolf A, Schüth F, *Angew. Chem.*, 111:2971, 1999.
33. Senkan S, Krantz K, Ozturk S, Zengin V, Onal I, *Angew. Chem.*, 111:2965, 1999.
34. Domínguez JM, HT Synthesis & Secondary Screening of NewCatlysts Libraries by Combnatorial Methods, *Combicat 2002*, The Catalyst Group Resources Inc., Lisbon, April 2002.

35. Schueth F, Busch O, Hoffmann C, Johann T, Kiener C, Demuth D, Klein J, Schunk S, Strehlau W, Zech T, *Topics in Catal.*, 21:55, 2002.
36. U.S. Patent No. 6149882 (2000), U.S. Patent No. 6410331 (2002) and Eur. Patent No. 1001846 (2002).
37. Mittasch A, Geschichte der Ammoniaksynthese, *Verlag Chemie.*, Weinheim, 113, 1951.
38. Mittasch A, *Adv. Catal.*, 2:81, 1950.
39. Maier WF, Combinatorial Chemistry—Challenge and chance for the development of new catalysts and materials, *Angew. Chem. Int. Ed.*, 38:9, 1999.
40. Domínguez JM, Terrés E, Montoya A, Armendariz H, Libraries design and synthesis of high surface area materials at IMP, presented at *ACS Div. Petr. Chem. Inc., 221st Natl. Meet. ACS*, San Diego, CA, April 1–5, p. 51, 2001.
41. Danielson E, Devenney M, Giaquinta EM, Golden JH, Haushalter RC, McFarland EW, Poojary DM, Reaves CM, Weinberg WH, *Science*, 279:837, 1998.
42. Cong P, Doolen RD, Fan Q, Giaquinta DM, Guan S, McFarland EW, Poojary DM, Self K, Turner HW, Weinberg WH, *Angew. Chem. Int. Ed.* 38:484, 1999.
43. Weinberg WH, Jandeleit B, Self K and Turner H, *Curr. Opin. Solid State Mater. Sci.*, 3:104, 1998.
44. Marshall S, *Res. Develop.*, 40:32, 1998.
45. Reddington E, Sapienza A, Guraou B, Viswanathan R, Sarangapani S, Smotkin ES, Mallouk TE, *Science*, 280:1735, 1998.
46. Taylor SJ, Morken JP, *Science*, 280:267, 1998.
47. Haap WJ, Walk TB, Jung G, *Angew. Chem.*, 110:3506, 1998.
48. US Patent No. 5959297 (1999).
49. U.S. Patent No. 6149882 (2000) and U.S. Patent 6149692 (1002).
50. Klein J, Lehmann CW, Schmidt HW, Meier W, *Angew. Chem. Int. Ed.*, 37(24):3369, 1998.
51. Bergh S, Cong P, Ehnebuske B, Guan S, Hagemeyer A, Lin H, Liu Y, Lugmair CG, Turner HW, Volpe AF Jr, Weinberg WH, Woo L, Zysk J, *Topics in catal.*, 23:65–79, 2003.
52. Werner H, Timpe O, Herein D, Uchida Y, Pfänder N, Wild U, Schlögl R, *Catal. Lett.*, 44:153, 1997.
53. Muhler M, Schlögl R, Ertl G, *J. Catal.*, 138:413, 1995.
54. Dautzenberg FM, in Ten Guidelines for Catalysts Testing, ACS Symposium Series, No. 411, *Characterization and Catalysts Development*, 1989, Bradley SA, Galhuso MJ, Bertolacini RJ, Eds., Los Angeles, CA.
55. Hendrix CD, *Chem. Technology*, 9(3):167, 1980.
56. Draper NR, Smith H, in *Applied Regresion Analysis*, John Wiley & Sons, New York, 1981.
57. Dautzenberg FM, Quality principles for catalyst testing during process development, *Combinatorial Catalysis and High Throughput Catalyst Design and Testing*, Ed. Kluwer Academic Publishers, The Netherlands, pp. 61–98, 2000.
58. Newsam JM, Freeman CM, Yao T, High Throughput Experimentation: The Role of Simulation, *Chemistry Today*, 31–37, 1998.
59. Dorsett Jr. DR, 222th *Nat. Meet. ACS*, Chicago USA, IL, Aug. 26–30, 2001.
60. Newsam JM, Design of catalysts and catalysts libraries: Computational techniques in HT experimentation for catalysis, in *Combinatorial Catalysis and HT Catalyst Design and Testing*, Derouane *et al.*, Eds., Kluwer Academic Publishers, The Netherlands, pp. 301–335, 2000.

16 Biological Libraries

Maria Dani

CONTENTS

I. INTRODUCTION

A biological display library consists of a pool of microorganisms expressing different polypeptides on their surfaces. Each microorganism displays only one type of peptide sequence and represents a clone. Each clone of the library can be replicated many times, and the progeny will express the same polypeptide on the surface. The polypeptide is coded by a specific sequence of DNA inserted into the microorganism during library construction (either into its genome or into a plasmid) (Figure 16.1). Because the polypeptide must be expressed on the surface of the microorganism, it must be fused to certain specific proteins of the outer cell membrane or the external envelope. If the fusion is successful, each polypeptide of the library will then be exposed to the medium.

A random polypeptide library construction starts with chemical synthesis of degenerated oligonucleotides. The pool of single-stranded oligonucleotides is then made double stranded, ligated into appropriate linearized vectors, and inserted into host cells for replication and display. A biological display library can also be constructed from cDNA, genomic DNA fragments,

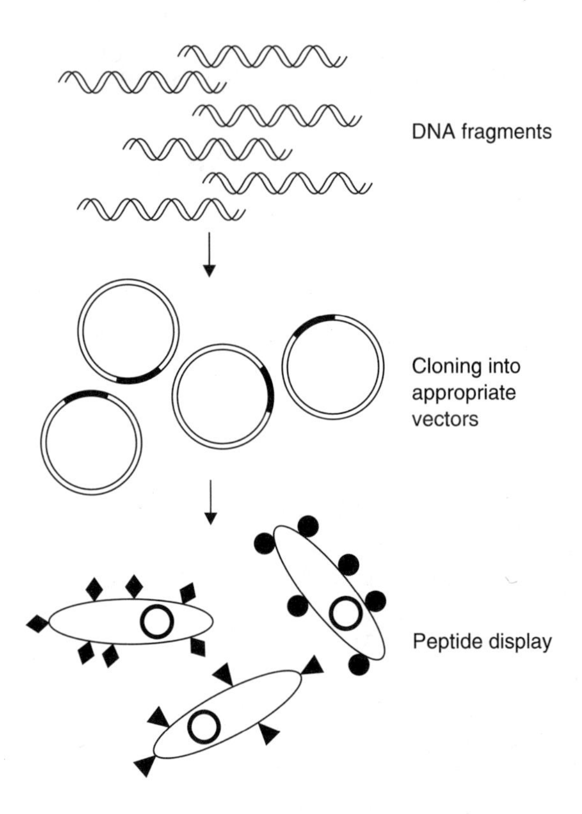

FIGURE 16.1 Biological display library construction. DNA fragments for coding for different peptides or proteins are cloned into appropriate vectors and inserted into specific organisms that synthesize and display the peptides on their surface.

or mutagenized specific gene fragments. Depending on the system used, the library will be constituted by viral particles or by cells.

Once the library has been constructed or acquired, the screening, called *biopanning*, is relatively simple (Figure 16.2). In about a week, the entire biological library can be screened against the target molecule, simply by incubating the microorganisms with the target bound to a solid support. The unbound microorganisms are washed away and the bound ones can be eluted by several methods, such as low pH, high concentration of free target, or direct infection of cells by bound phages. The eluted microorganisms are grown and reselected over the target molecule two or more times. At the end, they are grown at low density on solid medium and single clones are isolated. The DNA of about 30–50 clones is extracted and the portion coding for the polypeptides is put in sequence. The DNA sequences are translated into peptides and compared. If a consensus sequence is identified, the screening must have been successful. In this case it was necessary to test each peptide for its binding capacity, in order to confirm the results and identify the best candidate (Figure 16.3). This technique was first published

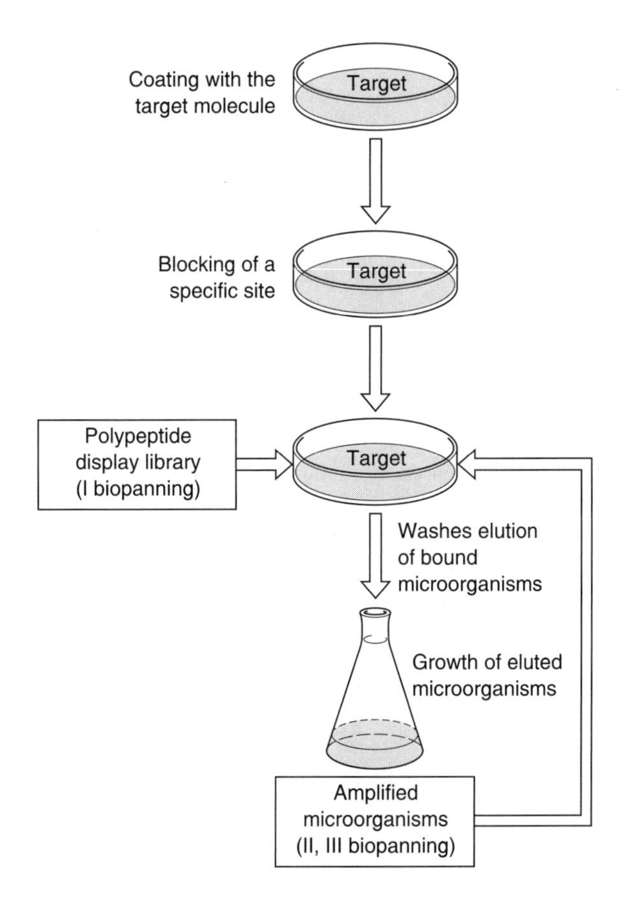

FIGURE 16.2 Biopanning. In a typical experiment the target molecule is bound to a solid support and aspecific sites are saturated. The display library is then incubated with the target and the unbound clones are eluted, amplified, and used for two or more rounds of selection.

in 1985 by Smith [1]. He demonstrated that foreign fragments could be inserted into a bacteriophage gene to create a fusion protein that is displayed on the surface of the virion. He was able to enrich more than 1000-fold the "fusion phage" over ordinary phages by using, as a target, an antibody directed against the foreign DNA fragment product. Starting from the beginning of the 1990s, this technique has spread to different fields of research and is now used widely for many different applications.

Biological libraries can display random polypeptides, protein fragments, mutagenized portions of a protein, entire proteins, and antibody fragments. The use of biological display libraries allows one to quickly design and engineer stable affinity ligands capable of discriminating and selectively binding a target protein in the presence of even distantly related impurities. This makes it ideally suited for use in production related chromatographic applications such as the purification of proteins. Other applications are study of protein–protein

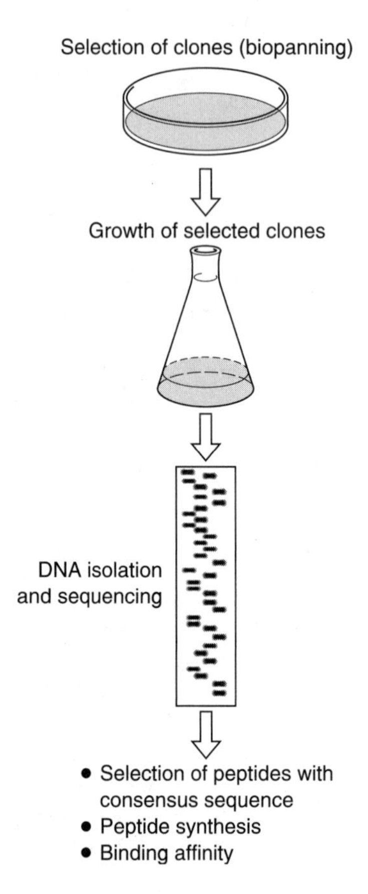

FIGURE 16.3 Analysis of biopanning selected clones.

interaction, receptor-binding site identification, isolation of targeting molecules, and drugs discovery. Antibody display libraries are used to complement hybridoma technology and represent a powerful tool for the selection and cloning of antibody fragments, either as Fabs or as single-chain Fvs. antibody fragments directed against a variety of target antigens that have been selected. Some of them are potentially useful therapeutic molecules for pathological conditions in humans. New applications of biological display libraries are emerging, such as expression of antigens on the surfaces of nonvirulent microorganisms for live vaccines, expression of antibodies or receptors for analytical applications, and bioseparations [2].

Several microorganisms can be used for surface display: bacteriophages, Gram-positive and Gram-negative bacteria, and yeasts. The most widely used microorganisms for peptide display libraries are *Escherichia coli* filamentous bacteriophages of the Ff class. During their normal life cycle, these phages infect *E. coli* by injecting their single-stranded genome into the bacterial cell and start to replicate their DNA. The DNA is packaged into phage particles across the bacterial membrane and secreted from the cell. Several coat proteins have been used for the display of peptides and proteins.

The lambda bacteriophage is a more appropriate vector for the display of proteins that fold up in the cytoplasm. Here virus secretion is not required for phage display. Indeed, lambda is a virus that, in its lytic cycle, assembles the viral particles intracellularly. Both Gram-negative and Gram-positive bacteria have been used for protein display on the cell surface, but only the Gram-negative *E. coli* has been used for polypeptide libraries. A recent alternative is the use of yeast cells, which are well suited for the display of mammalian proteins that require specific post-translational processing.

Biological libraries have several advantages and disadvantages over chemically synthesized libraries. One advantage is the possibility of regenerating the library simply by growing the microorganisms. However, this cannot be done too many times, because there can be a selective pressure on some of the clones that replicate either faster or slower. Furthermore, some of the displayed peptides or proteins may be toxic or may be translated less efficiently. A biological library can be made of up to 10^9 independent clones, compared to 10^6–10^7 in chemical peptide libraries. If a particular in-vivo recombination system is used, the size can even reach 10^{10}. The limit is due to the transformation efficiency of the microorganism used and the volume of cells that can be handled. In chemical libraries the diversity is 20^n (20 is the number of the different amino acids and n is the number of randomized positions). For example, a complete library constituted of five amino acids peptides will have 3.2×10^6 different molecules. In biological libraries, because of the codon degeneracy, some amino acids are more represented than others, while in chemical libraries there is no bias toward specific amino acids. The result is that not all the amino acids are evenly distributed as in chemical libraries, and more clones are necessary to have a complete library. A biological random peptide display library can be made of long random polypeptides. If the randomized positions total more than 7, the library is not complete. In most cases the binding region is limited to a few residues. Since a long variable peptide (e.g. 15 amino acids) contains within its sequence several short (e.g. 5 amino acids) peptide fractions, the total number of different short peptide sequences are higher than the number of different polypeptides that constitute the library. Furthermore, long random peptides allow affinity selection of peptide ligands that require the interaction of few small structural elements.

This chapter focuses on different microorganisms used for the construction of biological libraries and the applications of this powerful technique.

II. TYPES OF DISPLAY

A. Phage Display Libraries

The *E. coli* filamentous bacteriophages, such as M13, fd, and fl, are the most commonly used viruses in biological display systems. Their genomes are small and easily used as vectors for the construction of large display libraries. Unlike most other bacteriophages, they do not produce cell lysis. The infection begins with the attachment of the phage particle to the f pilus of a male *E. coli* cell mediated by the pIII viral protein. As the phage enters the cell, its coat protein is removed and the DNA starts to replicate using the host system. The DNA is then packaged into new phage particles, which are extruded through the cell wall into the medium without disrupting the host cell growth.

The filamentous bacteriophages M 13 coat for 10 different proteins. The single-stranded DNA genome is packaged by viral coat proteins (Figure 16.4A). Two types of proteins are usually used for phage display—pIII and pVIII. The pIII coat protein is present in 3–5 copies incorporated into one tip of the rod-shaped virion. It is 406 amino acids long and folds into two domains: the carboxy-terminal domain interacting with viral coat proteins and is required for viral assembly; the amino-terminal domain of the protein projects away from the virion and is responsible for the male *E. coli* infection by binding to F pili. The pIII protein can tolerate the insertion of large peptides without affecting the infectivity.

pVIII constitutes the most abundant tubular coat protein and is present in about 2700 copies. It is 50 residues long, and the amino-terminal end protrudes into the medium. The pVIII protein tolerates very short (up to 6 amino acids) inserts. Larger inserts are tolerated in the two-gene system. In spite of the lower copy number, it was demonstrated that more single-chain Fv (scFv) can be expressed with pIII than with pVIII fusion [3].

In filamentous phages two systems are used—the polyvalent display (or one gene system) and the monovalent display (or two-gene system [4]). In the

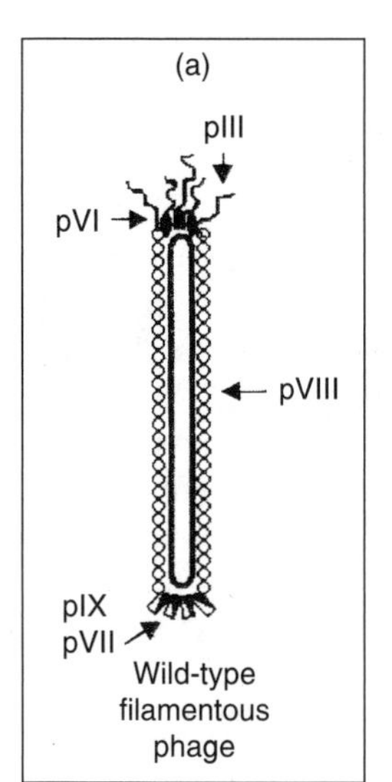

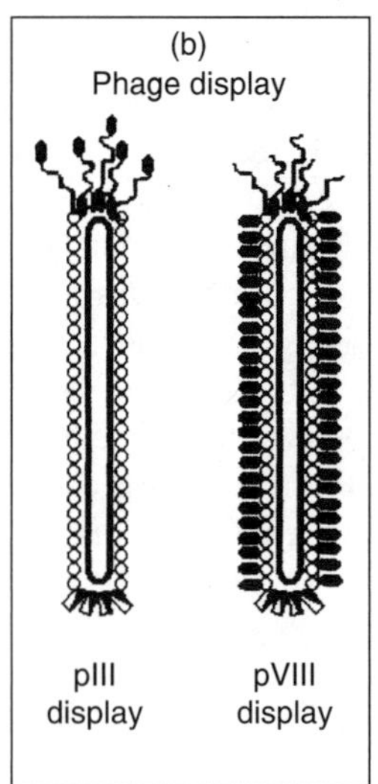

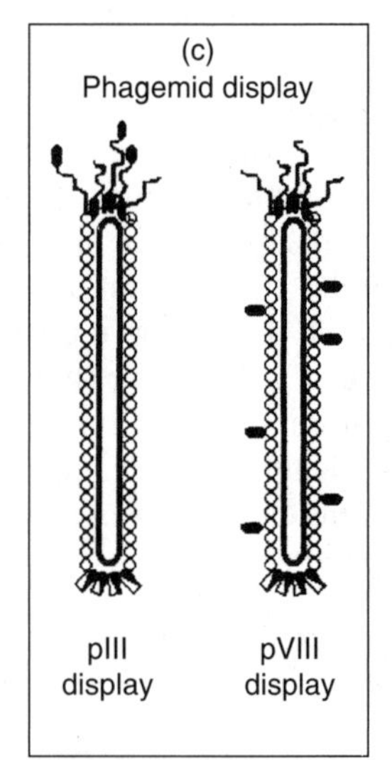

FIGURE 16.4 Schematic representation of phage (b) and phagemid (c) display of polypeptides fused to pIII and pVIII coat proteins, compared to wild-type phage (a).

polyvalent display (Figure 16.4B), the DNA fragments coding for the peptides are inserted into the phage vector, usually between a particular coat protein gene and its single peptide. All the coat protein molecules produced are fused to the peptide; consequently, the number of displayed peptides correspond to the number of coat proteins. In the monovalent display (Figure 16.4C), the phage genome is modified and the defective phage is called "phagemid," being something in between a phage and a plasmid. A phagemid contains the sequences needed for packaging into virions, but does not encode viral genes. When a cell harboring a phagemid is infected by a filamentous helper phage, virions are produced, because the helper phage supplies the defective genes. The virions produced display a mixture of recombinant coat proteins coded by the phagemid gene and the corresponding wild-type protein coded by the helper phage gene.

The one-gene system display (phage) is a good system for the display of small peptides, which do not alter the structure of the coat protein and do not prevent its assembly into viral particles. For the display of larger peptides the two-gene system (phagemid) is preferred or sometimes required, as for pVIII fusion.

The phage system is ideal for the construction of random peptide libraries, and many examples are found in the literature (Table 16.1). The libraries are easily constructed and propagated. It is possible to obtain cultures with high titers and also easily concentrate the viral particles by precipitation. This allows the possibility of performing the screening in small volumes even with very large libraries ($>10^9$ independent clones) and have all the library representation in a volume as small as 1–10 μL. Furthermore, the viral particles are stable, allowing elution of the phages bound to the target molecules in several ways, including low pH (2.2). Filamentous phages were also used for the display of various proteins or protein fragments, such as receptors, enzymes, growth factors, cytochines, and antibodies.

The lambda vector is more appropriate for the expression on its surface of foreign proteins that fold up in the cytoplasm, and secretion is not required. In fact, the bacteriophage lambda in the lytic cycle assembles intracellularly prior to release of viral particles from *E. coli* cells (Figure 16.5). The proteins used for the display are pV and D (decoration). pV is a major tail protein arranged in 32 disks each formed by 6 subunits. The C-terminal of this protein is not very essential and can be replaced by a foreign protein without altering phage assembly and infectivity. An example of pV display is the λfoo vector [5]. The 11-kDa D protein is essential for phage head morphogenesis and phage prohead while it fills with DNA. The D protein is assembled as trimers. Stenberg and Hoess [6] constructed a λ vector which is able to display peptides and proteins on its surface, fused to the amino-terminus of the D protein of the capsid.

Other bacteriophages have been tested for polypeptide display. The minor fibrous protein fibritin of phage T4 has been fused at the C-terminus with a polypeptide of 53 residues [7]. An antigenic peptide has been fused

TABLE 16.1
Examples of display peptide library screening on specific targets

Target	Length or type of random polypeptide libraries	References
Calveolin-scaffolding domain	15-mer	75
	10-mer	
NMDA glutamate receptor	10-mer	76
	15-mer	
Src-homology 2 domain of Grb2	Phosphopeptides	77
Enzyme I of the bacterial phosphoenolpyruvate-sugar phosphotransferase system	15-mer 10-mer 6-mer	78
Taq DNA polymerase Human insulin Apolipoprotein A-1	Randomization of 13 residues of α-helical bacterial receptor domain Z	81
Mouse anti-DNA antibody	10-mer	80
Phosphatase-1		82
α-Bungarotoxin	15-mer	79
Toxic shock syndrome toxin-1	15-mer	83
Mouse organs	Constrained pools	43
Human type I interleukin I receptor	8-mer, 9-mer, 10-mer, 12-mer	84
Human plasma kallikrein and human trombin	Variants of the first Kunitz domain of human lipoprotein-associated coagulation inhibitor	85
Human plasmin		107
mAb TL4 against Ap4A receptor	6-mer	86
Erytropoietin receptor	Cyclic 8-mer	87
Tyrosine kinase Itk/Tsk homology 3 domain	6-mer, 5-mer + fixed sequences	88
	10-mer	
Streptavidin	Cyclic 8-. 7-, 6-mer	89
Fibronectin	Cyclic 8-mer	90
Integrins	Cyclic pools 5-, 6-, 7-mers	91
	6-mer, 15-mer	92
	6-mer	93
Mouse CD1	22-mer	94
DNA	23-mer	95
	6-mer	96
Core antigen of hepatitis b virus	6-mer	97
Src homology 3 domain	15-mer	98
	44-, 25-mer	99
Surface immunoglobulin receptor of human B-cell lymphoma	8-, 12-mer	100
Tumor suppressor protein p53	6-, 12-, 20-mer	101
mAb 5.5 against the ligand-binding site ot the nicotinic acetylcholine receptor	6-mer	102
Bovine pancreatic RNase fragment	6-mer	103
concanavalin A	6-mer	104
	8-mer	105
Platelet glycoprotein IIb/IIIa	Cyclic 6-mer	106

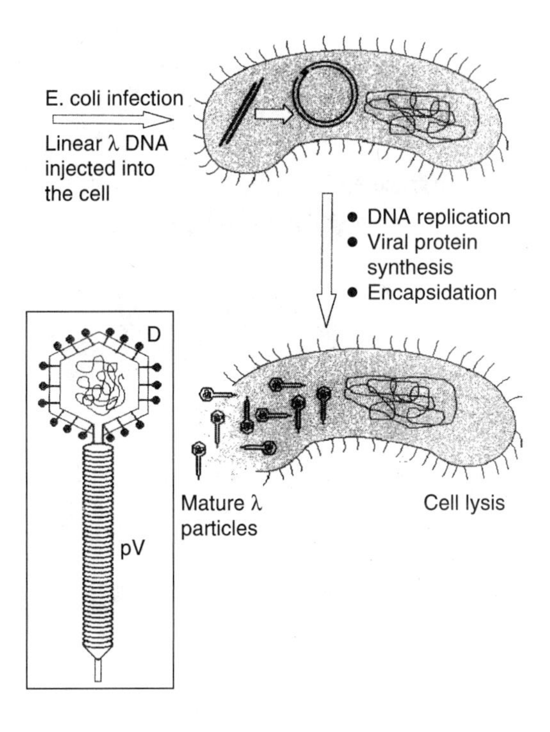

FIGURE 16.5 Lytic cycle of bacteriophage λ. D: decoration protein; pV: tail protein V.

to the C-terminus of the tailspike protein of *Salmonella typhimurium P22* bacteriophage [8].

B. Bacterial Display Libraries

Surface expression has been accomplished in Gram-positive and Gram-negative bacteria [9].

The Gram-negative bacteria has two membranes separated by the periplasmic space (Figure 16.6). Short peptides can be inserted into some surface exposed loops of the Outer Membrane Proteins (OMPs). The most used proteins are the maltoporin LamB, the outer membrane protein OmpA, and the phosphate inducible porin PhoE. LamB is an *E. coli* protein. Some permissive sites have been identified [10, 11], but the insertion of peptides longer than 60 amino acids perturbs its conformation and assembly. Another protein used is OmpA, which tolerates larger inserts [12, 13]. Lipoproteins (Lpp, TraT, and PAL) have also been used [14–16], as well as proteins present in the filamentous structures present on Gram-negative bacteria (FimA [17], FimH [18], P fibrillin [19], and flagellin [20]). Among the Gram-positive bacteria, nonpathogenic *Staphylococci* are used for the display, as well as the commensal bacterium *Streptococcus gordinii*, an attenuated mycobacterium, and *Bacillus subtilis*.

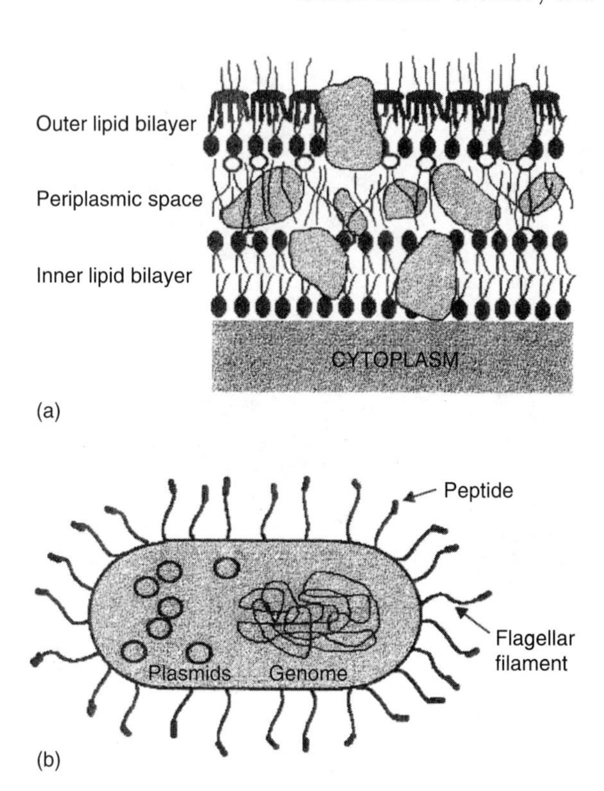

FIGURE 16.6 Bacteria display: (a) schematic representation of Gram-negative membrane; (b) example of flagellin thioredoxin display of polypeptides [20].

Bacterial display is used for antibodies and protein fragments rather than random peptide libraries. An exception is the thioredoxin-flagellin fusion display protein, in which random 12 amino acid peptides were inserted [20].

Despite the multivalent binding of the bacteria to the immobilized ligand and the large size of the cells, it is possible to isolate high-affinity peptides. One of the advantages of cell display is the possibility of using Fluorescence-Activated Cell Sorting (FACS) [21, 22]. Cells are incubated with fluorescently labeled target molecules, and those able to bind to the target can be separated. Cell sorting can highly enrich the positive clones and can discriminate between clones of different affinity and specificity. Furthermore, it allows screening with the target molecule in solution. In this way no elution is required, avoiding isolation of clones that bind nonspecifically to the solid support and elution problem of very tightly binding clones.

Display on bacteria can have various applications, such as the generation of recombinant bacterial vaccines, the screening of peptide libraries, and the use of cells as a source of immobilized proteins for purification or for enzymatic processes [2].

C. Yeast Display Libraries

Yeast display has been proposed recently as an alternative way to display mammalian proteins. Because some eukaryotic proteins expressed in *E. coli* are not in the soluble form, they cannot be incorporated into phage particles. Furthermore, phage display may sometimes select for reduced host toxicity or higher infectivity rather than increased affinity. Yeast display should alleviate expression biases present in *E. coli* and, as bacteria, have the advantage of high expression levels of displayed fusion polypeptides and selection through flow cytometric cell sorting.

Yeast cells are well suited for the display of mammalian surface and secreted proteins that require endoplasmic reticulum-specific post-translational processing in order to have the appropriate folding. In fact, the yeast *Saccharomyces cerevisiae* has a secretory and folding system very similar to that of mammalian cells. Two cell-surface mating adhesion receptors have been used for the display: α-agglutinin and a-agglutinin. Fusion to the C-terminal portion of α-agglutinin has been used for immobilizing enzymes and viral antigens on the surface of yeasts [23]. Boder [24] has displayed a functional antifluorescein scFv and c-myc epitope tag by fusion to a-agglutinin. Besides the isolation of high-affinity antibodies, yeast display can be used for the display of receptors, an important class of therapeutic drug target [25].

III. APPLICATIONS OF BIOLOGICAL DISPLAY LIBRARIES

A. Affinity Ligand Identification

Purification processes require several steps to obtain a commercially pure product. Affinity chromatography is therefore extremely useful. Until a few years ago, it was limited to antibodies produced by the immune systems of laboratory animals. However, antibodies often are unable to discriminate between closely related impurities. In addition, the drastic sanitation conditions used in the production of therapeutic products may denature the antibodies. Phage display technology allows the isolation of affinity ligands with the required physical and chemical properties. This technique also discriminates between the target and closely related impurities. Small peptides bound to resins are well suited for use in purification of proteins that are normally used as drugs. Once a peptide with a good affinity has been selected, it is possible to improve its binding affinity by modifying the amino acid composition at specific sites. These variants can be screened again based on conditions used for the purification and hence identify a commercially valuable ligand.

Phage display is the best system for random peptide library construction used for affinity ligand selection. Several examples are reported in Table 16.1. Libraries of 6- to 15-mers are usually used, and peptides are isolated to a variety of targets including receptors, enzymes, antibodies, DNA, toxins, lectines, and even organs.

B. Surface Display of Antibodies

Antibodies effect humoral immunity in vertebrates. They act as surface receptors on B cells which mature into plasma cells and secrete antibodies on binding with a specific antigen. The antigen-binding site is formed by two variable domains of about 110 amino acids, one from the heavy (V_H) and one from the light (V_L) chain [26]. The remainders of the heavy and light chains are constant regions.

Mammals can generate a vast array of antibodies against any antigen. This is possible because genes that encode antibodies' variable regions are assembled from multiple germ-line segments, which undergo somatic mutation and recombination [27] and consequently a continuous generation of novel clones are produced. Furthermore, the combinatorial assortment of different segments of the variable region and of the light and heavy chains contributes to the generation of a nearly limitless variety of antibodies. After antigen activation of a particular specific clone, this will proliferate, mature into antibody-producing cells, and then into memory cells. Antibodies are extremely important tools. They are used in many applications, including research, affinity purification, and diagnostic and therapeutic agents. Variable domains can function as the entire molecule and retain the antigen-binding capacity. Variable domains can be displayed on phages, phagemids, and bacteria, either as scFv fragments in which V_H and V_L chains are covalently linked, or as Fab fragments in which the two chains interact in a noncovalent way. Three types of antibody libraries can be constructed; synthetic, naïve, and antigen-biased [28] (Figure 16.7). The sources of synthetic libraries are the

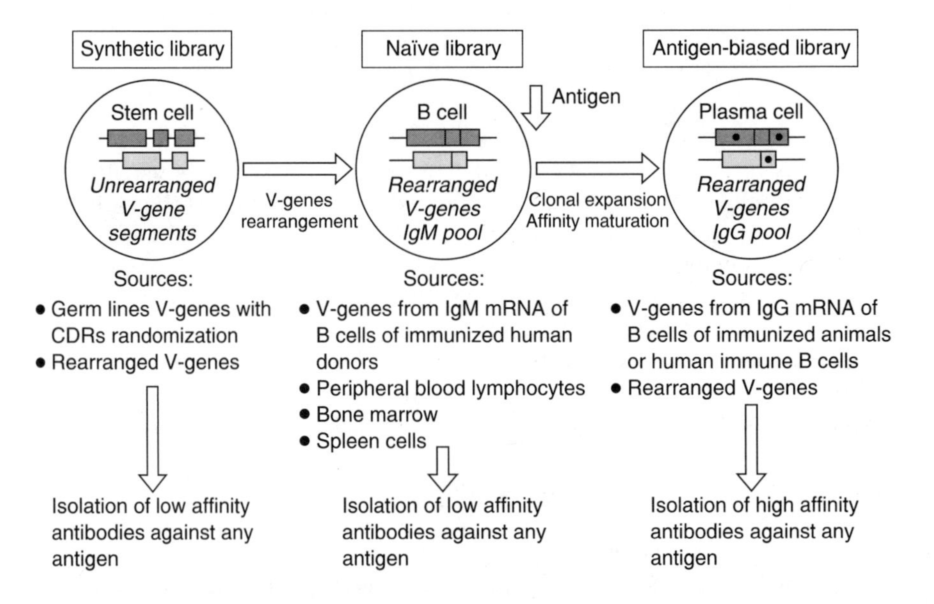

FIGURE 16.7 Different types of antibodies display libraries.

germ-line V genes from stem cells with randomization made in vitro by PCR of the variable CDRs (complementarity-determining regions). This library allows the isolation of low-affinity antibodies against any antigen. The source of naive library is B-cell mRNA of rearranged V genes. The antibodies isolated from a naive library are low-affinity antibodies against any antigen. In the antigen-biased library, the sources are rearranged V genes from immunoglobin G (IgG) mRNA of plasma B cells after clonal expansion and affinity maturation.

The antibodies isolated from antigen-biased libraries are high-affinity antibodies against specific immunogens. Phage antibody libraries are enriched for antigen-binding clones by subjecting the phages to multiple rounds of selection as with random peptide libraries. This technique is an alternative way to obtain monoclonal antibodies against any antigen without using laboratory animals or hybridomas [29]. Antibodies have been elicited against a variety of different molecules, such as haptens, proteins, polysaccharides, and DNA [30–33]. Using mutagenesis techniques of isolated antibodies, it was possible to identify variants with affinity in the picomolar range [34, 35], demonstrating the possibility of creating antibodies with an affinity and specificity not previously envisaged.

C. Pharmaceutical Applications

There are many therapeutic targets for intractable diseases, and new biological targets are being discovered. Pharmaceutical companies are interested in new compounds that can interact with these targets. The peptides isolated by screening display libraries against specific target molecules can be potentially used as drugs. However, a major drawback is the degradation of the peptides by proteolytic enzymes. Naturally synthesized peptides, such as those displayed in biological libraries, are in the L form. It has been demonstrated that it is possible to identify D-peptide ligands through mirror-image phage display [36]. The library is screened against the target molecule synthesized in the D-amino acid configuration. Once the L-peptide is isolated, the corresponding D-peptide is synthesized. For reasons of symmetry, the D-peptide will bind to the natural L-target protein.

The display of proteins, or protein fragments, can be used for live vaccines. In fact, it has been demonstrated that a surface recombinant protein (such as those displayed on bacteria or phages) is able to stimulate a good immune response in most cases through either parenteral or oral administration [37–42].

Peptides or protein fragments that display libraries have also been used for the selection of peptides (epitopes) which are able to bind to antibodies present in human serum from individuals immunized against a given antigen [43]. This is particularly useful when the antigen is not known. The serum-selected epitopes can be used as immunogens or as diagnostic reagents.

Another interesting and new application of phage display libraries is *in vivo* specific tissue targeting. This may be useful in targeting cells, drugs, and genes into selected tissue. In 1996 Pasqualini and Ruoslahti [44] injected phage libraries intravenously into mice and identified phages able to bind specific organs through repeated cycles of *in vivo* selection. In 1997 it was demonstrated [45] that a phage displaying an Arg-Gly-Asp (RGD)-containing peptide with high affinity for αv integrins targeted tumor tissues when injected into mice. Another paper [46] showed that a phage displaying a single-chain antibody fragment that recognizes a splice variant of fibronectin, can be used for tumor blood vessel targeting. Arap *et al.* [47] coupled the anticancer doxorubicin to peptides able to bind αv integrin and demonstrated the enhanced efficacy and reduced toxicity of the drug against human breast cancer xenografts in nude mice.

D. Display of cDNA or Genomic DNA

The importance of the surface expression of cDNA (or genomic DNA fragments) is the possibility to isolate unknown genes when a ligand molecule is available. The ligand-binding domain of a prokaryotic receptor gene was isolated by expressing chromosomal DNA fragments from *Staphylococcus aureus* into a phagemid vector [48, 49]. The screening was done by panning the phage against the immobilized ligands protein A and fibronectin. Recently, a new vector for direct cloning of polymerase chain reaction (PCR)-amplified fragments for surface display on bacteriophage M13 was constructed [50]. Other vectors for the display of cDNA libraries are also constructed [51, 52].

Random peptides, mutants of a single protein, or antibodies are relatively easy to display because they have well-defined 5′ and 3′ ends. On the other hand, genomic DNA fragments or cDNA have indeterminate 5′ and 3′ and potential translational stop codons. When the inserted DNA is fused to the amino-terminal end of phage coat proteins, it may disrupt the reading frame and consequently the viral protein correct synthesis. To overcome this problem, Light and co-workers [53] have recently developed an improved system based on the use of two antiparallel leucine zippers, one attached to the expressed cDNA proteins and one attached to the phage coat protein. In this system the cDNA proteins are expressed separately from the coat protein and are fused to the C-terminus of one zipper, with no possibility of disruption to the coding sequence.

E. Display of Proteins

Successful display of proteins requires several events to occur. Normally proteins require the correct expression, the translocation to the periplasm, the correct folding, and incorporation into the microorganism surface. Sometimes post-translational modifications are required. In spite of these limitations, however, many different proteins have been displayed on phages and bacteria. Some examples of phage display are protease inhibitors [4, 54, 55], receptor domains [56–59], protein A [60, 61], enzymes [62–66], cytochines and growth factors [67, 4, 67–70], and more. Examples of bacterial display can be found in both Gram-negative and Gram-positive bacteria, in which various antigens are displayed [2, 9]. Phage display of proteins is more appropriate for protein-protein interaction studies, such as display of mutated target regions that interact with the ligand, display of randomized proteins to select higher affinity variants, or display of variants to redirect the specificity. Bacterial display has been used mainly as live recombinant vaccines. The display of enzymes on the surface of bacteria could be used as immobilized cell biocatalysts.

F. Other Applications

Microbial cells displaying foreign proteins can be considered as microscopic protein matrices and be used in those applications that require protein immobilization. The display of enzymes such as β-lactamase [71] and cellulase [16] on the surface of *E. coli* has demonstrated the possibility of creating immobilized cell biocatalysts.

O'Brien *et al.* [72] have expressed in *E. coli* a recombinant histidine tailed T4 lysozyme fusion protein. They were able to purify the fusion protein by using particles charged with Cu^{2+}. Bacterial display of poly-histidine peptides could be used to remove contaminants such as metal ions from waste water. Cells expressing surface antibodies could be used in analytical applications as solid-phase immuno-reagents [73]. In fact, it is possible to chemically fix and stabilize *E. coli* cells displaying genetically engineered cell surface proteins by irreversibly adsorbing cells on treated chitosan particles, avoiding the slow leakage of cellular proteins [74].

REFERENCES

1. Smith GP, Filamentous fusion phage: Novel expression vectors that display cloned antigens on the virion surface, *Science*, 228:1315–1317, 1985.
2. Georgiou G, Stathopoulos C, Daugherty PS, Nayak AR, Iverson BL, Curtiss R III, Display of heterologous proteins on the surface of microorganisms: From the screening of combinatorial libraries to live recombinant vaccines, *Nat. Biotechnol.*, 15:29–34, 1997.
3. Kretzschmar T, Geiser M, Evaluation of antibodies fused to minor coat protein III and major coat protein VIII of bacteriophage M13, *Gene*, 155:61–65, 1995.
4. Clackson T, Wells JA, *In vitro* selection from protein and peptide libraries, *Trends Biotechnol.*, 12:173–184, 1994.
5. Maruyama IN, Maruyama HI, Brenner S, λfoo: A phage vector for the expression of foreign proteins, *Proc. Natl. Acad. Sci. USA*, 91:8273–8277, 1994.
6. Stemberg N, Hoess H, Display of peptides and proteins on the surface of bacteriophage λ, *Proc. Natl. Acad. Sci. USA*, 92:1609–1613, 1995.
7. Efimov VP, Nepluev IV, Mesyanzhinov VV, Bacteriophage T4 as a surface display vector, *Virus Genes*, 10:173–177, 1995.
8. Carbonell X, Villaverde A, Peptide display on functional tailspike protein of bacteriophage P22, *Gene*, 176:225–229, 1996.
9. Stahl S, Uhlen M, Bacterial surface display: Trends and progress, *Trends Biotechnol.*, 15:185–192, 1997.

10. Steidler L, Remaut E, Fiers W, LamB as a carrier molecule for the *functional* exposition of IgG-binding domains of the *Staphylococcus aureus* Protein A at the surface of *Escherichia coli* K12, *Mot. Gen. Genet.*, 236:187–192, 1993.
11. Charbit A, Molla A, Saurin W, Hofnung M, Versatility of a vector for expressing foreign polypeptides at the surface of Gram-negative bacteria, *Gene*, 70:181–189, 1988.
12. Hobom G, Arnold N, Ruppert A, OmpA fusion proteins for presentation of foreign antigens on the bacterial outer membrane, *Dev. Biol. Stand.*, 84:255–262, 1995.
13. Freudl R, Maclntyre S, Degen M, Henning U, Cell surface exposure of the outer membrane protein OmpA of *Escherichia coli* K-12, *J. Mol. Biol.*, 188:491–494, 1986.
14. Harrison JL, Taylor IM, O'Connor CD, Presentation of foreign antigenic determinants at the bacterial cell surface using the TraT lipoprotein, *Res. Microbiol.*, 141:1009–1012, 1990.
15. Fuchs P, Breitling F, Diibel S, Seehaus T, Little M, Targeting recombinant antibodies to the surface of *E. coli:* fusion to a peptidoglycan associated lipoprotein, *Biotechnology*, 9:1369–1372, 1991.
16. Francisco JA, Stathopoulos C, Warren RAJ, Kilburn DG, Georgiou G, Specific adhesion and hydrolysis of cellulose by intact *Escherichia coli* expressing surface anchored cellulase or cellulose binding domains, *Biotechnology*, 11:491–495, 1993.
17. Hedegaard K, Klemm P. Type 1 fimbriae of *Escherichia coil* as carriers of heterologous antigenic sequences, *Gene*, 85:115–124, 1989.
18. Pallesen L, Poulsen LK, Christiansen G, Klemm P, Chimeric. FimH. Adhesion. of type 1 fimbriae: A bacterial surface display system for heterologous sequences, *Microbiology*, 141:2839–2848, 1995.
19. Van Die I, Wauben M, Van Megen I, Bergmans H, Riegman N, Hoekstra W, Pouwels P, Enger-Valk B, Genetic manipulation of major P-fimbrial subunits and consequences for formation of fimbriae, *J. Bacteriol.*, 170:5870–5876, 1988.
20. Lu Z, Murray KS, Van Cleave V, laVallie ER, Stahl ML, McCoy JM, Expression of thioredoxin random peptide libraries on the *Escherichia coil* cell surface as functional fusions to flagellin: A system designed for exploring protein-protein interactions, *Biotechnology*, 13:366–372, 1995.
21. Leary JF., Strategies for rare cell detection and isolation, Method, *Cell. Biol.*, 42:331–358, 1994.
22. Fuchs P, Weichel, Dubel S, Breitling F, Little M, Separation of *E. coli* expressing functional cell-wall antibody fragments by FACS, *Immunotechnology*, 2:97–102, 1996.
23. Schreuder V, Mooren ATA, Toschka HY, Verrips CT, Kls FM, Immobilizing proteins on the surface of yeast cells, *Trends Biotechnol.*, 14:115–120, 1996.
24. Boder ET, Wittrup D, Yeast surface display for screening combinatorial polypeptide libraries, *Nat. Biotechnol.*, 15:553–557, 1997.
25. Pausch MH, G-protein-coupled receptors in *Saccharomyces cerevisiae:* High-throughput screening assays for drug discovery, *Trends Biotechnol.*, 15:487–494, 1997.
26. Poljak RJ, Amzel LM, Chen BL, Phizackerley RP, Saul F, The three-dimensional structure of the fab' fragment of a human myeloma immunoglobulin at 2.0-angstrom resolution, *Proc. Natl. Acad. Sci. USA*, 71:3440–3444, 1974.

27. Tonegawa S, Somatic generation of antibody diversity, *Nature*, 302:575–581, 1983.
28. Hoogenboom HR, Designing and optimizing library selection strategies for generating high-affinity antibodies, *Trends Biotechnol.*, 15:62–70, 1997.
29. Winter G, Griffiths AD, Hawkins RE, Hoogenboon HR, Making antibodies by phage display technology, *Ann. Rev. Immunol.*, 12:433–455, 1994.
30. Kruif J De, Terstappen L, Boel E, Logtenberg T, Rapid selection of cell subpopulation-specific human monoclonal antibodies from a synthetic phage antibody library, *Proc. Natl. Acad. Sci. USA.*, 92:3938–3942, 1995.
31. Tanha J, Forsyth G, Schorr P, Crosby W, Lee JS, Sequence and structure specific antibodies from phage display libraries, *Mol. Immunol.*, 34:109–113, 1997.
32. Chowdhury PS, Chang K, Pastan I, Isolation of anti-mesothelin antibodies from a phage display library, *Mol. Immunol.*, 34:9–20, 1997.
33. Chan SW, Bye JM, Jackson P, Allain JP, Human recombinant antibodies specific for hepatitis C virus core and envelope E2 peptides from an immune phage display library, *J. Gen. Virol.*, 77:2531–2539, 1996.
34. Yang WP, Green K, Pinz-Sweeney S, Briones ATB, Barbas CF, CDR walking mutagenesis for the affinity maturation of a potent human anti-HIV 1 antibody into the picomolar range, *J. Mol. Biol.*, 254:392–403, 1995.
35. Schier L, McCall A, Adams GP, Malmqvist M, Weiner LM, Marks D, Isolation of picomolar affinity anti-c-erbB2 single-chain Fv by molecular evolution of the complementarity determining regions in the center of the antibody binding site, *J. Mol. Biol.*, 263:551–567, 1996.
36. Schumacher TNM, Mayr LM, Minor DL, Milhollen MA, Burgess MW, Kim PS, Identification of o-peptide ligands through mirror-image phage display, *Science*, 271:1854–1857, 1996.
37. Leclerc C, Charbit A, Martineau P, Deriaud E, Hofnung M, The cellular location of a foreign B cell epitope expressed by recombinant bacteria determines its T cell-independent or T cell-dependent characteristics, *J. Immunol.*, 147:3545–3552, 1991.
38. Nguyen TN, Hansson M, Stahl S, Bachi T, Robert A, Domzig W, Binz H, Uhlen M, Cell-surface display of heterologous epitopes on *Staphylococcus xylosus* as a potential delivery system for oral vaccination, *Gene*, 128:89–94, 1993.
39. Wu JY, Newton S, Judd A, Stocker B, Robinson WS, Expression of immunogenic epitopes of hepatitis B surface antigen with hybrid flagellin proteins by a vaccine strain of *Salmonella, Proc. Natl. Acad. Sci. USA*, 86:4726–4730, 1989.
40. Newton SMC, Jacob CO, Stocker BAD, Immune response to cholera toxin epitope inserted in *Salmonella* flaggelin, *Science*, 244:70–72, 1989.
41. Poirier TP, Kehoe MA, Beaehey EH, Protective immunity evoked by oral administration of attenuated aroA *Salmonella typhimurium* expressing cloned streptococcal M protein, *J. Exp. Med.*, 1968:25–32, 1988.
42. Schorr J, Knapp B, Hundt E, Kupper HA, Amann E, Surface expression of malarial antigens *in Salmonella typhimurium:* Induction of serum antibody response upon oral vaccination of mice, *Vaccine*, 9:675–681, 1991.
43. Folgori A, Tafi R, Meola A, Felici F, Galfre G, Cortese R, Monaci P, Nicosia A, A general strategy to identify mimotopes of pathological antigens using only random peptide libraries and human sera, *EMBO J.*, 13:2236–2243, 1994.

44. Pasqualini R, Ruoslahti E, Organ targeting *in viva* using phage display peptide libraries. *Nature*, 380:364–366, 1996.
45. Pasqualini R, Koivunen E, Ruoslahti. av Integrins as receptors for tumor targeting by circulating ligands, *Nat. Biotechnol.*, 15:542–546, 1997.
46. Neri D, Carnemolla B, Nissim A, Leprini A, Querze G, Balza E, Pini A, Tarli L, Halin C, Neri P, Zandi L, Winter G, Targeting by affinity-matured recombinant antibody fragments of an angiogenesis associated fibronectin isoform, *Nat. Biotechnol.*, 15:1271–1275, 1997.
47. Arap W, Pasqualini R, Ruoslahti E, Cancer treatment by targeted drug delivery to tumor vasculature in a mouse model, *Science*, 279:377–380, 1998.
48. Jacobsson K, Frykberg L, Cloning of ligand-binding domains of bacterial receptors by phage display, *BioTechniques*, 18:878–885, 1995.
49. Jacobsson K, Frykberg L, Phage display shot-gun cloning of ligand binding domains of prokaryotic receptors approaches 100% correct clones, *BioTechniques*, 20:1070–1081, 1996.
50. Sampath A, Abrol S, Chaudhary VK, Versatile vectors for direct cloning and ligation-independent cloning of PCR-amplified fragments for surface display on filamentous bacteriophages, *Gene*, 190:5–10, 1997.
51. Crameri R, Hemmann S, Blaser K, pJuFo: A phagemid for display of cDNA libraries on phage surface suitable for selective isolation of clones expressing allergens, *Adv. Exp. Med. Biol.*, 409:103–110, 1996.
52. Jaspers LS, De Keyser A, Stanssens PE, Lambda ZLG6: A phage lambda vector for high-efficiency cloning and surface expression of cDNA libraries on filamentous phage, *Gene*, 173:179–181, 1996.
53. Light J, Maki R, Assa-Munt N, Expression cloning of cDNA by phage display selection, *Nucleic. Acids. Res.*, 21:4367–4368, 1996.
54. Roberts BL, Markland W, Siranosian K, Saxena, Guterman SK, Ladner RC, Protease inhibitor display M13 phage: selection of high-affinity neutrophil, elastase inhibitors, *Gene*, 121:9–15, 1992.
55. Pannekoek H, van Meijer M, Schleef RR, Loskutoff DJ, Barbas CD, Functional display of human plasminogen-activator inhibitor I (*PAI-1*) on phages: Novel perspectives for structure-function analysis by error-prone DNA synthesis, *Gene*, 128:135–140, 1993.
56. Abrol S, Sampath A, Arora K, Chaudhary VK, Construction and characterization of M13 bacteriophages displaying gp120 binding domains of human CD4, *Indian J. Biochem. Biophys.*, 31:302–309, 1994.
57. Chiswell DJ, McCafferty I, Phage antibodies: Will new coliclonal antibodies replace monoclonal antibodies? *Trends Biotechnol.*, 10:80–84, 1992.
58. Robertson MW, Phage and *Escherichia coli* expression of the human high affinity immunoglobulin E receptor alpha-subunit ectodomain, Domain localization of the IgE-binding site, *J. Biol. Chem.*, 268:12736–12743, 1993.
59. Scarselli E, Esposito G, Traboni C, Receptor phage. Display of functional domains of the human high affinity IgE receptor on the M13 phage surface, *FEBS Lett.*, 329:223–226, 1993.
60. Djojonegoro BM, Benedik MJ, Willson RC, Bacteriophage surface display of an immunoglobulin-binding domain of *Staphylococcus aureus* protein A, *BioTechnology*, 12:169–172, 1994.
61. Kushwaha A, Chowdhury PS, Arora K, Abrol S, Chaudhary VK, Construction and characterization of M13 bacteriophages displaying functional IgG-binding domains of staphylococcal protein A, *Gene*, 151:45–51, 1994.

62. Corey DR, Shiau AK, Yang Q, Janowski BA, Craik CS, Trypsin display on the surface of bacteriophage, *Gene*, 128:129–134, 1993.
63. Crameri R, Suter M, Display of biologically active proteins on the surface of filamentous phages: A cDNA cloning system for selection of functional gene products linked to the genetic information responsible for their production, *Gene*, 137:69–75, 1993.
64. McCafferty J, Jackson RH, Chiswell DJ, Phage-enzymes: expression and affinity chromatography of functional alkaline phosphatase on the surface of bacteriophage, *Protein Eng.*, 4:955–961, 1992.
65. Soumillon P, Jespers L, Bouchet M, Marchand-Btynaert J, Winter G, Fastrez I, Selection of beta-lactamase on filamentous bacteriophage by catalytic activity, *I. Mol. Biol.*, 237:415–422, 1994.
66. Light J, Lerner RA, Random mutagenesis of staphylococcal nuclease and phage display selection, *Bioorg. Med. Chem.*, 3:955–967, 1995.
67. Cabibbo A, Sporeno E, Toniatti C, Altamura S, Savino R, Paonessa G, Ciliberto G, Monovalent phage display of human interleukin (hIL)-6: Selection of superbinder variants from a complex molecular repertoire on the hIL-6 D-helix, *Gene*, 167:41–47, 1995.
68. Gram H, Strittmatter U, Lorenz M, Gluck D, Zenke G, Phage display as a rapid gene expression system: Production of bioactive cytokine-phage and generation of neutralizing monoclonal antibodies, *I. Immunol. Meth.*, 161:169–176, 1993.
69. Bass S, Greene R, Wells JA, Hormone phage: An enrichment method for variant proteins with altered binding properties, *Proteins*, 8:309–314, 1990.
70. Lowman HB, Wells JA, Affinity maturation of human growth hormone by monovalent phage display, *J. Mol. Biol.*, 234:564–578, 1993.
71. Francisco JA, Earhart CF, Georgiou G, Transport and anchoring of (3-lactamase to the external surface of *Escherichia coli, Proc. Natl. Acad. Sci. USA*, 89:2713–2717, 1992.
72. O'Brien SM, Sloane RP, Thomas OR, Dunnill P, Characterization of non porous magnetic chelator supports and their use to recover polyhistidine-tailed T4 lysozyme from a crude *E. coli* extract, *I. Biotechnol.*, 54:53–67, 1997.
73. Chen G, Cloud J, Georgiou G, Iverson BL, A quantitative immunoassay utilizing *Escherichia coli* cells possessing surface exposed single chain Fv molecules, *Biotechnol. Progr.*, 12:572–574, 1969.
74. Freeman A, Abramov S, Georgiou G, Fixation and stabilization of *E. coli* cells displaying genetically engineered cell surface proteins, *Biotechnol. Bioeng.*, 52:625–630, 1996.
75. Couet J, Li S, Okamoto T, Ikezu T, Lisanti MP, Identification of peptide and protein ligands for the caveolin-scaffolding domain, *J. Biol. Chem.*, 272:6525–6533, 1997.
76. Li M, Use of a modified bacteriophage to probe the interactions between peptides and ion channel receptors in mammalian cells, *Nat. Biotechnol.*, 15:559–563, 1997.
77. Gram H, Schmitz R, Zuber JF, Baumann G, Identification of phosphopeptide ligands for the Sre-homology 2 (SH2) domain of Grb 2 by phage display, *Eur. J. Biochem.*, 246:633–637, 1997.
78. Mukhija S, Erni B, Phage display selection of peptides against enzyme I of the phosphoenolpyruvate-sugar phosphotransferase system, *Mol. Microbiol.*, 25:1159–1166, 1997.

79. Balass M, Heldman Y, Cabilly S, Givol D, Katchalski-Katzir E, Fuchs S, Identification of a hexapeptide that mimics a conformation-dependent binding site of acetylcholine receptor by use of a phage-epitope library, *Proc. Natl. Acad. Sci. USA*, 90:10638–10642, 1993.
80. Gaynor B, Putterman C, Valdon P, Spatz L, Scharff MD, Diamond B, Peptide inhibition of glomerular deposition of an anti-DNA antibody, *Proc. Natl. Acad. Sci. USA*, 94:1955–1960, 1997.
81. Nord K, Gunneriusson E, Ringdahi J, Stahl S, Uhlen M, Nygren P, Binding proteins selected from combinatorial libraries of an (α-helical bacterial receptor domain, *Nat. Biotechnol.*, 15:772–777, 1997.
82. Zhao S., Lee EY, A protein phosphatase-l-binding motif identified by the panning of a random peptide display library, *J. Biol. Chem.*, 272:28368–28372, 1997.
83. Sato A, Ida N, Fukuyama M, Miwa K, Kazami J, Nakamura H, Identification from a phage display library of peptides that bind to toxic shock syndrome toxin-1 and that inhibit its binding to major histocompatibility complex (MHC) class II molecules, *Biochemistry*, 35:10441–10447, 1996.
84. Yanofsky SD, Baldwin DN, Butler JH, *et al.*, High affinity type I interleukin 1 receptor antagonists discovered by screening recombinant peptide libraries, *Proc. Natl. Acad. Sci. USA*, 93:7381–7386, 1996.
85. Markland W, Ley AC, Ladner RC. Iterative optimization of high-affinity protease inhibitors using phage display, 2. Plasma kallikrein and thrombin, *Biochemistry*, 35:8058–8067, 1996.
86. Liu G, Bryant RT, Hilderman RH, Isolation of a tripeptide from a random phage peptide library that inhibits P1, P4-diadenosine 5'-tetraphosphate binding to its receptor, *Biochemistry*, 35:197–201, 1996.
87. Wrighton NC, Farrel FX, Chang R, *et al.*, Small peptides as potent mimetics of the protein hormone erythropoietin, *Science*, 273:458–463, 1996.
88. Bunnel SC, Henry PA, Kolluri R, Kirchhausen T, Riekies RJ, Berg LI, Identification of ItklTsk Src Homology 3 domain ligands, *J. Biol. Chem.*, 271:25646–25656, 1996.
89. Giebel LB, Cass RT, Milligan DL, Young DC, Arze R, Johnson CR, Screening of cyclic peptide phage libraries identifies ligands that bind streptavidin with high affinities, *Biochemistry*, 34:15430–15435, 1995.
90. Pasqualini R, Koivunen E, Ruoslahti E, A peptide isolated from phage display libraries is a structural and functional mimic of an RGD-binding site on integrins, *J. Cell. Biol.*, 130:1189–1196, 1995.
91. Koivunen E, Wang B, Ruoslahti E, Phage libraries displaying cyclic peptides with different ring sizes: ligand specificities of the RGD-directed integrins, *Biotechnology*, 13:265–270, 1995.
92. Healy JM, Murayama O, Maeda T, Yoshino K, Sekiguchi K, Kikuchi M, Peptide ligands for integrin av(l3 selected from random phage display libraries, *Biochemistry*, 34:3948–3955, 1995.
93. Koivunen E, Gay DA, Ruoslahti E, Selection of peptides binding to the $\alpha 5\beta 1$ integrin from phage display library, *J. Biol. Chem.*, 268(27): 20205–20210, 1993.
94. Castano AR, Tangri S, Miller JEW, *et al.*, Peptide binding and presentation by mouse CD1, *Science*, 269:223–226, 1995.
95. Cheng X, Julian RL, Identification of a biologically significant DNA-binding peptide motif by use of a random phage display library, *Gene*, 171:1–8, 1996.

96. Krook M, Mosbach K, Lindbladh C, Selection of peptides with affinity for singlestranded DNA using a phage display library, *Biochem. Biophys. Res. Commun.*, 204(2): 849–854, 1994.
97. Dyson MR, Murray K, Selection of peptide inhibitors of interactions involved in complex protein assemblies: Association of the core and surface antigens of hepatitis B virus, *Proc. Natl. Acad. Sci. USA*, 92:2194–2198, 1995.
98. Cheadle C, Ivashchenko Y, South V, *et al.*, Identification of a Sre SH3 domain binding motif by screening a random phage display library, *J. Biol. Chem.*, 269:24034–24039, 1994.
99. Sparks AB, Quilliam LA, Thorn JM, Der CJ, Kay BK, Identification and characterization of Src SH3 ligands from phage-displayed random peptide libraries, *J. Biol. Chem.*, 269(39): 23853–23856, 1994.
100. Renschler MF, Bhatt RR, Dower WJ, Levy R, Synthetic peptide ligands of the antigen binding receptor induce programmed cell death in a human B cell lynphoma, *Proc. Natl. Acad. Sci. USA*, 91:3623–3267, 1994.
101. Daniels DA, Lane DP, The characterization of p53 binding phage isolated from phage peptide display libraries, *J. Mol. Biol.*, 243:639–652, 1994.
102. Balass M, Katchalski-Katzir E, Fuchs S, The α-bungarotoxin binding site on the nicotinic acetylcholine receptor: analysis using a phage-epitope library, *Proc. Natl. Acad. Sci. USA*, 94:6054–6058, 1997.
103. Smith GP, Schultz DA, Ladbury JE, A ribonuclease S-peptide antagonist discovered with a bacteriophage display library, *Gene*, 128:37–42, 1993.
104. Scott JK, Loganathan, Easley RB, Gong X, Goldstein IJ, A family of concanavalin A-binding peptides from a hexapeptide epitope library, *Proc. Natl. Acad. Sci. USA*, 89:5398–5402, 1992.
105. Oldemburg KR, Loganathan D, Goldstein IJ, Schultz PG, Gallop MA, Peptide ligands for sugar-binding protein isolated from a random peptide library, *Proc. Natl. Acad. Sci. USA*, 89:5393–5397, 1992.
106. O'Neil KT, Hoess RH, Jackson SA, Ramachandran NS, Mousa SA, DeGrado WF, Identification of novel peptide antagonists for GPIIb/IIIa from a conformationally constrained phage peptide library, Proteins: Structure, Function *Genet.*, 14:509–515, 1992.
107. Markland W, Ley AC, Lee SW, Ladner RC, Iterative optimization of high-affinity protease inhibitors using phage display, *I. Biochemistry*, 35:8045–8057, 1996.

17 Peptide Display Libraries: Design and Construction

Maria Dani

CONTENTS

I. INTRODUCTION

The biological display of peptides require the introduction of the genetic information (DNA) that codes for the peptides into a microorganism (Figure 17.1). Once inside the microorganism, the DNA is transcribed and translated into amino acids, according to the genetic code. The DNA fragment coding for the peptides are inserted into specific DNA molecules called vectors. In phage display the vector is constituted by the viral genome, while in bacterial display an extra-chromosomal circular DNA, called plasmid, is used. Vectors are usually genetically modified to allow the insertion of DNA fragments at specific sites and, sometimes, genes or regulatory regions are deleted or inserted.

Each DNA fragment of a library contains a different sequence. In order to obtain all the possible sequences it is necessary to synthesize degenerated oligonucleotides chemically. In fully degenerated oligonucleotides there is the possibility of introduction of stop codons that interrupt protein synthesis, and consequently the display. Hence, alternate methods are used. During the synthesis, these methods utilize different mixtures of nucleotides, especially in the third position of each triplet, lowering the probability of introduction of a stop codon. The pool of degenerated oligonucleotides are then made

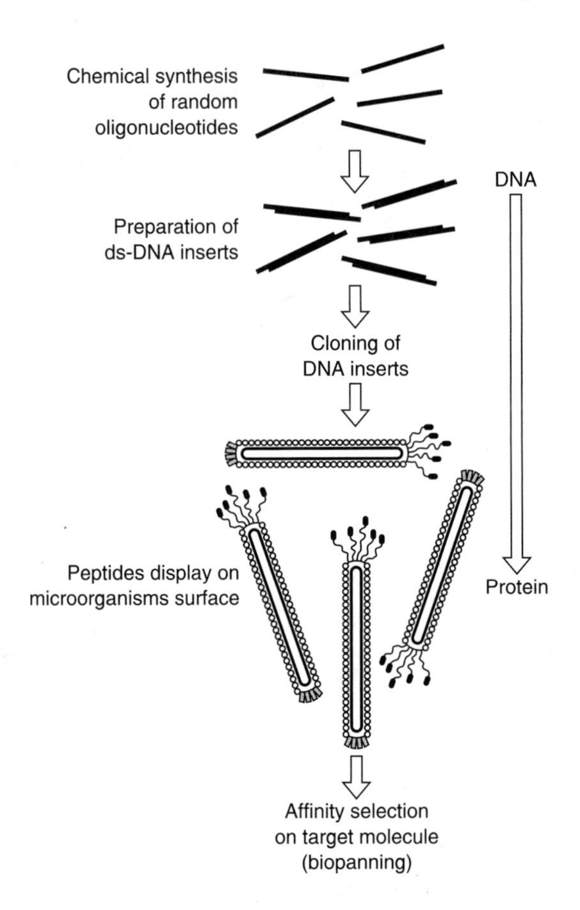

FIGURE 17.1 Schematic representation of a peptide display library. A pool of DNA fragments is synthesized, cloned into an appropriate vector, and expressed on the surface of microorganisms (e.g. bacteriophages).

double stranded by enzymatic reactions and, after cleavage of the ends with appropriate restriction enzymes, the DNA fragments are inserted into a linearized vector and used to transform the appropriate microorganism.

Because the peptides are expressed on the surface of the microorganism, the DNA has to be inserted into a gene coding for a protein to expose the peptide to the medium. The insertion should be correct, so that the coding sequence of the fusion protein is not disrupted. Furthermore, the foreign peptide should not change the conformation of important domains. Phages and bacteria have proteins on their surface, and for some of them it has been possible to find permissive sites to insert foreign peptides. For phages, coat proteins and proteins responsible for the infection are used. For bacteria, the most commonly used proteins are those that are found in the outer membrane. The site and the size of insertion are very important, especially when large peptides or entire proteins are displayed. Random peptide libraries range usually from 6 to 15 amino acids and pose fewer display problems.

Once the library is constructed, it is necessary to verify the quality of the library by sequencing some of the DNA fragments and comparing the observed amino acids frequencies with those that are expected.

Although different types of protein display libraries, such as antibody or cDNA libraries, can be created, this chapter focuses on construction of random polypeptide libraries.

II. LIBRARY CONSTRUCTION

A. Vectors

A vector is constituted of circular DNA (filamentous phages, bacterial plasmids) or linear DNA (lambda phage) which are able to replicate autonomously once it is inside the *Escherichia coli* cells. Usually, phage vectors consist of the entire virus genome, into which specific modifications are introduced. A phage vector carries the genetic information necessary for virus propagation and survival. A plasmid is a circular, self-replicating double-stranded DNA (dsDNA) which is naturally present in some bacterial strains and usually carries antibiotic resistance. For the construction of display peptide libraries, the DNA fragment coding for the peptides are inserted into the bacteriophage genome or into a bacterial plasmid DNA.

The wild-type vector is modified prior to the cloning of random DNA fragments. For the insertion of foreign DNA, the vector is cut at the site of insertion. This is done by restriction enzymes that recognize specific sequences of DNA and usually leave, after the cut, few protruding nucleotides at the ends. The insert to be cloned is cut with the same restriction enzyme to create complementary ends, able to hybridize to the vector. Because it is not always possible to find a unique restriction site at the insertion point, the vector has to be genetically modified. Furthermore, it is very often necessary to introduce new genes or regulatory regions, or to mutate the existing genes.

Constructing a new display vector requires specific skills, but using an existing one is relatively simple. Since display libraries are first constructed [1], several vectors are engineered for the display of random polypeptides (see Table 17.1). The most widely used are those that are derived from the Ff class of *E. coli* filamentous bacteriophage. This class of phages infects bacteria harboring the F episome and includes wild-type strains M13, fl and fd. The three strains are independent isolates of the same phage and differ by only a few nucleotides. The genome consists of a single-stranded, circular DNA molecule, 6407 nucleotides long (Figure 17.2). It codes for 10 proteins and contains an Intergenic Space (IG) between genes 4 and 2. The intergenic space contains the origins for (+) and (−) strand DNA synthesis, a transcription terminator, and a signal for packaging the (+) strand into phage particles. Filamentous phages are used extensively in molecular biology as cloning vectors. They do not have a size limit for DNA insertion—if the genome is longer, the DNA is packaged into longer phage particles. Other advantages of these vectors are the easy construction of large libraries, the

TABLE 17.1
Examples of vectors and display proteins used for the construction of biological display libraries

Microorganism	Vector	Display protein	References
fl phage	—	PIII	1
fl phage	—	PIII	70
fd-tet phage	fUSE1, fUSE2	PIII	4
M13 phage	M13LP67	PIII	19
fd-tet phage	fAFF1	PIII	72
fd-tet phage	fUSE5	PIII	5
fd phage	Fd-CAT1	PIII	73
fd phage and phagemid	fdH, pKfdH	PVIII	15
M13 phage	M13 mp18	PVIII	75
fl phage	p89, fag19	PVIII	76
fd phage	fdDOGl	PIII	77
M13 phage	M13PL-6	pill	78
Pasmid	pMCS	*lac* repressor	79
Pagemid	pSEX	PIII	74
Pagemid	pCBAK8	PVIII	16
Pagemid	pDONG6	PVI	21
Pagemid	pComb3	PIII	82
Pagemid	pSK4	PIII	80
lambda phage	λZapII	pet B leader	71
lambda phage	λfoo	V	9
lambda phage	λpRH814 λpRH804	D	8
E. coli	PFliTrx	flagellin-thioredoxin	10

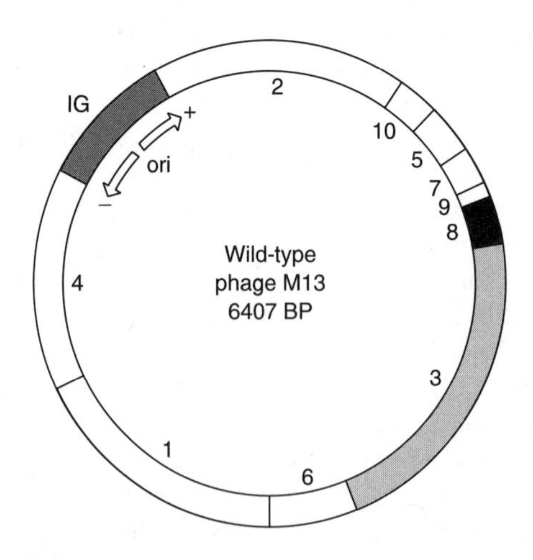

FIGURE 17.2 Genome of wild-type filamentous bacteriophage M13. Numbers correspond to different genes coding for 10 different viral proteins. IG: intergenic space; *ori:* origins of replication for the + and – strands.

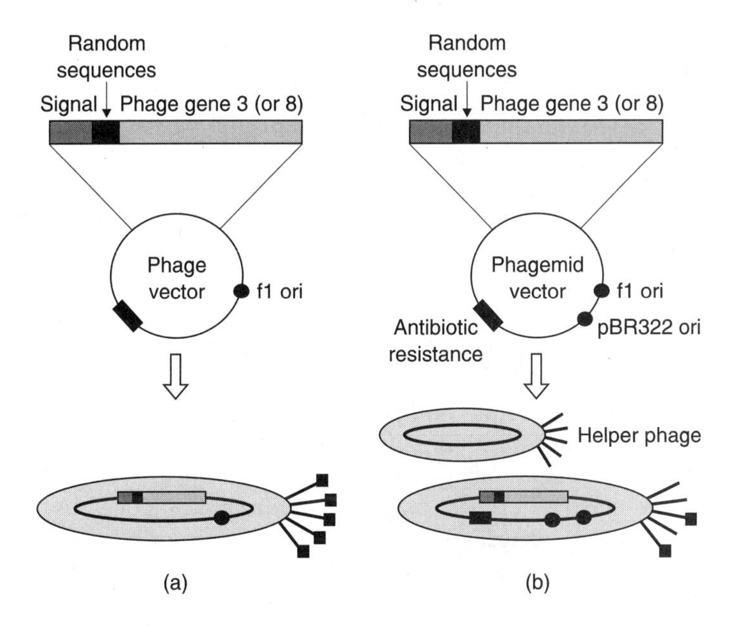

FIGURE 17.3 Differences between phage and phagemid. An example is presented. A phage vector retains all the functions and capability of forming viable viral particles. A phagemid vector is a defective phage that usually contains an antibiotic resistance gene and the *E. coli* origin of replication; it replicates in *E. coli* as a plasmid and forms viable viral particles only in the presence of a helper phage. See text for more details.

accumulation of viral particles at high titers, and the stability to extreme conditions such as low pH (pH 2) and high temperature (70°C), and allowing recovery of the affinity-bound phages by breaking noncovalent bonds. Furthermore, phage vectors allow the rapid isolation of single-stranded DNA for dideoxy sequencing [2] or oligonucleotide-directed mutagenesis [3]. The most widely used coat proteins for peptide display are pIII and pVIII. Fusion of foreign peptides at the amino terminus of these proteins generate viable phage particles that expose the foreign sequence on the virion surface. Figure 17.3a shows a typical phage vector with essential features only. Examples of more genetically engineered filamentous bacteriophage vectors are those of the fUSE series [4]. In particular, fUSE5 [5] has several interesting features. The vector is fd-tet, a derivative of fd with tetracycline resistance determinant spliced into the origin of minus-strand synthesis. The phage can be propagated like a plasmid in medium containing tetracycline. Figure 17.6a, below, shows the site of insertion. The fusion to polypeptides does not significantly debilitate pIII. Furthermore, only a single infection event is sufficient because, in the presence of tetracycline, the phage is propagated as a plasmid without the need for further infection. During the library construction the reading frame can be restored by inserts of the correct length, since the gene III is out of frame; in this case only phage-bearing gene III inserts are able to infect the bacteria, allowing the selection against phages without inserts. In fUSE5 vector the random peptide is flanked by few amino acids without a definite structure to minimize the influence of pIII on the peptide.

The filamentous phage genome can also be modified to obtain a phagemid, which is a hybrid vector that combines the advantages of phages and plasmids [6] (Figure 17.3b). A phagemid carries both filamentous phage (single-stranded) and plasmid (double-stranded) origins of replication. Phagemids can be grown as plasmids or as phages. Since the phagemid lacks few protein genes, cells transformed with a phagemid are infected with a helper phage to produce complete phage particles. The presence of a resistance marker on the phagemid permits selection of transformed cells on the antibiotic-containing medium. Other resistance marker present in the helper phage allows the selection of *E. coli* phagemid-containing cells that are also infected with the helper phage. Since the helper phage is packaged less efficiently, the majority of viral particles contain phagemid DNA. There are several examples of phagemid vectors in the literature (see Table 1); the majority of them have been used for the display of antibodies or other protein fragments that may disrupt viral protein structure. The advantage of phagemid relies on the low percentage of recombinant fusion proteins over native proteins. A commercial phagemid vector is pCANTAB 5 (Amersham Pharmacia Biotech), especially designed for the display of antibody variable regions. In this vector, the site of insertion is between gene 3 signal peptide and pIII protein. The expression of the antibody-g3p gene is controlled by an inducible *lac* promoter, which in turn is regulated by a *lac* repressor.

Another phage used for display is lambda (λ), though to a much lesser extent. The genome of bacteriophage lambda is a double-stranded DNA molecule, 50 kilobases long and packaged into phage particles as a linear molecule with single-stranded cohesive termini (Figure 17.4). The phage

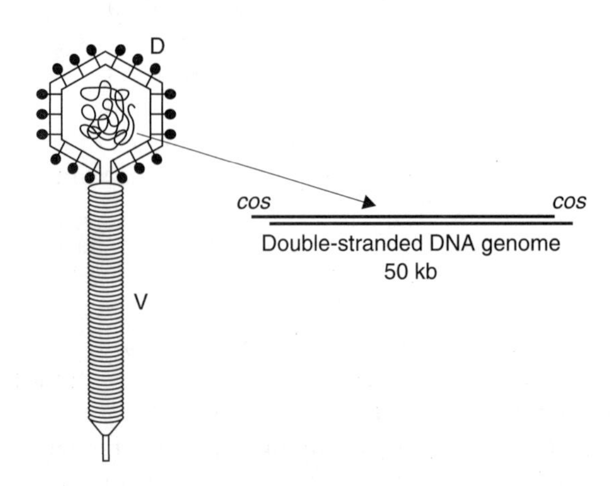

FIGURE 17.4 Bacteriophage lambda. D, decoration protein; V, tail protein; *cos*, cohesive ends.

adsorbs receptors of the outer membrane of *E. coli* coded by the *lamB* gene. Once inside the bacterium, the cohesive termini associate to form a circular DNA molecule. During the lytic cycle, the circular DNA is replicated many times, the gene products are synthesized and the viral particles are assembled intracellularly and released when the cell lyses. The structural proteins that constitute the envelope and their pathway are well known [7]. Two proteins are used for phage display—D or "decoration" protein and V protein. Peptides and proteins are fused successfully to the amino terminus of the D protein [8] by using the Cre-loxP site-specific recombination system *in vivo*, which allows the construction of large display libraries. Cells containing D fusion plasmids are infected with a A,Dminus phage containing a *loxP* site. During infection, phage and plasmid recombine at *loxP* sites by Cre recombinase and the plasmid is co-integrated into the lambda genome. In the λfoo vector [9], foreign proteins or peptides are fused to the C-terminus of a truncated pV protein.

The λ vector can also be used as an intermediate vector for construction of peptide libraries, which are then excised *in vivo* and the peptides displayed on phagemids. An example is the commercial vector Surf ZAP (Stratagene). The high ligation and packaging efficiencies of lambda result in the construction of very large primary libraries bypassing the lower transformation efficiency of bacteria. The vector derived from the lambda ZAP II contains the termination and initiation sequences of f1 origin of replication. Surf ZAP has been obtained by insertion of a ribosome-binding site, pe1B leader sequence, the foreign peptide cloning site, a glycine spacer and a portion of pIII. Proteins are expressed on fusion with amino acids 198–406 of pIII. When *E. coli* is co-infected with the lambda phage vector and the helper phage, the helper phage proteins recognize the initiation sequence of fl origin of replication, which are inserted into the lambda vector, nick one strand, and initiate DNA synthesis. The replication continues until it encounters the termination signal. The single-stranded DNA is then circularized by a gene II product of the helper phage and packaged. After secretion, lambda phages and cells are killed selectively by heat treatment. The library consists of phagemid particles.

The only bacterial vector used for random display libraries is pFliTrx [10, Invitrogen]. This vector (Figure 17.5) is useful when high level of expression and conformationally constrained peptides are desirable. The vector is used for the construction of random dodecapeptide libraries displayed on the surface of *E. coli* using the major bacterial flagellar protein (FIiC) and thioredoxin (TrxA). The random peptides are inserted into the thioredoxin active-site loop, a location that is highly permissive for the insertion of peptide sequences [11] and is known to protrude into the medium. The inserted sequences are conformationally constrained. Because thioredoxin is a cytoplasmic protein, its DNA was cloned into the fliC gene, replacing a large, medium-exposed nonessential domain [12, 13]. The fusion protein is exported to the cell surface, where it assembles into partially functional flagella.

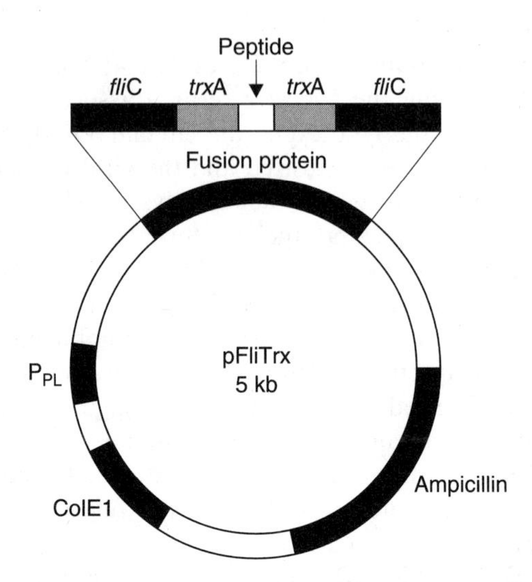

FIGURE 17.5 Bacterial plasmid vector FliTrx used for random peptide display library [10]. FliC, flagellin gene; trxA, thyoredoxin; P_{PL}, P_L, promoter; Co1E1, *E. coli* origin of replication; ampicillin, gene for ampicillin resistance.

B. Fusion Proteins

In order to express the peptides on the surface of microorganisms, the DNA is inserted into a gene coding for a coat or membrane protein, which is able to expose the peptide. Several such proteins are used.

The filamentous phage minor coat protein pIII is the most widely used display protein and is present at 3–5 copies per virion [14]. It is synthesized with amino-terminal signal peptide, which is cleaved during the translocation through the inner membrane. The carboxy terminal is in the cytosol, the amino terminal and the foreign peptide are in the periplasm. A single membrane-spanning domain anchors the protein to the membrane. As for the other viral proteins, pIII is assembled into the nascent virion as it emerges through the membrane. The mature pIII protein is 406 amino acids long and folds into two domains—the carboxy-terminal domain interacting with viral coat proteins and required for viral assembly, and the amino-terminal two-thirds of the protein project away from the virion and responsible for the binding of the F pilus and infection of male bacteria. The amino-terminal domain can tolerate even 100 amino acids long insertions, without greatly affecting the infectivity.

The product of gene 8 is the major capsid protein of the filamentous phage. The protein is synthesized as a precursor with an N-terminal extension of 23 amino acids, which is necessary for the insertion of the protein into the bacterial cytoplasmic membrane where the leader peptide is cleaved. During the process of phage assembly, about 2700 copies aggregate around the virus DNA to form a helical array with the amino terminus exposed to the medium. The mature protein is 50 amino acids long. The amino terminus has a predominance of acidic residues, where in the central part is hydrophobic, and the carboxy terminus is characterized by four lysine residues and interacts with the phage single-strand DNA. Up to six amino acids can be inserted into the pVIII without disrupting the protein coat assembly [15]. For longer peptide insertion it is necessary to use a phagemid vector. In this case only 1–10% of the pVIII will display the peptide; with this system up to 50 kDa antibody Fab fragment was displayed [16]. The N-terminal end of mature pVIII is not locked in a particular conformation [17]; consequently, short peptides fused to the N terminus would probably assume more than one conformation unless constrained by two flanking cysteines [18, 19].

The protein pVI is one of the minor coat proteins present at five copies per phage in one tip of the virion [20]. Its function is to connect the phage body to pIII protein and expose the C terminus to the solvent. Jespers and co-workers [21] have constructed a new vector in which they inserted cDNA fragments fused to the gene 6.

The 11-kDa D protein of bacteriophage λ plays a role in stabilizing the prohead of the phage as it fills with DNA [22, 23] and is essential for head morphogenesis. D protein is assembled as trimers and appears as a protrusion on the capsid surface [24]. Stemberg and Hoess [8] have fused peptides and protein domains to the amino terminus of the D protein. They have demonstrated that the fusion protein assembles into the viral capsid and the fusion peptide is recognized by a specific monoclonal antibody. The V protein is a major tail protein of the λ virion that forms the tubular part of the tail [25, 26]. The tail tube consists of 32 disks, each containing six subunits of pV. The C-terminal domain of the protein is on the outside. Since it was demonstrated that mutants lacking a portion of the C-terminal domain are still functioning [27], Maruyama *et al.* [9] have constructed a display λ vector (λfoo) with pV as a fusion protein and demonstrated that foreign proteins or peptides can be fused to the C terminus of the truncated pV by a peptide linker.

Bacteria are used mainly for protein display. The fusion proteins used are outer membrane proteins and the insertion site is in externally exposed loops. Display in bacteria require a leader sequence [28, 29], one or more hydrophobic membrane-spanning regions, and a signal peptide that directs the protein to the outer membrane. The used proteins most often are lamB and Omp A. LamB is a trimeric maltose channel; permissive sites are identified in the externally exposed loops. Omp A is used for larger protein fragments. For the expression of random peptide display libraries on bacteria, the fusion flagellin-thioredoxin is used [10]. Flagellin is the major structural protein found in *E. coli* flagella. This protein contains a large, nonessential domain exposed to the medium. Deletions of this region, or of the entire domain, are well tolerated, and some flagellar function is retained [30]. In the system used, the entire thioredoxin molecule is inserted into the nonessential domain of flagellin. The tertiary structure of thioredoxin reveals that its active-site sequence, CGPC, forms a tight, disulfide-constrained loop on the protein's surface [31]. This loop is shown as highly permissive for the insertion of different peptide sequences [11] without compromising thioredoxin folding. The random peptides inserted are constrained by a disulfide bond and a tight and stable tertiary fold of thioredoxin.

C. Insert Design

DNA consists of a chemically linked sequence of four nucleotides. Each trinucleotide sequence is coded for a corresponding amino acid. Because of the codon degeneracy, some amino acids are coded by more than one triplet. Since a library of random peptides should contain all the possible sequences, the corresponding DNA fragments would have to be degenerated oligonucleotides, which are chemically synthesized as a mixture of sequences.

For a fully degenerated oligonucleotide, each triplet would code for all 20 amino acids with no bias beyond what is due to the unequal degeneracy of the genetic code. At each coupling reaction an equal mixture of all four nucleotides (N) would be used for the first and second positions of each triplet. The third position would have a mixture of dC and dG or dG and T (NNK or NNS) [5, 32]. In this way, the mixture would contain only 32 triplets instead of 64, but all 20 aminoacids would be represented. More importantly, only one termination codon (amber) would be included, and this could be suppressed in special bacterial strains. Among the natural 64 triplets, three are signal terminations of the peptide chain. If a termination codon is present in the DNA fragment, the peptide synthesis would stop; and if the peptide is fused at the NH2 terminal, the fusion protein would not be synthesized and consequently the peptide would not be displayed.

Another way to design randomized oligonucleotides is to synthesize triplets as NNY and RNN, where Y represents dC or T and R represents dA or dG [33], all in equimolar mixtures. This design eliminates stop codons, but does not contain two of the 20 amino acids, and hence limiting the diversity.

In 1993 [34], a different method was published. This method minimizes stop codons and matches amino acid frequencies observed in 207 natural proteins. Three mixtures of nucleotides are designed, each corresponding to one of the three positions in the codon. The method, which is based on the manual use of a spreadsheet or a refining-grid search algorithm, minimizes termination codons to not more than 1%, codes for all 20 amino acids, balances internal vs external side chains, maintains a net charge near zero, and matches the individual amino acids as nearly as possible to the target values. This method produces longer polypeptides with desired balance of amino acids and greater sequence diversity.

Sometimes it is necessary to build a library starting from a particular sequence and introduce random substitutions. In that type of a case, the codons are biased toward a particular amino acid. Each nucleotide of the codon to be modified is a mixture of one nucleotide at a higher percentage (for example, at 60%) with an equimolar mixture of the other three (for example, at 40%). This results in a nucleotide substitution rate that is dependent on the molar ratio used (in the example: 30%) [32].

A typical random peptide library displays polypeptides usually from 6 to 15 amino acids long. The length of the peptides is chosen based on the type of ligand one wants to select, the number of randomized positions, and the size of the library. For example, ligands for affinity purification should be short in order to diminish the cost of peptide synthesis. A longer peptide would probably work better as a live vaccine. A larger number of randomized positions would increase sequence diversity and the size of the complete library.

The number of randomized positions determines the size of the library. A library of 7 randomized amino acids would be constituted by 1.3×10^9 different peptides. A complete library with 90% confidence of containing all possible amino acid sequences would need 7.9×10^{10} transformants [35]. The randomization of more positions would give incomplete libraries because the upper limit is set by the transformation efficiency of *E. coli*, which is usually around 10^9–10^{10}. Consequently, the display of long peptides may not allow the construction of a complete library. However, a long variable region would contain a larger fraction of shorter peptides than a library of the same size but made up of short peptides [36]. In most cases, in fact, the binding region is limited to a few residues and consequently a several-hundred-million 15-amino-acid library would have a larger diversity of short peptides than a library of 6 amino acids composed of the same number of clones.

The oligonucleotides chemically synthesized as a mixture are in the form of single-stranded DNA. In order to clone all different fragments into the appropriate vector, they must be in a double-stranded form, at least at both ends. Furthermore, since the fragments are to be inserted in specific restriction sites, they have to contain at both ends, the DNA sequences recognized by the appropriate restriction enzyme. Usually, an oligonucleotide mixture is designed to have the following features: (1) A restriction enzyme site is present at the 5′ end, preceded by 4–6 nucleotides to allow the restriction enzyme to bind and cut the DNA efficiently; the restriction site is usually designed not to disrupt the reading frame of signal peptide or display protein in which the random region is to be inserted; (2) The restriction site may be followed by other sequences coding for some amino acids of the signal peptide (if the random peptide is inserted between the signal peptide and the display protein) or a stretch of the fusion protein (if the random peptide is inserted in the coding region of the display protein); (3) The random region, which is formed by a variable number of nucleotides triplets corresponding to the same number of amino acids; (4) A constant region constituted by a glycine spacer, and/or sequences coding for some amino acids of the fusion display protein, and/or the restriction site at the 3′ end of the fragments plus 4–6 oligonucleotides for efficient restriction of enzyme digestion. Since a biological library is usually constituted of 10^8–10^9 independent clones, it is not practical to create a double-stranded degenerate insert simply by annealing complementary degenerate oligonucleotides, because, for each oligo, the complementary sequence would be present in only a few copies or sometimes not at all. Furthermore, the annealing can occur between highly homologous sequences, one of which would be repaired inside of the microorganism with a loss of library variability or deletions of some sequences. The best way to transform a single-strand degenerate oligonucleotide into a double-stranded fragment is to use a shorter oligo, able to anneal to its 3′ end. This assumes the presence of a stretch of nondegenerate sequences at the 3′ end complementary to the short oligo to be used as a primer. After the annealing of the primer to the degenerated oligo, the complementary strand is elongated by polymerase chain reaction (PCR) amplification or by DNA polymerase. An example of this strategy is shown in Figure 17.6a. Another way is to create gapped molecules (Figure 17.6b) by annealing two oligonucleotides to both sites of the random region. If the appropriate restriction site is included in the cut form, the resulting duplexes are ready to be cloned. The missing fragment of DNA would be completed *in vivo* after the transformation. However, if the gapped degenerate region is long, the incidence of rearrangements increases and the first method is preferable.

It is possible to constrain peptides by introducing two codons for cysteine in both sites of the random region. Constraining a peptide in its active conformation is an advantage, but if it is constrained into an incorrect conformation, binding to the target molecule can be prevented. Several peptides have been isolated from cyclic libraries. The screening of pools

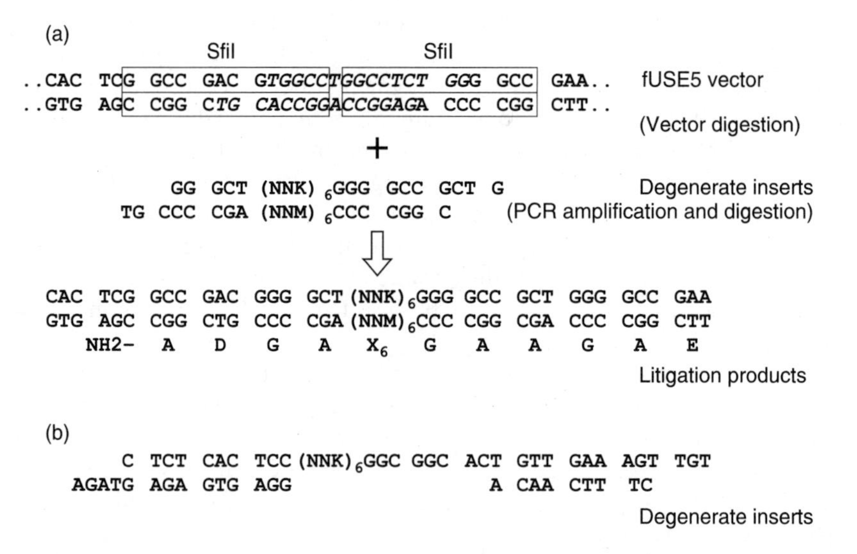

FIGURE 17.6 (a) Example of degenerate insert preparation by PCR amplification. DNA sequences of vector fUSE5 [5] at the site of insertion. Boxed sequences, SfiI restriction site. The shadowed sequences of the vector are removed during SfiI digestion and substituted by the degenerate inserts to form the ligation products as shown. (b) Example of degenerate insert preparation by gapped fragments: two short oligonucleotides are annealed at both sides of the random sequence. The protruding 3′ ends are compatible with the vector restriction site BstXI [72].

of libraries of peptides, flanked by a cysteine residue on each side, CX_5C, CX_6C, CX_7C, or only on one side, CX_9, resulted in the isolation of ligands to several integrins [81]. Cyclic and linear peptides libraries are also used to screen for peptides which are able to bind streptavidin. The analysis of the binding peptides demonstrated that the conformationally constrained cyclic peptides bound streptavidin with affinities up to three orders of magnitude higher than linear peptides [37]. From a library of cyclic 8-mers, a predominant cyclic peptide was isolated against fibronectin [38]. Also, by using 14-amino acid disulfide-bonded cyclic peptides, a consensus sequence could be identified with agonist-binding properties to the erytropoietin receptor [39]. Peptides from constrained library pools were also selected for their ability to bind to brain and kidney blood vessels [40]. But if some constrained peptides are isolated, many more linear peptides are able to bind many different targets [41–60].

The amount of double-stranded DNA fragments required depends on the number of randomized nucleotides and on the expectation of how many times a unique sequence combination should be represented in the pool. Each extension of the randomized sequence by one nucleotide requires a fourfold increase of nucleic acid molecules [61]. Another important parameter is the transformation efficiency of the system used.

When it is necessary to improve the affinity of the already-isolated peptide ligands, it is possible to construct a secondary library by introducing either targeted or random mutations. Different biological methods like cassette mutagenesis [62], regional mutagenesis [63,64], combinatorial mutagenesis [65], "spiked oligonucleotides" [66] can be used. In cassette mutagenesis, the target region is substituted by a synthetic DNA duplex with the desired mutations. In regional mutagenesis, mutations are introduced by chemical or enzymatic treatments at a controlled rate of alterations per nucleotide, and then the DNA is cloned. Combinatorial mutagenesis replaces several amino acids per protein, usually using the cassette method. The spiked oligonucleotides are synthesized by adding at predetermined positions a particular amount of a mixture of different bases in order to "spike" the wild-type base. An analysis of these methods are done, and optimal mutagenesis rates for specific conditions can be found at the amino acid level rather than at the nucleotide level [67]. In fact, the genetic code poses specific constraints in the creation of protein variants, being in between the nucleotide (mutagenesis unit) and the amino acid (protein-variation unit).

D. Cloning

The introduction of DNA fragments into an appropriate vector and the *E. coli* transformation (Figure 17.7) require basic molecular biology techniques [68]. However, particular attention must be paid to maximizing the cloning efficiency, especially for the construction of large libraries.

Since the cloning of a DNA fragment requires compatible ends with the vector, the two DNAs are cut with the same restriction enzyme or an enzyme

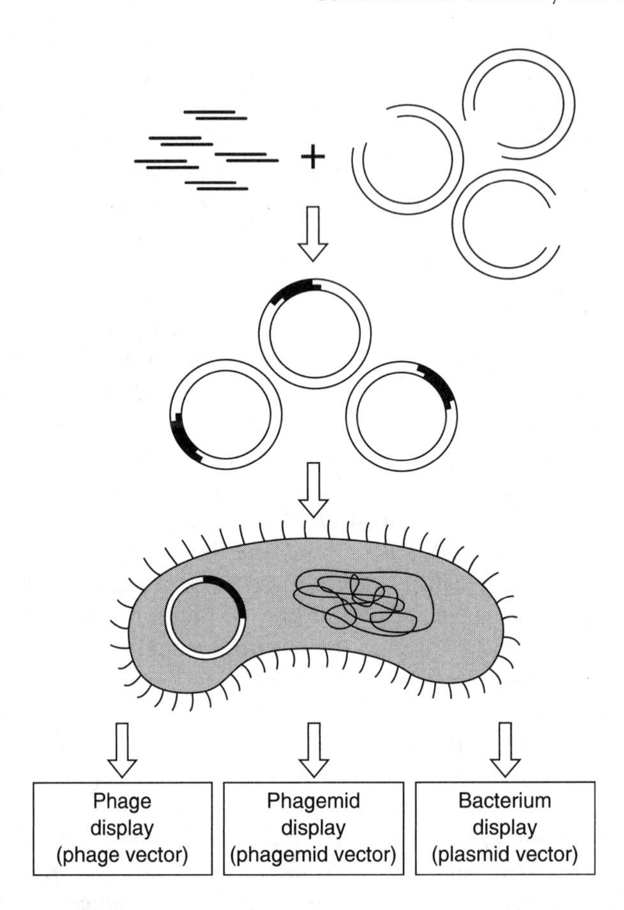

FIGURE 17.7 Schematic representation of a random peptide library construction. Random double-stranded DNA fragments are inserted into linearized vector molecules. Ligated molecules are used to transform *E. coli* cells. Phage vectors will generate phage particles; phagemid vectors will generate phage particles in the presence of a helper phage; plasmid vectors will display the peptides on the bacterium surface.

that produces compatible ends. The vector DNA is linearized and purified. A ligation mix is set up in which DNA degenerate fragments are mixed at a twofold molar excess with the vector and are ligated together with the enzyme T4 DNA ligase. The necessary amount of vector DNA is calculated based on the library complexity (number of independent clones) and on the transformation efficiency (number of clones obtained with 1 μg of vector DNA). It is therefore necessary to calculate the size of the library based on the randomized positions and to verify the transformation efficiency of the *E. coli* strain used. Usually a good transformation efficiency is obtained with electroporation and is around 10^9 transformants/μg of supercoiled vector DNA, while the efficiency of a cut-and-religated vector (as in a library construction) is about 10–100 times less. The ligation mix is used to transform competent *E. coli* cells in several separated electroporations. The bacterial strain is chosen based on the vector type. The transformed cells are then grown on solid medium and an aliquot counted for the calculation of the library complexity. Alternatively, the library can be grown on liquid medium. If the vector used is a phage, the library will appear as clear plaques on a lawn of bacteria. Each plaque will be formed by 10^6–10^7 phage particles derived from a single transformation event and therefore containing the same random peptide sequence. If the vector is a phagemid, the library would grow as bacteria colonies and each colony would constitute a clone with genetic information for just one random peptide; the peptide would be displayed on virions only after infection of the helper phage. If the vector is a plasmid, the library would be constituted by bacterial colonies.

At the time of writing this chapter, three commercial random peptide display libraries are available. Two phage Ph.D. libraries (New England Biolabs) are constituted of random 7-mers and random 12-mers fused to pIII protein of the filamentous phage M13. The FliTrx™ (Invitrogen) is an *E. coli* display library with constrained random dodeca-peptides inserted into flagellin-thioredoxin fusion protein [10].

E. Library Quality Control

If the size of the library is satisfactory, a certain number of clones must be verified for the correct insertion of the random oligonucleotides and for gross clonal bias and sequence abnormalities. Usually about 50 single clones are isolated and their DNA extracted. Many different methods are available for the extraction, including a vast number of commercial kits. The inserted fragments of DNA and the sequences surrounding the site of insertion are sequenced and translated into amino acids. The observed frequency of all amino acids is calculated and compared to the expected frequency, which is obtained by dividing the number of codons for that amino acid by the number of different codons represented in the degenerated oligonucleotides, multiplied by 100. For example, if a library has been constructed by using degenerate oligonucleotides with NNK triplets (N = G, A, T, C; K = G, T), the genetic code would be reduced to 32 codons for all 20 amino acids. In this library, the amino acid serine will be coded by TCG, TCT, and AGT. The expected frequency for serine will then be $3 \div 32 \times 100 = 9.4\%$. The observed frequency for each amino acid residue is calculated by dividing the number of observed codons for that amino acid by the total number of codons sequenced. The probability P that a library contains k clones with a given amino acid motif is calculated using the Poisson distribution $P(k) = e^{-\lambda}\lambda^k/k!$. For $k = 0$, $P(0) = e^{-\lambda} = e^{-np}$. For $k > 0$, $P(k > 0) = 1 - P(0) = 1 - e^{-np}$. n represents the complexity of the library (number of independent clones obtained after transformation). The probability p of a given motif being present in the library is obtained by multiplying all the observed frequencies for each amino acid of that motif (expressed as decimal values). If the given motif is smaller than the length of the library peptides, it is necessary to multiply p by the number of times the given motif appears within the peptide window. For example, in a hepta-peptide library a given penta-peptide motif could occupy positions 1–5, 2–6, or 3–7, and therefore p must be multiplied by 3.

Christian *et al.* [69] evaluated the genetic diversity of their dodeca-peptide library by a modified colony hybridization protocol. This protocol allows the discrimination of perfect matches by hybridizing 17 degenerated oligonucleotide probes to the library clones and washing by incrementing the temperature until the perfect matches alone are able to hybridize. The observed number of positive clones for each oligonucleotide probe is correlated with the theorized number.

If the quality is satisfactory, the library, constituted by alive phages or bacteria, can be stored frozen in aliquots and used when needed for bioplanning experiments.

REFERENCES

1. Smith GP, Filamentous fusion phage: Novel expression vectors that display cloned antigens on the virion surface, *Science*, 228:1315–1317, 1985.
2. Sanger F, DNA sequencing with chain-terminating inhibitors, *Proc. Natl. Acad. Sci. USA*, 74:5463–5467, 1977.
3. Kunkel TA, Roberts JD, Zakour RA, Rapid and efficient site-specific mutagenesis without phenotypic selection, *Meth. Enzymol.*, 154:367–382, 1987.
4. Parmley SF, Smith GP, Antibody-selectable filamentous fd phage vectors: affinity purification of target genes, *Gene*, 73:305–318, 1988.
5. Scott JK, Smith GP, Searching for peptide ligands with an epitope library, *Science*, 249:386–390, 1990.
6. Vieira J, Messing J, Production of single-stranded plasmid DNA, *Meth. Enzymol.*, 153:3–11, 1987.
7. Feiss M, Becker A, In Hendrix RW, Roberts JW, Stahl FW, Weisberg RW, eds., *Lambda* II, Laboratory Press Cold Spring Harbor Lab Press, New York, 1983, pp. 279–304.
8. Stemberg N, Hoess RH, Display of peptides and proteins on the surface of bacteriophage λ *Proc. Natl. Acad. Sci. USA*, 92:1609–1613, 1995.
9. Maruyama IN, Maruyama HI, Brenner S, λfoo: a λ phage vector for the expression of foreign proteins, *Proc. Natl. Acad. Sci. USA*, 91:8273–8277, 1994.
10. Lu Z, Murray KS, Van Cleave V, LaVallie ER, Stahl ML, McCoy JM, Expression of thioredoxin random peptide libraries on the surface as functional fusion to flagellin: A system designed for exploring protein-protein interactions, *BioTechnology*, 13:366–372, 1995.
11. La Vallie ER, Diblasio EA, Kovacic S, Grant KL, Schendel PF, McCoy JM, A thioredoxin gene fusion expression system that circumvents inclusion body formation in the *E. coli* cytoplasm, *BioTechnology*, 11:1187–1193, 1993.
12. La Vallie ER, Stahl ML, Cloning of the flagellin gene from *Bacillus subtilis* and complementation studies of an *in vitro*-derived deletion mutation, *J. Bacteriol.*, 171:3085–3094, 1989.
13. Wilson AR, Beveridge TJ, Bacterial flagellar filaments and their component flagellins, *Can. J. Microbiol*, 39:451–472, 1993.
14. Model P, Russet M, Filamentous bacteriophage, In Calendar R, ed., *The Bacteriophages*, New York, Plenum Press, 1988, pp. 375–456.

15. Greenwood J, Willis AE, Perham RN, Multiple display foreign peptides on a filamentous bacteriophage: Peptides from *Plasmodium falciparum* circumsporozoite protein as antigens, *J. Mol. Biol.*, 220:821–827, 1991.
16. Kang AS, Barbas CF, Janda KD, Benkovic SJ, Lerner RA, Linkage of recognition and replication functions by assembling combinatorial antibody Fab libraries along phage surfaces, *Proc. Natl. Acad. Sci. USA*, 88:4363–4366, 1991.
17. Banner DW, Nave C, Marvin DA, Structure of the protein and DNA in fd filamentous bacterial virus, *Nature*, 289:814–816, 1981.
18. Luzzago A, Felici F, Tramontano A, Pessi A, Cortese R, Mimicking of discontinuous epitopes by phage-displayed peptides, I, Epitope mapping of human H ferritin using a phage library of constrained peptides, *Gene*, 128:51–57, 1993.
19. Devlin JJ, Panganiban LC, Devlin PE, Random peptide libraries: A source of specific protein binding molecules, *Science*, 249:404–406, 1990.
20. Simons GFM, Konings RNH, Schoenmakers JGG, Genes VI, VII and IX of phage M13 code for minor capsid proteins of the virion, *Proc. Natl. Acad. Sci., USA*, 78:4194–4198, 1981.
21. Jespers LS, Messens JH, De Keyser A, Eeckhout D, Van den Brande I, Gansemans YG, Lauwereys MJ, Viasuk GP, Stanssens PE, Surface expression and ligand-based selection of cDNAs fused to filamentous phage gene VI, *Biotechnology*, 13:378–382, 1995.
22. Stemberg N, Weisberg R, Packaging of coliphage lambda DNA, 11, The role of the gene D protein, *J. Mol. Biol.*, 117:733–759, 1977.
23. Imher R, Tsugita A, Wurtz M, Hohn T, Outer surface protein of bacteriophage lambda, *J. Mol. Biol.*, 139:277–295, 1980.
24. Dokland T, Murialdo H, Structural transitions during maturation of bacteriophage lambda capsids, *J. Mol. Biol.*, 233:682–694, 1993.
25. Buchwald M, Murialdo H, Siminovitch L, The morphogenesis of bacteriophage lambda, II, Identification of the principal structural proteins, *Virology*, 42:390–400, 1970.
26. Casjens SR, Hendrix RW, Locations and amounts of major structural proteins in bacteriophage lambda, *J. Mol. Biol.*, 88:535–545, 1974.
27. Katsura I, Structure and function of the major tail protein of bacteriophage lambda, Mutants having small major tail protein molecules in their virion, *J. Mol. Biol.*, 146:493–512, 1981.
28. Von Heijne G, Patterns of amino acids near signal sequence cleavage sites, *Eur. J. Biochem.*, 133:17–21, 1983.
29. Von Heijne G, A new method for predicting signal sequence cleavage sites, *Nucleic Acids. Res.*, 14:4683–4690, 1986.
30. Kuwajima G, Construction of a minimum-size functional flagellin of *E. coli, J. Bacteriol.*, 170:3305–3309, 1988.
31. Katti SK, LeMaster DM, Eklund H, Crystal structure of thioredoxin from *E. coli* at 1.68 angstroms resolution, *J. Mol. Biol.*, 212:167–184, 1990.
32. Smith GP, Scott JK, Libraries of peptides and proteins displayed on filamentous phage, *Meth. Enzymol.*, 217:228–257, 1993.
33. Mandecki W, A method for construction of long randomized open reading frames and polypeptides, *Protein Eng.*, 3:221–226, 1990.
34. LaBean TH, Kauffman SA, Design of synthetic gene libraries encoding random sequence proteins with desired ensemble characteristics, *Protein Sci.*, 2:1249–1254, 1993.

35. Clackson T, Wells JA, *In vitro* selection from protein and peptide libraries, *Trends Biotechnol.*, 12:173–184, 1994.
36. Scott JK, Discovering peptide ligands using epitope libraries, *Trends Biotechnol.*, 17:241–245, 1992.
37. Giebel LB, Cass RT, Milligan DL, Young DC, Arze R, Johnson CR, Screening of cyclic peptide phage libraries identifies ligands that bind streptavidin with high affinities, *Biochemistry*, 34:15430–15435, 1995.
38. Pasqualini R, Koivunen E, Ruoslahti E, A peptide isolated from phage display libraries is a structural and functional mimic of an RGD-binding site on integrins, *J. Cell. Biol.*, 130(5):1189–1196, 1995.
39. Wrighton NC, Farrel FX, Chang R, Kashyap AK, Barbone FP, Mulcahy LS, Johnson DL, Barrett RW, Jolliffe LK, Dower WJ, Small peptides as potent mimetics of the protein hormone erythropoietin, *Science*, 273:458–463, 1996.
40. Pasqualini R, Ruoslahti E, Organ targeting *in vivo* using phage display peptide libraries, *Nature*, 380:364–366, 1996.
41. Couet J, Li S, Okamoto T, Ikezu T, Lisanti MP, Identification of peptide and protein ligands for the caveolin-scaffolding domain, *J. Biol. Chem.*, 272:6525–6533, 1997.
42. Li M, Use of a modified bacteriophage to probe the interactions between peptides and ion channel receptors in mammalian cells, *Nat. Biotechnol.*, 15:559–563, 1997.
43. Mukhija S, Erni B, Phage display selection of peptides against enzyme I of the phosphoenolpyruvate-sugar phosphotransferase system, *Mol. Microbiol.*, 25:1159–1166, 1997.
44. Balass M, Heldman Y, Cabilly S, Givol D, Katchalski-Katzir E, Fuchs S, Identification of a hexapeptide that mimics a conformation-dependent binding site of acetylcholine receptor by use of a phage-epitope library, *Proc. Natl. Acad. Sci. USA*, 90:10638–10642, 1993.
45. Gaynor B, Putterman C, Valdon P, Spatz L, Scharff MD, Diamond B, Peptide inhibition of glomerular deposition of an anti-DNA antibody, *Proc. Natl. Acad. Sci. USA*, 94:1955–1960, 1997.
46. Nord K, Gunneriusson E, Ringdahi J, Stahl S, Uhlen M, Nygren P, Binding proteins selected from combinatorial libraries of an alpha-helical bacterial receptor domain, *Nat. Biotechnol.*, 15:772–777, 1997.
47. Sato A, Ida N, Fukuyama M, Miwa K, Kazami J, Nakamura H, Identification from a phage display library of peptides that bind to toxic shock syndrome toxin-I and that inhibit its binding to major histocompatibility complex (MHC) class 11 molecules, *Biochemistry*, 35:10441–10447, 1996.
48. Yanofsky SD, Baldwin DN, Butler JH, *et al.*, High affinity type I interleukin 1 receptor antagonists discovered by screening recombinant peptide libraries, *Proc. Natl. Acad. Sci. USA*, 93:7381–7386, 1996.
49. Markland W, Ley AC, Ladner RC, Iterative optimization of high-affinity protease inhibitors using phage display, 2. Plasma kallikrein and thrombin, *Biochemistry*, 35:8058–8067, 1996.
50. Castaño AR, Tangari S, Miller JEW, Holocombe HR, Jackson HR, Huse DW, Kronenberg M, Peterson PA, Peptide binding and presentation by mouse CD1 *Science*, 269:223–226, 1995.
51. Cheng X, Julian RL. Identification of a biologically significant DNA-binding peptide motif by use of a random phage display library, *Gene*, 171:1–8, 1996.

52. Krook M, Mosbach K, Lindbladh C, Selection of peptides with affinity for single stranded DNA using a phage display library, *Biochem. Biophys. Res. Commun.*, 204(2):849–854, 1994.
53. Dyson MR, Murray K, Selection of peptide inhibitors of interactions involved in complex protein assemblies: Association of the core and surface antigens of hepatitis B virus, *Proc. Natl. Acad. Sci. USA*, 92:2194–2198, 1995.
54. Cheadle C, Ivashechenko Y, South V, Searfoss G, French S, Howk R, Ricca G, Jaye M, Identification of a Src SH3 domain binding motif by screening a random phage display library, *J. Biol. Chem.*, 269:24034–24039, 1994,
55. Sparks AB, Quilliam LA, Thorn JM, Der CJ, Kay BK, Identification and characterization of Src SH3 ligands from phage-displayed random peptide libraries, *J. Biol. Chem.*, 269(39):23853–23856, 1994.
56. Renschler MF, Bhatt RR, Dower WJ, Levy R, Synthetic peptide ligands of the antigen binding receptor induce programmed cell death in a human B cell lynphoma, *Proc. Natl. Acad. Sci. USA*, 91:3623–3627, 1994.
57. Daniels DA, Lane DP, The characterization of p53 binding phage isolated from phage peptide display libraries, *J. Mol. Biol.*, 243:639–652, 1994.
58. Balass M, Katchalski-Katzir E, Fuchs S, The α-bungarotoxin binding site on the nicotinic acetylcholine receptor: Analysis using a phage-epitope library, *Proc. Natl. Acad. Sci. USA*, 94:6054–6058, 1997.
59. Scott JK, Loganathan D, Easley RB, Gong X, Goldstein IJ, A family of concanavalin A-binding peptides from a hexapeptide epitope library, *Proc. Natl. Acad. Sci. USA*, 89:5398–5402, 1992.
60. Oldemburg KR, Loganathan D, Goldstein IJ, Schultz PG, Gallop MA, Peptide ligands for sugar-binding protein isolated from a random peptide library, *Proc. Natl. Acad. Sci. USA*, 89:5393–5397, 1992.
61. Tablet M, Benos P, Dorr M, Representation of unique sequences in libraries of randomized nucleic acids, *Nucleic Acids Res.*, 24:3437–3438, 1996.
62. Reidhaar-Olson JF, Sauer RT, Combinatorial cassette mutagenesis as a probe of the informational content of protein sequences, *Science*, 241:53–57, 1988.
63. Chen K, Arnold FH, Enzyme engineering for nonaqueous solvents: random mutagenesis to enhance activity of subtilisin E in polar organic media, *Biotechnology*, 9:1073–1077, 1991.
64. Holm L, Koivula A, Lehtovaara P, Hemminki A, Knowles J, Random mutagenesis used to probe the structure and function of *Bacillus stearothermophilus* alpha-amylase, *Protein. Eng.*, 3:181–191, 1990.
65. Merino E, Osuna F, Bolívar F, Sóberon X, A general PCR-based method for single or combinatorial oligonucleotide-directed mutagenesis on pUC-M13 vectors, *Biotechniques*, 12:508–510, 1992.
66. Hermes JD, Parekh SM, Blacklow SC, Koster H, Knowles JR, A reliable method for random mutagenesis: The generation of mutant libraries using spiked oligodeoxyribonucleotide primers, *Gene*, 84:143–151, 1989.
67. Del Río G, Osuna J, Soberón X, Combinatorial Libraries of proteins: Analysis of efficiency of muthagenesis techniques, *Biotechniques*, 17:1132–1138, 1994.
68. Sambrook J, Fritsch EF, Maniatis T, In Nolan C, ed., *Molecular Cloning*, Cold Spring Harbor, NY, Cold Spring Harbor Lab Press, 1989.
69. Christian RB, Zuckermann RN, Kerr JM, Wang L, Malcom BA, Simplified methods for construction, assessment and rapid screening of peptide libraries in bacteriophage, *J. Mol. Biol.*, 227:711–718, 1992.

70. de la Cruz VF, Lal AA, McCutchan TF, Immunogenicity and epitope mapping of foreign sequences *via* genetically engineered filamentous phage, *J. Biol. Chem.*, 263:4318–4322, 1988.
71. Huse WD, Sastry L, Iverson SA, Kang AS, Alting-Mees M, Burton DR, Benkovic SJ, Lerner RA, Generation of a large combinatorial library of the immunoglobulin repertoire in phage lambda, *Science*, 246:1275–1281, 1989.
72. Cwirla SE, Peters EA, Barrett RW, Dower WJ, Peptides on phage: A vast library of peptides for identifying ligands, *Proc. Natl. Acad. Sci. USA*, 87:6378–6382, 1990.
73. McCafferty, Griffiths AD, Winter G, Chiswell DJ, Phage antibodies: Filamentous phage displaying antibody variable domains, *Nature*, 348:552–554, 1990.
74. Breitling F, Dubel S, Seehaus T, Klewinghaus I, Little M, A surface expression vector for antibodies screening, *Gene*, 104:147–153, 1991.
75. Markland W, Roberts BL, Saxena MJ, Guterman SK, Ladner RC, Design, construction and function of a multicopy display vector using fusion to the major coat protein of bacteriophage M13, *Gene*, 109:13–19, 1991.
76. Felici F, Castagnoli L, Mustacchio A, Jappelli R, Cesareni G, Selection of antibody ligands from a large library of oligopeptides expressed on multivalent exposition vector, *J. Mol. Biol.*, 222:301–310, 1991.
77. Clackson T, Hoogenboom HR, Griffiths AD, Winter G, Making antibody fragments using phage display libraries, *Nature*, 352:624–628, 1991.
78. O'Neil KT, Hoess RH, Jackson SA, Ramachandran NS, Mousa SA, DeGrado WF, Identification of novel peptide antagonists for GPIIb/IIIa from a conformationally constrained phage peptide library, Proteins: Structure, Function, *Genet.*, 14:509–515, 1992.
79. Cull MG, Miller JF, Schatz PT, Screening for receptor ligands using large libraries of peptides linked to the C terminus of the lac repressor, *Biochemistry*, 89:1865–1869, 1992.
80. Guzman LM, Belin D, Carson MJ, Beckwith J, Tight regulation, modulation, and high-level expression by vectors containing the arabinose PBAD promoter, *J. Bacteriol.*, 177:4121–4130, 1995.
81. Koivunen E, Wang B, Ruoslahti E, Phage libraries displaying cyclic peptides with different ring sizes: Ligand specificities of the RGD-directed Integrins, *Biotechnology*, 13:265–270, 1995.
82. Barbas CF III, Kang AS, Lerner RA, Benkovic SJ, Assembly of combinatorial antibody libraries on phage surfaces: The gene III site, *Proc. Natl. Acad. Sci. USA*, 88:7978–7982, 1991.

18 Making and Selecting from Phage Antibody Libraries

Andrew Bradbury, Daniele Sblattero, Hennie R. Hoogenboom and Simon Hufton

CONTENTS

I. INTRODUCTION

The use of phage display to select antibodies that recognize specific antigens is arguably the most successful use of phage display. The concepts behind such selection are identical to those used for any of the other biological molecular diversity techniques described in this book and illustrated in Figure 18.1—the creation of diversity, followed by a series of recursive cycles of selection on antigens, each of which involves binding, washing, elution, and amplification, and finally analysis of selected clones.

Antibodies with affinities comparable to those obtained using traditional hybridoma technology can be selected from large naive antibody libraries, and the affinity of these can be further increased, to levels unobtainable in the immune system, by using the selected antibodies as the basis for subsequent libraries and selection.

In general, two kinds of libraries can be used: naive or immune. The naive libraries are derived from natural, unimmunized human rearranged V genes [1, 2], or synthetic human V genes [3–5]. Although the immune libraries are created from V genes from immunized humans [6–8] or mice [9–11], such libraries are biased only toward antibodies of a certain specificity, and have also been used to select antibodies against antigens which were not used in the immunization [7]. In general, the affinity of the antibodies selected is proportional to the size of the library, with K_u's ranging from $10^{-6/7}$ for the smaller libraries [2, 3] to 10^{-9} for the larger ones [1, 4], a finding which is in line with theoretical considerations [12]. Antibodies selected from immunized

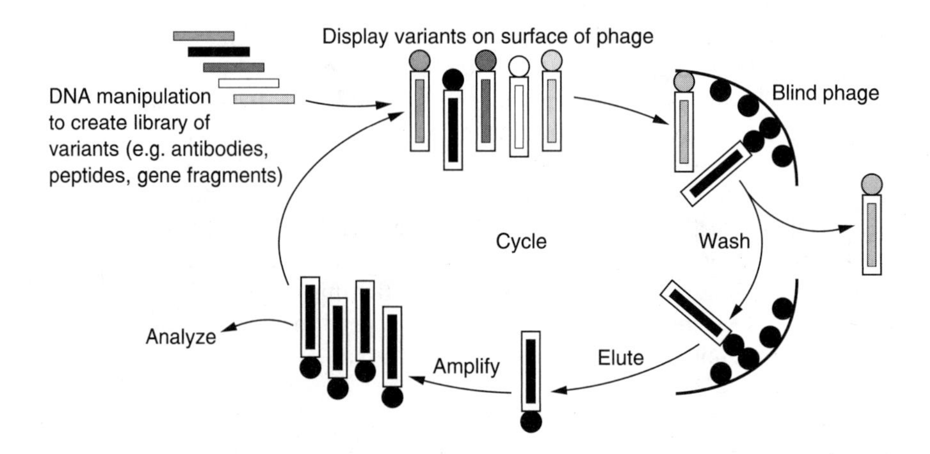

FIGURE 18.1 The phage display cycle.

libraries tend to have higher affinities for the antigen used for immunization from an equivalent library size.

Although a number of different phage (mid) antibody libraries have been published (see Table 18.1), the number which are "naive" is relatively small. To date, all phage antibody libraries have been made on the filamentous phage, Ff (which comprises fd, f1, and M13). In the creation of such libraries a number of different choices can be made:

1. The display protein used (p3, p6, or p8)
2. Phage or phagemid
3. Antibody form [single-chain Fv (scFv) or Fab]
4. How to assemble the V regions
5. For scFvs, the linker and the order of V regions
6. The source of V-region diversity (natural or synthetic)
7. Primers used to amplify the V genes

II. THE DISPLAY PROTEIN

The phage protein p3 is involved in bacterial infection and is present in three to five copies. It is responsible for binding to the F pilus and has a tripartite structure composed of three domains separated by glycine-rich regions. The N-terminal domain is important in penetration of the phage into the bacteria, while the second domain is responsible for binding to the F piles. Although peptides have been displayed between the two N-terminal domains, the site at the N terminus, after the leader, has become the site of choice for display of foreign proteins.

The protein p8 is the major coat protein, found in 2700 copies per phage. It is a small protein (50 amino acids) which is not very tolerant to large insertions. The protein p6 is a minor coat protein found at the same end of the phage as p3. It is not known to be involved in infection and has the characteristic that the C terminus rather than the N terminus is exposed.

Although p3, p6, and p8 have all been used to display proteins (see Table 18.2), p3 is the display protein *par excellence*, having been used to display large numbers of different proteins. The general characteristics of the three proteins when used as display proteins are shown in Table 18.2. Where a direct comparison has been made between the display efficiency of p3 and p8, it was found that p3 displayed more copies of a Fab than p8, notwithstanding its far lower copy number [13]. For these reasons, p3 has been the display protein used for all published antibody libraries (see Table 18.1).

III. PHAGE OR PHAGEMID?

Both phage and phagemid vectors have been used to display antibodies and other proteins. When proteins are displayed on phage, the gene encoding the recombinant display protein is included in the phage genome and as a result

TABLE 18.1
Published antibody phage libraries

Library	Theoretical diversity	Construction	V-gene source	Antibody form	Vector	Best affinity
Marks 91	3×10^7	Natural naïve PCR μVH assembled with PCR Vλ & Vκ	Human PBL	ScFv	Phagemid	10^{-6}–10^{-1}
Hoogenboom 92	2×10^7	Synthetic 49 VH, CD3 5 or 8, 1 JH 1 Vλ	Synthetic human V genes	ScFv	Phagemid	7×10^6
Orum 93	5×10^6	Natural immunized (factor VII) PCR fd, PCR VL, assemble	Mouse spleen	Fab	Phagemid	10^{-1}–10^{-9}
Barbas 93	10^7	Natural immunized (HIV) PCR td, PCR Vie, assemble	Human BM	Fab	Phagemid	?
Williamson 93	3×10^6	Natural immunized PCR fd, PCR Vκ, assemble	Human BM	Fab	Phagemid	?
Griffiths 94	6×10^{10}	Synthetic 49 VH, CDR 3 4–12, 1 JH 26 Vκ CDR3 8–10 (1–3) 21 Vκ CDR3 8–13 (0–5)	Synthetic human V genes	Fab	Phage	3.8×10^{-9}
Nissim 94	3×10^8	Synthetic 49 VH, CDR3 6–15, 1 JH 1λ	Synthetic human V genes	scFv	Phagemid	10^{-6}–10^{-7}
Kruif 95	3×10^8	Synthetic 49 VH, CDR3, 6–15, 1 JH CDR3 conservatively randomized for CDR >94 Vκ3Vλ	Synthetic genes	scFv	Phagemid	10^{-7}
Sheets 98	7×10^9	Natural Ig-specific primers, PCR V regions, clone VH and VL repertoires and assemble	Human spleen PBL	scFv	Phagemid	7×10^{-9}
Vaughan 96	1.4×10^{10}	Natural PCR VH (from hexamer primed, therefore all isotypes) cloned (10^8) PCR Vκ and Vλ cloned with upstream GS linker (10^7), both reamplified and PCR assembled	Human PBL-15, BM-24, tonsil-4	scfv	Phagemid	3×10^{-10}

TABLE 18.2
Comparisons between the different display proteins

p3	p6	p8
Antibodies have been displayed	CDNAs have been displayed	Antibodies have been displayed
Well tested for many proteins	Not well tested	Not well tested
Proteins displayed at high levels	Generally lower levels of expression	Generally lower levels of expression
Infection selections strategies possible	Not possible	Not possible
Phage or phagemid forms possible		Only phage forms containing both recombinant and wt g8p, or phagemids, possible for all but the smallest peptides
Very good for peptides of many lengths		Very good for peptides up to 8 amino acids
N-terminal fusions, or C-terminal fusions if Jun/Fos systems used	C-terminal fusions	N-terminal fusions
1–5 copies per phage (mid)	1–5 copies per phage (mid)	Up to 2700 copies per phage (mid) for peptides up to 8 amino acids

all phage (excluding loss due to proteolysis) will display the recombinant protein (see Figure 18.2), and all phage will contain only the recombinant phage genome.

In the case of phagemid, the recombinant g3p fusion protein is encoded on a plasmid (phagemid) which also contains the packaging signal of Ff. This phagemid makes large amounts of the recombinant display protein, but is unable to make phage unless the bacteria carrying the phagemid also contains a helper phage, which supplies all the other proteins required to make functional phage. Helper phages are essentially normal Ff phages with a couple of modifications: their packaging signal is disabled, and they usually carry antibiotic resistance genes. The disabled packaging signal does not prevent the helper phage from making phage when alone in a bacterium, but in the presence of a phagemid, which contains an optimal packaging signal, the phagemid will be packaged in preference to the helper phage. Phagemid pHEN1 (see Figure 18.3) is a typical example of a phagemid display vector.

As a result, phagemid preparations are both phenotypically and genotypically heterogeneous (see Figure 18.4, in which the two different genotypes are indicated by different phage lengths). The gene 3 may be either wild type (derived from the helper phage) or recombinant (derived from the phagemid), and the contained genome may be either phage or phagemid. The antibiotic resistance genes carried by phage and phagemid allow one to select for bacteria that contain both the phagemid and the helper phage.

From the point of view of phage antibody libraries, the choice of phagemid or phage has a number of considerations (see Table 18.3). From a practical point of view it is easier to work with phagemids than with phage (preparing DNA, transfection efficiency), and as a result it is far easier to create large libraries in phagemid than phage. This advantage is offset, however, by the slightly greater difficulty in using phagemid libraries in selection (the cycles of recovery are longer because of the need for helper phage superinfection). The fact that phage carry 3–5 copies of the recombinant protein, whereas with phagemid 1–10% of particles contain a single copy of the displayed protein [14],

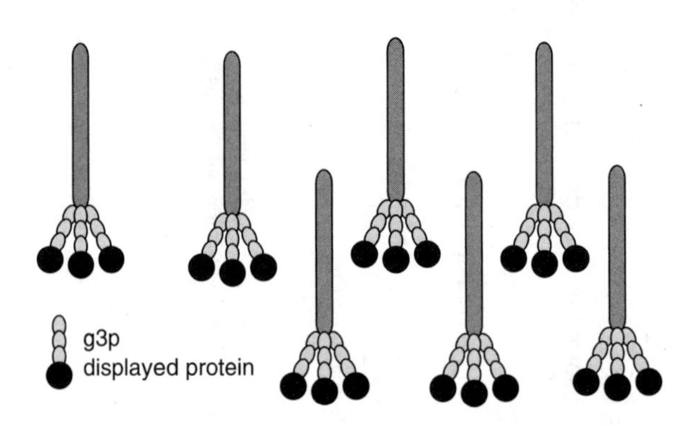

FIGURE 18.2 Phage homogeneity.

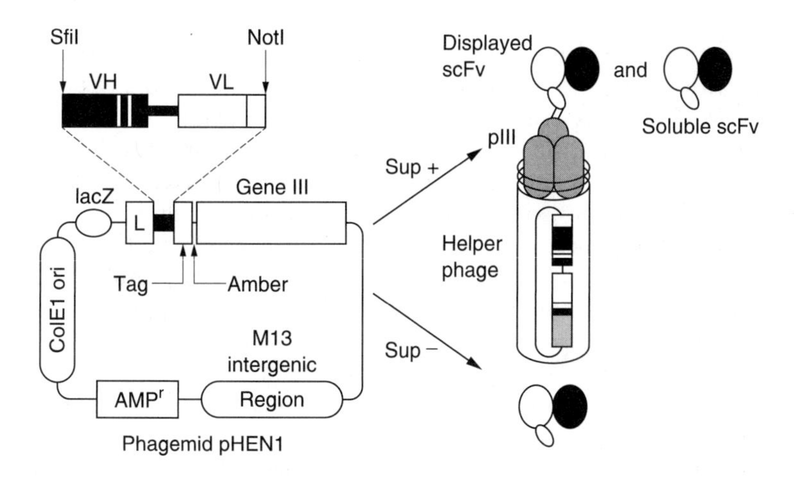

FIGURE 18.3. Phagemid pHEN1.

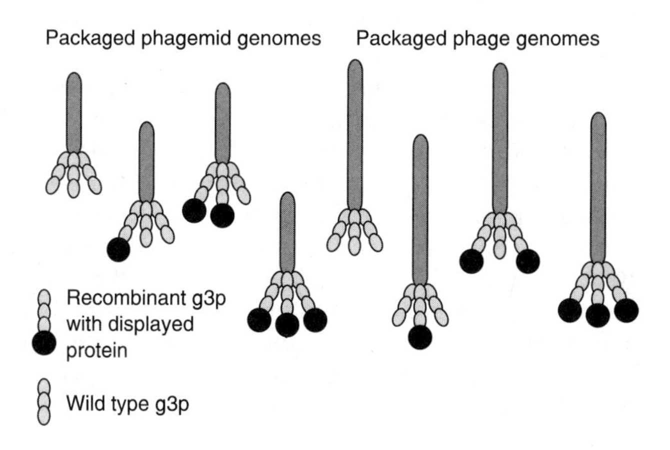

FIGURE 18.4 Phagemid heterogeneity.

TABLE 18.3
Phage or phagemid

Phage	Phagemid
3–5 copies of antibody per phage	Only 1–10% of phagemids have 1 copy of the displayed antobody
Difficult to transfect and make DNA	Easy to handle
Faster to select	Slower to select (helper phage needed)
Must subclone to make soluble Ab	Soluble Ab made directly
Phenotypic and genotypic homogeneity	Soluble Ab made directly
Genetically less stable (deletions)	More stable genetically
Infection selection strategies possible	Infection selection strategies unlikely

means that when selecting with phage, avidity effects allow the selection of particles with lower affinities. It also means that it is more difficult to improve the affinity of proteins displayed on phage, as such avidity effects overcome any small changes in affinity brought about by mutation. Another practical advantage of phagemid libraries is that it is easier to produce soluble proteins if an amber stop codon is inserted between the displayed protein and p3 (see Figure 18.3) [15]. Such soluble protein is secreted into the periplasmic space, but can also be isolated from the culture medium as a result of bacterial lysis. The ratio of soluble to displayed protein in a suppressor strain is always in favor of the production of soluble protein and will depend on the strength of the suppressor (TG1 is stronger than DH5α, F), varying from 50% to 90% of all the protein produced. Although soluble protein can also be made in phage display libraries using a similar genetic arrangement, the low copy number of the vector and the weakness of the p3 promoter and ribosome-binding site results in levels of soluble protein which are too low for most practical purposes.

Another advantage of phagemid vectors concerns their relative resistance to deletions of extraneous genetic material. Filamentous phage vectors in general have a tendency to delete unneeded DNA. This occurs as a result of the selective growth advantage a smaller phase has over a larger one. Phagemids suffer far less from this disadvantage and as a result are more stable.

One advantage of phage display libraries is that selection strategies which employ infection are more likely to be successful, since it seems that more than one functional p3 is required for infection. This has been recently exploited in two model systems. In the first, an antigen is attached to the C terminal of the first two domains of g3p and antibodies are attached to the N terminus of the last domain of g3p [16, 17]. Libraries for this kind of selection need to be specifically prepared. In the second case, an antigen is displayed within the context of the pilus and the antibody is displayed at the N terminus of p3 [18]. Selections using these methods are discussed later.

IV. ANTIBODY FORM

The choice of which form of antibody fragment to use in a phage antibody library (Figure 18.5) is on the whole dictated by practical considerations (Table 18.4).

Fabs consist of two chains, the VH+CH1 and the VL+CL, which need to assemble. scFvs, on the other hand, are single proteins with the two V regions joined by a flexible linker [19].

In general, Fabs are more difficult to assemble, are more likely to be degraded, have lower yields as soluble fragments, and are more likely to cause problems of DNA instability in the phage due to the larger size of the encoding DNA. All these problems arise from the fact that Fabs contain two protein chains which need to assemble, each of which is the same size as the single scFv chain. However, Fabs also have advantages: large libraries have been

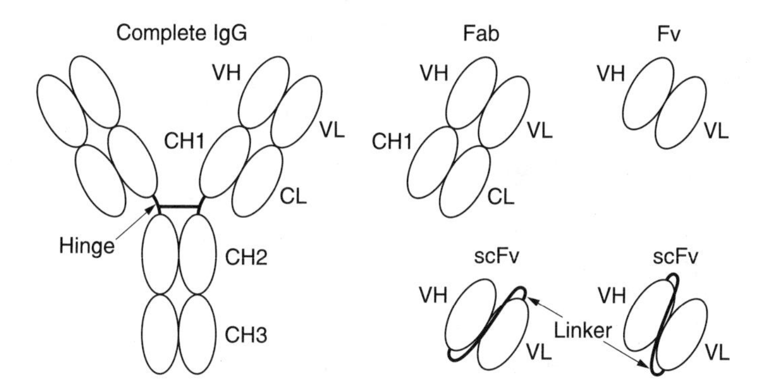

FIGURE 18.5 Structure of antibody derivatives.

TABLE 18.4
scFv or Fab

Better tolerated by bacteria	More difficult to synthesize
Less likely to be degraded	More likely to be degraded
Can form dimers (diabodies)	No dimerization
Single protein molecule	Two protein molecules
$(Gly_4Ser)_3$ linker can shorten	No such problems
DNA insert 700 bp	DNA insert >1500bp
In vivo recombination published, scFv libraries possible in principle	In vivo recombination used to create large libraries

made using *in vivo* recombination (see later) which uses an infection step, so avoiding the need for large numbers of electroporations, and Fabs do not appear to suffer from the problem of dimerization which afflicts scFvs. Recently, the use of a recombination signal (loxP) as a scFv linker and the use of the cre/lox recombination system to switch variable regions indicates that, at least in principle, it should be possible to make libraries using recombination with scFvs as well [20].

Although scFvs do not suffer from many of the problems described above, they can form dimers in which the VH of one scFv interacts with the VL of another, this problem being aggravated by shorter linkers. The formation of such dimeric antibodies has been exploited to create dimeric antibodies (termed diabodies) in which each member of the dimer expresses a different antibody [21]. Although this problem can be reduced by increasing the length of the linker to more than 20 amino acids, it has also been used to create phage antibody libraries in which each phage carries two antibody specificities [22].

V. ASSEMBLING V GENES

To create phage antibody libraries, VH regions need to be coexpressed with VL regions, a goal which involves placing both genes on a single plasmid. This is true if the antibody form used for selection is either scFv or Fab. Although the discussion below refers specifically to the creation of scFvs, it applies equally to the creation of Fabs. In each case the region between the V genes is that which is involved in the assembly. With scFvs, this region is a polypeptide linker which joins the two V genes covalently, whereas in the case of Fabs this region will be a piece of DNA which must contain a ribosome-binding site and a leader sequence (see Figure 18.6).

V regions may be assembled in a number of different ways (Table 18.5).

In PCR assembly methods (Figure 18.7), V regions are amplified with regions of overlap, either to a separately amplified linker region, or to each other, in such a way that mixing the two V regions recreates a linker

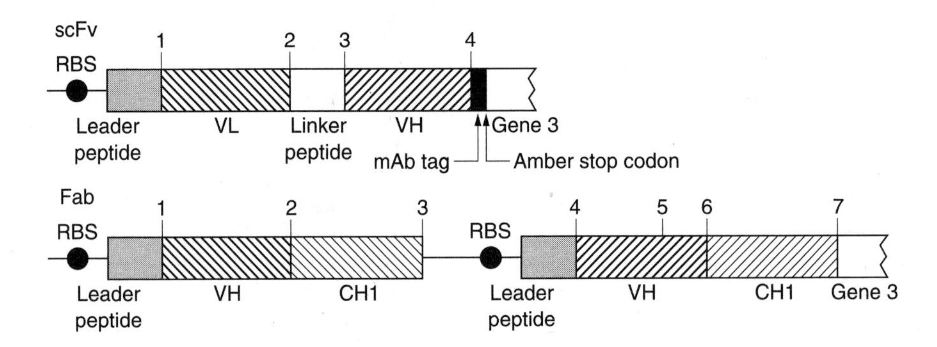

FIGURE 18.6 The genetic structure of scFv and Fabs used in phage display.

TABLE 18.5
How to link VH to VL

Method	Comments
2- or 3-fragment PCR assembly	Can be difficult to get the PCR to work.
	No easy way to assess degree of recombination.
	Single cloning step.
	Many transfections required to get a diverse library.
Cloning	More cloning steps.
	Better idea of degree of recombination.
	Many transfections required to get a diverse library
In vivo recombination	Methodologically more difficult.
	Can produce unwanted by-products.
	Will produce the highest diversity with the least work, as infection rather than transfection is used

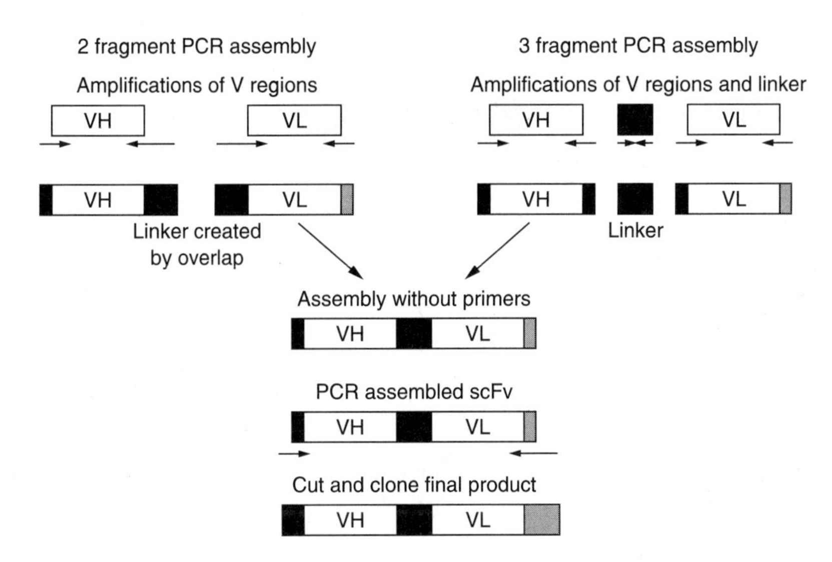

FIGURE 18.7 PCR assembly of V genes.

region joining the two V genes. A number of amplification cycles without the addition of external primers are first performed. These involve an initial annealing of the regions of overlap, following by an extension. In this way VH regions are joined to VL and the complete scFv can then be amplified by subsequent PCR in which external primers are added. This method can be difficult and is sometimes unsuccessful, requiring a careful titration of the amounts of VH, VL, and linker regions. Furthermore, it is impossible to assess the degree of recombination which has occurred, since residual VH and VL fragments are usually observed. It has the advantage, however, that the cloning involves the single scFv fragment rather than separate VH and VL fragments.

An alternative method involves the assembly of the library by cloning. To do this one must use restriction enzymes which do not cut frequently within V regions (see Table 18.6 for a list of enzymes based on VBASE [23]).

Cloning involves the separate amplification and cloning of VH and VL regions into separate vectors and the joining of these by cloning steps (Figure 18.8). It is important that the enzymes used do not cut within the V genes, otherwise those V regions will be unclonable and diversity will be lost. This has the advantage that it is relatively straightforward to create separate VH and VL libraries with diversities of 10^{7-8}, this providing a readily accessible source of V regions for further PCR or cloning.

The third method of joining heavy and light chains involves the use of recombination systems. One phage library in the Fab format has been constructed in this fashion [4]. Separate VH+CH1 and VL+CL libraries were made in different vectors. The VH+CH1 library was cloned into a plasmid vector with the VH+CHI domains flanked by two recombination signals

TABLE 18.6
Frequency of rare restriction sites in human germline V genes

Enzyme	Recognition site	337 VH	80 D/JH	168 VL	12 JL
Af*l*II	C/TTAAG	0	0	2	0
AgeI	A/CCGGT	0	0	8	0
AscI	GG/CGCGCC	0	0	0	0
*BspE*I	T/CCGGA	12	0	2	0
*BssH*II	G/CGCGC	0	0	0	0
BstBI	TT/CGAA	1	0	0	0
*Miu*I	A/CGCGT	1	0	0	0
Nhel	G/CTAGC	0	0	0	0
Nrul	TCG/CGA	0	0	0	0
Pact	TTAAT/TAA	0	0	0	0
PmlI	CAC/GTG	1	0	0	0
Sail	G/TCGAC	0	0	0	0
SfiI	GGCCN4/NGGCC	0	0	0	0
SnaBI	TAC/GTA	0	0	0	0
Spa	C/GTACG	2	0	1	0
SrfI	GCCC/GGGC	0	0	1	0
Xhol	C/TCGAG	0	0	2	0

Source: IM Tomlinson, SC Williams, SJ Corbett, JPL Cox, G Winger, V BASE sequence directory. Cambridge, UK: MRC Centre for Protein Engineering, 1996.

(lox P wild type and lox P 511), while the light-chain library was made in a phage containing a dummy VH+CH1 flanked by the same recombination signals.

These recombination signals are only able to recombine in a homologous fashion, i.e. loxP with loxP and loxP5I I with loxP511. Bacteria containing the heavy-chain library were infected with phage containing the light-chain library and the heavy chain was linked with the light chain by the expression of cre recombinase (see Figure 18.9). This was originally done by infecting bacteria with phage P1, which carries the cre recombinase, but is more easily done by inducing the expression of cre with an inducible promoter. As this method involves infection (which approaches 100% efficiency), it is theoretically far easier to make large libraries.

Although requiring far fewer transfections than the other methods, this system does suffer from a number of problems: (1) the recombination system is reversible, and as a result the library will be "contaminated" with the starting phage (containing dummy VH chains) and plasmids (containing the VH+CH1 library or the dummy VH+CHI) at relatively high levels; (2) there are by-products of the recombination process which consist of double plasmids; (3) it has been found that most plasmids, the pUCI9-based plasmid used here included, although they do not contain an Ff origin, can nevertheless be

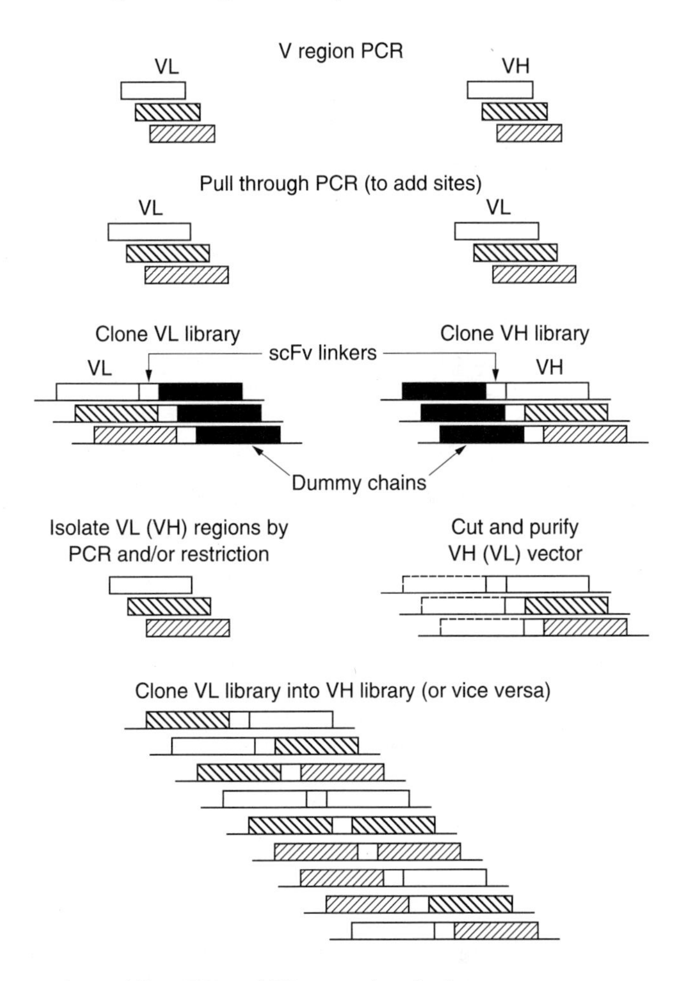

FIGURE 18.8 Assembling VH and VL genes by cloning.

incorporated into phage with variable efficiencies. As a result, the final library contains a mixture of many different genetic elements, so reducing the effective functional diversity; (4) the use of a phage to produce the final genotypic and phenotypic diversity results in genetic instability.

VI. VARIABLES TO BE CONSIDERED WHEN USING scFvs

A. Linkers

Many different linkers have been used in single antibodies expressed as scFv (Table 18.7). However, although some comparisons have been made for single scFvs, no direct comparison has been made between these different linkers in the creation of phage antibody libraries. In fact, the only linker which has been used extensively in scFv libraries is the $(Gly_4\text{-}Ser)_3$ linker. This has a number of

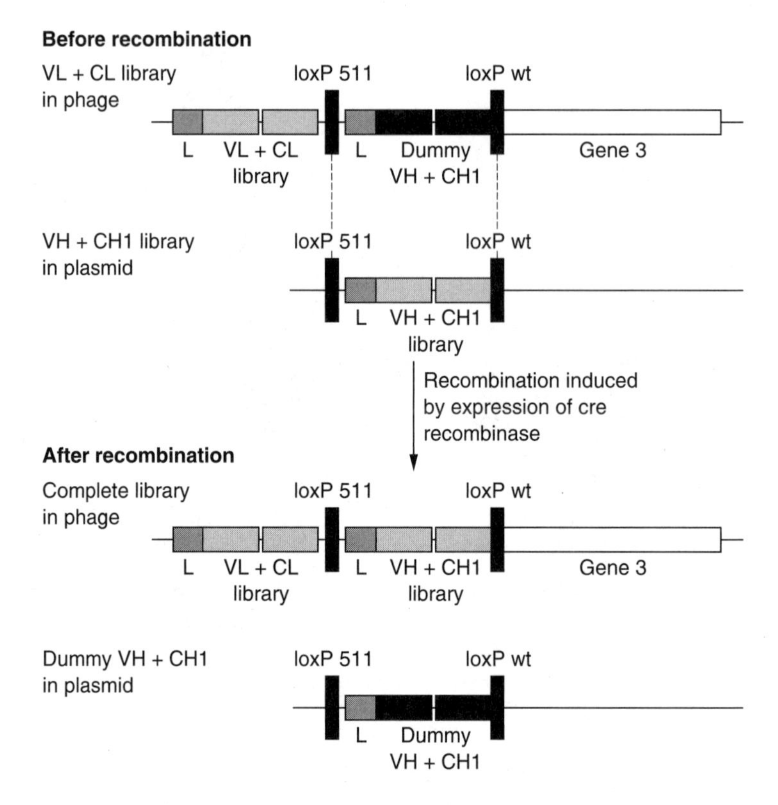

FIGURE 18.9 Creating libraries by recombination.

known disadvantages, including the tendency to shorten and form dimeric scFvs. One recent development which may be useful is the use of a loxP site as an scFv linker, which allows the use of the Cre recombinase to make large scFv libraries [20], although some of the problems described above will still apply.

B. V-Region Order

The scFvs have two possible conformations: VH-VL or VL-VH. The first scFvs and all libraries published to date have been made using the VH-VL format. Although few direct comparisons have been made between these two formats, one theoretical consideration suggests that the VL-VH format may be better: VL chains are generally more soluble than VH chains, and as such may be more likely to induce the formation of soluble scFv-g3p fusion protein if they enter the periplasmic space before the VH chain. Furthermore, the use of the VL-VH format has allowed the use of a very short tag sequence, based on the FLAG epitope, at the N terminus of the light chain [24]. This would be particularly useful to visualize full-length scFvs in Western blots, although it has not been systematically shown that the addition of this tag at the

TABLE 18.7
scFvs linkers

Linkers used in scFvs	Length	Order	Comments
GGGGSGGGGSGGGGS	15	VH/NL	Linker used in many scFv libraries so far (VH/NL format). Has a tendency to make scFv dimers.
GGGGSGGGGSGGGGSGGGGS	20	VL/VH	Longer GS linker which has been shown to reduce dimerization in the VLNH format
GSTSGSGKPGSGEGSSKG	18	VL/VH	218: based on the original GS linker and claimed to be less susceptible to Proteolysis and aggregation.
SGGSTSGSGKPGSGEGSSGS	20	VL/VH	220: based on the 218 linker but 2 as longer to reduce dimer formation.
PGGNRGTTTTRRPATTTGS SPGPTQSHY	28	VH/VL	Natural linker from fungal cellulase, forms monomers, but contains an internal protease site.
ASTSSGGGGSITSYSIHYTK LSGGGGSEL	29	VH/VL	loxP wild-type recombination site, flanked by extra amino acids, used to make a single functional scFv.
	18	VL/VH	Each of these has been used for isolated scFvs. On the whole they tend to be short and so will probably suffer from problems of dimerization.
GQPKSSPSVTLFPPSSNG	14	VL/VH	
GSTSGSGKSSEGKG	15	VL/NH	
GGSGSGGSGSGGSGS	15	VH/NL	
GGGGTGGGGTGGGGT EGKSSGSGSESKEF	14	VL/VH	
GGSGGSGGSGGSGG	14	VL/NH	

N terminus of VL, at a site very close to the antibody-binding site, does not alter antibody reactivity.

C. The Source of V-Region Diversity

Some scFv and Fab libraries have been made from both natural and synthetic V regions (Table 18.8). Natural V regions can be considered to be those found in lymphocytes (see later for a discussion of sources), which may or may not have undergone antigen stimulation. As a result, some of the V genes amplified will be rearranged germline, whereas others will contain mutations induced by the encounter of the B cell with antigen. Such genes are amplified using primers which recognise the 5′ end of the V genes and the 3′ end of the J genes or the 5′ end of the CL or CH1 domain (Figure 18.10). In general, the majority of V regions which are amplified from natural sources are functional and do not contain stop codons or frame shifts. This is because the transcription of such nonfunctional V regions is suppressed in B cells. It should be noted that the same is not true of hybridomas, in which nonfunctional V regions are often encountered [25].

TABLE 18.8
Synthetic or natural

Synthetic	Natural
May create "unnatural" loops	"Tested" over millions of years
Usually 1 JH, and few VL, although recent unpublished libraries have used multiple VH and VL genes	Theoretically, all JH and VLs
No somatic mutation	Somatic mutation of other parts of V possible
Stops may be inserted	Nonproductive V regions may occasionally
V genes can be limited to those which are well expressed in bacteria	All V genes, well expressed and not, are used

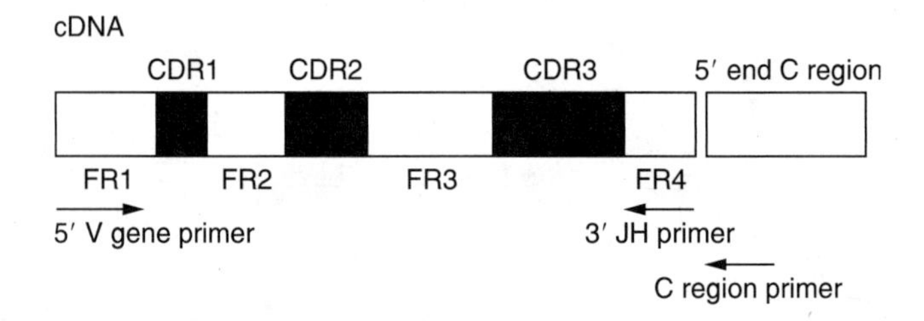

FIGURE 18.10 Cloning natural V regions. FR: Framework; CDR: Complementarity determining region.

Synthetic V regions, on the other hand, have usually been made by PCR and involve the addition of random CDR3's and framework region 4 to the 3′ ends of cloned immunoglobulin genes (Figure 18.11) using long oligonucleotide primers [4, 26]. As such they are based on natural sequences but have synthetic elements within the CDR3. The CDR3 can be made with a wide variety of lengths, depending on the oligonucleotides used for amplification.

A comparison of the best affinities achieved from antibody libraries made in different ways (Table 18.1), although not strictly valid for a number of reasons (different antigens, some affinities are against protein antigens, others against haptens, different selection conditions, and different vectors) reveals that libraries made from natural antibodies tend to give antibodies of higher affinities than those obtained from similarly sized libraries made from synthetic V genes. This may be due to the use of CDR residues, which are either not tolerated by the antibody structure, or from the creation of structures which are not able to bind at high affinity. The use of PCR to create synthetic V regions in this way has also been found to result in PCR errors, which may cause frameshifts, as well as stop codons in the region of variability.

The problem of stop codons within the regions of variability can be overcome by the use of trinucleotides instead of single bases to create the oligonucleotides [27]. However, this technology is expensive and not yet generally available. An alternative method to ensure that the V regions used are functional involves the initial creation of separate heavy- and light-chain libraries which are then selected on protein A and protein L, respectively (Tomlinson I unpublished). In this case the library is limited to VH and VL genes known to bind these proteins. One advantage of the use of synthetic V regions, rather than natural V regions, is that V genes can be chosen or engineered which are known to express well in bacteria. However, this advantage can be undermined by the fact that single amino acid changes within CDRs can cause great differences in bacterial expression levels. A number of unpublished libraries have been made using synthetic techniques which allow synthetic sequences to be inserted into any of the CDRs using flanking restriction sites; these may be particularly useful for subsequent affinity maturation of the antibody.

As the structural features of V regions become further elucidated [28], it is possible that rules describing the amino acids, or amino acid combinations,

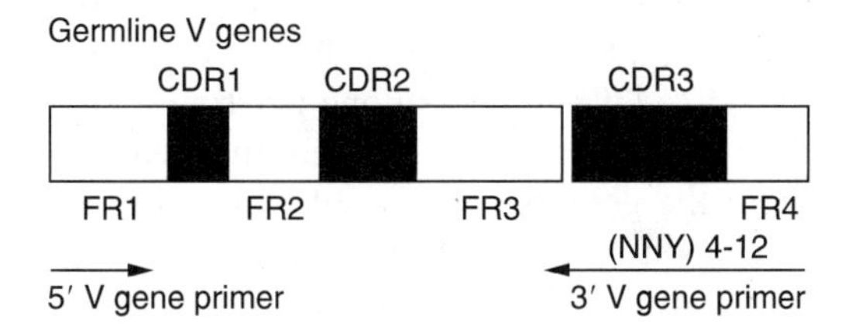

FIGURE 18.11 Creating synthetic V genes.

permitted or frequently present at different CDR positions, may be derived. For such rules to be useful, they are likely to be more complicated than a statistical analysis of amino acid frequency at different CDR positions, since it is possible that the presence of particular amino acids at particular positions may preclude the presence of other amino acids at other sites. Neural networks are capable of understanding relationships of this kind. Such rules, in association with trinucleotide chemistry and the use of well-expressed V genes, may lead to the creation of synthetic libraries which are better than those presently derived from natural sources.

D. Sources of Natural V Genes

When making a phage antibody library from natural V genes, a number of different sources of B cells (and hence V genes) can be considered. None of these has been systematically examined for its advantages in phage antibody libraries, and hence the following comments are based more on immunological theory than practical knowledge. In general, the choice can be considered to be between V genes which have not undergone somatic mutation (naive and generally IgM) and those which have undergone somatic mutation (immune and generally IgG), there being a gradient of naivity going from IgM to IgG.

Early experiments with libraries of diversities less than 10^8 suggested that IgM-derived V genes were better than IgG-derived V genes [29], as antigen-binding phage antibodies were only isolated from IgM-derived libraries and not from IgG-derived libraries. However, a far larger library which has been subsequently made [1] using unfractionated V genes from bone marrow, peripheral blood, and tonsil, appears to be very diverse and has been used successfully to isolate antibodies against a wide source of antigens. Analysis of the V genes derived from this library shows that most of the antigen-binding V genes selected contained many mutations, suggesting that they were derived either from IgG or from the mutated IgM compartment mentioned below. Other large libraries which have been made [4] (Sheets *et al.* [93]; Hoogenboom *et al.*, unpublished) have invariably used IgM-derived V genes.

Although it is still unclear whether IgM- or IgG-derived B cells provide better V-gene sources, it is true to say that most V genes selected from natural phage antibody libraries are mutated when compared to the germline genes from which they were derived. B-cell sources which have been used include bone marrow cells, peripheral blood lymphocytes, spleen cells, and tonsil cells. These tissues all have different cellular properties. Naive B cells are produced (and over 75% killed) in the bone marrow. From there they enter the blood and migrate to the spleen. When they encounter antigen they proliferate (usually in the germinal centers of lymphoid organs-lymph nodes, spleen, gut-associated lymphoid tissue) and their V regions acquire somatic mutations. Following stimulation they can become either plasma cells or memory cells. Plasma cells are found in the secondary lymphoid organs (especially spleen) and bone marrow, and memory cells are found in the circulation. It is clear from this brief tour of B-cell development that all tissues used contain B cells at various stages of development, containing V regions with different degrees of mutation. In general, cells from bone marrow are the most naive (although they also contain plasma cells), and those from tonsil are the most mutated. As tonsils are usually removed from patients suffering from tonsillitis, one would expect that the V genes found there would be heavily biased toward recognition of the agent causing the tonsillitis, perhaps not the best theoretical choice of V genes for a diverse one-pot library. However, the large library described below [1] did include a small percentage (10%) of V genes from tonsil and does appear to be very diverse. Similar arguments may apply to V genes from PBLs obtained from patients suffering infectious diseases [7, 30]. In persons not suffering from infections, the V genes in PBLs or spleen would be expected to be a mixture of those found in memory cells (and hence recognizing past infectious agents) and naive ones which have yet to encounter antigen. Peripheral-blood B cells can be divided into three groups: naive IgM, mutated IgM, and mutated chain switched (mainly IgG). Although the proportions of cells in each of these populations are 75%, 10%, and 15%, the mutated cells contain 7–11 times more mRNA than the naive cells, and as a result most V regions isolated from PBLs will be mutated rather than germline [31].

This discussion has focused on the desire to create a one-pot naive phage antibody library. However, phage display has also been used very effectively to isolate binding antibodies when libraries are made from immunized experimental animals [32], or from naturally immunized humans [7, 30]; in these cases the V-gene source being either spleen or peripheral-blood lymphocytes.

Despite the differences between the "naivety" of different lymphocyte sources, it is still not clear which source (more or less naive) provides the best V genes to make single-pot phage antibody libraries.

E. Primers for the Amplification of Natural V Genes

An ideal set of primers to amplify V genes would pick up all V genes, equally efficiently, with the minimum number of primers. Recently, all human V, D, and J genes have been cataloged in the VBASE database maintained by Tomlinson *et al.*, at the MRC in Cambridge. This can be most easily accessed and downloaded via the World Wide Web (http://www.mrc-cpe.cam.ac.uk/imt-doc/pub-lic/INTRO.html). A set of primers able to amplify all V genes contained in VBASE is now found within this site. These primers have a degeneracy no greater than twofold. A smaller set of primers, based on the same database, but with greater degeneracy, has also been published [33]. These two sets of primers supersede those published before the availability of VBASE [2, 29, 34–37], although very good libraries have been produced using some of these primer sets.

Mouse antibodies can also be displayed on phage, and a series of primers which will pick up almost all mouse V genes has been described [38]. These are likely to be very useful for the cloning of hybridoma V regions and for making antibody libraries from immunized mice.

F. Phage Rescue

After the construction of the phage (mid) antibody library, phage-displaying antibodies representing the whole library need to be produced. In the case of phagemid libraries, rescue with helper phage is required; whereas in the case of phage libraries, helper phage is unnecessary, and it is sufficient to grow the bacteria, phage being produced during growth.

In the phagemid vectors most often used (such as pHEN1 and pCANTAB5/6) [15], the expression of the antibody fragment is driven by the lacZ promoter, although other promoters have also been used. Before phage rescue, bacteria containing the library are grown to logarithmic phase in the presence of glucose. This prevents any bias induced by possibly lethal antibody expression, and, most important, minimizes the lacZ transcription level of the pIII product below a certain threshold, making infection by the helper phage feasible. This is because the presence of p3 within the bacteria inhibits the formation of the pilus and so infection by more than one phage. If glucose is omitted, the efficiency of infection drops dramatically, sharply reducing the size of the library. After infection by helper phage, the glucose needs to be removed, to permit p3, and hence library, expression. p3 expression is then driven by the low basal level of the lac promoter. Inducing this promoter with IPTG is counterproductive and leads to very low display levels, due to the toxicity of the p3/antibody fusion protein. For scFv and Fab libraries, typically 30°C is chosen for the overnight growth phage, as this seems to result in better antibody display levels. Before the selection, phages are prepared from the supernatant of the culture by PEG-precipitation. This removes soluble antibody fragments, arising from nonsuppression or proteolytic cleavage, which can compete with the phages for binding, as well as any other contaminants which may interfere with the selection.

VII. SELECTING FROM PHAGE ANTIBODY LIBRARIES

Phage antibody libraries are enriched for antigen-binding clones by subjecting the phage to recursive rounds of selection. This is illustrated in its simplest format in Figure 18.1, but the same principles apply for all selection methods. Ideally, only one round of selection should be required, but the aspecific binding of phage limits the enrichment that can be achieved per round; therefore in practice 2–5 rounds are necessary. An extensive collection of methodologies is available to separate binding clones from nonbinding clones. These include biopanning on immobilized antigen coated onto solid supports, columns, or BIAcore sensorchips [4, 39], selection using biotinylated antigen [40], panning on fixed prokayotic cells [41] and on mammalian cells [42], subtractive selection using sorting procedures such as FACS [43] and MACS [44], enrichment on tissue selections or pieces of tissue [45] and, in principle, selections using living animals, as reported for peptide phage libraries [46]. More complex selection methods are "Pathfinder" [47] and methods based on the restoration of phage infectivity via the antibodyantigen interaction [17]. In general, selection methods can be divided into two broad classes: those which attempt to isolate antibodies against known antigens, and those which use phage antibodies as a research tool to target unknown antigens (e.g. against cell surface antigens).

A. The Selection Step

When designing phage selection regimes, it should be born in mind that retrieval of the desired antibody will be most efficient if purified antigen is available. If this is the case, it is crucial to have a method for its immobilization or labeling. This will facilitate the separation of antigen-bound phage from the large excess of nonbinding particles. Although small quantities of antigen are required for selection, far larger amounts are needed for screening. In general, about 500 μg–1 mg of antigen is enough to complete a generous series of selections and screenings, although it can be done with as little as 100 μg.

It is important to perform the selection in a suitable blocking agent which inhibits to some extent nonspecific phage binding. Semiskimmed milk powder, gelatine, BSA, and caseine have been used. The addition of Tween-20 also helps to reduce nonspecific phage binding. Typically, 10^{13} phage (defined as colony-forming units) are used for the first round of selection, when the initial diversity is 10^{8}–10^{10}. This will ensure that each phage antibody clone will be present in the starting repertoire. The number of phage used may be decreased in later rounds of selection. It is important to note that, with a display efficiency of less than 10% [14], there will be only 10 to 100 copies of each phage antibody at the start of selection. Therefore, it is crucial that this first round is not too stringent, to ensure that all binding antibodies are recovered for subsequent amplification. There are many different strategies for the enrichment of specific phage antibodies; Table 18.9 and Figure 18.12 show the most frequently used methods. Antigen may be adsorbed onto Petri dishes or onto plastic tubes [2], or immobilized on Sepharose and loaded onto columns [48]. A particular problem with these physical methods of selection is that the epitope may be partially denatured after immobilization, which may lead to selection of antibodies which do not recognize the native antigen. Although this is not a common occurrence, it can be avoided by indirect coating, using another antigen-specific antibody. When using impure antigen preparations or targeting defined epitopes, specific elution can be used. This may be achieved using the antigen itself or other reagents that bind the antigen.

With panning or column methods, the amount of native antigen captured onto a solid support and therefore available for selection is difficult to determine. This can be more easily done when selections are carried out using

TABLE 18.9
Selection methods

1. Panning
2. Biotinylated antigen
3. Peptides
4. Natural cell (prokaryotic and eukaryotic) surfaces
5. Cloned proteins expressed on cell (prokaryotic and eukaryotic) surfaces
6. Fluorescence-activated cell sorting (FACS)
7. Magnetic-activated cell sorting (MACS)
8. Selection on pieces of tissue
9. "*In vivo*" selections
10. "Pathfinder" selections
11. Selection methods based on the pill–pilus interaction

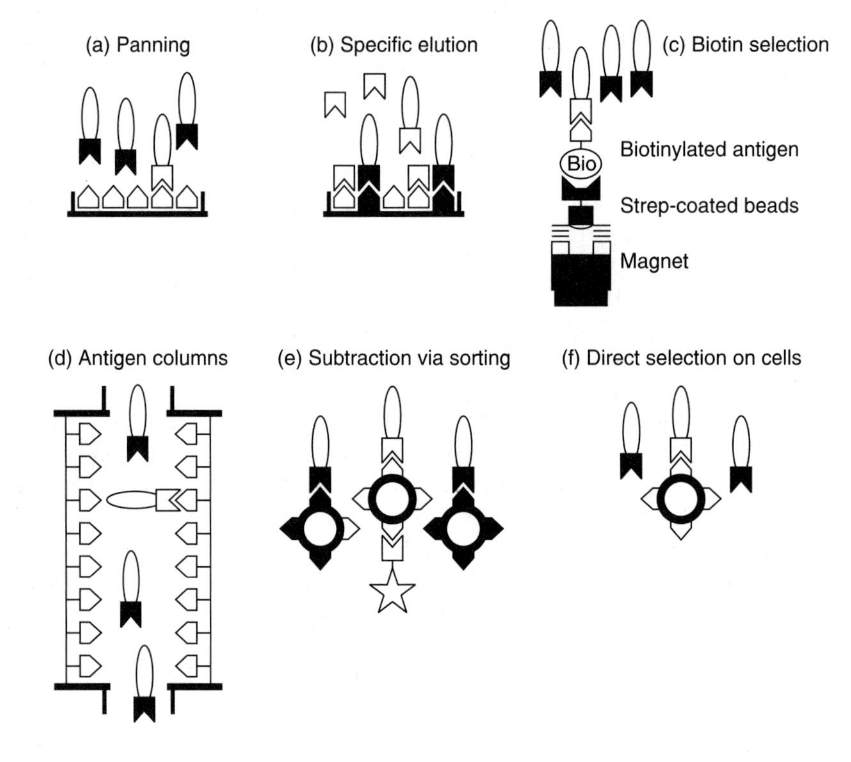

FIGURE 18.12 Selection methods.

labeled or tagged antigen in solution [40]. (Biotinylation is the method of choice.) After incubation of the phage library with the biotinylated antigen, phage bound to the antigen-biotin complex are retrieved with streptavidin-coated paramagnetic beads and a magnet. A potential problem with using biotinylated antigen is that antibodies to streptavidin are also isolated, in particular with nonimmune libraries. When the number of streptavidin binders is high, a depletion step with streptavidin coated onto paramagnetic beads will reduce their frequency, or an alternative method is to elute binding phages with DTT, at least when biotin is linked to antigen by a disulfide bridge [49].

During the selection of the primary phage libraries, the antigen concentration may be decreased to favor the enrichment of antibodies with a higher affinity in a similar fashion to the *in vivo* selection of B cells. The selection regime may be chosen to favor affinity or kinetic parameters such as off-rate. This hinges on the use of limited and decreasing amounts of antigen and on performing the selections in solution rather than by avidity-prone panning on coated antigen [40]. Typically, for large unbiased antibody libraries, a starting antigen concentration of 300–500 nM is used, and this is reduced fivefold with each subsequent round of selection. When selecting from a secondary phage library (see improving antibody affinities), the antigen concentration is typically reduced below the K_d of the parent clone to allow preferential selection of higher-affinity mutants [40]. This may be monitored by titrating the input and output phage from each round of selection.

B. Factors Governing Selection

The actual number of rounds of selection required is almost impossible to predict in advance and is dependent on the degree of enrichment afforded by the selection method and the relative affinities of the antibodies in the population. The degree of enrichment, (i.e. the ratio of binders to nonbinders before and after selection) typically varies from 5- to 1000-fold at each round, and more rounds of selection are needed to resolve antibodies with similar affinities than for those with widely differing affinity constants. All selection methods favor, to differing extents, those variants with stronger binding, since these are automatically best suited to survive the selection process. These antibodies will have either highest affinity or greatest avidity. The highest-affinity antibodies within a population are those which, at a set concentration, bind the greatest proportion of antigen at the lowest antigen concentrations. Selection using low concentrations of antigen can be used to preferentially select those antibodies of highest affinity. Selection may also favor variants that form multimers, particularly when the antigen is polymeric or present at high density. In this case, a low-affinity antibody with a tendency to multimerize may be preferentially selected by virtue of the slower rate of dissociation. This can be calculated as the product of the off-rates of each component of the multimer, although thc true dissociation rate tends to be lower [50].

The selection efficiency is also dependent on the percentage of phage(mid) particles which display a particular antibody. This usually varies between 1% and 10%, although it can be far lower [14]. The selection of two antibodies with similar affinities but different display levels will result in the preferential recovery of the antibody with the highest display level. The level of display is the result of a number of different parameters, including (1) antibody folding within the periplasm, (2) resistance to proteolysis, (3) resistance to aggregation, and (4) toxicity to the bacterial expression host.

C. Washing, Elution, and Reinfection Steps

When selection is carried out by panning, washing is performed by simply rinsing between 10 and 40 times in a suitable solution, usually PBS-containing detergent (e.g. 0.1% Tween 20) in PBS. Washing after the first cycle should be less stringent, as it is important to recover all phage-binding antigens, including those present in very few copies. After the first cycle, such phages are present at much higher levels, and washing can be more stringent. When using biotinylated antigen and streptavidin-paramagnetic beads, washing is carried out by using a magnet to pull the beads with bound phages to one side of the tube between washings. For panning cells, the washing procedure is a simple repetition of gentle centrifugation, removal of the supernatant, and resuspension of the cell pellet in fresh buffer. In all cases, care must be taken to avoid contamination of the phage to be eluted (total titer 10^6–10^9) with minute quantities of the starting phage preparation (10^{9-10}/μL). Phage antibodies bound to antigen can be eluted in different ways: with acidic solutions such as 0.1 M HCl or 0.2 M glycine pH 2.5 [51], with basic solutions such as 0.1 M triethylamine [2], with 0.1 M DTT when biotin is linked to antigen by a disulfide bridge [49], by enzymatic cleavage of a protease site (e.g. factor X or genenase) engineered between the antibody and gene III [52], or by competition with excess antigen [32] or antibodies to the antigen [53]. Finally, bacteria may be incubated directly with the antigenphage antibody mix, without disrupting the antibody-antigen interaction; this is effective for antibodies with affinities up to 10 nM (Roovers and Hoogenboom, unpublished); however, there are data indicating that ultrahigh-affinity antibodies will not be able to infect unless the antigen-antibody interaction is broken. In such extreme cases 100 mM HCI, urea, or 3 M guanidium hydrochloride followed by dialysis before reinfection can also be used; alternatively, DNA can be extracted with phenol prior to PCR or transfection. Similar conditions may be used for elution from cells: acid or alkaline elution which lyses the cells is very efficient.

Eluted phage are then incubated with bacteria and grown on solid media containing glucose and ampicillin. Amplification in liquid medium may introduce growth biases due to differences in the toxicities of individual antibodies and should therefore be avoided. A number of different host bacteria have been used, including TG1, DH5αF′, XLI-Blue for making libraries and rescuing phages. The selected repertoire is then rescued using helper phage, usually M13K07, and purified phages are then ready to start the next round of selection.

D. Following Selection: When Do You Know It Is Working?

The phage titers should change between rounds of selection: the phage recovered should increase with each selection cycle, as should the output-to-input ratio. Without this increase, selection has usually failed, although there are exceptions. It is also possible that this increase can also be seen without an accompanying enrichment of genuine antigen binders; this is seen particularly with selection on complex antigen mixtures or cells, and where the selection is overtaken by deletion mutants which have growth advantages.

When starting selections on a new antigen, it is advisable first to determine the experimental conditions that give the best enrichment and yield, before using very large phage antibody libraries. This may be done by using mixes of a binding phage carrying, for example, the beta-lactamase gene in its genome, and a nonbinding phage carrying an alternative antibiotic resistance marker (i.e. tetracycline resistance). After selection, the mix is reinfected into bacteria and the ratio of ampicillin- versus tetracycline-resistant clones is determined before and after selection. This rapid method allows the experimental determination of the ideal incubation, washing, and elution conditions. This is particularly helpful when selecting on complex antigen mixtures or on cells. When no model antibody is available, the simplest and fastest approach is to perform a few rounds of panning and/or biotin selections with a large antibody library, and use the selected antibodies to assist in designing an optimum selection strategy.

E. Advanced Selection Strategies

Selections on impure antigen preparations are significantly more difficult, due to the problems of enriching phage antibodies that are specific for nonrelevant antigens. Examples include antigens that cannot be easily purified from contaminants with similar properties, or cell surface antigens whose native conformation is dependent on folding across a lipid bilayer, such as many of the seven transmembrane receptor family.

Selection from phage antibody libraries provides a new tool for the isolation of novel self-antigens, such as tumor- or disease-associated antigens. Both the de novo combined V-domain pairs in naive and synthetic antibody libraries (but particularly the latter) are not shaped by the constraints of the immune system, and for the first time avoid library bias by *in vivo* tolerance mechanisms. Therefore, the chances of detecting unique self-antigens and self-epitopes are dramatically improved compared to traditional immunization-dependent hybridoma-based antigen screening. However, success is dependent on powerful cell sorting or subtraction methods.

By developing careful selection and screening strategies, antibodies have been made to traditionally difficult antigens, such as MHC-peptide complexes [54], and human blood group antigens [55]. Also, selection experiments probing lymphocyte and tumor cell surfaces [43, 43] have yielded a number of promising, but as yet uncloned, cell-type specific antigens. The lack of reactivity of the selected phage antibodies in immunostaining, their low affinity, or the low abundance of antigen, however, complicates their full characterization.

F. Selection on Cell Surfaces

Direct panning on cell surfaces carrying the antigen may be carried out on adherent cells grown in monolayers, or on intact cells in suspension [42, 55, 56]. Selections are usually performed at room temperature, in the presence of sodium azide to prevent phage internalization, and in the absence of detergents which will cause cell lysis. Selections have been carried out on many different tumor cell lines, transfected CHO cells, primary dendritic cells, and on fragile endothelial cells. The conditions for selection need to be determined empirically for each cell line. For the most robust cells such as CHO cells, selections may be carried out with cells in suspension. Washes are performed by careful centrifugation to spin down cells with bound phage particles and then resuspension. An alternative to selecting on cells in solution or in monolayers is the use of fluorescence-activated cell sorting (FACS) [43] or magnetic-activated cell sorting (MACS) [44]. Using FACS, cells are labeled with a fluorescent antibody that recognizes a cell subpopulation of interest, incubated with the phage antibody library and then sorted for the fluorescent antibody. Cells (with attached phage) recognized by the fluorescent antibody are then lysed and the bound phage are rescued. This method has the advantage that complex cell mixtures can be used, providing antibodies recognizing the cell type of interest [43], and has been used to isolate new antibody specificities recognizing plasma cells, activated polymorphonuclear cells, and tonsil B cells. It should be noted that, in many cases, antigens will be present at very low densities on the cell surface, and antigen concentrations during selections will reach values much lower than the K_d of any antibody in the library. Even when antigen concentration is sufficient for antibody binding and retrieval, antigen inaccessibility through steric hindrance caused by the presence of other proteins or glycosylation may prevent the selection of antibodies.

The use of FACS to sort cells and their bound phage has the problem of being possibly too stringent, especially for early selection rounds: every first selection leads to a set of approximately 1000 phage in the first round (versus 10^{6-8} when using panning), from which a couple of binders are isolated in the second round. Since enrichment factors per round are never higher than 10^{4-5} the use of large libraries will automatically yield new and different binders in every new selection. MACS (magnetic-activated cell sorting) appears to be less stringent, allowing sampling of the complete repertoire in one selection (Kenter and Hoogenboom, unpublished). It remains to be seen whether any subtraction method may be powerful enough to yield new antigens. Systematic studies to compare the various enrichment methods are essential to understand and design appropriate cell-surface selection strategies of phage antibodies.

Antigens can also be displayed on the surface of bacteria by cloning epitopes or protein antigens within the context of membrane proteins such as LamB and OmpA. Such bacteria can then be used to construct antigen columns and to select antibody phage recognizing the displayed antigen [41]. Such selections are similar in principle to those performed using tissue culture cells, with the exception that bacteria are less delicate than tissue culture cells, and can also be fixed prior to use and stored for prolonged periods at 4°C.

In *in vivo* selections, in which phage repertoires are injected directly into animals, binding peptides localized to certain organs are then rescued [46]. This may be extended to the in-vivo selection of phage-displayed antibody repertoires to isolate, for example, phage antibodies specific for organ-specific endothelial cell antigens. It is also possible to select on pieces of tissue, or histological sections [45], a method which is likely to be particularly useful for the generation of antibody reagents used for pathological and diagnostic-based applications.

G. "Pathfinder" Selections

A recently described method, termed "Pathfinder" selection, may overcome several difficulties associated with phage antibody selections on complex antigens such as cell surfaces or the problems associated with dominant epitopes [47]. This method allows the isolation of antibodies which bind closely to a known ligand. A horseradish peroxidase (HRP)-labeled ligand is incubated with a target (which can be cells, tissue, or purified protein) and the phage antibody library. After washing, phage antibodies binding closely (within 25 nm) to the HRP-labeled ligand are biotinylated using biotin tyramine. The biotinylated phage are then easily separated from other unlabeled phage using magnetic avidin beads. The procedure was exemplified by selecting antibodies to TGF-1β, CEA, and a cell surface receptor, CC-CKR5. This technique has the advantage that antibodies can be isolated which are specific for the antigen in its native configuration and environment.

H. Selection Methods Based on the PIII–Pilus Interaction

Methods that use the interaction between the phage and the pilus of the bacterium to mediate the selection process have been developed by several academic groups [16, 57, 58], and are reviewed in [17]. The method is based on the principles that infection by Ff phage requires a complete pill protein. If part of pill is removed, phage are noninfective. However, infectivity can be restored (albeit at low levels) by reconnecting the two separate parts of pIII by a noncovalent interaction, such as an antigen/antibody interaction. A library of antibodies displayed on phage which have the N terminus domain of pill deleted can be selected by interacting with an antigen attached to the deleted N-terminal domain. Those phage antibodies that recognize the antigen are then able to reassemble pill, such that phage can infect via the pilus and can be enriched from irrelevant phage antibodies which are unable to reassemble pill. The fusion product may be supplied *in vitro* or produced *in vivo*, where both antibody and antigen are encoded on the same phage genome. Both affinity and folding properties have been shown to influence the efficiency of selection of antibodies with the SIP methodology [59], and selection on the basis of both binding and also kinetics has been reported. Potential problems are (1) the low efficiency of reestablishing the infection event, which negates the advantage of using larger repertoires; (2) the occurrence of infection events independent of the antibody-antigen interaction, for example, selection of antibodies binding directly to the pilus or to the soluble part of gIII, or to the bacteria themselves; and (3) the need to make specific antibody libraries lacking the N-terminal domain of p3.

A more recent procedure has involved the display of antigen on bacterial surfaces. Bacteria expressing an antigenic peptide within the context of the bacterial F pilus were shown to be infected by phage only when displaying an antibody specific for the antigenic peptide [18]. It remains to be seen whether these selection systems will be more efficient than standard procedures in isolating antibodies from very large naive or synthetic antibody libraries. A better understanding of the infection process itself [60, 61] may eventually lead to more generic application of these procedures in isolating interacting protein-ligand pairs, which will ultimately lead to the co-selection of repertoires of both antibody and antigen.

VIII. SCREENING ASSAYS

A. Designing Screening Assays for Binding

After the selection procedure, a fast and robust screening assay for binding is required; for example, a simple ELISA with coated antigen or with whole cells, FACS, or immunocyto- or histochemistry. Screening procedures employing a 96-well format such as ELISA allow many clones to be screened in parallel. Even if the end application of the antibody is not directly compatible with such simple ELISAs, a crude 96-well screen can still be used to focus on the best candidates for further detailed analyses. Although the antigen used in selections needs to be as pure as possible, that used in screening need not be as pure if availability is limited. To speed up screening procedures, phagemid vectors that incorporate a dual purpose have been developed. These vectors allow both monovalent display of antibody fragments and the production of soluble antibody fragments for screening without the necessity to subclone the antibody V genes. This is achieved by the insertion of an amber codon between the antibody and pIII genes [15]. A variety of tags have been described that can be appended to the antibody fragment for detection, including the myc-derived tag recognized by the antibody 9E10 [2], and the Flag sequence [62, 63]. This setup will allow the use of unpurified phage antibodies or antibody fragments, present in crude supernatant or periplasmic extracts, for screening assays. After reintroduction into TGI or DH5αF$'$ (suppressor +) or HB2151 (suppressor −) bacteria, individual clones are either rescued to produce monoclonal phage antibodies, or soluble scFv/Fab fragments are induced by the addition of IPTG. In both cases the expression of the antibody is driven by the lacZ promotor. Display on phage is dependent on the leaky expression in the absence of glucose, while IPTG is used for the induction of soluble antibody fragments. Finally, it is possible to insert (for example, between the antibody and gIII), a histidine-encoding tag which allows the purification of soluble antibodies using immobilized metal affinity chromatography [64, 65].

It is important to check both the antigen specificity of the clones and their diversity, by sampling clones from all rounds of selection. Nonspecific or polyreactive antibodies may predominate in later rounds of selection in the absence of a high-affinity specific antibody to compete. Moreover, sampling early selection rounds ensures adequate diversity, since the number of different antibodies tends to decrease with further rounds of selection, and if selections are continued for too long, specific binding antibodies may be lost to deletion phage which possess selective growth advantages.

One drawback for analysis after the *first* screen using ELISA is that the antibody expression level in *E. coli* is dependent on the primary sequence of the individual antibody, and can be extremely variable (from 10 μg to 100 mg/L). Unless expression is at sufficiently high levels, consideration should be given to recloning the antibody into another expression system (for review, see [66]). To reformat selected antibodies, fast recloning methods are needed in those cases where large numbers of clones need to be screened. Recently, eukaryotic expression vectors were described that may be used for one-step recloning of V genes derived from any phage repertoire, and cloned for expression as Fab fragments or whole antibody [67], as well as for targeting to different intracellular compartments [68]. This permits facile and rapid one-step cloning of antibody genes for either transient or stable expression in mammalian cells. By carefully choosing restriction sites that are rare in human V genes, the immunoglobulin genes of selected populations may be batch-cloned into these expression vectors. All the important elements in the vectors (promoter, lead sequence, constant domains, and selectable markers) are flanked by unique restriction sites, allowing simple substitution of elements and further engineering. By design of the correct promoter cassette, ribosome-binding site, "consensus" signal sequences, and by using "intron space" appropriately, it should be possible to make vectors that mediate both phage display of antibody fragments in prokaryotic hosts, as well as expression of antibody fragments or whole antibodies from eukaryotic hosts (A.B. and H.R.H., unpublished). Such shuttle vectors would be suitable to link selection of panels of binding antibodies with a screening assay based on a particular format of the antibody. In addition, it would allow the combination of different methods of display or combinatorial library screening. For example, a preselection for binding from a very large bacteriophage display library, followed by a fine-tuned affinity selection by means of flow cytometry of the medium-sized library, using yeast [69] or bacterial surface display [70], could be envisaged.

B. Designing Screening Assays for Function

Where the influence of affinity or kinetic behavior is unclear or less important than the precise epitope recognized, for example, in virus or cytokine neutralization, or in receptor triggering, fast read-out systems independent of binding tests are required.

With large libraries at hand, we may go beyond the in-vitro binding interaction itself, and select for a particular function. For example, provided reporter systems with sufficient sensitivity are used, it may eventually be possible to sort cells which have been triggered by a phage particle displaying an (ant)agonistic ligand. Such sorting procedures could allow the direct selection of phage particles with agonist or antagonist activity for a given receptor directly from the phage library. With new reporter genes and sensitive fluorescent read-out methods under development (for review, see [71]), we envisage that such "functional selection" schemes will be useful tools for drug discovery. Such methods may be used to identify peptide ligands for orphan receptors (such as the many related opioid receptors), for which a function but not a natural ligand is known. An example of an already-demonstrated functional selection method is found in retrieving catalytic antibodies. As yet underexplored routes are selection for cell internalization [72], cell survival or killing (induction of apoptosis) upon ligand binding, cell transfection, specific inhibition of certain cell surface molecules such as drug pumps, (inhibition of) viral entry, and, finally, receptor cross-linking or triggering.

IX. IMPROVING ANTIBODY AFFINITIES

The affinity of antibodies selected from naive, synthetic, or even immune phage libraries is typically sufficient for use as a research reagent, but too low for some specific therapeutic applications such as viral neutralization or tumor imaging. For many applications, affinity maturation can be bypassed completely by the construction of multivalent molecules. If this is not sufficient selected antibody clones may be affinity matured (see Figure 18.13).

In vivo antibodies are affinity matured in a stepwise fashion, by gradually incorporating mutations that cause small incremental improvements in the affinity which provide selective advantages. From the structure of affinity-matured antibodies it appears that many mutations affect affinity indirectly by influencing the positioning of side chains contacting the antigen [73]. Mutant

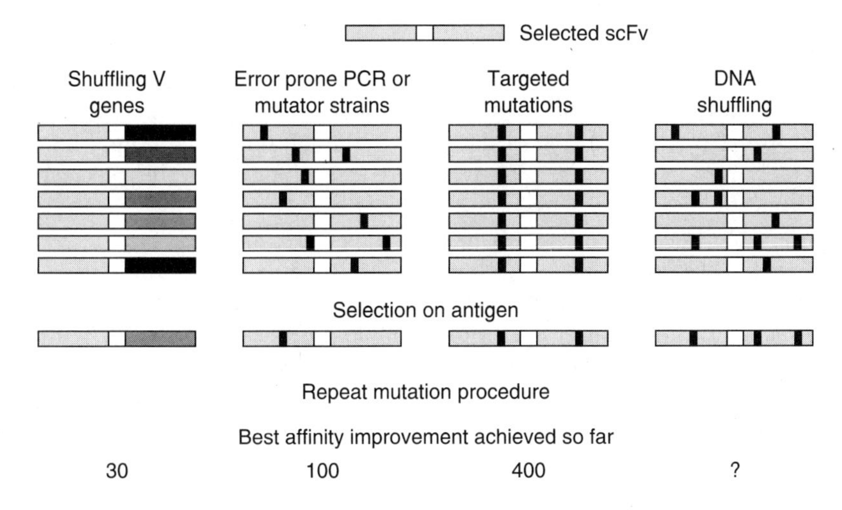

FIGURE 18.13 Improving phage antibody affinity.

residues may also increase the affinity by providing new contact residues (particularly when they are located in or near the center of the antigen-combining site), or by replacing "repulsive" or low-affinity contact residues with contact residues with more favorable energetics. Several methods have been used to in-vitro affinity mature antibodies derived from "primary" libraries, or from antibodies cloned from hybridomas. The process essentially involves three steps: (1) introduction of diversity in the antibody V genes of the candidate antibody for affinity maturation, so creating a "secondary" library; (2) selection of the higher-affinity from the low-affinity variants; and (3) screening to allow discrimination of antibody variants with differences in affinity or kinetics of binding.

Diversity in the antibody genes may be introduced using a variety of methods, which are either random or localized. Methods include using mutator strains [74, 75], error-prone PCR [40], chain shuffling [32, 76], and DNA shuffling [77], or directed to defined residues or regions of the V genes using codon-based mutagenesis, oligonucleotide-directed mutagenesis, and PCR techniques [27, 78–82].

Error-prone PCR relies on the natural error rate of Taq polymerase (which lacks a proofreading function) to incorporate mutations into a PCR-amplified segment of DNA. This has been used in conjunction with chemical mutagenesis to increase the affinity of an anti-carbohydrate scFv by 10-fold [82]. This is the same order of affinity improvement obtained by using mutator strains (30-fold) and chain shuffling (30-fold), in which the VH gene from the best phage antibody isolated is kept constant and displayed in association with a library of light chains, the procedure then being reversed with a library of VH genes for the best new light-chain V. DNA shuffling allows PCR errors from different amplified molecules to combine at random. This has been applied to antibodies; however, under the experimental system used (scFv display on phage) [77]), there was an apparent increase in affinity of the phage scFv which was not seen in the soluble scFv. This may have been due to induced shortening of the scFv linker with a subsequent dimerization or higher-order oligomerization of the scFv. Although this method appears to be extremely powerful, its usefulness applied to antibodies awaits experimental confirmation.

The nondirected approaches have been used to mature some antibodies with relatively low starting affinity [75, 76, 82]. Once antibodies with nanomolar affinities are used as starting leads, it appears that CDR-directed approaches are more successful. For example, residues that modulate affinity may be randomized, ideally 4–6 residues at a time to allow efficient sampling of the sequence space. Such residues that contact the antigen or influence other residues contacting the antigen may be defined experimentally by chain shuffling [83], by alanine scanning of the CDR regions [84], or by parsimonious mutagenesis [85, 86]. Targeting CDRs in parallel has been carried out [81], but additive effects of mutants are frequently unpredictable. The most successful approaches report improvements of affinity to below 100 pM, by saturation mutagenesis and affinity selection of CDR3 of heavy and light chain [80, 81]. A detailed study of the sequence diversity of human antibodies created in the primary and secondary immune responses also suggests other key residues.

In contrast to the *in vivo* situation, the in-vitro selection may be modified to allow selection for chosen kinetic parameters [40] such as off-rate or affinity. This hinges on the use of limited and decreasing amounts of antigen, on performing the selections in solution rather than by avidity-prone panning, and on having sensitive and accurate screening methods, such as BIAcore-based affinity screening, available. For example, antibodies with the highest affinity can be preferentially selected by using the antigen concentration at or below the desired dissociation constant [40]. Antigen is added at the desired K_d, and in molar excess over phage: for example, to select antibodies of 10 nM or higher affinity, antigen is used at 10 nM or lower concentration to select 10^{11} or fewer phage (10^{11} phage/mL is 1 nM, etc.) in a 1-mL volume. Alternatively, antibodies with improved off-rate may be selected; the rate of dissociation from the antigen often distinguishes primary from affinity-mature antibodies. Off-rate selections can be performed as described in detail by Hawkins *et al.* (1992). In brief, phage are equilibrated with biotinylated antigen, then a molar excess of unlabeled antigen added, such that phage dissociating from the biotinylated antigen will most likely rebind to nonbiotinylated antigen and are therefore lost. Slower off-rate antibodies will remain bound to biotinylated antigen for longer and are isolated when phage are captured at later time points. The point of maximum discrimination has to be determined experimentally, since background binding eventually reduces discrimination. The "selective infection of phage" procedure may also be used to enrich antibodies for certain kinetic parameters [87]; but it remains to be seen if this technique will be generally applicable.

The main complication while selecting for high-affinity clones is avidity, which is caused by two or more copies of the antibody on the phage head interacting with adjacent epitopes. Most scFvs will dimerize to a certain extent, and it is important to bear in mind that the selection method may favor poor antibodies with a tendency to dimerize. The presentation of the antigen is the most obvious way to control this. If the antigen is a hapten coupled to a carrier, derivatize at a low ratio of hapten to carrier. If the antigen is a repeating polymer, then selection on fragments or synthetic versions containing fewer repeat units can be used. For soluble proteins, selection methods incorporating solution capture are recommended. All published examples of affinity maturation use monovalent rather than multivalent phage display.

X. NEW DISPLAY APPROACHES FOR ANTIBODY AFFINITY MATURATION

The recently described method of "ribosome display" has been proposed as a method to affinity-mature antibodies. This approach involves the translation of proteins *in vitro* and their selection while attached to ribosomes [88, 89]. This can occur if the mRNA lacks a stop codon. In this case, the ribosome is not able to detach, and mRNA and encoding protein remain attached together and can be selected on a relevant ligand. After selection, the mRNA from selected polysomes is converted into cDNA, amplified by PCR, and used for the next transcription, translation, and selection round. "Ribosome display" has been used for the display [89, 90] and evolution of a scFv antibody *in vitro* [89], but no improvement in affinity was reported. However, the combination of this display technique (which potentially allows access to unlimited genetic diversity, unlike phage display) with methods such as DNA shuffling will see greater affinity improvements for selected antibodies. While this technique seems very promising, it remains to be seen whether it will be applicable to all antibodies, or whether it may find a niche in the selection and affinity maturation of antibodies for intracellular immunization [91].

Finally, the use of yeast display to affinity-mature antibodies has also been reported [69]. A randomly mutated scFv library was displayed on the surface of yeast and selected using flow cytometry; however, only a modest decrease in the k_{off} rate was reported.

XI. DOWNSTREAM USE OF ANTIBODIES

The isolation of an antibody is an extremely useful first step toward understanding the function of the protein to which it binds. scFvs derived from phage antibody libraries have been used in immunofluorescence, immunoprecipitation, fluorescence-activated cell sorting, Western blotting, and inhibition of function studies, both *in vivo*, in tissue culture cells, and *in vitro*. In this sense, they can essentially be used in the same way as conventional hybridoma-derived antibodies. They have the advantage, however, that the genes for the variable regions are cloned simultaneously with selection. This allows the fusion of functional elements, such as dimerization domains, effector or detector functions to selected scFvs [38], the re-creation of complete functional antibodies from selected scFvs [67], or the expression of intracellular scFvs in different cellular compartments [68], an exciting technology still under development that may give insights into protein function by the creation of cellular "knockouts", in which the function of the protein recognized is inhibited in situ [91, 92].

REFERENCES

1. Vaughan TJ, Williams AJ, Pritchard K, Osbourn JK, Pope AR, Earnshaw JC, McCafferty J, Hodits RA, Wilton J, Johnson KS, Human antibodies with sub-nanomolar affinities isolated from a large non-immunised phage display library, *Nat. Biotechnol.*, 14:309–314, 1996.
2. Marks JD, Hoogenboom HR, Bonnert TP, McCafferty J, Griffiths AD, Winter G. By-passing immunization-human antibodies from V-gene libraries displayed on phage, *J. Mol. Biol.*, 222:581–597, 1991.
3. Nissim A, Hoogenboom HR, Tomlinson IM, Flynn G, Midgley C, Lane D, Winter G, Antibody fragments from a single pot phage display library as immunochemical reagents, *EMBO. J.*, 13:692–698, 1994.
4. Griffiths AD, Williams SC, Hartley O, Isolation of high affinity human antibodies directly from large synthetic repertoires, EMBO *J.*, 13:3245–3260, 1994.
5. Kruif J de, Boel E, Logtenberg T, Selection and application of human single chain Fv antibody fragments from a semi-synthetic phage antibody display library with designed CDR3 regions, *J. Mol. Biol.*, 248:97–105, 1995.
6. Barbas CF, Kang AS, Lerner RA, Benkovic SJ, Assembly of combinatorial antibody libraries on phage surfaces: the gene III site, *Proc. Natl. Acad. Sci. USA*, 88:7978–7982, 1991.
7. Williamson RA, Burioni R, Sanna PP, Partridge LJ, Barbas CF, Burton DR, Human monoclonal antibodies against a plethora of viral pathogens from single combinatorial libraries, *Proc. Natl. Acad. Sci. USA*, 90:4141–4145, 1993.
8. Zebedee SL, Barbas CF, Hom Y, Human combinatorial antibody libraries to hepatitis B surface antigen, *Proc. Natl. Acad. Sci. USA*, 89:3175–3179, 1992.
9. Orum H, Andersen PS, Oster A, Johansen LK, Riise E, Bjornvad M, Svendsen I, Engberg J, Efficient method for constructing comprehensive marine Fab antibody libraries displayed on phage, *Nucleic Acids Res.*, 21:4491–4498, 1993.
10. Ames RS, Tornetta MA, Jones CS, Tsui P, Isolation of neutralizing anti-C5a monoclonal antibodies from a filamentous phage monovalent Fab display library, *J. Immunol.*, 152:4572–4581, 1994.
11. Ames RS, Tometta MA, McMillan LJ, Neutralizing murine monoclonal antibodies to human IL-5 isolated from hybridomas and a filamentous phage Fab display library, *J. Immunol.*, 1545:6355–6364, 1995.
12. Perelson AS, Oster GF, Theoretical studies of clonal selection: minimal antibody repertoire size and reliability of self non-self discrimination, *J. Theor. Biol.*, 81:645–670, 1979.
13. Kretzschmar T, Geiser M, Evaluation of antibodies fused to minor coat protein III and major coat protein VIII of bacteriophage M13, *Gene*, 155:61–65, 1995.
14. Clackson T, Wells JA, *In vitro* selection from protein and peptide libraries, *TIBTECH.*, 12:173–184, 1994.

15. Hoogenboom HR, Griffiths AD, Johnson KS, Chiswell DJ, Hudson P, Winter G, Multi-subunit proteins on the surface of filamentous phage: methodologies for displaying antibody (Fab) heavy and light chains, *Nucleic Acids. Res.*, 19:4133–4137, 1991.
16. Krebber C, Spada S, Desplancq D, Pluckthun A, Co-selection of cognate antibody antigen pairs by selectively-infective phages, *FEBS Lett.*, 377:227–231, 1995.
17. Spada S, Pluckthun A, Selectively infective phage (SIP) technology: a novel method for the *in vivo* selection of interacting protein-ligand pairs, *Nat. Med.*, 6:694–696, 1997.
18. Malmborg A-C, Soderlind E, Frost L, Borrebaeck CAK, Selective phage infection mediated by epitope expression on F pilus, *J. Mol. Biol.*, 273:544–551, 1997.
19. Bird RE, Walker BW, Single chain antibody variable regions, *Trends. Biotech.*, 9:132–138, 1991.
20. Tsurushita N, Fu H, Warren C, Phage display vectors for *in vivo* recombination of immunoglobulin heavy and light chain genes to make large combinatorial libraries, *Gene.*, 172:59–63, 1996.
21. Holliger P, Prospero T, Winter G, Diabodies, small bivalent and biospecific antibody fragments, *Proc. Natl. Acad. Sci. USA.*, 90:6444–6448, 1993.
22. McGuinness BT, Walter G, FitzGerald K, Schuler P, Mahoney W, Duncan AR, Hoogenboom HR, Phage diabody repertoires for selection of large numbers of bispecific antibody fragments, *Nat. Biotech.*, 14:1149–1154, 1996.
23. Tomlinson IM, Williams SC, Corbett SJ, Cox JPL, Winter G, BASE V, Sequence Directory, Cambridge, UK. MRC Centre for Protein Engineering, 1996. (www.mrc-cpe.cam.ac.uk)
24. Knappik A, Pluckthun A, An improved affinity tax based on the FLAG peptide for the detection and purification of recombinant antibody fragments, *Biotechniques*, 17:754–761, 1994.
25. Ruberti F, Cattaneo A, Bradbury A, The use of the RACE method to clone hybri-doma cDNA when V region primers fail, *J. Immunol. Meth.*, 173:33–39, 1994.
26. Hoogenboom HR, Winter G, Bypassing immunisation: human antibodies from synthetic repertoires of germ line VH-gene segments rearranged *in vitro*, *J. Mol. Biol.*, 227:381–388, 1992.
27. Virnekas B, Ge , Pluckthun A, Schneider KC, Wellnhofer G, Moroney SE, Trinucleotide phophoramidites: ideal reagents for the synthesis of mixed oligonucleotides for random mutagenesis, *Nucleic. Acids. Res.*, 22:5600–5607, 1994.
28. Morea V, Tramontano A, Rustici M, Chothia C, Lesk AM, Antibody structure, prediction and redesign, *Biophys. Chem.*, 68:9–16, 1997.
29. Marks JD, Tristem, Karpas A, Winter G, Oligonucleotide primers for polymerase chain reaction amplification of human immunoglobulin variable genes and design of family-specific oligonucleotide probes, *Eur. J. Immunol.*, 21:985–991, 1991.
30. Bender E, Pilkington GR, Burton DR, Human monoclonal Fab fragments from a combinatorial library prepared from an individual with a low serum titer to a virus, *Hum. Antibodies Hybridomas*, 5:3–8, 1994.
31. Klein U, Küppers R, Rajewsky K, Evidence for a large compartment of IgM-expressing memory B cells in humans, *Blood*, 89:1288–1298, 1997.

32. Clackson T, Hoogenboom HR, Griffiths AD, Winter G, Making antibody fragments using phage display libraries, *Nature*, 352:624–628, 1991.
33. Sblattero D, Bradbury A, A definitive set of oligonucleotide primers for amplifying human V regions, *Immunotechnology*, 3:271–278, 1998.
34. Larrick JW, Danielsson L, Brenner CA, Wallace EF, Abrahmson M, Fry KR, Barrebaeck CAK, Polymerase chain reaction using mixed primers: cloning of human monoclonal antibody variable region genes from single hybridoma cells, *Biotechnology*, 7:934–938, 1989.
35. Watkins BA, Davis AE, Fiorentini S, Reitz Jr MS, V-region and class specific RT-PCR amplification of human immunoglobulin heavy and light chain genes from Bcell lines. *Scand. J. Immunol.*, 42:442–448, 1995.
36. De Boer M, Chang S-Y, Eichinger G, Wong HC, De sing and analysis of PCR primers for the amplification and cloning of human immunoglobulin Fab fragments, *Hum. Antibodies Hybridomas*, 5:57–64, 1994.
37. Welschof M, Terness P, Kolbinger F, Zewe M, Dubel S, Dorsam H, Hain C, Amino acid sequence based PCR primers for amplification of rearranged human heavy and light chain immunoglobulin variable region genes, *J. Immunol. Meth.*, 179:203–214, 1995.
38. Krebber A, Bornhauser S, Burmester J, Honeggar A, Willuda J, HR, B, Pluckthun A, Reliable cloning of functional antibody variable domains from hybridomas and spleen cell repertoires employing a reengineered phage display system, *J. Immunol. Meth.*, 201:35–55, 1997.
39. Malmborg AC, Duenas M, Ohlin M, Soderlind E, Borrebaeck C, Selection of binders from phage displayed antibody libraries using the BIAcore biosensor, *J. Immunol. Meth.*, 198:51–57, 1996.
40. Hawkins RE, Russell SJ, Winter G, Selection of phage antibodies by binding affinity: mimicking affinity maturation, *J. Mol. Biol.*, 226:889–896, 1992.
41. Bradbury A, Persic L, Werge T, Cattaneo A, From gene to antibody: the use of living columns to select specific phage antibodies, *BioTechnology*, 11:1565–1569, 1993.
42. Cai X, Garen A, Anti-melanoma antibodies from melanoma patients immunized with genetically modified autologous tumor cells, selection of specific antibodies from single-chain Fv fusion phage libraries, *Proc. Natl. Acad. Sci. USA*, 92:6537–6541, 19.
43. Kruif J de, Terstappen L, Boel E, Logtenberg T, Rapid selection of cell subpopulation-specific human monoclonal antibodies from a synthetic phage antibody library, *Proc. Natl. Acad. Sci. USA*, 92:3938–3942, 1995.
44. Siegel DL, Chang TY, Russell SL, Bunya VY, Isolation of cell surface-specific human monoclonal antibodies using phage display and magnetically-activated cell sorting: applications in immunohematology, *J. Immunol. Meth.*, 206:73–85, 1997.
45. Van Ewijk W, Kruif J de, Germeraad WTV, Berenedes P, Ropke C, Platenburg PP, Logtenberg T, Subtractive isolation of phage displayed single-chain antibodies to thymic stromal cells by using intact thymic fragments, *Proc. Natl. Acad. Sci. USA*, 94:3903–3908, 1997.
46. Pasqualini R, Ruoslahti E, Organ targeting *in vivo* using phage display peptide libraries, *Nature.*, 380:364–366, 1996.
47. Osbourn JK, Derbyshire EJ, Vaughan TJ, Field AW, Johnson KS, Pathfinder selection: in situ isolation of novel antibodies, *Immunotech.*, 3:293–302, 1998.

48. McCafferty J, Griffiths AD, Winter G, Chiswell DJ, Phage antibodies, filamentous phage displaying antibody variable domains, *Nature*, 348:552–554, 1990.
49. Griffiths AD, Malmqvist M, Marks JD, Bye JM, Embleton MJ, McCafferty J, Baier M, Holliger KP, Garick BD, Hughes-Jones NC, Hoogenboom MR, Winter G, Human anti-self antibodies with high specificity from phage display libraries, *EMBO. J.*, 12:725–734, 1993.
50. Neri D, Momo M, Prospero T, Winter G, High-affinity antigen binding by chelating recombinant antibodies (CRAbs), *J. Mol. Biol.*, 246:367–373, 1993.
51. Kang AS, Barbas CF, Janda KD, Benkovic SJ, Lerner RA, Linkage of recognition and replication functions by assembling combinatorial antibody Fab libraries along phage surfaces, *Proc. Natl. Acad. Sci. USA.*, 88:4363–4366, 1991.
52. Ward RL, Clark MA, Lees I, Hawkins NJ, Retrieval of human antibodies from phage-display libraries using enzymatic cleavage, *J. Immunol. Meth.*, 189:73–82, 1996.
53. Meulemans EV, Slobbe R, Wasterval P, Ramaekers FC, van Eys GJ, Selection of phage-displayed antibodies specific for a cytoskeletal antigen by competitive elution with a monoclonal antibody, *J. Mol. Biol.*, 244:353–360, 1994.
54. Andersen PS, Stryhn A, Hansen BE, Fugger L, Engberg J, Buns S, A recombinant antibody with the antigen-specific, major histocompatibility complex-restricted specificity of T cells, *Proc. Natl. Acad. Sci. USA.*, 93:1820–1824, 1996.
55. Marks JD, Ouwehand WH, Bye JM, Finnern R, Gorick BD, Voak D, Thorpe S, Hughes-Jones NC, Winter G, Human antibody fragments specific for human blood group antigens from a phage display library, *Biotechnology*, 11:1145–1149, 1993.
56. Palmer DB, George AL, Ritter MA, Selection of antibodies to cell surface determiants on mouse thymic epithelial cells using a phage display library, *Immunology*, 91:473–478, 1998.
57. Duenas M, Borrabaeck CA, Clonal selection and amplification of phage displayed antibodies by linking antigen recognition and phage replication, *Biotechnology*, 12:999–1002, 1994.
58. Gramatikoff K, Georgiev O, Schaffner W, Direct interaction rescue, a novel filamentous phage technique to study protein-protein interactions, *Nucleic. Acids. Res.*, 22:5761–5762, 1994.
59. Pedrazzi G, Schweseinger F, Honegger A, Krebber C, Pluckthun A, Affinity and folding properties both influence the selection of antibodies with the selectively infective phage (SIP) methodology, *FEBS. Lett.*, 415:289–293, 1997.
60. Holliger P, Riechmann L, A conserved infection pathway for filamentous bacteriophages is suggested by the structure of the membrane penetration domain of the minor coat protein g3p from phage fd, *Structure*, 5:265–275, 1997.
61. Riechmann L, Holliger P, The C terminal domain of ToIA is the coreceptor for filamentous phage infection of *E. coli*, *Cell*, 90:351–360, 1997.
62. Lindner P, Bauer K, Krebber A, Nieba L, Kremmer E, Krebber C, Honegger A, Klinger B, Mocikat R, Plückthun A, Specific detection of his-tagged proteins with recombinant anti-His tag scFv-phosphatase or scFv-phage fusions, *Biotechniques*, 22:140–149, 1997.

63. Lah M, Goldstraw A, White FJ, Dolezal O, Malby R, Hudson PJ, Phage surface presentation and secretion of antibody fragments using an adaptable phagemid vector, *Hum. Antibodies. Hybridomas.*, 5:48–56, 1994.
64. McCafferty J, Fitzgerald KJ, Earnshaw J, Chiswell DJ, Link J, Smith R, Selection and rapid purification of murine antibody fragments that bind a transition-state analog by phage display, *Appl. Biochem. Biotechnol.*, 47:157–171, 1994.
65. Hochuli E, Bannwarth W, Döbeli H, Gentz R, Stüber D, Genetic approach to facilitate purification of recombinant proteins with a novel metal chelate adsorbent, *Biotechnology*, 6:1321–1325, 1998.
66. Pluckthun A, Pack P, New protein engineering approaches to multivalent and bispecific antibody fragments, *Immunotechnology*, 3:83–105, 1997.
67. Persic L, Roberts A, Wilton J, Cattaneo A, Bradbury A, Hoogenboom H, An integrated vector system for the eukaryotic expression of antibodies or their fragments after selection from phage display libraries, *Gene*, 187:9–18, 1997.
68. Persic L, Righi M, Roberts A, Hoogenboom HR, Cattaneo A, Bradbury A, Targeting vectors for intracellular immunisation, *Gene*, 187:1–8, 1997.
69. Boder ET, Wittrup KD, Yeast surface dispay for screening combinatorial polypeptide libraries, *Nat. Biotechnology*, 15:553–557, 1997.
70. Georgiou G, Stathopoulos C, Daugherty PS, Nayak AR, Iverson BL, Curtiss R, 3rd, Display of heterologous proteins on the surface of microorganisms, from the screening of combinatorial libraries to live recombinant vaccines, *Nat. Biotechnol.*, 15:29–34, 1997.
71. Broach JR, Thorner J, High-throughput screening for drug discovery, *Nature*, 384:14–16, 1996.
72. Hart SL, Knight AM, Hartbottle RP, Mistry A, Hunger HD, Cutler DF, Williamson R, Coutelle C, Cell binding and internalization by filamentous phage displaying a cyclic Arg-Gly-Asp-containing peptide, *J. Biol. Chem.*, 269:12468–12474, 1994.
73. Hawkins RE, Russell SJ, Baier M, Winter G, The contribution of contact and noncontact residues of antibody in the affinity of binding to antigen. The interaction of mutant D1.3 antibodies with lysozyme, *J. Mol. Biol.*, 234:958–964, 1993.
74. Irving RA, Kortt AA, Hudson PJ, Affinity maturation of recombinant antibodies using *E. coli* mutator cells, *Imunotechnology*, 2:127–143, 1996.
75. Low NM, Holliger PH, Winter G, Mimicking somatic hypermutation: affinity maturation of antibodies displayed on bacteriophage using a bacterial mutator strain, *J. Mol. Biol.*, 260:359–368, 1996.
76. Marks JD, Hoogenboom HR, Griffiths AD, Winter G, Molecular evolution of proteins on filamentous phage: mimicking the strategy of the immune system, *J. Biol. Chem.*, 267:16007–16010, 1992.
77. Crameri A, Cwirla S, Stemmer WP, Construction and evolution of antibody-phage libraries by DNA shuffling, *Nat. Med.*, 2:100–102, 1996.
78. Glaser SM, Yelton DE, Huse WD, Antibody engineering by codon-based mutagenesis in a filamentous phage vector system, *J. Immunol.*, 149:3903–3913, 1992.
79. Jackson JR, Sathe G, Rosenberg M, Sweet R, *In vitro* antibody maturation. Improvement of a high affinity, neutralizing antibody against IL-1 beta, *J. Immunol.*, 154:3310–3319, 1995.

80. Schier R, Bye J, Apell G, *et al.*, Isolation of high-affinity monomeric human anti-cerbB-2 single chain Fv using affinity-driven selection, *J. Mol. Biol.*, 255:28–43, 1996.
81. Yang WP, Green K, Pinz-Sweeney S, Briones AT, Burton DR, Barbas CFR, CDR walking mutagenesis for the affinity maturation of a potent human anti-HIV-1 antibody into the picomolar range, *J. Mol. Biol.*, 254:392–403, 1995.
82. Deng SJ, MacKenzie CR, Sadowska J, Michniewicz J, Young NM, Bundle DR, Narang SA, Selection of antibody single-chain variable fragments with improved carbohydrate binding by phage display, *J. Biol. Chem.*, 269:9533–9538, 1994.
83. Thompson J, Pope T, Tung JS, Chan C, Hollis G, Mark G, Johnson KS, Affinity maturation of a high-affinity human monoclonal antibody against the third hypervariable loop of human immunodeficiency virus: use of phage display to improve affinity and broaden strain reactivity, *J. Mol. Biol.*, 256:77–88, 1996.
84. Schier R, McCall A, Adams GP, *et al.*, Isolation of picomolar affinity anti-c-erB-2 single-chain Fv by molecular evolution of the complementarity determining regions in the center of the antibody binding site, *J. Mol. Biol.*, 263:551–567, 1996.
85. Balint RF, Larrick JW, Antibody entering by parsimonious mutagenesis, *Gene*, 137:109–118, 1996.
86. Schier R, Balint RF, McCall A, Apell G, Larrick JW, Marks JD, Identification of functional and structural amino-acid residues by parsimonious mutagenesis, *Gene*, 169:147–155, 1996.
87. Denas M, Malmborg AC, Casalvilla R, Ohlin M, Borrebaeck CA, Selection of phage displayed antibodies based on kinetic constants, *Mol. Immunol.*, 33:279–285, 1996.
88. Mattheakis LC, Bhatt RR, Dower WJ, An *in vitro* polysome display system for identifying ligands from very large peptide libraries, *Proc. Natl. Acad. Sci. USA*, 91:9022–9026, 1994.
89. Hanes J, Plückthun A, *In vitro* selection and evolution of functional proteins using ribosome display, *Proc. Natl. Acad. Sci. USA*, 94:4937–4942, 1997.
90. He M, Taussig MJ, Antibody-ribosome-mRNA (ARM) complexes as efficient selection particles for *in vitro* display and evolution of antibody combining sites, *Nucleic. Acids. Res.*, 25:5132–5134, 1997.
91. Biocca S, Cattaneo A, Intracellular immunization: antibody targeting to subcellular compartments, *Trends. Cell. Biol.*, 5:248–252, 1995.
92. Biocca S, Cattaneo A, Intracellular antibodies: developments and applications, Austin, TX, *Landes Bioscience*, distributed by Springer Verlag, 1997.
93. Sheets MD, *et al.*, Efficient construction of a large non-immune phage antibody library: the production of panels of high affinity human single-chain antibodies to protein antigens, *Proc. Natl. Acad. Sci. USA.*, 95:6157–6162, 1998.

19 Biopanning

Giovanna Palombo

CONTENTS

I. INTRODUCTION

The ability to identify new potential ligands against a variety of targets of interest has been drastically improved through the use of random peptide libraries. Diverse repertoires of molecules have been generated and, using genetic engineering techniques, peptides and/or antibodies have been displayed on the surface of filamentous bacteriophages [1–4] or expressed on the surface of bacteria [5].

The great power of biological libraries lies in the possibility to make use of "biopanning" [6, 7], which represents an *in vitro* selection process that has been employed to isolate target-binding sequences from a large pool of different molecules [8, 9]. The procedure, pioneered by Smith and others [10, 11], is performed by iterative cycles which include the binding of the library to an immobilized target, washing steps to remove aspecific interaction, elution of retained phage particles, and amplification in *Escherichia coli* [6].

Since its introduction, this technique has gained more and more importance in the combinatorial field. The main advantage of biological libraries, in comparison to chemical libraries, depends on the possibility of increasing remarkably the number of different molecules that can be screened at the same time. Furthermore, once the phage displaying the active sequence has been selected, it is able to replicate itself by infecting host bacterial cells; successive selection and amplification cycles result in an exponential enrichment of the active peptide sequence.

On the basis of these considerations, this technique has become a widely used selection method to identify new binding molecules [12–17], mimotopes [18–23], or to characterize protein–ligand and/or protein–protein interactions [24–37]. Moreover, attempts to optimize the selection process have resulted in the setting up of new biopanning procedures, which have been performed not only on purified targets [24–37], cells [38, 39], and tissues [40] (*in vitro* selection), but also on living animals [41] (*in vivo* selection).

All these applications have enormously increased the usage of this methodology in both research and diagnostics. The selection of phage antibody libraries has provided a powerful method both in the production of monoclonal antibodies [3, 4] and in the identification of cell surface antigens previously unknown [42, 43]; on the other hand, biopanning selection carried out on cells or on intact tissues allowed the identification of cell- and/or tissue-specific ligands that could be used for the development of new receptor-mediated gene delivery systems [44, 45].

The power of this approach is not limited to the search for peptides and/or antibodies fragments acting as ligands [46] or inhibitors [47] of biological molecules. From this point of view, biopanning experiments performed *in vivo* [41] may enlarge the horizons of molecular medicine, allowing the localization and identification of new receptors involved in pathological processes.

II. *IN VITRO* SELECTION

A. Biopanning on Plate

This biopanning procedure, which represents the most frequently used *in vitro* selection method [48–59], is represented schematically in Figure 19.1.

Polystyrene microtiter plates are coated with purified target, washed to remove the excess of the protein, then saturated with a specific reagent, such as BSA, dried milk, or gelatin, to block the uncoated plastic surface.

After removal of blocking solution, approximately 10^{11}–10^{12} virions of a random display library are incubated for a variable time. During this step, phages displaying active sequences are retained on the surface coated with the target, while nonadherent phages are washed away.

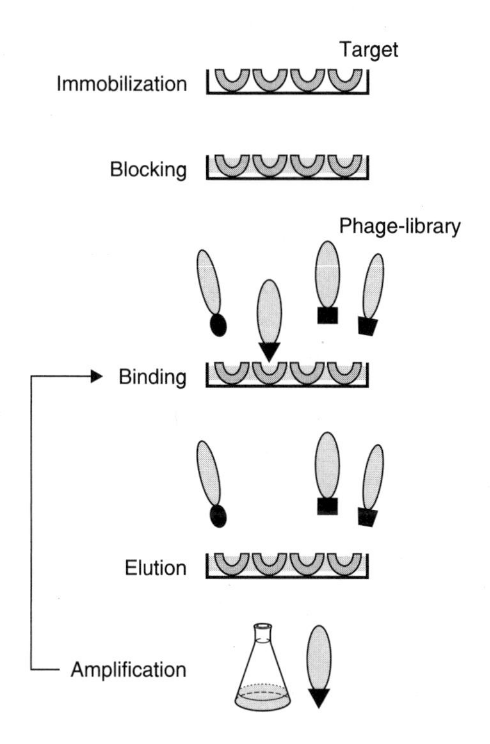

FIGURE 19.1 Biopanning on plates. The target, immobilized on a Petri dish, is incubated with the library in solution. Phages displaying the active sequence are eluted, amplified, and submitted to subsequent biopanning cycles.

Bound phages can be recovered by low pH buffers [48–50, 52, 53, 55, 56, 58, 59] or by competitive elution [12, 15], amplified into bacteria, and submitted for additional rounds of selection.

1. Stringency of Selection

The eluted pool is a heterogeneous population which contains phages that bind with different affinities as well as phages adsorbed aspecifically, that are not removed even through extensive washing steps. As a consequence, in order to isolate the binding peptides, it is necessary to increase the stringency of selection.

Generally, nonstringent conditions (long incubation time, nonextensive washings, excess of ligand immobilized) are used early in a selection process to ensure that even molecules with low binding activity are not going to be lost. In subsequent rounds, a higher stringency of selection can be reached by

Increasing the number and length of washings steps [59]
Changing the ionic strength of the washing buffer to remove weakly or aspecifically bound phages

Increasing the number of biopanning cycles [58]
Decreasing the protein coating concentration in order to select for higher-affinity ligands [50, 52].

In addition, aspecific phages interaction can be avoided by performing either of the following procedures.

"Subtractive biopanning" (Figure 19.2a), in which the library is first incubated in a Petri dish coated with the blocking solution. Nonadherent phages are removed and added to the protein-coated plate. This procedure is useful to remove, from the entire population, molecules that bind the blocking reagent, and selects for molecules that specifically bind the target [49, 51].

"Competitive biopanning" (Figure 19.2b), performed in the presence of a soluble competitor (usually a protein with similar chemical and physical properties to the target), which binds specific phages that will not be available for interaction with the target. This approach allows the selection of phage-bearing peptides with high selectivity and specificity [56, 57].

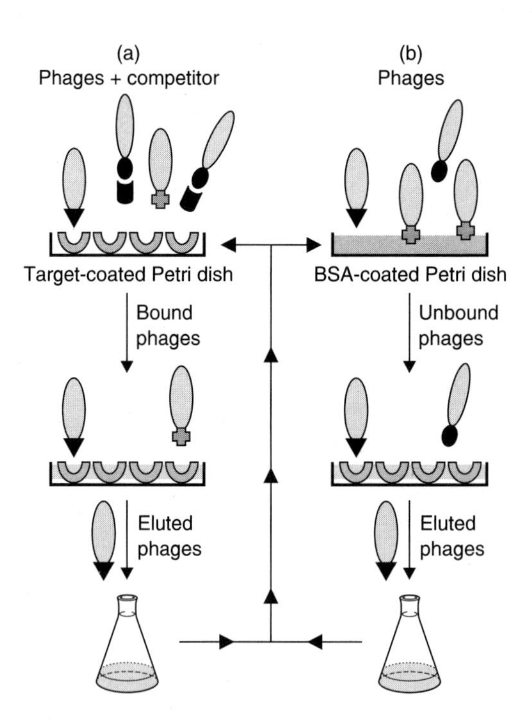

FIGURE 19.2 (a): Biopanning on plate performed in the presence of a competitor in solution. Competitor-binding phages do not interact with the target and are removed by washing. Bound phages are eluted and amplified in *E. coli* cells. (b): subtractive biopanning on plate. The phage library is added to a Petri dish saturated with blocking reagent. Aspecifically bound phages are removed, while unbound phages are incubated in the presence of the immobilized target. The retained phages are then recovered by elution, amplified, and submitted to iterative cycles of selection.

Biopanning on plates has been applied either to select new monoclonal antibody fragments [55, 58], or to identify peptide ligands against cell surface receptors [50, 51], nonreceptor proteins [48], or tumor suppressor proteins such as p53, which is frequently implicated as a target for genetic mutation in the progress of human carcinogenesis [59]. This selection method requires the availability of purified target and its consequent immobilization on the solid support. In some instances, however, the coating procedure leads to partial destruction of the three-dimensional structure of the protein, with the risk of selecting for sequences against other epitopes. This problem can be avoided by capturing the target on a solid support using a monoclonal or a polyclonal antibody [25, 60, 61]. Moreover, the power of this procedure relies on the possibility of selecting peptide-ligands against specific epitopes from a crude extract [62] (Figure 19.3). The advantage of using such a selection strategy becomes evident when the pure target is not available and the purification procedure, laborious and time consuming, may not preserve the conformational integrity of the final product. In addition, the fact that the antigens and/or the proteins are presented in the panning in their native structure allows the isolation of antibodies and/or peptides that recognize conformationally structured epitopes.

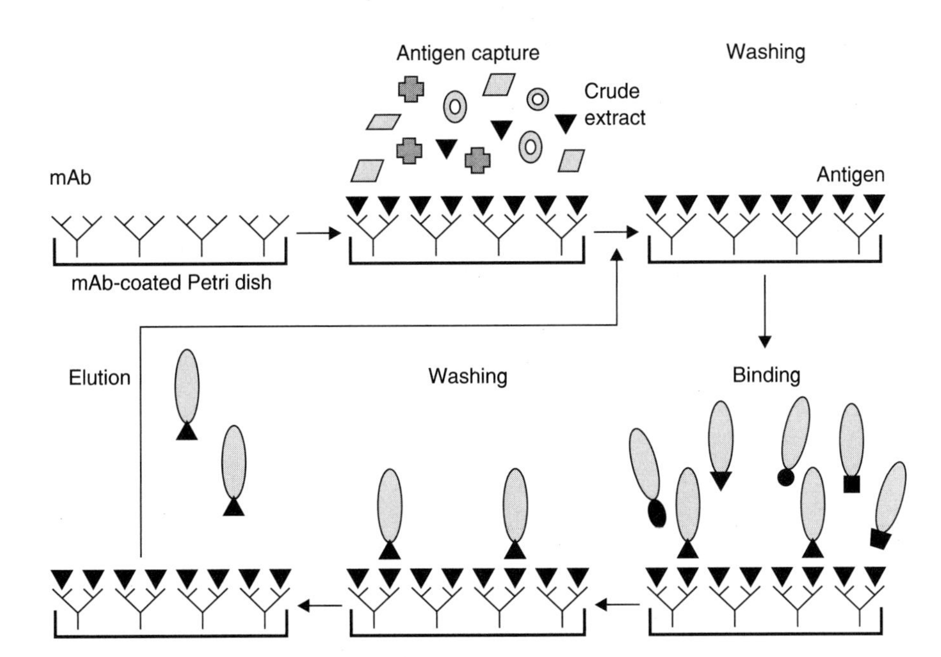

FIGURE 19.3 Selection of antigen-binding peptide, through a "capture-biopanning" procedure. In the first step, a monoclonal antibody, immobilized on solid supports, is used to capture the corresponding antigen in a crude extract. The plate is washed extensively to remove nonspecific interactions, then incubated with a phage library. During this incubation, phages displaying the antigen-binding peptides are selected and recovered by elution.

Alternatively, biopanning on plates is based on the strong biotin-streptavidin interaction to isolate target-binding phages [6]. Figure 19.4 shows a schematic representation of the procedure.

The library is allowed to react with biotinylated target in solution, usually a MAB, then added to a streptavidin-coated Petri dish. Phages displaying peptides binding to the antibody are adsorbed to the plastic surface through biotin-streptavidin bonds.

Unbound phages are removed by extensive washings.
Bound phages are eluted by a low-pH buffer and used to infect *E. coli* cells [6].

The first round of biopanning is critical to achieve success. Ordinarily, each clone is represented by only 100 infectious units (I.U.) in the original library.

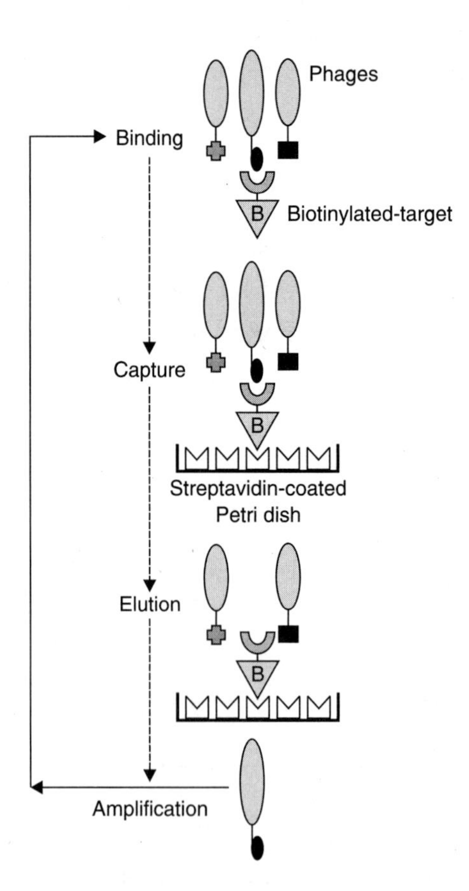

FIGURE 19.4 Biopanning selection through biotin–streptavidin interaction. The library is incubated with the biotinylated target in solution; phages displaying the active binding peptide are captured on a streptavidin-coated Petri dish, eluted, and amplified.

Because biopanning gives only a 1% yield with strongly binding phages, many binding clones will be represented by a single T.U. As a consequence, a large amount of biotinylated target is used in the first round, to maximize the yield of ligand phage even at the cost of reducing discrimination [6]. The decrease in ligate concentration in successive cycles allows the selection of higher-affinity epitopes [6].

This selection procedure has been employed to identify DNA-binding peptides [63] as well as novel protein-binding ligands [27, 28, 34] and, above all, to characterize the antigen–antibody interaction by identifying new epitopes for MAB-directed against different targets [8, 23, 64].

B. Biopanning on Magnetic Particles

An alternative way of selecting is to use magnetic particles as solid support for isolating active binding molecules. The procedure includes binding the library with the biotinylated target, followed by capture on streptavidin-coated magnetic beads and selection of bound phages using a magnetic field (Figure 19.5).

Hawkins *et al.* [65] have utilized this biopanning procedure to select high-affinity antigen-ligands by screening of a phage display antibody library. More recently, Gaynor *et al.* [66] reported a successful use of this biopanning procedure to identify peptide-ligands that bind pathogenic anti-DNA antibodies. The identification of such peptides may be a valid contribution to characterizing the binding site of anti-DNA antibodies; on the other hand, the identification of ligands that bind in or near the DNA-binding site could be helpful from a therapeutic point of view, to avoid antibody-mediated diseases.

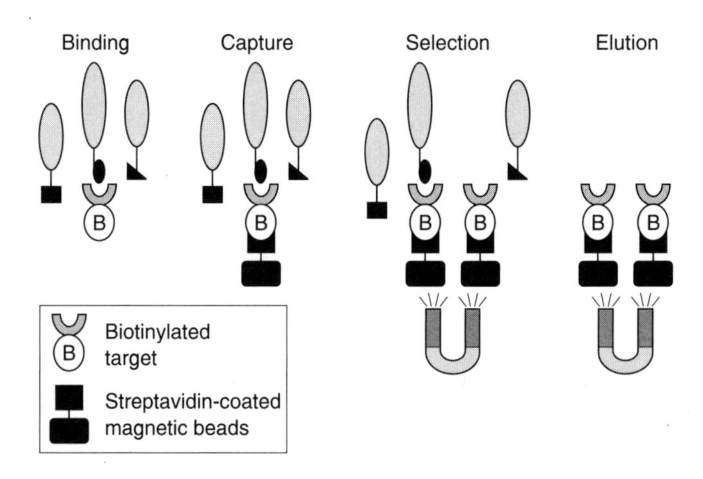

FIGURE 19.5 Biopanning on magnetic beads. The binding of the library to biotinylated target in solution is followed by capture of phages on streptavidin-coated magnetic beads and by selection of bound phages through a magnetic field.

C. Biopanning on a Column

Biopanning on a column is illustrated in Figure 19.6.

The target is covalently immobilized on a chromatographic support such as sepharose [3, 4, 29, 30, 67, 68] or agarose beads [37, 69].

The library is applied to the column, which has previously been equilibrated with buffer at a defined pH and ionic strength.

After the phages are bound to the target, extensive washing steps are required to remove matrix-binding particles.

Adsorbed phages can be dissociated by low pH buffers [29, 30, 37] or by competitive elution using a high concentration of the target or of the real ligand if it is available.

This biopanning technique has been employed to select peptide ligands that bind proteins [29, 68, 69], antibodies [30], and nucleic acid molecules [67] with moderate affinity. However, the fact that the column also retains sequences with low affinity compromises the utility of using this affinity selection to discriminate between tight or moderately binding sequences.

1. Stringency of Selection

A variety of experimental parameters can be changed to increase the stringency of the selection in order to identify high-affinity ligands.

(1) Washing. Aspecifically or weakly bound molecules can be eliminated by extensive washing steps, or by changing the ionic strength of the washing buffer. For instance, buffers with high ionic strength may

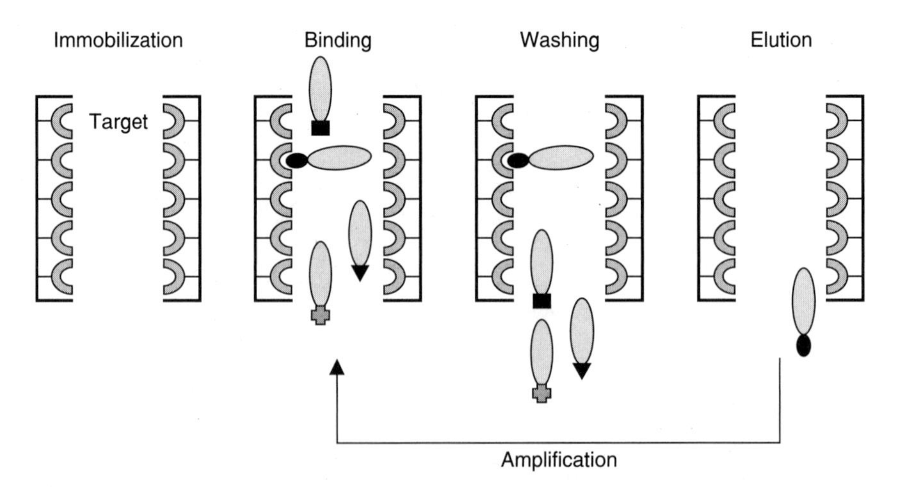

FIGURE 19.6 Biopanning on a column. The target is covalently immobilized on a chromatographic support. The phage library is then applied to the column and, after an extensive washing step, bound phages are recovered by elution, amplified, and submitted to a successive selection cycle.

release some weakly bound molecules that form hydrogen bounds or salt bridges with the ligand.

(2) Alternative Biopannings. One of the most common artifacts occurring with this biopanning procedure is the selection of matrix-binding clones that are difficult to remove even through extensive washings. These aspecific molecules will be amplified during the successive cycles of selection and amplification, leading to the preferential enrichment of nonbinding phages and, consequently, high background yield. To circumvent these problems, comparative biopanning can be performed on a control column containing the same matrix without the immobilized target. The sequence analysis of the phages recovered from the two different selections will distinguish matrix-binding peptides from those that bind the target specifically [30, 68].

Alternatively, competitive biopanning can be performed by sorting the library against an immobilized target in the presence of a competitor in solution. This procedure removes competitor-binding phages from the population and allows the isolation of peptides that recognize the target specifically (Figure 19.7).

D. Biopanning on Cells

So far, selection of ligand for cell surface receptors or antigens has required the availability of the target receptor in its purified form [15–17, 23–28]. In some

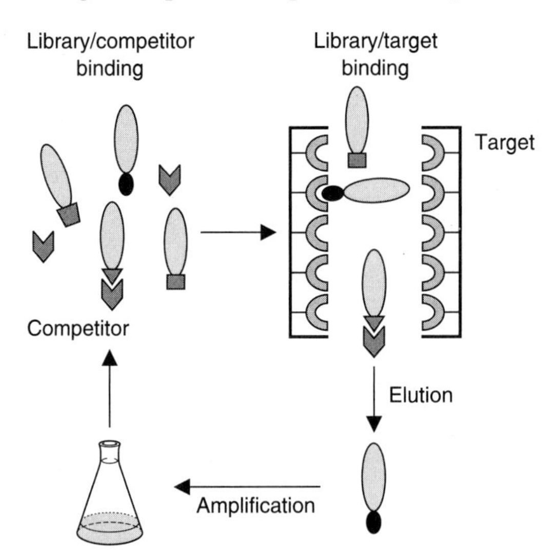

FIGURE 19.7 Competitive biopanning on a column. The library in incubated with a competitor in solution, then applied to the column. Competitor-bound phages are not available to interact with the immobilized target, allowing the isolation of specific binding phages.

instances, however, the purified target is not available and the purification procedures, which are laborious and time expensive, may not preserve the conformational structure of the final product. Further, it could be interesting to identify new cell surface receptors, involved in important cellular functions, without any structural or molecular knowledge of the target. When these situations occur, an alternative and useful approach is to select cell-type-specific ligands using whole cells as an affinity matrix [39] (Figure 19.8).

This approach offers a number of significant advantages. Since whole cells are used as the affinity support, the receptors are likely to be in their native conformation, allowing the isolation of peptide ligands directed against conformationally structured epitopes [70, 71]. Moreover, the technique leads to the identification of cell surface markers which distinguish otherwise similar cell types such as antigens, epitopes, or receptors, without the need to either identify or purify a particular receptor in advance [38, 42, 43, 72].

The advantage of this application becomes more evident considering that ligand–receptor interactions, which influence cellular activities, represent essential tools to understand and control cellular processes. Generally, the analysis of receptor–ligand interaction requires biological assays that can be difficult to perform. For example, a receptor-based screening assay requires identification of the receptor responsible for the cellular function under examination. Once the receptor has been identified, setting up the assay can be a problematic process, particularly if the receptor requires an active cell metabolism or comprises multiple subunits, each necessary to explicate the biological activity. From this point of view, phage display technology represents a powerful approach, which opens new perspectives in the

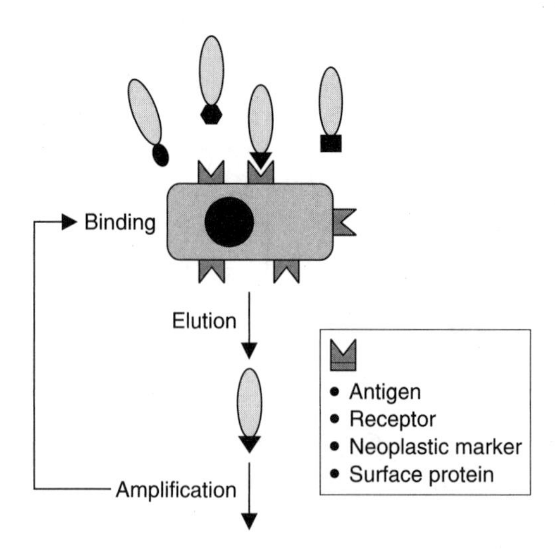

FIGURE 19.8 Biopanning procedure on cells. The binding of the phage library, followed by elution and amplification steps, allows the selection of ligands directed against specific cell markers.

identification of new cell surface proteins [42, 43, 72]. In the first step, peptides that bind the target specifically, by screening whole cells, are identified; in the second step, the binding peptides are tested individually in biological assays, avoiding the need to identify a particular receptor. If the ligand demonstrates activity in cell-based assays, it can be used as an affinity matrix to purify the previously unknown receptors.

E. Phage Display Antibodies: An Alternative to Hybridoma Technology?

The construction of libraries of antibodies fragments expressed on the surface of filamentous bacteriophages, [3, 4] and the selection of phage antibodies by binding to antigen [7], represent a powerful means for generating new biomolecules for research and clinical applications. From this point of view, the phage display approach represents an alternative to conventional hybridoma technology for the production of monoclonal antibodies [3, 4]. In many instances, in fact, the pure antigen is not available as immunogen, particularly cell surface antigens or integral membrane proteins that lose their conformation during the purification steps. Moreover, immunization with intact human cells may result in the production of antibodies against irrelevant epitopes. With phage display, the antibodies can be made completely *in vitro*, bypassing the immune system and immunization procedures. In this approach, the DNA encoding the portion of antibodies in the form of single-chain Fv (scFv), a disulfide-stabilized Fv (dsFv), or Fab is fused in frame with gene III, which encodes the minor surface protein gIIIp of the filamentous phage. When the phage population propagates in a bacterial host, each phage particle displays the antibody fragment as a fusion to the gulp surface protein. The result is a phage population displaying a library of antibody fragments on its surface, which can be subjected for binding to a particular antigen, or a cell of interest [73–75].

The most common selection procedure for antibody libraries has been performed on purified antigen coated directly on the plastic surface [4, 49, 52, 55, 56, 58]. Although antibodies obtained through this procedure perform well in ELISA, their effectiveness in other biological assays is generally limited. This consideration can be explained by the assumption that the coating procedure can lead to the alteration and/or partial destruction of the three-dimensional structure of the antigen, with the possibility of isolating antibody fragments directed against epitopes that are structurally and functionally different from those exposed on the native antigen. This hypothesis has been extensively supported by comparative biopanning experiments performed on purified target and on antigen-positive cells [76]. The results indicated that the antibodies obtained through the two different procedures did not present any structural homology or sequence similarity [76]. The choice of an adequate strategy selection depends on the research goals and on the final applications of the isolated molecule. Obviously, if an antibody has been designed for final use on biological cellular assays or for diagnostic applications, a better strategy would be to pan the phage library on antigen-positive cells. Successive elution and amplification of the bound phage particles allows the recovery of ligands directed against cell surface epitopes [43]. In addition, the possibility of obtaining phage libraries constructed from blood of human donors with autoimmune diseases [77] allows the identification of molecules that can interact with targets previously inaccessible to conventional methodologies, such as toxic substances or self-antigens [78, 79]. The advantage of this approach is particularly evident considering that, in the field of biotechnology and molecular medicine, antibodies play a key role in both diagnostic and therapeutic applications. In the field of cancer immunology, efforts to isolate antibodies that react selectively with human tumoral cells represent the main objective of researchers and scientists. The most common approach aims to generate a large pool of MAB from mice immunized with human neoplastic cells and to screen the antibodies for reactivity against the tumor.

Phage display technology represents an alternative approach not only to isolate specific antibodies but also to identify new neoplastic-associated antigens previously unknown [42, 80]. Moreover, selection procedures on target cells can be extended by a subtractive step in which the library is incubated with adsorbing cells to remove undesired aspecificities. Selection of phage for binding to melanoma cells, followed by adsorption of eluted phages on normal melanocytes, yielded antibodies that recognized the neoplastic cells exclusively [81].

F. FACS Analysis: Applications in Phage Display Technology

FACS analysis represents a powerful approach to identify and select specific cell types related to antigens or epitopes expressed on their surface. When a live stained-cell suspension is put through a fluorescence-activated cell sorter, the machine measures the flourescence intensity of each cell that is separated according to its particular fluorescent brightness [82, 83]. This selection method, in combination with phage display technology, has provided success in isolating antibodies directed against specific cell surface markers [84]. The library is incubated with a cell suspension in which the subpopulation of interest has been previously stained with the appropriate fluorochrome-labeled MAB. The recovery of phages bound to the cells sorted in this matter allows the identification of antibodies and/or peptide fragments against either known or new surface antigens expressed on defined subpopulation of cells [84, 85].

G. Biopanning on Cells: A New Approach for Cell-Targeting Gene Therapy Vectors

Over the past decade, the development of methods for delivering genes to mammalian cells has stimulated great interest in the possibility of treating human diseases by gene-based therapies. However, the transfection of appropriate target cells and the correct delivery of genes still represent critical steps in gene therapy. The main limitation is represented by the vectors utilized, which have features that may limit their applicability, particularly with regard to *in vivo* applications. For instance, retroviral vectors penetrate cells in a manner that is absolutely dependent on the presence of the appropriate viral receptor on the target cells. Since the identities of most retroviral receptors are unknown, it has not been possible to determine their distribution in different cell types. Moreover, such a system may have deleterious effects associated with the expression of some viral proteins which may be toxic to cells, leading to the induction of a deleterious immune response. From this point of view, strategies for gene therapy require either the development of new cell targeting methods or the use of a new entry system for gene delivery. Attempts to generate cell targeting molecules have been focused primarily on monoclonal antibodies. Despite substantial progress, this approach presents several limitations, including the complexity of isolating the appropriate monoclonal antibody and its immunogenicity [86].

Along this line, phage display technology may offer an alternative strategy to generate cell targeting ligands useful for gene therapy vectors. Peptides have been selected against platelets, indicating the potential for using phage libraries to identify ligands against a variety of cell types [72]. Cell-binding peptides selected in this manner could be linked by physical or genetic manipulation to gene therapy vectors that mediate their own endocytosis (for example, adenovirus) [44].

More recently, phages that display known integrin-binding proteins have been shown not only to bind to mammalian cells but also to mediate cellular internalization through a mechanism of receptor-mediated endocytosis [87].

These observations demonstrate the applicability of phage display technology for the identification of ligands that can be used for cell targeting and entry. Targeted filamentous phages will probably not be used as gene transfer vectors themselves, as they have a single-stranded, circular DNA genome which is expressed poorly or absent in mammalian cells. But the ligand itself, given the advantage of its small size, could be useful for the development of new receptor-mediated gene delivery systems, by generating chimerae with gene therapy vectors that do not have the inherent capacity to enter the cell (such as polycationic complex); further, it could also provide a means of targeting other bioactive agents, such as drugs, to particular cells *in vivo* [88].

H. Biopanning on Tissues

Preparations of cells for phage selection may require some procedures that can damage or modify the morphology of the final product. Treatment of tissues with proteolytic enzymes can lead to the loss of some phenotypic characteristic of the target cells. Moreover, upon isolation and culture, these cells, removed from their natural environment, could lose some phenotypic and/or structural characteristics necessary to perform a given biological activity.

To avoid these problems, recently the possibility of using intact fragments or whole tissues as targets for phage selection has been described [40]. An interesting application of this methodology has been recently reported for the isolation of single-chain antibodies to thymic stromal cells which form a microenvironment that controls different steps in *T*-cell differentiation [40]. Intact murine thymic tissue fragments were used as the target, and a subtractive panning was performed, preadsorbing the library with thymocytes and spleen cells to remove aspecifically bound molecules and generate monoclonal antibodies directed against specific subpopulations of thymic stromal cells. The mild fixation of the tissue seems to be an advantage in this selection procedure, since it either preserves the antigenicity of epitopes expressed on the cell surface or prevents the internalization of antigens recognized by monoclonal antibodies during the incubation of the tissue with the phage library. Finally, fixation inhibits the degradation effect caused by proteolytic enzymes present in the intact tissue.

This technique represents an innovative method to isolate antibodies against novel membrane molecules and epitopes expressed in their native conformation and allows the identification of new tissue-specific targeting molecules. From this point of view, this approach opens a new perspective in phage display technology and represents the first key step toward *in vivo* application.

III. *IN VIVO* SELECTION

A. Biopanning *In Vivo*

Biopanning technology is not limited to *in vitro* selection. Phages able to recognize specific target organs may be selected by injecting a phage population into living animals [41] (Figure 19.9). After a few minutes, phages that have been retained in specific organs are eluted, amplified, and reinjected for successive panning cycles. Analysis of rescued phages allows either the identification of new peptide motifs responsible for organ specificity [41], or the tissue-specific localization of known receptors, whose importance in some human disorders has been already clarified [89]. A classical example is represented by tumor vasculature that occurs during angiogenesis, which expresses specific markers on the endothelial surface including certain receptors for vascular growth factors, such as various VEGF receptors [90] and the $\alpha v\beta 3$ integrin [91]. Since many integrins recognize an Arg-Gly-Asp (RDG) sequence as the critical determinant in their ligand [92], phages displaying the RDG peptide were injected intravenously in tumor-bearing mice; the analysis of rescued phages showed a specific localization into tumoral tissues, such as malignant melanomas and breast carcinoma [89].

The main objective of this technique, in the near future, will be the identification of unknown receptors involved in pathological diseases, and the broadest challenge will be to transfer the process across species, and finally in humans. Although the technique is still not feasible as currently performed, a preliminary approach could be to select peptides in mice and subsequently analyze them for appropriate binding of immunohistochemical staining of human tissue.

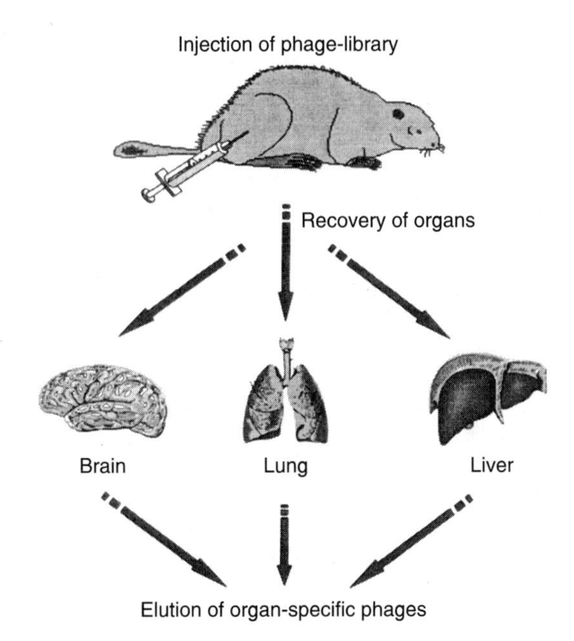

FIGURE 19.9 Biopanning *in vivo*. The phage library is injected intravenously into mice, allowing the identification of phage-peptide motifs that have been retained in specific organs.

REFERENCES

1. Smith GP, Filamentous fusion phage: Novel expression vectors that display cloned antigens on the virion surface, *Science*, 228:1315–1317, 1985.
2. Smith GP, Surface presentation of protein epitopes using bacteriophage expression system, *Curr. Opin. Biotech.*, 2:668–673, 1991.
3. Clackson T, Hoogenboom HR, Griffiths AD, Winter G, Making antibodies fragments using phage display library, *Nature*, 352:624–628, 1991.
4. McCafferty J, Griffiths AD, Winter G, Chiswell DJ, Phage antibodies: filamentous phage displaying antibody variable domains, *Nature*, 348:552–554, 1990.
5. Lu Z, Murray KS, Cleave VV, La Valtie ER, Stahl ML, McCoy JM, Expression of thioredoxin random peptide libraries on the *Escherichia coli* cell surface as functional fusions to flagellin: a system designed for exploring protein–protein interaction, *Biotechnology*, 13:366–372, 1995.
6. Scott JK, Smith GP, Searching for peptide ligands with an epitope library, *Science*, 249:386–390, 1990.
7. Hoogenboom HR, Designing and optimizing library selection strategies for generating high-affinity antibodies, *Trends Biotechnol.*, 15:62–70, 1997.
8. Cwirla SE, Peters EA, Barrett RW, Dower WJ, Peptide on phage: a vast library of peptides for identifying ligands, *Proc. Natl. Acad. Sci. USA*, 87(16): 6378–6382, 1990.
9. McGregor D, Selection of proteins and peptides from libraries displayed on filamentous bacteriophage, *Mol. Biotechnol.*, 6(2):155–162, 1996.

10. Parmley SF, Smith GP, Antibody-selectable filamentous fd phage vectors: affinity purification of target genes, *Gene*, 73(2):305–318, 1988.
11. Scott JK, Discovering peptide ligands using epitopes libraries, *Trends Biochem. Sci.*, 17(7):241–245, 1992.
12. Choi SJ, Ahn M, Lee JS, Jung WJ, Selection of high affinity angiogenin-binding peptide from a peptide library displayed on phage coat protein, *Mol. Cells*, 7(5):575–581, 1997.
13. Felici F, Castagnoli L, Musacchio A, Jappelli R, Cesareni G, Selection of antibody ligands from a large library of oligopeptides expressed on a multivalent exposition vector, *J. Mol. Biol.*, 222(2):301–310, 1991.
14. Krook M, Mosbach K, Lindbladh C, Selection of peptides with high affinity for single stranded DNA using a phage display library, *Biochem. Biophys. Res. Commun.*, 204(2):849–854, 1994.
15. Healy JM, Murayama O, Maeda T, Yoshino K, Sekiguchi K, Kikuchi M, Peptide ligands for integrin $\alpha_v\beta_3$ selected from random phage display libraries, *Biochemistry*, 34:3948–3955, 1995.
16. Giebel LB, Cass RT, Milligan DL, Young DC, Arze R, Johnson CR, Screening of cyclic peptide phage libraries identifies ligands that bind streptavidin with high affinities, *Biochemistry*, 34:15430–15435, 1995.
17. Gram H, Schmitz R, Zuber JF, Baumann G, Identification of phosphopeptide ligands for the Src-homology 2 (SH2) domain of Grb2 by phage display, *Eur. J. Biochem.*, 246(3):633–637, 1997.
18. Folgori A, Tafi R, Meola A, Felici F, Galfre G, Cortese R, Monaci P, Nicosia A, A general strategy to identify mimotopes of pathological antigens using only random peptide libraries and human sera, *EMBO J.*, 13:2236–2243, 1994.
19. Cortese R, Felici F, Galfre G, Luzzago A, Monaci P, Nicosia A, Epitope discovery using peptide libraries displayed on phage, *Trends Biotechnol.*, 12:262–267, 1994.
20. Prezzi C, Nuzzo M, Meola A, Delmastro P, Galfre G, Cortese R, Nicosia A, Monaci P, Selection of antigenic and immunogenic mimics of hepatitis C virus using sera from patients, *J. Immunol.*, 156:4504–4513, 1996.
21. Tafi R, Bandi R, Prezzi C, Mordelli MU, Cortese R, Monaci P, Nicosia A, Identification of HCV core mimotopes: Improved method for the selection and use of disease-related phage-displayed peptides, *Biol. Chem.*, 378:495–502, 1997.
22. Cortese R, Monaci P, Luzzago A, Santini C, Bartoli F, Cortese I, Fortugno P, Galfre G, Nicosia A, Felici F, Selection of biologically active peptides by phage display of random peptide libraries, *Curr. Opin. Biotech.*, 7:616–621, 1997.
23. Barchan D, Balass M, Souroujon MC, Katchalski-Katzir E, Fuchs S, Identification of epitopes within a highly immunogenic region of acetylcholine receptor by phage epitope library, *J. Immunol.*, 155:4264–4269, 1995.
24. Cesareni G, Peptide display on filamentous phage capsids, A new powerful tool to study protein–ligand interaction, *FEBS Lett.*, 307(1):66–70, 1992.
25. Wrighton NC, Farrell FX, Chang R, Kashyap AK, Barbone FP, Mulchay LS, Johnson DL, Barrett RW, Jolliffe LK, Dower WJ, Small peptides as potent mimetics of the protein hormone erythropoietin, *Science*, 273:458–463, 1996.
26. van Meijer M, Roelofs Y, Neels I, Horrevoets AL, van Zonneveld AJ, Pannekoek H, Selective screening of a large phage display library of

plasminogen activator inhibitor 1 mutants to localize interaction sites with either thrombin or the variable region 1 of tissue-type plasminogen activator, *J. Biol. Chem.*, 271(13):7423–7428, 1996.

27. Oldenberg KR, Loganathan D, Goldstein IJ, Schultz PG, Gallop MA, Peptide ligand for a sugar-binding protein isolated from a random peptide library, *Proc. Natl. Acad. Sci. USA*, 89:5393–5397, 1992.
28. Scott JK, Loganathan D, Easley RB, Gong X, Goldstein IJ, A family of concavalin A-binding peptides from a hexapeptide epitope library, *Proc. Natl. Acad. Sci. USA*, 89:5398–5402, 1992.
29. DeLeo FR, Yu L, Burritt JB, Bond CW, Jesaitis AJ, Quinn MT, Mapping sites of interaction of p47-phox and flavocytochrome b with random-sequence peptide phage display libraries, *Proc. Natl. Acad. Sci. USA*, 92:7110–7114, 1995.
30. Burritt JB, Quinn MT, Jutila MA, Bond CW, Jesaitis AJ, Topological mapping of neutrophil cytochrome b epitopes with phage-display libraries, *J. Biol. Chem.*, 270(28):16974–16980, 1995.
31. Horn IR, Moestrup SK, van den Berg BMM, Pannekoek H, Nielsen MS, van Zonneveld AJ, Analysis of the binding of pro-urokinase and urokinase-plasminogen activator inhibitor-I complex to the low density lipoprotein receptor-related protein using a Fab fragment selected from a phage-displayed Fab library, *J. Biol. Chem.*, 270(20):11770–11775, 1995.
32. Dyson MR, Murray K, Selection of peptide inhibitors of interactions involved in complex protein assemblies: association of the core and surface antigens of hepatitis B virus, *Proc. Natl. Acad. Sci. USA*, 92:2194–2198, 1995.
33. Onda T, Laface D, Baier G, Brunner T, Honma N, Mikayama T, Altman A, Green DR, A phage display system for detection of T cell receptor-antigen interactions, *Mol. Immunol.*, 32(17/18):1387–1397, 1995.
34. Balass M, Katchalski-Katzir E, Fuchs S, The alpha-bungarotoxin binding site on the acetylcholine receptor: analysis using a phage-epitope library, *Proc. Natl. Acad. Sci. USA*, 94(12):6054–6058, 1997.
35. Linn H, Ermekova KS, Rentschler S, Sparks AB, Kay BK, Sudol M, Using molecular repertoires to identify high-affinity peptide ligands of the WW domain of human and mouse YAP, *Biol. Chem.*, 378(6):531–537, 1997.
36. Cheadle C, Ivashchenko YC, South V, Identification of a Src SH3 domain binding modif by screening a random phage display library, *J. Biol. Chem.*, 269(39):24034–24039, 1994.
37. Couet J, Li S, Okamoto T, Ikezu T, Lisanti MP, Identification of peptide and protein ligands for the calveolin-scaffold domain, *J. Biol. Chem.*, 272(10):6525–6533, 1997.
38. Marks JD, Ouwehand WH, Bye JM, Finnern R, Gorick BD, Voak D, Thorpe SJ, Hughes-Jones NC, Winter G, Human antibody fragments specific for human blood group antigens from a phage display library, *Biotechnology*, 11:1145–1149, 1993.
39. Watters JM, Telleman P, Junghans RP, An optimized method for cell-based phage display panning, *Immunotechnology*, 3(1):21–29, 1997.
40. van Ewijk W, de Kruif J, Germeraad WTV, *et al.*, Subtractive isolation of phage displayed single-chain antibodies to thymic stromal cells by using intact thymic fragments, *Proc. Natl. Acad. Sci. USA*, 94:3903–3908, 1997.
41. Pasqualini R, Rouslahti E, Organ targeting *in vivo* using phage display peptide libraries, *Nature*, 380:364–366, 1996.

42. Pereira S, Maruyama H, Siegel D, Van Belle P, Elder D, Curtis P, Herlyn D, A model system for detection and isolation of a tumor cell surface antigen using antibody phage display, *J. Immunol. Meth.*, 203(1):11–24, 1997.
43. Palmer DB, George AJ, Ritter MA, Selection of antibodies to cell surface determinants on mouse thymic epithelial cells using a phage display library, *Immunology*, 91(3):473–478, 1997.
44. Barry MA, Dower WJ, Johnston SA, Toward cell-targeting gene therapy vectors: selection of cell-binding peptides from random peptide-presenting phage libraries, *Nat. Med.*, 2(3):299–305, 1996.
45. Feero WG, Rosenblatt JD, Sirianni N, Morgan JE, Partridge TA, Huang L, Hoffman EP, Selection and use of ligands for receptor-mediated delivery to myogenic cells, *Gene. Ther.*, 4(7):664–674, 1997.
46. Pasqualini R, Koivunen E, Rouslahti E, A peptide isolated from phage display libraries is a structural and functional mimic of an RGD-binding site on integrins, *J. Cell. Biol.*, 130(5):1189–1196, 1995.
47. Liu G, Bryant RT, Hilderman RH, Isolation of a tripeptide from a random phage peptide library that inhibits P',P^4-diadenosine 5'-tetraphosphate binding to its receptor, *Biochemistry*, 35:197–201, 1996.
48. Sparks AB, Quilliam LA, Thom JM, Der CJ, Kay BK, Identification and characterization of Src SH3 ligands from phage-displayed random peptide libraries, *J. Biol. Chem.*, 269(39):23853–23856, 1994.
49. Sato A, Ida N, Fukuyama M, Miwa K, Kazami J, Nakamura H, Identification from a phage display library of peptides that bind to toxic shock syndrome toxin-1 that inhibits its binding to major histocompatibility complex (MHC) class II molecules, *Biochemistry*, 35:10441–10447, 1996.
50. Koivunen E, Wang B, Rouslahti E, Phage libraries displaying cyclic peptides with different ring sizes: ligand specificities of the RDG-directed integrins, *Biotechnology*, 13:265–270, 1995.
51. Koivunen E, Gay DA, Rouslahti E, Selection of peptides binding to the $\alpha_5\beta_1$ integrin from phage display library, *J. Biol. Chem.*, 268(27):20205–20210, 1993.
52. Fu Y, Shearing LN, Haynes S, Crewther P, Tilley L, Anders RF, Foley M, Isolation from phage display libraries of single chain variable fragment antibodies that recognize conformational epitopes in the malaria vaccine candidate, apical membrane antigen-1, *J. Biol. Chem.*, 272(41):25678–25684, 1997.
53. Renschler MF, Bhatt RR, Dower WJ, Levy R, Synthetic peptide ligands of the antigen binding receptor induce programmed cell death in a human B-cell lymphoma, *Proc. Natl. Acad. Sci. USA*, 91:3623–3267, 1994.
54. O'Neil KT, Hoess RH, Jackson SA, Swamy Ramachandran N, Mousa SA, De Grade WF, Identification of novel peptide antagonists for GPIIb/IIIa from a conformationally constrained phage peptide library, Proteins: Structure, Function, *Genet.*, 14:509–515, 1992.
55. Yamanaka HI, Inoue T, Ikeda-Tanaka O, Chicken monoclonal antibody isolated by a phage display system, *J. Immunol.*, 157:1156–1162, 1996.
56. Ames RS, Tornetta MA, Jones CS, Tsui P, Isolation of neutralizing anti-C5a monoclonal antibodies from a filamentous phage monovalent Fab display library, *J. Immunol.*, 152:4572–4581, 1994.
57. Smith JW, Hu D, Satterthwait A, Pinz-Sweeney S, Barbas III CF, Building synthetic antibodies as adesive ligands for integrins, *J. Biol. Chem.*, 269(52):32788–32795, 1994.

58. Baca M, Presta LG, O'Connors SJ, Wells JA, Antibody humanization using phage display, *J. Biol. Chem.*, 272(16):10678–10684, 1997.
59. Daniels DA, Lane DP, The characterization of p53 binding phage isolated from phage peptide display libraries, *J. Mol. Biol.*, 243:639–652, 1994.
60. Yanofsky SD, Baldwin DN, Butler JH, High affinity type I interleukin 1 receptor antagonists discovered by screening recombinant peptide libraries, *Proc. Natl. Acad. Sci. USA*, 93:7381–7386, 1996.
61. Li M, Yu W, Chen CH, Cwirea S, Whitehorn E, Tate E, Raab R, Bremer H, Dower B, *In vitro* selection of peptides acting at a new site of NMDA glutamate receptors, *Nat. Biotechnol.*, 14:986–991, 1996.
62. Sanna PP, Williamson RA, De Logu A, Directed selection of recombinant human monoclonal antibodies to herpes simplex virus glycoproteins from phage display libraries, *Proc. Natl. Acad. Sci. USA*, 92:6439–6443, 1995.
63. Rebar EJ, Pabo CO, Zinc finger phage: affinity selection of fingers with new DNAbinding specificities. *Science*, 263:671–673, 1994.
64. Balass M, Heldman Y, Cabilly S, Givol D, Katchalski-Katzir E, Fuchs S, Identification of a hexapeptide that mimics a conformation-dependent binding site of acetylcholine receptor by use of a phage-epitope library, *Proc. Natl. Acad. Sci. USA*, 90:10638–10642, 1993.
65. Hawkins RE, Russell SJ, Winter G, Selection of phage antibodies by binding: affinity mimicking affinity maturation, *J. Mol. Biol.*, 226:889–896, 1992.
66. Gaynor B, Putterman C, Valadon P, Spatz L, Scharff MD, Diamond B, Peptide inhibition of glomerular deposition of an anti-DNA antibody, *Proc. Natl. Acad. Sci. USA*, 94:1955–1960, 1997.
67. Wang B, Dickinson LA, Koivunen E, Rouslahti E, Kohwi-Shigematsu T, A novel matrix attachment region DNA binding motif identified using a random phage peptide library, *J. Biol. Chem.*, 270(40):23239–23242, 1995.
68. De Leo FR, Ulman KV, Davis AR, Jutila KL, Quinn MT, Assembly of the human neutrophil NADPH oxidase involves binding of $p67^{phox}$ and flavocytochrome b to a common functional domain in $p47^{phox}$, *J. Biol. Chem.*, 271(29):17013–17020, 1996.
69. Markland W, Charles Ley A, Lee SW, Charles Ladner R, Iterative optimization of high-affinity protease inhibitors using phage display, *Plasmin Biochemistry*, 35:8045–8057, 1996.
70. Goodson RJ, Doyle MV, Kaufman SE, Rosenberg S, High-affinity urokinase receptor antagonists identified with bacteriophage peptide display, *Proc. Natl. Acad. Sci. USA*, 91:7129–7133, 1994.
71. Szardenings M, Törnroth S, Mutulis F, Muceniece R, Keinanen K, Koosinen A, Wikberg JE, Phage display selection on whole cells yields a peptide specific for melanocortin receptor, *J. Biol. Chem.*, 272(44):27943–27948, 1997.
72. Fong S, Doyle LV, Devlin JJ, Doyle MV, Scanning whole cells with phage-display libraries: identification of peptide ligands that modulate cell function, *Drug Develop Res.*, 33:64–70, 1994.
73. Marks JD, Hoogenboom HR, Bonnert TP, McCafferty J, Griffiths AD, Winter G, By-passing immunization, Human antibodies from V-gene libraries displayed on phage, *J. Mol. Biol.*, 222(3):581–597, 1991.
74. Marks JD, Griffiths AD, Malmqvist M, Clackson TP, Bye JM, Winter G, By-passing immunization: building high affinity human antibodies by chain shuffling, *Biotechnology*, 10:779–783, 1992.

75. Hoogenboom HR, Winter G, By-passing immunization, Human antibodies from synthetic repertoires of germline VH gene segments rearranged *in vitro*, *J. Mol. Biol.*, 227(2):381–388, 1992.
76. Chowdhury PS, Chang K, Pastan I, Isolation of anti-mesothelin antibodies from a phage display library, *Mol. Immunol.*, 34(l):9–20, 1997.
77. Finnern R, Pedrollo E, Fisch I, Human autoimmune anti-proteinase 3 scFv from a phage display library, *Clin. Exp. Immunol.*, 107(2):269–281, 1997.
78. Finnern R, Bye JM, Dolman KM, Zhao MM, Short A, Marks JD, Lockwood MC, Ouwehand WH, Molecular characteristic of anti-self antibody fragments against neutrophil cytoplasmic antigens from human V gene phage display libraries, *Clin. Exp. Immunol.*, 102(3):566–574, 1995.
79. Griffiths AD, Malmqvist M, Marks JD, Bye JM, Embleton MJ, McCafferty J, Baier M, Holliger, KP, Govick BD, Hughes-Jones NC, Human anti-self antibodies with high specificity from phage display libraries, *EMBO J.*, 12(2):725–734, 1993.
80. Pereira S, Van Belle P, Elder D, Maruyama H, Jacob L, Sivanandhan M, Wallack M, Siegel D, Herlyn D, Combinatorial antibodies against human malignant melanoma, *Hybridoma*, 16(l):11–16, 1997.
81. Cai X, Garen A, Anti-melanoma antibodies from melanoma patients immunized with genetically modified autologous tumor cells: selection of specific antibodies from single-chain Fv fusion phage libraries, *Proc. Natl. Acad. Sci. USA*, 92:6537–6541, 1995.
82. Davey HM, Kell DB, Flow cytometry and cell sorting of heterogeneous microbial population: the importance of single-cell analysis, *Microbiol. Rev.*, 60(4):641–696, 1996.
83. Orfao A, Ruiz-Arguelles A, General concept about cell sorting techniques, *Clin. Biochem.*, 29(1):5–9, 1996.
84. De Kruif J, Terstappen L, Boel E, Logtenberg T, Rapid selection of cell subpopulation-specific human monoclonal antibodies from a synthetic phage antibody library, *Proc. Nail. Acad. Sci. USA*, 92:3938–3942, 1995.
85. Li M, Use of a modified bacteriophage to probe the interaction between peptides and ion channel receptors in mammalian cells, *Nat. Biotechnol.*, 15(6): 559–563, 1997.
86. Pietersz GA, McKenzie IFC, Antibody conjugates for the treatment of cancer, *Immunol. Rev.*, 129:57–80, 1992.
87. Hart SL, Knight AM, Harbottle RP, *et al.*, Cell binding and internalization by filamentous phage displaying a cyclic Arg-Gly-Asp-containing peptide, *J. Biol. Chem.*, 269(17):12468–12474, 1994.
88. Arap W, Pasqualini R, Rouslahti E, Cancer treatment by targeted drug delivery to tumor vasculature in a mouse model, *Science*, 279:377–380, 1998.
89. Pasqualini R, Koivunen E, Rouslahti E, av Integrins as receptors for tumor targeting by circulating ligands, *Nat. Biotechnol.*, 15:542–546, 1997.
90. Martiny-Baron G, Marme D, VEGF-mediated tumor angiogenesis: a new target for cancer therapy, *Curr. Opin. Biotechnol.*, 6:675–680, 1995.
91. Brooks PC, Clark RA, Cheresh DA, Requirement of vascular integrin $\alpha v\beta 3$ for angiogenesis, *Science*, 264:569–571, 1994.
92. Rouslahti E, Pierschbacher MD, New perspectives in cell adhesion: RGD and integrins, *Science*, 238:491–497, 1987.

20 Globular Oligonucleotide Screening via the SELEX Process: Aptamers as High-Affinity, High-Specificity Compounds for Drug Development and Proteomic Diagnostics

Larry Gold

CONTENTS

I. INTRODUCTION

Craig Tuerk and I published a paper in *Science* in August 1990 entitled "Selection of Ligands by Exponential Enrichment: RNA Ligands to Bacteriophage T4 DNA Polymerase" [1]. We were studying the regulation of translation of bacteriophage T4 DNA polymerase, a protein that represses its own translation by binding to an RNA site that overlaps the ribosome-binding site in its own mRNA [2]. We showed that randomizing the eight nucleotides in a hairpin loop in that domain (so as to provide 65,536 different octamer sequences in the loop), and then selecting for tight binders with purified T4 DNA polymerase, led to the isolation of two octamers—the wild-type sequence and a quadruple mutation of that sequence (that is, four changes within eight nucleotides). We used amplification (RT-PCR) of the binding subset between rounds, thus creating a winnowing or culling of winners from the bulk of the RNA sequences as the SELEX process proceeded. In the abstract of the paper we wrote, "These protocols with minimal modification can yield high-affinity ligands for any protein that binds nucleic acids as part of its function; high-affinity ligands could conceivably be developed for any target molecule." That bold statement regarding the potential of the SELEX process (the products of the SELEX process are called "aptamers," based on a suggestion made by Ellington and Szostak in a paper published just after ours [3]) was reiterated in the paper to make the point: we believed that we had found something as powerful as antibodies for measuring or inactivating/activating therapeutically interesting proteins [4]. Craig and I said that the "products of SELEX can affect the activity of the protein to which they have been fit." We implied that aptamers could be used to agonize or antagonize protein targets, and thus might become therapeutic and/or diagnostic agents. This review article and many lovely scientific papers from academic labs and NeXagen/NeXstar Pharmaceuticals suggest strongly that Craig Tuerk and I had seen correctly the power and potential of oligonucleotide aptamers.

Probably a few hundred SELEX experiments have been aimed at protein and other target molecules. Extremely high-affinity aptamers (with low pM monovalent K_d's) have been isolated repeatedly, and more than 10 aptamer antagonists have been tried in preclinical efficacy models. These nuclease-resistant aptamers perform in animals in a dose-dependent manner, just like other therapeutic agents (Aptamers obviously are delivered to animals parenterally—the average mass of an aptamer is 10,000 without PEG or other adducts that are used for improved pharmacokinetics and biodistribution.) The first aptamer to reach clinical trials, NX1838 (an antagonist of vascular endothelial growth factor, VEGF), is being tested for age-related macular degeneration by NeXstar Pharmaceuticals. The toxicity package for NX1838 is very promising: the aptamer is neither toxic nor immunogenic at high doses. Recent internal development of high-throughput SELEX machines at NeXstar suggests that aptamers might be a first class of therapeutic compound to be used in so-called target validation experiments, as the many protein targets uncovered through genomics are filtered into potentially useful targets for the treatment of disease. Over and over we are struck by the high affinity and specificity of aptamers [5]; in fact, aptamers already represent a class of compounds with performance features at least as powerful as monoclonal monovalent or divalent antibodies. A recent paper from Tom Steitz's lab (in which the co-crystal of an aptamer with its target—HIV reverse transcriptase—was presented [6]) showed dramatically why aptamers are such remarkable compounds and why people at NeXstar have fallen in love with them.

Other interesting combinatorial chemistry methods have been developed over the last few years in response to the unmet needs of the pharmaceutical industry. The rising costs of drug discovery and drug development and the slow rate of approval of "new chemical entities" (NCEs) are at the center of these unmet needs. Combinatorial chemistry may offer the industry faster and cheaper drug discovery paradigms that will increase the rate at which successful compounds reach the patient.

The unmet needs of the pharmaceutical industry are centered on the difficulties in identifying high-potency compounds for therapeutic use, and the even greater difficulties inherent to compound improvement with further organic chemistry. Organic chemists now collaborate with computational "chemists" in attempts to do virtual screening of vast libraries *in silico*, hoping that a well-designed virtual library will lead to the rapid synthesis and testing of compounds that actually do what they are meant to do. Actual screening at very high throughput along, with massive parallel chemical syntheses, may provide number of tested compounds in the range of only 100,000 or so.

An equally robust platform, somewhat less favored in the present pharmaceutical industry, would be the synthesis and screening of enormous mixtures of compounds followed by deconvolution and iterative resynthesis and rescreening of the best compounds from the previous tests. It seems obvious that the number of compounds tested might be higher with these methods. [The first and most commonly used systems that screen mixtures were aimed at peptides, proteins, or antibodies that are displayed on cells, phages, or even (today) ribosomes. Several chapters in this volume describe these methods. In general, libraries of 10^6 to perhaps 10^9 compounds can be screened efficiently with these techniques; ribosome display promises to raise these numbers further.] SELEX libraries contain about 10^{15} different oligonucleotide sequences which are meant to fold into intramolecular globular structures, some of which (by chance) are able to interact with a target protein with remarkable affinity and specificity.

II. THE STATUS OF APTAMERS—AN EXAMPLE

An aptamer that binds to vascular endothelial growth factor (VEGF), one of many identified since 1990, has been identified and studied [7]. The aptamer binds tightly to VEGF and in so doing blocks the binding of VEGF to both the Flt-1 receptor and the KDR receptor. The epitope within VEGF to which the aptamer binds includes peptide residues from within exon 7 sequences, the domain encoding the critical residues within VEGF responsible for very high-affinity binding to both VEGF receptors, heparin, and the aptamer. A cross-link between one part of the aptamer and cysteine 137 in the VEGF protein has been identified [7].

The VEGF aptamer has been used *in vivo* to block corneal angiogenesis and to slow xenograft tumor growth. The compound is now in the clinic for the

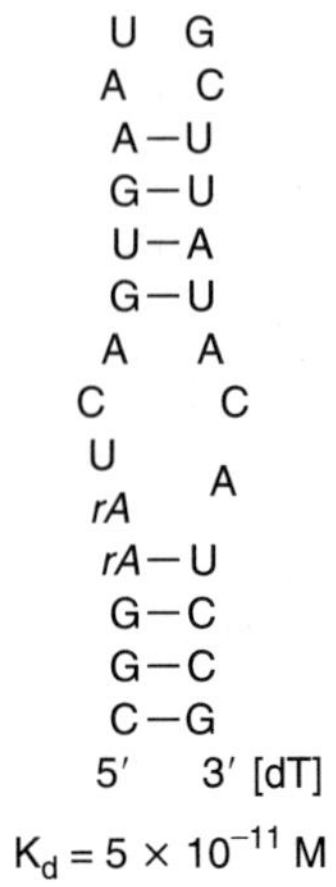

```
     U  G
    A    C
     A—U
     G—U
     U—A
     G—U
    A    A
   C      C
   U
           A
   rA
    rA—U
     G—C
     G—C
     C—G
    5'    3' [dT]
```

$K_d = 5 \times 10^{-11}$ M

FIGURE 20.1 The VEGF Aptamer (NX1838).

treatment of age-related macular degeneration (ARMD). Angiogenesis, the process of new blood vessel elaboration that is a major event in opthalmic pathologies and cancer, might be slowed in human diseases by the VEGF aptamer.

The affinity of the aptamer for VEGF protein is about 50 pM. The VEGF aptamer (NX1838) is shown in Figure 20.1. The aptamer is a single-stranded oligonucleotide comprised of 27 nucleotides. The molecular weight of this aptamer is around 9,000 D. The pyrimidines contain 2′-fluorine and the purines mostly contain 2′-OCH_3; only A's at positions 4 and 5 contain 2′-OH. The compound has been formulated for appropriate human pharmacokinetics and biodistribution. The VEGF aptamer behaved well in animals with a number of model neovascular pathologies. The purpose of this first example is to make clear to the reader what qualities of aptamers might be exciting to the pharmaceutical and diagnostics industry-high potency and *in vivo* efficacy.

The aptamer binds tightly and specifically to the intended target protein and to few other proteins. The idea of an (oligonucleotide) aptamer behaving like a protein antibody is at odds with the history of molecular biology, a history delineated in *The Eighth Day of Creation* [8]. The "central dogma" of molecular biology stated that the information in genes (DNA) was transcribed into mRNA, which was subsequently translated into protein. Nucleic acids were conceived to be linear "tapes" possessing four letters (the nucleotides), while proteins were "shapes" built as globular molecules from 20 monomeric letters (the natural amino acids). An aptamer behaving like an antibody is at odds with this simple idea. That is, while globular proteins and enzymes are standard, globular single-stranded oligonucleotides have been thought to be less common [9]; rather, most single-stranded oligonucleotides (such as messenger RNAs) have been thought to be largely unstructured. The side groups on amino acids have varied chemistries, while the four bases of RNA seem to have limited recognition potentials (aimed largely at the

"Watson-Crick" complementary partners). When (globular) single-stranded ribozymes were described, the surprise was sufficient to warrant (appropriately) the award of a Nobel Prize for Chemistry in 1990 [10].

Old paradigms, especially useful ones like the concept of the central dogma, change slowly. In fact, in spite of strong structural work on RNA molecules that have shape nuances beyond Watson-Crick pairing, textbooks continue to depict double-helical DNA, single-stranded RNA, and proteins in much the same manner as they were described by Crick when he elaborated the central dogma of molecular biology. In spite of the wonderful intellectual and experimental history of DNA and RNA molecules as (largely) conveyors of primary sequence information, the driving idea behind aptamers as high-affinity/high-specificity binding reagents is that oligonucleotide secondary structures create loops and other irregularities in the shape and electronic distributions within the oligonucleotide, to afford a tight fit between an aptamer and a target protein or target small molecule [11]. This idea is supported by a large number of structures of aptamers and other interesting oligonucleotides (from X-ray and NMR studies) that have accumulated in the literature over the last 10 years.

III. THE SELEX PROCESS—A BRIEF DESCRIPTION

The few paragraphs that follow briefly summarize an extensive description of the methods published by Fitzwater and Polisky [12]. First, a library of double-stranded DNA sequences is prepared by a mixture of chemical and enzymatic steps. Using standard solid-phase DNA synthesis, a mixture of single-stranded DNA sequences is made with defined sequences on the 5′ and 3′ ends of each molecule. The interior sequences of each molecule are randomized by condensing mononucleotides from a mixture of all four chemical substrates, such that each individual sequence has roughly a 0.25 likelihood of stochastically coupling with A, C, G, or T. The randomized domain length is typically 40 or so nucleotides (see below). The defined sequence on the 5′ ends (fixed A) contains (typically) a promoter for bacteriophage T7 RNA polymerase, while the defined sequence on the 3′ ends (fixed B) is chosen to be low in potential intramolecular secondary structures and (usually, but not in some specific cases in the literature aimed at special aptamer structures) in sequences complementary to the sequence that comprises fixed A. The library of single-stranded DNA sequences is then made double-stranded by filling in the single-stranded library after annealing a primer (fixed B′) to the 3′ ends of the library; the library can be amplified by the polymerase chain reaction with primers fixed B′ and fixed A. Such libraries are stored for the next step in the SELEX process. Typically, 10^{14}–10^{15} molecules are made synthetically prior to the amplification step; thus most sequences are missing from a library with 40 randomized positions (4^{40} is about 10^{24}; libraries of the same length made twice will have very few identical molecules).

Second, the double-stranded DNA is transcribed with T7 RNA polymerase to yield a library of single-stranded RNA or modified RNA (or replicated asymmetrically to yield a library of single-stranded DNA or modified DNA). The choice of chemistry is determined by the intended use of the aptamer: if one intends to use aptamers as the intracellular products following transcription of synthetic genes *in vivo* (after transformation of an appropriate cell with a synthetic gene that encodes the aptamer), one would select a normal RNA aptamer. If one wanted (as is often the case at NeXstar Pharmaceuticals) a drug candidate aimed at an extracellular protein target (such as VEGF, above), the SELEX RNA library might be "front-loaded" with modified triphosphates (typically normal rATP and rGTP, along with 2′-F CTP and UTP) to provide serum stability against nucleases. If one wanted (as has been the case occasionally) a DNA aptamer for use in affinity chromatography to purify a protein on a large scale, one might use single-stranded DNA libraries so that the aptamer column could be reused after alkaline "cleaning in place." In any event, the key parameter is that the single-stranded library is free to form (unknown and not predicted) secondary and tertiary structures that lead to high-affinity aptamers for any intended target molecule.

Third, the single-stranded library is challenged with the intended target molecule (or mixtures of molecules or even crude cell lysates [13]). The challenge step might be as simple as simply mixing the library with an appropriate amount of the intended target in solution, or it might entail passing the library through a column which contains the target substance attached to an appropriate matrix material. The idea is to achieve partitioning of the members of the library that bind more tightly to the target molecule than the bulk of the library. Much work has gone into calculating and simulating the partitioning step [14, 15], so that rare winning molecules are not lost in this first round of the SELEX process; obviously, if the best aptamers are infrequent [11], one must take care not to lose them during the first partitioning step, since nothing one does after the first round can recover those molecules that are lost.

Fourth, the selected (partitioned) molecules are reverse transcribed (if the library was RNA based) and then amplified by PCR using the primers fixed B′ and fixed A; for DNA libraries PCR is sufficient. Thus, the subset of sequences that bound to the intended target becomes the pool for the second round of the SELEX process. Several more rounds of the SELEX process are done until the library complexity drops from 10^{15} sequences to (perhaps) a hundred, after which they are cloned and sequenced and tested for the desired properties. For protein targets, the SELEX process usually has taken from 7 to 15 rounds, although a robotic SELEX process will take fewer rounds (see Figure 20.2). The entire (front-loaded) SELEX process is diagrammed in Figure 20.2.

The cloned sequences often (but not always) have obvious secondary structures in common, leading to "visual" truncations by which the fixed sequences and other unnecessary sequences from the random regions are eliminated from the aptamer, often resulting in slightly increased affinity over the starting full-length aptamer. When "visual" truncation is not possible, experimental truncation is done instead. After truncation the aptamer may be further altered by substituting nuclease-resistant purines for the normal

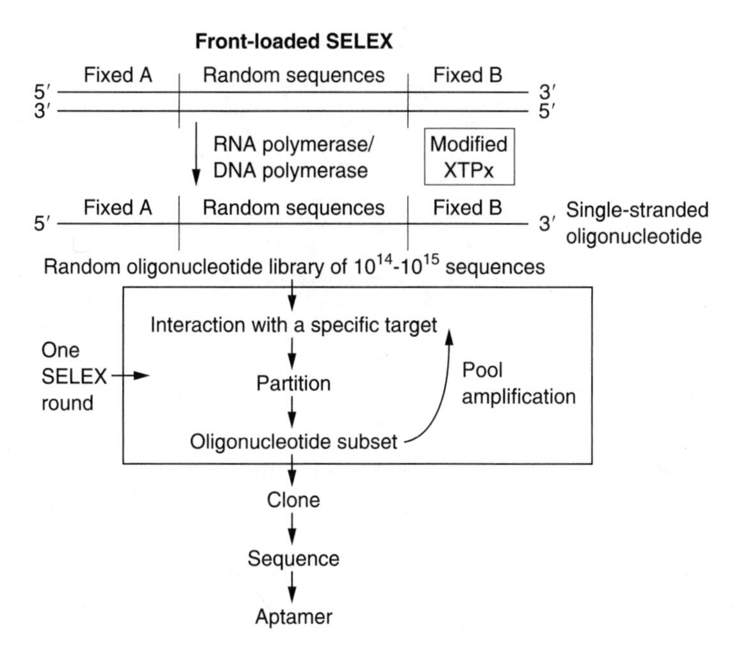

FIGURE 20.2 The SELEX process.

ribo-purines in the truncated aptamer. Many purines in a specific aptamer will be easily changed to, for example, 2′-OCH_3 purines, without any impact on the aptamer affinity for the target; other purines will remain 2′-OH because that 2′ position is critical for aptamer binding. The purine "scan" can be accomplished (for a short aptamer) in a few weeks.

In overview, the SELEX process bears substantial similarities to other *in vitro* and *in vivo* selection technologies such as phage display, ribosome display, or even display of proteins on bacterial surfaces. The real differences are that the very molecule that undergoes selection is amplifiable directly rather than through a nucleic acid intermediary, because the selected molecules in the SELEX process are nucleic acids. The major differences between selections of proteins or peptides and oligonucleotides are: oligonucleotide libraries may be screened at vast sizes compared to protein or peptide libraries ($\sim 10^6$ times more oligonucleotides than proteins or peptides); oligonucleotide libraries may be "front-loaded" (as above) to provide substantial stability against serum/blood enzymatic attack; and, most important, aptamers have binding properties equal to or better than anything that has emerged from peptide and protein (including antibody) selections.

IV. APTAMERS—SOME GENERALIZATIONS

The single most surprising biochemical attribute of aptamers is their extraordinary binding affinities and specificities for their intended targets.

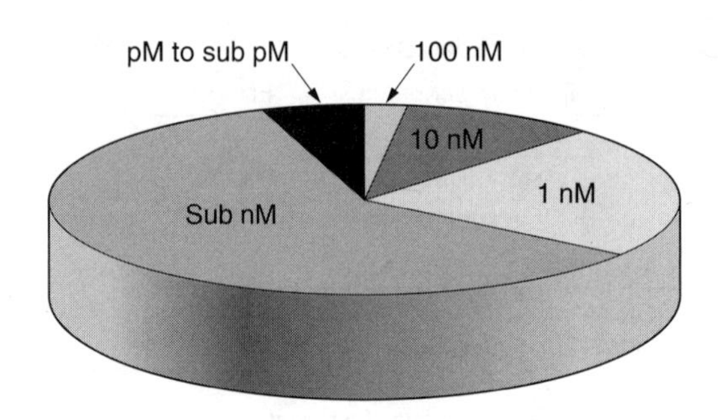

FIGURE 20.3 K_d values from the first 100 SELEXs aimed at proteins.

A large collection of aptamers (probably more than 200 at this time) aimed at protein targets has been made and studied. The affinities have been measured (usually in solution) as monovalent binding interactions between a unique aptamer sequence and the protein; the measured binding constants (k_d's) are shown in Figure 20.3 for the first 100 or so aptamers isolated at NeXstar Pharmaceuticals and at the University of Colorado in Boulder.

Rarely is a selected aptamer bound weakly by its intended protein target (worse than 100 nM K_d), while *most* of the time the aptamer binds to its intended target with a K_d of 1 nM or better. If one reads the literature for other kinds of "combinatorial libraries" (including the chapters in this book), aptamers stand alone as the most avidly bound compounds to flow quickly from selection of winners from large libraries. Every paradigm about the glory of proteins, every simple biochemical notion (where is the histidine in a nucleic acid, where is the tyrosine, where is the lysine/arginine? At least people never ask, "where are the aspartates and glutamates...") points toward peptides and proteins as better binding reagents than would be provided by single-stranded oligonucleotides, and yet the data seem clear: monovalent interactions between an aptamer and a protein target provide higher affinities that those provided by most monovalent antibodies aimed at similar proteins. The SELEX process has power and value because of this empirical and unexpected finding.

The data for specificity are equally surprising. As shown in Table 20.1, an aptamer aimed at basic fibroblast growth factor (bFGF) does not bind tightly to other members of the FGF family or to other proteins known to interact with an acidic substance, heparin (these data were reported in [5]). Similar data exist for aptamers aimed at several reverse transcriptases, serine proteases, P- and L-selectin, and cytokines such as VEGF and PDGF. Aptamers *in vitro* show extreme specificity for their intended targets.

With a compound discovery method in place that achieves high potency and specificity, all that remains for aptamers to have utility as drugs would be stability, pharmacokinetics, toxicity, and low immunogenicity of aptamers *in vivo*. Because the literature was clear about the major human plasma endonucleases that destroy RNA molecules, one early event in the NeXstar

TABLE 20.1
The bFGF aptamer high specificity

Protein	Relative binding
BFGF	$100 = 0.5\,nM\,K_d$
Denatured bFGF	0.05
FGF-1	0.02
FGF-4	0.03
FGF-5	2
FGF-6	0.03
FGF-7	0.04
PDGF	0.1
VEGF	0.05
IL-8	<0.0005
Thrombin	<0.0005
ATIII	<0.0005

development of aptamers was to include pyrimidine 2′ ribose modifications that blocked pyrimidine-specific attack. Both 2′-fluoro-UTP and -CTP are incorporated by bacteriophage T7 RNA polymerase during transcription and provide well-behaved RNAs. The plasma half-lives of RNAs with only this modification is reasonable. Oligonucleotides can be further modified by doing post-SELEX purine modification with 2′O-methyl-sugars, using a protocol that does not require incorporation of 2′-O-methyl-ATP-or -GTP during transcription (see above). Individual aptamers allow variable purine modification as a function of the idiosyncratic requirements for specific purine 2′ OH groups in the aptamer structure. A second approach, so-called mirror-image SELEX, utilizes target proteins and peptides prepared with D-amino acids so that the normal/biological/unstable RNA aptamer can be synthesized with L-ribose and provide complete stability to serum nucleases [16]. This is a good idea, suffering only from the issues of target protein preparation [17].

The pharmacokinetics of aptamers has been largely predictable. At average molecular weights of 10,000, aptamers are cleared rapidly through the kidneys. Experiments have been done with PEG adducts, lipid adducts, and presentation of aptamers on the outside of liposomes, and each of these formulations has provided a long lifetime in animals. Aptamers have been provided parenterally by IV, sub-Q, footpad injection, IP, and intramuscularly, and all such deliveries have provided sustained systemic circulation. Aptamers with, for example, 20,000 MW PEG adducts, distribute to tissues, and so aptamers are not restricted to vessel targets. Lastly, direct injection into the vitreous (the proposed means by which an aptamer aimed at VEGF may be useful for the treatment of macular degeneration) leads to high local and low systemic concentration.

Aptamers have not shown toxicity in rodents. Furthermore, aptamers have not been immunogenic in mice, even after repeated injections of nuclease-resistant aptamers, formulated so as to provide very long lifetimes in the vasculature. Aptamers have been injected with adjuvants and foreign basic proteins, as suggested by the literature for the preparation of antibodies against DNA, and no IgG's have been raised. More work will be required in the clinical trials of aptamers, but so far the data are promising.

Finally, the most serious issue for the therapeutic use of aptamers may have been solved. Large-scale production of oligonucleotides has been studied intensively (because of the anti-sense paradigm for drug development), yet the costs of production using solid-phase synthetic means remain high. Recently, a new approach to oligonucleotide synthesis has been developed by Pieken, Eaton, and their colleagues [18]; the approach, called PASS (for Product-Anchored Sequential Synthesis), allows solution-phase coupling of monomers to a growing chain, followed by capture of the chain that successfully received that monomer and purification away from the "failure" sequence and the remaining monomer. The purified chain is then released into solution for the addition of the next monomer. PASS provides (in principle) inexpensive purification of the growing oligonucleotide after every coupling plus the capacity to scale-up the solution reactions to large volumes. The PASS process is diagrammed in Figure 20.4.

Thus, the efficacy of aptamers *in vivo* has not been described (in this review), with the exception of the VEGF aptamer, NXI 838 (above). At this point we have evaluated 10 aptamers, seven for efficacy and three (including one of the first set of seven aptamers tested for efficacy) for diagnostic imaging. In one set of published experiments with an aptamer aimed at human L-selectin, the IC_{50} *in vivo* was not very much higher than the K_d *in vitro*, as though not too many proteins in serum/blood bound the aptamer and lowered its effective free concentration. Every experiment with each of the aptamers showed *in vivo* activity (Table 20.2).

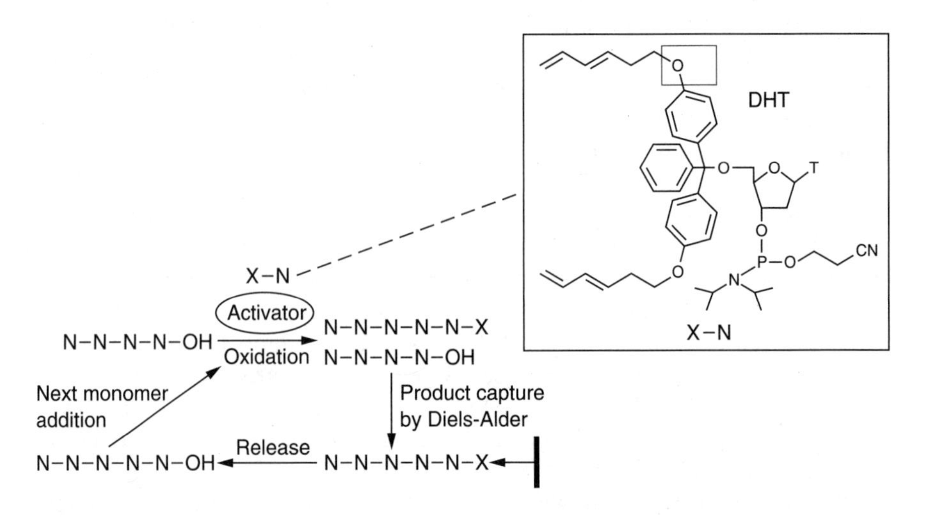

FIGURE 20.4 The PASS process.

TABLE 20.2
***In vivo* activity of aptamers**

Aptamers Ten (of ten tried) work *in vivo*
Seven for seven for efficacy (various preclinical models):
Anti-VEGF, anti-bFGF, anti-PDGF, anti-L-selectin, anti-P-selectin, anti-neutrophil elastase, anti-thrombin
Three for three for *in vivo* imaging
Blood clots, restenotic lesions, tumors

Finally, the SELEX process can be automated, as is clear from inspection of Figure 20.2. No single step in the SELEX process is difficult, and all steps can be and have been accomplished with robotic pipetting stations. At NeXstar prototypes have been built that accomplish the various steps of the SELEX process, using as targets proteins that have been immobilized on solid supports (within 96 well plates). When SELEX is performed with immobilized targets, the number of rounds required to reach the winning pool is a bit lower than in "classic" SELEX, although more experiments will be required to verify this early observation and explain it (probably the more rapid culling of nonbinders reflects the capacity to lower nonspecific binding during the partitioning steps [14, 15]). The point is that the SELEX process will be done on multiple pure protein targets at the same time, and one such "machine" should generate winning aptamer pools for (say) 96 target proteins in a very short time. The idea of lots of aptamers aimed at lots of proteins ("Keeping Pace with Genomics" [19]) seems to be within reach.

V. PROTEOMICS—THE USE OF APTAMERS FOR DIAGNOSTICS

We spent a substantial amount of time over the last few years trying to develop aptamers as replacements for antibodies for *in vitro* diagnostics (both cell sorting [20] and protein assays [21]). These experiments were successful, yet they missed what might be a substantial value of aptamers, especially if the SELEX process is placed within a robotics context, so that aptamers are generated effortlessly against any number of protein targets. Recently, we have been developing suitable platforms for the high-density presentation of aptamers (arrays of aptamers) for the simultaneous measurement of large numbers of proteins in blood, urine, or other relevant human fluids. The simple idea is that the present list of "markers" measured by reference labs, while useful, is inadequate for a complete status report on a patient. Extensive measurements of a range of proteins over the years of a patient's lifetime (my definition of "proteomics"—other definitions exist) offer that patient (as, say, a portion of a yearly physical examination) a "wellness" check, an early warning regarding the beginning of pathology, a measure of the

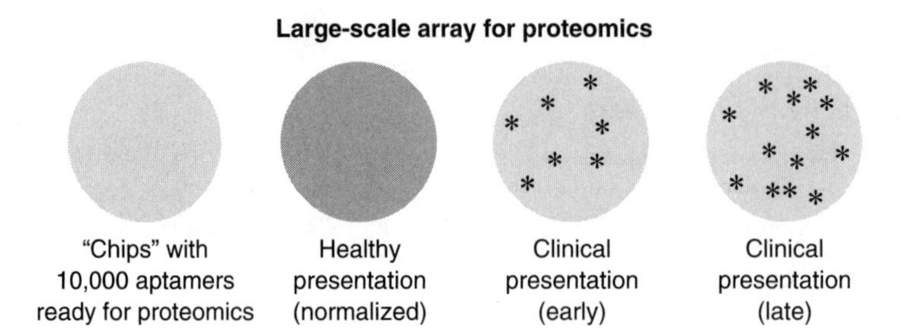

FIGURE 20.5 A proteomics "chip." * = proteins expressed differentially.

appropriateness of particular drug interventions, and/or the surveillance of disease progression during treatment.

Aptamers placed on a matrix bind specific proteins from dilute solution in the context of many other proteins [22]. Thus, an array of different aptamers on a "chip" could be used to capture different proteins (from blood, for example) at specific locations (in X, Y space) for subsequent (protein) quantitative measurements and informatics (Figure 20.5):

Aptamers within an array may offer the health-care community a new paradigm for patient monitoring, during both periods of health and periods of disease. This is an exciting possibility, equivalent to a larger bank of antibody-based ELISAs than is presently available, or even the technically more challenging notion of antibodies on chips [23].

VI. SUMMARY AND FUTURE PROSPECTS

To contrast the SELEX process with other combinatorial chemistry paradigms (which are discussed in detail within this book, and which have enthusiastic and intelligent proponents) is unnecessary. There will remain many useful ways to discover drugs, and the SELEX process will be but one of them (at least we hope that aptamers will be an important set of parenteral drugs). At NeXstar the driver continues to be rapid assessment of the highest-affinity aptamers for both therapeutic and diagnostic use. The major power of SELEX and aptamers resides in the potency and specificity of the identified aptamers along with the speed with which aptamers can be identified. As such, aptamers have an obvious role to play in validating protein targets associated (through functional genomics) with various pathologies; the major limitation toward this end appears to be the present focus on extracellular targets (to obviate the need for cellular uptake). The true value of the SELEX technology will become apparent as more *in vivo* experiments are reported using aptamers as "drugs." The use of aptamers as reagents for proteomic diagnostics seems straightforward and is under intense investigation.

The major liability of aptamers as drugs may reside in their need for parenteral delivery and the corresponding unwillingness of pharmaceutical companies and perhaps patients to use them (although examples abound of parenteral compounds that are widely used). Successful experimental efforts toward simple delivery systems would allow aptamers to be exploited fully, thus taking advantage of the high potency and specificity that flow from the large surface interactions between aptamers and their targets.

ACKNOWLEDGMENTS

I thank Craig Tuerk (once again) for being a strong proponent of the "aptamer paradigm" early, when the formal objections to aptamer development seemed overwhelming. In addition, I thank a huge group of 60 people at NeXstar Pharmaceuticals, including but not limited to Barry Polisky, Ed Brody, Bruce Eaton, Ray Bendele, Nebojsa Janjic, David Emerson, Stan Gill, Dom Zichi, Drew Smith, Nikos Pagratis, David Parma, Judy Ruckman, Louis Green, Tim Fitzwater, Rob Jennison, Dan Drolet, and Sumedha Jayasena. This review does not attempt to cite all relevant papers, many of which may be found by author searches on the people listed above, or which can be obtained by contact with the author.

REFERENCES

1. Tuerk C, Gold L, Systematic evolution of ligands by exponential enrichment: RNA ligands to bacteriophage T4 DNA polymerase, *Science*, 249:505–510, 1990.
2. Tuerk C, Eddy S, Parma D, Gold L, The autogenous translational operator recognized by bacteriophage T4 DNA polymerase, *J. Mol. Biol.*, 213:749–761, 1990.
3. Ellington AD, Szostak JW, *In vitro* selection of RNA molecules that bind specific ligands, *Nature*, 346:818–822, 1990.
4. U.S. Patents, 5,270,163; 5-12/14/93; 5,475,096-12/12/95; 5,696,249-12/9/97; 5,670,637-9/23/97.
5. Eaton BE, Gold L, Zichi DA, Let's get specific: The relationship between specificity and affinity, *Chem. Biol.*, 2:633–638, 1995.
6. Jaeger J, Restle T, Steitz TA, The structure of HIV-I reverse transcriptase complexed with an RNA pseudoknot inhibitor, *EMBO. J.*, 17:4535–4542, 1998.
7. Ruckman JL, Green S, Beeson J, Waugh S, Gillette WL, Henninger DD, Claesson-Welsh L, Januic N, 1998; 2′-Flouropyrimidine RNA-based aptamers to the 165-amino acid form of vascular endothelia growth factor ($VEGF_{165}$), *J. Biol. Chem.*, 273(32):20556–20567.
8. Judson HF, *The Eighth Day of Creation: The Makers of the Revolution in Biology*, Cold Spring Harbor, NY, Cold Spring Harbor Laboratory, 1996.
9. Brawerman G, Gold L, Eisenstadt J, A ribonucleic acid fraction from rat liver with template activity, *Proc. Natl. Acad. Sci. USA*, 50:630–638, 1963.
10. Gold L, Catalytic RNA: a Nobel Prize for small village science, *New Biologist*, 2:1–4, 1990.
11. Gold L, Polisky B, Uhlenbeck O, Yarns M, Diversity of oligonucleotide functions, *Ann. Rev. Biochem.*, 64:763–797, 1995.

12. Fitzwater T, Polisky B, A SELEX primer, In: Methods in Enzymology, *Combinatorial Chemistry Volume*, 267, San Diego, Academic Press, 1996.
13. Morris KN, Jensen KB, Julin CM, Weil M, Gold L, High affinity ligands from *in vitro* selection: complex targets, *Proc. Natl. Acad. Sci. USA*, 95:2902–2907, 1998.
14. Irvine D, Tuerk C, Gold L, Selexion: systematic evolution of ligands by exponential enrichment with integrated optimization by nonlinear analysis, *J. Mol. Biol.*, 222:739–761, 1991.
15. Vant-Hull B, Payano-Baez A, Davis RH, Gold L, The mathematics of SELEX against complex targets, *J. Mol. Biol.*, 278:579–597, 1998.
16. Klubmann S, Nolte A, Bald R, Erdman VA, Fiirste JP, Mirror-image RNA that binds D-adenosine, *Nat. Biotech.*, 14:1112–1115, 1986.
17. Gold L, Reflections on mirrors, *Nat. Biotech.*, 14:1080, 1996.
18. Pieken W, McGee D, Settle A, Zhai Y, Huang JP, Method for solution phase synthesis of oligonucleotides, WO 97/14706.
19. Gold L, Alper J, Keeping pace with genomics through combinatorial chemistry, *Nat. Biotech.*, 15:297, 1997.
20. Davis KA, Lin Y, Abrams B, Jayasena SD, Staining of cell surface human CD4 with 2′-F-pyrimidine-containing RNA aptamers for flow cytometry, *Nucleic Acid Res.*, 26:3915–3924, 1998.
21. Drolet DW, Moon-McDermott L, Romig TS, An enzyme-linked oligonucleotide assay, *Nat. Biotech.*, 14:1021–1025, 1996.
22. Romig TS, Bell C, Drolet DW, Aptamer affinity chromatography; combinatorial chemistry applied to protein purification, *J. Chromatogr B Biomed. Sci. Appl.*, 731:275–284, 1999.
23. Zipkin L, Technology focus: proteomics, *BioCentury*, 6(60):A1–A4, 1998.

21 Applications of RNA and DNA Aptamers in Basic Science, Diagnostics and Therapy

Henning Ulrich

CONTENTS

I. INTRODUCTION

The SELEX technology (Systematic Evolution of Ligands by Exponential enrichment) was introduced by Larry Gold and Jack Szostak [1, 2] and provides a powerful tool for the *in vitro* selection of nucleic acids (aptamers) from combinatorial DNA or RNA libraries against a target molecule.

The SELEX technique involves reiterative selection rounds with target presentation during which target-binding oligonucleotides are amplified until the original random oligonucleotide library is purified to a few molecules that bind to their target with high affinity. During the last decade the SELEX technology was used extensively to isolate high-affinity ligands for a wide variety of proteins and other molecules of therapeutic importance, such as growth factors and neuropeptides [3–8], antibodies [9], enzymes such as HIV reverse transcriptase and non-structural hepatitis C virus protease [10, 11], cell-surface antigens such as selectins and the prostate-specific membrane antigen [12–14], as well as for the discovery of new ribozymes [15, 16]. Recently the use of the SELEX method has been extended to find ligands that bind to complex targets such as erythrocyte ghosts [17], the membrane-bound nicotinic acetylcholine receptor [18, 19], the neurotensin receptor [20], rat tumor brain vessels [21] and live trypanosomes [22, 23]. These ligands have dissociation constants in the picomolar to low micromolar range for their protein targets. Nowadays aptamers can be selected against nearly every target. The SELEX technique can be used to evolve ligands that act on extracellular or intracellular targets. Aptamers can be expressed in cells in order to continuously act on their intracellular target proteins.

Figure 21.1 shows a tentative flow chart for application of aptamers in basic science, diagnostics and therapy, taking into account the latest technological developments. The next step following identification of potential targets, based on knowledge from genomics and proteomics, is to determine

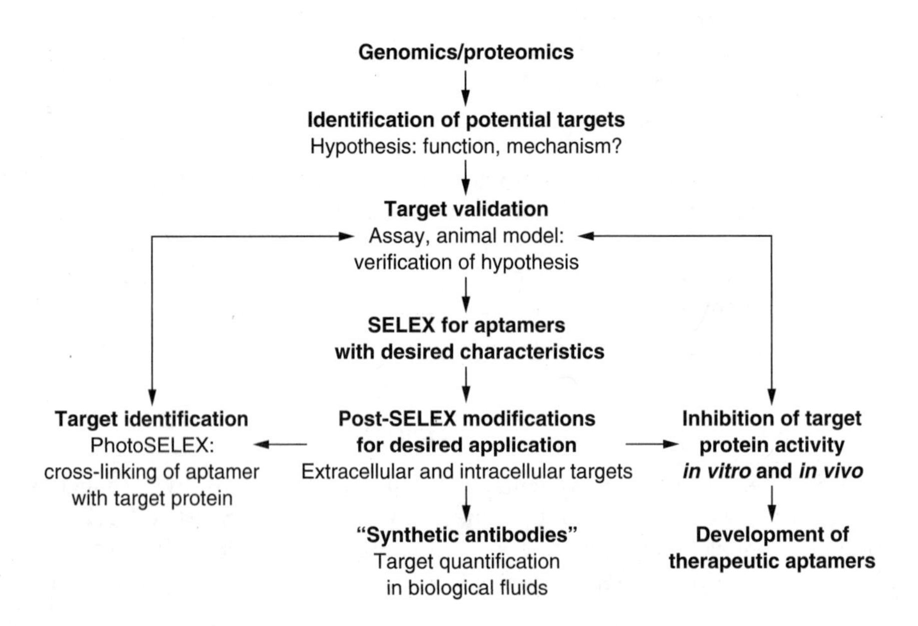

FIGURE 21.1 Flow chart for selection and subsequent optimization of oligonucleotide aptamers.

the importance and mechanism of function of these targets in a biological process of interest. Once a target is validated, aptamers are selected, identified and then modified for a specific application.

II. APPLICATIONS OF SELEX

A. RNA and DNA Aptamers as Tools for Target Validation and for Studying the Mechanism of Protein Function *In Vitro*

RNA and DNA aptamers have been found to be effective in studying the function and mechanism of activation and inactivation of receptor proteins, both in extracellular and intracellular environments. Aptamers can influence receptor–ligand interactions and enzyme actions by either binding to the ligand of a receptor, or to a substrate of an enzyme, or by displacing the ligand, or substrate from a receptor-binding site, or from a catalytic site of an enzyme. Aptamers that alter the activity of a target protein can bind to an allosteric site on the receptor or enzyme and thereby compete with the ligand and may therefore act either as agonists or inhibitors. Aptamers that act through a non-competitive mechanism on their target, bind to regulative sites different from the ligand site and may act as agonists or antagonists by stabilizing either active or inactive protein forms.

The mechanism of aptamer action (competitive or non-competitive) is predicted by the choice of the agonist or antagonist for the displacement of high-affinity RNA or DNA binders to a receptor during the SELEX process.

The first example will demonstrate how RNA or DNA aptamers are used to provide proof of mechanism for the inhibition of a neurotransmitter receptor.

1. The Isolation of RNA Aptamers that Alleviate Inhibition of the Nicotinic Acetylcholine Receptor (nAChR) Caused by Cocaine

Understanding the mechanism of the nicotinic acetylcholine receptor (n AChR) and its inhibition is a long-standing problem, with major implications for drug addiction, especially for cocaine abuse [24]. Steady-state kinetic and electrophysiological studies indicated a simple and generally accepted mechanism in which cocaine enters the open channel and sterically inhibits the flow of ions [25]. Novel rapid kinetic techniques of electrophysiology were developed that suggested that cocaine binds to a regulatory site on the closed receptor and induces a change of the protein conformation that prevents the receptor channel from opening. This hypothesis suggested the possibility of finding compounds that bind to the inhibitory site of the receptor, without inducing the subsequent change of protein conformation that leads to inactive receptor forms. Since not all RNA molecules induce a change in the conformation of the protein they bind to, SELEX can be used to select for RNA molecules that displace cocaine from its binding site on the nAChR but do not inhibit the receptor function by themselves. Following selection and identification of cocaine-RNA aptamers (explained in Figure 21.2), two classes of RNA aptamers, denominated class I and class II, with different consensus regions were identified [18]. Secondary structure predictions using free energy minimization [26] revealed that consensus motifs in the previous random regions are located in stem-loop structures that are responsible for biological activity. Two aptamers from each class, denominated aptamer I-14 and aptamer II-3, were selected for further characterization (Figure 21.3). These aptamers served as tools to prove the suggested mechanism of inhibition of the nAChR [19]. Aptamer I-14 was a potent inhibitor of receptor function, and bound with higher affinities to the closed channel form of the receptor than to the open channel, and thereby shifted the equilibrium between open and closed channels towards the closed, inactive receptor. Aptamer II-3 did not affect the receptor activity by itself, but did displace cocaine from the receptor. The ability of aptamer II-3 to protect the nAChR against

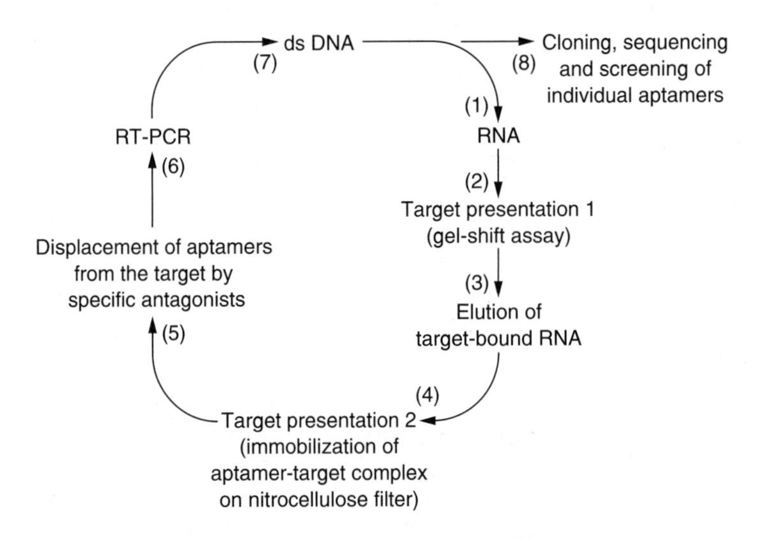

FIGURE 21.2 SELEX for cocaine-displaceable aptamers bound to the nicotinic acetylcholine Receptor (nAChR). (1) The original RNA pool containing 10^{13} different sequences and structures is obtained by *in vitro* transcription from a partial randomized DNA. (2) The RNA pool is incubated with cell membranes enriched in nAChR protein. Receptor-bound RNA is separated from free RNA by a gel-shift assay. (3) Receptor-bound RNA is purified from the protein-RNA complex and (4) again presented to the target that has been immobilized to a nitrocellulose membrane. (5) RNA bound to the cocaine-binding site of the receptor is displaced by a cocaine analogue. (6) The cocaine-displaceable RNA molecules are amplified by RT-PCR to restore the double-stranded DNA pool and (7) again *in vitro* transcribed to give a RNA pool enriched in molecules with the desired properties. Reiterative SELEX cycles are carried until no further increase in binding affinity is obtained. (8) Individual aptamers are obtained by cloning the final pool into a bacterial plasmid and sequencing of the inserts.

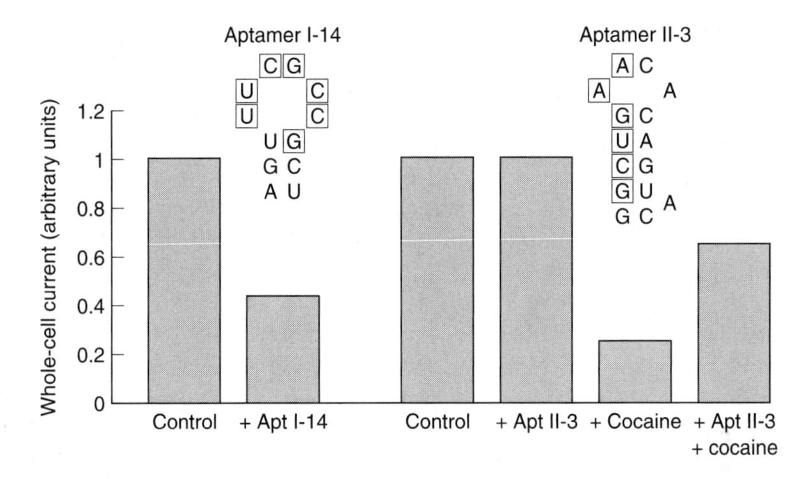

FIGURE 21.3 Action of class I and II cocaine-displaceable aptamers: The effect of cocaine-displaceable aptamers I-14 and II-3 on carbamoylcholine (100 μM)-induced stimulation of acetylcholine receptor activity in the absence (control) and presence of cocaine was determined *in vitro* by fast kinetic electrophysiology. Aptamer I-14 (0.5 μM) inhibited receptor activity (left panel); aptamer II-3 (5 μM) does not inhibit receptor activity (middle panel); aptamer II-3 (3 μM) alleviates inhibition of receptor function by cocaine (150 μM) (right panel). Prediction of the secondary structure [26] revealed the localization of consensus motifs of class-I and -II aptamers in stem-loop regions (letters in boxes). These stem loops are believed to be necessary for aptamer action. The figure reveals the data published in references [18, 19, 74].

inhibition by cocaine was determined *in vitro* by fast kinetic electrophysiology, with BC_3H1 cells expressing the muscle-type nAChR. Cells were stimulated by an agonist and the resulting whole-cell is determined in the absence or presence of cocaine or cocaine and protecting aptamer II-3. The aptamer had a dose-dependent effect in alleviating cocaine inhibition ($K_A = 700$ nM). This is taken as proof that cocaine is not a sterical channel blocker, but binds to a regulatory site different from the channel pore. The developed RNA aptamers demonstrated the feasibility of the approach, and based on the obtained results cocaine analogues were screened and a molecule (RTI 4229–70) [27] was found that acts exactly like aptamer II-3. For *in vivo* target validation and pharmacological evaluation of cocaine-displaceable aptamers, post-SELEX modifications are needed that include the stabilization and minimization of the aptamer and the attachment of linkers that increase its half-time in the plasma.

The second example discusses approaches to regulate the activity of intracellular expressed aptamers.

2. Controlling the Activity of Intracellularly Expressed Aptamers

Permanently and transiently intracellular expressed RNA aptamers were evolved against different targets that include pathogens such as hepatitis C virus, HIV-1 virus reverse transcriptase [11, 28] and intracellular proteins involved in signal transduction events [29, 30]. In order to study the involvement of targeted proteins during signal intracellular transduction events, it may be interesting to express an intracellular aptamer whose activity can be controlled and switched on or off at a desired point of time. The expression of aptamers, but not their activity, can be induced at a certain time point by placing the aptamer expression under the control of an inducible promoter [31].

Direct control over aptamer activity is achieved by expressing the aptamer in the cell and then activating or inactivating the aptamer upon demand. This goal can be achieved by selecting aptamers that bind with high-affinity to their intracellular protein target but also contain a second binding site for a membrane-permeable, small ligand molecule (principle of allosteric aptamer activation and inactivation reviewed in [32]). The binding to this ligand molecule changes the conformation and activity of the aptamers (Figure 21.4). These ligand-regulated aptamers can be controlled in their activity when they already interact with their target protein. In the absence of its regulator that binds to a different site on the aptamer than its target protein, an aptamer-protein complex is formed and the target protein is inhibited. Following addition of the regulator, the binding of the regulator changes the aptamer conformation and dissociates the aptamer-protein complex, thereby abolishing the inhibition of the protein.

The expression of RNA aptamers needs to be at sufficiently high levels within cells in order to influence their phenotype and alter target protein function in its cellular environment. One potential disadvantage of intracellularly expressed aptamers is that once the aptamer is expressed in the cell and the target protein inhibited, there is no further possibility of precise control of the aptamer's action [33]. Temporal regulation of the aptamer's activity may be crucial if the inhibition of a target protein is desired

(a) Allosteric inhibition of aptamer activity:

$A + R \rightleftharpoons \underline{A}\,R$ (aptamer active and target inhibited)

$L + \underline{A}\,R \rightleftharpoons L\,A + R$ (aptamer inactive and target not interfered)

(b) Allosteric activation of aptamer activity:

$A + R \rightleftharpoons A + R$ (no complex formation and target not interfered)

$L + A + R \rightleftharpoons L\,\underline{A}\,R$ (aptamer active and target inhibited)

FIGURE 21.4 Possible mechanism for the action of ligand-regulated aptamers. (a) Addition of the ligand inhibits aptamer function and promotes the dissociation of the aptamer-receptor complex. (b) Addition of the ligand induces formation of the aptamer-receptor complex and activates aptamer function. Inhibitory activity of aptamers is assumed in both cases. L: Ligand that controls aptamer activity; A: inactive aptamer; $\underline{A}$: active aptamer; R: receptor (aptamer target).

during certain stages in the cell cycle or development. Precise regulation of aptamer action during a time course is achieved by including a small organic molecule that penetrates the cell and disrupts the RNA-protein complex. As a proof of principle, aptamers that target the DNA repair enzyme formamido-pyrimidine glycosylase were isolated. Neomycin controls aptamer activity and binds to the inhibitory aptamer bound to the enzyme, thereby dissociating the aptamer-enzyme complex [33].

Another example of ligand-induced intracellular aptamer regulation was given by Suess and coworkers [34]. The authors used a conditional gene expression system in which direct RNA-metabolite interactions are exploited as a mechanism of genetic control. Tetracycline-binding aptamers were inserted into the 5′-Untranslated Region (5′-UTR) of a Green-Fluorescent Protein (GFP) encoding mRNA. Aptamer insertion generally reduced GFP expression. Inhibition of GFP expression was enhanced by tetracycline addition. The formation of the aptamer-ligand complex increased the thermodynamic stability and activity of the aptamer. In contrast to the first example, where addition of the ligand resulted in inactivation of the aptamer, in this case the ligand is needed for the aptamer's activity.

In summary, as aptamers can virtually recognize every target, they are ideal tools to interfere with the biological activity of extra- and intracellular proteins and to study the mechanism of conformational changes during activation and inactivation, as shown for the acetylcholine receptor. Ligand-regulated aptamers allow the study of activation and inactivation of proteins in a cellular system, allowing transient functional target protein knock-outs or knock-ins upon aptamer activation or inactivation by a ligand. Table 21.1 summarizes recent accomplishments of aptamers selected against extra- and intracellular targets.

B. Aptamers as "Synthetic Antibodies" for Target Validation and Quantification in Biological Fluids

As aptamers can be evolved against virtually any target and their specificity of target recognition is similar to that of antibodies [35], aptamers have also been termed as "synthetic antibodies" (reviewed in [36]). In fact, aptamers evolved by *in vitro* selection have some advantages over antibodies from animal sources. For instance, toxins and targets that do not elicit immune responses in animals can be targeted. Aptamers can be protected against enzymatic degradation in biological systems and can be easily modified by attaching reporter molecules at the researcher's will. These observations make aptamers promising tools for *in vitro* diagnostics and *in vivo* imaging. Examples of the use of aptamers as diagnostics are given by the work of Fredriksson *et al.* (2002) [37], who developed a sandwich assay for the quantification of Platelet-Derived Growth Factor (PDGF) in biological fluids. The assay system consists of two DNA aptamers binding to the target—the first immobilizing PDGF to a support and the second aptamer, which is attached to a fluorescent reporter, quantifies the concentration of the target protein.

TABLE 21.1
Accomplishments of aptamers targeting cell surface receptor activity and intracellular targets

Receptor ligands targeted	Studied effect	Accomplishments	Ref. (No.)
Vasopressin	Inhibition of receptor activation	Inhibition of vasopressin-induced cAMP formation *in vitro*.	4
VEGF	Inhibition of receptor binding and angiogenesis	Aptamer preventing neovascularization tested in clinical tests; reduction of size of Wilms tumor in animal model.	7, 49–51
PDGF	Inhibition of receptor activation	Antagonism in experimental glomerulonephrities in animal model. Transcapillary transports in tumors *in vivo* enhanced.	8, 48, 53
L/P-selectin	Inhibition of cell-adhesion	Inhibition of lymphocyte trafficking *in vivo*.	12, 13, 57, 68
Interferon-γ (IFN-γ)	Inhibition of IFN-receptor binding	Inhibition of IFN-γ-induced regulation *in vitro*.	59, 75
Receptors as targets			
Neurotensin (NTS-1) Receptor	Specific binding to receptor	Neurotension receptor-specific aptamer. Interference with receptor activity not demonstrated.	20
Nicotinic acetylcholine receptor (nAChR)	Displacement of cocaine	One class of aptamers that inhibits the nACHR like cocaine and a second class of aptamers that protect against inhibition by cocaine.	18,19
Intracellular targets			
HIV reverse transcriptase	Inhibition of HIV virus replication	Intracellular expressed RNA aptamers suppresses HIV virus particle release *in vitro* and *in vivo*.	10, 28, 76
HCV NS3 (hepatitis C virus) protease	Inhibition of NS3 protease	In cells expressed RNA ribozyme-aptamer repeated in tandem ensure high intracellular aptamer dosage. Aptamers inhibit virus protease action *in vitro* and *in vivo*.	11, 77
Intracellular domain of the β-integrin LFA-1	Binding to functional subdomain of LFA-1	Cell cultures expressing intracellular aptamers show reduced binding to Intercellular Adhesion Molecule I (ICAM-I).	29
Sec 7 domain of Cytohesin-1 (ADP-Ribosylation factor)	Inhibition of guanine-nucleotide-exchange-factor activity	Intracellular RNA expression inhibits binding of cells to ICAM-I and induces reorganization of actin distribution *in vitro*.	30

Aptamers attached to a fluorescent reporter are useful for quantification of a cell surface antigen on living cells by flow cytometry (see references [21, 38]).

The suitability of aptamers as *in vivo* imaging agents is demonstrated in references [39–41], who used anti-neutrophil elastase and anti-thrombin aptamers for imaging of inflammation and thrombus *in vivo*, respectively.

C. Aptamers as Tools for Target Purification and Identification

Aptamers can be selected against target proteins within a complex mixture of potential targets and then be used for a ligand-mediated target purification. One of the advantages of combinatorial library approaches over structural drug design is that no knowledge of the mechanism of interaction between a target protein and its antagonist is required. In many cases, the binding site of the antagonists or agonists used for the elution of high-affinity aptamers are not determined. In other cases, the receptor for the antagonists or agonists used for displacement of the aptamer during the SELEX process is not even known. The work of Ulrich and coworkers [23] can be cited where RNA aptamers compete with host-cell matrix molecules for their binding sites on unknown receptors on the cell surface of *Trypanosoma cruzi*. Larry Gold and coworkers [17] evolved aptamers against erythrocyte membranes that bind specifically to prior unidentified target proteins. These studies indicate the possibility of purifying any protein target for which a high-affinity nucleic acid ligand is identified. In order to generate covalent cross-links of selected aptamers with their targets, pyrimidines of the oligonucleotides are modified at the C5 position of the ribose by attaching an iodine or bromine function. The aptamers can either be selected from oligonucleotide libraries that contain photo-cross linkable pyrimidines or already selected aptamers can be modified posterior to the SELEX process [42, 7].

These modified oligonucleotides, that are attached to a reporter molecule, generate reactive groups that, when irradiated by UV-light, can form covalent linkages with another molecule in close proximity [43, 44]. The target-linked aptamer can then be used for a ligand-mediated purification of the protein target.

D. Therapeutic Approaches

The potential utility of aptamers as therapeutic agents is considerably enhanced by chemical modifications that lend resistance to nuclease attack [38, 45–47]. Important target applications against those aptamers were selected with possible clinical applications that include combating growth factor induced pathophysiology and tumor progression [7, 8, 49–53], targeting of virus proteins and inhibition of virus-replication and propagation [10, 11, 28, 54, 55], blocking of cell-adhesion events affected in disease [56, 57] or cell adhesion receptors used by pathogens for cell infection [23], the blockade of immunoglobulins and cell surface antigens involved in inflammation, allergy and autoimmune disease [58–62], and the protection of neurotransmitter receptors against inhibition by abused drugs and toxins [18, 19] (see Table 21.1 for examples of aptamers with possible future applications in therapy).

Approaches are evolved using aptamers by the SELEX technique to block defective signal transduction events caused by growth factor receptor overactivity, and induce tumor proliferation by including the inhibition of interactions between Vascular Endothelial Growth Factor (VEGF) [7], Nerve Growth Factor (NGF) [63], basic Fibroblast Growth Factor (bFGF) [6], Platelet-Derived Growth Factor (PDGF) [8], vasopressin [4], substance P [64] and their respective receptors. Pharmacologically the best-characterized aptamer is the anti-VEGF aptamer, patented and named as NX1838 [65, 66]. This aptamer inhibited angiogenesis *in vivo* and tumor growth in its presence was dramatically reduced. As an inhibitor of neovascularization, the aptamer is being clinically tested for treatment of age-related vascular disease [67]. The possible therapeutic use of aptamers in blocking tumor proliferation is not restricted to blocking growth factor actions. As cell adhesion events, for example those mediated by selectins, aptamers are necessary prerequisites for the attachment of metastatic cells to blood vessel endothelia. Aptamers directed against L- and P-selectins [12, 13] that have shown biological activity *in vivo* [57, 68] also may become important for cancer therapy. Nuclease-resistant aptamers against tenascin-C, a protein expressed in high concentration on tumor cells, may have the same effects by preventing tenascin-C binding to its integrin receptor and blocking tenascin-C-induced angiogenesis [56].

The therapeutical potential of aptamers depends on many of the factors that apply to convential pharmaceuticals. Binding their targets with high affinity and specificity is the first requirement. In this regard, the large surface area of aptamers permits potentially more interactions with their target molecules than small molecules share with their receptors. Repeated selection rounds of a huge excess of possible ligands over their binding sites ensure the amplification of the best fitting molecules. This results in binding affinities of aptamers to their targets up to low picomolar (comparable to monoclonal antibodies), in many cases much higher than the binding affinity of the natural ligands to their targets. Aptamers are capable of distinguishing between isozymes of the same protein with 96 percent homology [69] and when small molecules are targeted, aptamers can even distinguish between the presence and absence of an hydroxyl group [70].

The industrial and pharmacological use of aptamers was limited in the past, due to the lack of nuclease-resistance, short half time in the plasma, and problems in delivering aptamers into target cells. The coupling of large linkers such as polyethyleneglycol and liposome anchors to the aptamers [65, 71] promises prolonged half times in the plasma and enhance the aptamers' aridity of interaction with their targets on tumor cells. Since oligoribonucleotides are particularly prone to the attack of nucleases present both in serum and inside cells, the development of chemical modifications is essential for their biological applications. Usual modifications include 2′amino- or 2′-fluoro-pyrimidines that are incorporated enzymatically before the selection process (see Figure 21.5). The incorporation of these modified nucleotides into RNAs results in an enhanced resistance of the RNA against nuclease attack, but the modified aptamers lack the 2′-hydroxyl group, which is shown to be essential for RNA folding and, consequently, fundamental for the development of modified RNA like ribozymes, ribozyme inhibitors and RNA aptamers. In the particular case of ribozymes, the presence of the 2′-hydroxyl group in defined positions has proved to be indispensable for maintaining the activity. For better thermal stabilization and protection against RNA activity, sugar-modified analogues, 2′-*C*-methyl-nucleotides, can be used in post-SELEX modifications that provide stability of the RNA and possess a 2′-hydroxyl group able to make similar interactions as those displayed by the natural moieties [72].

The lack of effectiveness of aptamers in some experimental studies, such as in the inhibition of PDGF-induced brain tumor cells *in vitro*, may be explained by the fact that, in contrast to classical tyrosine kinase inhibitors that enter the tumor cell and inhibit receptor autophosphorylation, the anti-PDGF aptamers do not recognize their binding motif on PDGF when PDGF is bound to its receptor and are therefore less active [73].

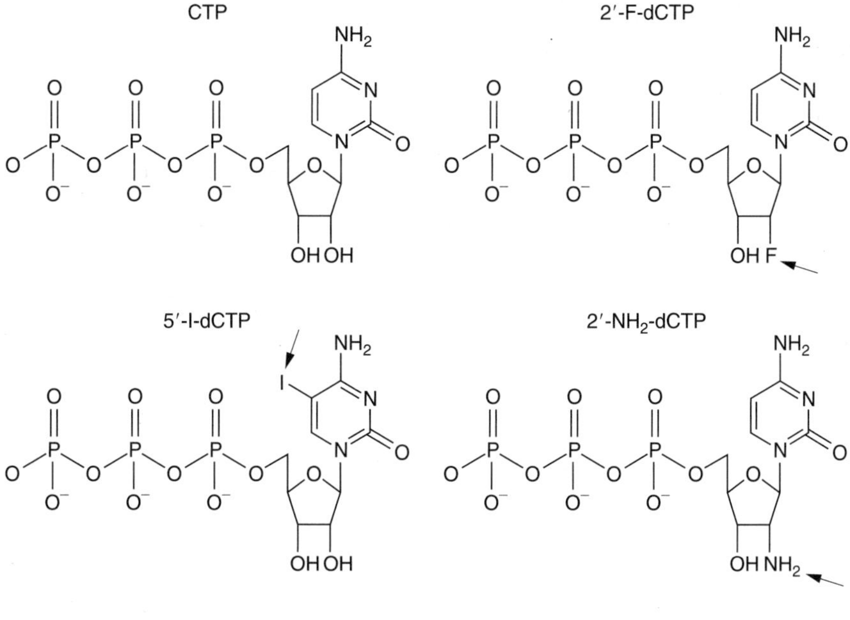

CTP

2′-F-dCTP

5′-I-dCTP

2′-NH_2-dCTP

FIGURE 21.5 Chemical structures of modified nucleotides used in the SELEX procedure. Aptamers containing 2′F- and 2′NH_2-modified nucleotides are resistant against degradation by nucleases and therefore suitable for *in vitro* and *in vivo* applications. Aptamers containing modifications (I or Br) at the C-5 position of pyrimidine can be photo-cross-linked to their protein targets. Arrows indicate the respective chemical modifications.

Recent modifications of nucleotides that increase uptake rates into cells and the intracellular expression of RNA aptamers [11, 28–30] promise accessibility of intracellular targets that are involved in virus replication and tumor proliferation. Cytoplasmic expression of aptamers has been used to combat virus replication and modulate intracellular signal transduction pathways of integrin-induced adhesion events and to dissect single steps of intracellular events [29, 30]. For instance, tumor cells may, therefore, be transfected with expression vectors coding for aptamers that bind to intracellular domains of growth factor receptors and inhibit their activity. Controlling the activity of intracellularly expressed aptamer by cell-permeable ligands permits activation and inactivation of aptamer function and growth factor receptor inhibition at desired time points.

In summary, as DNA and RNA aptamers recognize their targets with similar specificity as antibodies and can be selected against almost every target, they are suitable for *in vitro* and *in vivo* experiments [65, 68], diagnostics and drug development, such as the anti-VEGF-aptamers that are undergoing human clinical testing [66, 67].

ACKNOWLEDGMENTS

I thank FAPESP (Fundação de Amparo à Pesquisa do Estado de São Paulo) and CNPQ (Conselho Nacional de Desenvolvimento Cientifico e Técnologico), Brazil, for financial support.

REFERENCES

1. Tuerk C, Gold L, Systematic evolution of ligands by exponential enrichment: RNA ligands to bacteriophage T4 DNA polymerase, *Science*, 249:505–510, 1990.
2. Ellington AD, Szostak JW, *In vitro* selection of RNA molecules that bind specific ligands, *Nature*, 346:818–822, 1990.
3. Nieuwlandt D, Wecker M, Gold L, *In vitro* selection of RNA ligands to substance P, *Biochemistry*, 34:5651–5659, 1995.
4. Williams KP, Liu XH, Schumacher TN, Lin HY, Ausiello DA, Kim PS, Bartel DP, Bioactive and nuclease resistant L-DNA ligand of vassopressin, *Proc. Natl. Acad. Sci. USA*, 94:11285–11290, 1997.
5. Proske D, Höfliger M, Söll RM, Beck-Sickinger AG, Farmulok M, A Y2 receptor-mimetic aptamer directed against neuropeptide Y, *J. Biol. Chem.*, 277:11416–1646, 2002.
6. Jellinek D, Lynott CK, Rifkin DB, Janjic N, High-affinity RNA ligands to basic fibroblast growth factor inhibit receptor binding, *Proc. Natl. Acad. Sci. USA*, 90:11227–11231, 1993.
7. Ruckman J, Green LS, Beeson J, Waugh S, Gillette WL, Henninger DD, Claesson-Welsh L, Janjic N, 2′-Fluoropyrimidine RNA-based aptamers to the 165-amino acid form of vascular endothelial growth factor ($VEGF_{165}$), *J. Biol. Chem.*, 273:20556–20567, 1998.
8. Green LS, Jellinek D, Jenison R, Eldin CH, Janjic N, Inhibitory DNA ligands to platelet-derived growth factor B-chain, *Biochemistry*, 35:14413–14424, 1996.

9. Hamm J, Characterisation of antibody-binding RNAs selected from structurally constrained libraries, *Nucleic Acids Res.*, 24:2220–2227, 1996.
10. Tuerk C, Macdougal S, Gold L, RNA pseudoknots that inhibit human immunodeficiency virus type-1 reverse transcriptase, *Proc. Natl. Acad. Sci. USA*, 89:6988–6992, 1992.
11. Nishikawa F, Kakiuchi N, Funaji K, Fukuda K, Sekiya S, Nishikawa S, Inhibition of HCV NS3 protease by RNA aptamers in cells, *Nucleic Acids Res.*, 31:1935–1943, 2003.
12. O'Connell D, Koenig A, Jennings S, Hicke B, Han H-L, Fitzwater T, Chang Y-F, Varki N, Parma D, Varki A, Calcium-dependent oligonucleotide antagonists against L-selectin, *Proc. Natl. Acad. Sci. USA*, 93:5883–5887, 1996.
13. Jenison RD, Jennings SD, Walker DW, Bargatze RF, Oligonucleotide inhibitors of P-selectin-dependent neutrophil-platelet adhesion, *Antisense Nucleic Acid Drug Dev*, 8:265–279, 1998.
14. Lupold SE, Hicke BJ, Lin Y, Coffey DS, Identification and characterization of nuclease-stabilized RNA molecules that bind human prostate cancer cells *via* the prostate-specific membrane antigen, *Cancer Res.*, 60:5237–5243, 2000.
15. Bartel DP, Szostak JW, Isolation of new ribozymes from a large pool of random sequences, *Science*, 261:1411–1418, 1993.
16. Robertson MP, Ellington AD, Design and optimization of effector-activated ribozyme ligases, *Nucleic Acids Res.*, 28:1751–1759, 2000.
17. Morris KN, Jensen KB, Julin CM, Weil M, Gold L, High affinity ligands from *in vitro* selection: Complex targets, *Proc. Natl. Acad. Sci. USA*, 95:2902–2907, 1998.
18. Ulrich H, Ippolito JE, Pagan OR, Eterovic VE, Hann RM, Shi H, Lis JT, Eldefrawi ME, Hess GP, *In vitro* selection of RNA molecules that displace cocaine from the nicotinic acetylcholine receptor, *Proc. Natl. Acad. Sci. USA*, 95:14051–14056, 1998.
19. Hess GP, Ulrich H, Breitinger H-G, Niu L, Gameiro AM, Grewer C, Srivastava S, Ippolito JE, Lee SM, Jayaraman V, Coombs SE, Mechanism-based discovery of ligands that prevent inhibition of the nicotinic acetylcholine receptor, *Proc. Natl. Acad. Sci. USA*, 97:13895–13900, 2000.
20. Daniels DA, Sohal AW, Rees S, Grisshammer R, Generation of RNA aptamers to the G-protein-coupled receptor for Neurotensin, NTS-1, *Anal. Biochem*, 305:214–226, 2002.
21. Blank M, Weinschenk T, Priemer M, Schluesener H, Systematic evolution of a DNA aptamer binding to rat tumor brain microvessels. Selective targeting of endothelial regulatory protein pigpen, *J. Biol. Chem.*, 276:16464–16468, 2001.
22. Homann M, Göringer H, Combinatorial selection of high affinity RNA ligands to live African trypanosomes, *Nucleic Acids Res.*, 27:2006–2014, 1999.
23. Ulrich H, Magdesian MH, Alves MJM, Colli W, *In vitro* selection of RNA aptamers that bind to cell surface receptors of *Trypanosoma cruzi* and inhibit cell invasion, *J. Biol. Chem.*, 277:20756–62, 2002.
24. Carroll FI, Howell LL, Kuhar MJ, Pharmacotherapies for treatment of cocaine abuse: preclinical aspects, *J. Med. Chem.*, 42:2721–2736, 1999.
25. Lena C, Changeux JP, Allosteric modulations of the nicotinic acetylcholine receptor, *Trends Neurosci.*, 16:181–186, 1993.
26. Zuker M, Mathews DH, Turner DH, Algorithms and Thermodynamics for RNA Secondary Structure Prediction: A Practical Guide. In: J Barciszewski &

BFC Clark, eds, *RNA Biochemistry and Biotechnology*, NATO ASI Series, Kluwer Academic Publishers, pp. 11–43, 1999.

27. Hess GP, Gameiro AM, Schoenfeld RC, Chen Y, Ulrich H, Nye JA, Caroll FI, and Ganem B, Reversing the action of non-competitive inhibitors (MK-801 and cocaine) on a protein (nicotinic acetylcholine receptor) mediated reaction, *Biochemistry*, 42:6106–6114, 2003.
28. Chaloin L, Lehmann MJ, Sczakiel G, Restle T, Endogeneous expression of a high-affinity pseudoknot RNA aptamer suppresses replication of HIV-1, *Nucleic Acids Res.*, 30:4001–4008, 2002.
29. Blind M, Kolanus W, Famulok M, Cytoplasmatic RNA modulators of an inside-out signal transduction cascade, *Proc. Natl. Acad. Sci. USA*, 96:3606–3610, 1999.
30. Mayer G, Blind M, Nagel W, Bohm T, Knorr T, Jackson CL, Kolanus W, Famulok M, Controlling small guanine-nucleotide exchange factor function through cytoplasmatic RNA intramers, *Proc. Natl. Acad. Sci. U.S.A.*, 98:4961–4965, 2000.
31. Shi H, Hoffman BE, Lis JT, RNA aptamers as effective protein antagonists in a multicellular organism, *Proc. Natl. Acad. Sci. USA*, 96:10033–10038, 1999.
32. Soukup GA, Breaker RR, Nucleic acid molecular switches, *Trends Biotechnol.*, 17:469–476, 1999.
33. Vuyisich M, Beal PA, Controlling protein activity with ligand-regulated RNA aptamers, *Chem. Biol.*, 9:907–913, 2002.
34. Suess B, Hanson S, Berens C, Fink B, Schroeder R, Hillen W, Conditional gene expression by controlling translation with tetracycline-binding aptamers, *Nucleic Acids Res.*, 31:1853–1858, 2003.
35. Xu W, Ellington AD, Anti-peptide aptamers recognize amino acid sequence and bind a protein epitope, *Proc. Natl. Acad. Sci. USA*, 83:7475–7480, 1996.
36. Jayasena SD, Aptamers, an emerging class of molecules that rival antibodies in diagnostics, *Clin. Chem.*, 45:1628–1650, 1999.
37. Fredriksson S, Gullberg M, Jarvius J, Olsson C, Pietras K, Gustafsdottir SM, Ostman A, Landegren U, Protein detection using proximity-dependent DNA ligation assays, *Nat. Biotechnol.*, 20:473–477, 2002.
38. Davis KA, Lin Y, Abrams B, Jayasena SD, Staining of cell surface CD4 with 2′-F-pyrimidine-containing RNA aptamers for flow cytometry, *Nucleic Acids Res.*, 26:3915–324, 1998.
39. Charlton J, Sennello J, Smith D, *In vivo* imaging using an aptamer inhibitor of human neutrophil elastase, *Chem. Biol.*, 4:809–816, 1997.
40. Dougan H, Lyster DM, Vo CV, Stafford A, Weitz JI, Hobbs JB, Extending the lifetime of anticoagulant oligodeoxynucleotide aptamers in blood, *Nucl. Med. Biol.*, 27:289–297, 2000.
41. Dougan H, Weitz JI, Stafford AR, Gillespie KD, Klement P, Hobbs JB, Lyster DM, Evaluation of DNA aptamers directed to thrombin as potential thrombus imaging agents, *Nucl. Med. Biol.*, 30:61–72, 2003.
42. Golden MC, Collins BD, Willis MC, Koch TH, Diagnostic potential of photoSELEX-evolved ssDNA aptamers, *J. Biotechnol.*, 81:167–178, 2000.
43. Willis MC, Hicke BJ, Uhlenbeck OC, Cech TR, Koch TH, Photocrosslinking of 5-iodouracil-substituted RNA and DNA to proteins, *Science*, 262:255–227, 1993.

44. Meisenheimer KM, Meisenheimer PL, Willis MC, Koch TH, High yield photocrosslinking of a 5-iodocytidine (IC) substituted RNA to its associated protein, *Nucleic Acids Res.*, 24:981–982, 1996.
45. Ito Y, Teramoto N, Kawazoe N, Inada K, Imanishi Y, Modified nucleic acid for systematic evolution of RNA ligands by exponential enrichment, *J. Bioactive Compatible Polymers*, 13:114–123, 1998.
46. Kusser W, Chemically modified nucleic acid aptamers for *in vitro* selections, evolving evolution, *J. Biotechnol.*, 74:27–38, 2000.
47. Faria M, Ulrich H, The use of synthetic oligonucleotides as protein inhibitors and anticode drugs in cancer therapy: accomplishments and limits, *Current Cancer Drug Targets*, 2:355–36, 2002.
48. Floege J, Ostendorf T, Janssen U, Burg M, Radeke HH, Vargeese C, Gill SC, Green LS, Janjic N, Novel approach to specific growth factor inhibition *in vivo*, Antagonism of platelet-derived growth factor in glomerulonephritis by aptamers, *Am. J. Pathol.*, 154:169–179, 1999.
49. Carrasquillo KG, Ricker JA, Rigas IK, Miller JW, Gragoudas ES, Adamis AP, Controlled delivery of the anti-VEGF aptamer EYE001 with poly (lactic-o-glycolic) acid microspheres, *Invest Ophtalmol. Vis. Sci.*, 44:290–299, 2003.
50. Green LS, Jellinek D, Bell C, Beebe LA, Feistner BD, Gill SC, Jucker FM, Janjic N, Nuclease-resistant nucleic acid ligands to vascular permeability factor/vascular endothelial growth factor, *Chem. Biol.*, 2:683–695, 1995.
51. Huang J, Moore J, Soffer S, Kim E, Rowe D, Manley CA, O'Toole K, Middlesworth W, Stolar C, Yamashiro D, Kandel J, Highly specific antiangiogenic therapy is effective in suppressing growth of experimental Wilms tumors, *J. Pediatr. Surg.*, 36:357–361, 2001.
52. Kim ES, Serur A, Huang J, Manley CA, McCrudden KW, Frischer JS, Soffer SZ, Ring L, New T, Zabski S, Rudge JS, Holash J, Yancopoulos GD, Kandel JJ, Yamashiro DJ, Potent VEGF blockade causes regression of coopted vessels in a model of neuroblastoma, *Proc. Natl. Acad. Sci. USA*, 99(17):11399–404, 2002.
53. Pietras K, Ostman A, Sjoquist M, Buchdunger E, Reed RK, Heldin CH, Rubin K, Inhibition of platelet-derived growth factor receptors reduces interstitial hypertension and increases transcapillary transport in tumors, *Cancer Res.*, 61:2929–2934, 2001.
54. Giver L, Bartel DP, Zapp ML, Green MR, Ellington AD, Selection and design of high-affinity RNA ligands for HIV-1 Rev, *Gene*, 137:19–24, 1993.
55. Allen P, Worland S, Gold L, Isolation of high-affinity RNA ligands to HIV-integrase from a random pool, *Virology*, 209:327–336, 1995.
56. Hicke BJ, Marion C, Chang YF, Gould T, Lynott CK, Parma D, Schmidt PG, Warren S, Tenascin-C aptamers are generated using tumor cells and purified protein, *J. Biol. Chem.*, 276:48644–48654, 2001.
57. Hicke BJ, Watson SR, Koenig A, Lynott CK, Bargatze RF, Chang YF, Ringquist S, Moon-McDermott L, Jennings S, Fitzwater T, Han HL, Varki N, Albinan I, Willis MC, Parma D, DNA aptamers block L-selectin function *in vivo*, Inhibition of human lymphocyte trafficking in SCID mice, *J. Clin. Invest*, 98:2688–2692, 1996.
58. Wiegand TW, Williams PB, Dreskin SC, Jouvin MH, Kinet JP, Tasset DM, High-affinity oligonucleotide ligands to human IgE inhibit binding to Fc epsilon receptor, *J. Immunol.*, 157:221–230, 1996.

59. Kubik MF, Bell C, Fitzwater T, Watson SR, Tasset DM, Isolation and characterization of 2′fluoro, 2′amino, and 2′fluoro/amino-modified RNA ligands to human IFN gamma that inhibit receptor binding, *J. Immunol.*, 159:259–267, 1997.
60. Kraus E, James W, Barclay AN, Cutting edge: Novel RNA ligands able to bind CD4 antigen and inhibit CD4+ T lymphocyte function, *J. Immunol.*, 160:5209–5212, 1998.
61. Lee SW, Sullenger BA, Isolation of a nuclease-resistant decoy RNA that can protect human acetylcholine receptors from myasthenic antibodies, *Nature Biotechnol.*, 15:41–45, 1997.
62. Hwang B, Lee SW, Improvement of RNA aptamer activity against myasthenic autoantibodies by extended sequence selection, *Biochem. Biophys. Res. Commun.*, 290:656–662, 2002.
63. Binkley J, Allen P, Brown DM, Green L, Tuerk C, Gold L, RNA ligands to human nerve growth factor, *Nucleic Acids Res.*, 23:3198–3205, 1995.
64. Nieuwlandt D, Wecker M, Gold L (1995), *In vitro* selection of RNA ligands to substance P, *Biochemistry*, 34:5651–5659, 1995.
65. Tucker CE, Chen LS, Judkins MB, Farmer JA, Gill SC, Drolet DW, Detection and pharmacokinetics of an anti-vascular endothelial growth factor oligonucleotide-aptamer (NX1838) in rhesus monkeys, *J. Chromatogr. B. Biomed. Sci. Appl.*, 732:203–212, 1999.
66. Sun S, Technology evaluation: SELEX, Gilead Sciences Inc, *Curr. Opin. Mol. Ther.*, 2:100–105, 2000.
67. Eyetech Study Group, Preclinical and phase 1A clinical evaluation of an anti-VEGF pegylated aptamer (EYE001) for the treatment of exudative age-related macular degeneration, *Retina*, 22:143–152, 2002
68. Watson SR, Chang YF, O'Connell D, Weigand L, Ringquist S, Parma DH, Anti-L-selectin aptamers: Binding characteristics, pharmacokinetic parameters, and activity against an intravascular target *in vivo*, *Antisense Nucleic Acid Drug Dev.*, 10:63–75, 2000.
69. Conrad R, Keranen LM, Ellington AD, Newton AC, Isozyme-specific inhibition of protein kinase C by RNA aptamers, *J. Biol. Chem.*, 269:32051–32054, 1994.
70. Sassanfar M, Szostak JW, An RNA motif that binds ATP, *Nature*, 364:550–553, 1993.
71. Willis MC, Collins BD, Zhang T, Green LS, Sebesta DP, Bell C, Kellogg E, Gill SC, Magallanez A, Knauer S, Bendele RA, Janjic N, Liposome-anchored vascular endothelial growth factor aptamers, *Bioconj. Chem.*, 9:573–582, 1998.
72. Gallo M, Monteagudo E, Cicero DA, Torres HN, Iribarren AM, M, 2′-C-methyluridine phosphoramidite: A new building block for the preparation of RNA analogues carrying the 2′-hydroxyl group, *Tetrahedron*, 57:5707–5713, 2001.
73. Uhrbom L, Hesselager G, Ostman A, Nister M, Westermark B, Dependence of autocrine growth factor stimulation in platelet-derived growth factor-B-induced mouse brain tumor cells, *Int. J. Cancer*, 85:398–406, 2000.
74. Niu L, Abood LG, Hess GP, Cocaine: Mechanism of inhibition of a muscle acetylcholine receptor studied by a laser-pulse photolysis technique, *Proc. Natl. Acad. Sci. USA*, 92:12008–12012, 1995.

75. Balasubramanian V, Nguyen LT, Balasubramanian SV, Ramanathan M, Interferon-gamma-inhibitory oligodeoxynucleotides alter the conformation of interferon-gamma, *Mol. Pharmacol.*, 53:926–932, 1998.
76. Bai J, Banda N, Lee NS, Rosi J, Akkina R, RNA based anti-HIV-1 gene therapeutic constructs in SCID-hu mouse model, *Mol. Ther.*, 6:770–782, 2002.
77. Fukuda K, Vishnuvardan D, Sekiya S, Hwang J, Kakiuchi N, Taira K, Shimotohno K, Kumar PKR, Nishikawa S, Isolation and characterization of RNA aptamers specific for the hepatitis C virus NS3 protease, *Eur. J. Biochem.*, 267:3685–3694, 2000.

22 Combinatorial Proteomic

Maria Marino, Luca Beneduce, Daniela Palomba, Antonio Fiordelisi and Giorgio Fassina

CONTENTS

I. INTRODUCTION

During the past century, the clinical behavior of human cancer has been predicted using its microscopic appearance. This has been a useful approach because cancer has hundreds of "faces" under the microscope, but unfortunately "looks" alone can only predict general categories of biological behavior [1]. Tumors with a similar histological appearance can follow significantly different clinical courses and show different responses to therapy [2]. For example, prostate carcinoma arising in two patients may look virtually identical under the microscope, but each patient may have a different clinical outcome. Given the wide diversity of tumors, even for those derived from the same tissue, additional methods of classification are urgently required.

The widespread application of serum markers for cancer screening, like prostate-specific antigen (PSA) for prostate cancer and Carcino-Embryonic Antigen (CEA) for colon cancer, has led to the perception that specific and accurate markers, yet to be discovered, will be the answer to early diagnosis and prognosis.

The identification and development of tumor biomarkers have challenged the research for over 150 years. Although there have been some successes, a biomarker that is useful in the early diagnosis of disease in otherwise asymptomatic individuals has proved to be an elusive goal. The majority of the existing tumor markers are most useful in educating a clinical decision-making process after an initial suspicion has been raised by more conventional means [3].

An ideal tumor marker would be a molecule specific to one type of tumor (100% specificity, i.e. no false positives) and detectable right from the initial stage of the disease (100% sensitivity, i.e. no false negatives). It would be undetectable in healthy subjects, and enable the screening and diagnosis of cancer. The tumor marker level should correlate closely with tumor size, contribute to the initial extension of the profile and evaluation of therapeutic efficacy, as well as the early detection of recurrent diseases.

The pursuit of the ideal tumor marker has generated many tests for use in the diagnosis and management of cancer, several of which are now widely available [4], but no tumor marker meets all of those requirements, hence no test is highly confident.

Listed in Table 22.1 are different tumor markers for certain types of cancer, with their primary applications and the techniques related with their discovery process. Among them the alpha 1 fetoprotein and CEA are the most frequently exploited.

Alpha 1 FetoProtein (AFP) is normally produced during fetal and neonatal development by the liver, yolksac, and in small concentrations by the gastrointestinal tract. After birth, serum AFP concentration decreases rapidly, and by the second year of life and thereafter only trace amounts are normally detected in serum.

The significance of elevated serum alpha-fetoprotein in testicular seminoma was discovered in the 1980s by serial serum determinations and cellular localization of the protein utilizing radioimmunoassay and immunocytochemical techniques [5]. In the case of nonseminomatous testicular cancer, a direct relationship has been observed between the incidence of elevated AFP levels and the stage of disease [6]. Elevation of serum AFP to abnormally high values also occurs in hepatocellular carcinoma (HCC), but does not appear to play an important role in the diagnosis of HCC; its use, in fact, is limited because at least one-third of small HCC and 10% of the advanced HCC are missed [7]. Elevated serum AFP concentrations have been measured in patients with other noncancerous diseases, including ataxia telangiectasia [8], hereditary tyrosinemia [9], neonatal hyperbilirubinemia [10], acute viral hepatitis [11], chronic active hepatitis [12], and cirrhosis [12]; additionally elevated serum AFP concentrations are also

TABLE 22.1
Tumor biomarkers

Tumor marker	Primary cancer site	Primary application	Discovery year	False Positives
Alpha-fetoprotein (AFP)	Hepatocellular, germ cell tumors (testis)	Diagnosis, monitoring	1980 [5]	Cirrhosis, hepatitis
Carcinoembryonic antigen (CEA)	Gastrointestinal, breast, other adenocarcinomas	Monitoring	1965 [14]	Pancreatitis, hepatitis, inflammatory bowel disease, smoking (*About 5% of population has above normal CEA*)
CA 125	Ovarian	Monitoring, prognosis	1983 [22]	Menstruation, peritonitis, pregnancy
Prostate specific antigen (PSA)	Prostate	Screening, diagnosis, monitoring	1979 [23]	Prostatitis, benign prostatic hypertrophy
CA 19-9	Pancreatic	Monitoring	1975 [29]	Pancreatitis, ulcerative colitis

Adapted from Ref. [30]

observed in pregnant women [13]. This allows a limited use of AFP as molecular marker.

CEA was one of the first oncofetal antigens to be described and exploited clinically. This was first identified in 1965 by Gold P. *et al.* [14] in colon cancer, but elevated levels are subsequently found in many other cancers including pancreatic [15], gastric [16], lung [17] and breast [18]. It is also detected in benign conditions including cirrhosis, inflammatory bowel disease [19], chronic lung disease [20], and pancreatitis [21]. Although CEA has been the subject of interest for many investigators for over 30 years, many questions still remain unanswered concerning the CEA molecule and especially its clinical potential as a tumor marker related to its specificity, sensitivity and distribution.

Cancer Antigen (CA) 125 is a tumor-associated antigen present on 80 percent of nonmucinous ovarian carcinoma [22]. It is used for the follow-up of epithelial ovarian cancer [23]; with surgical resection or chemotherapy, the level correlates with patient response. Other malignant pathologies including endometrial, pancreatic [15], lung, breast, and colon cancer, as well as benign pathologies such as cardiovascular and chronic liver disease [24], are diagnosed in patients with increased CA 125 levels. This supports the opinion that CA-125 lacks utility as a marker for malignancy, but could rather have a role in the follow-up of cardiovascular, hepatic and tumoral diseases with serosal involvement [24].

Between tumor markers the Prostate-Specific Antigen (PSA) seems to be the only one to have the capability of achieving at least one of the characteristics of an ideal tumor marker: the tissue specificity [25]. Actually, PSA testing is used for: 1, the evaluation of men at risk for prostate cancer [26, 27], 2, assistance in pretreatment staging [27], and 3, the post-treatment monitoring and management of men with this disease [27]. Due to its low sensitivity (false negatives) and specificity (false positives), PSA-based screening for cancer remains a controversial issue because it is not yet known if the process actually saves lives [28].

All markers listed in Table 22.1 have been discovered by immunometric techniques by using antibodies raised against cell lines established from human cancerous tissues (i.e. CA 125) [22, 23], or against cells derived from cancer (i.e. CEA) [14], or against cells from normal tissue (PSA) [25]. They in turn have generated several products which are commercially available in kits for blood determination of antigens, such as CEA and AFP that have generated more than twenty kits, or only four in the case of PSA free fraction determination.

II. NEW TRENDS IN BIOMARKERS DISCOVERY

Advances in genomics and proteomics methodologies, directed at identifying those genes expressed at elevated levels in tumors compared to normal tissues, have led and accelerated the identification process of many potentially useful tumor biomarkers for the diagnosis and therapy.

Nowadays the genomics-based techniques mainly employed include microarrays [31], serial analysis of gene expression (SAGE) [32] and circulating nucleic acids isolated from plasma or serum (CNAPS) [33].

A. MICROARRAYS

DNA microarray technology provides a method for monitoring the RNA expression levels of many thousands of genes simultaneously in primary tumors and cell lines. In principle, the emerging of such technology could make possible the development of better methods for the diagnosis and treatment of cancer, allowing researchers to design treatment aimed at the specific mutations present in each patient. The rationale behind this proposition is that the overall behavior of a cancer must be determined by the expression of the genes within it. It should, therefore, be possible to identify sets of genes whose expression or lack of expression defines each individual property of a tumor or, in general, of a disease, including its precise diagnosis and clinical behavior [31, 34, 35].

B. SERIAL ANALYSIS OF GENE EXPRESSION (SAGE)

SAGE is a powerful genetic profiling technology that enables qualitative and quantitative assessment of gene transcript populations. It can be applied to any cell line or tissue. Comparison of gene expression levels in neoplastic tissues with those seen in nonneoplastic tissues can, in turn, identify novel tumor markers. Several potential tumor markers have been identified in a wide variety of organ system by SAGE analysis [32, 36–40]. An example is the Prostate Stem Cell Antigen (PSCA) that has been discovered by SAGE analysis; it has been demonstrated to be overexpressed in prostatic carcinoma where it is correlated with high stage of this cancer, and in pancreatic carcinoma where a role as biomarker has been postulated [36].

C. Circulating Nucleic Acids Isolated from Plasma or Serum (CNAPS)

The recent discovery that cell-free DNA can be shed into the bloodstream as a result of tumor cell death has generated great interest. The *2nd International Symposium on Circulating Nucleic acids in Plasma and Serum*, held in Hong Kong in 2001, presented numerous studies on tumor-specific alterations in DNA recovered from plasma or serum of patients with various malignancies, providing reports on the microsatellite alterations, tumor associated viral presence and detection and quantification of mRNA associated with malignant diseases [33]. The implication related to this technique is that tumor-derived nucleic acids of human or viral origin can be retrieved from blood by a minimally invasive procedure, and used for molecular diagnosis and prognosis [41].

D. Proteomic

Significant improvements in the technologies of high-resolution two-dimensional Polyacrylamide Gel Electrophoresis (2-D PAGE) and Mass Spectrometry (MS) have marked the start of proteome analysis. Proteomics permits the analysis of thousand of proteins simultaneously, and have the potential to identify markers for early detection, classification and prognosis of diseases, as well as pinpointing targets for improved treatment outcomes [42].

The need for technologies that allow highly parallel quantization of specific proteins in a rapid, low-cost and low-sample volume format has led to the implementation of strategies for identification and quantification of proteins from biological samples [43–45]. Antibody and protein arrays are now widely used; beside them a new platform technology is emerging: the Combinatorial Proteomic™.

III. COMBINATORIAL PROTEOMIC™

In the postgenomic era, the greatest effort is to face the unprecedented challenge of assigning molecular and cellular functions to thousands of newly predicted gene products and explaining how these products cooperate in complex physiological processes.

To address this problem, recently a new strategy for proteome analysis has emerged. This technology, named Combinatorial Proteomic™, uses antibody libraries as probes to profile the expression and function of protein families in complex proteomes. The use of antibodies allows the detection of iper- and ipo-expressed proteins, even if they are at pico-quantity level, overcoming one of the proteomic limitations: of difficulty in detecting low abundance proteins [46, 47].

Combinatorial Proteomic™ shares with Proteomic the goals of developing and applying technologies for the global analysis of protein expression and function. Their greatest therapeutic value lies in the comparison of cells from normal tissue with those representing a disease state. Such comparisons enable the identification of disease-specific biomarkers that could be used for diagnostic tests, or target proteins that have the potential for drug intervention.

The striking feature of Combinatorial Proteomic™ is to exploit the natural antibody diversity in order to select specific antibodies with high avidity for selected protein families.

A single individual may produce a population of antibody specificities, an antibody repertoire, which is a reflection of all the B cell clones (lymphocyte repertoire) capable of immunoglobulin (Ig) synthesis and secretion in response to antigenic stimulation, but also in absence of exposure to environmental pathogens [48, 49]. Natural antibodies, in fact, are germ line-encoded molecules produced by a distinct population of peritoneal B cells, bearing the cell surface marker CD5 and are present in the sera and interstitial fluids of healthy individuals [49–51].

The majority of natural antibodies are immunoglobulin M isotype; they are polyreactive, with various affinities for multiple antigens, including pathogens and toxins [52, 53]. Although their contribution to immune defense has only recently been appreciated, natural antibodies are able to kill bacteria *in vitro* [53] and help clearance of lipopolysaccarides *in vivo* [54, 55]. In addition they facilitate uptake, processing and presentation of antigen by B cells [52, 56] and may help localize pathogens and their antigens to lymphoid organs [50, 51, 53]. In fact, several findings suggest a role for natural antibodies in limiting the initial pathogen burden prior to the development of adaptive immune response [53].

Antibodies from both natural or adaptive immunity are characterized by an extraordinary diversity of structure, considering there are more than 1×10^7 and perhaps as many as 10^9 structurally different antibody there are molecules present in every individual, each with unique amino acid sequences in their antigen-combining sites [57, 58].

This structure diversity, genetically defined, accounts for the extraordinary specificity of antibodies for antigens, because each amino acid difference may produce a difference in antigen binding [59], since such extensive sequence diversity is confined to hypervariable regions.

Two parameters describe the physicochemical characteristics of antigen-antibody binding: affinity and avidity. In a simplified system consisting of an antigen that has only one determinant per molecule and a population of identical antibody molecules specific for this determinant, the affinity may be described as a measure of the strength of binding, and is defined as the concentration of antigen that allows one half of the antibodies to be in complex with antigen and leave one half free.

In a natural serum there is a mixture of antibodies specific for natural antigens, with more than one determinant interacting with different antibody-combining site. In fact, unless inhibited by steric constraints, a single antibody molecule may be able to attach to a single multivalent antigen by more than one binding site. For IgG and IgE, this attachment can involve, at most, two binding sites because there are only two combining regions per antibody molecule, one on each Fab. For IgM (Figure 22.1), however, a single antibody may bind at up to ten different sites. Although the affinity of any one site is unchanged, the overall strength of attachment must take into account binding at all of the sites. This overall strength of attachment is called avidity and is much stronger than the affinity of any given site. In this way a low-affinity IgM molecule can still bind very tightly to a multivalent antigen because many low-affinity interactions can produce a single high-avidity interaction.

Such a feature of IgM makes it the most suitable molecule to be used in the Combinatorial Proteomic™ technology.

FIGURE 22.1 Pentameric structure of IgM: IgM is composed of a pentamer of the basic four chain structure held together by inter H chain disulfide bonds. The heavy chain (μ) has four constant domains. A small, cysteine rich protein called J chain initiates cross linking of C3 and C4 of five IgM monomers to make the circulating, pentameric form of IgM.

IV. COMBINATORIAL PROTEOMIC™ OPERATIONAL PROCEDURE

Combinatorial Proteomic™ is a process with five main phases, shown in Figure 22.2:

1. *in silico* analysis of human proteome using the pattern matcher PatScan and Prosite database
2. combinatorial synthesis of selected signature libraries by Fmoc chemistry and Mix & Split methods
3. immobilization on solid support of selected signature libraries
4. purification of signature-specific IgM antibodies by affinity chromatography experiments
5. differential analysis of cellular protein expression levels

1. *In Silico* Analysis of Human Proteome

The human proteome refers to the whole array of proteins that are found in widely varying amounts in different human cells and tissues. The number of different proteins is enormous, perhaps as many as 1,000,000 in humans. The Prosite database groups these different proteins, on the basis of similarities in their sequences into a limited number of families sharing functional domains, assigning them specific signatures [60]. A further comparative characterization of conserved protein regions by PatScan software allows obtaining undersized signatures characteristic for a given family. The undersized signatures should match almost the same members of original families and identify synthesizable peptide libraries [61]. An example of undersized signatures is reported in Table 22.2.

In the example described, the human proteome has been divided in three distinct groups—membrane proteins, cytosolic proteins and nuclear proteins. Each group can be furthermore described in subfamily by the undersized signatures in silico identified as reported for membrane proteins in Table 22.3, where it is shown that 26 Prosite signatures are derived from 519 membrane proteins, and 33 undersized signatures are obtained by PatScan analysis which identifies 80% of the original proteins.

2. Combinatorial Synthesis of Selected Signature Libraries

The selected signature libraries could be obtained by using Mix and Split solid phase combinatorial method [62] or Pre-mix method [63–65]; both allow to rapidly synthesize thousands of combinations of molecules.

The mix and split method involves building libraries by randomly combining various amino acids in a step-by-step fashion until the appropriate peptide length is synthesized. Different combinations are obtained by splitting collections of beads and mixing them. The achieved collection is

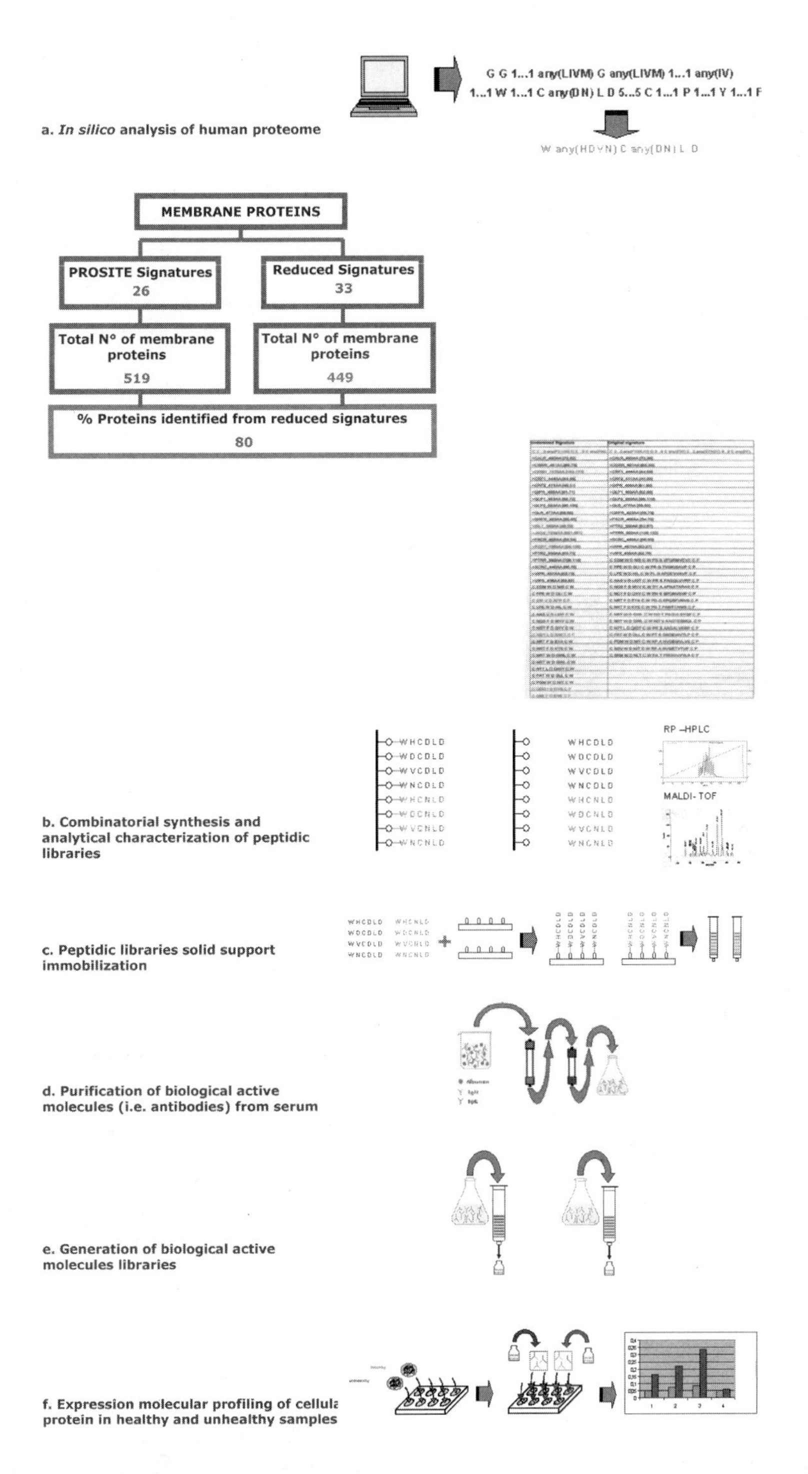

FIGURE 22.2 Schematic representation of steps involved in Combinatorial Proteomic®.

TABLE 22.2
Undersized signatures for ATP P2X receptors

	ATP P2X receptors **P2X PURINORECEPTORS 1–7**
SIGNATURE	G G 1...1 any(LIVM) G any(LIVM) 1...1 any(IV) 1...1 W 1...1 C any(DN) L D 5...5 C 1...1 P 1...1 Y 1...1 F
	>P2X1_399AA:[250, 276]
	G G V V G I T I D W H C D L D WHVRH C R P I Y E F
	>P2X2_471AA:[259, 285]
	G G V I G V I I N W D C D L D LPASE C N P K Y S F
	>P2X3_397AA:[236, 262]
	G G V L G I K I G W V C D L D KAWDQ C I P K Y S F
	>P2X4_388AA:[250, 276]
	G G I M G I Q V N W D C N L D RAASL C L P R Y S F
	>P2X5_421AA:[251, 277]
	G G V I G I N I E W N C D L D KAASE C H P H Y S F
	>P2X6_431AA:[251, 277]
	G G S VG I R V H W D C D L D TGDSG C W P H Y S F
	>P2X7_595AA:[249, 275]
	G G I M G I E I Y W D C N L D RWFHH C R P K Y S F
UNDERSIZED SIGNATURE	W any(HDVN) C any(DN) L D
	>P2X1_399AA:[259, 264]
	W H C D L D
	>P2X2_471AA:[268, 273]
	W D C D L D
	>P2X3_397AA:[245, 250]
	W V C D L D
	>P2X4_388AA:[259, 264]
	W D C N L D
	>P2X5_421AA:[260, 265]
	W N C D L D
	>P2X6_431AA:[260, 265]
	W D C D L D
	>P2X7_595AA:[258, 263]
	W D C N L D

composed of n^x equimolar peptides, where n is the number of monomers used and x is the length of the library.

3. Immobilization on Solid Support of Selected Signature Libraries

Selected signature libraries may be immobilized on a solid matrix such as activated silica resin, cellulose microporous modified membranes [66], Sepharose®, magnetic beads based on MagaPhase™ technology. The affinity support obtained is used for IgM antibodies parting.

TABLE 22.3
Proteome analysis of membrane proteins

G-protein coupled receptors family 1	
1.Signature Prosite	any(GSTALIVMFYWC) any(GSTANCPDE) notany(EDPKRH) 2...2any(LIVMNQGA)2...2any(LIVMFT) any(GSTANC)any(LIVMFYWSTAC)any(DENH)Rany(FYWCSH)2...2any(LIVM)
Undersized Signature1	D R Y any(AWRVIYLFTCM) any(ASLGRCTN) any(IVL)
Undersized Signature2	any(ILVAYFST) D R any(CFHSW) any(ILVAQTRY) any(CSAQVL) any(IV)
G-protein coupled receptors family 2	
2.Signature Prosite	C 3...3 any(FYWLIV) D 3...4 C any(FW) 2...2 any(STAGV) 8...9 C any(PF)
Undersized Signature1	C 3...3 any(FWLIV) D 3...4 C any(FW)
3.Signature Prosite	Q G any(LMFCA) any(LIVMFT) any(LIV) 1...1 any(LIVFST) any(LIF) any(VFYH) C any(LFY)1..1 N 2..2 V
Undersized Signature1	Q G any(FAL) any(VFLIM) any(VLI) any(YASF) any(TLVI) any(ILF)
G-protein coupled receptors family 3	
4.Signature Prosite	any(LV)1..1N any(LIVM)any(LIVM)1..1LF1..1 I any(PA)Qany(LIVM)any(STA)1..1any(STA)any(STAN)
Undersized Signature1	V any(AQ) N any(ILV) L any(GQR) L
5.Signature Prosite	C C any(FYW) 1...1 C 2...2 C 4...4 any(FYW) 2...4 any(DN)2...2any(STAH) C 2...2 C
Undersized Signature1	C C any(WF) any(HTILE) C any(EVTI)
6.Signature Prosite	F N E any(STA) K 1...1 I any(STAG) F any(ST) M
Undersized Signature1	F N E A K any(FYP) I
Undersized Signature1	K any(FYP) I any(TAG) F any(ST) M
Neurotransmitter-gated ion-channels	
7.Signature Prosite	C 1...1 any(LIVMFQ) 1...1 any(LIVMF) 2...2 any(FY) P 1...1 D 3...3 C
Undersized Signature1	C 1...1 any(LIVMFQ) 1...1 any(LIVMF) 2...2 any(FY) P1...1 D
Receptor tyrosine kinase class II	
8.Signature Prosite	any(DN) any(LIV) Y 3...3 Y Y R
Undersized Signature1	Y any(ASERK) any(TNAG) any(DS) Y Y R
Receptor tyrosine kinase class III	
9.Signature Prosite	G 1...1 H 1...1 N any(LIVM) V N L L G A C T
Undersized Signature1	V N L L G A
Receptor tyrosine kinase class V	
10.Signature Prosite	F 1...1 any(DN) 1...1 any(GAW) any(GA) C any(LIVM) any(SA) any(LIVM)any(LIVM)any(SA)any(LV) any(KRHQ)any(LIVA) 3...3any(KR)any(PSAW)
Undersized Signature1	F any(HQ) any(ND) any(QYIVP) G any(GA) C
Undersized Signature2	C any(LIVM) any(SA) any(LI) any(LIV) any(SA) any(LV) any(KRH) any(LIVA)
11.Signature Prosite	C 2...2 any(DE) G any(DEQKRG) W 2...3 any(PAQ) any(LIVMT) any(GT) 1...1 C 1...1 C 2...2 G any(HFY) any(EQ)
Undersized Signature1	any(DE) G any(EQK) W 2...3 any(PA) any(IV) any(GT)
Undersized Signature2	G any(EQGK) W any(LMA) any(EVW) any(PAQ)
Guanylate cyclase coupled receptors	
12.Signature Prosite	G P 1...1 C 1...1 Y 1...1 A A 1...1 V 1...1 R 3...3 H W
Undersized Signature1	C any(VE) Y any(APA) A A any(PS) V
Undersized Signature2	G P any(GV) C any(VE) Y any(AP)
Long hematopoietin receptor, gp130 family	
13.Signature Prosite	N 4...4 S 28...35 any(LIVM) 1...1 W 0...3 P 5...9 any(FY) 1...2 any(VILM) 1...1 W
Undersized Signature1	any(GTEKMDY) any(YF) any(VIC) any(ILV) any(KE) W
Long hematopoietin receptor, soluble alpha chains family	
14.Signature Prosite	any(LIV) 1...1 P D P P 2...2 any(LIV) 8...11 any(LV) 3...3 W 2...2 P 1...1 any(ST) W 4...6 any(FY) 1...1 L 1...1 any(FY) 1...1 any(LVI)
Undersized Signature1	P D P P any(AE) N any(VI)
Short hematopoietin receptor family 1	
15.Signature Prosite	any(LIV) 9...9 any(LIV) R 9...20 W S 1...1 W S 4...4 any(FYW)
Undersized Signature1	any(IHTFQ) W S any(EPD) W S any(TQPMEH)
Undersized Signature2	W S any(EPD) W S any(TQPMEH) any(PSA)
Short hematopoietin receptor family 2	
16.Signature Prosite	any(LIVM) 1...1 C 1...1 W 2...2 G 5...5 D 2...2 Y 1...1 any(LIVM) 10...14 C
Undersized Signature1	C any(ST) W any(ALK) any(RPV) G any(PRIT)
ABC transporters family	
17.Signature Prosite	any(LIVMFYC) any(SA) any(SAPGLVFYKQH) G any(DENQMW) any(KRQASPCLIMFW) any(KRNQSTAVM) any(KRACLVM) any(LIVMFYPAN) not any(PHY) any(LIVMFW) any(SAGCLIVP) not any(FYWHP) not any(KRHP) any(LIVMFYWSTA)
Undersized Signature1	L S G G any(MEKQW) any(KRMQ) any(RQMKA)
LDL-receptor class A (LDLRA) domain	
18.Signature Prosite	C any(VILMA) 5...5 C any(DNH) 3...3 any(DENQHT)C 3...4 any(STADE) any(DEH) any(DE) 1...5 C
Undersized Signature1	S D E any(AL) any(HDNAS) C
ATP P2X receptors	
19.Signature Prosite	G G 1..1 any(LIVM) G any(LIVM) 1..1 any(IV)1..1 W 1..1 C any(DN) L D 5...5 C 1...1 P 1...1 Y 1...1 F
Undersized Signature1	W any(HDVN) C any(DN) L D
Amiloride-sensitive sodium channels	
20.Signature Prosite	Y 2..2 any(EQTF) 1..1 C 2..2 any(GSTDNL)C 1..1 any(QT)2..2 any(LIVMT) any(LIVMS) 2..2 C 1..1C
Undersized Signature1	Y any(ST) any(IQRL) any(TQ) any(AVI) C any(LRI)
TNFR/NGFR family cysteine-rich region signature	
21.Signature Prosite	C 4...6 any(FYH) 5...10 C 0...2 C 2...3 C 7...11 C 4...6 any(DNEQSKP) 2...2 C
Undersized Signature1	C C any(RTSNH) any(GKRLE) C any(PHQSER) any(AKP)
Undersized Signature2	C any(QETLKRDAF) any(LASPKHRQIN) C any(RVSTDQ) any(PVTSKWHREMQAL) C
Integrins alpha chain	
22.Signature Prosite	any(FYWS) any(RK) 1...1 G F F 1...1 R
Undersized Signature1	any(KR) any(CAILMV) G F F any(KDR) R
Integrins beta chain cysteine-rich domain	
23.Signature Prosite	C 1...1 any(GNQ) 1...3 G 1...1 C 1...1 C 2...2 C 1...1 C
Undersized Signature1	C S any(GN) any(NRH) G any(ERKDV) C
Transmembrane 4 family	
24.Signature Prosite	G 3...3 any(LIVMF) 2...2 any(GSA) any(LIVMFT) any(LIVMF)G C 1...1 any(GA) any(STAP) 2...2 any(EG) 2...2 any(CWN)any(LIVMG) any(LIVM)
Undersized Signature1	any(GA) any(FLV) any(VLIF) G C any(LCYIVMF) any(GA)
ATP1G1 / PLM / MAT8 family	
25.Signature Prosite	any(DNS) 1...1 F 1...1 Y any(DN) 2...2 any(ST) any(LIVM) any(RQ) 2...2 G
Undersized Signature1	any(DS) P F any(FHTY) Y any(DN) any(EYW)
E1-E2 ATPases phosphorylation	
26.Signature Prosite	D K T G T any(LIVM) any(TI)
Undersized Signature1	D K T G T any(LI) T

4. Purification of Signature-Specific IgM Antibodies

Prior to fractionation on columns prepared by the immobilization of signature libraries, IgM are prepared from crude serum by gel filtration or other affinity fractionation techniques. Purified IgM are then loaded on to different signature libraries columns to get the corresponding signature specific IgM fractions.

5. Differential Analysis of Cellular Proteins

Molecular analysis of cells in their native tissue environment provides the most accurate picture of the *in vivo* disease state. This goal is accomplished by protein level analysis. Creating a portrait of the protein patterns may be a more reliable indicator of specific conditions than the gene expression. Results reported from many studies [67] suggest that scientists would soon be able to monitor these patterns routinely. Although several new technologies are introduced for high-throughput protein characterization and discovery [68–70], the bulk of protein identification continues to be made by two-dimensional (2-D) gel electrophoresis. 2-D gels have traditionally required a large amount of protein starting material and generally are not practical for rapid profiling.

Comparative analysis of protein and peptide levels can be conducted by differential ELISA assay on biological sample from affected and healthy individuals. Theoretically, all diseases are studied by using molecular profiling. Specific clinical goals derived from such studies include:

Screening tests for early disease detection
Improved diagnostic markers
Improved prognostic markers
New therapeutic targets
Markers to evaluate therapeutic efficacy
New approaches and technologies for less invasive screening and diagnosis

REFERENCES

1. Holland JF, Frei E, eds. *Cancer Medicine*, 5th ed., Section 1 Williams and Wilkins, New York, 1999.
2. Alizadeh AA, Eisen MB, Davis RE, Ma C, Lossos IS, Rosenwald A, Boldrick JC, Sabet H, Tran T, Yu X, Powell JI, Yang L, Marti GE, Moore T, Houdson J, Lu L, Lewis DB, Tibshirani R, Sherlock G, Chan WC, Greiner TC, Weisenburger DD, Armitage JO, Warnke R, Levy R, Wilson W, Grever MR, Byrd JC, Botstein D, Staudt LM, Distinct types of diffuse large B-cell lymphoma identified by gene expression profiling, *Nature*, 403:503–511, 2000.
3. Nicolette CA and GA Miller, The identification of clinically relevant markers and therapeutic targets, *Drug Discov. Today*, 8:31–38, 2003.

4. Bates SE, Clinical application of serum tumor markers, *Ann. Intern. Med.*, 115:623–638, 1991.
5. Javadpour N, Significance of Elevated Serum Alpha-Fetoprotein (AFP) in Seminoma, *Cancer*, 45:2166–8, 1980.
6. Calaminus G, Schneider DT, Bokkerink JP, Gadner H, Harms D, Willers R, Gobel U, Prognostic value of tumor size, metastases, extention into bone, and increased tumor marker in children with malignant Sacrococcygeal germ cell tumors: A prospective evaluation of 71 patients treated in the German cooperative protocols Maligne Keimzelltumoren (MAKEI) 83/86 and MAKEI 89, *J. Clin. Oncol.*, 21:781–786, 2003.
7. Sherlock S Dooley, Disease of the Liver J and Biliary System, Blackwell Scientific, Oxford, pp. 503–531, 1993.
8. Jason JM, Gelfand EW, Diagnostic considerations in ataxia-telangiectasia, *Arch. Dis. Child.*, 54:682–686, 1979.
9. Pitkanen S, Salo MK, Kuusela P, Holmberg C, Simell O, Heikinheimo M, Serum levels of oncofetal markers CA 125, CA 19–9, and alpha-fetoprotein in children with hereditary tyrosinemia type I, *Pediatr. Res.*, 35:205–208, 1994.
10. Lee PI, Chang MH, Chen DS, Hsu HC, Lee CY, Prognostic implications of serum alpha-fetoprotein levels in neonatal hepatitis, *J. Pediatr. Gastroenterol. Nutr.*, 11:27–31, 1990.
11. Francioni S, M Pastore, Alpha-fetoprotein and acute viral hepatitis type B, *J. Nucl. Med. Allied. Sci.*, 33:103–6, 1989.
12. Stein DF, Myaing M, Normalization of markedly elevated alpha-fetoprotein in a virologic nonresponder with HCV-related cirrhosis, *Dig. Dis. Sci.*, 47:2686–2690, 2002.
13. Jansen MW, Brandenburg H, Wildschut HI, Martens AC, Hagenaars AM, Wladimiroff JW, Veld PA, The effect of chorionic villus sampling on the number of fetal cells isolated from maternal blood and on maternal serum alpha-fetoprotein levels, *Prenat. Diagn.*, 17:953–959, 1997.
14. Gold P, Freedman SO, Demonstration of tumor-specific antigens in human colonic carcinomata by immunological tolerance and absorption techniques, *Journal of Experimental Medicine*, 122:467–481, 1965.
15. Bassi C, Salvia R, Gumbs AA, Butturini G Falconi M, Pederzoli P, The value of standard serum tumor markers in differentiating mucinous from serous cystic tumors of the pancreas: CEA, Ca 19–9, Ca 125, Ca 15–3, *Langenbecks Arch. Surg.*, 387:281–285, 2002.
16. Fujii S, Kitayama J, Kaisaki S, Sasaki S, Seto Y, Tominaga O, Tsuno N, Umetani N, Yokota H, Kitamura K, Tsuruo T, Nagawa H, Carcinoembryonic antigen mRNA in abdominal cavity as a useful predictor of peritoneal recurrence of gastric cancer with serosal exposure, *J. Exp. Clin. Cancer Res.*, 21:547–553, 2002.
17. Buccheri G, Ferrigno D, Identifying patients at risk of early postoperative recurrence of lung cancer: A new use of the old CEA test, *Ann. Thorac. Surg.*, 75:973–980, 2003.
18. Dorvillius M, Garambois V, Pourquier D, Gutowski M, Rouanet P, Mani JC, Pugniere M, Hynes NE, Pelegrin A, Targeting of Human Breast Cancer by a Bispecific Antibody Directed against Two Tumour-Associated Antigens: ErbB-2 and Carcinoembryonic Antigen, *Tumour Biol.*, 23:337–347, 2002.

19. Radovic S, Selak I, Babic M, Pasic F, Carcinoembryonic antigen (CEA) in colonic inflammatory-regenerative and dysplastic epithelial lesions, *Croat. Med. J.*, 39:15–18, 1998.
20. Stockley RA, Shaw J, Whitfield AG, Whitehead TP, Clarke CA, Burnett D, Effect of cigarette smoking, pulmonary inflammation, and lung disease on concentrations of carcinoembryonic antigen in serum and secretions, *Thorac.*, 41:17–24, 1986.
21. Hamori J, Arkosy P, Lenkey A, Sapy P, The role of different tumor markers in the early diagnosis and prognosis of pancreatic carcinoma and chronic pancreatitis, *Acta. Chir. Hung.*, 36:125–127, 1997.
22. Bast RC Jr, Klug TL, St John E, Jenison E, Niloff JM, Lazarus H, Berkowitz RS, Leavitt T, Griffiths CT, Parker L, Jr Zurawski VR, Knapp RC, A radioimmunoassay using a monoclonal antibody to monitor the course of epithelial ovarian cancer, *N. Engl. J. Med.*, 309:883–7, 1983.
23. Canney PA, Moore M, Wilkinson PM, James RD, Ovarian cancer antigen CA 125: a prospective clinical assessment of its role as a tumor marker, *Br. J. Cancer*, 50:765–769, 1984.
24. Miralles C, Orea M, Espana P, Provencio M, Sanchez A, Cantos B, Cubedo R, Carcereny E, Bonilla F, Gea T, Cancer antigen 125 associated with multiple benign and malignant pathologies, *Ann. Surg. Oncol.*, 10:150–4, 2003.
25. Wang MC, Valenzuela LA, Murphy GP, Chu TM, Purification of a human prostate specific antigen, *Invest. Urol.*, 17:159–63, 1979.
26. Ito K, Yamamoto T, Ohi M, Kurokawa K, Suzuki K, Yamanaka H, Free/total PSA ratio is a powerful predictor of future prostate cancer morbidity in men with initial PSA levels of 4.1 to 10.0 ng/mL. *Urology*, 61:760–764, 2003.
27. Prostate-specific antigen (PSA) best practice policy, American Urological Association (AUA), *Oncology (Huntingt)*, 14:267–72, 277–8, 280 passim, 2000.
28. Perron L, Moore L, Bairati I, Bernard P-M, Meyer F, PSA screening and prostate cancer mortality, *CMAJ*, 166:586–591, 2002.
29. Koprowski H, Steplewski Z, Mitchell K, Herlyn M, Herlyn D, Fuhrer P, Colorectal carcinoma antigens detected by hybridoma antibodies, *Somatic Cell Genetics*, 5:957–972, 1979.
30. Thomas CMG, Sweep CGJ, Serum tumor markers: Past, state of the art, and future, *Int. J. Biol. Markers*, 16:73, 2001.
31. Sanchez-Carbayo, Use of high-throughput DNA microarrays to identify biomarkers for bladder cancer, *Clin. Chem.*, 49:23–31, 2003.
32. Valculescu VE, Zhang L, Vogelstein B, Kinzler KW, Serial analysis of gene expression, *Science*, 270:484–7, 1995.
33. Fleischacker M, The 2nd International Symposium on Circulating Nucleic Acids in Plasma and Serum (CNAPS-2), Hong Kong, February 20–21, 2001, *Eur. J. Med. Res.*, 6:364–368, 2001.
34. Cooper CS, Application of microarray technology in breast cancer research, *Breast Cancer Res.*, 3:158–175, 2001.
35. Li X, Gu W, Mohan S, Baylink DJ, DNA microarrays: Their use and misuse, *Microcirculation*, 9:13–22, 2002.
36. Argani P, Rosty C, Reiter RE, Wilentz RE, Murugesan SR, Leach SD, Ryu B, Skinner HG, Goggins M, Jaffee EM, Yeo CJ, Cameron JL, Kern SE, Hruban RH, Discovery of new markers of cancer through serial analysis of gene

expression: Prostate stem cell antigen is overexpressed in pancreatic adenocarcinoma, *Cancer Res.*, 61:4320–24, 2001.

37. Porter DA, Krop IE, Nasser S, Sgroi D, Kaelin CM, Marks JR, Riggins G, Polyak K, A SAGE (Serial analysis of Gene Expression) view of breast tumor progression, *Cancer Res.*, 61:5697–5702, 2001.
38. Buckhaults P, Rago C, St Croix B, Romans KE, Saha S, Zhang L, Vogelstein B, Kinzler KW, Secreted and cell surface genes expressed in benign and malignant colorectal tumors, *Cancer Res.*, 61:6996–7001, 2001.
39. Zhang L, Zhou W, Velculescu VE, Kern SE, Hruban RH, Hamilton SR, Vogelstein B, Kinzler KW, Gene expression profiles in normal and cancer cells, *Science*, 276:1268–72, 1997.
40. Boon K, Osorio EC, Greenhut SF, Schaefer CF, Shoemaker J, Polyak K, Morin PJ, Buetow KH, Strausberg RL, De Souza SJ, Riggins GJ, An anatomy of normal and malignant gene expression, *Proc. Natl. Acad. Sci. USA*, 17:11287–11292, 2002.
41. Ziegler A, Zangemeister-Wittke U, Stahel RA, Circulating DNA: A new diagnostic gold mine? *Cancer Treat. Rev.*, 28:255–271, 2002.
42. Poon TC, Johnson PJ, Proteome analysis and its impact on the discovery of serological tumor markers, *Clin. Chim. Acta.*, 313:231–239, 2001.
43. Haab BB, Dunham MJ, Brown PO, Protein microarrays for highly parallel detection and quantitation of specific proteins and antibodies in complex solutions, *Genome Biology*, 2:1–13, 2001.
44. de Wildt RM, Mundy CR, Gorick BD, Tomlinson IM, Antibody arrays for High-throughput screening of antibody-antigen interactions, *Nat. Biotechnol.*, 18:989–994, 2000.
45. Miller JC, Zhou H, Kwekel J, Cavallo R, Burke J, Butler EB, Teh BS, Haab BB, Antibody microarray profiling of human prostate cancer sera: Antibody screening and identification of potential biomarkers, *Proteomics*, 3:56–63, 2003.
46. Corthals GL, Wasinger VC, Hochstrasser DF, Sanchez JC, The dynamic range of protein expression: a challenge for proteomic research, *Electrophoresis*, 21:1104–1115, 2000.
47. Lopez MF, Better approaches to finding the needle in a haystack: optimizing proteome analysis through automation, *Electrophoresis*, 21:1082–1093, 2000.
48. Casali P and Schettino EW, Structure and function of natural antibodies, *Curr. Top. Microbiol. Immunol.*, 210:167–179, 1996.
49. Kasaian MT, Ikematsu H and Casali P, Identification and analysis of a novel human surface CD5-B lymphocyte subset producing natural antibodies, *J. Immunol.*, 148:2690–2702, 1992.
50. Greenberg AH, Antibodies and natural immunity, *Biomed. Pharmacother.*, 39:4–6, 1985.
51. Kasaian MT, Casali P, Autoimmunity-prone B-1 (CD5 B) cells, natural antibodies and self recognition, *Autoimmunity*, 15:315–329, 1993.
52. Carroll MC, Prodeus AP, Linkage of innate and adaptive immunity, *Curr. Opin. Immunol.*, 10:36–40, 1998.
53. Ochsenbein AF, Fehr T, Lutz C, Suter M, Brombacher F, Hengartner H, Zinkernagel RM, Control of early viral and bacterial distribution and disease by natural antibodies, *Science*, 286:2156–2159, 1999.
54. Moss RB, Hsu YP, Van Eede PH, Van Leeuwen AM, Lewiston NJ, De Lange G, Altered antibody isotype in cystic fibrtosis: impaired natural antibody response to polysaccharide antigens, *Pediatr. Res.*, 22:708–713, 1987.

55. Reid RR, Prodeus AP, Khan W, Hsu T, Rosen FS, Carroll MC, Endotoxin shock in antibody-deficient mice: unraveling the role of natural antibody and complement in the clearance of lipopolysaccharide, *J. Immunol.*, 159:970–975, 1997.
56. Thornton BP, Vetvicka V, Ross GD, Natural antibody and complement-mediated antigen processing and presentation by B lymphocytes, *J. Immunol.*, 152:733–740, 1994.
57. Tonegawa S, Hozumi N, Matthyssens G, Schuller R, Somatic changes in the context of immunoglobulin genes, Cold Spring Harbor Symposium on quantitative, *Biology*, 41:877–888, 1975.
58. Weigert M, Cesari I, Yonkovitch S, Cohn M, Variability in the light chain sequences of mouse antibody, *Nature*, 228:1045–1047, 1970.
59. Lantto J, Ohlin M, Functional consequences of insertions and deletions in the complementary-determining regions of human antibodies, *J. Biol. Chem.*, 277:45108–14, 2002.
60. Sigrist CJ, Cerutti L, Hulo N, Gattiker A, Falquet L, Pagni M, Bairoch A, Bucher P, PROSITE: A documented database using patterns and profiles as motif descriptors, *Brief Bioinform*, 3:265–274, 2002.
61. Dsouza M, Larsen N, Overbeek R, Searching for patterns in genomic data, *Trends Genet*, 13(12):497–8, 1997.
62. Furka A, Sebestyen M, Asgedom M, Dibo G, General method for rapid synthesis of multicomponent peptide mixture, *Int. J. Peptide Protein Res.*, 37:487, 1991.
63. Eichler J, Houghten RA, Identification of substrate-analog trypsin inhibitors through the screening of synthetic peptide combinatorial libraries, *Biochemistry*, 32(41):11035–41, 1993.
64. Pinilla C, Appel J, Blondelle S, Dooley C, Dorner B, Eichler J, Ostresh J, Houghten RA, A review of the utility of soluble peptide combinatorial libraries, *Biopolymers*, 37(3):221–240, 1995.
65. Pinilla C, Appel JR, Blanc P, Houghten RA, Rapid identification of high affinity peptide ligands using positional scanning synthetic peptide combinatorial libraries, 13(3):412–421, 1992.
66. Cattoli F, Sarti GC, Separation of MBP fusion proteins through affinity membranes, *Biotechnol. Prog.*, 18:94–100, 2002.
67. Haab BH, Dunham JM, Brown OP, Protein microarrays for highly parallel detection and quantitation of specific proteins and antibodies in complex solutions, *Genome Biology*, 1(6):1–22, 2000.
68. Paweletz CP, Gillespie JW, Ornstein DK, Simone NL, Brown MR, Cole KA, Wang QH, Huang J, Hu N, Yip TT, Rich WE, Kohn EC, Linehan WH, Weber T, Taylor P, Emmert-Buck MR, Liotta LA, Petricoin III EF, Biomarker profiling of stages of cancer progression directly from human tissue using a protein biochip, *Drug. Develop. Res.*, 49:34–42, 2000.
69. Gygi SP, Rist B, Gerber S, Turecek F, Gelb M, Aebersold R, Quantitative analysis of complex protein mixtures using isotope-coded affinity tags, *Nature Biotechnol.*, 17:994–999, 1999.
70. Buckholz RG, Simmons CA, Stuart JM, Weiner MP, Automation of yeast two-hybrid screening, *J. Mol. Microbiol. Biotechnol.*, 1:135–140, 1999.

23 A Combinatorial Approach to Gene Expression Analysis: DNA Microarrays

Concetta Ambrosino, Luigi Cicatiello, Claudio Scafoglio, Lucia Altucci and Alessandro Weisz

CONTENTS

I. INTRODUCTION

The microarray technology is based on analytical tools that parallelize the quantitative and qualitative analysis of nucleic acids, proteins and tissue sections, one of its more recent applications. By miniaturizing the size of the reaction and sensing area, microarrays allow assessment of the activity of thousands of genes in a given tissue or cell line at once in a rapid and quantitative way, and to carry out serial comparative tests in multiple samples. These tools, that stem from the innovations resulting from the technological improvements and knowledge arising from the genome sequencing projects, can be considered as a combinatorial technique that can rapidly provide significant information about complex cellular pathways and processes within one or few "mass scale" and comprehensive testing of a biological sample's composition.

DNA microarrays, the focus of this review, deal predominantly with the genome-wide gene expression analysis (gene expression profiling). They are formed by a planar support, usually a glass microscopy slide, allowing the binding of nucleic acids (cDNA or oligonucleotides) or, in other cases, proteins and oligopeptides. The detector molecules immobilized on the surface of the slide are defined "probes" and the mixtures being interrogated are defined "targets" [1]. The slides, each containing up to several thousand probes arranged in ordered arrays, are used to analyze labeled samples, generally prepared by fluorescent tagging of nucleic acids (DNA or RNA) extracted from a cellular or tissue sample under investigation. Specific binding of the unique components, of the tested sample, mix to its complementary probe, immobilized on the solid support, leading to the appearance of "spots", the glow of which is proportional to the activity of the expressed gene.

The microarray technology was developed at the Stanford University in the early 1990s [2]. From the beginning, it was clear that this technique could have the same impact in biomedical and biotechnological research that the "polymerase chain reaction" (PCR) had in the 1980s. PCR reactions are now extensively used in microarray manufacturing.

The microarray technology is unique, as no other analytical approach allows the exploration to such an extent of the biochemical complexity of biological samples and combines expertise from many different disciplines such as biology, chemistry, physics, engineering, mathematics, and computer science. The role of the recombinant DNA technology, developed in the 1970s, was important, not only for the discovery of the enzymatic tools used in the microarray technology, such as RNA and DNA polymerases, but also for the tools and techniques it made available, in particular the cDNA libraries and nucleic acids hybridization protocols. The microarray technology required a modification of the hybridization techniques to make them suitable for a glass support. The first hybridization on glass was performed in the early 1990s. Rapid, efficient and cost effective chemical synthesis of natural and derivatized polynucleotides, is one more domain whose progress made possible the advent of the "microarray era." This also required the full development of the fluorescent microscopy technology. Since 1970, fluorescent dyes are used for cell biology and microscopy studies, some of which are later being adapted for nucleic acid labeling. Microscopic analysis of chromosome structure by florescence *in situ* hybridization (FISH) is an example. The evolution of more sophisticated fluorescence microscopy devices, such as the confocal microscopes in the 1990s, was necessary for the development of efficient microarray reading tools (see below). The above mentioned know-how, combined with the development of combinatorial oligonucleotides synthesis, the improvement of linker and surface synthesis technologies and detection methods led to the development of the first microarray assay in 1995. Furthermore, all this and the work that introduced robotization in microarrays manufacturing paved the way to the present success and diffusion of DNA microarray applications for gene expression profiling.

A. The DNA Microarray: Components and Characteristics

A DNA microarray, also known as DNA chip, gene chip or more generically biochip, is a microscopic slide on which multiple DNA samples are deposited ("spotted") in predefined positions to constitute an ordered array of probe elements. The chemical nature of these probes in the arrays used for quantitative gene expression analysis can be different, e.g. DNA, PNA or RNA, although in most cases they are represented by cDNAs or chemically synthesized oligonucleotides. The amount of probes to be spotted (optimal probe concentration) as well as the number of spots for unit of area (optimal probe density) is first evaluated experimentally, as these parameters greatly depend upon the detection protocol and device to be implemented and the nature of the experimental test. The microarray surface plays an important role in determining the probe binding efficiency and specificity as well as the sensitivity of the detection step, which greatly affects the quality of the data generated. To be used as analytical devices for genome-wide gene expression studies, arrays of probes which are planar, microscopic and specific are to be put in order. The array elements ("spots") are ordered in rows and columns so that the columns cross the rows in a perpendicular manner. The ordered elements have, as much as possible, the same size, spacing and a unique location on the array, this facilitates manufacture of the slide, as well as design and application of microarray reading devices and software for image and data analysis. Regularity of spot spacing is a prerequisite for correct data analysis, enabling the use of standard analysis templates, while the uniformity of the spot size is required for quantitation and assay precision, as this ensures that the same amount of probe is spotted in each location. Quantitation templates are grids superimposed to the graphical image generated by the scanner, necessary to define the borders of each element and to calculate, for every one of them, its signal intensity and the relative statistics ("shape"), considering the pixels included within the area delimited by these borders. The presence of microscopic spots on the slide enables the examination of a large number of genes, up to an entire genome, with a single test and is a necessary prerequisite for automation of the whole process. The slide surface has to be planar since a planar support allows an accurate scanning and imaging, due to the uniform detection distance between the optical element and the microarray surface. Furthermore, this surface needs to be impermeable to liquids, allowing the use of a small reaction (hybridization) volume. Specific recognition between the probes on the array and the target is the key criterion of microarray-based nucleic acids analysis, as it allows precise quantitation of the amount of molecules present in the sample. This aspect is also important for the statistical evaluation of the data (reproducibility, precision, confidence level, etc.).

B. The DNA Microarray Technology

Modern microarray technology interfaces biology, engineering and physics. Five steps are necessary to perform a microarray experiment: microarray manufacture, probe labeling, hybridization, detection and data analysis (Figure 23.1). In this section the general concepts of this technique are briefly discussed. An important prerequisite to what follows is the concept that the biological question addressed is a key determinant in the design of a microarray experiment, as its formulation dictates not only the technological platform to be selected but also other experimental parameters, such as positive and negative controls, reference and the strategies to apply for computational analysis of the resulting data.

Several criteria need to be kept into account for the choice of a microarray manufacture procedure, including the need to be able to use the technology in a given laboratory setting, the time required for its production or its analysis, the probe content (the total amount of DNA probes delivered on the slide) and density (number of target spots per unit area), the spot size, the purity and the reactivity of the element spotted and the overall costs. The microarray manufacturing technologies are in continuous evolution and they already ensure mass production of biochips as well as assay automation, leading to improved quality, dissemination and reproducibility of the experiments, and cost effectiveness, in the near future. As more extensively reported in other sections that follow, DNA chip manufacturing technology follows two different approaches, known as "delivery" and "synthesis." In the first case (Figure 23.1a) the probes, usually DNAs generated by PCR amplification of cloned cDNAs or synthetic oligonucleotides, are transferred to the slide surface by automatic robotic platforms. This was the way used for the preparation of the first microarray ever. The synthesis approach, currently applicable only to oligonucleotide arrays, foresees that all probes are synthesized *in situ* by light-driven chemical reactions, one base at a time, till the desired oligonucleotide is synthesized in each array position (Affymetrix® technology: Figure 23.1b).

The target for microarray experiments can be of different nature. Nucleic acids are more commonly used nowdays. Diverse techniques can be used for target labeling. Selection in each case depends upon the molecular and biological nature of the target. Binding of the labeled target nucleic acids to

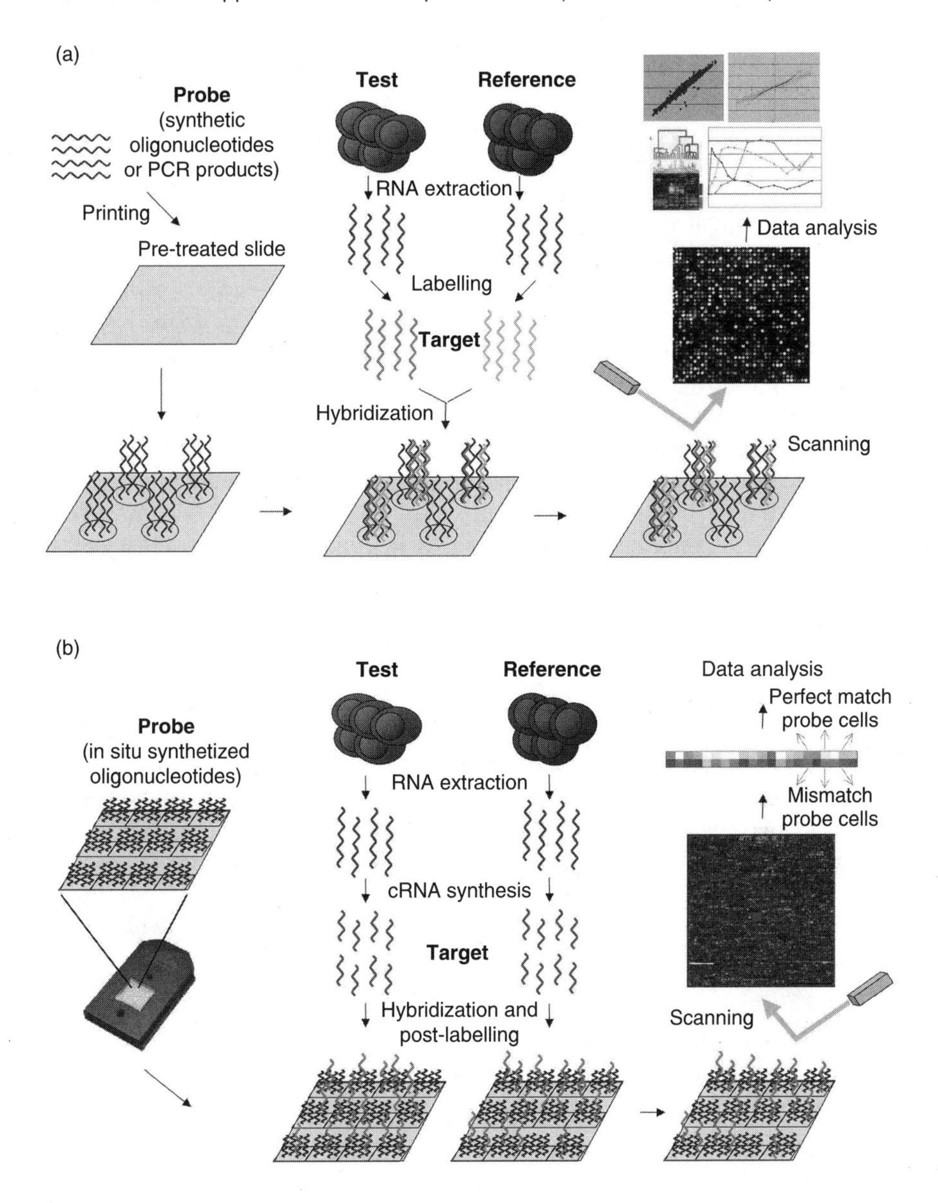

FIGURE 23.1 Schematic representation of genome-scale gene expression analysis with DNA microarrays. (a) DNA microarrays produced by probe deposition; (b) Oligonucleotide microarrays produced by *in situ* probe synthesis (Affymetrix® technology).

the probe onto the microarray occurs by molecular hybridization. Conditions for hybridization, and slide washing to remove targets unspecifically bound to the slide and/or to the probe, are experimentally determined, as they relate to the nature of the biochip and can be differentiated in order to achieve the degree of specificity desired. Once hybridized, the microarray is scanned with a detection device in an automated manner. Fluorescent detection schemes, to detect very low concentrations of label, are generally used. The devices to detect the fluorescent signal on such slides are either (laser-powered) scanners or imagers. In both cases, the fluorescent dyes are excited and the emitted fluorescence is read by converting its stream of photons into an electrical current that, in turn, is transformed in to digital values that can be stored and analyzed by a computer. The processes by which numerical values are obtained from the data files in a format that can provide information helpful to determine each target's concentration in the original mixture, is known as quantitation. Bioinformatics and computational approaches are then used for data normalization, mining and modeling. In this respect, availability of comprehensive electronic libraries and databases containing various functional and structural information on genes or proteins, provide a great advantage in the interpretation of gene expression data. This is to be discussed in detail later.

Each step of microarray technology is still evolving at a high pace and several modifications of the original, basic technology described here, are being continuously introduced, while ever increasing new applications and improvements are further expected [1, 3–5].

II. A CLOSER LOOK AT THE DNA MICROARRAY TECHNOLOGY

A typical gene expression profiling experiment takes place in five separate processes. They are (i) microarray fabrication, (ii) purification and labeling of the target material, (iii) hybridization, (iv) detection and (v) data analysis. The characteristics of each step were briefly discussed in the introduction. A closer look at each of these steps is the object of this section. Here we mainly refer to biochips where the probe is constituted by nucleic acids (DNA microarrays).

A. Microarray Fabrication: Solid Support, Probe Synthesis and Immobilization Techniques

The microarray assay is based on hybridization reactions between labeled single stranded molecules in solution (target) and complementary molecules immobilized on the flat surface of the slide (probe). The fabrication of a microarray requires the synthesis of the target and its deposition on the slide surface (deposition technology). Alternatively, a different approach involving the synthesis of the target directly onto the surface can also be employed.

The slide surface has to be planar, uniform, inert and accessible. Several materials can provide slides with these characteristics, but the glass surface is the most commonly used for fluorescent labeling, since most plastics do not permit the use of fluorescent dyes. Different glasses are available that are suitable for slide preparation (borosilicate, fused silica, etc.) and all are stable materials with low intrinsic fluorescence and reflectivity and an efficient transmission throughout the visible range. In order to allow the efficient attachment of nucleic acids to the surface of the glass, various chemical pre-treatments of the slide are applied, including the most used: (i) poly-lysine covered glasses, with positive charged surface where unmodified nucleic acids can be bound and fixed by UV cross-linking, (ii) aldehyde covered slides, that attach DNAs carrying amino-modified end nucleotides and (iii) amine covered slides that also provide a positive charged surface. The choice between these options will depend on the experimental procedures.

The first step of microarray fabrication process is the synthesis of the probe, commonly constituted by PCR products or oligonucleotides.

1. Printing Microarrays with PCR Products

Usually, primer pairs which are gene-specific or optimized to anneal in the vector sequence, are used to amplify any cDNA or express sequence tags (EST) from an available library. The amplicons are then deposited (printed) into the slide surface at pre-defined positions. Good quality PCR products are to be used for microarray construction and hence stringent QC procedures are applied, including analysis of each amplification product by agarose gel or capillary electrophoresis, to control amplification efficiency and quality of its products and to exclude production pipeline samples presenting a poor amplification or nonspecific bands and smears. The length of the PCR products used to generate microarrays varies generally between 300 and 800 nucleotides.

The main advantages of this approach are that it is not necessary to know the full sequence of the starting DNA clone (although it is preferable!) and the signals obtained are generally strong, and hence the array is easier to implement and quality of the data generated is higher. The main disadvantage is the denaturation step has to be introduced in to the procedure, due to the presence of a double stranded target on the microarray, but also other cross-hybridization problems are possible.

2. Printing with Oligonucleotides

For this strategy to be practised 50 to 70 mer oligonucleotides are generally used. The major benefits here with respect to cDNA probes are: (i) higher specificity, since the sequences are optimised to minimize cross-hybridization, (ii) the possibility to design several oligonucleotides for different parts of the same gene, and monitor the specificity of hybridization and detect alternative splicing products (different alleles) and (iii) the possibility to normalize the hybridization conditions by construction of oligonucleotide sets having similar melting-temperature (Tm).

In both the cases decribed above, the second step in microarray fabrication involves ordered deposition of the probes on the surface (spotting). The purified DNA probes (PCR fragment or oligonucleotides) are spotted on a modified glass slide in an ordered grid, by means of array spotters and robotic instruments that allow for precise deposition of a few nanoliters of DNA solution, with accuracy and reproducibility leading to replicate uniformity. Two main spotting technologies are currently employed. They are "the contact deposition" and "the noncontact deposition". The "contact deposition" uses particular pins that aspirate by capillarity a small volume of pre-made target solutions and deposit a drop of it onto the slide surface on physical contact. The "noncontact deposition" distributes sub-nanoliters of the target solutions to specified locations. Two "noncontact deposition" technologies are mainly used: (i) the piezoelectric technology, which uses electricity in order to modify the morphology of a piezoelectric crystal that encircles a capillary containing nucleic acid solutions, resulting in a squeezing of the capillary and delivery jet onto the surface; (ii) the syringe-solenoid deposition, in which a positive pressure, produced from a syringe and regulated from a solenoid valve, allows the deposition of micro-volumes of the target solutions onto the slide when the valve is opened. A third option, recently introduced, uses an adaptation of an ink jet printing technology to deliver sub-nanoliter droplets of DNA solution onto the glass surface.

All these approaches allow high-density spotting and are easy to implement at low costs.

3. Printing Microarrays by *In Situ* Probe Synthesis

DNA targets are synthesized *in situ* by a modified photo-lithography procedure, one base at a time for several cycles, until the desired sequence is obtained in each element of the array. The technology for semiconductor production is combined with photolithography in the Affymetrix® method, which uses ultraviolet light and solid-phase chemical synthesis for solid-state polynucleotide synthesis. In photo-lithography, the glass slide is modified with a surface providing the reactive amine groups, modified with a photo-protecting group to control their reactivity. The amine group is activated by ultraviolet light. A predefined mask (photo mask) is applied to select the sites that have to be activated during each photo-activation step. In the de-protected regions, modified phosphoramidite nucleotides can then be covalently bound. The cycle of removing the photo-protecting group by UV-light and subsequently the coupling step facilitates oligonucleotide synthesis.

The advantages of this approach are similar to those involving delivered oligonucleotides. All the steps regarding sample production, handling and storage are eliminated and the oligonucleotides are produced directly using sequences from the databases. The use of "perfect match" vs "mismatch" probe pairs is a unique concept introduced in this case. For each probe, perfectly complementary to a target sequence (the "perfect match" probe; PM), an associate probe that carries a single base mismatch in its 13th position is also synthesized in the same array (the "mismatch" probe; MM). This system allows the subtraction of the signals due to nonspecific cross-hybridization to the "MM" probe and provides a key information for signal specificity. Moreover, the chip-to-chip variations are significantly reduced. These kind of chips contain high numbers of targets (up to 400,000 oligonucleotide cells within 1.6 cm^2).

The main disadvantage of this approach is the short length of the *in situ* synthesized oligonucleotides (less then 30 nts) and the high manufacturing costs, although it might be the least expensive way to produce chips covering the whole mammalian genome.

B. Target Preparation and Labeling Procedures

The method of labeling a target depends on its molecular nature, and on the microarray technology implemented. Here we shall describe the labeling method used for a DNA chip with an RNA as the target molecule, which is used for gene expression profiling. Other target preparation protocols are found in Refs. 3–4.

The purity and integrity of the RNA isolated from tissues or cell lines are critical for microarray experiments. The most common method for total RNA isolation involves organic extraction of RNA from homogenized samples like guanidinium isothiocyanate or guanidinium hydrochloride. The RNA sample should be devoid of carbohydrates, DNA, lipids and proteins, as the presence of these contaminants may affect the performance of the sample in the downstream procedures. Several commercially available methods and buffers are currently used. The RNA is significantly more labile than DNA because it is readily susceptible to degradation by endogenous and contaminating ribonucleases, which are stable and ubiquitous enzymes. To obtain a high quality RNA and to maintain its integrity during the subsequent procedures, several precautions are taken, which include: (i) processing the sample as soon as possible, and (ii) avoiding RNase contamination using disposable gloves and RNase-free glassware, plasticware, water and salt solutions, obtained by autoclaving or treatment with DEPC. The purified RNA is stored at –70°C.

Total RNA or (polyA+)RNA (mRNA) can be used for experiments. In the latter case, a purification step is necessary, as mRNA is isolated from total cellular RNA by affinity chromatography on oligo-dT immobilized to a solid support. The amount of the purified RNA is determined by its dual wavelength absorbance at 260 nm and 280 nm and the quality checked by agarose gel or capillary electrophoresis.

In the most recent microarray experiments, multicolor fluorescence labeling is used for simultaneous analysis of two or more samples in a single assay. For this, total RNA or mRNA are labelled with fluorescent nucleotides by a reverse transcription reaction. Cyanines Cy3 and Cy5 are used for dual color analysis.

Various RNA labeling strategies involving direct or indirect labeling exist. Direct labeling is the most diffused method. The RNA template is converted to fluorochrome-labeled first-strand cDNA by a reverse transcription reaction (RT). Reverse transcriptase synthesizes cDNA using the RNA as a template, in the presence of oligo (dT) primers that hybridize with the poly-A tail of the mRNA, incorporating at the same time modified Cy3- or Cy5-conjugated deoxy-nucleotide triphosphates. This method is fast and simple. The main drawback is that cyanine-labeled nucleotides are not efficiently incorporated in the polymerizing step, due to steric hindrance caused by the large fluorophores. After this step, the template is removed by RnaseH1 digestion or NaOH treatment and the free nucleotides are eliminated by gel filtration on dedicated columns. Generally, direct DNA labeling does not produce long cDNA molecules, and for this reason probes complementary to the nucleotide sequence near the poly-A plus tail of the target are recommended. In the indirect labeling, the 5-(3-amino-allyl)-2′-deoxynucleotide 5′-triphosphate (aa-dNTP) modified nucleotides are used. They are incorporated with the same efficiency of the unmodified nucleotides in the first-strand cDNA synthesis. After removal of the RNA template and purification of the amine-modified cDNA, the coupling reactions with N-hydroxysuccinimide-esters (NHS-esters) of Cy3 or Cy5 are performed to produce uniformly labeled probes. The labeled cDNA requires a re-purification step to remove the unincorporated free Cy-Dye.

When the amount of the RNA sample is low (for example in tissues from biopsies), an RNA amplification step is needed, that involves labeling through a linear RNA amplification method. For this, the mRNA population is converted into a double-strand cDNA containing a strong promoter sequence from viral RNA polymerases, such as T3 or T7 phage promoters, by using an oligo(dT) primer including a 5′ extension including the viral promoter for first strand cDNA synthesis, followed by complementary cDNA strand synthesis with DNA polymerases. Several RNA copies are synthesized from each template of double-strand cDNA in the presence of Cy-Dye ribonucleotides and appropriate DNA-dependent RNA polymerases. If biotinylated ribonucleotides are used in the transcription amplifications to generate biotinylated cRNA, a post-labelling reaction is performed following target hybridization to the array and the washing steps, by staining with streptavidin-phycoerythrin conjugates (see http://www.affymetrix.com/technology/ge_analysis/index.affx for more details) or CyDye streptavidin conjugates (see: http://www4.amershambiosciences.com/aptrix/upp01077.nsf/Content/codelink_bioarray_system). Both technologies use short oligonucleotide probe microarrays (25mer- and 30mer-long, respectively). Fragmentation of cRNA before hybridization is necessary to avoid secondary structure of RNA interfering with the target-probe annealing.

A very recent labeling method, instead of fluorescence, uses Resonance Light Scattering (RLS) technologies, based on the optical light scattering properties of nano-sized metal colloidal particles. In this technique, biotinylated nucleotides are used in the first-strand cDNA synthesis and the coupling steps are performed with anti-biotin-coated gold or silver particles. The main advantages of this technology are the high sensitivity and the absence of signals due to photochemical bleaching.

C. Hybridization and Washing

In the hybridization step, the ability of labeled targets to bind immobilized probes are tested. Referring again to the cases where DNA microarrays are used, the labeled targets are first annealed with their complementary probes immobilized onto the slide surface. Subsequently, serial steps of post-hybridization washing are carried out in order to remove unbound materials, to improve the signal-to-noise ratio and to minimize cross-hybridization between labeled targets and probes. Hybridization and washing steps are performed in the dark, to avoid signal loss due to photo-bleaching of the fluorescent dyes. Several parameters are taken into account in order to obtain successful results, linked to the two main experimental variables: temperature of hybridization and composition of the hybridization solution. The base composition of the nucleic acids involved in the annealing reaction have a large effect on duplex yield of the solvents normally used for hybridization, due to the different stability of A : T vs G : C base pairs. Short oligonucleotides have extreme biases in composition and, hence, correspondingly show large differences in melting temperature ($T_{m,}$ the temperature at which 50% of the target is denatured). As a rule-of-thumb, the addition of an A : T base pair increases the T_m by 2°C, compared with 4°C for a G : C pair. This difference is minimized by adding TMAC (tetra-methyl-ammonium chloride) to the hybridization mix. The optimum range of hybridization temperature in aqueous solutions is about 65–75°C. The degradation of the target due to high hybridization temperature is minimized by the use of 50% formamide, that allows the reduction of the hybridization temperature of about 25°C, thereby protecting the targets from degradation. Usually, in the presence of formamide, hybridization is carried out at 42°C. The other parameters to be considered in this process are the pH (neutral value promotes hydrogen bonding between base-pairs), and salt concentration (improves the hybridization efficiency by shielding negative phosphate groups of the nucleic acid and minimizing electrostatic repulsion). The post-hybridization washings eliminate unbound labeled target. Usually these washes are performed in SSC/SDS solutions, progressively diminishing the concentration of salt. To improve the reproducibility of the microarray analysis by reducing the variability due to hybridization and washing steps, automatic hybridization-washing stations are developed, but their high cost is still a deterrent for their widespread diffusion. After the slides are washed, they are dried by blowing nitrogen gas or by low speed centrifugation at, and quickly passed, to the subsequent processing steps.

D. Target Detection

In microarray analysis, detection is the step in which the signal for each spot is revealed and quantified, giving an image that is like photography of the microarray. In the following paragraphs we shall provide basic information on the currently used fluorescence detection systems for microarrays and the characteristics of corresponding detection devices.

1. Fluorescent Dyes

Fluorescence is the light emission process in which a fluorophore, a molecule able to adsorb light, after reaching an excited state releases light (photons) with

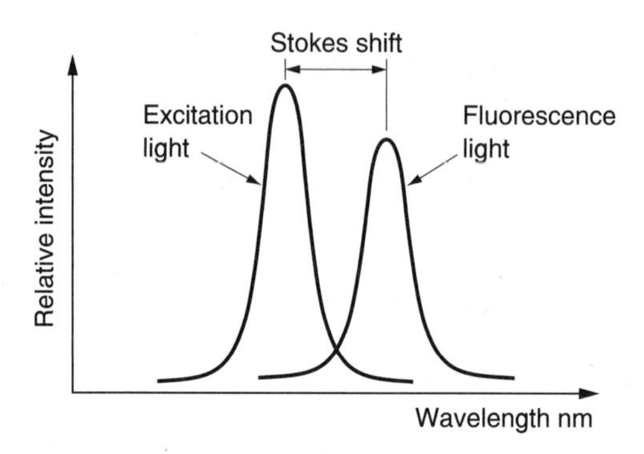

FIGURE 23.2 Stokes shift diagram depicting optimal relationships between excitation and fluorescence light of a fluorophore suitable for microarray applications.

less energy and, consequently, a longer wavelength than the exciting light (Stokes shift; Figure 23.2). In fluorophores the distance between excitation and emission wavelengths are greater and more suitable for microarray application. Part of the excitation energy is emitted in processes different from fluorescence (nonproductive fates), whose values have to be evaluated to choose the optimal fluorescent label and detection device. When a moderate excitation light is applied, a single fluorescent molecule can be cyclically excited and detected, increasing consistently the sensitivity of the assay. On the other hand, the signal emitted is reduced when the exciting light applied is too intense, a phenomenon causing permanent loss of fluorescence signal (photo-bleaching).

Several fluorescent labeling dyes are commercially available (Alexa series, Oregon green, Rhodamine and Cyanine dyes, etc.). They are all characterized by the presence of double bonds on every other carbon atom of a cyclic structure, containing the electron that once on excitation emitted fluorescent light. The cyanine dyes are the most widely used at the moment. They are bright, easily added to the nucleotides, stable to photo-bleaching and with a Stokes shift value of about 20 nm. The cyanine dyes used in microarray analysis are Cy3 (absorption at 550 nm and emission at 570 nm) and Cy5 (absorption at 649 nm and emission at 670 nm) that are already available as phosphoramidite derivatives.

The use of fluorescence in microarray technology has several advantages. Indeed, fluorescence significantly increases the sensitivity and the speed of the assay and enables the collection of very large amounts of data in automated fashion. Moreover, the fluorescent dyes are not too toxic, stable, and much safer than radioactivity. Their major advantage, however, is the spatial resolution they provide, as this allows correct assessment of both weak and strong signals, even when they are emitted by elements located adjacent to one another on the array grid, as any signal spreading effect is avoided. This property allows the construction of high-density arrays. Two or more fluorescent dyes are often used in conjunction in microarray experiments, to increase the reliability of the comparative analyses and to permit the analysis of several differently labelled samples on the same microarray.

2. The Microarray Reading Devices

The two main detection systems currently used for microarray reading are the scanners and the imagers. The microarray scanners acquire the entire image by moving the array or the optic system back and forth in small increments (10 μm). The imagers are able to collect data from larger areas (1 cm^2) without any movement. Both devices have to include functions necessary to detect a microarray image, in particular: (i) a source of excitation light with a precisely defined wavelength, provided by a laser or a lamp, requires the addition of filters to provide beams of selected wavelengths; (ii) a system of optic lens collecting the fluorescent light in three dimensions, the numerical aperture, and a parameter defining the collection angle, e.g. collection efficiency of 50% – a lens with a numerical aperture of 1 collects the light over an entire hemisphere; (iii) the spatial addressing, that refers to fluorescence detection from a defined area of the glass slide, usually divided in pixels, smaller than the element size to reveal artifacts due to spotting problems or to dust on the slide; (iv) the ability to discriminate between excitation and emission light, the last being of much lower intensity than the first one; and (v) the ability to convert the low level light into electrical signals.

The microarray detection devices, scanners and imagers, are sophisticated instruments that require components specifically designed to achieve the specifications listed above. Confocal scanners are widely used as microarray detection devices. Differing from a normal scanner, these systems have two focal points configured to limit the field of view in three dimensions, driving data acquisition one pixel at a time. In the basic design, a laser light is directed to an excitation filter allowing passage only to light of the wavelength of interest, usually corresponding to the excitation peak of the dye. The laser light is monochromatic (single color), coherent (the photons in the beam have the same phase) and collimated (highly parallel), implying that no other optic systems are required to get a beam that is really intense and precisely reflected. The device has to include multiple gas or solid-phase lasers, one for each wavelength required. The emitted light is reflected through the microscope objective by a beam-splitter, an optic filter that separates out the returning excitation beam. The main function of the splitter is to reject most of the reflected laser light (4% of the input), while allowing passage to most of the fluorescent light. The excitation of the fluorescent dye molecules results in the emission of fluorescent light in several directions, collected by the objective. The return beam goes to the beam-splitter, which in turns transmits only the fluorescence beam to a mirror directing it to the detector. This emission filter allows the passage of a selected narrow band of fluorescence, while rejecting the reflected laser light. At this point, the fluorescence beam is directed to the detector lens, that in turns focuses it on a detector device able to convert the beam into electrical signals. The most common detectors are photo-multiplier tube (PTM), which transforms the photons to an amplified electrical signal. The fluorescence beam is directed onto a light-sensitive surface of the PTM, the photo-cathode, that then releases electrons. These will jump onto a charged electrode, inducing the release of a higher number of electrons (up to about 1 million-fold amplification) and are finally collected on the anode there by sending out of the PTM the electrical current that is easily recorded and measured. The recorded signal is converted to digital data, and stored as image (TIFF) files that are further computed.

The microarray imagers, on the other hand, are detection instruments able to capture images from a larger portion of the microarray in a single detection step. Conceptually, an imager is organized as a scanner but several technical details are different. The imagers have a white light source (polychromatic) that is directed to the optic filters to obtain the monochromatic beam, necessary to excite the fluorescent dye. The fluorescent beam is directed to a beam-splitter to remove the reflected light. A light sensor, also known as the detector, located in the camera, captures the fluorescence light. The detector consists of a checkerboard matrix of light sensitive pixels, like for example in the charged-coupled devices (CCDs) in which a pixel is coupled to each photo-sensor, allowing the charge accumulation of the photo-sensor to be transferred and amplified across the matrix. The sensing region, defining the CCD-chip, contains around 1,000 pixels in each direction. The smaller the pixels the higher is the image resolution. Only smaller images are captured each time as less charge is stored in the device. The amplified electrical signal is recorded, measured and transformed to digital data, as is the case with scanners.

E. Analysis of DNA Microarray Data

The raw data, generally fluorescence measurements extracted from the TIFF-format images generated by the scanning devices, require first scaling and normalization, to eliminate the systematic sources of variation between samples as well as the different intrinsic fluorescence labeling or hybridization efficiency among the two dyes analyzed in parallel, the unequal spreading of the hybridization mix on the array surface, and the variations in image analysis (laser power fluctuations, photo-multiplier gain adjustments, etc.). Indeed, raw analysis of data relative to replicate experiments reveal a high variability present, when comparing the same RNA labeled with two different dyes on a single array (self–self differential hybridization). As a consequence, at least three technical replicates (i.e. three different hybridization reactions for each sample-control pair) are required to allow an efficient analysis of variance and correction of systematic errors (biases) and variability due to stochastic events (variation).

The data analysis process, starting from normalization to eliminate casual sources of variability within and among arrays, proceeds through statistical analysis, aiming at identifying those genes whose expression is significantly different in the two (or more) investigated samples. The identified genes are then subjected to further bio-informatic analysis, to group them according to their expression patterns, functional role, etc., or to test their predictive value with respect to biological hypotheses [6].

We shall describe here some of the approaches used for computation of data from "two-dye" comparative gene expression profiling analyses carried out with cDNA arrays.

The initial step proceeds through "within-array normalization," aimed at correcting biases within the data sets due to intrinsic dye fluorescence, intensity-dependent or local (sub-array) variation. The more immediate and simplest way to address the problem is the total intensity normalization, which is based on two postulates: (i) the slide containing a large number of gene probes is likely to show the same activity in both test and reference samples, (ii) the total mass is same for the two samples that are hybridized competitively to the array, so that the total fluorescence is same among them, even if some RNA species are over-represented in one sample and vice versa in the other. As a result, the data are adjusted so that either the sum, the mean or the median of the measured intensities, are equal for the two fluorescence channel readings. A variation of this normalization step is to scale all data according to reference genes whose expression levels are assumed to be constant, based on biological and functional considerations (such as the so called "housekeeping" genes, for example). Probes for these reference genes are spotted on various regions of the array to correct sub-array variation and used as an alternative means to normalize data from replicate experiments. In any case, this type of scaling does not correct intensity-dependent variation, as standard deviation data often varies with the signal intensity because casual fluctuation affects signal detection more incisively at the lower end of the fluorescence scale than at its higher end.

To adjust this source of variation, a locally weighted linear regression ("lowess": locally weighted scatter-plot smoothing, [7], that computes an intensity-dependent normalization factor for each gene, should be carried out [8]). In this way, however, local differences within the array (sub-array variation) are not addressed. The differences due to spatial location of the spots are in regard to those slides on which different arrays are spotted through different pins (the so-called "print tip groups," "pen groups" or "sub-grids"). The variance is due to slight differences in the geometry of pins, deformation of the pins after a long activity, or to unequal distribution of the hybridization efficiency over the slide surface. The same variability is observed among replicate slides. A possible solution is the execution of different "lowess" analyses for the various sub-grids, and a scaling of the obtained data similar to across-array normalization (see below).

A particular type of "within-array" analysis is the so called "self–self" hybridization [9], in which two dyes are used to label the same RNA species, so that the fluorescence values acquired by the scanner for each gene is supposed to be the same for the two channels. This approach allows the identification of the variability which depends only on systematic biases or on stochastic processes. Some authors suggest the performance of some "self–self hybridization" for each experiment, to establish an error model used to correct data derived from experimental measurements.

The second step is the "across-array normalization." The simplest way to compare replicate arrays is to scale their intensities according to a total fluorescence method, computing the standard deviation of replicates measurement for each gene, and excluding from the analysis those genes whose variance is too high. A complex approach to correct stochastic sources of variability is the "variance regularization" [10], which implies the adjustment of data relative to the different slides (or sub-grids) in order to center the fold-change distribution around zero (normalization step), and then multiplication of each element for the respective scaling factor, computed for each array by dividing its variance for the geometric mean of all the variances.

A different type of across-arrays data comparison is the "flip-dye" analysis, based on the inversion of the dyes used for labeling the test and reference samples in at least one replicate [dye swap, 11]. The comparison of at least two dye-swapped hybridization reactions reveals the presence of differences in fluorescence values which are not due to effective changes in RNA levels, but instead to casual fluctuations; the genes for which the ratio between the fold-changes of the swapped arrays is far from one and should be excluded from further analyses. Therefore, in order to eliminate as many variability sources as possible, correct planning of a gene expression profiling test with microarrays should include not only replicate hybridizations but also dye swapping. It has even been suggested that each sample should be subjected to balanced hybridization, carrying out as many labeling with Cy3 or Cy5.

The last step of this initial data analysis phase is the selection of significant genes whose expressions are different in the two compared samples. The first and simplest method used is based on computing fold-change differences for each gene, by averaging replicate results and choosing the first cut-off value that defines differentially expressed genes. Generally, a gene is considered differentially expressed in two samples when the differences in mRNA detection among them by microarray hybridization are at least two-fold. If the data relative to many replicates are consistent, a lower cut-off (down to ±1.5-fold change) is acceptable. However, a cut-off value has to be selected based on statistical significance, and for this reason a great number of computational approaches are introduced to compute the level of confidence associated with the selection of truly differentially expressed genes. The most used among those methods go from the standard t-test [12], to a Significance Analysis of Microarrays method [SAM, 13], to analysis of variance [ANOVA, 14] or application of Bayesian mathematics [15], to the "maximum likelihood" method [16].

The t-test analysis computes for each gene the probability that the difference between the mean fluorescence intensities of the test and reference samples is falsely called significant (p-value), by theoretical t-distribution or permutation test.

SAM involves a modified t-test and computes a "False Discovery Rate" (FDR, representing the expected incidence of false positives) for each chosen differential expression (significance) cut-off.

ANOVA takes in to account the different sources of variation, dependent on the arrays themselves (the different slides are spotted and hybridized under slightly different conditions), the dyes, (one dye is often brighter than the other) the samples (their concentrations can be slightly different) the genes (individual probes can show different efficiency of hybridization) the microarray elements (a complete control over the amount of DNA immobilized on the slide is not possible). Numerous other mathematical methods are proposed, but their complete listing is beyond the scope of this review.

A different algorithm for data normalization and selection of significantly expressed genes is used for the Affymetrix®-type oligonucleotide arrays, in which each gene is represented by a probe set of 16–20 perfect match (PM) oligonucleotides, each of them paired with a single-based mutant (MM) to allow computing the quote of non specific annealing reaction (see above). The Affymetrix® algorithm first discriminates the genes effectively expressed from those whose levels are similar to MM, by executing a t-test for each probe set. Then the fluorescence value relative to the probe set is computed by averaging the intensities of the perfect matches subtracted of mismatch and, finally, a t-test is performed to compare the test and reference samples, hybridized to distinct biochips [http://www.affymetrix.com, 17].

The list of positive, differentially expressed genes obtained by either one of the above mentioned procedures, are then subjected to other investigations to gather further insights on its biological meaning. A first analysis can be based on gene "clustering," where grouping of the genes according to similarities of their expression patterns in each sample is done. A basic principle of genome-wide expression analysis is that genes linked by similar expression profiles respond in a similar fashion to the environmental and internal signals reflecting the functional state of the cell, while the products they encode are likely to act in concert toward achieving a cellular phenotype. According to this view, data clustering is the first step for interpretation of microarray data toward identification of the biologically relevant processes they underscore. Different clustering algorithms have been proposed, among which the most used are: hierarchical clustering, K-means clustering, self-organizing maps, supervised clustering and Best Score Clustering [BSC, 18].

Hierarchical clustering [19] consists of computing the distances between each couple of genes of the studied list, thus constructing a distance matrix in which the distances of each gene from all the others is reported; the smallest distances computed will form the first cluster (composed by gene pairs). Another distance matrix is then constructed, considering now groups of genes (or "objects") instead of single genes, and the process of classification is repeated until a single group or cluster remains. The similarity hierarchy so computed can be represented, likewise a phylogenetic tree, whose branch length is proportional to the correlation between the elements connected (expressed as Pearson's score). Besides clustering genes according to their expression pattern (gene clustering), samples can also be grouped according to their gene expression profile (array/sample clustering). This clustering is effective in identifying the main groups of similar expression, but the hierarchic tree may be too rigid to represent the combinatorial complexity of gene expression data. Moreover, this method yields a large number of clusters in the tree-like structure, making it difficult to link the expression patterns to biological processes.

K-means clustering [19] is based on the assumption that a certain number of classes must be identified in the data set. The genes are first randomly assigned to these classes, and then rearranged in the clusters through successive steps, involving computation of the distance of each gene from the mean of each of the selected groups and shuffling it to the nearest class. This approach is used when *a priori* hypothesis concerning the number of expected clusters is formulated in advance.

Self-organizing maps [20] are also based on the establishment of a certain number of nodes in a k-dimensional gene expression diagram, followed by iterative mapping of the nodes in the space according to their distance from points corresponding to the gene expression values determined experimentally. The advantages of this method are the flexible structure of the clusters and their easy visualization and interpretation. The spatial representation of clusters better reflects the multiple distinct ways in which gene expression patterns can relate to each other.

Despite the many approaches proposed, one is still far from linking clusters to biologically relevant groups, as sufficient information about the biological role of the genes and the classes they group in are still missing. To address this problem, supervised clustering methods [21] are proposed, in which genes and other notions of interest are associated with labels that provide information about a pre-existing classification. The information used to drive the analysis may include knowledge of gene function or regulation, disease subtype or tissue origin of a cell type. The methods comprise a training phase (supervised learning), in which the expression profiles associated with each class are defined by using a set of informative genes, and a test phase, in which new genes are classified according to their similarity to the pre-defined classes.

More complex approaches to this problem involve the use of artificial neural networks [22], Bayesian networks [23] and support vector machines [24], which in turn are based on the same principle of supervised learning [25].

In parallel with the cluster analysis, a functional classification has to be carried out in order to identify groups of co-expressed or co-regulated genes that play a common or complementary role in the cellular homeostasis or in the response to external stimuli. An international effort, the Gene Ontology (GO) consortium [http://www.geneontology.org, 26], is currently under way to establish precise and univocal definitions of the involvement of all genes from various species in biological processes, including the molecular functions and cellular localizations to their products. The GO dictionary is organized in a hierarchical structure, in the form of Directed Acyclic Graphs (DAGs), in which each term belongs to a parent class and has in turn a certain number of child terms, going from the broader to the narrowest category [27]. The resource is public and available online through different browsers: AmiGo (http://www.godatabase.org/cgi-bin/go.cgi), MGI (http://www.informatics.jax.org/userdocs/GO_help.shtml), QuickGo (http://www.ebi.ac.uk/ego/). Some data bases of GO annotations (i.e. matching single genes to the GO terms) have been compiled and are periodically updated (http://www.geneontology.org), while instruments are introduced to evaluate the statistical significance of the number of genes belonging to a given GO functional class identified in a microarray experiment [OntoExpress, http://vortex.cs.wayne.edu/Projects.html, 28]. Although the simple number of genes can be irrelevant for activation of a biological process, the aggregation of many elements in a functional or spatial group may be a useful tool to drive further research about the biological meaning of the observed gene expression patterns.

In conclusion, microarray data analysis is a complex process, due to either the very large amount of information yielded by each single experiment or the high frequency of systematic and stochastic errors. Any of the different normalization and transformation methods described here can substantially modify a DNA microarray data set, so that a strenuous work of optimization and standardization of all steps of data gathering and analysis is required to gain the highest reproducibility and to allow direct comparison of data generated from multiple microarray analyses, especially when it is from different laboratories. Furthermore, for functional data analysis, harmonization and integration of available databases [29] are required. These are the challenges that allow a capillary diffusion of this technology and the fulfillment of the expectations raised by its potentials.

III. APPLICATIONS OF THE MICROARRAY TECHNOLOGY FOR ASSESSMENT OF GENOME ACTIVITY IN NORMAL AND PATHOLOGIC CELLS AND TISSUES

The DNA microarray technology has several applications. In the beginning it was applied for gene expression monitoring and then for mutation detection, mapping and evolutionary studies. Some of these aspects are discussed in this section.

Gene expression analysis through DNA microarrays is a powerful means to study the global profile of gene activity of any cell type or tissue, allowing many applications, to include molecular profiling of different tissues or stages of embryonic development, diseases classification according to gene expression signatures of pathologic specimens, identification of transcriptional modifications induced by drugs (pharmaco-genomics) and dynamic description of gene expression changes triggered by a particular stimulus in the cell, through single determinations or time-course analyses.

Sequencing of the entire genome of various species (Yeast, *C. elegans*, *A. thaliana*, *D. Melanogaster*, the house mouse and *H. Sapiens*) paved the way for a dynamic analysis of the genetic material of each cell type in the various differentiation and functional states (post-genomic or post-sequencing studies). Once all the genes present in the cells are identified the scientific community needs to verify which of them is effectively activated in each given cell type. This leads to a functional genomic classification of all the tissues and organisms in all stages of differentiation and functional activation; each of these conditions are univocally defined by a specific combination of activated and/or repressed genes. DNA microarrays nowadays represent the analytical technology that best fulfills this need, and for this reason is currently applied at an increasing rate. This involves the concept that massive accumulation of gene expression data would occur quite rapidly and all this needs to be made rapidly and effectively available to all laboratories throughout the world. To this aim, publicly available gene expression data banks are organized, including Gene Expression Omnibus [www.ncbi.nlm.nih.gov/geo, 30], the Stanford University Microarray Database [genome-www5.stanford.edu/MicroArray/SMD/, 31] and the EMBL database [www.ebi.ac.uk/arrayexpress/, 32]. Furthermore, to make data from different laboratories directly and effectively comparable, a common scheme to standardize microarray data presentation is being studied (Minimum Information About Microarrays Experiments, [33–34]).

These databases are constructed to include gene expression profiles, not only of normal tissues but also of pathologic ones, as virtually all diseases are studied through DNA microarrays. The identification of pathologic gene expression patterns are useful for different aims, including: (i) definition of pathogenic alterations that underlie the disease, through the reconstruction of cellular pathways hyper-activated, or impaired, in the pathologic tissues; (ii) identification of gene expression patterns associated with the pathology, to be used for diagnostic applications; (iii) extraction of expression profiles useful for the prognostic evaluation; and (iv) identification of new therapeutic targets through reconstruction of cellular pathways implicated in disease pathogenesis. Particular attention is given to molecular classification of cancer, as early diagnostic tools and more effective prognostic factors are actually required for most forms of malignant neoplasia. The complexity and wideness of microarray analysis provide a useful tool to investigate the heterogeneity of neoplastic diseases. In fact, large scale gene expression analyses show that each tumor has its own pattern of gene expression, which is different from that of other tumors derived from the same histological type [35]. The observation of specific expression profiles in many different tumors have suggested that the gene activation pattern (the so-called molecular signature, or portrait) is the result of a complex network of factors, including the genetic background of the patient, the tissue of origin, the grade of de-differentiation of the tumor, the clonal genetic alterations characterizing the neoplasia, the proliferation rate and the different cellular types that form the tumor mass. Therefore, different subsets of genes are identified in each tumor, reflecting the various components of its genetic background. Some genes are common to virtually all tumors or characterize the pathologic tissues from the normal counterparts. These genes are generally involved in cell-cycle control, adhesion and motility, apoptosis and angiogenesis [36–37]. On the other hand, supervised clustering methods show that other genes identified by microarray analysis allow us to distinguish tumors according to their tissue of origin [38], or to discriminate the grade of differentiation of neoplasia derived from the same tissue. Some studies show that gene expression analysis can even permit the distinction of functional subclasses among histologically homogeneous tumor samples, with different grades of malignancy and, therefore, quite different clinical outcomes [39–41]. This is possible since the gene expression patterns reflect some biological properties of the tumor, that influence its ability to infiltrate the surrounding tissues, to give rise to metastasis [42–43] and to respond to therapy [44].

IV. APPLICATION OF MICROARRAY-BASED GENE EXPRESSION PROFILING ANALYSIS TO THE CHARACTERIZATION OF THE HORMONE-RESPONSIVE PHENOTYPE OF BREAST CANCER

Breast cancer is the most frequent malignant neoplasm in women, and is the best example of hormone-dependent cancer, defining in this way a tumor that needs a hormonal stimulus to grow and expand. Estrogens are the female sexual hormones and represent the main endogenous factor promoting breast cancer cells proliferation [45]. To increase our knowledge in the biological pathways involved in estrogen-dependent growth of breast cancer cells, gene expression analysis with cDNA microarrays was carried out on a hormone-dependent breast cancer cell line (ZR-75.1) before and after stimulation with a mitogenic dose of the natural estrogen 17β-estradiol. Time-course analysis was carried out in hormone-stimulated cells to provide a kinetic view of the gene responses to the hormone throughout a whole mitotic cycle (32 hrs in these cells). mRNA extracted from treated cells was used to synthesize cDNA labeled with the fluorescent dye Cy5, that was then mixed with an equal amount of a common reference target consisting of Cy3-labelled cDNA, extracted from hormone-deprived cultures and hybridized to a glass array including 9,182 cDNA elements, representing 8,372 randomly selected unique gene/ESTs clusters. Three independent hybridization assays were performed for each sample pair and dye swapping (see above) was included in the protocol. 6,080 genes were selected as informative, or expressed in this cell line at detectable levels, by SAM statistics [13]; among these genes, 344 showed significant changes in activity in estrogen-treated cells [46–47]. We grouped estrogen-responsive genes through an unsupervised hierarchical clustering algorithm according to similarities in their inhibition or activation profiles in hormone stimulated vs control cells. Eight main clusters summarize the main patterns of gene expression changes detectable in estrogen growth-stimulated breast cancer cells (Figure 23.3). The first three clusters (1–3) group all down-regulated genes, with different kinetics of decrease in mRNA expression: significant down regulation occurring already 1 to 4 hrs into estrogen stimulation for the genes belonging to cluster 1, after 6 to 8 hrs for those of cluster 2 or after ≥12 hrs for genes of cluster 3. Clusters 4 to 6, instead, comprise activated genes whose transient expression patterns appear to be linked to cell cycle phasing, while the last two (clusters 7 and 8) include genes showing persistent activation

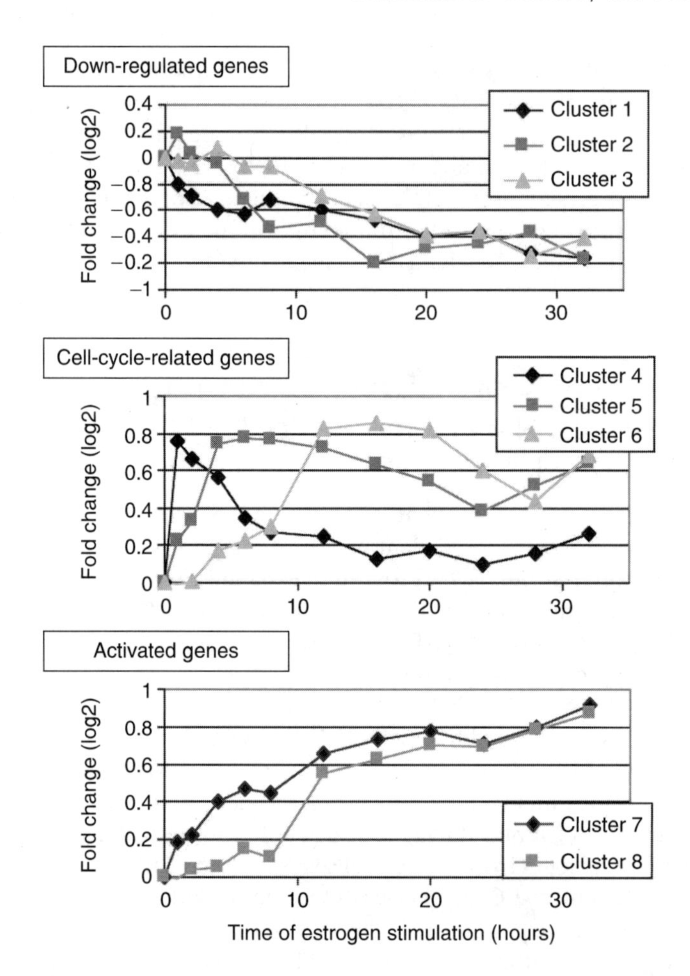

FIGURE 23.3 Example of the gene regulation patterns induced by a biological stimulus in responsive cells, in this case human breast cancer cells stimulated with a mitogenic dose of estrogens.

by the hormone for up to 32 hrs, with RNAs starting to increase within the first 1 to 6 hrs of stimulation (cluster 7) or only after 8hrs (cluster 8).

A functional classification of these genes according to Gene Ontology reveals that the cell cycle gene clusters (clusters 4, 5 and 6) comprise some important genes involved in cell cycle regulation (as c-*fos*, c-*jun*, c-*myc*, cyclin D1), while clusters 7 and 8 encompass a larger number of genes, with a general activation of some metabolic pathways, including glycolysis, nucleotide and cholesterol biosynthesis, revealing a clear activation of anabolic processes in these time windows.

This set of estrogen-regulated genes were then used to classify both breast cancer cell lines and specimens, whose expression data were publicly available in on-line databases (the NCI60 gene expression database for the molecular pharmacology of cancer [http://genome-www.stanford.edu/nci60/] and the Stanford University "Molecular Portraits of Human Breast Tumours" [48] web site [http://genome-www.stanford.edu/breast_cancer/molecularportraits/], and the possibility of identifying different subclasses with prognostic significance in an apparently homogeneous population of tumors was tested. A subset of 49 genes was able to distinguish between estrogen receptor (ER) positive and negative breast cancer cell lines or tumor specimens, thus confirming the predictive value of these genes [Figure 23.4 and Refs. 49–50]. This is fundamental for prognostic evaluation of these tumors, as ER expression is among the main prognostic factors for breast cancer, and reflects the disease responsiveness to hormonal therapy. These results confirm the possibility to discriminate between biologically different forms of cancers through the study of gene expression signatures related to relevant physiological or pathological stimuli.

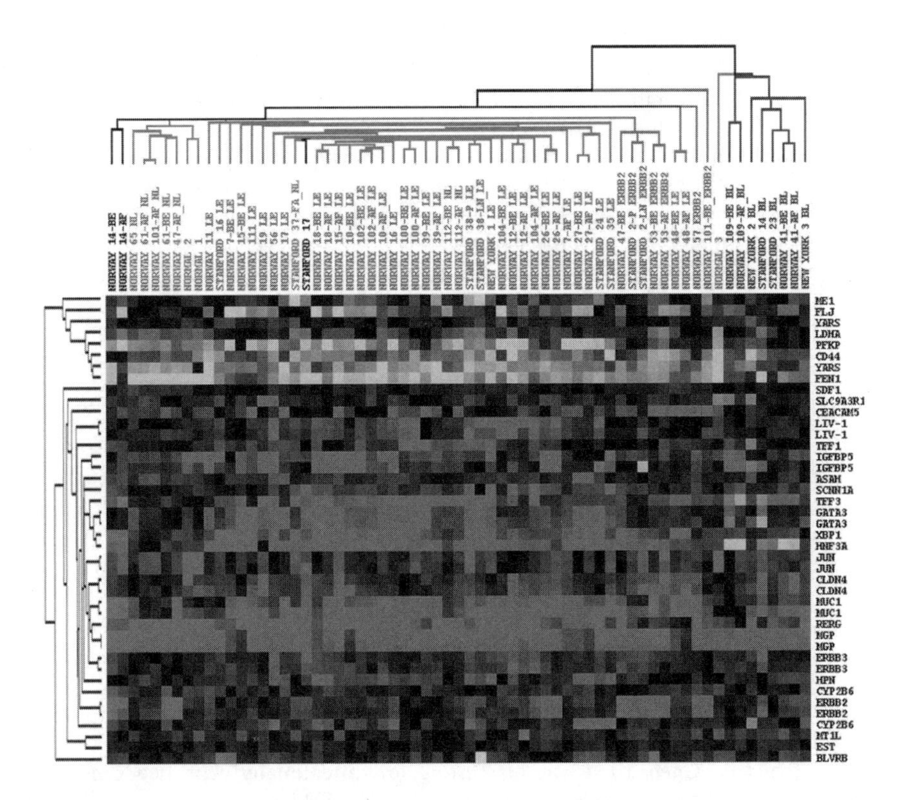

FIGURE 23.4 Clustering of breast cancer specimen according to gene expression profiling of a defined set of estrogen-responsive genes. Cluster analysis of 62 breast tumor surgical specimens and 3 normal mammary gland biopsies, based on expression of a subset of 27 estrogen responsive genes identified in BC cell lines and 4 molecular markers of ESR1 (ERα) positive breast tumors. Expression data were from Perou *et al.* [48]; sample denomination has been maintained the same as in the original study.

ACKNOWLEDGMENTS

Preparation of this review and of the experimental data reported therein were supported by research grants from: Associazione Italiana per la Ricerca sul Cancro, Seconda Università degli Studi di Napoli (Fondi per la Ricerca di Ateneo e per Assegni e Dottorati di Ricerca), Ministero dell'Istruzione, dell'Università e della Ricerca (PRIN 2002067514, 2004067020, 2004055579 and FIRB RBNE0157EH) and the European Union (Contracts BMH4-CT98-3433, QLG1-CT-2000-01935 and QLK3-CT-2002-02029).

REFERENCES

1. The Chipping forecast II, Nature Genetics, 32 (supplement), 462–552, 2002.
2. Schena M, Shalon D, Davis RW, Brown PO, Quantitative monitoring of gene expression pattern with a complementary DNA microarray, *Science*, 270:467–470, 1995.
3. Schena M, ed., *Microarray analysis*, Wiley-Liss, Hoboken New Jersey, USA, 2003.
4. Bowtell D, Sambrook J, eds., *DNA mycroarrays—A molecular cloning manual*, Cold Spring Harbor Laboratory Press, Cold Spring Harbor, New York, USA, 2003.
5. Schena M, Heller RA, Theriault TP, Konrad K, Lachenmeier E, Davis RW, Microarray: Biotechnology's discovery platform for functional genomics, *Trends Biotechnol.*, 16:301–306, 1998.
6. Cleveland WS, Robust locally weighted regression and smoothing scatterplots, *J. Amer. Stat. Assoc.*, 74:829–836, 1979.
7. Knudsen S, *A Biologists's Guide to Analysis of DNA Microarray Data*, New York, Wiley-Liss, 2002.
8. Workman C, Jensen LJ, Jarmer H, Berka R, Gautier L, Nielser HB, Saxild HH, Nielsen C, Brunak S, Knudsen S, A new nonlinear normalization method for reducing variability in DNA microarray experiments, *Genome Biol.*, 3:research 0048, 2002.
9. Yang IV, Chen E, Hasseman JP, Liang W, Frank BC, Wang S, Sharov V, Saeed AI, White J, Li J, Lee NH, Yeatman TJ, Quackenbush J, Within the fold: Assessing differential expression measures and reproducibility in microarray assays, *Genome Biol.*, 3:research 0062, 2002.
10. Yang YH, Dudoit S, Luu P, Lin DM, Peng V, Ngai J, Speed TP, Normalization for cDNA microarray data: A robust composite method addressing single and multiple slide systematic variation, *Nucleic Acids Res.*, 30: e15, 2002.
11. Churchill GA, Fundamentals of experimental design for cDNA microarrays, *Nat. Genet.*, 32 Suppl:490–495, 2002.
12. Tsai CA, Chen YJ, Chen JJ, Testing for differentially expressed genes with microarray data, *Nucleic Acids Res.*, 31:E52, 2003
13. Tusher VG, Tibshirani R, Chu G, Significance analysis of microarrays applied to the ionizing radiation response, *Proc. Natl. Acad. Sci. USA.*, 98:5116–5121, 2001.
14. Kerr MK, Martin M, Churcill GA, Analysis of variance of gene expression microarray data, *J. Comput. Biol.*, 7:819–837, 2000.

15. Baldi P, Long AD, A Bayesian framework for the analysis of microarray expression data: Regularized t-test and statistical inferences of gene changes, *Bioinformatics*, 17:509–519, 2001.
16. Ideker T, Thorsson V, Siegel AF, Hood LE, Testing for differentially-expressed genes by maximum-likelihood analysis of microarray data, *J. Comput. Biol.*, 7:805–817, 2000.
17. Liu G, Loraine AE, Shigeta R, Cline M, Cheng J, Valmeekam V, Sun S, Kulp D, Siani-Rose MA, NetAffx: Affymetrix probe sets and annotations, *Nucleic Acids Res.*, 31:82–86, 2003.
18. Iazzetti G, Calabrò V, Saviozzi S, Weisz A, Lania L, Calogero R, BSC: A clustering program for DNA array expression data, *Proceedings of Biocomp, 2001*, Siena, 2001. http://obelix.bio.uniroma2.it/www/abstr_2001.html#iazzetti
19. Eisen MB, Spellman PT, Brown PO, Botstein D, Cluster analysis and display of genome-wide expression patterns, *Proc. Natl. Acad. Sci. USA.*, 95:14863–14868, 1998.
20. Tamayo P, Slonim D, Mesirov J, Zhu Q, Kitareewan S, Dmitrovsky E, Lander ES, Golub TR, Interpreting patterns of gene expression with self-organizing maps: Methods and application to hematopoietic differentiation, *Proc. Natl. Acad. Sci. USA.*, 96:2907–2912, 1999.
21. Dettling M, Buhlmann P, Supervised clustering of genes, *Genome. Biol.*, 3: Research, 0069, 2002.
22. Mateos A, Herrero J, Tamames J, Dopazo J, Supervised neural networks for clustering conditions in DNA array data after reducing noise by clustering gene expression profiles, In: Lin SM, Johnson KF, eds., *Methods of Microarray Data Analysis II*, Boston, Kluwer Academic Publ, pp. 91–103, 2002.
23. Friedman N, Linial M, Nachman I, Pe'er D, Using Bayesian networks to analyze expression data, *FEBS Lett.*, 451:142–161, 1999.
24. Brown MP, Grundy WN, Lin D, Cristianini N, Sugnet CW, Furey TS Jr, Ares M, Haussler D, Knowledge-based analysis of microarray gene expression data by using support vector machines, *Proc. Natl. Acad. Sci. USA*, 97:262–267, 2000.
25. Tobler JB, Molla MN, Nuwaysir EF, Green RD, Shavlik JW, Evaluating machine learning approaches for aiding probe selection for gene-expression arrays, *Bioinformatics 18 Suppl.*, 1:S164–S171, 2002.
26. Ashburner M, Ball CA, Blake JA, Botstein D, Butler H, Cherry JM, Davis AP, Dolinski K, Dwight SS, Eppig JT, Harris MA, Hill DP, Issel-Tarver L, Kasarskis A, Lewis S, Matese JC, Richardson JE, Ringwald M, Rubin GM, Sherlock G, Gene ontology: Tool for the unification of biology, the gene ontology consortium, *Nat. Genet.*, 25:25–29, 2000.
27. Ashburner M, Ball CA, Blake JA, Butler H, Cherry JM, Corradi J, Dolinski K, Eppig JT, Harris MA, Hill DP, Lewis S, Marshall B, Mungall C, Reiser L, Rhee S, Richardson JE, Richter J, Ringwald M, Rubin GM, Sherlock G, Yoon J, Creating the gene ontology resource: Design and implementation, *Genome Res.*, 11:1425–1433, 2001.
28. Facchiano A, A Weisz Internet tools for the analysis of gene expression by database integration, *Proceedings of the NETTAB 2001 Workshop: "CORBA and XML—Toward a bioinformatics integrated network environment"*, Italy, 2001, pp. 99–102.

29. Khatri P, Draghici S, Ostermeier GC, Krawetz SA, Profiling gene expression using onto-express, *Genomics.*, 79:266–270, 2002.
30. Edgar R, Domrachev M, Lash AE, Gene Expression Omnibus: NCBI gene expression and hybridization array data repository, *Nucleic Acids Res.*, 30:207–210, 2002.
31. Sherlock G, Hernandez-Boussard T, Kasarskis A, Binkley G, Matese JC, Dwight SS, Kaloper M, Weng S, Jin H, Ball CA, Eisen MB, Spellman PT, Brown PO, Botstein D, Cherry JM, The Stanford Microarray Database, *Nucleic Acids Res.*, 29:152–155, 2001.
32. Brazma A, Parkinson H, Sarkans U, Shojatalab M, Vilo J, Abeygunawardena N, Holloway E, Kapushesky M, Kemmeren P, Lara GG, Oezcimen A, Rocca P -Serra, Sansone SA, ArrayExpress—a public repository for microarray gene expression data at the EBI, *Nucleic Acids Res.*, 31:68–71, 2003.
33. Brazma A, Hingamp P, Quackenbush J, Sherlock G, Spellman P, Stoeckert C, Aach J, Ansorge W, Ball CA, Causton HC, Gaasterland T, Glenisson P, Holstege FC, Kim IF, Markowitz V, Matese JC, Parkinson H, Robinson A, Sarkans U, Schulze-Kremer S, Stewart J, Taylor R, Vilo J, Vingron M, Minimum information about a microarray experiment (MIAME)-toward standards for microarray data, *Nat. Genet.*, 29:365–371, 2001.
34. Spellman PT, Miller M, Stewart J, Troup C, Sarkans U, Chervitz S, Bernhart D, Sherlock G, Ball C, Lepage M, Swiatek M, Marks WL, Goncalves J, Markel S, Iordan D, Shojatalab M, Pizarro A, White J, Hubley R, Deutsch E, Senger M, Aronow BJ, Robinson A, Bassett D, Jr Stoeckert CJ, Brazma A, Design and implementation of microarray gene expression markup language (MAGE-ML), *Genome Biol.*, 3:Research 0046, 2002.
35. Chung CH, Bernard PS, Perou CM, Molecular portraits and the family tree of cancer, *Nat. Genet.*, 32 Suppl:533–540, 2002.
36. Hanahan D, Weinberg RA, The hallmarks of cancer, *Cell*, 100:57–70, 2000.
37. Khan J, Wei JS, Ringner M, Saal LH, Ladanyi M, Westermann F, Berthold F, Schwab M, Antonescu CR, Peterson C, Meltzer PS, Classification and diagnostic prediction of cancers using gene expression profiling and artificial neural networks, *Nat. Med.*, 7:673–679, 2001.
38. Ramaswamy S, Tamayo P, Rifkin R, Mukherjee S, Yeang CH, Angelo M, Ladd C, Reich M, Latulippe E, Mesirov JP, Poggio T, Gerald W, Loda M, Lander ES, Golub TR, Multiclass cancer diagnosis using tumor gene expression signatures, *Proc. Natl. Acad. Sci. USA.*, 98:15149–15154, 2001.
39. Weiss MM, Kuipers EJ, Postma C, Snijders AM, Siccama I, Pinkel D, Westerga J, Meuwissen SG, Albertson DG, Meijer GA, Genomic profiling of gastric cancer predicts lymph node status and survival, *Oncogene*, 22:1872–1879, 2003.
40. Sorlie T, Perou CM, Tibshirani R, Aas T, Geisler S, Johnsen H, Hastie T, Eisen MB, van de Rijn M, Jeffrey SS, Thorsen T, Quist H, Matese JC, Brown PO, Botstein D, Eystein Lonning P, Borresen-Dale AL, Gene expression patterns of breast carcinomas distinguish tumor subclasses with clinical implications, *Proc. Natl. Acad. Sci. USA*, 98:10869–10874, 2001.
41. Beer DG, Kardia SL, Huang CC, Giordano TJ, Levin AM, Misek DE, Lin L, Chen G, Gharib TG, Thomas DG, Lizyness ML, Kuick R, Hayasaka S, Taylor JM, Iannettoni MD, Orringer MB, Hanash S, Gene-expression profiles predict survival of patients with lung adenocarcinoma, *Nat. Med.*, 8:816–824, 2002.

42. van't Veer LJ, Dai H, van de Vijver MJ, He YD, Hart AA, Mao M, Peterse HL, van der Kooy K, Marton MJ, Witteveen AT, Schreiber GJ, Kerkhoven RM, Roberts C, Linsley PS, Bernards R, Friend SH, Gene expression profiling predicts clinical outcome of breast cancer, *Nature*, 415:530–536, 2002.
43. Kikuchi T, Daigo Y, Katagiri T, Tsunoda T, Okada K, Kakiuchi S, Zembutsu H, Furukawa Y, Kawamura M, Kobayashi K, Imai K, Nakamura Y, Expression profiles of nonsmall cell lung cancers on cDNA microarrays: Identification of genes for prediction of lymph-node metastasis and sensitivity to anti-cancer drugs, *Oncogene*, 22:2192–2205, 2003.
44. Okutsu J, Tsunoda T, Kaneta Y, Katagiri T, Kitahara O, Zembutsu H, Yanagawa R, Miyawaki S, Kuriyama K, Kubota N, Kimura Y, Kubo K, Yagasaki F, Higa T, Taguchi H, Tobita T, Akiyama H, Takeshita A, Wang YH, Motoji T, Ohno R, Nakamura Y, Prediction of chemosensitivity for patients with acute myeloid leukemia, according to expression levels of 28 genes selected by genome-wide complementary DNA microarray analysis, *Mol. Cancer Ther.*, 1:1035–1042, 2002.
45. Weisz A, Estrogen regulated genes, In Oettel M, Schillinger E, eds., *Handbook of Experimental Pharmacology*, Vol. 135/I: Estrogens and Antiestrogens I. Berlin-Heidelberg-New York, Springer Verlag, 1999, pp. 127–151.
46. Cicatiello L, Facchiano A, Calogero R, De Bortoli M, Bresciani F, Weisz A, Gene expression monitoring in hormone-responsive human breast cancer cells during estrogen-induced cell cycle progression, *Proceedings of AACR/ Nature Genetics Joint Conference: Oncogenomics: Dissecting Cancer through Genome Research*, Tucson, AZ, 2001. http://www.nature.com/cgi-taf/DynaPage.taf?file=/ng/journal/v27/n4s/full/ng0401supp_95a.html&_UserReference=C0A804EC4650B9B7E02E44E479153B0D3D72
47. Cicatiello L, Scafoglio C, Altucci L, Cancemi M, Natoli G, Facchiano A, Iazzetti G, Calogero R, Biglia N, De Bortoli M, Sfiligoi C, Sismondi P, Bresciani F, Weisz A, A genomic view of estrogen actions in human breast cancer cells by expression profiling of the hormone-responsive transcriptase, *J. Mol. Endocrinol*, 32:719–775, 2004.
48. Perou CM, Sorlie T, Eisen MB, van de Rijn M, Jeffrey SS, Rees CA, Pollack JR, Ross DT, Johnsen H, Akslen LA, Fluge O, Pergamenschikov A, Williams C, Zhu SX, Lonning PE, Borresen-Dale AL, Brown PO, Botstein D, Molecular portraits of human breast tumours, *Nature*, 406:747–752, 2000.
49. Weisz A, Identification of the gene expression signature which characterizes human breast cancer cells response to estrogen, *Proceedings of Second International Cancer Congress: Translational Research in Cancer*, Rovigo, p. 83, 2001.
50. Weisz A, Basile W, Scafoglio C, Natoli G, Altucci L, Bresciani F, Facchiano A, Sismondi P, Cicatiello L, De Bortoli M, Molecular identification of ERalpha-positive breast cancer cells by the expression profile of an intrinsic set of hormone regulated genes, *J. Cell. Physiol*, 200:440–450, 2004.

24 Economics of Combinatorial Chemistry and Combinatorial Technologies

Aris Persidis

CONTENTS

I. INTRODUCTION

The biotechnology industry as a whole, and the human therapeutics area in particular, are the subjects of much empirical analysis and description, whose aim is to describe and discuss economic characteristics and overall patterns of success and failure for participating companies. Recently, the advent and recognition of the importance of so-called platform biotechnologies has caused the participants and observers of biotechnology to pay very close attention to various factors of the business arena that affect the development of these platforms. Platform biotechnologies can be defined as specific technical developments that can be applied to multiple areas and that also spawn supporting mini-industries, and they include novel drug-generation methods and therapeutic modalities which can be applied to multiple diseases. This chapter reviews the historical evolution of combinatorial chemistry as a platform business, examines the economics displayed by this sector, and describes how this platform has influenced and been influenced by the business development of participating companies.

II. THE IMPORTANCE OF COMBINATORIAL CHEMISTRY

The advent of molecular medicine, by which our understanding of disease has shifted from the large, macro scale to the molecular pathway level, has changed the way that new drugs are discovered and optimized. Nowadays, once a potential molecular target is identified in a particular disease setting, the race is afoot to find appropriate drug leads that can exert a specific effect on the particular target, such that the end result is a therapy.

The large-scale sequencing of the human genome, advances in our understanding of cellular events, including programmed cell death (or apoptosis), signal transduction, chromosome telomere control, cytoskeletal development, upregulation and downregulation of stress-related proteins, cell surface display of antigens by the major histocompatibility complex molecules and other molecular scenarios, are filling the potential target pipeline.

In a sense, given these and other advances, finding good potential targets that are known to be participating in the development and progression of specific diseases is no longer a bottleneck, although it is well known that validating such targets in a clinically relevant setting is not a trivial task. Discovering and optimizing potential drug leads that are directed toward the newly defined targets is where combinatorial chemistry has made and will continue to exert its largest impact. Table 24.1 shows that it takes one medicinal chemist one month to generate four compounds directed against a particular target, at a total cost of about $30,000, or $7500 per compound.

TABLE 24.1
The power of combinatorial chemistry

	Traditional chemistry	Combinatorial chemistry
Compounds per one chemist	4	3300
Total cost	$30,000	$40,000
Cost per compound	$7500	$12

Combinatorial chemistry enables the synthesis *of a* vast number *of* test compounds at a fraction *of* the cost and time spent per compound, compared to traditional chemistry methods.

Source: Booz, Allen & Hamilton, 1996 Survey.

By comparison, combinatorial chemistry applied by one chemist over one month can produce 3300 compounds for $40,000, or about $12 per compound.

This, then, is the immense power and potential of combinatorial chemistry. In quantitative terms, both financial and chemistry ones, it is capable of producing almost 1000 times more compounds than traditional methods, and for about 600 times less cost per compound, in the same time period.

These overall performance characteristics of combinatorial chemistry are very significant when one considers that it takes on average 10 years and approximately $350 million to get a new drug to the market. In addition, and based on historical data, for every drug that has made it to the market, 4000 compounds have been made by traditional chemistry methods and have been screened [1]. By using the data shown in Table 24.1, it would thus take 100 chemists 10 months to make as many compounds for testing by traditional methods, or it would take one combinatorial chemist just 5 weeks to make the same number of compounds. Although this calculation makes several assumptions about the types of molecules made, it nevertheless serves the purpose of demonstrating unequivocally that combinatorial chemistry is a drug lead-generation method that has enormous potential as a cost-saving device. This cost-saving characteristic, then, is the reason why so many companies are involved in the field, as will be discussed later on.

Furthermore, it is possible today to enter the field of combinatorial chemistry relatively easily, and achieve the immense economies of scale shown in Table 24.1. For example, it is possible to buy ready-made combinatorial libraries built around specific molecular themes and consisting of many thousands of compounds, and it is also possible to create a high-throughput screening system for that library using automated, off-the-shelf instrumentation. This may enable one to be using combinatorial chemistry very quickly, and is one of the key characteristics of the field today: it can be considered a commodity, available to anyone for a modest capital cost.

III. BRIEF HISTORICAL OVERVIEW OF THE BUSINESS DEVELOPMENT OF COMBINATORIAL CHEMISTRY

To see how combinatorial chemistry has become one of the major platforms of modern biotechnology, it is necessary first to have a basic understanding of how the relevant science developed and eventually merged with business interests.

The experimental discipline that is now known as combinatorial chemistry—although it is also known by such evocative terms as test tube or in-vitro evolution—arose almost 30 years ago. At that time, RNA evolution experiments showed that it was possible to create selection-pressure conditions in an experimental setting that caused certain species of RNA, but not others, to survive from one generation to the next [2]. This work was complemented by advances of theories on how biopolymers organize themselves and change over time under specific conditions [3]. About 13 years ago, the use of evolution and Darwinian selection was proposed for RNA [4] and for proteins [5] in an experimental setting, which would permit such biopolymers to be subjected to mutation and selection pressures (hence the term test-tube evolution). Finally, the development of solid-phase peptide synthesis in the mid 1980s enabled the automated, large-scale synthesis of libraries of molecules [6–8]. This allowed the large-scale, practical implementation of the theories of biopolymer organization and launched the practical and business side of the field. Since that time, test-tube evolution has steadily developed novel proteins and nucleic acids, in addition to peptidomimentics, peptide nucleic acids, and carbohydrate-based oligomers, catalytic RNAs (ribozymes) and combinatorially generated catalytic antibodies (abzymes) being particularly notable products, as well as a variety of small-molecule drug leads directed against a multitude of targets.

It must be noted that it was the advent of automated nucleic acid and peptide synthesizers/analyzers that spawned the practical development of the field. As a result, it is not surprising that first-generation combinatorial technologies focused largely on peptides and oligonucleotides. The first companies dedicated or founded around particular aspects of combinatorial chemistry started to appear in the late 1980s and early 1990s. They included companies such as Affymax (Palo Alto, CA), Aptein Corp. (Seattle, WA), Darwin Molecular (Seattle, WA), Gilead Sciences (Foster City, CA), Houghten Pharmaceuticals (San Diego, CA, now Trega), Isis Pharmaceuticals (Carlsbad, CA), Ixsys, Inc. (San Diego, CA), NeXstar Pharmaceuticals (originally NeXagen, Boulder CO), Protein Engineering Corp. (Cambridge, MA, now Dyax), Selectide Corp. (Tucson, AZ), and Sphinx Pharmaceuticals (Durham, NC).

Ultimately, the question always remained whether combinatorial chemistry would be just another fad in the drug discovery world. This possibility was dispelled in the mid 1990s, when the field announced its arrival as a valid technology by passing the acid test of mergers and acquisitions. In 1994, Eli Lilly & Co. (Indianapolis, IN) bought Sphinx for $80 million. This was followed by Marion Merrell Dow (Kansas City, MO), spending $58 million for Selectide in 1995. The last—and also one of the biggest in all of biotech's history—major acquisition was Glaxo's (London, UK) investment of $533 million in 1995 for Affymax.

IV. FINANCIAL PERFORMANCE OF COMBINATORIAL CHEMISTRY

Having seen how combinatorial chemistry emerged as a science and business, it is important to review the field's overall business performance. One of the key questions that participants, observers, and strategists involved directly with combinatorial chemistry have to answer has to do with the overall financial health of combinatorial chemistry. How is the field doing financially? Is it a good investment? Is it a field worth entering? Is it a field worth maintaining and even increasing one's position in? Is it a competitive field?

To answer these questions, two good metrics of the industry are presented here that focus on macro factors, rather than micro business conditions and circumstances. The first such indicator is the ability of combinatorial chemistry to raise capital in the public and private markets, excluding any contributions from strategic alliances and collaborations. Thought of in a different way, this metric is a direct reflection of the short-, medium-, and long-term confidence that combinatorial chemistry can instill in sophisticated and critical investors, either in private placements or during Initial Public Offerings (IPOs) and post-IPO financings.

To understand the significance of this metric, it is important to understand first that investors need to be convinced to invest in particular biotechnologies based on a number of criteria. These criteria include—can the technology produce more than one product, or more than one kind of product?

Are the products themselves valuable and useful?
Do the products address unmet needs and economically attractive markets?
Can the technology's products generate real sales relatively quickly?
Does the technology interact and fit well with other established technologies?
Does the technology have a unique proprietary position, which will prevent large-scale copying?
Can the technology evolve with time, and ensure the continuous generation of new products?
Is the technology based on quality, solid science?

If the answers to the above questions are satisfactory in the eyes of investors, then they will support a particular technology by investing in it in various placements, IPOs and post-IPOs and they will maintain their investment and support for extended periods of time. The approach of quantifying investor confidence through actual patterns of investment is a more accurate indicator of an industry's health and its perception in the capital markets than simple stock performance, which may be affected by short-term exogenous considerations. The results for combinatorial chemistry are shown in Figure 24.1, which plots the public and private money raised by about 40 combinatorial companies from the Biovista™ Database source, excluding money raised through alliances and collaborations.

The data show that there has been a steady increase in the capital raised during the 1990s, which have seen to date an overall increase of about 65-fold, although the trend has been tapering off in the last three years. Part of the reason for the steady increase is, of course, the arrival of new entrants. However, the trend shows that there is growing confidence in the field's overall development. This is indicative of the fact that combinatorial chemistry is well received in the public and private capital markets, which do not hesitate to invest.

In terms of the criteria used by investors to decide whether or not to invest in a particular technology, as listed above, here is how combinatorial chemistry is perceived as an investment.

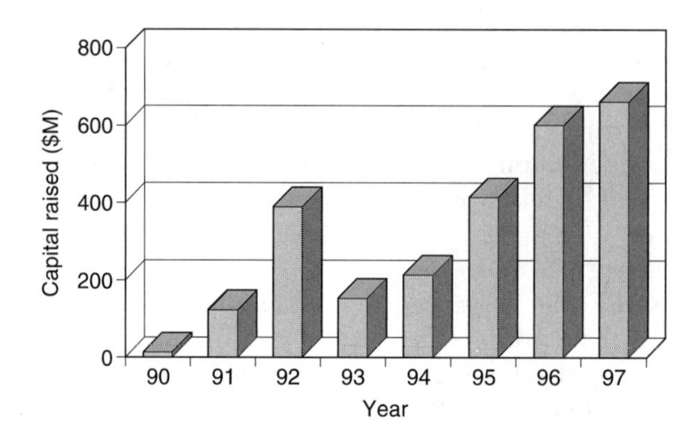

FIGURE 24.1 Capital raised by combinatorial chemistry companies. Ever since the field of combinatorial chemistry made its first corporate appearance in 1990, its ability to raise capital in the public and private markets has increased steadily over the years. By 1997, the field had raised 100 times more capital than in 1990. *Source:* Biovistar^{TM} (www.biovista.com).

Combinatorial chemistry can produce many different products and also many different kinds of products. For example, it can produce peptide drug leads against multiple disease targets, or drug leads based on entirely different molecular scaffolds against multiple diseases, or even high-affinity ligands for purification purposes, in addition to products that are used to make and screen libraries. Combinatorial chemistry thus produces a vast array of different products.

The products of combinatorial chemistry, namely, drug leads, instrumentation software for library data analysis, affinity ligands, etc., are all valuable and useful, because they are geared toward specific applications. For example, entire libraries have value as starting points for drug-lead screening purposes, and leads themselves are very valuable if they show desired effects in screening assays.

The products of combinatorial chemistry address vast markets, such as cancer, infections, auto-immune and neurodegenerative diseases, in the human, veterinary and also agro sectors. An investment in combinatorial chemistry, therefore, may produce products that have an essentially guaranteed market need that they fill.

Combinatorial chemistry can produce products that reach the marketplace relatively quickly. Although the drug leads produced by combinatorial chemistry are subject to the same long development times as leads generated by other methods, the technology also delivers products, such as affinity ligands, and the dedicated instrumentation and software that achieve sales immediately.

Combinatorial chemistry is a platform technology that is integrating very well with other technologies. For example, functional genomics and proteomics have and are producing numerous important drug targets, which are used by combinatorial chemistry to focus new lead generation.

Combinatorial chemistry has an unusual patent history. No single company or individual has control of any one strategic patent and there is no single strategic patent that defines the field. Many companies have patents that enable them to pursue unique chemistries, and this provides investors with many choices. On the other side, exclusivity is reduced.

Combinatorial chemistry is a very dynamic technology that will always be changing and improving, as new chemistries are developed, new building blocks for libraries are created, and automation is improved. Its value will increase with time because of this feature.

Combinatorial chemistry is the result of some of the most powerful and elegant science in recent times. Its fundamental concepts are simple, although the chemistries and details are very complex. The field has a very solid scientific footing and is practised by some of the best scientists in the world, which generates a lot of confidence in the eyes of investors.

Given the excellent investment profile outlined above, it is not surprising that combinatorial chemistry attracts consistent investment, as shown in Figure 24.1.

The second metric used to assess the overall health of combinatorial chemistry as a business is related directly to the performance of participating companies in the public capital markets. Specifically, it plots the aggregate market capitalization of combinatorial chemistry companies over time. Ultimately, excluding seasonality factors and microconditions which affect the stock performance of individual companies, a good measure of relative success of a field is simply to look at market capitalization over time for as large a collection or representative companies as is possible. This is shown in Figure 24.2, for a total of nine companies that are combinatorial chemistry specific and are also publicly traded. With an aggregate market value of more

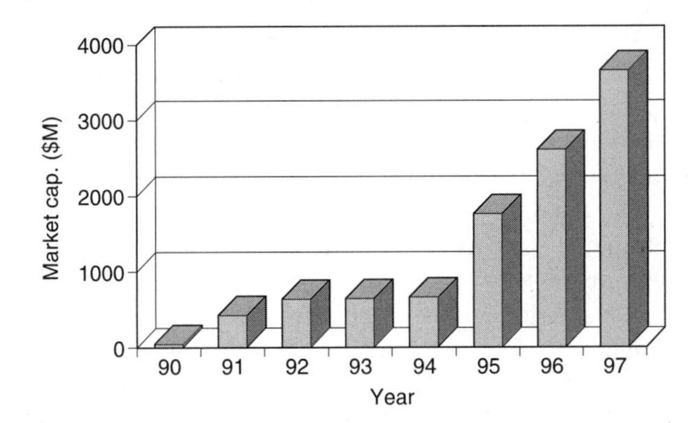

FIGURE 24.2 Aggregate market capitalization of public combinatorial chemistry companies. Just as the ability to raise capital in combinatorial chemistry has increased steadily during the 1990s, so has the combined market capitalization of companies dedicated to the field that are also traded publicly in stock exchanges. Specifically, the combined market value of these companies has increased by a factor of 65 during the 1990s. *Source:* Biovista^{TM} (www.biovista.com).

than $3 billion in 1997, which is 100 times more than what it was in 1990, companies that are involved in combinatorial chemistry and are publicly traded are doing well, showing a significant appreciation trend over time.

These two metrics, then, namely, capital-raising ability and also aggregate market capitalization, show that combinatorial chemistry is perceived as a solid investment over time, worthy of steady infusion of funds for further development, and also that it has returned substantial growth in the capital marketplace to its investors.

V. STRUCTURE OF THE INDUSTRY

In reviewing the overall economics of combinatorial chemistry as a business, it is important to understand the fundamental structure of the industry. Is combinatorial chemistry an integrated industry, or is it fragmented? Does any single company have a more significant market share than the competitors? What types of companies actually make up the industry of combinatorial chemistry?

The answer to the last question has changed with time during the development of the combinatorial chemistry industry. Initially, the pioneer companies focused on developing chemistries that allowed them to perform combinatorial chemistry on peptides and nucleic acids. These were the original library makers, and a significant number of such companies now have proprietary chemistries that enable them to generate libraries on demand. The library scaffolds and building blocks have increased dramatically since the days of nucleic acids and peptides and include carbohydrates, peptide nucleic acids, peptidomimentics, aromatic monomers, steroid derivatives, and whole families of small-molecule derivatives. Companies that are pure suppliers of the chemical building blocks and relevant reagents for combinatorial chemistry are the commoditizers of the industry, as they have enabled essentially anyone with a basic chemistry laboratory and a modest budget to use libraries in target screening programs.

Coupled to the development of novel chemistries and building blocks for creating libraries are developments in instrumentation and information management. On the hardware side, several companies have been created that make automated systems for the creation and manipulation of libraries, either as new start-ups or as spin-offs of larger companies. They include companies such as Argonaut, CombiChem, Irori, Ontogen, Affymax, and others. Combinatorial chemistry and library manipulation, both during the creation and also during the deconvolution stages, generate vast amounts of information that must be collected, archived, condensed and reported appropriately. This has led to the development of a significant number of software platforms and companies that make, use and/or sell these, including Tripos, Molecular Simulations, 3-D Pharmaceuticals and others.

A third group of companies consists of high-throughput screening companies. These companies have developed large-scale, automated methods of screening combinatorial libraries against large numbers of targets.

TABLE 24.2
Types of combinatorial chemistry companies

Company type	Number
Library commoditizers	6
Library value-adders	14
Commodity users	145
Hardware/software developers	16

The field of combinatorial chemistry is fragmented. Depending on whether a company simply makes or uses libraries, or whether it makes the hardware or software for synthesizing libraries or analyzing the data, the field consists of several companies in each category. Several companies, of course, cross over the four categories given here, but they are not counted twice, for clarity.

Source: Biovista™ (www.biovista.com).

Such companies typically have 100-plus *in vitro* tests for lead compounds, which they test against collections of receptors and intracellular molecules of pharmacological interest. They will also offer their screening capabilities to partners and such companies include Aurora, Diversomer, Orchid, Pharmacopeia and others. These companies can be considered as library value-adders, because they take libraries whose value is meaningless without identified leads and add value by filtering those libraries through appropriate screening assays.

Finally, the fourth type of company involved in combinatorial chemistry is the library users, who do not make or screen libraries but who provide targets for high-throughput screening and library generation. Although these companies represent the majority of participants in the combinatorial chemistry universe, it is likely that they have some internal capabilities to make and screen libraries, although on the whole they prefer to partner this need with dedicated companies that specialize in these tasks. Table 24.2 shows how many companies are involved in each of the categories listed above.

Table 24.2 also shows that the combinatorial chemistry industry is highly fragmented, consisting of four major types of companies that make different contributions to the field and have different requirements. Within each of these categories it may be possible to find market leaders, although, because of the fragmented nature of the industry, it is impossible and meaningless to search for an overall leader in the field of combinatorial chemistry.

VI. COMBINATORIAL CHEMISTRY ALLIANCE OVERVIEW

Combinatorial chemistry is a platform technology that is characterized by a very high degree of partnering and strategic alliances among its participants. For example, it has been estimated that alliances among approximately 30 companies for which financial data were available exceed $2.6 billion and that at least 130 companies worldwide are engaged in combinatorial chemistry-driven alliances, out of a total of about 180 companies involved in the field [9]. Given this tendency among companies participating in this sector to be particularly active in alliances, it is useful to examine the typical structure of these alliances and also their financial elements.

In terms of overall positioning along the drug development chain, from lead discovery and optimization to clinical trials, alliances in combinatorial chemistry tend to focus on the upstream portion of drug discovery. Specifically, these alliances have their objective, the sourcing of compound libraries and/or the generation of new leads against specific targets and/or the optimization of already-identified leads for particular targets. That alliances in combinatorial chemistry should have this upstream drug discovery focus is not surprising, given the nature of the drug discovery process today. Drug discovery and development consist initially of two components. First is the identification of suitable targets that are clinically relevant. Identifying particular molecules, such as receptors, signal transduction molecules, genes, genetic control elements, etc., in the context of a particular disease pathway is the starting point for drug discovery and is driven by a molecular understanding of the basis of disease. The second step is to generate leads against these particular targets, and this is exactly where combinatorial chemistry makes its contributions. Given clinically relevant targets, combinatorial chemistry is capable of generating vast numbers of leads for testing. Alliances therefore, often focus on this upstream component of drug development.

For example, many alliances focus currently on outright procurement or sourcing of libraries for screening. Here, pharmaceutical companies and biotechnology companies enter into agreements with pure suppliers of libraries of compounds that they can use to screen against their own targets. Examples of such alliances are those made by Arris (now Axys), ArQule, Pharmacopeia and others. The need for libraries of compounds to be used in high-throughput screening has resulted in companies being created that do nothing but make libraries for sale.

Another group of alliances is centered on lead generation, where specific targets are known and it is desired to generate new drug leads against those targets through combinatorial chemistry. These alliances are based on the premise that the company who will be doing the lead generation has a particular molecular scaffold and a proprietary ability to manipulate that scaffold to develop new leads in a focused manner. Examples of alliances in this category are those of Chiron, ISIS, Cadus and others.

The third group of alliances focuses on lead optimization. Here, a particular lead is known, such as a small peptide and it is desired to optimize the activity of that lead. In the case of peptides this could be accomplished through phage display and/or peptidomimetic combinatorial chemistry. Companies involved in alliances of this type have proprietary technologies that enable them to generate usually small-molecule libraries built around particular molecular themes, for example, steroid cores. Examples include alliances of Ontogen, Dyax, ArQule, etc.

VII. FINANCIAL CHARACTERISTICS OF COMBINATORIAL CHEMISTRY ALLIANCES

Having examined the three principal focus points of combinatorial chemistry alliances, it is illuminating to examine the financial elements of these agreements. Here, scarcity of good data is restricting, because many companies do not wish to reveal the specific financial terms of alliances they have entered into. However, in some cases, publicly traded companies are obligated under stock market rules to reveal such elements, enabling the analysis that is presented next and which is based on data from about 25 alliances. The data are displayed in Table 24.3.

It can be seen from the table that the greatest variation in the terms of these alliances occurs in the actual R&D investment, which ranges from $1 to $47M. This amount, however, shows a direct positive correlation with the number of targets on which the alliance is focusing. For example, the most significant commitment is the $47M paid by American Home Products to Affymax, which was to develop leads against a total of 11 specific targets in the first instance.

Another significant observation from the table is that there is a considerable monetary commitment that is guaranteed upon signing and it averages $12M through a combination of equity and up-front payment. This suggests that there is significant confidence on the part of the sponsoring party toward the combinatorial chemistry companies and the sponsor is, therefore, prepared to pay up front for the lead discovery program. This is not the norm, given that alliances now are generally not 'front-loaded,' but have payment terms that are contingent on the achievement of defined goals and milestones.

Finally, another important observation from the table is that investments made in the form of equity were done at a significant premium over the market price for equity that was raised in the last round of financing. On average, this

TABLE 24.3
Financial elements of combinatorial chemistry alliances

Alliance term	Average size
Duration	3 yr (0.5–5 yr range)
R&D investment	$3.3M p.a. (1$M–$47M range)
Guranteed commitment	$12M (equity+up-front payment)
Payment upon signing	$8M (equity +up-front payment)
Equity participation	$6.4M
Equity premium	26% + over last round
Sponsor termination	0.5–4 yr, in 50% of cases

Alliances and collaborations driven by combinatorial chemistry can be very profitable financially for the library company. Significant money is guaranteed up front, and even more significant milestones await companies that succeed in doing what they say they can do.

Source: Recombinant Capital (www.recap.com); Biovista^{TM} (www.biovista.com).

was a 26% premium, demonstrating again that sponsor companies acknowledged the value of the alliance and the potential of the companies in which they were investing.

It is important to remember that the very high level of strategic alliance activity observed between companies involved in combinatorial chemistry is completely understandable—biotechnology companies can survive only if they participate in collaborative projects with other companies. The requirement for strategic alliances, joint ventures and collaborations makes absolute sense, given the high cost and associated risk of failure of developing a drug. Alliances occur between biotechnology and pharmaceutical firms of all sizes and also between biotechnology or pharmaceutical firms themselves. They provide the innovator—usually the smaller firm—with research capital, with access to complementary technologies, and with access to established marketing and production infrastructures. The larger firm obtains a window into a promising technological advance, which would otherwise be too expensive to develop internally [10–13].

VIII. COMBINATORIAL CHEMISTRY BUSINESS FORMATS

Modern drug discovery and development requires efficient methods to find and optimize drug leads. As discussed previously, drug development has three key upstream components, namely, lead generation, target identification and, finally, lead validation against those targets. Lead generation was traditionally accomplished through natural products screening and also more recently through processes of rational drug design, which incorporated prior knowledge of specific targets into design regimes that would yield drugs with prespecified properties. Vertex is a good example of a company that pioneered this process of rational drug design. Combinatorial chemistry complements these approaches to lead generation by virtue of its brute-force ability to supply huge numbers of potential leads that then have to be screened. This complementarity with the more traditional methods of drug design has led to the vast proliferation of strategic alliances in the field.

As discussed elsewhere, the combinatorial chemistry industry is fragmented, consisting of four major types of companies, separated essentially by whether they make or use libraries. This has resulted in large numbers of alliances between these companies, and the characteristics of these alliances as business formats are developed elsewhere in this chapter.

Apart from simple bilateral strategic alliances, however, which are the norms in biotechnology as a whole, combinatorial chemistry has also caused vertical integration to be adopted as a business format. For example, NeXagen merged with Vestar in 1995, to form NeXstar Pharmaceuticals. This consolidation combined NeXagen's combinatorial expertise with Vestar's liposome drug delivery technology. It enables the resulting company not only to generate a rich pipeline of drug leads, but also to have a platform for the pharmaceutical delivery of those leads in a clinical setting.

Another two examples of integration efforts are shown by the recent acquisition by ChiroScience's in the UK of the US company Darwin Molecular Corp. in 1996 for $120 million and of ChromaXome by Trega, both in the United States. These two cases represent integration of drug discovery platforms.

In addition, it must be remembered that, because of the fragmented nature of the combinatorics industry, one of the business challenges that participants, especially the smaller ones, face is that such companies are both producers and users of their technology. As such, value to them exists in both their proprietary combinatorial approach (proprietary monomers, novel encoding, novel software and automated instrumentation), and also in the new "hits" or drug leads they develop. The challenge is to decide which avenue to pursue, whether to invest resources in developing specific leads that arise from combinatorial chemistry, or whether to develop the actual tools used in the practice of combinatorial chemistry itself. Several companies take the dual approach of incorporating both their "tool" and their "target" capabilities in their commercialization strategies, such as NeXstar's SELEX™, the biochips of Affymax, ArQule's DirectedArrays™ and many others. This business format enables companies to be more flexible in their approach, for example, by generating revenue through the licensing of their "tool" technology to other companies, and then using this revenue to drive the generation of novel drug leads in their own internal R&D programs.

Finally, a very interesting business format that has appeared in combinatorial chemistry is the consortium approach. As shown in Table 24.4, two consortia have emerged in recent years. They both intend to enable their partners to share resources and progress in combinatorial chemistry through sharing and resource-pooling strategies, which it is hoped will enable them to achieve goals they would otherwise not have been able to. It is thus hoped that the drug discovery process will be made more efficient and cost-effective.

As is argued elsewhere [14], creation of a consortium is an important response of participating companies that realize they cannot dominate the field, and actually need access to complementary technology protected by patents owned by others.

IX. CLINICAL RELEVANCE

Although it is widely believed that combinatorial chemistry will prove to be very efficient and cost-effective in new-drug discovery, at present there are no drugs on the market which were discovered by the application of this method. However, a large number of leads are in various stages of preclinical development. This number is in excess of 350, and is based on the 180+ companies that are active in the field of combinatorial chemistry and also leads that they report directly that they are developing through the application of combinatorial chemistry.

Also, Eli Lilly & Co. (Indianapolis, IN) now has a small-molecule agent for the central nervous system in human Phase II/III trials, which took less than

TABLE 24.4
The consortium business format in combinatorial chemistry

The Diversity Biotechnology Consortium (founded in 1994; directed out of the Santa Fe Institute)	The Combinatorial Chemistry Consortium (founded in 1996; directed out of Molecular Simulations, Inc.)
Consortium partners	
Affymax Research Institute (Palo Alto, CA)	Bayer (Pittsburgh, PA)
CalTech (Pasadena, CA)	FMC (Chicago, IL)
Cytogen Corp. (Princeton, NJ)	Glaxo Wellcome (London, UK)
Department of Biology, University of North Carolina	Molecular Simulations, Inc. (San Diego, CA)
(Chapel Hill, NC)	
Department of Chemistry, Indiana University (Bloomington, IN)	Procept, Inc. (Cambridge, MA)
Departments of Biochemistry and Microbiology, Duke University	SmithKline Beecham (Philadelphia, PA)
Medical Center (Durham, NC)	
Glaxo Research Institute (Research Triangle Park, NC)	Teijin (Tokyo, Japan)
Institute for Molecular Biotechnology (Jena, Germany)	The University of Queensland (Australia)
Isis Pharmaceuticals, Inc. (Carlsbad, CA)	Vertex Pharmaceuticals (Cambridge, MA)
Ixsys, Inc. (San Diego, CA)	
Los Alamos National Laboratory (Los Alamos, NM)	
The Santa Fe Institute (Santa Fe, NM)	

Combinatorial chemistry has led to the formation of two large consortia. Although organized around different structures, they share their ultimate objective, which is to spread resources and risk among the participants.

Source: Biovista (www.biovista.com).

two years to identify and enter into the trials. In addition, Magainin Pharmaceuticals (Exton, PA) has developed a novel antibacterial drug lead through the application of combinatorial chemistry that is in Phase I/II trials.

X. COMBINATORIAL CHEMISTRY RELATIVE TO OTHER PLATFORM BIOTECHNOLOGIES

Having examined the principal features of combinatorial chemistry as an ongoing business concern, it is instructive and useful to compare and contrast the field with other major platform biotechnologies that are shaping modern drug discovery and development. In particular, combinatorial chemistry will be compared with gene therapy and also with the business that has emerged as a result of the invention of the Polymerase Chain Reaction (PCR).

Like combinatorial chemistry, gene therapy is a platform biotechnology that relies on several key technical contributions that have arisen in several academic and company laboratories. By contrast, PCR is a single invention, unlike combinatorial chemistry or gene therapy, but, like the other two platforms used as examples here, it has also been able to completely revolutionize the process of drug discovery.

Combinatorial drug discovery, gene therapy and the PCR platform have several distinct features in common. First, they are all enabling technologies that transcend several disciplines. Not only do they combine disciplines, such as various fields of chemistry, biochemistry and molecular biology, but their end result, namely, novel drug leads, can be generated for any disease target. Second, all three platforms elicit intense commercial interest. Their platform nature has a very reassuring business effect, in that it spreads investment risk in case any one particular application fails. The breadth of potential applications causes investors to consider the three platforms to have both short-term and also long-term growth prospects. Third, all three platforms can be used as tools for novel drug discovery or value can be derived from their end products themselves, namely, the specific drug leads obtained. This again has the effect of spreading risk and multiple-level value enhancement, which are two very important components of truly successful platform technologies.

However, despite their common features, combinatorial drug discovery, gene therapy and the PCR platform have a key difference between them, which has affected their business development profoundly. PCR developed as a business from a very strong early patent position, which survived and gained from an extensive court battle. It originated with one company (Cetus Corporation, Emeryville, CA), which owned all the relevant patents [15]. Combinatorial drug discovery and gene therapy, unlike PCR, are characterized by a diffuse lineage of intellectual property, distributed across several individuals and companies. This feature is one of the primary reasons why there are a multitude of players in combinatorial drug discovery and gene therapy today, with more than 100 companies in each of the areas listing these technologies as part of their R&D portfolio [16]. If nobody has a proprietary position which inhibits others from entering the field, if entry barriers are relatively low and this is certainly the case for several technical subsegments of combinatorial drug discovery and gene therapy, and if the promise of a field is great, as is also the case for combinatorial drug discovery and gene therapy, then the combined effect of these factors is to attract a large number of players who want to be part of a new industry "bandwagon" [17].

XI. CONCLUSIONS

During its 25-year history to date, the human therapeutics biotechnology industry has created about 35 major therapeutics currently on sale, with combined annual sales exceeding U.S. $7 billion [18]. These medicines are used against cystic fibrosis, heart attacks, several cancers, infectious diseases such as

AIDS, hepatitis, blood disorders, multiple sclerosis and others. At present, another 300 potential therapeutics are in various human clinical trials, although it is impossible to estimate how many of those will actually reach the marketplace. Combinatorial chemistry promises to revolutionize the rate of new-drug development by accelerating initial lead generation. Already, in its approximately 8-year corporate history, combinatorial chemistry has produced about 350 leads in various stages of preclinical development.

Practising combinatorial chemistry enables companies to produce thousands of potential leads for a fraction of the cost of producing the same number of leads by traditional chemistry. This is probably why so many companies, in excess of 180, are involved in the field. These companies can be divided into four major categories, depending on their use of combinatorial chemistry—library makers, library value-adders, library users and finally hardware/software developers. The industry is, therefore, fragmented with no clear leadership position enjoyed by any single company. This also probably accounts for why the field is so rich in alliances and collaborations, whose value exceeds a few billion dollars and which occur among at least 130 of the 180 companies involved in combinatorial chemistry.

Finally, the field has enjoyed excellent growth during the 1990s, showing a 65-fold increase in aggregate market capitalization for nine dedicated companies involved and also a 100-fold increase in the amount of capital raised in various public and private rounds of financing. These figures show that combinatorial chemistry is very well perceived by the biotechnology investment community and, as technological advances continue to improve its range of applications and efficiency, it is very likely that the field will continue to grow and prosper.

REFERENCES

1. Source: Pharmaceutical Manufacturers Association.
2. Spiegelman S, An approach to the experimental analysis of precellular evolution, *Quart. Rev. Biophys.*, 4:213–253, 1971.
3. Eigen M, Self-organization of matter and the evolution of biological macromolecules, *Naturwissenschaften*, 58:465–523, 1971.
4. Eigen M, Gardiner W, Evolutionary molecular engineering based on RNA replication, *Pure. Appl. Chem.*, 56:967–978, 1984.
5. Kauffman SA, Autocatalytic sets of proteins, *J. Theor. Biol.*, 119:1–24, 1986.
6. Merrifield RB, Solid phase synthesis (Nobel Lecture), *Angew. Chem.*, 97:801, 1985.
7. Geysen HM, Meloen RH, Bartelig SJ, Use of peptide synthesis to probe viral antigens for epitopes to a resolution of a single amino acid, *Proc. Natl. Acad. Sci. USA*, 81:3998–4002, 1984.
8. Houghten RA, General method for the rapid solid-phase synthesis of large numbers of peptides: Specificity of antigen-antibody interaction at the level of individual amino acids, *Proc. Natl. Acad. Sci. USA*, 82:5131–5135, 1985.
9. Persidis A, Biovista^{TM} Industry Review & Company Database: Combinatorics (www.biovista.com), 1998.

10. Stone D, Drug firms need biotech drug discovery, *Biotechnology*, 13:208–209, 1995.
11. Pisano GP, The governance of innovation: Vertical integration and collaborative arrangements in the biotechnology industry, *Res. Policy.*, 20:237–249, 1991.
12. Rotman D, Biotech industry forges new strategies, increases revenues, *Chem. Week*, 15:16, 1992.
13. Hamilton WF, Vila J, Dibner MD, Patterns of strategic choice in emerging firms: positioning for innovation in biotechnology, *Calif Management Rev.*, 32:73–86, 1990.
14. Persidis A, Persidis A, Biotechnology consortia versus multifirm alliances: Paradigm shift at work? *Nat. Biotechnol.* 14:1657–1660, 1996.
15. Daniell E, PCR in the marketplace, In Mullis KB, Ferré F, Gibbs RA, eds., *The Polymerase Chain Reaction*. Boston, Birkhauser, 1994, pp. 421–426.
16. Persidis A, Tomczyk M, Critical issues in gene therapy commercialization, *Nat. Biotechnol.*, 15:689–690, 1997.
17. Persidis A, Enabling technologies and the business of science, *Biotechnology*, 13:1172–1176, 1995.
18. Lee K, Burrill S, *Biotech 97*: Alignment (Ernst & Young 11th Industry Annual Report), 1997.

Index

D

E

T

U

V

X